Lecture Notes in Computer Science   10531

Commenced Publication in 1973
Founding and Former Series Editors:
Gerhard Goos, Juris Hartmanis, and Jan van Leeuwen

More information about this series at http://www.springer.com/series/7411

Olga Galinina · Sergey Andreev
Sergey Balandin · Yevgeni Koucheryavy (Eds.)

Internet of Things, Smart Spaces, and Next Generation Networks and Systems

17th International Conference, NEW2AN 2017
10th Conference, ruSMART 2017, Third Workshop NsCC 2017
St. Petersburg, Russia, August 28–30, 2017
Proceedings

 Springer

Editors
Olga Galinina (iD)
Electronics and Communication Engineering
Tampere University of Technology
Tampere
Finland

Sergey Andreev (iD)
Electronics and Communication Engineering
Tampere University of Technology
Tampere
Finland

Sergey Balandin
FRUCT Oy
Helsinki
Finland

Yevgeni Koucheryavy (iD)
Tampere University of Technology
Tampere
Finland

ISSN 0302-9743   ISSN 1611-3349 (electronic)
Lecture Notes in Computer Science
ISBN 978-3-319-67379-0   ISBN 978-3-319-67380-6 (eBook)
DOI 10.1007/978-3-319-67380-6

Library of Congress Control Number: 2017952841

LNCS Sublibrary: SL5 – Computer Communication Networks and Telecommunications

Printed on acid-free paper

This Springer imprint is published by Springer Nature
The registered company is Springer International Publishing AG
The registered company address is: Gewerbestrasse 11, 6330 Cham, Switzerland

Preface

We welcome you to the joint proceedings of the 17th NEW2AN conference (Next Generation Teletraffic and Wired/Wireless Advanced Networks and Systems) and the 10th conference on Internet of Things and Smart Spaces, ruSMART (Are You Smart), held in St. Petersburg, Russia, on August 28–30, 2017.

Originally, the NEW2AN conference was launched by ITC (International Teletraffic Congress) in St. Petersburg in June 1993 as an ITC-Sponsored Regional International Teletraffic Seminar. The first event was entitled "Traffic Management and Routing in SDH Networks" and held by R&D LONIIS. In 2002, the event received its current name, the NEW2AN. In 2008, NEW2AN acquired a new companion in Smart Spaces, ruSMART, hence boosting interaction between researchers, practitioners, and engineers across different areas of ICT. From 2012, the scope of the ruSMART conference has been extended to cover the Internet of Things and related aspects.

NEW2AN and ruSMART are now well-established conferences with a unique cross-disciplinary mixture of telecommunications-related research and science. They are accompanied by outstanding keynotes from universities and companies across Europe, the USA, and Russia.

The 17th NEW2AN technical program addresses various aspects of next-generation data networks. This year, special attention was given to advanced wireless networking and applications as well as to lower-layer communication enablers. In particular, the authors demonstrated novel and innovative approaches to performance and efficiency analysis of ad hoc and machine-type systems, employed game-theoretical formulations, Markov chain models, and advanced queuing theory. It is also worth mentioning the rich coverage of graphene and other emerging materials, photonics and optics, generation and processing of signals, as well as business aspects.

The 10th conference on Internet of Things and Smart Spaces ruSMART 2017 provided a forum for academic and industrial researchers to discuss new ideas and trends in the emerging areas of the Internet of Things and Smart Spaces that create new opportunities for fully-customized applications and services. The conference brought together leading experts from top affiliations around the world. This year, the event attracted a high level of participation from representatives of various players in the field, including academic teams and industrial world-leader companies, particularly representatives of Russian R&D centers, which have a good reputation for high-quality research and business in innovative service creation and applications development.

The 3rd International Workshop on Nano-scale Computing and Communications (NsCC 2017) aims to foster the advanced development of nanotechnologies through communication and networking at the nanoscale. Via the communication and networking process, nanonetworks have been formed between nanomachines. The interconnection of nanonetworks to the wider Internet is also envisioned for the future, leading to new paradigms known as the Internet of Nano Things or Internet of Bio-Nano Things. However, unlike traditional communication systems, where devices have sufficient processing capabilities, this will be a major challenge at the nanoscale. Therefore, this brings along a new set of challenges, as well as new molecular and nanoscale communication paradigms. This year, the workshop included manuscripts that captured the current state of the art in the field of molecular and nanoscale communications, e.g., information, communication and network theoretical analysis of molecular and nanonetworks, mobility in molecular and nanonetworks, novel and practical communication protocols, routing schemes and architectures, design/engineering/evaluation of molecular and nanoscale communication systems, as well as their potential applications and interconnection to the Internet (e.g., Internet of Nano Things). O. Akan, I. Balasingham, S. Balasubramaniam, M. Barros, and C. Han have made the NsCC 2017 a successful event.

We would like to thank the Technical Program Committee members of all three events, as well as the associated reviewers, for their hard work and important contribution. This year, the papers presented met the highest quality criteria with an acceptance ratio of around 35%.

The conferences and workshop were organized in cooperation with National Instruments, IEEE Communications Society Russia Northwest Chapter, Radiozavod im. A.S. Popova, YL-Verkot OY, Open Innovations Association FRUCT, Tampere University of Technology, St. Petersburg State Polytechnical University, Peoples' Friendship University of Russia (RUDN University), The National Research University Higher School of Economics (HSE), St. Petersburg State University of Telecommunications, and the Popov Society.

We also wish to thank all those who contributed to the organization of the events. In particular, we are grateful to Aleksandr Ometov for his substantial work on supporting the conference website and his excellent job on the compilation of camera-ready papers and interaction with Springer.

We believe that the 17th NEW2AN, 10th ruSMART, and 3rd NsCC conferences delivered an informative, high-quality, and up-to-date scientific program. We also hope that participants enjoyed both technical and social conference components, the Russian hospitality, and the beautiful city of St. Petersburg. The conference is holding in the framework of the RUDN University Competitiveness Enhancement Program "5-100".

August 2017

Olga Galinina
Sergey Andreev
Sergey Balandin
Yevgeni Koucheryavy

Organization

NEW2AN International Advisory Committee

Igor Faynberg	Stargazers Consulting, LLC; Stevens Institute of Technology, USA
Jarmo Harju	Tampere University of Technology, Finland
Villy B. Iversen	Technical University of Denmark, Denmark
Andrey Koucheryavy	State University of Telecommunications, Russia
Kyu Ouk Lee	ETRI, South Korea
Sergey Makarov	St. Petersburg State Polytechnical University, Russia
Svetlana V. Maltseva	National Research University Higher School of Economics, Russia
Mohammad S. Obaidat	Monmouth University, USA
Andrey I. Rudskoy	St. Petersburg State Polytechnical University, Russia
Konstantin Samouylov	Peoples' Friendship University of Russia, Russia
Manfred Sneps-Sneppe	Ventspils University College, Latvia
Michael Smirnov	Fraunhofer FOKUS, Germany
Sergey Stepanov	MTUCI, Russia

NsCC International Advisory Committee

Ozgur Akan	University of Cambridge, UK
Ilangko Balasingham	Norwegian University of Science and Technology, Norway
Sasitharan Balasubramaniam	Tampere University of Technology, Finland and Telecommunication Software and Systems Group, Waterford Institute of Technology, Ireland
Michael Taynnan Barros	Telecommunication Software and Systems Group, Waterford Institute of Technology, Ireland
Chong Han	Shanghai Jiao Tong University, China

NEW2AN, ruSMART and NsCC Technical Program Committee

Naveed Abbasi	Koc University, Turkey
Bayram Akdeniz	Bogazici University, Turkey
Hassen Alsafi	IIUM, Malaysia
Baris Atakan	Izmir Institute of Technology, Turkey
Konstantin Avrachenkov	Inria Sophia Antipolis, France
Sergey Balandin	FRUCT Oy, Finland
Michael Barros	Waterford Institute of Technology, Ireland
Kalil Bispo	Federal University of Sergipe, Brazil
Jose Carrera	University of Bern, Switzerland

Paulo Carvalho	Centro Algoritmi, Universidade do Minho, Portugal
Oktay Cetinkaya	Koc University, Turkey
Youssef Chahibi	Georgia Institute of Technology, USA
Wei Koong Chai	Bournemouth University, UK
Ji-Woong Choi	DGIST, South Korea
Chrysostomos Chrysostomou	Frederick University, Cyprus
Meltem Civas	Koc University, Turkey
Gianpaolo Cugola	Politecnico di Milano, Italy
Bruno Dias	Universidade do Minho, Portugal
Roman Dunaytsev	The Bonch-Bruevich Saint-Petersburg State University of Telecommunications, Russia
Lyndon Fawcett	Lancaster University, UK
António Fernandes	Centro Algoritmi, Universidade do Minho, Portugal
Dieter Fiems	Ghent University, Belgium
Ivan Ganchev	University of Limerick, Ireland
Margarita Gapeyenko	Tampere University of Technology, Finland
Mikhail Gerasimenko	Tampere University of Technology, Finland
Wolfgang Gerstacker	University of Erlangen-Nuernberg, Germany
Regina Gumenyuk	TUT, Finland
Mustafa Gursoy	Bogazici University, Turkey
Chong Han	Shanghai Jiao Tong University, China
Matthew Higgins	University of Warwick, UK
Philipp Hurni	University of Bern, Switzerland
Pedram Johari	University at Buffalo (SUNY), USA
Eirini Kalogeiton	University of Bern, Switzerland
Mostafa Karimzadeh	University of Bern, Switzerland
Alexey Kashevnik	SPIIRAS, Russia
Andreas J. Kassler	Karlstad University, Sweden
Tooba Khan	Koc University, Turkey
Geun-Hyung Kim	Dong Eui University, South Korea
Mikhail Komarov	National Research University Higher School of Economics, Russia
Alexey Koren	State University of Aerospace Instrumentation, Russia
Andrey Koucheryavy	SPbSUT, Russia
Kirill Krinkin	St.-Petersburg State Electrotechnical University LETI, Russia
Aravindh Krishnamoorthy	Friedrich-Alexander-Universität Erlangen-Nürnberg, Germany
Mark Leeson	University of Warwick, UK
Elena Simona Lohan	Tampere University of Technology, Finland
Nuno Lopes	University of Minho, Portugal
Valeria Loscrì	Inria Lille-Nord Europe, France
Ninoslav Marina	Princeton University, USA
Ricardo Martini	University of Minho, Portugal
Daniel Martins	Waterford Institute of Technology, Ireland

Pavel Masek	Brno University of Technology, Czech Republic
Diomidis Michalopoulos	Nokia Bell Labs, Germany
Dmitri Moltchanov	Tampere University of Technology, Finland
Edmundo Monteiro	University of Coimbra, Portugal
Tadashi Nakano	Osaka University, Japan
Shuai Nie	Georgia Institute of Technology, USA
Aleksandr Ometov	Tampere University of Technology, Finland
Antonino Orsino	Oy L M Ericsson AB, Finland
Mustafa Ozger	Koc University, Turkey
Michele Pagano	University of Pisa, Italy
David Perez Abreu	University of Coimbra, Portugal
Dmitry Petrov	Nokia, Finland
Vitaly Petrov	Tampere University of Technology, Finland
Edison Pignaton de Freitas	Federal University of Rio Grande do Sul, Brazil
Alexander Pyattaev	YL Verkot, Finland
Nicholas Race	Lancaster University, UK
Giuseppe Raffa	Intel Corporation, USA
Josè Carlos Ramalho	Centro Algoritmi, Universidade do Minho, Portugal
Gianluca Reali	University of Perugia, Italy
Simon Pietro Romano	University of Napoli Federico II, Italy
Andrey Samuylov	Tampere University of Technology, Finland
Robert Schober	Friedrich-Alexander University Erlangen-Nuremberg, Germany
Nikolay Shilov	SPIIRAS, Russia
João Marco Silva	HASLab, INESC TEC & Universidade do Minho, Portugal
Alexander Smirnov	SPIIRAS, Russia
Dmitrii Solomitckii	Tampere University of Technology, Finland
Ales Svigelj	Jozef Stefan Institute, Slovenia
Takeshi Takahashi	National Institute of Information and Communications Technology, Japan
Jouni Tervonen	University of Oulu, Finland
Bige Deniz Unluturk	Georgia Institute of Technology, USA
Anna Maria Vegni	Università Roma 3, Italy
Karima Velasquez	University of Coimbra, Portugal
Chao-Chao Wang	Zhejiang University of Technology, China
Wei Wei	Xi'an University of Technology, China
Xin-Wei Yao	Zhejiang University of Technology, China
Qussai Yaseen	Jordan University of Science and Technology, Jordan
Haiyang Zhang	University of Limerick, Ireland

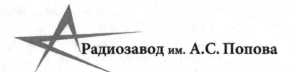

Contents

New Generation of Smart Services (ruSMART 2017)

Developing of Emerging Internet Applications for Home Healthcare   3
  Konstantin G. Korotkov, Konstantin P. Semenov, Victor I. Malyugin,
  and Dmitry V. Kiesewetter

Digital Business Model and SMART Economy Sectoral Development
Trajectories Substantiation   13
  L.A. Ismagilova, T.A. Gileva, M.P. Galimova, and V.V. Glukhov

Ontology Matching for Socio-Cyberphysical Systems:
An Approach Based on Background Knowledge   29
  Alexander Smirnov, Nikolay Teslya, Sergey Savosin, and Nikolay Shilov

Battery Monitoring Within Industry 4.0 Landscape:
Solution as a Service (SaaS) for Industrial Power Unit Systems   40
  Mathieu Devos and Pavel Masek

Opportunistic Data Collection for IoT-Based Indoor
Air Quality Monitoring   53
  Aigerim Zhalgasbekova, Arkady Zaslavsky, Saguna Saguna,
  Karan Mitra, and Prem Prakash Jayaraman

The IoT Identification Procedure Based on the Degraded Flash
Memory Sector   66
  Sergey Vladimirov and Ruslan Kirichek

DisCPAQ: Distributed Context Acquisition and Reasoning for Personalized
Indoor Air Quality Monitoring in IoT-Based Systems   75
  Tamara Belyakhina, Arkady Zaslavsky, Karan Mitra, Saguna Saguna,
  and Prem Prakash Jayaraman

Cloud Computing Solution for Investment Efficiency Measurement
in Biomedicine   87
  Petra Maresova and Vladimir Sobeslav

Time Series Distributed Analysis in IoT with ETL and Data Mining
Technologies   97
  Ivan Kholod, Maria Efimova, Andrey Rukavitsyn, and Shorov Andrey

Exploring SDN to Deploy Flexible Sampling-Based Network Monitoring . . .   109
  Catarina Pires da Silva, Solange Rito Lima, and João Marco Silva

VNF Orchestration and Modeling with ETSI MANO
Compliant Frameworks   121
   Ales Komarek, Jakub Pavlik, Lubos Mercl, and Vladimir Sobeslav

Performance Evaluation of OpenFlow Enabled Commodity and Raspberry
pi Wireless Routers   132
   Muhammad Zeeshan Asghar, M. Ahsan Habib, and Timo Hämäläinen

Design Issues of Information and Communication Systems for New
Generation Industrial Enterprises   142
   Valery Leventsov, Anton Radaev, and Nikolay Nikolaevskiy

SWM-PnR: Ontology-Based Context-Driven Knowledge Representation
for IoT-Enabled Waste Management   151
   Inna Sosunova, Arkady Zaslavsky, Theodoros Anagnostopoulos,
   Petr Fedchenkov, Oleg Sadov, and Alexey Medvedev

Supporting Data Communications in IoT-Enabled Waste Management   163
   Petr Fedchenkov, Arkady Zaslavsky, Alexey Medvedev,
   Theodoros Anagnostopoulos, Inna Sosunova, and Oleg Sadov

**International Workshop on Nano-Scale Computing
and Communications (NsCC 2017)**

Fiber-Optic Transmission System for the Testing of Active Phased Antenna
Arrays in an Anechoic Chamber   177
   Roman V. Davydov, Ivan K. Saveleiv, Vladimir A. Lenets,
   Margarita Yu. Tarasenko, Tatiana R. Yalunina, Vadim V. Davydov,
   and Vasily Yu. Rud'

Advanced Materials for Fiber Communication Systems   184
   Victor A. Klinkov, Alexandr V. Semencha, and Evgenia A. Tsimerman

Dynamic Data Packaging Protocol for Real-Time Medical Applications
of Nanonetworks.   196
   Rustam Pirmagomedov, Mikhail Blinnikov, Ruslan Glushakov,
   Ammar Muthanna, Ruslan Kirichek, and Andrey Koucheryavy

Nano Communication Device with an Embedded Molecular Film:
Electromagnetic Signals Integration with Dynamic
Operation Photodetector   206
   Dmitrii Dyubo and Oleg Yu. Tsybin

A Formal Definition for Nanorobots and Nanonetworks.   214
   Florian Büther, Florian-Lennert Lau, Marc Stelzner,
   and Sebastian Ebers

Features of Use Direct and External Modulation in Fiber Optical Simulators
of a False Target for Testing Radar Station   227
  Margarita Yu. Tarasenko, Vadim V. Davydov, Vladimir A. Lenets,
  Natalya V. Akulich, and Tatyana R. Yalunina

**Next Generation Wired/Wireless Advanced Networks and Systems
(NEW2AN 2017)**

On Detection of Network-Based Co-residence Verification Attacks
in SDN-Driven Clouds.   235
  Mikhail Zolotukhin, Elena Ivannikova, and Timo Hämäläinen

Health-Care Pervasive Environments: A CLA Based Trust Management   247
  Omid Bushehrian and Shayeste Esmail Nejad

Physical-Layer Security for DF Two-Way Dual-Hop Cooperative Wireless
Networks over Nakagami-m Fading Channels.   258
  Islam M. Tanash, Mamoun F. Al-Mistarihi, and Amer M. Magableh

Physical-Layer Security for DF Two-Way Full-Duplex Cooperative
Wireless Networks over Rayleigh Fading Channels   270
  Islam M. Tanash, Amer M. Magableh, and Mamoun F. Al-Mistarihi

DNS Tunneling Detection Techniques – Classification, and Theoretical
Comparison in Case of a Real APT Campaign   280
  Viivi Nuojua, Gil David, and Timo Hämäläinen

Digital Watermarking Method Based on Image Compression Algorithms   292
  Sergey Bezzateev and Natalia Voloshina

Development of the Credit Risk Assessment Mechanism of Investment
Projects in Telecommunications   300
  Sergei Grishunin and Svetlana Suloeva

High-Tech Sector in the Conditions of Institutionalization of the Smart
Economy (on the Example of the Telecommunication Industry)   315
  N.V. Vasilenko, A.J. Linkov, and V.V. Glukhov

Virtual Telecommunication Enterprises and Their Risk Assessment   326
  N.V. Apatova, O.V. Boychenko, Tatyana P. Nekrasova, and S.V. Malkov

Peculiarities of Creation of Information System at the Enterprises
of Telecommunication Branch   337
  Ye. Yu. Vinogradova, A.I. Galimova, Natalya V. Mukhanova,
  and S.L. Andreeva

A Game-Theoretic Model for Investments
in the Telecommunications Industry   351
  Sergey A. Chernogorskiy and K.V. Shvetsov

Business Perception Based on Sentiment Analysis Through Deep Neuronal
Networks for Natural Language Processing   365
  Mónica Pineda Vargas, Octavio José Salcedo Parra,
  and Miguel José Espitia Rico

Cognitive Models for Access Network Management   375
  Vladimir Akishin, Alex Goldstein, and Boris Goldstein

Bargaining in a Dual Radar and Communication System
Using Radar-Prioritized OFDM Waveforms   382
  Andrey Garnaev, Wade Trappe, and Athina Petropulu

Discrete Time Bulk Service Queue for Analyzing LTE Packet Scheduling
for V2X Communications   395
  Vitalii Beschastnyi, Valeriy Naumov, Pasquale Scopelliti,
  Irina Gudkova, Claudia Campolo, Giuseppe Araniti, Iliya Dzantiev,
  and Konstantin Samouylov

Structure Analysis of an Explanatory Dictionary Ontological Graph   408
  Yu. N. Orlov and Yu. A. Parfenova

Performance Optimization of a Clustering Adaptive Gravitational Search
Scheme for Wireless Sensor Networks.   420
  Elham Pourabdollah, Reza Mohammadi Asl, and Theodore Tsiligiridis

A Retrial Queueing System with Preemptive Priority and Randomized
Push-Out Mechanism.   432
  Alexander Ilyashenko, Oleg Zayats, Maria Korenevskaya,
  and Vladimir Muliukha

NGN/IMS and post-NGN Management Model   441
  Alex Goldstein

The Principles of Antennas Constructive Synthesis in Dissipative Media   455
  Roman U. Borodulin, Boris V. Sosunov, and Sergey B. Makarov

Amplitude and Phase Stability Measurements of Multistage
Microwave Receiver   466
  Yuriy V. Vekshin and Alexander P. Lavrov

Evolving Toward Virtualized Mobile Access Platform
for Service Flexibility   473
  Seung-Que Lee and Jinup Kim

Model of Photonic Beamformer for Microwave Phased Array Antenna   482
  Sergey I. Ivanov, Alexander P. Lavrov, and Igor I. Saenko

Nanosecond Miniature Transmitters for Pulsed Optical Radars   490
  Alexey V. Filimonov, Valery E. Zemlyakov, Vladimir I. Egorkin,
  Andrey V. Maslevtsov, Marc Christopher Wurz,
  and Sergey N. Vainshtein

Fog Computing for Telemetry Gathering from Moving Objects.   498
  Ivan Kholod, Nikolai Plokhoi, and Andrey Shorov

Stability and Delay of Algorithms of Random Access
with Successive Interference Cancellation.   510
  Nikolay Apanasenko, Nikolay Matveev, and Andrey Turlikov

Throughput Analysis of Adaptive ALOHA Algorithm Using Hybrid-ARQ
with Chase Combining in AWGN Channel   519
  Artem Burkov, Nikolay Kuropatkin, and Nikolay Matveev

Characterizing Time-Dependent Variance and Coefficient
of Variation of SIR in D2D Connectivity.   526
  Anastasia Ivchenko, Yuri Orlov, Andrey Samouylov, Dmitri Molchanov,
  and Yuliya Gaidamaka

Analysis of Admission Control Schemes Models for Wireless Network
Under Licensed Shared Access Framework   536
  Ekaterina Markova, Dmitry Poluektov, Darya Ostrikova, Irina Gudkova,
  Iliya Dzantiev, Konstantin Samouylov, and Vsevolod Shorgin

Low-Complexity Iterative MIMO Detection Based
on Turbo-MMSE Algorithm.   550
  Mikhail Bakulin, Vitaly Kreyndelin, Andrey Rog, Dmitry Petrov,
  and Sergei Melnik

Rubidium Atomic Clock with Improved Metrological Characteristics
for Satellite Communication System   561
  Alexander A. Petrov, Vadim V. Davydov, Nikita S. Myazin,
  and Vladislav E. Kaganovskiy

Provision of Connectivity for (Heterogeneous) Self-organizing Network
Using UAVs   569
  Alexander Paramonov, Ilhom Nurilloev, and Andrey Koucheryavy

Performance Evaluation of COPE-like Network Coding in Flying Ad Hoc
Networks: Simulation-Based Study   577
  Danil S. Vasiliev, Irina A. Kaysina, and Albert Abilov

Communication Technologies in the Space Experiment "Kontur-2"   587
Vladimir Muliukha, Vladimir Zaborovsky, Alexander Ilyashenko,
and Yuri Podgurski

Accuracy of Secondary Surveillance Radar System Remote
Analysis Station   598
Igor A. Tsikin and Ekaterina S. Poklonskaya

Spectral and Energy Efficiency of Optimal Signals with Increased Duration,
Providing Overcoming "Nyquist Barrier"   607
Anna S. Ovsyannikova, Sergey V. Zavjalov, Sergey B. Makarov,
Sergey V. Volvenko, and Trinh Luong Quang

Choosing Parameters of Optimal Signals with Restriction
on Correlation Coefficient   619
Anna S. Ovsyannikova, Sergey V. Zavjalov, Sergey B. Makarov,
and Sergey V. Volvenko

The Filtration of Composite Signals from Interference by the Maximum
Likelihood Method   629
Kseniia V. Vlasova, Valerii A. Pakhotin, Evgenii V. Korotey,
and Sergey B. Makarov

Direct Signal Processing for GNSS Integrity Monitoring   635
Igor A. Tsikin and Antonina P. Melikhova

An Intentional Introduction of ISI Combined with Signal Constellation Size
Increase for Extra Gain in Bandwidth Efficiency   644
Van Phe Nguyen, Anton Gorlov, and Aleksandr Gelgor

Foreground Detection Using Region of Interest Analysis
Based on Feature Points Processing   653
Nikita Ustyuzhanin and Marat Gilmutdinov

Modified EM-Algorithm for Motion Field Refinement in Motion
Compensated Frame Interpoliation   662
Nikolay Nemcev and Marat Gilmutdinov

The Models of Moving Users and IoT Devices Density Investigation
for Augmented Reality Applications   671
M. Makolkina, A. Koucheryavy, and A. Paramonov

Quality of Experience Estimation for Video Service Delivery
Based on SDN Core Network   683
Maria Makolkina, Ammar Muthanna, and Steve Manariyo

LTE MCS Cell Range and Downlink Throughput Measurement
and Analysis in Urban Area   693
    Yi Hua Chen, Kai Jen Chen, and Jyun Jhih Yang

Transfer of Multimedia Data via LoRa   708
    Ruslan Kirichek, Van-Dai Pham, Aleksey Kolechkin,
    Mahmood Al-Bahri, and Alexander Paramonov

Practical Results of WLAN Traffic Analysis   721
    A. Paramonov, A. Vikulov, and S. Scherbakov

Wi-Fi Based Indoor Positioning System Using Inertial Measurements   734
    Mstislav Sivers, Grigoriy Fokin, Pavel Dmitriev, Artem Kireev,
    Dmitry Volgushev, and Al-odhari Abdulwahab Hussein Ali

Power Allocation in Cognitive Radio with Distributed Antenna System   745
    Jerzy Martyna

Multi-level Cluster Based Device-to-Device (D2D) Communication
Protocol for the Base Station Failure Situation   755
    Abdelhamied A. Ateya, Ammar Muthanna, Anastasia Vybornova,
    and Andrey Koucheryavy

Author Index   767

New Generation of Smart Services
(ruSMART 2017)

Developing of Emerging Internet Applications for Home Healthcare

Konstantin G. Korotkov[1(✉)], Konstantin P. Semenov[1],
Victor I. Malyugin[2], and Dmitry V. Kiesewetter[2]

[1] Saint Petersburg Federal University of Information Technologies,
Mechanics and Optics, Saint Petersburg, Russia
korotkov2000@gmail.com
[2] Peter the Great Saint Petersburg Polytechnic University,
Saint Petersburg, Russia

Abstract. The system of collection, transmission and processing of data for wearable sensors for personal health care using the heart rate monitoring, skin temperature and electrophotonic imaging (EIP) sensors is described. EPI instrument based on the stimulation of photon and electron emissions from the surface of the object is used. The paper describes the interaction of the system components, the methods and hardware implementation of data transfer, information about computer software for working with database servers and client terminals as well as components of a remote patient monitoring system based on cloud architecture. Examples of information presented to the patients and doctors are presented.

Keywords: Data acquisition · Cloud processing · Data transmission · Internet · Electrophotonic imaging sensor · Smart environments · Remote health monitoring

1 Introduction

Recent years have seen a rising interest in wearable sensors and today several devices are commercially available [1–3] for personal health care, fitness, and activity awareness. In addition to the niche recreational fitness arena catered to by current devices, researchers have also considered applications of such technologies in clinical applications in remote health monitoring systems for long term recording, management and clinical access to patient's physiological information [4–8]. In home health care applications the sensors would several times during the day record signals correlated with person's key physiological parameters and relay the resulting data to a database linked with his/her health records. The doctor would has available not only conventional clinic/lab-test based on static measurements of person's physiological and metabolic state, but also the much richer longitudinal record provided by the sensors. Using the available data, and aided by decision-support systems that also have access to a large corpus of observation data for other individuals, the doctor can make a much better prognosis for patient health and recommend treatment, early intervention, and life-style choices that are particularly effective in improving the quality of health. Such

O. Galinina et al. (Eds.): NEW2AN/ruSMART/NsCC 2017, LNCS 10531, pp. 3–12, 2017.
DOI: 10.1007/978-3-319-67380-6_1

a disruptive technology could have a transformative impact on global healthcare systems and drastically reduce healthcare costs and improve speed and accuracy for diagnoses [9].

Technologically, the vision presented in the preceding paragraph has been feasible for a few years now. Yet, wearable sensors have, thus far, had little influence on the current clinical practice of medicine. In this paper, we focus particularly on developed by our team approach to developing home appliances for individual healthcare monitoring based on Internet data storage and processing.

2 Demands to System Architecture

Most proposed frameworks for remote health monitoring leverage a three tier architecture: a Wireless Body Area Network (WBAN) consisting of wearable sensors as the data acquisition unit, communication and networking and the service layer [4, 7, 8, 10]. Most of currently developed devices propose a system that recruits wearable sensors to measure various physiological parameters such as blood pressure and body temperature [11]. Sensors transmit the gathered information to a gateway server through a Bluetooth connection. The gateway server turns the data into an Observation and Measurement file and stores it on a remote server for later retrieval by patient or clinicians through the Internet. Utilizing a similar cloud based medical data storage a health monitoring system is presented in [12] in which medical staff can access the stored data online through content service application. Targeting a specific medical application, an end to end remote health monitoring and analytics system is presented for supervision of patients with high risk of heart failure [13]. In addition to the technology for data gathering, storage and access, data analysis and visualization are critical components of remote health monitoring systems. Accurate diagnoses and monitoring of patient's medical condition relies on analysis of medical records containing various physiological characteristics over a long period of time. The use of data mining and visualization techniques had previously been addressed as a solution to the aforementioned challenge [14, 15]; these methods have only recently gained attention in remote health monitoring systems [16, 17].

Figure 1 illustrates the system architecture for a remote health monitoring system, whose major components we describe next.

Data Acquisition is performed by multiple wearable sensors that usually measure physiological biomarkers, such as heart rate, skin temperature, respiratory rate, EMG (muscle activity), and gait (posture). In the nearest future such important parameters as sweating, blood pressure, blood Glucose and oxygenation may be added.

The sensors connect to the network though an intermediate data aggregator or concentrator, which is typically a smart phone located in the vicinity of the patient.

The Data Transmission components of the system are responsible for conveying recordings of the patient from the patient's house (or any remote location) to the Internet storage with assured security and privacy, ideally in near real-time. Typically, the sensory acquisition platform is equipped with a short range radio such as Zigbee or low-power Bluetooth, which it uses to transfer sensor data to the concentrator.

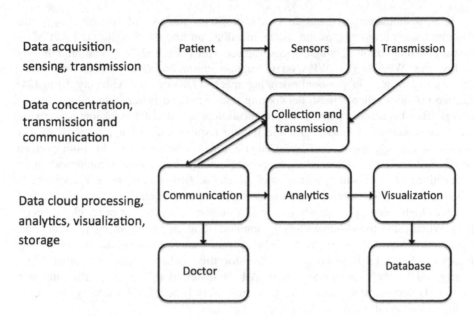

Fig. 1. Components of a remote patient monitoring system based on Cloud architecture.

Sensors in the data acquisition part form an Internet of Things (IoT)-based architecture as each individual sensor's data can be accessed through the Internet via the concentrator [18, 19].

Cloud Processing has three distinct components: storage, analytics, and visualization. The system is designed for long term storage of patient's biomedical information as well assisting health professionals with diagnostic information. Cloud based medical data storage and the upfront challenges have been extensively addressed in the literature [20, 21]. Additionally, visualizationis a key requirement for any such system because it is impractical to ask physicians to pore over the voluminous data or analyses from wearable sensors. Visualization methods that make the data and analyses accessible to them in a readily digestible format are essential if the wearable sensors are to impact clinical practice.

Physiological data is acquired by wearable devices that combine miniature sensors capable of measuring various physiological parameters, minor preprocessing hardware and a communications platform for transmitting the measured data. The wearability requirement, poses physical limitations on the design of the sensors. The sensors must be light, small, and should not hinder a patient's movements and mobility. Also, because they need to operate on small batteries included in the wearable package, they need to be energy efficient. Though the battery may be rechargeable or replaceable, for convenience it is highly desirable that they provide extended durations of continuous operation without requiring charging or replacement. Recent designs [5, 22, 23] of flexible sensors that can be placed in contact with the skin in different body parts are particularly attractive for medical applications because, compared to alternatives, the close contact with the skin allows measurement of more physiological parameters and with greater accuracy.

Energy limitation of these devices necessitates the use of suitable low power communication protocols, as the communication can account a significant part of the power consumption in sensing devices. ZigBee over IEEE 802.15.4 is commonly used in low rate WPANs (LR-WPANs) to support communication between low power devices that operate in personal operating space (POS) of approximately 10 m [24]. ZigBee provides reliable mesh networking with extended battery life. Bluetooth low energy (BLE) is another wireless communication protocol suitable for low power short range communication suitable for the unique requirements of applications such as health monitoring, sports, and home entertainment. The original Bluetooth protocol (IEEE 802.15.1) was designed to support relatively short range communication for applications of a streaming nature, such as audio. BLE modifies the framework by utilizing much longer sleep intervals to decrease the overall energy consumption. BLE achieves higher energy efficiency in terms of number of bytes sent per Joule of energy [25]. When using the aforementioned communication protocols, an intermediate node (data concentrator) is necessary to make sensors data and control accessible through Internet. Most logical solution is to use for this mobile phones or tablets. Data aggregated by the concentrator needs to be transferred to the cloud for long term storage. Offloading data storage to the cloud offers benefits of scalability and accessibility on demand, both by patients and clinicians.

Secure Data Storage in the Cloud: Privacy is of tremendous importance when storing individual's electronic medical records on the cloud. According to the terms defined by Health Insurance Portability and Accountability Act (HIPAA), the confidential part of medical records has to be protected from disclosure. When the medical records are outsourced to the cloud for storage, appropriate privacy preserving measures need to be taken to prevent unauthorized parties from accessing the information. Secure cloud storage frameworks have therefore been proposed for use with sensitive medical records [26, 27]. Standard encryption techniques can be employed to ensure security in such settings [28].

It is impractical to ask physicians to pore over the voluminous data or analyses from IoT-based sensors. To be useful in clinical practice, the results from the analytics data processing need to be presented to physicians in an intuitive format where they can readily comprehend the inter-relations between quantities and eventually start using the sensory data in their clinical practice. Visualization is recognized as an independent and important research area with a wide array of applications in both science and day to day life [29]. Of course, format of the data and diagnostic conclusions presented to physician and to the patient, should be different.

3 Practical Implementation

We have developed a system of home health care application based on combination of several sensors using presented above methodology. At the moment for the first prototype we have implemented the heart rate monitoring, skin temperature and Electrophotonic imaging (EPI) sensors. Heart rate monitoring allows applying Heart Rate Variability analysis [30, 31] which gives valuable information on the activity of the autonomic nervous system. Monitoring of skin temperature during the day allows to follow up circadian rhythm of a person's physiology, and EPI analysis provides complex information on psycho-physiological condition of a person.

EPI technique is a novel Russian technology which have been developed in Saint Petersburg Polytechnical University from 1980-th [32–35]. EPI instrument (previous name is gas discharge visualization - GDV) is based on the stimulation of photon and electron emissions from the surface of the object. This process is called 'photo-electron emissions' and it has been thoroughly studied with physical electronic methods. The emitted particles accelerate in the electromagnetic field, generating electronic avalanches on the surface of the dielectric (glass) plate. This process is called 'sliding gas discharge'. The discharge causes glow from the excitement of molecules in the surrounding gas, and this glow is what is being measured by the EPI instrument. Voltage pulses stimulate optoelectronic emission, while intensifying this emission in the gas discharge. Electrophotonic is a promising direction in the construction of new non-invasive automatic procedures that assess the body's condition during minimum interventions into vital functions. Electrophotonic method investigates human functional states, by assessing electro-optical parameters of the skin that are based on the registration of physical processes emerging from electron components of tissue conductivity. This technique allows one to capture the image of plasma glow around human fingertips and extract information about sympathetic and parasympathetic activities. The biophysical concept of the principles of measurements is based on the ideas of quantum biophysics [36].

Electrophotonic technique is being used in different areas [37–39]; in particular: in medicine [40–42]; in monitoring energy levels during different treatments [43, 44]; in evaluation of cognitive functions [45–47]; and study of different materials [40–50]. The latest EPI instrument named "Bio-Well" [51–53] is actively used in preparation of the Russian Olympic and Paralympic teams [54, 55]. It was shown that evaluation of psycho-physiological condition of elite athletes' at various stages of preparation and in international competition was conducted. It was shown that computer analysis of EPI information offers a fast, highly precise, non-invasive method to assess an athlete's level of readiness during both training and at the time of competition. To the moment more than 3000 specialists worldwide are using Electrophotonic technique.

All sensors transmit data in real time mode via Bluetooth low energy communication protocol to the mobile device – phone, tablet or computer. This testing version is using Windows platform, but we plan to expand to MacOS and Android operational systems. Data aggregated by these devices transferred to the cloud for processing and long term storage.

For cloud processing we are using Amazon server, which provides sufficient communication speed and practically unlimited storage facilities. The program is built on network architecture, on the three-level principle, where services inside the cloud consist of three broad categories: storage, data processing, and communication with the customer [56]. It has the following components:

- client programing (terminal);
- program for the connection to the server;
- program for the connection to the database server.

The client is an interface graphical component, representing the first level, the application for the end user. This level has no direct links with the database and does not store application states. It has simple business logic: login interface, encryption, verification of input values for validity and compliance with the format, simple operations (sorting, grouping, etc.) with the data already downloaded to the terminal, as well as receiving data from the sending device to the server and data printing. The Application Server is a second level, where the bulk of business logic and programming interfaces are stored, connecting the client components with the application database logic. At the third level, the database server provides data storage (standard relational database management system). The application server and the database server in this case is physically located on the same server running CentOS with installed Apache web server and PHP 5.X. (the so-called LAMP). The application server is divided into two parts, which are responsible for different functions:

- Parametric modules.
- Modules that implement business logic, as well as network sharing with clients.

Owing to the fact that the calculation of the parameters is quite difficult from a computational point of view, modules that implement this task require maximum performance. The modules are written in standard C++, and gathered in the executable files. Each module takes as arguments of the command-line some of the parameters and outputs the result to the specified folder. Source texts are provided with the description for cmake utilities [57], which creates the necessary project files for the selected platform and compiler. All Bio-grams are stored on the server as PNG images, so as a part of the source code libraries LIBPNG [58] and ZLIB [59] are included. Each module takes as a parameter the path to the folder, wherein are stored raw data from the sensors and some other technical parameters, and forms an output as ini-file in the specified directory. The whole complex of the modules allows client applications to interact with the system through a set of exported functions. Interaction with client applications is done via SOAP protocol [60], while the description of the interface itself is implemented using WSDL language.

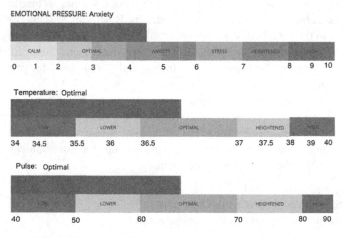

Fig. 2. Example of information presented to the patient.

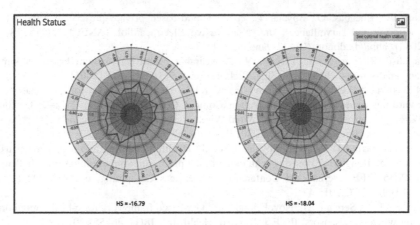

Fig. 3. Example of information presented to the doctor.

Program allows to the patient to see only limited information, such as heart rate, temperature, stress level and energy level (Fig. 2), while doctor have access to a big array of information stored in the Internet Database (Fig. 3). System is being tested in test mode and after approval will be available to the users.

4   Conclusion

In this paper, we reviewed the current state and projected future directions for integration of remote health monitoring technologies into the clinical practice of medicine and presented our developed approach in this field. Wearable sensors offer attractive options for enabling observation and recording of data in home and work environments, over much longer durations than are currently done at office and laboratory visits. This treasure trove of data, when analyzed and presented to physicians in easy-to-assimilate visualizations has the potential for radically improving healthcare and reducing costs.

References

1. Jawbone Inc.: Jawbone fitness trackers. https://jawbone.com/up/trackers. Accessed Apr 2015
2. FitBit Inc.: Flex: Wireless activity + sleep wristband. https://www.fitbit.com/flex. Accessed Apr 2015
3. Apple Inc.: Apple watch. https://www.apple.com/watch/. Accessed Apr 2015
4. Pantelopoulosand, A., Bourbakis, N.: A survey on wearable sensor-based systems for health monitoring and prognosis. IEEE Trans. Syst. Man Cybern. Part C Appl. Rev. **40**(1), 1–12 (2010)
5. Son, D., Lee, J., Qiao, S., et al.: Multifunctional wearable devices for diagnosis and therapy of movement disorders. Nat. Nanotechnol. **9**(5), 1–8 (2014)

6. Page, A., Kocabas, O., Soyata, T., et al.: Cloud-based privacy-preserving remote ECG monitoring and surveillance. Ann. Noninvasive Electrocardiol. (ANEC) **20**(4), 328–337 (2014). http://dx.doi.org/10.1111/anec.12204
7. Paradiso, R., Loriga, G., Taccini, N.: A wearable health care system based on knitted integrated sensors. IEEE Trans. Inf. Tech. Biomed. **9**(3), 337–344 (2005)
8. Milenkovi, A., Otto, C., Jovanov, E.: Wireless sensor networks for personal health monitoring: issues and an implementation. Comput. Commun. **29**(1314), 2521–2533 (2006). (Wireless Sensors Networks and Wired/Wireless Internet Communications. http://www.sciencedirect.com/science/article/pii/S0140366406000508)
9. Hassanalieragh, M., Page, A., Soyata, T., et al.: Health monitoring and management using Internet-of-Things (IoT) sensing with cloud-based processing: opportunities and challenges. In: 2015 IEEE International Conference on Services Computing, pp. 285–292 (2015). doi:10.1109/SCC.2015.47
10. Benharref, A., Serhani, M.: Novel cloud and SOA-based framework for e-health monitoring using wireless biosensors. IEEE J. Biomed. Health Inf. **18**(1), 46–55 (2014)
11. Babu, S., Chandini, M., Lavanya, P., Ganapathy, K., Vaidehi, V.: Cloud-enabled remote health monitoring system. In: International Conference Recent Trends Information Technology (ICRTIT), pp. 702–707 (2013)
12. Rolim, C., Koch, F., Westphall, C., Werner, J., Fracalossi, A., Sal-vador, G.: A cloud computing solution for patient's data collection in health care institutions. In: Second International Conference eHealth Telemedicine Society Medical, ETELEMED 2010, pp. 95–99 (2010)
13. Lan, M., Samy, L., Alshurafa, N., Suh, M.-K. et al.: Wanda: an end-to-end remote health monitoring and analytics system for heart failure patients. In: Proceedings of Conference Wireless Health Services, WH 2012, pp. 9:1–9:8. ACM, New York (2012)
14. Wei, L., Kumar, N., Lolla, V., Keogh, E., et al.: A practical tool for visualizing and data mining medical time series. In: Proceedings of 18th IEEE Symposium on Computer-Based Medical System, pp. 341–346 (2005)
15. Mao, Y., Chen, Y., Hackmann, G., et al.: Medical data mining for early deterioration warning in general hospital wards. In: IEEE 11th International Conference on Data Mining Workshops (ICDMW), pp. 1042–1049 (2011)
16. Ukis, V., Rajamani, S., Balachandran, B., Friese, T.: Architecture of cloud-based advanced medical image visualization solution. In: IEEE International Conference on Cloud Computing in Emerging Markets (CCEM), pp. 1–5, October 2013
17. Rao, B.: The role of medical data analytics in reducing health fraud and improving clinical and financial outcomes. In: 2013 IEEE 26th International Symposium on Computer-Based Medical Systems (CBMS), p. 3, June 2013
18. Zhao, W., Wang, C., Nakahira, Y.: Medical application on Internet of Things. In: IET International Conference on Communication Technology and Application, ICCTA 2011, pp. 660–665, October 2011
19. Hu, F., Xie, D., Shen, S.: On the application of the Internet of Things in the field of medical and health care. In: IEEE International Conference on Green Computing and Communications (GreenCom) and IEEE Cyber, Physical and Social Computing, (iThings/CPSCom), pp. 2053–2058, August 2013
20. Nalinipriya, G., Aswin Kumar, R.: Extensive medical data storage with prominent symmetric algorithms on cloud - a protected framework. In: IEEE International Conference on Smart Structures and Systems (ICSSS), pp. 171–177, March 2013
21. Hani, A.F.M., Paputungan, I.V., Hassan, M.F., et al.: Development of private cloud storage for medical image research data. In: International Conference on Computer and Information Sciences (ICCOINS), pp. 1–6, June 2014

22. Xu, S., Zhang, Y., Jia, L., et al.: Soft microfluidic assemblies of sensors, circuits, and radios for the skin. Science **344**, 70–74 (2014)
23. Kim, D.-H., Ghaffari, R., Lu, N., Rogers, J.A.: Flexible and stretchable electronics for biointegrated devices. Ann. Rev. Biomed. Eng. 113–128 (2012)
24. Lee, J.-S., Su, Y.-W., Shen, C.-C.: A comparative study of wireless protocols: Bluetooth, UWB, ZigBee, and Wi-Fi. In: 33rd Annual Conference of the IEEE Industrial Electronics Society, IECON 2007, pp. 46–51, November 2007
25. Siekkinen, M., Hiienkari, M., Nurminen, J., Nieminen, J.: How low energy is Bluetooth low energy? Comparative measurements with Zig-Bee/802.15.4. In: Wireless Communications and Networking Conference Workshops (WCNCW), pp. 232–237. IEEE, April 2012
26. Li, M., Yu, S., Zheng, Y., Ren, K., Lou, W.: Scalable and secure sharing of personal health records in cloud computing using attribute-based encryption. IEEE Trans. Parallel Distrib. Syst. **24**(1), 131–143 (2013)
27. Ruj, S., Stojmenovic, M., Nayak, A.: Privacy preserving access control with authentication for securing data in clouds. In: Proceedings of the 2012 12th IEEE/ACM International Symposium on Cluster, Cloud and Grid Computing, CCGRID 2012, pp. 556–563. IEEE Computer Society, Washington, DC (2012)
28. National Institute of Standards and Technology: Advanced encryption standard (AES), November 2001. fIPS-197
29. Tufte, E.R., Graves-Morris, P.: The Visual Display of Quantitative Information, vol. 2. Graphics Press, Cheshire (1983)
30. Heart Rate Variability: Standards of measurement, physiological interpretation, and clinical use. Eur. Heart J. **17**, 354–381 (1996)
31. Brosschot, J.F., Van Dijk, E., Thayer, J.F.: Daily worry is related to low heart rate variability during waking and the subsequent nocturnal sleep period. Int. J. Psychophysiol. **63**(1), 39–47 (2007). doi:10.1016/j.ijpsycho.2006.07.016. PMID 17020787
32. Korotkov, K.G.: The study of the properties of the discharge during the formation of the gas discharge images of the surface. Trans. LPI **371**, 51–54 (1980). (in Russian)
33. Bankovsky, N.G., Korotkov, K.G.: The study of the physics of the process gas discharge visualization. JTP Lett. **8**(4), 216–300 (1982). (in Russian)
34. Bankovsky, N.G., Korotkov, K.G., Petrov, N.N.: Physical processes of imaging in gas-discharge visualization (the Kirlian effect). Radiotech. Electron. **31**(4), 625–642 (1986). (in Russian)
35. Korotkov, K.G., Bankovsky, N.G.: Experimental investigation of discharge characteristics in a narrow gap dielectric. Trans. LPI **412**, 64–68 (1985)
36. Korotkov, K., Williams, B., Wisneski, L.: Biophysical energy transfer mechanisms in living systems: the basis of life processes. J. Altern. Complement. Med. **10**(1), 49–57 (2004)
37. Jakovleva, E.G., Korotkov, K.G.: Electrophotonic Applications in Medicine. EPI Bioelectrography Research. Amazon.com Publishing, Seattle (2013)
38. Korotkov, K.G., Matravers, P., Orlov, D.V., et al.: Application of electrophoton capture analysis based on gas discharge visualization technique in medicine: a systematic review. J. Altern. Complement. Med. **16**(1), 13–25 (2010)
39. Hossu, M., Rupert, R.: Quantum events of biophoton emission associated with complementary and alternative medicine therapies. J. Altern. Complement. Med. **12**(2), 119–124 (2006)
40. Alexandrova, R., Fedoseev, B., Korotkov, K.G., et al.: Analysis of the bioelectrograms of bronchial asthma patients. In: Proceedings of Conference "Measuring the Human Energy Field: State of the Science", pp. 70–81. National Institute of Health,   Baltimore, MD (2003)

41. Aleksandrova, E., Zarubina, T., Kovelkova, M., et al.: Gas discharge visualization (EPI) technique is new in diagnostics for patient with arterial hypertension. J. New Med. Technol. **17**(1), 122–125 (2010)
42. Yakovleva, E.G., Buntseva, O.A., Belonosov, S.S., et al.: Identifying patients with colon neoplasias with gas discharge visualization technique. J. Altern. Complement. Med. **21**(11), 720–724 (2015)
43. Polushin, J., Levshankov, A., Shirokov, D., Korotkov, K.G.: Monitoring energy levels during treatment with EPI technique. J. Sci. Heal. Outcome **2**(5), 5–15 (2009)
44. Korotkov, K.G., Shevtsov, A., Mohov, D., et al.: Stress reduction with osteopathy assessed with EPI electro-photonic imaging: effects of osteopathy treatment. J. Altern. Complement. Med. **18**(3), 251–257 (2012)
45. Rgeusskaja, G.V., Listopadov, U.I.: Medical technology of electrophotonics in evaluation of cognitive functions. J. Sci. Heal. Outcome **2**(5), 16–19 (2009)
46. Bundzen, P.V., Korotkov, K.G., Unestahl, L.E.: Altered states of consciousness: review of experimental data obtained with a multiple techniques approach. J. Altern. Complement. Med. **8**(2), 153–165 (2002)
47. Kostyuk, N., Meghanathan, N., Isokpehi, R.D., et al.: Biometric evaluation of anxiety in learning English as a second language. Int. J. Comput. Sci. Netw. Secur. **10**(1), 220–229 (2010)
48. Szadkowska, I., Masajtis, J., Gosh, J.H.: Images of corona discharges in patients with cardiovascular diseases as a preliminary analysis for research of the influence of textiles on images of corona discharges in textiles' users. Autex Res. J. **1**(10), 26–30 (2001)
49. Korotkin, D.A., Korotkov, K.G.: Concentration dependence of gas discharge around drops of inorganic electrolytes. J. Appl. Phys. **89**(9), 4732–4736 (2001)
50. Korotkov, K.G., Orlov, D.V.: Analysis of stimulated electrophotonic glow of liquids. http://www.waterjournal.org/volume-2/korotkov. Accessed 29 Jan 2016
51. www.bio-well.com. Accessed Mar 2017
52. Korotkov, K.G., Yusubov, R.R.: Method for determining the condition of a biological object and device for making same. US Patent 8,321,010 B2, 27 Nov 2012
53. Korotkov, K.G.: Method for determining the anxiety level of a human being. US Patent 7,869,636 B2, 11 Jan 2011
54. Drozdovsky, A.K., Korotkov, K.G., Evseev, S.P.: Psychophysiological factors that contributed to the successful performance of skiers and biathlon at the Paralympic games Sochi-2014. Adapt. Phys. Cult. **5**(58), 13–15 (2014). (in Russian)
55. Drozdovski, A., Gromova, I., Korotkov, K., Shelkov, O., Akinnagbe, F.: Express-evaluation of the psycho-physiological condition of Paralympic athletes. Open Access J. Sports Med. **3**, 215–222 (2012). http://www.dovepress.com/express-evaluation-of-the-psycho-physiological-condition-of-paralympic-peer-reviewed-article-OAJSM
56. Kuzmin, V.V.: The urgency and complexity of the application of "cloud computing", vol. 1, pp. 93–96. Herald Dmitrovgradskogo Engineering Institute (2013)
57. http://cmake.org/. Assessed Mar 2017
58. http://libpng.sourceforge.net/index.html. Assessed Mar 2017
59. http://zlib.net/. Assessed Mar 2017
60. http://en.wikipedia.org/wiki/SOAP. Assessed Mar 2017

Digital Business Model and SMART Economy Sectoral Development Trajectories Substantiation

L.A. Ismagilova[1(✉)], T.A. Gileva[1], M.P. Galimova[1], and V.V. Glukhov[2]

[1] Ufa State Aviation Tecnological University, Ufa, Russia
{ismagilova_ugatu, t-gileva, polli66}@mail.ru
[2] Peter the Great St. Petersburg Polytechnic University, Saint Petersburg, Russia
vicerector.me@spbstu.ru

Abstract. Tendencies and perspectives of sectoral development in SMART economy have been shown. Digitalization and IoT development levels have been given. The necessity of precedence development in sectors that create a base for telecommunications industry has been substantiated. Original schema of substantiation of sectoral development trajectories in SMART economy based on identification of gaps between current and required levels of sectoral development and the usage of complex of transformable business models as a tool for overcoming this gaps have been proposed. The analysis of current state of electronic components manufacturing industry in Republic Bashkortostan has been conducted; low level of compliance with the requirements of SMART economy and sectoral target indexes has been identified. The digital business model of electronic components manufacturing industry development oriented on creation of integrated production manufacturers maintaining all stages of creation, production, selling the product (Integrated Device Manufacturers, IDM), and realizing as virtual integrator technology (VIDM) has been formed. The matrix of sectoral development substantiation in SMART economy based on criteria of production-technological and market readiness and digitalization levels has been built. Three basic trajectories of sectoral development (vertical, horizontal and mixed) based on different schemas of transformations of business models depending on available resources and government support level have been identified.

Keywords: SMART economy · Telecommunications and electronic components manufacturing industry · Digital business model · Development trajectory

1 Sectoral Development Trends in SMART Economy

Nowadays, the SMART or the digital economy is the key trend that defines the prospects for sectoral developement. The digital technologies evolution [2]:

- allows to significantly reduce costs of economic and social transactions for enterprises, individuals and the public sector;

O. Galinina et al. (Eds.): NEW2AN/ruSMART/NsCC 2017, LNCS 10531, pp. 13–28, 2017.
DOI: 10.1007/978-3-319-67380-6_2

– rapidly improves efficiency – existing activities and services are becoming cheaper, faster, more convenient:
– stimulates the searching for new solutions, since the maximum level of efficiency can be reached when transactions are performed automatically without the personal participation (specific search engines, e-commerce, online payments, crowd funding platforms, new supply models etc.);
– allows individuals to get access to previously inaccessible services.

According to McKinsey data [15] the real prospects of the digital economy development are:

– increase in labor productivity in technical professions due to various labor automation aspects up to 45–55% more;
– equipment idle time – 30–50% less;
– maintenance costs reduction on 10–40%;
– stocks storage costs reduction on 20–50%;
– increase in forecasting accuracy up to 85 ± 5%;
– reduction of time before the product accesses the market – on 20–50%.

This can be achieved because of [9, 16]: digitalization of products and services; digitalization and integration of vertical and horizontal value creation chains; application of new digital business models improving client communication and access, client service process based on serving complex solutions in single digital ecosystem.

IoT (Internet of Things) which can be described as a system of integrated telecommunication networks and physical world objects (Things) packed with sensors and software for collecting and exchanging their data with a remote and automated [3] control ability is the technological basis of the digital economy. IoT technologies are widely used in such industry segments as «smart houses», «smart transport», «smart city». IoT spread analysis in different industry segments done by Capgemini Consulting и MIT Sloan School of Management [19] is showing that high-tech, banking and retail industries are having the highest profits from the «digitalization»; after them with a little lag there are telecommunications and hospitality industries, after - power production and housing industries. The lowest positions in this rating are hold by pharmacy, industrial production and consumer goods production.

According to IDC (International Data Corporation), company specialising on IT markets researches, IoT worldwide expenditure by the end of 2016 are estimated in $737 billion. This amount includes infrastructure and equipment costs, software and services costs, communication costs etc. IDC forecasts IoT market growth up to $1.7 trillion by 2020. Annual average growth rate during 2015–2020 is estimated at 15.6%. Machina Research and Cisco have given the more «forward» forecast – according to them, IoT market value in 2025 estimated at $4.3 trillion; also according to Accenture analysts IIot (Industrial IoT) market value will top $14 trillion [3] by 2030.

Thereby, despite the differences between the mentioned forecasts, their results and the results of many other researchers are showing the growth of digital maturity as a stable trend of sectoral development.

Being the one of leaders in digitalisation, telecommunication industry requires the specific material base from electronic industry to proceed it's development. Therefore, the lag of industrial manufacturing in modern Russia can cause serious problems and lead to irreversible lagging in SMART economy. According to experts from Boston Consulting Group [10], nowadays Russia exports about 90% of hardware and 60% of software. High export and investment stoppage led to the stagnation of the digital economy share in GDP since 2014.

In 2015 the share of digital economy was 2.1% of GDP, that is 1.3 times higher than 5 years before but 3–4 times lower than the leaders. The lag of Russia in general digitalization rating from the leaders is estimated at the level of 5–8 years. Without the adders-based stimulation of digital economy this lag estimation will grow to the level of 15–20 years in next 5 years [10].

The process of decreasing a current gap in condition of limited resources requires to define the sectoral development priorities correctly. Due to the impossibility of SMART economy «flagship» branches like IT and Telecommunications development without the appropriate material base procurement, special attention should be payed to various industrial sectors the progress in which is necessary for building a reliable foundation for digital economy in Russia. The example of that kind of industrial sectors is electronics manufacturing. It's specificity in SMART economy is that:

1. this industry provides the technological foundation of digital economy;
2. it's level of digital maturity and abilities to operate in new conditions are poor in Russia in general or in specified federal regions and do not satisfy requirements to complete the task;
3. the progress in this field should be conducted not only by the manufacturing technologies, but also the management practices.

Therefore, the feasibility studies in field of industries that provide the basis of digital economy, demand to develop new methodological approaches and analytical and decision-making tools.

2 Research Design

The relevance of developing new management instruments designed for completing tasks specific to sectoral development in SMART economy has been proven by researches [20]. According to the results of analysis of more than 400 companies from different industries four types were identified: digital leaders (Digirati), followers of «fashion» (Digital Fashionistas), beginners (Digital Beginners) and conservatives (Digital Conservatives). Companies that actively use digital technologies and new management methods (Digirati) get avg. 26% of income than their competitors. More conservative companies (Digital Conservatives) that perform only management upgrades increase their income on 9%. Digital Fashionistas, investing a lot in digital technologies while paying less attention to management, are not able to get the synergetic effect and create significant additional value based on digital applications; they get 11% lower in the financial indicators. Finally, companies that use digital technologies potential and management potential insufficiently get avg. 24% less income.

In this research the focused analysis of sectoral gaps that was chosen to become a basis for sectoral development trajectory rationalisation; it presumes (Fig. 1):

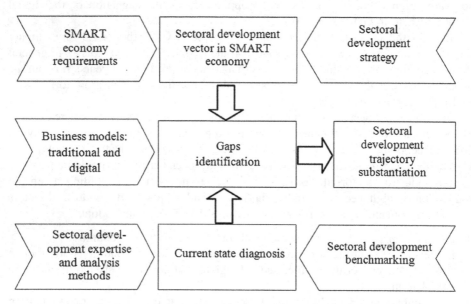

Fig. 1. Sector development in SMART economy rationale scheme

1. current state of sector diagnosis based on complex expertise and analytical scoring [6], including resources, processes and results of the chosen sector development. Strategy analysis and expertise scoring methods, matrix and factor analysis, sectoral development benchmarking – all this tools and methods should be used to complete diagnostic activities;
2. researches on sector development vector based on matching the strategy (or program) and the SMART economy requirements;
3. identification of strategic gaps and determination of sector development. Also, the correlation-regression analysis is used for the determination of prioritized vectors and projects of sector development.

Step-by-step trajectory is forming for the sake of overcoming detected sectoral development gaps. As a toolset for growth management we consider business models [6, 11, 14] by which we mean the way of creation customer value and get profit. According to experts [10] digital economy has an enormous influence on conservative business models; there are quite many researches on IoT business models or digital business models. However the necessity of counting of special aspects and maturity level requires to clarify contents of the model in case of electronic components manufacturing.

3 Sectoral Development Diagnostics and Gap Identification: Electronic Components Manufacturing

Nowadays, manufacturers located in Saint Petersburg and Moscow are leading in electronic parts manufacturing. Their share in overall amount of this industry production – 12.5% and 15.8% respectively. After the mentioned there go manufacturers from Sverdlovskaya and Samarovskaya regions – 4.6% and 3.2%.

Let us describe the state and perspectives of electronic components manufacturers on the example of Republic Bashkortostan (RB).

Industry Development Results in RB. Today contribution of companies located in RB to overall amount of sector production is quite humble – about 1.17% [7, 8]. The share of «electronic equipment, electronics and optical equipment» in manufacturing industry of RB equals 1.75%; of overall amount of shipped by RB manufacturing – 1.34%. This indexes are lower than in Russia or other world where the share of this high-tech manufacturing grows permanently and reaches 5–7% and 10–15% accordingly [7].

Sector Resources Analysis. Condition of the material base is characterised by the following marks: average age of equipment and it's deterioration are higher than same in overall industrial manufacturing (25.6 and 23 years accordingly), capital productivity of fixed assets is lower (2.15 RUB/RUB and 2.56 RUB/RUB) due to high levels of wearout (>50%) and incomplete usage of production capacity (about 60%).

Labor productivity in considered sector is losing to labor productivity in overall industrial manufacturing (1126 thous RUB/person and 1700 thous RUB/person accordingly). Value of this index is 1.5–2 times lower then in Russia, 4 times lower than in EU and 6 times lower than in USA; also, the growth rate of labor productivity is falling.

Processes Analysis. From the point of sectoral development the key processes are investing and investment implementation. Main investing vectors are: modernisation and renewal of production assets – more then 90%, investment projects on prior scientific and technical directions – less than 5%, others – 5%. The character of investments of sector's companies is chaotic, «patchy». The main emphasis is on the existing resource base and it's simple reproduction. All this reveals the lack of wide-scale targeted investment projects and will not lead to the transition to SMART economy.

Innovational product share in the sector was less than 1.8%. The sector does not accomplish the «high-tech» status.

On the other side, the government program «Electronics and electronic components manufacturing industry development program for 2014–2015» (approved by Russian Government decree at 15.04.2014, № 329) defines the following targeted markers for the sector:

– financial income increase 5.7 times in 2025 from the level of 2014;
– increase Russian-made electronic components share in the domestic market from 19% in 2014 to 30% in 2020 and 35% in 2025;

- sector production export level increase to 2025 on more than 3.5 times from the level of 2014;
- increase of production per 1 sector worker 4.4 times to 2025 from the level of 2014; the salary should become 2.6 times bigger;
- creation of 30 thou. new high-tech workplaces.

Due to overcome detected gaps in quantitive indicators and growth quality and fit the demands of SMART economy we need new development trajectories based on current business model development, the last's adaptation to digital.

4 Sectoral Development Digital Business Model Design

4.1 Traditional Sector Development Business Models Analysis

Choosing a complex development model focused on professional segments is justified in a mentioned above «Electronic components industry development strategy in Russia to 2030». The priority status goes to special purpose electronic parts manufacturing and import substitution of key special hardware and equipment with a decrease of customer electronics share.

In the context of sector development in Republic Bashkortostan, this exact vector should be considered as a sector development focus because companies located there are working mainly in the fields of professional and special purpose electronics [6].

In the worldwide management practices, the following business models support a basis in the segments of professional electronics and differ by coverage of value creation chain:

- integrated production manufacturer supporting all three stages of creating, manufacturing and selling products (Integrated Device Manufacturers);
- contract developers (Fabless) – doing all the RnD cycle and stand as customers of contract manufacturers;
- contract manufacturers (Foundry, Original Equipment Manufacturer, Original Design Manufacturer) - differ by scopes of contract work, by manufactured production type and by originality and licensing;
- contract Assembly and QA (testers).

Table 1 gives a brief comparative characteristic of mentioned business models considering their role in scientific-and-technical and manufacturing chain.

The existent sector business model in Republic Bashkortostan is the model with a prevalence of company types M2, M3 and M4. It can be described as an incorporation in sectoral and global production chains. Today more than 70% RB companies are integrated in vertical-integrated holdings. Besides, the analysis of traditional business models (Table 2) shows that they do not fit the tasks of region development and Russia in SMART economy and occur transitional.

The key criteria for choosing a sectoral development business model is it's ability to ensure the accomplishing of strategic targets. As seen in Table 2, the complete realisation of all objectives can be provided only on IDM model exploitation, the modification of which in digital economy is a virtual integrator model (VIDM).

Table 1. Traditional sectoral development business models characteristics

Indexing criteria	IDM (M1)	Fabless (M2)	Foundry (M3)	Assembly (M4)
1. Fullness of value creation chain coverage	All stages	Development	Production	Assembly
2. RnD investments level and RnD reserve value	High	High	Low/medium	Low
3. Production-technological potential	High	Low	Medium	Low/medium
4. Existence of competences through the whole value creation chain	High	Low	Medium	Low
5. Market potential	High	Low	Medium	Low
6. Investment capacity	High	Low	Medium	Low
7. Costs level	High	High	Medium	Low
8. Production and innovational infrastructure levels	High	Low	Low/medium	Low
9. Added value level	High	Low	Medium	Low

Table 2. Comparative analysis of sectoral development business models

Comparsion criteria	IDM (M1)	Fabless (M2)	Foundry (M3)	Assembly (M4)
1. Ability to ensure high sectoral (current sector + «customer» sectors) growth rates	+	−	−	−
2. Sufficiency of existent production-technological potential for the development	+	−	+/−	+/−
3. Ability for production capacity upbuilding with a considering of abilities of innovational and digital economy	+	+	+	+/−
4. Ability of model realization in current investment and financial conditions	−	+	+/−	+
5. Ability to overcome sectoral technological lags from world tendencies	+	+/−	+/−	−
6. Maneuvering and adaptation to unstable outer conditions abilities	+	−	+/−	−

In the process of forming the development trajectories the following must be considered:

1. sectoral development resource contradiction. On the one hand, the sector potential does not allow to realize integrated models, on the other hand, exactly this type of business models is needed for accomplishing strategic targets and multiplied growth;
2. separate companies cannot achieve technical-economical level that can be achieved with a cooperative functioning of participants with resource and competence cooperation [1, 5].

4.2   Electronic Parts Manufacturing Digital Business Model Design

Today there are many publications on researching of business models transformation in IoT ecosystem [12, 13, 17, 21, 22]. The main attention is given to interaction dynamics between business model components and necessity of fundamental revision of value creation and value capture principles. The ecosystem business model designing approach based on value network is being developed [13, 17, 21]. Realization of business models on this level is determined by the abilities of SMART economy technological platforms.

The basis of forming a digital business model for electronic components manu-facturing sector is built on Osterwalder-Pigneur model [18]. The model is represented by two interconnected aspects: customer (Table 3) and industrial and technological (Table 4). Informative part of this model is transformed considering the requirements of SMART environment and key targets of sector development. All in all, full business model is realizing the VIDM concept.

Key business model transformations within the customer aspect in SMART economy:

1. key trend in SMART economy is a time-based competition which requires high speed of decision making and significant decrease of product's time-to-market type;
2. intellectual property, information and competences are the key resources and products;
3. radical changes in end-product structure: increase of intellectual property or rights on using it, research results share to 60–90% of overall profit and corresponding decrease of product share to 10–40% what leads to loss-profit structure changes;
4. expansion of SMART tools, instruments and client interaction mechanisms: digital communication channels, platform-style solutions, crowd-based technologies etc.;
5. border erasure between «key partners» and «key customers». Customer becomes a participant and investor of production process. Partners become a part of customer concept.

Key transformations within the industrial and technological aspect:

1. accents shifting to RnD activities (RnD, engineering, scientific consulting);
2. radical changes in industrial processes' principles: speed and agility become the key aspects;
3. disaggregation tendencies in industrial sector, reorientation to medium and small enterprises. Small business is building on a basis of existing development centres, which integrate «niche development» into global solutions and push them to the market;
4. value creation chain defragments to separate levels, growth of outsourcing approach;
5. resource dispositions damping with assist from virtual companies and virtual industrial clusters;
6. rate of innovation generating and it's commercialisation increasing due to dynamically growing clusters and business accelerators; changes in cluster's structure and character. In SMART economy clusters are represented as VIC - virtual industrial clusters, with an aim to provide access to remote unique resources.

Table 3. Customer aspect of VIDM model

Value for customer (VP)	Product (P)/Customer segments (CS)	Customer relations (CR)/Sales channels (C)	Revenue (RS)
Complex personified offering and integrated maintenance VP1-Product provisioning time (speed) VP2-High-Tech Competitive Product VP3-Providing high income to the consumer for a long time VP4-Providing long-term technological advantages VP5-Long life cycle with short stages of recession and curtailment of production VP6-Ability to modify and develop the product VP7-Comprehensive solution of the problem «on a turn-key basis» - accompaniment, maintenance, consulting VP8-Low cost of ownership (low operating costs) VP9-High level of customization VP10-Exclusive ownership of IP, embodied in the product VP11-Impossibility of copying by competitors VP12-The opportunity to participate in product development	**Products** P1-Professional electronics P2-Components, electronic base P3-Special technological equipment P4-Licenses P5-Rights (IP) P6-Engineering P7-Franchise P8-Competences P9-Staff leasing P10-RnD results **Customer segments** CS1-Final product manufacturers in RB, Russia, EU, countries of BRICS, SCO CS2-Cooperation chain partners CS3-Contractors CS4-Developers CS5-Enterprises implementing individual stages of the production cycle CS6-Engineering companies CS7-Electronic base developers CS8-Cluster members CS9-Subcontractors CS10-Outsourcers	**Customer relations** CR1-Virtual industrial cluster (VIC) CR2-Digital platforms: crowdfunding, technological exchange markets CR3-Virtual enterprise (corporation) CR4-Cooperative design and development CR5-Joint venture CR6-Digital order processes **Sale channels** C1-Automatic product transfer channels C2-Digital channels C3-Digital platforms C4-Internet platforms and communities C5-Wholesale C6-Order-based C7-Joint networks C8-Cluster channels C9-Government orders	RS1-From the sale of products, components, equipment RS2-From engineering RS3-From outsourcing RS4-From outstaffing RS5-From license sales RS6-From rights sales RS7-From participation in the authorized capital as owner of IP RS8-From scientific consulting RS9-From RnD RS10-From sales of competences RS11-From expertise RS12-From whole business selling RS13-From securities, crowd investment RS14-From crowdfunding platforms RS15-From e-commerce RS16-From franchise RS17-From advertising

Only the virtual integrator business model (VIDM) fully corresponds to characteristics given in Tables 3 and 4. Other models described in Table 1 corresponds this characteristics partly only. The accordance of business models to SMART economy requirements is shown in Table 5.

Table 4. Industrial and technical aspect of VIDM model

Key processes	Key partners	Key resources	Expenses
KA1-RnD	**KP1**-Design	**KR1**-	**CS1**-For production
KA2-Production	centers	Intellectual	**CS2**-For product placement
KA3-Sales	**KP2**-Crowdfunding	property	**CS3**-For buying licenses and
KA4-Service	platforms	**KR2**-Unique	IP rights
KA5-Supply	**KP3**-Internet	competences	**CS4**-For creating and
chain	platforms	**KR3**-	maintaining of Internet
management	**KP4**-Contract	Organizational	platforms
KA6-Production	manufacturers	competences	**CS5**-For creation and
processes	(foundry)	**KR4**-Products	maintenance of a database
management	**KP5**-Contract	and technologies	and knowledge base
KA7-Internet	assemblers and	**KR5**-Production	**CS6**-For RnD
commerce	testers (assembly)	areas (ability to	**CS7**-For creation of new
KA8-Internet	**KP6**-Service	access)	enterprises
platforms	companies		**CS8**-For technologies transfer
administration	**KP7**-QA		**CS9**-Networking costs
KA9-Investor	organizations		**CS10**-For project expertise
attraction	**KP8**-ICT providers		**CS11**-For securing and
KA10-Stock	**KP9**-Component		protecting IP rights
market work	Providers		**CS12**-Credits
KA11-PR-GR	**KP10**-Patent		**CS13**-Competences rent
management	attorneys		**CS14**-Personnel rent
KA12-Security	**KP11**-Banks		**CS15**-For freelancing
and protection of	**KP12**-Private		
IP	investors		
KA13-New	**KP13**-Government		
productions	investors		
designing	**KP14**-Cluster		
KA 4-Offset	members		
transactions	**KP15**-Virtual bank		
KA15-Startups	**KP16**-Competence		
acquisition	providers		
	KP17-Developers		

So, the models M2–M4 are considered as transitional steps (transitional business models) of sectoral «ascending» to a new level, fulfilling the SMART economy requirements.

Table 5. Accordance of business models to SMART economy requirements

Components	M2 (fabless)	M3 (foundry)	M4 (assembly)
Customer aspect			
Value for customer (**VP**)	VP1-VP2-VP3-VP4-VP5-VP6-VP10-VP11-VP12	VP1-VP6-VP8-VP9-VP12	VP1-VP8
Product (**P**)	P1-P2-P3-P4-P5-P6-P8-P10	P1-P2-P3-P7-P8-P9	P1
Customer segments (**CS**)	CS1-CS2-CS3-CS5-CS8-CS9	CS1-CS2-CS3-CS4-CS6-CS7-CS10	CS1-CS3-CS5-CS8-CS10
Customer relations (**CR**)	CR1-CR2-CR3-CR4-CR5-CR6	CR1-CR2-CR3-CR4-CR5-CR6	CR1-CR3-CR5-CR6
Sales channels (**C**)	C1-C2-C3-C4-C6-C7-C8-C9	C1-C2-C3-C4-C5-C6-C7-C8-C9	C5-C6-C7-C8-C9
Income (**RS**)	RS1-RS2-RS5-RS6-RS7-RS8-RS9-RS10-RS11-RS12-RS17	RS1-RS2-RS3-RS4-RS7-RS10-RS12	RS1-RS7-RS10-RS12
Production-technological aspect			
Key processes (**KA**)	KA1-KA3-KA7-KA8-KA9-KA11-KA12-KA14	KA2-KA3-KA4-KA5-KA6-KA7-KA8-KA9-KA10-KA11-KA13-KA14	KA2-KA3-KA6
Key partners (**KP**)	KP1-KP2-KP4-KP5-KP6-KP7-KP8-KP10-KP11-KP12-KP13-KP14-KP15-KP16	KP3-KP4-KP5-KP6-KP7-KP8-KP9-KP11-KP12-KP13-KP14-KP15-KP16	KP4-KP6-KP7-KP9-KP14-KP16
Key resources (**KR**)	KR1-KR2	KR3-KR4-KR5	KR3-KR4-KR5
Expenses (**CS**)	CS3-CS5-CS6-CS8-CS9-CS10-CS11-CS12-CS13-CS15	CS1-CS2-CS3-CS4-CS5-CS7-CS8-CS9-CS12-CS14	CS1-CS9

5 «Sectoral Development Trajectory in SMART Economy» Matrix Rationale

The analysis of current condition of electronic components parts manufacturing industry in Republic Bashkortostan showed that the potential of region does not allow to realize fully the IDM model. Due to rationale of sectoral development trajectory the positioning matrix was developed (Table 6).

Combining of business models signifies the consistent integration of a company into vertical and horizontal value creation chains or in realization of the co-ordinated partnership model.

As it was shown in [20], effective functioning of a company depends on two constituents: digital intensity characterising investing in digital initiatives with a purpose to change company's operating activities and the deepness of management intensity transformations. Therefore the development of company defining the sectoral progress must be considered with both digital and management transformations.

Table 6. «Sectoral development trajectory in SMART economy» matrix rationale

	High	M2&M3 or M2&M3&M4	M1 (VIDM)	M1 (VIDM)
	Medium	M2	M2&M3 or M2&M3&M4	M1 (IDM)
	Low	M2 or M4	M3 or M4	M2&M3 or M2&M3&M4
		Low	Medium	High
	The level of industrial-technological and market sectoral readiness (of enterprises)			

(Left vertical axis label: Sector digitalization level)

According to the matrix (Table 6) there are three types of sectoral development trajectories:

1. Vertical trajectory (revolutionary scenario) – gap overcoming by using a combination of digital and management tools and instruments with a scope on digital. Visually it can be described as a direction from lower left quadrant to upper right: (M2 or M4) → (M2&M3 or M2&M3&M4) → (M1). Assumes development of unique competences through the whole value creation chain. Resource potential grows due to usage of partners' resources.

   Digital transformations and tools: virtual enterprises, network communities, virtual industrial clusters, crowdfunding platforms, IoT technology stack. Management transformation tools: technological stock markets, technology transfer systems, business acceleration.

   To make this scenario real the government support is needed strongly in fields like: investments in RnD stimulation, education development, building an infrastructure for RnD results commercialisation, special tax credits etc.

2. Horizontal trajectory (evolutionary scenario) – gap overcoming provided by a combination of digital and management tools with a scope on management. Assumes the prime upbuilding of management competences with a progressive transition on digital technologies: (M2 or M4) → (M3 or M4) → (M2&M3 or M2&M3&M4) → (M1).

   Management transformations and it's realisation tools are: manufacturing and technological modernisation, production processes automation, production models, mass customisation models, building of agile and adaptive production and competence centres, private-public partnerships, technological stock markets. Digital transformations: virtual industrial clusters, crowd-investing platforms.

3. Mixed transformation (base scenario)

   Gap overcoming due to combination of digital and management tools with an advance in digital technologies. Except the investments in production and technical modernisation, investments in human capital or filling the deficit a.o. virtual cluster or virtual production personnel. Development of production powers can be also achieved with using a cooperation potential, licensed production development and outsourcing [4].

Digital transformations and tools: network communities, virtual industrial clusters, crowd funding and crowd-investment platforms. Management transformations and tools: outstaffing, personnel lease, outsourcing, license purchasing, licensed production creation.

6 Conclusion

Currently we can observe a strong unevenness of digitalization as in different countries and regions, as in different economy sectors. However for a development of SMART economy's «flagships» – telecommunications and IT – things like in-advance development of such base sectors as, at first, electronic components manufacturing are necessary for the whole sector development.

SMART economy development requires to only the wide digital technologies implementation, but the changes in management tools and instruments. Traditional business models should be changed fundamentally in the aspect of radical revision of value creation and appropriation principles, «value network» formation. As a digital business model the integrated device manufacturers or IDM model should be considered, supporting all stages of creation, manufacturing and selling the production, embodying the technology of virtual integrator (VIDM).

Current development level of electronic components manufacturing industry in most Russian regions does not meet the requirements of SMART economy.

The choice of sectoral development trajectory in SMART economy depends on three factor groups: current industrial-technical and market readiness levels, sector digitalization level and amount of government support. To substantiate a choice, the sectoral positioning matrix was developed, according to that the transition from current to objective state is carried out on the basis of consistent transformations of traditional business models into digital. Transformation process is oriented at the industrial-technological potential upbuilding and/or digitalization level depending on a chosen development focus (digital or management [20]), available resources and government support level.

There are three basic development trajectories:

- vertical trajectory (revolutionary scenario) – gaps overcoming by using a combination of digital and management tools and instruments with focus on digital. Presumes the development of unique competences through all the stages of value creation chain. Provides maximal results in the shortest time, but can not be realized without the government support and investments in RnD stimulation, education development, RnD results commercialisation infrastructure creation, special tax credits etc.;
- horizontal trajectory (evolutionary scenario) – gaps overcoming by using a combination of digital and management tools and instruments with focus on management. Presumes the prime accent on production-technical potential development

and limited by enterprise's investing capability. Possible to integrate in current chains or enterprises' entry in integrated sectoral structures due to overcome the technological and management competences deficit;

– mixed trajectory – presumes a higher priority of intellectual competences linked to production RnD, maximisation of digital technologies and virtual structure forming platforms (including integrated) usage. In order to overcome the deficit in management competences staff from a virtual enterprise or virtual cluster can be used. Productive capacity development can be also provided by using a cooperation potential, licensed manufacturing and outsourcing. Realisation difficulties of this trajectory are highly connected to the necessity of changes in management style and methods, risks linked to it and lack of appropriate management competences.

The advancing in sectoral development can be possible only on a vertical development trajectory basis based on complexed government support and investments in RnD stimulation, education development, creation of RnD results commercialisation infrastructure, special tax credits etc.

According to the results of current and expected conditions of the electronic components manufacturing industry the conclusion was made: in current conditions the only development trajectory that can be realised is horizontal, which implies way less growth rates and will not allow to achieve targeted development index values within the planned period.

Within the chosen development trajectory following recommendations on electronic components manufacturing industry in Republic Bashkortostan are formed:

– creation of cooperative enterprises as with CIS countries, as within the SCO and BRICS arrangements; mastering new sectoral directions with a scope on professional electronics segment.
– incorporation in technology chains as a supplier of unique competences (production, engineering, project design) as on regional, as on a global level;
– «soft integration» with intangible assets to technological chains of corporations and holdings;
– licensed production development, franchised technologies (buying and selling franchises);
– development of small business zones around big investor companies (i.e.g. CG Rostech)
– creation of a closed exchange market for technologies and patent bases interchange between enterprises.

References

1. Kleyner, G., Babkin, A.: Forming a telecommunication cluster based on a virtual enterprise. In: Balandin, S., Andreev, S., Koucheryavy, Y. (eds.) ruSMART 2015. LNCS, pp. 567–572, vol. 9247. Springer, Cham (2015). doi:10.1007/978-3-319-23126-6_50
2. Doklad o mirovom razvitii Tsifrovyie dividendyi. Obzor. Vsemir-nyiy bank, Mezhdunarodnyiy bank rekonstruktsii i razvitiya, 46 (2016)

3. Internet veschey, IoT, M2M (mirovoy ryinok). http://www.tadviser.ru/index.php/Statya: Internet_veschey,_IoT,_M2M_(mirovoy_ryinok)#.D0.9F.D1.80.D0.BE.D0.B3.D0.BD.D0. BE.D0.B7_Gartner

4. Ismagilova, L.A., Galimova, M.P., Gileva, T.A.: Vyibor promyishlennogo autsorsera na osnove metoda strukturirovaniya funktsii kachestva. Vestnik Kazanskogo gosudarstvennogo tehnicheskogo universiteta im. A.N. Tupoleva # 5, 97–103 (2015)

5. Ismagilova, L.A., Gileva, T.A., Galimova, M.P.: Organizatsionno-ekonomicheskie aspektyi formirovaniya tsentrov tehnologicheskih kompetentsiy. Nauchno-tehnicheskie vedomosti Sankt-Peterburgskogo politehnicheskogo uni-versiteta. Ekonomicheskie nauki, # 5, 125–132 (2013)

6. Mashinostroitelnyiy kompleks regiona: diagnostika, konkurentospo-sobnost, strategicheskie prioritetyi (na primere Respubliki Bashkortostan). L.A. Ismagilova i dr. – M.: Izdatelstvo «Innovatsionnoe mashinostroenie», 417 (2016)

7. Promyishlennost Respubliki Bashkortostan: statisticheskiy sbornik. – Ufa: Bashkortostanstat (2014)

8. Promyishlennost Rossiyskoy Federatsii: statisticheskiy sbornik – M: Rocstat (2014)

9. Pshenichnikov, V.V., Babkin, A.V.: Elektronnyie dengi kak faktor razvitiya tsifrovoy ekonomiki. Nauchno-tehnicheskie vedomosti Sankt-Peterburgskogo politehnicheskogo universiteta. Ekonomicheskie nauki. # 1, 32–42 (2017)

10. Rossiya onlayn? Dognat nelzya otstat. In: Banke, B., Butenko, V., Kotov, I., Rubin, G., Tushen, Sh., Syicheva, E.: The Boston Consulting Group (2016). http://www.bcg.ru/ documents/file210280.pdf

11. Tretyak, O.A., Klimanov, D.E.: Novyiy podhod k analizu biznes-modeley. Rossiyskiy zhurnal menedzhmenta. Tom 14, # 1, 115–130 (2016)

12. Fleisch, E., Weinberger, M., Wortmann, F.: Business models and the internet of things. Bosch IoT Lab. White Paper, August 2014. http://cocoa.ethz.ch/downloads/2014/10/2090_ EN_Bosch%20Lab%20White%20Paper%20GM%20im%20IOT%201_2.pdf

13. Chan, H.: Internet of things business models. J. Serv. Sci. Manag. **8**, 552–568 (2015). http:// file.scirp.org/pdf/JSSM_2015081214471082.pdf

14. Geissbauer, R., Vedso, J., Schrauf, S.: A Strategist's Guide to Industry 4.0. Strategy + Business, issue 83, Summer (2016). https://www.strategy-business.com/article/A-Strategists-Guide-to-Industry-4.0?gko=7c4cf

15. Industry 4.0 at McKinsey's model factories. Get ready for the disruptive wave. McKinsey digital (2016). https://capability-center.mckinsey.com/files/downloads/2016/digital4. 0modelfactoriesbrochure_0.pdf

16. Industry 4.0: Building the digital enterprise. PwC (2016). https://www.pwc.com/gx/en/ industries/industries-4.0/landing-page/industry-4.0-building-your-digital-enterprise-april-2016.pdf

17. IoT Business Models Framework. H2020 – UNIFY-IoT Project (2016). http://www.internet-of-things-research.eu/pdf/D02_01_WP02_H2020_UNIFY-IoT_Final.pdf

18. Osterwalder, A., Pigneur, Y.: Business Model Generation. Wiley, Hoboken (2010)

19. Strategy, Not Technology, Drives Digital Transformation. In: Kane, G.C., Palmer, D., Phillips, A.N., Kiron, D., Buckley, N.: MIT Sloan Management Review and Deloitte University Press, July 2015. https://dupress.deloitte.com/content/dam/dup-us-en/articles/ digital-transformation-strategy-digitally-mature/15-MIT-DD-Strategy_small.pdf

20. The Digital Advantage: How digital leaders outperform their peers in every industry. Capgemini Consulting, MIT Sloan Management (2012). https://www.capgemini.com/ resource-file-access/resource/pdf/The_Digital_Advantage__How_Digital_Leaders_ Outperform_their_Peers_in_Every_Industry.pdf

21. Weiller, C., Neely, A.: Business Model Design in an Ecosystem Context. Cambridge Service Alliance, June 2013. http://cambridgeservicealliance.eng.cam.ac.uk/resources/Downloads/Monthly%20Papers/2013JunepaperBusinessModelDesigninEcosystemContext.pdf
22. Westerlund, M., Leminen, S., Rajahonka, M.: Designing business models for the internet of things. Technol. Innov. Manag. Rev. **4**(7), 5–14 (2014). http://timreview.ca/article/807

Ontology Matching for Socio-Cyberphysical Systems: An Approach Based on Background Knowledge

Alexander Smirnov[1,2], Nikolay Teslya[1,2(✉)] ⓘ, Sergey Savosin[1],
and Nikolay Shilov[1,2]

[1] SPIIRAS, 14 line, 39, 199178 St. Petersburg, Russia
{smir, teslya, svsavosin, nick}@iias.spb.su
[2] ITMO University, 49, Kronverksky Pr., 197101 St. Petersburg, Russia

Abstract. Nowadays there are exist many ontology matching methods mostly based on using context and content in various combinations. The accuracy of existing methods can be increased by using the background knowledge – an external reference knowledge that can facilitate the matching process. The paper proposes an approach for automated matching of elements (fragments) of the ontologies based on the combination of existed ontology matching methods (pattern and context-based), neural network matching and additional control by community-driven matching. All of them are using the background knowledge to get additional information that helps to find more precise correspondence between ontologies concepts. Also, the background knowledge helps to explain result of ontology matching to the user, which is going to check the correspondence manually. In addition, the paper proposes an architecture of ontology matching service that is built based on the presented approach. The service is used for providing semantic interoperability in socio-cyberphysical systems.

Keywords: Ontology matching · Combination · Background knowledge · Socio-cyberphysical system

1 Introduction

The socio-cyberphysical system (SCPS) unites different entities from employees to entire enterprises in single complex system. One of the ways to support semantic-based interoperability in a SCPS is application of the ontology management technology. Development of efficient methods of ontology matching could significantly improve the knowledge processing and interoperability between them. Nowadays there are many ontology matching methods mostly based on using context and content [1] in various combinations.

The accuracy of existing methods can be increased by using the background knowledge [2, 3]. The background knowledge can be understood as "any external reference knowledge that can facilitate the matching process" [4]. Shvaiko and Euzenat [5, 6] declares the following challenges for ontology matching that can be addressed to background knowledge usage: (i) discovering specific context of background knowledge

© Springer International Publishing AG 2017
O. Galinina et al. (Eds.): NEW2AN/ruSMART/NsCC 2017, LNCS 10531, pp. 29–39, 2017.
DOI: 10.1007/978-3-319-67380-6_3

for the ontologies, (ii) explanation of matching results. The last challenge can also be addressed to cognitive support due to the using of the background knowledge in user-driven matching to provide additional information about matched concepts during the process that helps to make decision about matching.

The paper addresses the challenges presented above and proposes an approach for automated matching of elements (fragments) of the ontologies based on the combination of existed methods built around ontology patterns and context with additional control by community-driven matching. These methods should utilize the background knowledge in their work as well as provide explanation of their decision to the community-based method. The methods based on patterns and context could be implemented using the neural networks. The requirement of matching explanation is important because most of existing methods, especially methods based on the neural networks, cannot provide it in a simple, clear, and precise way to the user in order to facilitate the correction of proposed matching.

The rest of the paper is structured as follows. Section 2 describes existing approaches of using the background knowledge in ontology matching and systems that provides cognitive support during the matching process. Section 3 provides the description and classification of background knowledge sources. Section 4 provides the approach for usage the background knowledge in ontology matching. Section 5 shows the architecture of ontology matching service with usage of background knowledge for socio-cyberphysical system. Section 6 provides the conclusion and future work.

2   State-of-the-Art

Searching and processing background knowledge is normally carried out automatically. The approach presented in paper [7] describes an iterative semantic matching with usage of background knowledge. Iterations are conducted from the root of the ontology tree to the branches. As a result, if a discrepancy is found at one of the levels the further comparison is not possible. When a critical point is found, in which additional information about ontology concepts is required, the search of background knowledge is conducted. For this reason, WordNet thesaurus is accessed to acquire the boundary values of the semantic connectivity of the ontology concepts.

Papers [8, 9] describe the scheme of the method based on the use of the problem area upper ontology as a source of background knowledge for ontology matching. The method includes two steps: (1) anchoring the source and target concepts of the matched ontologies to the background knowledge ontology; (2) defining if the relationship between the source and target concepts exists by searching for a similar relationship in the background knowledge ontology. Paper [10] extends the results of previous works of the authors, considering the use of several ontologies as sources of background knowledge.

There are also studies aimed at analyzing the requirements for cognitive support for the task of comparing ontologies [11]. The authors note a lack of research related to cognitive support for ontology matching with a large number of different matching techniques and provide 13 requirements related to simplifying the interaction of the user with the cognitive assistant and reducing the cognitive burden on the user. The work also presents an extension for PROMT ontology matching plugin, intended for the ontology matching in Protégé, in which most of the provided requirements are implemented. The architecture of the cognitive support system for ontology matching is proposed in [12]. Based on the proposed architecture the authors developed a tool for cognitive support. For this purpose, an analysis of user interaction with two plugins for the comparison of ontologies, COMA++ and PROMT, was carried out. This allowed to identify the issues that cause particular difficulty for users. The developed tool provides the user with an interface in which the auxiliary information associated with the result of the mapping is visually presented.

Paper [13] focuses on the development of an ontology matching system, in which the concepts are modeled on the basis of cognitive knowledge units that contain objects, attributes, and relationships. The system defines semantically related concepts by collecting the attributes' similarity degree, the structural similarity degree and the degree of semantic similarity of concepts with the use of the concept algebra, which is an extension of formal analysis of concepts. The concept algebra includes a set of formal relations to find out how widely the concepts of ontologies are semantically presented and how strongly they are related to one another. Reducing the dimension of the matching space for concepts is achieved by dividing ontologies into clusters using the idea of a network of concepts. Clusters include semantically related concepts, which allows matching only within them.

To solve the task of comparing ontologies, neural networks can also be used. In [14], a neural network is proposed for ontology matching that expands the ability to represent and display complex relationships in a neural network for identical elements searching (Identical Elements Neural Network). The network can learn the high-level properties of concepts in various tasks and use them to find a correspondence between tasks. This allows to train the network to find correspondence for specific concepts between different problem areas.

The transformation of the task of mapping ontology concept instances into the problem of binary classification and its solution using machine learning methods is proposed in [15]. To reach this goal, a similarity vector was developed, independent of the presence of coinciding properties in the instances of concepts. The training of neural network to classify pairs of concepts for the sets of matches and mismatches is carried out based on this vector. To improve performance, it is suggested to use information about already existing alignments. As a classification mechanism, a random decision tree was chosen to construct weak classifiers, which are combined using the AdaBoost algorithm.

3 Background Knowledge Sources

The section provides description and classification of background knowledge sources that can be used in ontology matching. The following two main classes can be distinguished in background knowledge sources based on the way of knowledge extraction type: *explicit* and *implicit*.

Explicit sources can be understood as sources that do not need any additional processing of provided data. The choice of explicit sources can be explained by the need to determine the semantic correspondence of individual concepts of ontologies without their connections within the ontologies under consideration. It can also be noted that the use of background knowledge makes it possible to compensate for the inadequacy of the structure and lexical overlapping of the concepts of the ontologies being compared, as well as to significantly increase the speed of comparison [8, 9]. The examples of explicit sources are the following:

- Linked data. It can be used as upper ontology for context-based matching. The examples of the linked data sources are DBpedia, YAGO (Yet Another Great Ontology). Usually, linked data sources are linked between each other. Therefore, it is enough to use only one source.
- Domain specific or general-purpose ontologies. This class can also be used as a source of upper ontology for the context-based matching. The examples of domain specific ontologies are BMO (an e-Business Model Ontology) and Friend of a Friend (an ontology for describing persons). The examples of general ontologies are BabelNet and Basic Formal Ontology (BFO).
- Dictionaries and thesauri. These are used to get definition of word and linked words like synonyms, homonyms, antonyms for syntax and semantic-based matching. The examples of such type of sources are "Keyword AAA"[1] (covers terminology common to most organisations) and WordNet.
- Lexical databases. They combine dictionary and thesauri in a single database, allow to get lexical category of ontology concept, semantics, synonyms, etc., and provide built-in functions to calculate similarity of words. The examples are WordNet, DANTE (http://www.webdante.com/).

The *implicit* sources require data processing before they can be used in the matching process. Ontology design patterns and ontology alignment patterns could also be viewed as implicit sources of background knowledge. Ontology design patterns can be defined using data analysis and can be further applied when searching for correspondences to take into account the relations between ontology concepts. The examples of this type of sources are common ontology design patterns, profiles and templates of specific ontology creators. An analysis and existing libraries of existing ontology design patterns, such as W3C, ontologydesignpatterns.org portal and Manchester ODP catalog, are presented in [16]. Usage of ontology design patterns in the WebProtégé ontology development environment is described in [17]. It offers an extension of the environment, allowing to search, specialize and integrate templates.

Ontology matching patterns are also developed in addition to ontology design patterns. The authors of [18] propose creating of ontology matching patterns that can be reused to refine the matching. Such a pattern includes the concept context and rules of matching that are executed when a specified context is detected.

Papers [19–21] introduce ontology matching patterns, which are proposed to be used as reusable patterns of repeated correspondences. Based on a detailed analysis of frequent ontology discrepancies, a library of common templates was developed. Each template is a general solution to a specific mapping task.

---

[1] https://www.records.nsw.gov.au/recordkeeping/resources/keyword-products/keyword-aaa.

4 Ontology Matching Approach Based on Background Knowledge

The methodology used in this research is based on integration of the organizational and individual contexts of the entities of socio-cyberphysical systems that make it possible to include the variability of the constraints used for decision making (taking into account the roles of the users – SCPS members' decision makers). At the same time, the ontological models of the contexts are described at two levels [22]:

1. Abstract, defining the conceptual description of a typical task solved by a decision maker as a fragment of the SCPS ontology
2. Operational, representing knowledge and information about the particular parameters of the task being solved at the moment, obtained from the information environment of the SCPS.

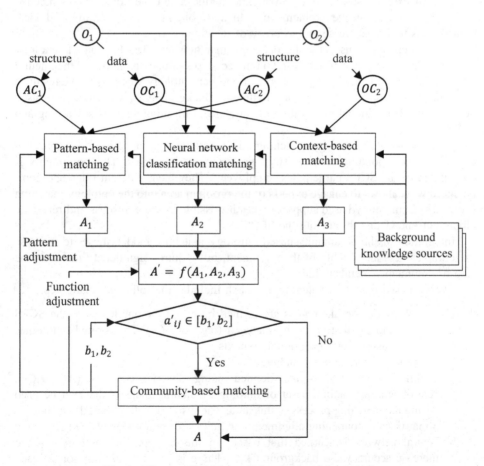

Fig. 1. The overall scheme of the multimodel approach

The proposed approach (Fig. 1) is based on the following models:

- Pattern-based matching;
- Context-based matching;
- Neural network classification matching;
- Community-driven matching.

The matching patterns [1, 20] are aligned groups of linked ontology elements (ontology fragments). These patterns are used as templates to find more precise alignments between separate elements or fragments. In particular, patterns can be related to particular classes, to classes and relationships between them or to complex clusters of classes including relationships [23]. The proposed approach provides possibilities to create new and adapt existing ontology matching patterns depending on the problem area.

The context-based ontology matching model assumes considering the environment of the matched ontologies. This environment can be an upper level ontology that is used as an external source of common knowledge about the concepts of matched ontologies, as well as the problem area. In addition, information from the lexical databases is used to increase the accuracy of matching.

Neural network is used for the ontology matching by classification of ontology concepts. Learning is performed based on the concepts of one ontology and then neural network is trying to classify the concepts of another ontology vice versa. The result of both matches composes in common alignment. Background knowledge can be used here to provide additional information about concept meaning during the learning and classification phases.

If the automated ontology matching for a certain problem area is not efficient enough, a corresponding community of professionals is formed. This community is constituted of the SCPS participants' employees. They have different roles and competences what allows to engage experts of the problem area into the ontology matching processes. Community-based support makes it possible to share, discuss and reuse the alignment knowledge throughout the SCPS.

Integration of the community-based ontology matching model allows to increase the completeness and quality of the alignment, but requires significant time from the side of community members [24].

The proposed ontology matching approach includes the following main steps.

1. The ontologies of problem areas are formed based on the ontologies of the SCPS and its participants that are to be matched (e.g., O_1, O_2). The formed ontologies include the abstract and operational contexts.
2. Calculation of intermediate matches:
    a. Specific problem areas are defined using abstract contexts (AC_1, AC_2). A problem area defines a set of ontology matching patterns that can be used during the matching process in this area. The result of pattern-based matching is a matrix (A_1) containing alignment coefficients of ontologies elements.
    b. Neural network is learned first based on the concepts of ontology O_1. To increase accuracy the background knowledge is provided in addition to each concept. After the learning phase6 neural network tries to classify concepts from the ontology O_2. Concepts that has been classified as concepts of ontology O_1

are marked as corresponding. Then the process repeated based on concepts from the ontology O_2. The result of the matching with neural network is alignment matrix A_2 that contains correspondences found by both classifications.

c. The context-based matching of operational contexts (OC_1, OC_2) is done. During this context-based matching, the instances of ontology elements constituting the contexts are matching. The result is matrixes (one matrix for each couple of operational contexts, A_3) that contain alignment coefficients for the matched instances.

3. The results of all autonomous matchers are used to refine the matching coefficients in the ontologies alignment matrix (A'). For example, the value of the alignment coefficient can be increased if the value of the alignment coefficient of instances is higher than the value of the element alignment coefficient. The calculation of matrix a is proceeded by the following formula:

$$A' = f(A_1, \ldots, A_n) = \sum_{i=1}^{n} \alpha_i A_i, \tag{1}$$

where α_i - weight coefficients to adjust impact of pattern-based, neural network and context-based matching results; $\sum_{i=1}^{n} \alpha_i = 1$.

4. The alignment coefficients $(a'_{ij} \in A')$ are checked for each problem area if their values fall into the predefined value ranges. The ranges define if the elements are aligned, not aligned or the alignment is ambiguous (the alignment cannot be defined automatically). The range boundaries are set by experts from the professional community (community members who have highest ratings among other members with same roles) before the matching process. The professional community resolves such ambiguous situations when the alignment coefficient belongs to the ranges of ambiguous alignment coefficient values (e.g., $[b_1, b_2]$). The coefficients of matching technique impact $\alpha_1, \alpha_2, \alpha_3$ can also be adjusted by professional community.

5. After passing the above steps, the resulting alignment (A) is defined.

The recommendation of the professional community can also provide an adjustment of the alignment coefficients of range boundaries, creation of the new matching patterns or modification (adaptation, adjustment) of existing matching patterns.

5 Architecture of Ontology Matching Service

To provide semantic interoperability between the entities of socio-cyberphysical system it is proposed to develop a special service for entities' ontology matching. Architecture of the service is presented on the Fig. 2. It describes a needed modules and links between them to provide the ontology matching.

Architecture can be conditionally divided into three major components related to the ontology matching in the socio-cyberphysical system. The first one is related to the entities of the SCPS, which needs to have support for semantic interoperability. Also, among these entities there are a *community of experts* who participate in the process of ontologies matching, providing their knowledge and experience to adjust the alignment obtained automatically. The entity of the system that request semantic interoperability

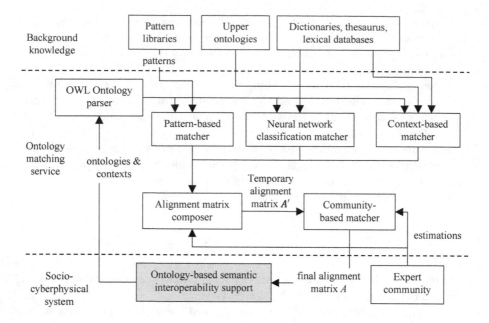

Fig. 2. Architecture of ontology matching service for socio-cyberphysical system

support provides ontologies in form of OWL and the corresponding context to the *ontology matching service*.

The received ontologies and context are divided into abstract and operational contexts in the ontology matching service, according to the approach described in the previous section. Abstract contexts are matched in the module *"Patter-based matcher"*, which uses background knowledge in form of ontology design and correspondence patterns from public *pattern libraries*. Operational contexts are used in the *"Context-based matcher"*, which, in addition to ontologies' own context, refers to sources of background knowledge for additional information while calculating the correspondence. These sources are upper level ontologies and lexical databases.

Also, ontology concepts are transferred to the input of the neural network, where the training and further classification is carried out. The results of the work of all matchers are combined according to the formula (1) in the module *"Alignment matrix composer"*. The intermediate result of the matching (temporary alignment matrix A') is transferred to the *"Community-based matcher"*, where it can be adjusted by the community of experts that are part of the socio-cyberphysical system. Decisions of experts can also influence to the work of *"Alignment matrix composer"*, by changing the coefficients in formula (1). The result is transferred to the entity of the socio-cyberphysical system, which initiated the process of ontology matching.

6 Conclusion

One of the ways to support semantic-based interoperability in a SCPS is the ontology management technology. Development of efficient methods of ontology matching could significantly improve the knowledge processing and interoperability between them.

The accuracy of existing methods can be increased by using the background knowledge. There are a lot of works aimed to utilize background knowledge for ontology matching. Two main classes of background knowledge have been distinguished in the paper based on the way of their gathering: explicit and implicit. Explicit sources can be accessed by simple query without any additional processing of provided data. The following types of this kind of sources have been proposed in the paper: linked data, domain specific or general-purpose ontologies, dictionaries and thesauri and lexical databases. The implicit sources require data processing before they can be used in the matching process. The examples of this kind of sources are ontology design and correspondence patterns.

To utilize background knowledge in ontology matching an approach has been proposed that combines four models of ontology matching: pattern-based, context-based, neural network classification, and community-driven matching. All of them can be configured to use background knowledge during the matching process. The results from first three methods are composed to temporal alignment matrix that is adjusted by community of socio-cyberphysical system.

The architecture of ontology matching service based on the proposed approach has been described in the paper. The architecture allows to change the number of matcher modules by adding or removing any of them and adjusting the function that calculates the temporal alignment matrix. The usage of community-based matcher allows to utilize knowledge and experience of the socio-cyberphysical system's community if autonomous matchers cannot find correspondences. It is expected that community-based matcher will be used in the cases of strong uncertainty, when the most of alignment coefficients calculated by autonomous matchers will belong to the ranges of ambiguous alignment coefficient values. Therefore, the slowest matcher will determine the overall speed of computations for the matching process, which is now the neural network-based matcher due to the difficult procedure of learning for classification of ontology concepts.

The future work is aimed at the evaluation of the proposed approach. The estimations of processing time, accuracy and precision will be conducted based on the ontologies that are provided by OAEI (Open Alignment Evaluation Initiative). Using the OAEI allows to compare the proposed approach with other approaches through usage of the same dataset. Three main metrics of the alignment quality will be calculated: precision, f-measure and recall.

Acknowledgments. This work was supported by projects funded by grants # 16-29-12866, 16-29-04349, 17-07-00327, 17-07-00328 of the Russian Foundation for Basic Research, programs # I.5 (# 0073-2015-0006) and # III.3 (# 0073-2015-0007) of the Russian Academy of Sciences. The work has been partially financially supported by Government of Russian Federation, Grant 074-U01.

References

1. Euzenat, J., Shvaiko, P.: Ontology Matching, 2nd edn. Springer, Heidelberg (2013)
2. Faria, D., Pesquita, C., Santos, E., Palmonari, M., Cruz, I.F., Couto, F.M.: The agreementmakerlight ontology matching system. In: Meersman, R., Panetto, H., Dillon, T., Eder, J., Bellahsene, Z., Ritter, N., De Leenheer, P., Dou, D. (eds.) ODBASE 2013. LNCS, vol. 8185, pp. 527–541. Springer, Heidelberg (2013)
3. Tigrine, A.N., Bellahsene, Z., Todorov, K.: Light-weight cross-lingual ontology matching with LYAM ++. In: Debruyne, C., Panetto, H., Meersman, R., Dillon, T., Weichhart, G., An, Y., Ardagna, C.A. (eds.) ODBASE. LNCS, vol. 9415, pp. 527–544. Springer, Cham (2015). doi:10.1007/978-3-319-26148-5_36
4. Tigrine, A.N., Bellahsene, Z., Todorov, K.: Selecting optimal background knowledge sources for the ontology matching task. In: Blomqvist, E., Ciancarini, P., Poggi, F., Vitali, F. (eds.) EKAW 2016. LNCS, vol. 10024, pp. 651–665. Springer, Cham (2016). doi:10.1007/978-3-319-49004-5_42
5. Shvaiko, P., Euzenat, J.: Ten challenges for ontology matching. In: Meersman, R., Tari, Z. (eds.) OTM 2008. LNCS, vol. 5332, pp. 1164–1182. Springer, Heidelberg (2008). doi:10.1007/978-3-540-88873-4_18
6. Shvaiko, P., Euzenat, J.: Ontology matching: state of the art and future challenges. IEEE Trans. Knowl. Data Eng. **25**(1), 158–176 (2013)
7. Giunchiglia, F., Shvaiko, P., Yatskevich, M.: Discovering missing background knowledge in ontology matching. In: 17th European Conference on Artificial Intelligence (ECAI 2006), vol. 1, pp. 382–386 (2006)
8. Aleksovski, Z., Ten Kate, W., Van Harmelen, F.: Exploiting the structure of background knowledge used in ontology matching. In: Proceedings of the 1st International Workshop on Ontology Matching (OM 2006), vol. 225, pp. 13–24 (2006)
9. Aleksovski, Z., ten Kate, W., van Harmelen, F.: Ontology matching using comprehensive ontology as background knowledge. In: Proceedings of the International Workshop on Ontology Matching at ISWC, pp. 13–24 (2006)
10. Aleksovski, Z., Ten Kate, W., Van Harmelen, F.: Using multiple ontologies as background knowledge in ontology matching. In: CEUR Workshop Proceedings, vol. 351, pp. 35–49 (2008)
11. Falconer, S.M., Noy, N.F., Storey, M.A.: Towards understanding the needs of cognitive support for ontology mapping. In: Proceedings of the 1st International Workshop on Ontology Matching (OM 2006), vol. 225, pp. 25–36 (2006)
12. Falconer, S.M., Storey, M.-A.: A cognitive support framework for ontology mapping. In: Aberer, K., et al. (eds.) ASWC/ISWC -2007. LNCS (LNAI, LNB), vol. 4825, pp. 114–127. Springer, Heidelberg (2007). doi:10.1007/978-3-540-76298-0_9
13. Saruladha, K., Ranjini, S.: COGOM: cognitive theory based ontology matching system. Procedia - Procedia Comput. Sci. **85**, 301–308 (2016)
14. Peng, Y., Munro, P., Mao, M.: Ontology mapping neural network: an approach to learning and inferring correspondences among ontologies. In: CEUR Workshop Proceedings, vol. 658, pp. 65–68 (2010)
15. Rong, S., Niu, X., Xiang, E.W., Wang, H., Yang, Q., Yu, Y.: A machine learning approach for instance matching based on similarity metrics. In: Cudré-Mauroux, P., et al. (eds.) ISWC 2012. LNCS, vol. 7649, pp. 460–475. Springer, Heidelberg (2012). doi:10.1007/978-3-642-35176-1_29

16. Svátek, V., Vacura, M., Homola, M., Kl'uka, J.: Mapping structural design patterns in OWL to ontological background models. In: Proceedings of the Seventh International Conference on Knowledge Capture - K-CAP 2013, pp. 117–120. ACM Press, New York (2013)
17. Hammar, K.: Ontology design patterns in webprotégé. In: 14th International Semantic Web Conference (ISWC-2015), Betlehem, pp. 1–4 (2015)
18. Hamdi, F., Reynaud, C., Safar, B.: Pattern-based mapping refinement. In: Cimiano, P., Pinto, H.S. (eds.) EKAW 2010. LNCS (LNAI, LNB), vol. 6317, pp. 1–15. Springer, Heidelberg (2010). doi:10.1007/978-3-642-16438-5_1
19. Scharffe, F., Euzenat, J., Fensel, D.: Towards design patterns for ontology alignment. In: Proceedings of the 2008 ACM Symposium on Applied Computing - SAC 2008, p. 2321. ACM Press, New York (2008)
20. Scharffe, F., Fensel, D.: Correspondence patterns for ontology alignment. In: Gangemi, A., Euzenat, J. (eds.) EKAW 2008. LNCS (LNAI, LNB), vol. 5268, pp. 83–92. Springer, Heidelberg (2008). doi:10.1007/978-3-540-87696-0_10
21. Scharffe, F., Zamazal, O., Fensel, D.: Ontology alignment design patterns. Knowl. Inf. Syst. 40(1), 1–28 (2014)
22. Smirnov, A., Levashova, T., Shilov, N.: Patterns for context-based knowledge fusion in decision support. Inf. Fusion 21, 114–129 (2015)
23. Scharffe, F.: Representing correspondence patterns with expressive ontology alignments. Ph.D. thesis, University of Innsbruck, Innsbruck, Austria (2008)
24. Zhdanova, A.V., Shvaiko, P.: Community-driven ontology matching. In: Sure, Y., Domingue, J. (eds.) The Semantic Web: Research and Applications. ESWC 2006. LNCS, vol. 4011, pp. 34–49. Springer, Heidelberg (2006). doi:10.1007/11762256_6

Battery Monitoring Within Industry 4.0 Landscape: Solution as a Service (SaaS) for Industrial Power Unit Systems

Mathieu Devos[1(✉)] and Pavel Masek[2,3]

[1] Tampere University of Technology, Tampere, Finland
mathieu.devos@tut.fi
[2] Brno University of Technology, Brno, Czech Republic
[3] Peoples Friendship University of Russia (RUDN University), Moscow, Russia

Abstract. The current globalization already faces the challenge of meeting the continuously growing demand for new consumer goods by simultaneously ensuring a sustainable evolution of human existence. The industrial value creation must be geared towards sustainability. In order to overcome this challenge, tightly coupling the production and its axiomatization processes is required in the paradigm of Industry 4.0. This technology bridges together a vast amount of new interconnected smart devices being mostly battery powered. Batteries are the heart of industrial motive power and electric energy storing solutions in the infrastructures of today. The charges related to the batteries are among the biggest cost (2.000–5.000 EUR per unit). Unfortunately, the batteries are not always treated properly and the badly managed ones lose their ability to store energy quickly. In this work, we present the developed modular Cloud solution utilizing Solution as a Service (SaaS) to monitor and manage industrial power unit systems. Modular approach is realized using simple miniature non-intrusive wireless sensors combined with cloud platform that provides the battery intelligence.

Keywords: Industry 4.0 · Internet of Things · Battery consumption · Industrial motive battery · IoT platform

1 Introduction

Today, modern cyber-physicals systems are facing the new phase in the transformation towards forthcoming interconnected industrial revolution [1]. Recent advancement has driven the dramatic increase for industrial productivity which in turn is accelerating the development in industrial automation platforms and devices [2]. Generally, the Industry 4.0 paradigm could be described as "the merger of digital with physical workflows", i.e., physical production phases are strongly integrated and accompanied by computer-based processes, as it is depicted in Fig. 1. A wide range of factory machinery, operator's equipment, and vehicles involved would be interconnected optimizing the production process deeply [3].

© Springer International Publishing AG 2017
O. Galinina et al. (Eds.): NEW2AN/ruSMART/NsCC 2017, LNCS 10531, pp. 40–52, 2017.
DOI: 10.1007/978-3-319-67380-6_4

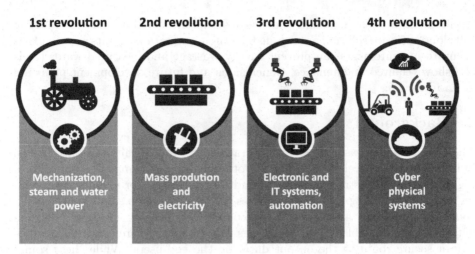

Fig. 1. Evolution towards Industry 4.0

The ongoing development towards Industry 4.0 has shown a tremendous influence on "goods" manufacturing industry in particular [4]. It is based on the establishment of smart factories, smart products, and smart services as part of the Internet of Things (IoT) coupled with already in the limelight Industrial Internet [5]. Indeed, the number of industrial-operating "machines" even today is vast and it is expected to grow further [6–8]. While a significant portion of those is constantly connected to the power supply, many are battery operated, and this trend is not expected to be changed[1] especially in case of Machine-Type Communications (MTC) [9].

Industrial motive power batteries, such as the ones used in forklifts, are often a large price factor in the costs concerning these devices [10,11]. Such batteries are either charged when thoroughly drained, or occasionally, the latter proving to be more wasteful for the overall life of the battery. Conventional forklifts, or other industrial-sized mobile appliances, usually measure their state of charge (SoC) in simplified methods which is no longer suitable in an Industry 4.0 scenario. The rise of Wireless Sensor Networks [12], their industrial offspring, Industrial Wireless Sensor Networks (IWSN) [13], and Cloud of Things [14] have both made beneficial advancements to energy consumption and the move towards the Industrial Automation [15] in Smart Cities [16,17]. The main goal of this manuscript is to present the developed integrated system providing cloud access to monitoring IoT devices, and to bring the reader's attention to the improper battery utilization issues.

The rest of the manuscript is organized as follows. Section 2 provides background information on the topic. In Sect. 3, the full system solution to the

---

[1] See "Five fundamental considerations for a battery-powered, wireless IoT sensor product", 2016: http://www.silabs.com/whitepapers/battery-life-in-connected-wireless-iot-devices.

requirement problem, including the device setup, measured parameters, and the full cloud software solution are detailed. In the next section, results showing that accurate measurements can prevent energy wastage and measured water levels can show required maintenance. Conclusions and future development plans are given in Sect. 5.

2 Background

The idea behind measuring a single parameter in industrial sized batteries is, indeed, not very novel [18,19]. On the other hand, the complete solution of a highly scalable Industry 4.0 is still a mind touching topic for manufactures willing to optimize their production expenses [20,21]. The main set of requirements could be formalized as follows: (i) to combine elementary measuring devices with a highly scalable Cloud of Things (CoT) middleware; and (ii) to provide a secure route to the output data for the end-users. While these sound plainly implementable, the details of the complete system reveal themselves more challenging.

The process of the sensing device design is constrained by the following set of requirements.

– **Size:** The device should be suitable to fit in the free space of the battery storage casing, while still the transmitter should be powerful enough to export the stored data through the IP-enabled wireless network.
– **Storage:** The corresponding measuring devices operate under an unreliable network connection, thus the need to store the recorded data is added in the sizing compartment, i.e., aggregating the data inside the device until an active network connection is established is a must.
– **Protection:** The harsh chemical and physical surroundings of the deployment have a significant impact on the design. Due to the sensor placement next to the lead-acid, water submerged batteries, the casing should be resistant to all of the aforementioned materials still allowing a signal to emerge from the casing, and penetrate through the often heavily plated industrial settings.
– **Execution efficiency:** The fuel cells are designed for simplicity and execution cost, because of this, all the measurements are required to be run by the device itself. This includes designing the individual sensors for their specific environments and measurements [22].
– **Security:** As the devices can become plentiful, distributed among different costumers, the network protocols should support security out of the box.
– **Interoperability:** Tied in with security, the sensors are required to be capable of running on top of state of the art, Internet of Things, application layer protocols [23–25], such as MQTP, SIP, CoAP, AMQP, WebSocket, and legacy HTTP.

The needs continue for the middleware and presentation entities to the end-users. Middleware [26] requirements and the corresponding specific role are often unclear to inexperienced developers [27].

The full stack solution should supplement one or more unified input gateways for the data to be received. These gateways require high scalability, providing for a massive number of devices as the Industrial Internet of Things (IIoT) [28] continues to grow exponentially. Moreover, the gateways must accept the Application Layer protocols, in addition to providing authentication, registration, and integrity for the connected devices. In short, the gateways receive the data, while providing means to do over a secure channel.

Following the information reception concept, the data is piped through the system to data manipulation and storage entity. The related flows are either part of a cloud solution or are to be self-implemented. The enablers for cloud solutions to be considered are Amazon AWS, Google IoT, Microsoft Azure, or other suppliers. On the other hand, they can be custom developed in systems such a self-developed *node.JS* environment. Similarly, the data storage can be either part of a cloud full stack solution, or any self-chosen high-scalability database. Big Data solutions are presented to the user custom implementation or Big Data cloud entities as part of a larger solution [29].

During the final execution steps, the collected and manipulated data is presented to the different sets of end-users [30,31]. The end-users can be either consumers, developers, clients, or any other custom type requesting to interpret the collected data. Moreover, connecting to this end system to review the selection should happen only over secure channels [32], which are often mandatory in state-of-the-art cloud solutions. Alternatively, a developer can choose to design their own security system, or use predefined protocols for the execution. The overview of the system is given in Fig. 2.

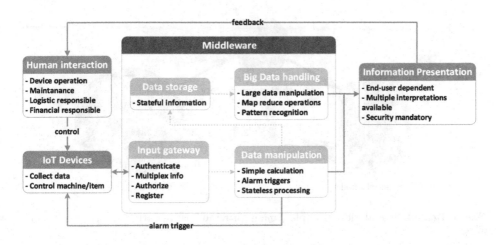

Fig. 2. Requirement framework overview

The consumers efficiently pay "as they go" using the SaaS business model [33], i.e., the initial cost is only limited to the amount of requested measuring devices. In earlier stages of the solution, installation and devices themselves may be provided without a cost. Beyond the installation, a simple thin client application, often a web browser, is used to access the end data. Delivering the business model becomes quite straightforward, as the end-user simply pays for the selected services, plus a static cost concerning the device and its installation [34].

3   Complete System Solution

The previous section surveyed the requirements and showed a general list of potential solutions applicable for the implementation of battery-powered Industry 4.0 machinery. In this section, we inspect the specific requirements for the system, and discuss the selected solution and its reasoning.

The selected data processing device is a straightforward Internet enabled system on chip solution with different external sensors. Protection for the device comes in the form of a plastic all-surrounding case. Connection to the Internet is provided through a IEEE 802.11n (a.k.a. WiFi) infrastructure connection [35,36] of the factory (or BLE for local monitoring [37]), i.e., the devices are required to connect to the costumers WiFi in order to forward the data to the middleware [38]. The developed solution is provided as system from the box and the example installation from Bamomas industrial trial is given in Fig. 3. Current version of the device is suitable for sensing the following set of measurements: (i) Shunt-voltage; (ii) Voltage; (iii) Temperature; and (iv) Water level (custom developed sensor).

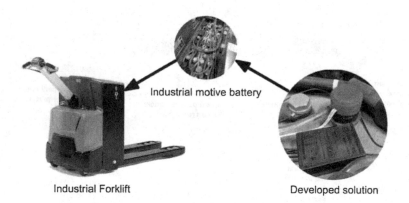

Fig. 3. Real-life installation example (Image provided as a courtesy by the Bamomas project.)

Further, the data is transferred to the middleware. The Amazon AWS IoT solution was selected mainly due to its scalability[2]. AWS IoT maintains a large

---

[2] See "AWS IoT – Amazon Web Services", 2017: https://aws.amazon.com/iot/.

portion of the full-scale middleware, specifically it handles the input gateway, and a potential minimal chunk of the data manipulation.

Additional data manipulation is allocated with the implementation of the AWS Lambda[3]. Data storage is available using Amazon DynamoDB.

The current representation to end-users is implemented through an AWS Lambda function which forwards the data securely to the IoT-Ticket service[4]. IoT-Ticket provides a secure connection for the users, and meaningful interface to display the collected data. A snapshot of the measured data for a single device is depicted in Fig. 4.

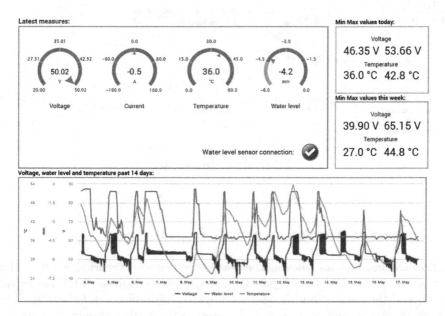

Fig. 4. End-user data presentation (Screenshot provided as a courtesy by the Bamomas project.)

The end-users are often not only people working on the project, but also customers willing to gain more information on their battery usage. End-users are thus responsible for the feedback loop to the human interaction with the device, in this case, the industrial motive battery. Since many of these batteries are only designed to be utilized for a specific amount of cycles, this IoT solution provides them with information as to when the energy storage unit becomes obsolete. Furthermore, preventing occasional charging may be beneficial for the battery in the long run, as occasional charging consumes a cycle. While this type of

---

[3] See "AWS Lambda – Serverless Compute – Amazon Web Services", 2017: https://aws.amazon.com/lambda/.

[4] See "IoT-Ticket, the Office Suite for Internet of things", 2017: https://iot-ticket.com/.

charging can be beneficial for the operation of the motive device, short charging times may "stress" the battery, and lead to a faster degradation. Given this complete solution, we present the updated full system solution with Bamomas elected implementations in Fig. 5.

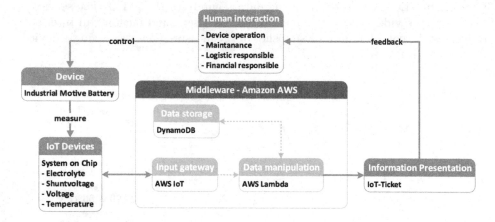

Fig. 5. Full system solution

4   Operation Results

To show the benefits of deploying the system we provide numerical analysis of a large set of measured data. We provide a step-by-step analysis of the numerical results, and their importance in moving towards an Industry 4.0 platform. The first set of data, presented in Fig. 6, shows a scatter plot containing the energy difference $\Delta E = E_{intake} - E_{out}$. To clarify the data, we define that every time the battery is connected to the charger, a new cycle is initialized. During this time, we measure the total connected time with the charger, presented on the X-axis in Fig. 6, and the shunt voltage. The shunt voltage is used to calculate the real current, and with the combination of these variables we can calculate the ΔE, measured in Ampere*hour (Ah). During the normal operation of the battery, we enumerate the total spent energy through similar measurements resulting in the total energy output. The difference between the energy intake and spent energy is thus corresponded as the delta energy (ΔE), given in Fig. 6. As such, some of the cycles contain the possibility of a larger energy expenditure compared to the intake, as such these values can be negative.

As it is depicted in the Fig. 6, the beginning of the curve has exponential behavior. The *slow start* is explained due to the water heating up when the battery is connected to the charger. The liquid unavoidable heating consumes a significant amount of the incoming energy after the slow charging phase is over, the battery charge status is affected tremendously. Further, we may observe the

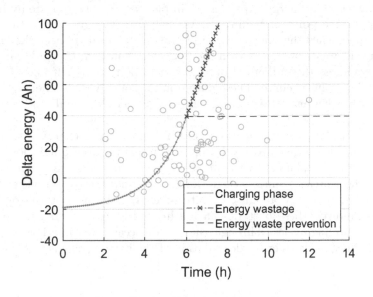

Fig. 6. Scatter plot ΔE versus time

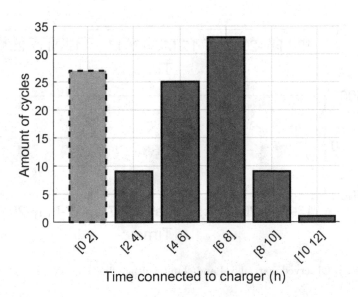

Fig. 7. Amount of cycles for a given charging time

linear section of the curve appearing at around 40 Ah, showing that the battery is fully charged. The energy expended is simply "wasted" by repeatedly heating the water, to the point of boiling it. As such, we strongly advice that when the data approaches the linear section of the curve, to disconnect the charger, preventing the waste of energy with the optimized solution. Closing remarks on this figure address the outliers of data in the section to the left and above the exponential and linear curve. These data points show a very high energy intake coupled with a low energy spending, thus being very wasteful, and are to be always avoided.

Figure 7 displays the amount of cycles for their given charging times. We observe that a large group of cycles only contains an exceedingly low charge time. This can be explained by the practice of occasional charging. An operator who has a short break, connects the forklift to the charger to ensure that he/she does not need to worry about the state of charge throughout the operation timespan. However, these batteries have a limited lifespan[5], often measured in amount of charging-cycles. Through the use of occasional charging, a cycle is consumed which leads to faster degradation of the battery, this is to be avoided.

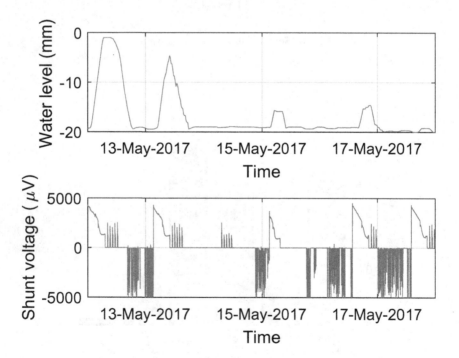

Fig. 8. Observed water level and shunt voltage fluctuations over time

---

[5] See "Industrial Batteries Motive Power Economical robust reliable EPzS series": http://www.sue.ba/sue/files/file/classic/Classic_EPzS_int.pdf.

In Fig. 8 we observe the water level fluctuation over time in conjunction with the shunt voltage over time. This clearly illustrates that charging the battery causes the water level to rise due to a mix of the expansion of the liquid and the partial boiling of the water. The charging status is given in Fig. 8 (bottom) plot, which shows a positive shunt voltage during the charge, zero during idle operations, and a negative during a discharge. However, the Fig. 8 also points to a problem in the battery – we note a lower maximum of the water level during the charging status over time. This indicates a too low water level, and should be conveyed to the operator of the battery to top the battery up with water preventing degradation of the capacitor. In current operation modus, this simply triggers an alarm informing that the item requires maintenance.

It is possible to anticipate the degradation of the batteries when their energy intake and outtake rapidly increase compared to nominal numbers. Furthermore, the maintenance of the water levels for the batteries could be predicted, and the optimal charging times for a certain energy expenditure could be delivered as well.

5 Conclusion and Future Work

In this work, we have presented the developed modular Cloud utilizing Solution as a Service (SaaS) to monitor and manage industrial battery systems for Industry 4.0 scenario. The proposed modular approach utilizes simple miniature non-intrusive wireless sensor combined with cloud platform that provides the battery intelligence. The integration of the proposed framework allows to obtain the information about upcoming problems before they really occur with the battery; the maintenance cycles could be organized before the battery actually faults, and the lifetime can be improved.

Future goals include data collection, aggregation, and deep analysis for longer periods of time, more specifically, the full analysis of the lifetime of battery. Here, we suggest testing the observed numerical values to the common knowledge of batteries and improve the alarms for degradation of said batteries. Furthermore, feedback can be coupled with the complete system using alarms to disconnect the charger automatically preventing the waste of energy after the battery has been fully charged.

Finally, supplementary measuring of the acidity of the battery, individual cell measurements, internal cell temperature, would provide us with more accurate measurements.

Acknowledgments. Research described in this paper was financed by the National Sustainability Program under grant LO1401. For the research, infrastructure of the SIX Center was used. This research was funded by the Technology Agency of Czech Republic project No. TF02000036.

We thank our colleagues from the Bamomas project for providing insight and expertise into the development of industrial motive sensors and batteries. The IoT devices and framework are developed by Bamomas. We acknowledge Pavel Marek and

Dr. Raimo Vuopionperä for assistance with the numerical data analysis and unique insights in the life of industrial motive battery units.

References

1. Hermann, M., Pentek, T., Otto, B.: Design principles for industrie 4.0 scenarios. In: Proceedings of 49th Hawaii International Conference on System Sciences (HICSS), pp. 3928–3937. IEEE (2016)
2. Rüßmann, M., Lorenz, M., Gerbert, P., Waldner, M., Justus, J., Engel, P., Harnisch, M.: Industry 4.0: the future of productivity and growth in manufacturing industries. Boston Consulting Group, p. 14 (2015)
3. Schmidt, R., Möhring, M., Härting, R.-C., Reichstein, C., Neumaier, P., Jozinović, P.: Industry 4.0 - potentials for creating smart products: empirical research results. In: Abramowicz, W. (ed.) BIS 2015. LNBIP, vol. 208, pp. 16–27. Springer, Cham (2015). doi:10.1007/978-3-319-19027-3_2
4. Stock, T., Seliger, G.: Opportunities of sustainable manufacturing in Industry 4.0. In: Proceedings of CIRP Conference, vol. 40, pp. 536–541 (2016)
5. Wan, J., Tang, S., Shu, Z., Li, D., Wang, S., Imran, M., Vasilakos, A.V.: Software-defined industrial Internet of Things in the context of Industry 4.0. IEEE Sens. J. **16**(20), 7373–7380 (2016)
6. VNI Cisco: Global mobile data traffic forecast 2016–2021. White Paper (2017)
7. Stusek, M., Masek, P., Kovac, D., Ometov, A., Hosek, J., Kröpfl, F., Andreev, S.: Remote management of intelligent devices: using TR-069 protocol in IoT. In: Proceedings of 39th International Conference on Telecommunications and Signal Processing (TSP), pp. 74–78. IEEE (2016)
8. Ometov, A., Bezzateev, S., Kannisto, J., Harju, J., Andreev, S., Koucheryavy, Y.: Facilitating the delegation of use for private devices in the era of the internet of wearable things. IEEE Internet Things J. (2017)
9. Gerasimenko, M., Petrov, V., Galinina, O., Andreev, S., Koucheryavy, Y.: Energy and delay analysis of LTE-advanced RACH performance under MTC overload. In: Proceedings of IEEE Globecom Workshops (GC Workshops), pp. 1632–1637. IEEE (2012)
10. Seitz, C.W.: Industrial battery technologies and markets. IEEE Aerospace Electron. Syst. Mag. **9**(5), 10–15 (1994)
11. Renquist, J.V., Dickman, B., Bradley, T.H.: Economic comparison of fuel cell powered forklifts to battery powered forklifts. Int. J. Hydrogen Energy **37**(17), 12054–12059 (2012)
12. Yick, J., Mukherjee, B., Ghosal, D.: Wireless sensor network survey. Comput. Netw. **52**(12), 2292–2330 (2008)
13. Gungor, V.C., Hancke, G.P.: Industrial wireless sensor networks: challenges, design principles, and technical approaches. IEEE Trans. Indus. Electron. **56**(10), 4258–4265 (2009)
14. Parwekar, P.: From Internet of Things towards cloud of things. In: Proceedings of 2nd International Conference on Computer and Communication Technology (ICCCT), pp. 329–333. IEEE (2011)
15. Shrouf, F., Ordieres, J., Miragliotta, G.: Smart factories in Industry 4.0: a review of the concept and of energy management approached in production based on the Internet of Things paradigm. In: Proceedings of International Conference on Industrial Engineering and Engineering Management (IEEM), pp. 697–701. IEEE (2014)

16. Masek, P., Masek, J., Frantik, P., Fujdiak, R., Ometov, A., Hosek, J., Andreev, S., Mlynek, P., Misurec, J.: A harmonized perspective on transportation management in smart cities: the novel IoT-driven environment for road traffic modeling. Sensors **16**(11), 1872 (2016)
17. Fujdiak, R., Masek, P., Mlynek, P., Misurec, J., Olshannikova, E.: Using genetic algorithm for advanced municipal waste collection in smart city. In: Proceedings of 10th International Symposium on Communication Systems, Networks and Digital Signal Processing (CSNDSP), pp. 1–6. IEEE (2016)
18. Ceraolo, M.: New dynamical models of lead-acid batteries. IEEE Trans. Power Syst. **15**(4), 1184–1190 (2000)
19. Sharaf, H.: Method of testing the capacity of a lead-acid battery. US Patent 3,808,522 (1974)
20. Lee, J., Bagheri, B., Kao, H.A.: A cyber-physical systems architecture for industry 4.0-based manufacturing systems. Manuf. Lett. **3**, 18–23 (2015)
21. Brettel, M., Friederichsen, N., Keller, M., Rosenberg, M.: How virtualization, decentralization and network building change the manufacturing landscape: an industry 4.0 perspective. Int. J. Mech. Indus. Sci. Eng. **8**(1), 37–44 (2014)
22. Grankin, M., Khavkina, E., Ometov, A.: Research of MEMS accelerometers features in mobile phone. In: Proceedings of the 12th Conference of Open Innovations Association FRUCT; Oulu, Finland, pp. 31–36 (2012)
23. Al-Fuqaha, A., Guizani, M., Mohammadi, M., Aledhari, M., Ayyash, M.: Internet of Things: a survey on enabling technologies, protocols, and applications. IEEE Commun. Surv. Tutorials **17**(4), 2347–2376 (2015)
24. Sheng, Z., Yang, S., Yu, Y., Vasilakos, A., Mccann, J., Leung, K.: A survey on the IETF protocol suite for the Internet of Things: standards, challenges, and opportunities. IEEE Wireless Commun. **20**(6), 91–98 (2013)
25. Masek, P., Hosek, J., Zeman, K., Stusek, M., Kovac, D., Cika, P., Masek, J., Andreev, S., Kröpfl, F.: Implementation of true IoT vision: survey on enabling protocols and hands-on experience. Int. J. Distrib. Sens. Netw. (2016)
26. Bandyopadhyay, S., Sengupta, M., Maiti, S., Dutta, S.: A survey of middleware for Internet of Things. In: Özcan, A., Zizka, J., Nagamalai, D. (eds.) Recent Trends in Wireless and Mobile Networks, vol. 162, pp. 288–296. Springer, Heidelberg (2011). doi:10.1007/978-3-642-21937-5_27
27. Bandyopadhyay, S., Sengupta, M., Maiti, S., Dutta, S.: Role of middleware for Internet of Things: a study. Int. J. Comput. Sci. Eng. Surv. **2**(3), 94–105 (2011)
28. Jeschke, S., Brecher, C., Song, H., Rawat, D.B.: Industrial Internet of Things. Springer, Cham (2017)
29. Olshannikova, E., Ometov, A., Koucheryavy, Y., Olsson, T.: Visualizing big data. In: Big Data Technologies and Applications, pp. 101–131. Springer, Cham (2016). doi:10.1007/978-3-319-44550-2_4
30. Olshannikova, E., Ometov, A., Koucheryavy, Y.: Towards big data visualization for augmented reality. In: Proceedings of 16th Conference on Business Informatics (CBI), vol. 2, pp. 33–37. IEEE (2014)
31. Olshannikova, E., Ometov, A., Koucheryavy, Y., Olsson, T.: Visualizing big data with augmented and virtual reality: challenges and research agenda. J. Big Data **2**(1), 22 (2015)
32. Heer, T., Garcia-Morchon, O., Hummen, R., Keoh, S.L., Kumar, S.S., Wehrle, K.: Security challenges in the IP-based Internet of Things. Wireless Pers. Commun. **61**(3), 527–542 (2011)
33. Buxmann, P., Hess, T., Lehmann, S.: Software as a service. Wirtschaftsinformatik **50**(6), 500–503 (2008)

34. Dubey, A., Wagle, D.: Delivering software as a service. McKinsey Q. **6**(2007), 2007 (2007)
35. Ometov, A.: Fairness characterization in contemporary IEEE 802.11 deployments with saturated traffic load. In: Proceedings of 15th Conference of Open Innovations Association FRUCT, pp. 99–104. IEEE (2014)
36. Condoluci, M., Militano, L., Orsino, A., Alonso-Zarate, J., Araniti, G.: LTE-direct vs. WiFi-direct for machine-type communications over LTE-A systems. In: Proceedings of 26th Annual International Symposium on Personal, Indoor, and Mobile Radio Communications (PIMRC), pp. 2298–2302. IEEE (2015)
37. Iera, A., Militano, L., Romeo, L.P., Scarcello, F.: Fair cost allocation in cellular-Bluetooth cooperation scenarios. IEEE Trans. Wireless Commun. **10**(8), 2566–2576 (2011)
38. Ometov, A.: Short-range communications within emerging wireless networks and architectures: a survey. In: Proceedings of 14th Conference of Open Innovations Association (FRUCT), pp. 83–89. IEEE (2013)

Opportunistic Data Collection for IoT-Based Indoor Air Quality Monitoring

Aigerim Zhalgasbekova[1(✉)], Arkady Zaslavsky[2], Saguna Saguna[1],
Karan Mitra[1], and Prem Prakash Jayaraman[3]

[1] Luleå University of Technology, Skellefteå, Sweden
aigzha-6@student.ltu.se
[2] CSIRO, Melbourne, Australia
[3] Swinburne University of Technology, Melbourne, Australia

Abstract. Opportunistic sensing advance methods of IoT data collection using the mobility of data mules, the proximity of transmitting sensor devices and cost efficiency to decide when, where, how and at what cost collect IoT data and deliver it to a sink. This paper proposes, develops, implements and evaluates the algorithm called CollMule which builds on and extends the 3D kNN approach to discover, negotiate, collect and deliver the sensed data in an energy- and cost-efficient manner. The developed CollMule software prototype uses Android platform to handle indoor air quality data from heterogeneous IoT devices. The CollMule evaluation is based on performing rate, power consumption and CPU usage of single algorithm cycle. The outcomes of these experiments prove the feasibility of CollMule use on mobile smart devices.

Keywords: Internet of Things (IoT) · Opportunistic Sensing (OppS) · Analytic Hierarchy Process (AHP) · Indoor Air Quality (IAQ) monitoring

1 Introduction

Internet of Things (IoT) is a massively distributed network of "things" connected to the Internet where they communicate and exchange data. It has been studied in numerous research papers such as [1–3]. The "things" or "objects" are broadly referred to the electronic devices producing data with the ability to compute, sense the environmental phenomenon and communicate with external world [1]. In this paper, we call them IoT or smart devices. Nowadays, smart buildings are being equipped with an increasing number of IoT devices. Most of these are powered with a battery that has a strictly limited life. Therefore, they need to be recharged periodically. The energy consumption of these large number of IoT devices is a challenging issue [2]. To extend the IoT devices' battery lifespan, there is a need to save the energy consumed by them. Therefore, its reduction is one of the main research challenges today [3].

For solving this challenge, another paradigm has arisen which is called opportunistic sensing (OppS) [4]. It enables techniques whereby sensors negotiate

© Springer International Publishing AG 2017
O. Galinina et al. (Eds.): NEW2AN/ruSMART/NsCC 2017, LNCS 10531, pp. 53–65, 2017.
DOI: 10.1007/978-3-319-67380-6_5

with smart devices while passing by and upload data which is delivered in an energy-efficient manner. OppS uses the opportunity of gathering data about an environment from sensing IoT devices (sensors) by mobile smart gadgets (e.g. smartphones, tablets, etc.) carried by people, without their direct involvement [5]. Thus, there is no need in predefined infrastructure and additional hardware. Moreover, as the devices and sensors directly communicate without mediator hops, it reduces the bandwidth usage and, consequently, energy consumption. However, OppS also brings other challenges like coverage [6], sensing quality, resource cost, privacy, security and data integrity [4]. This paper aims to develop an efficient sensor data collection system applied to indoor air quality (IAQ) monitoring. Its primary contributions are: (1) proposing a system to achieve the energy efficiency while gathering real-time data from IoT devices; and (2) realising and testing it in real life scenario.

The rest of the paper is organised as follows. The next section discusses the related work. Section 3 describes the proposed system namely CollMule. Section 4 presents the implementation of our system. Section 5 evaluates it in real time scenario. Section 6 concludes the paper.

2   Related Work

There are many research works addressing the OppS challenges, we review some of the recent ones [7–12], in this section. Most of these systems are intended for use in the outdoor environment. Only WiFi-Amber [7] can be utilised both outdoors and indoors. Its authors address privacy, security, resource utilisation and sensing effectiveness. SenseMyCity [8] estimates different participant engagement techniques that could enhance sensing quality. R-DA and DC-BSF algorithms [12] target to increase coverage and reduce data redundancy, thereby, improve the effectiveness of memory utilisation. The former divides the sensing area to regions and the latter estimates the frequency of data collection in each region. SCmules [10] priorities the data collected by mobile smart devices for decreasing the system's memory usage. The authors of [11] propose a model to estimate energy consumption of Bluetooth Low Energy devices. While [9] estimates mobile devices' resource usage using different communication protocols and standards. All of the described research works, except WiFi-Amber [7], validate their system through evaluation of its performance.

All of the above-mentioned frameworks directly or indirectly cope with energy constraints using OppS paradigm and proposed algorithms. However, only [12] proposes a method for collecting data from sensors to mobile device considering collection frequency. Though its authors do not estimate the energy efficiency of their method, it indirectly affects it. We believe that the effectiveness of data collection by mobile devices can be improved through eliminating needless connections to unreliable IoT devices. To realise this, the data can be gathered from the most cost-efficient sensing devices. Such method was proposed by 3DkNN algorithm [13]. It dynamically computes a k (number of neighbours in a set) size set of cost-efficient nearest sensors (enclosed by the sphere on Fig. 1) around a mule (the smartphone carried by a human on Fig. 1) within wireless sensor networks (WSNs). Where "mule" is a mobile device carried by a human or vehicle. Its authors consider three-dimensional (3D) space and metrics pointing out to sensor's efficiency such as Signal to Noise Ratio (SNR) and distance. However, the sensors were ranked by estimating a ratio between the metrics manually assigning their weights. Moreover, it was not implemented in real life scenario. In the next section, we propose our system that extends 3DkNN for IoT paradigm.

3   CollMule System

In this paper, we propose a system for efficient data collection from sensing IoT devices. This section presents a motivating scenario to show a need in our system. Further, it describes its architecture based on a basic structure of OppS systems and CollMule algorithm, which define an efficient way of interaction among the architecture's elements.

3.1   Motivating Scenario

This scenario presents how an OppS approach can be applied to IAQ monitoring. Let's assume that the monitoring system is deployed in some office building. Numerous IoT devices sensing different air characteristics are situated in 3D space (see Fig. 1) of the building.

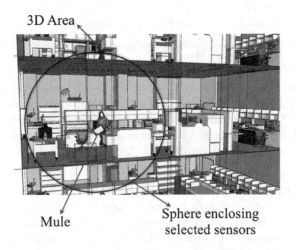

Fig. 1. Sensor selection in a three dimensional area.

For example, let's follow the possible path of a person named Marie who works there and suffers from asthma, who changes her location during the usual working day. Her smartphone (mule on Figure 1) directly gathers data from sensors around her (that are enclosed in a sphere on Figure 1). Thereby, Marie always knows the air quality of her current location. Moreover, she can share the data gathered by the smartphone with her colleagues who can also help Marie in finding the air quality in other parts of the building. However, they are concerned about their gadget's power consumption, because data gathering consumes a considerable amount of its energy. It can prevent them from using such a system. Therefore, there is a need to reduce the energy consumption of such a system so that it can attract people to participate in data gathering.

3.2 CollMule System Architecture

Considering the OppS paradigm, our system consists of a mobile smart device carried by a human ("mule"), sensing appliances and central repository. This paper is focused on the gathering data from the sensing devices and uploading data to mobile devices which are passing by. Therefore, we propose the architecture for each device.

Mule Architecture. Mule is represented as a software application on a mobile smart device that consists of different components (see Fig. 2a): *Sensors Detection Manager (SDM)* discovers sensor nodes that are accessible in the communicational range; *Node's Information Repository (NIR)* stores information about the nodes within a building; *Location Tracker (LT)* traces the device's position; *Query Manager (QM)* is responsible for organising a data collection order; *Communication Interface (CI)* is accountable for establishing a connect to the sensors and obtaining the sensed data, and sending it to the central repository when it is possible; *Data Manager (DM)* is intended to manage data storage until it is sent to the centre layer of the system architecture; *Controller* manages all processes involved in the functioning of the software.

The QM determines an order of most cost-efficient sensors towards the mule. For realising this, it uses information about discovered sensors and the device's location provided by SDM, NIR and LT. Further, the controller passes an ordered list of sensors to CI that connects to k number of them following its order and retrieves their data. The obtained information is then temporarily stored in DS till the centre layer becomes accessible.

Sensing IoT Device Architecture. The role of sensing IoT devices in our system is to measure environmental phenomenon and send the acquired data to mules whenever it is requested. In our system, each of them functions as follows: advertises its accessibility by broadcasting messages that contain information about the device (like RSSI, power level and accuracy) and available types of measurements (combined in a service) it provides; connects to the mule when it is requested; sends the current measures to it. A such device is a complex appliance containing two main parts: an IoT unit and sensors (see Fig. 2b). The former controls accomplishing tasks. It consists of: *Communication Interface (CI)* broadcast messages about availability, and realises connection with mules; *Power Level Estimator (PLE)* computes consumed power; *Sensor Data Extractor*

(SDE) retrieves sensed data from sensors; *Controller* processes requests and manages tasks within the IoT device.

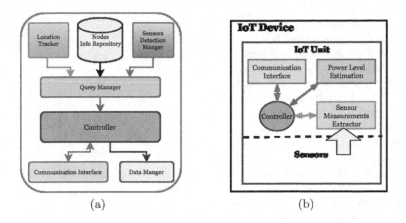

Fig. 2. Architecture of (a) Mule; and (b) Sensing IoT device.

PLE estimates the battery power level which is broadcast with other information about the device by CI. After establishing a connection and receiving a request for sensed data, SDE retrieves it from the sensors part and transfers it to Controller. That sends it to the querying end through CI. The second part of the IoT device contains several different sensors that detect environmental events and changes, which are returned to the IoT unit on demand.

3.3   CollMule Algorithm

This scenario presents how the proposed approach works in application to IAQ monitoring. CollMule extends 3DkNN algorithm [13] discussed in the previous section by introducing it to an IoT environment and an appropriate method for prioritising the IoT devices regarding their efficiency. Thereby, our algorithm provides a more robust computation of the cost-efficient devices in the context of IoT paradigm.

CollMule algorithm has three phases: (1) Discovering phase implies gathering initial information about sensing IoT devices; (2) Ranking phase - ordering the devices according to their efficiency; (3) Collecting phase - obtaining data from the k (number of) most cost-efficient devices.

During the first phase, a mule discovers surrounding IoT devices by listening to radio channels where they broadcast messages informing about their presence. This phase returns L_{avS} and L_{metric} that are lists of available IoT devices and their metrics respectively. The broadcast packets usually contain data about the device like name, signal strength, radio power, manufacturer data, services the device provides etc. In our algorithm, it is assumed that they include metrics pointing out on the device's efficiency such as location, distance, battery

---

Algorithm 1. CollMule Algorithm

Input: Set of IoT nodes with their coordinates L_S.
Output: Set of collected data L_{data}.

```
1  while true do
2  |   switch into scanning state
3  |   detect the presence of IoT devices from L_S
4  |   if IoT devices satisfies a certain condition then
5  |   |   add them into new list L_avS
6  |   |   foreach device d ∈ L_avS do
7  |   |   |   obtain its metrics into L_metric
8  |   |   end
9  |   end
10 |   return L_avS, L_metric;
11 |   obtain the mule's location Location;
12 |   compute AHP to rank the devices in L_avS according the metrics in L_metric
   |   and Location into L_ordered
13 |   foreach device d ∈ L_ordered do
14 |   |   if priority p ≥ certain value then
15 |   |   |   add d to L_kOrdered
16 |   |   end
17 |   end
18 |   return L_kOrdered;
19 |   foreach device d ∈ L_kOrdered do
20 |   |   connect to d & read its data into L_data
21 |   end
22 |   return L_data;
23 end
```

---

level, sensor accuracy, received signal strength indication (RSSI), link quality indication, packet delivery rate, packet error rate [14], data size and etc. Then, CollMule ranks the devices from the less to the most cost-efficient ones using the metrics retrieved from these messages. The ranking phase returns $L_{ordered}$ that is a list of ordered devices. The last phase determines $L_{kOrdered}$ that k size list of devices to collect data from. Further, the mule connects and reads the sensed data from them to L_{data} set. The optimal k size is found experimentally. Algorithm 1 represents a pseudo code of CollMule algorithm.

The ranking phase of CollMule algorithm has to order available IoT devices in communication radio range according to their cost-efficiency towards a mule considering multiple metrics. This process relates to Multiple Criteria Decision Making (MCDM) discipline [15] evaluating multiple incompatible criteria in making a decision. We use Analytic Hierarchy Process (AHP) [16], which is one of the widely used approaches to solving MCDM problems.

Figure 3 presents the IoT devices hierarchy based on different metrics. The first layer is a goal of CollMule's second phase, which targets to rank the available devices regarding their efficiency. The second layer contains efficiency attributes, which are the metrics. The bottom layer comprises available IoT devices for all the metrics in the second layer.

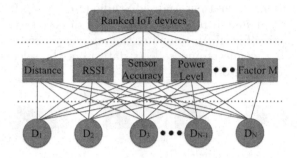

Fig. 3. AHP model for CollMule.

To achieve the goal, the ranking phase, firstly, needs to find the relative cost-efficiency of each IoT device. In a comparison of the two IoT devices, each metric and device's values needs to be assigned to weights taking into account their relative importance. CollMule assigns weight to each attribute using the unified scale from 1 to 9, as suggested in [16], that indicates the importance of one attribute over another. Thereby, it can calculate weights for all the metrics and device's values. For each metric, CollMule constructs a comparison matrix from the weighed devices' values. Then, CollMule calculates priority vectors of these matrices. On the second level, the averages of each metric's values are calculated and weighted. They are used to calculate priority vector of the metrics' comparison matrix. In order to check the consistency of each matrix, the Consistency Ratio (CR) is calculated, which must be less than 10%, according to the AHP theory [16]. Further, the overall device priority is computed from the sum of devices' priorities multiplied by relative metrics' priority.

4 Implementation

We implement our system in the real world scenario and evaluate it through conducting different experiments. The chosen application for the implementation is Indoor Air Quality Monitoring. It is one of the broadly used applications of the IoT systems. It measures air pollutants concentration within indoor environments. We assume that sensor nodes of our system are static. Thus, their addresses and locations are known. Also, we assume that the coordinates of a gathering device are known while the ranking phase. In our implementation, we consider distance, RSSI, power and accuracy level of IoT appliances in the second phase of CollMule algorithm.

Our testbed consists of: two mules - mule prototype implemented on a smartphone Sony Xperia M5 and Google Nexus 7; eight sensing IoT devices - IoT device prototype installed on eight sensing IoT devices composed of a Raspberry Pi, a Sensly Hat, a BME280 and Sharp Dust Sensor.

Mule Prototype Implementation. The application prototype is implemented in Java for Android platform using Android Studio (see Fig. 4). It follows mule's architecture described in the previous chapter. It discovers sensors available within radio range (see Fig. 4(a)), computes the ranking phase of CollMule algorithm (see Fig. 4(b)), connects and collects data from the selected most cost-efficient sensors (see Fig. 4(c)).

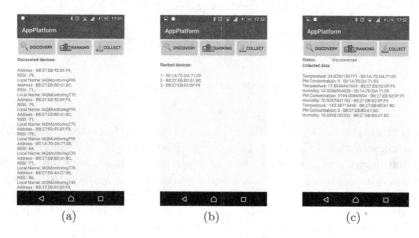

| (a) | (b) | (c) |

Fig. 4. (a) Discovering nearby devices; (b) ranking of available IoT devices; (c) collecting data from the most cost-efficient ones.

Earlier it has been decided to use BLE communication radio technology in our project. We have implemented the part of the interaction with the IoT devices using bluetooth and bluetooth.le packages of Android API. The ranking part of the CollMule algorithm is implemented using the linear package of Apache Commons Math 3.3 API.

Sensing IoT Device Prototype. The prototype running on the IoT devices is implemented using Golang[1] using its Gatt library[2] and Python. Moreover, we have updated BlueZ[3] version to 5.43 on Raspberry Pis as Raspbian Jessie package contains its old version. The prototype implementation corresponds to the sensing layer architecture described in the previous section.

5   System Evaluation

In this section, we run an experiment in real time scenario to demonstrate the system's performance and determining k most efficient IoT devices to collect data

---

[1] https://golang.org.

[2] https://github.com/paypal/gatt.

[3] Bluetooth protocol stack for Linux platform (http://www.bluez.org).

from. Then, we run an experiment with the different number of nearby sensors ("sample size K") varied from two to seven within one room to test the speed of ranking computation. Then, we run another experiment with different k sizes of sensors sets to measure the time taken to collect the data from them. Moreover, there are two different connection ways: sequential and simultaneous which are compared here. Within this experiment, we also measure Battery Power and CPU Load of the mobile application prototype while scanning, ranking and gathering data. For this purpose, we use an application called "Trepn Profiler[4]". In our experimental setup, the sensing devices are sparsely distributed because of their limited number.

5.1 Experiments

After each element of the system has been implemented, we deploy our system within Luleå University of Technology's[5] B building of Skellefteå Campus. Figure 5 demonstrates our experimental network setup on the plan of the building's second floor. The stars signed with the letters are positions of the collectors from which they gather data from sensors that are illustrated as blue circles on the plan. Also, the dotted blue circles represent the sensors situated on the staircase and the first floor. Red dotted lines show the user's path.

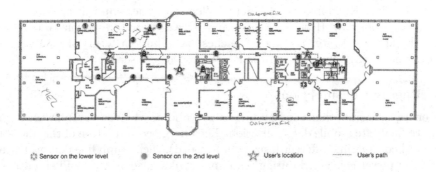

Fig. 5. Experimental setup on the second floor plan.

A basic experimental scenario is: the mule (user with his/her smart device) follows the path and collects the sensor data at location points that are shown on Fig. 5. The sensing IoT devices advertise their availability and service they provide. Following CollMule algorithm, mule collects the data from k number of selected sensors that are available nearby at that moment and location.

Therefore, the mobile smart device detects the nearby sensors via BLE scanning (listening radio channels for sensory advertisements). After that, the scan results are processed and the devices are ranked regarding the weights computed

---

[4] https://developer.qualcomm.com/software/trepn-power-profiler.
[5] https://ltu.se.

using AHP. According to the results, the number of nearby sensors for different positions varies from three to five. However, not all of the visible sensors have an appropriate signal strength for connection. Therefore, such devices are excluded from the ranking. The high weights define the most cost-efficient sensors which data should be gathered. In cases when only two devices are ordered the data can be gathered from both of them as their weights differ slightly. In other cases, we can exclude the sensors having the weight values less than 20%. As they perform at a lower level in terms of the metrics, that might cause an increase in power consumption of both devices (the collector and sensor). Thus, the k size differs for the different locations. Moreover, it is specific for this experimental setup.

5.2   Latency

The experiments have been conducted on the smartphone and tablet. We have done ten experiments measuring the computational latency for each sample size K and collection latency for each k size. The graphs on Fig. 6 show their average values.

The results show that the measurement of both computational and collection latencies on the Xperia M5 and Nexus 7 depends on the devices' capacity. Therefore, further experiments are conducted using only the device with higher capacity that is the smartphone. According to the results (see Fig. 6(a)), the computational latency does not exceed one second for both devices in case of all sample sizes K. Moreover, the data collection is accomplished faster using the simultaneous way. However, the latency is really close to sequential one (see Figs. 6(b) and 6(c)) after the k size reaches six. Thus, in terms of the latency, both methods can be used when the k size exceeds six sensors.

5.3   Energy Consumption

The energy consumption is a key metric to validate the use of CollMule algorithm to facilitate mobile data collection. The energy consumption of the mules is measured considering different k size. On Fig. 6(d), each graph line demonstrates maximum power consumed during scanning, ranking, sequential and simultaneous collections. Analysing the results we can observe that the utilisation during the scanning phase has decreased considerably from $k = 2$ to $k = 5$. While the ranking usage is stable which average equals to about $600\,mW$ (mile-Watts). According to Fig. 6(d), the consumption of both sequential and simultaneous collections fluctuates, though for the former it is more significant differences. Moreover, the latter type has lower battery usage values for most of the k sizes. Therefore, in terms of energy utilisation, the simultaneous connection should be used for CollMule's last phase.

5.4   CPU Usage

The smart device shares CPU cycles among all its running processes. Therefore, it is important to know the CPU usage of our application because it can consume too much processing time, thereby, causing a decrease in the processing

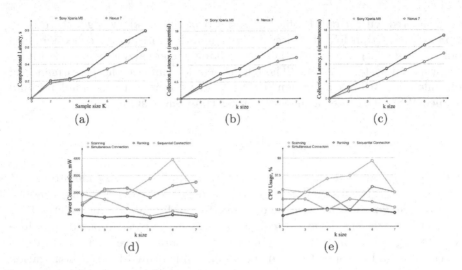

Fig. 6. (a) Computational Latency vs sample size K; Collection Latency of (b) sequential, and (c) simultaneous connections vs k size; (d) Maximum Energy Consumption; and (e) Maximum CPU usage of the prototype with different k size of sensor set.

speed of the entire device system that hosts it. The optimal maximum value of CPU load for an individual application is considered to be no more than 30% assuming a margin for sudden spikes in the load [17]. On Fig. 6(e), each graph line demonstrates the maximum values of CPU usage during scanning, ranking, sequential and simultaneous collections. Analysing the results we can observe that the CPU utilisation during all CollMule's phases follows the same trends as for power consumption. Moreover, the values exceed 30% only in case of sequential collection. Therefore, in terms of CPU utilisation, the simultaneous connection should be used for CollMule's last phase.

5.5   Discussion

This section evaluated how quickly the ranking computation and collection are performed by CollMule prototype on the mobile devices regarding sample size K; and how much energy and CPU are used by the mule prototype regarding to the k size. According to the obtained results, the computational and collection latencies are acceptable assuming that the mule moves with considerably slow velocity and stops for a few seconds during processing CollMule algorithm. Though, we should test the algorithm performance while a mule moves with different velocities and stop durations. In addition, the battery power consumption and CPU utilisation of the prototype on a gathering device can be considered reasonable for all CollMule phases: discovering, ranking and collecting (using simultaneous connection), compare to a commercial BLE connection application which consumes around 3000 mW and 35% for the maximum amount of these resources respectively during connection and data reading from one device (these were also measured with Trepn Profiler). To the best of our knowledge, there is no OppS system within IoT that uses any particular algorithm for collecting data from the cost-efficient sensors, thereby, making a comparison of our algorithm and proposed system difficult. Though, we could compare the performance of CollMule with 3DkNN algorithm. However, that approach was simulated and assumed WSNs that are significantly larger than our system.

6   Conclusion and Future Work

This paper proposed a data collection system applied to IAQ monitoring. It is intended to efficiently gather data from sensing IoT devices using a mobile smart device. For this purpose, we developed CollMule algorithm which selects the most cost-efficient devices to collect data from. This is supposed to reduce overall energy consumption as it prevents the smart device from attempts to connect unreliable sensors. CollMule has been implemented for a demonstration of its performance and appropriateness in real life scenario. The implemented system has been evaluated considering the computational and data collection latencies, energy consumption and CPU usage while running the algorithm on a smart device. The results show that it utilises a sufficiently efficient amount of the device's battery power and CPU for collecting data without overloading and slowing the device. In the future, to address security which is another serious issue faced by modern systems, it might be required to add authentication of the system user while connecting to the IoT devices.

Acknowledgements. The research reported here was supported and funded by the PERCCOM Erasmus Mundus Program of the European Union [18]. Part of this work has been carried out in the scope of the project bIoTope, which is co-funded by the European Commission under Horizon-2020 program, contract number H2020-ICT-2015/ 688203-bIoTope. The research has been carried out with the financial support of the Ministry of Education and Science of the Russian Federation under grant agreement RFMEFI58716X0031.

References

1. Guo, B., Zhang, D., Wang, Z., Yu, Z., Zhou, X.: Opportunistic IoT: exploring the harmonious interaction between human and the internet of things. J. Netw. Comput. Appl. **36**, 1531–1539 (2013)
2. Zhu, C., Leung, V.C.M., Shu, L., Ngai, E.C.-H.: Green Internet of Things for smart world. IEEE Access **3**, 2151–2162 (2015)
3. Atzori, L., Iera, A., Morabito, G.: Understanding the Internet of Things: definition, potentials, and societal role of a fast evolving paradigm. Ad Hoc Netw. **56**, 122–140 (2017)
4. Liu, J., Shen, H., Zhang, X.: A survey of mobile crowdsensing techniques: a critical component for the Internet of Things. In: 25th International Conference on Computer Communication and Networks (ICCCN) (2016)

5. Jayaraman, P.P., Perera, C., Georgakopoulos, D., Zaslavsky, A.: MOSDEN: a scalable mobile collaborative platform for opportunistic sensing applications. EAI Endorsed Transactions on Collaborative Computing 1 (2014)
6. Ahmed, A., Yasumoto, K., Yamauchi, Y., Ito, M.: Distance and time based node selection for probabilistic coverage in people-centric sensing. In: 8th Anual IEEE Communications Society Conference on Sensor, Mesh and Ad Hoc Communications and Networks (2011)
7. Song, S., Shin, S., Jang, Y., Lee, S., Choi, B.-Y.: Effective opportunistic crowd sensing IoT system for restoring missing objects. In: IEEE International Conference on Services Computing (2015)
8. Rodrigues, J.G.P., Aguiar, A., Queiros, C.: Opportunistic mobile crowdsensing for gathering mobility information: lessons learned. In: 19th International Conference on Intelligent Transportation Systems (ITSC) (2016)
9. Aloi, G., Caliciuri, G., Fortino, G., Gravina, R., Pace, P., Russo, W., Savaglio, C.: Enabling IoT interoperability through opportunistic smartphone-based mobile gateways. J. Netw. Comput. Appl. **81**, 74–84 (2017)
10. Tang, Z., Liu, A., Huang, C.: Social-aware data collection scheme through opportunistic communication in vehicular mobile networks. IEEE Access **4**, 6480–6502 (2016)
11. Aguilar, S., Vidal, R., Gomez, C.: Opportunistic sensor data collection with bluetooth low energy. Sensors **159**, 17 (2017)
12. Ma, Y., Zhang, S., Lin, C., Li, L.: A data collection method based on the region division in opportunistic networks. ACES J. **32**, 43–49 (2017)
13. Jayaraman, P.P., Zaslavsky, A., Delsing, J.: Intelligent mobile data mules for Cost-Efficient sensor data collection. Int. J. Artif. Intell. Neural Netw. Complex Probl.-Solving Technol., 225–234 (2010)
14. Jang, W.S., Healy, W.M.: Wireless sensor network performance metrics for building applications. Energy Buildings **6**, 862–868 (2010)
15. Figueira, J., Greco, S., Ehrogott, M.: Multiple Criteria Decision Analysis: State of the Art Surveys. Springer, New York (2005)
16. Saaty, T.L.: Decision making with the analytic hierarchy process. Int. J. Serv. Sci. **1**(1), 83 (2008)
17. Subramanian, K.: 4 things you need to know about CPU utilization of your Java application. http://karunsubramanian.com/java/4-things-you-need-to-know-about-cpu-utilization-of-your-java-application/. Accessed 05 May 2017
18. Klimova, A., Rondeau, E., Andersson, K., Porras, J., Rybin, A., Zaslavsky, A.: An international master's program in green ICT as a contribution to sustainable development. J. Cleaner Prod. **135**, 223–229 (2016)

The IoT Identification Procedure Based on the Degraded Flash Memory Sector

Sergey Vladimirov[1]($\boxtimes$) and Ruslan Kirichek[1,2]

[1] The Bonch-Bruevich State University of Telecommunications,
St. Petersburg, Russian Federation
vladimirovs@spbgut.ru, kirichek@sut.ru
[2] Peoples' Friendship University of Russia (RUDN University),
6 Miklukho-Maklaya St., Moscow, Russian Federation

Abstract. The article presents general principles of IoT identification on the basis of the degraded bucket of NOR flash memory used in network devices for storing microprograms. We proposed the identification procedure based on the sector of forcibly degraded flash memory and considered variants of its application for identifying network devices. The proposed identification method can be used both on the basis of chips already existing in network devices and on the basis of embedded units specially designed for identification.

Keywords: Internet of things · IoT · Identification · Information security · Flash memory

1 Introduction

The development of network technologies and, in particular, of the Internet of things and sensor networks leads to the widespread use of network devices. They are used in practically all areas of human activity—production, agriculture, education, medicine, and many others [1–3]. This level of usage allows to solve many tasks in these industries, for example, the automation task, but at the same time place engineers and developers before new challenges, among which the information security is on the top [1,4–6].

One of the main issues related to the information security of the Internet of things is the task of unambiguous identification of the device, which should ensure that this device can not be replaced by another—false—for the purpose of implementing some type of security threat—network attack, leakage or substitution of information and so on [1,5,7]. In application to automation task, this problem can be formulated differently. When it comes to managed and control devices, the managed device must uniquely determine that it receives commands from the correct control device, and vice versa, the control device must know that it sends commands and receives data from the desired managed device. Currently, the common identification of network devices bases on identifiers that the manufacturer writes to the network device's memory. A typical example of such

© Springer International Publishing AG 2017
O. Galinina et al. (Eds.): NEW2AN/ruSMART/NsCC 2017, LNCS 10531, pp. 66–74, 2017.
DOI: 10.1007/978-3-319-67380-6_6

identifiers are the widely used extended unique identifiers EUI-48 and EUI-64—MAC addresses of network devices, often used as the lower part of the IPv6 network address; digital identifiers of the object DOI; the CID (Communication Identifier) and Ecode (Entity Code) systems proposed by Chinese companies [8]. However, such identifiers can be changed programmatically or falsified by storing the identifier of another device in memory [9]. Another option for identification is the use of RFID tags [7,8]. However, they require special readers and also do not provide the necessary security level [7]. That is why recent years research has been conducted on methods of the unambiguous identification of network and IoT devices. Some methods are proposed for identifying devices by their physical characteristics or radio signal characteristics [10]. Another promising option is the identification of devices by the properties of the built-in storage device. For example, it is suggested to use the buil t-in flash memory as a hardware random number generator and a source of unique device identifiers [11,12]. Also, a method of using a degraded flash memory as a network device ID was proposed earlier [13]. In this paper, we consider the version of the last identification method, based on forced degradation of the flash memory bucket.

2 Background

Semiconductor flash memory is an array of elementary memory cells located on a matrix of conductors in the form of a two-dimensional (NOR-flash) or three-dimensional (NAND-flash) array. Each elementary memory cell can store from one to four bits of information (from two to sixteen charge levels) depending on the type of cell. During the program and erase operations in the memory cells, the stored charge level value changes. With that irreversible changes accumulate in the structure of the memory cells—the chip decays. It is gradually leads to the appearance of so-called "bad blocks"—memory cells that do not change their state during program and erase operations. For example, in NOR flash memory, the initial state of the bit cell is "1", and during programming information to memory, the desired cells are programmed with zeros. However, after a large number of overwrites, the bit cell can stop returning to the initial state and its value becomes always "0"—there is a "bad-block" [14,15]. Hereafter we will also consider NOR flash memory chips, because NOR flash memory with free access is used in network devices and their processors for storing firmware [15]. In the Fig. 1 there is an example of the page of the degraded NOR flash memory in comparison to the page of correctly working memory.

It is said that each memory chip and even each sector of memory cells array in a single chip is unique due to production features [11]. This leads to the fact that in each chip of flash memory, in case of degradation, "bad blocks" appear randomly. Therefore, if we take two degraded flash memory chips, then the "bad blocks" patterns in them will be unique. Thus, this pattern of "bad blocks" can be used as a unique identifier, which theoretically can not be repeated on another chip of flash memory [13].

From the point of view of architecture and the order of working with memory, the usual NOR-flash chip has a hierarchical structure, shown in Fig. 2. The

Page of correctly working NOR-flash after erasing

	00	01	02	03	04	05	06	07	08	09	0A	0B	0C	0D	0E	0F
00	FF	FF	FF	FF	FF	FF	FF	FF	FF	FF	FF	FF	FF	FF	FF	FF
01	FF	FF	FF	FF	FF	FF	FF	FF	FF	FF	FF	FF	FF	FF	FF	FF
02	FF	FF	FF	FF	FF	FF	FF	FF	FF	FF	FF	FF	FF	FF	FF	FF
03	FF	FF	FF	FF	FF	FF	FF	FF	FF	FF	FF	FF	FF	FF	FF	FF
04	FF	FF	FF	FF	FF	FF	FF	FF	FF	FF	FF	FF	FF	FF	FF	FF
05	FF	FF	FF	FF	FF	FF	FF	FF	FF	FF	FF	FF	FF	FF	FF	FF
06	FF	FF	FF	FF	FF	FF	FF	FF	FF	FF	FF	FF	FF	FF	FF	FF
07	FF	FF	FF	FF	FF	FF	FF	FF	FF	FF	FF	FF	FF	FF	FF	FF
08	FF	FF	FF	FF	FF	FF	FF	FF	FF	FF	FF	FF	FF	FF	FF	FF
09	FF	FF	FF	FF	FF	FF	FF	FF	FF	FF	FF	FF	FF	FF	FF	FF
0A	FF	FF	FF	FF	FF	FF	FF	FF	FF	FF	FF	FF	FF	FF	FF	FF
0B	FF	FF	FF	FF	FF	FF	FF	FF	FF	FF	FF	FF	FF	FF	FF	FF
0C	FF	FF	FF	FF	FF	FF	FF	FF	FF	FF	FF	FF	FF	FF	FF	FF
0D	FF	FF	FF	FF	FF	FF	FF	FF	FF	FF	FF	FF	FF	FF	FF	FF
0E	FF	FF	FF	FF	FF	FF	FF	FF	FF	FF	FF	FF	FF	FF	FF	FF
0F	FF	FF	FF	FF	FF	FF	FF	FF	FF	FF	FF	FF	FF	FF	FF	FF

Page of degraded NOR-flash after erasing

	00	01	02	03	04	05	06	07	08	09	0A	0B	0C	0D	0E	0F
00	FF	FF	FF	FF	FF	FE	FF	FF	3F	FF	FF	FF	EF	FF	FF	FF
01	FF	FF	FF	FF	FF	FF	FF	FF	DF	FF	FF	FF	FB	FF	DF	FF
02	FF	FF	FF	FF	FB	FB	FF	FF	FF	FF	FF	FF	FF	FF	FF	FF
03	FF	FF	FF	FF	7F	FF	F7	FF	EB	FF	FF	FF	FF	FF	FF	FF
04	FF	FF	FF	FF	FF	FF	FF	FF	FF	FF	EF	FF	FF	FF	F7	FF
05	FF	FF	FF	FF	FF	FF	7F	FF	FF	FF	BF	FF	FF	FF	FF	FF
06	FF	FF	DB	FF	FF	FF	FF	FF	FF	FF	FB	FF	FD	FF	FF	FF
07	BF	FF	FF	FF	FF	FF	7F	FF	FF	BF	FF	FF	FF	FF	FF	FF
08	FF	FF	FF	BF	FF	FF	FF	FF	DF	FF	FF	FF	FF	FF	FF	FF
09	FF	FF	FF	F4	FF	FF	FF	FF	FB	FF	FD	7F	FF	FD	FF	FF
0A	FB	FB	EF	FF	FF	FF	FF	FF	FF	FF	EF	FF	FF	FF	FD	FB
0B	FF	FE	FF	FF	FF	FF	FF	FF	FF	FF	BF	FF	FF	FF	FF	
0C	FF	F7	FF	FF	FF	FF	FF	FF	FF	FB	FF	FF	FF	FF	FF	
0D	FF	FF	FF	FF	FF	FF	FF	FF	F7	FF	FD	FB	FF	FF	FF	FF
0E	FF	FF	DF	FF	FF	FF	FF	BF	FF	FF	EF	FF	FF	EF	FF	FF
0F	DF	FF	FF	F7	DF	FF	FF	FF	FF	FF	FF	FF	FD	FF	FF	EF

Fig. 1. Comparison of pages of correctly working NOR flash memory and degraded NOR flash memory

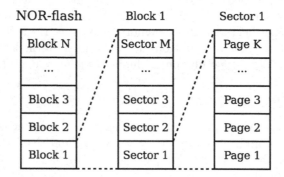

Fig. 2. Hierarchical structure of the usual NOR-flash chip

whole array of chip memory is divided into large sections—blocks, each of which is divided into sectors of smaller size, in turn consisting of minimal structural sections—pages. The minimum section for which the erase operation is available is the sector, and the program operation is done on a page-by-page basis. Accordingly, to change at least one information bit stored in the memory chip, you need to use the following algorithm [11,14,15]:

1. Read all information from the relevant sector.
2. Change the necessary information in the read data array.
3. Erase the entire relevant sector.
4. Page-by-page program the information to the erased sector.

Therefore, in the NOR-flash memory, the minimum bucket that can be degraded is the memory sector. Thus, since all modern network devices already have a flash chip for storing firmware, a single memory sector of this chip can be used as a unique identifier, carrying out the procedure of its forced degradation. However, this method can be used only if the degradation of this sector of memory does not affect neighboring sectors of memory. Otherwise, you can add a separate flash memory chip to the network device, intended only for identification of this device.

3   Method and Results of the Study

To test the identification method, we chose NOR chips of 16 Mbit (2 Mbyte) flash memory in the SO-8 case and a serial SPI interface. These chips are used in many network devices. The test was performed for EN25F16 chips manufactured by Eon Silicon Solution, Inc. and W25Q16V chips manufactured by Winbond Electronics Corporation. These chips have the same memory structure, consisting of $N = 32$ blocks of 64 KB each, which in turn contains $M = 16$ sectors of 4 KB (total 512 sectors per chip) consisting of $K = 16$ pages By 256 bytes (total 8192 pages per chip). The selected chips use a standard command system (instructions).

The experimental unit was built on the basis of the Raspberry Pi B microcomputer, containing the built-in SPI interface, and the standard scheme SPI programmer connected to RPis GPIO connectors. The scheme of the experimental unit is shown in Fig. 3.

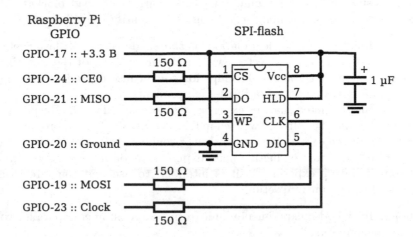

Fig. 3. The scheme of the experimental unit

To work with the memory chip, we wrote Python software using the py-spidev module, which defines the functions for working with the Linux kernel spidev driver, which provides low-level mechanisms for working with the SPI interface. The software performs the following functions:

1. Cyclic overwriting of one sector of the flash memory chip until the stable appearance of "bad blocks".
2. Saving an image (pattern of "bad blocks") of the degraded sector of the flash memory chip.
3. Comparing the sector of the flash memory chip connected to the experimental installation with the stored images and identifying the chip using the maximum likelihood method.

Cyclic overwriting is performed by erasing the flash memory sector with the appropriate command and then page-program the sector with an array of zeros, which ensures the equiprobability of the potential degradation of the elementary memory cells of the sector.

As a result of the experiment, the sectors of each studied memory chip were successfully degraded. Two main properties were identified:

1. The image of the "bad blocks" of the degraded 4 KB sectors of each chip studied is really different, which made it possible to identify the chips within the test sample. The number of "bad blocks" that coincide when comparing the erased identifying sector of a single memory chip with identifying sectors of other memory chips turned out to be two orders of magnitude less than the number of "bad blocks" that coincide when compared with the previously written in memory identifying "bad block" pattern of this sector.
2. Some distorted bits ("bad blocks") can sometimes be reset when erased. Thus, to get a complete image of "bad blocks", it is necessary to conduct several sector erasings in a row, accumulating the positions of "bad blocks". After this, the resulting sector image can be used to identify the chip.

The possibility of forced degradation of the single sector of flash memory without degradation of other sectors within the framework of the experiment was not confirmed, since one of the chips for some cause degraded completely. This assumption will be tested in further experiments.

It should also be noted that the experimental unit based on the Raspberry Pi microcomputer and the software written in the interpreted Python programming language, for all the ease of working with it, provides a chip degradation speed that is not sufficient for conducting mass experiments on large samples. In this connection, in future experiments it is intended to use and compare several variants of the experimental setup:

1. Extensively extended experimental unit from a large number of parallel working microcomputers.
2. Developed installation on the basis of parallel connected USB–SPI devices, control computer and the control software written in the C programming language.
3. An experimental unit in which the flash memory sector degradation mechanism is placed in a separate device based on fast FPGA.

4   Identification Procedure

Based on the obtained results, we formulated the network device identifying procedure using a degraded chip of flash memory.

1. Forced degradation of the flash memory chip by cyclic rewriting of one of its sectors (the first sector was used in the experiment) until the stable appearance of "bad blocks".

2. Creating of a unique array-identifier $\mathbf{S}_{\text{ID}}(c)$ of the memory chip c. For this, we need to initially form a zero array $\mathbf{S}_{\text{ID}}(c)$, corresponding to the size of the memory chip sector. Then, the degraded sector is erased ten times in a row. After each i erasure, the sector contents are read into the array $\mathbf{B}_i$. Next, the elementwise conjunction of $\mathbf{B}_i$ arrays is performed, with the result written in $\mathbf{S}_{\text{ID}}(c)$. As a result, after ten erasures the array $\mathbf{S}_{\text{ID}}(c)$ contains the unique identifier of the c memory chip. This operation can be described by the formula

$$\mathbf{S}_{\text{ID}}(c) = \prod_{i=1}^{10} \mathbf{B}_i,$$

where the symbol of the product corresponds to the operation of elementwise logical multiplication (conjunction).

3. The final array identifier $\mathbf{S}_{\text{ID}}(c)$ of each c memory chip is written into the identifier's storage.

4. To identify the network device containing the memory identification chip, the selected sector of the memory chip is erased, the sector contents are read after erasure and the result is compared with each identifier $\mathbf{S}_{\text{ID}}$ from the identifier's storage. The conformity of the identifier is determined by the majority principle.

The simplest version of this procedure is the usage of degraded flash memory chips as hardware identifiers, the contents of which are read directly by a special reader on the SPI interface. Such a reader performs the entire identification procedure: erasing the identifying sector, reading the unique array identifier $\mathbf{S}_{\text{ID}}(c)$, querying the identifier's storage (or searching the built-in storage in case of its small size), and displaying the search result.

The reliability of identifying a network device containing an identification memory chip directly depends on the size of the degraded memory block selected as an identifier. The larger the block size, the higher the reliability of identification. However, with the increase in the size of the identifier, the time of its processing and comparison with the contents of the identifier's storage grows. Thus, for simple identification, you can use a shortened identifier, for example, one or two pages of memory. We denote this shortened identifier as $\mathbf{P}_{\text{ID}}(c)$. To identify during important tasks, it is reasonable to use the block identifier $\mathbf{S}_{\text{ID}}(c)$ of a larger size, for example, the whole memory sector or several sectors, so-called full identifier. During the critical tasks the image of "bad blocks" for entire memory chip can be used as an identifier.

This principle can be applied to identification within a data transfer session between two network devices. Let's consider the case when identifiers verification is performed in a separate network device—identifier's storage—which stores known images of the full identifiers of network devices $\mathbf{S}_{\text{ID}}$ and compares them with network identifiers (network addresses) N_{ID}, forming an identification table. The network scheme for this example is shown in Fig. 4(1). The figure shows the identifier's storage IS and two network devices D1 and D2 that have identifiers $\mathbf{S}_{\text{ID}}(\text{D1}) \rightarrow N_{\text{ID}}(\text{D1})$ and $\mathbf{S}_{\text{ID}}(\text{D2}) \rightarrow N_{\text{ID}}(\text{D2})$, respectively.

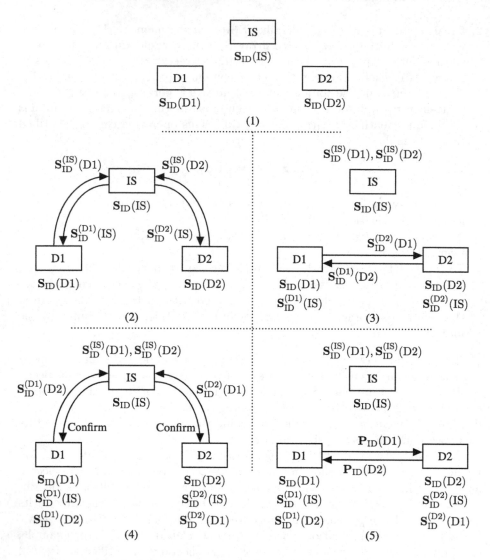

Fig. 4. A simplified network devices identifying scheme

Initially, images of the full identifiers $\mathbf{S}_{ID}$ of all network devices in the network must be written into the identifier's storage IS. We denote the identifier images for devices D1 and D2 in IS as $\mathbf{S}_{ID}^{(IS)}(D1)$ and $\mathbf{S}_{ID}^{(IS)}(D2)$. These images can be written into the IS by the network administrator or can be recorded automatically when the device is directly connected to the IS. The unique full identifier of the storage itself $\mathbf{S}_{ID}(IS)$ is stored in the memory of each device so that when accessing it the device can be sure that it is specifically IS. This procedure is shown in Fig. 4(2).

During the connection establishing, each device initializes the memory chip used for identification by erasing the identification sector. Then the devices read their full identifiers $\mathbf{S}_{\mathrm{ID}}$ and exchange them as shown in Fig. 4(3) to authenticate. After receiving the full ID of the interlocutor, the device D2 writes it to its memory in the form of the image $\mathbf{S}_{\mathrm{ID}}^{(D2)}(D1)$ and sends the request to the IS, which specifies the network $N_{\mathrm{ID}}(D1)$ and full $\mathbf{S}_{\mathrm{ID}}^{(D2)}(D1)$ identifiers of the interlocutor. The IS compares received identifier $\mathbf{S}_{\mathrm{ID}}^{(D2)}(D1)$ with the stored image $\mathbf{S}_{\mathrm{ID}}^{(IS)}(D1)$ by the majority principle. If the identifier to be checked and the stored control image are correlated, the IS sends a D2 confirmation in which it indicates its unique full identifie r $\mathbf{S}_{\mathrm{ID}}(IS)$, which serves as proof that the confirmation is received from a valid source. Having received the confirmation, D2 set the image $\mathbf{S}_{\mathrm{ID}}^{(D2)}(D1)$ as correct. In the future, this image is used to dynamic identification of devices during the communication session. Similar actions are performed by the device D1 with respect to the identifier $\mathbf{S}_{\mathrm{ID}}(D2)$. This procedure is shown in Fig. 4(4).

Further, in the data transfer process, we can use the shortened identifier $\mathbf{P}_{\mathrm{ID}}$ for periodic authentication, as shown in Fig. 4(5). In this case, we can use both a static shortened identifier, for example, only the first page of the full identifier, as well as various variants of dynamic identification, for example, alternating the shortened identifiers for one or another algorithm or using the cyclic shift of the shortened identifier within the full identifier. In this case, we can use both synchronous dynamic identification, when each device knows which part of the full identifier $\mathbf{S}_{\mathrm{ID}}$ it should receive as $\mathbf{P}_{\mathrm{ID}}$, and asynchronous dynamic identification, in which the correspondence $\mathbf{P}_{\mathrm{ID}}$ and $\mathbf{S}_{\mathrm{ID}}$ is determined by the ejection of the cross-correlating function

$$\mathrm{CCF}_i(\mathbf{P}_{\mathrm{ID}}, \mathbf{S}_{\mathrm{ID}}) = \sum_j \mathbf{P}_{\mathrm{ID}\,j} \oplus \mathbf{S}_{\mathrm{ID}\,i+j}.$$

At the end of the communication session, the saved images of the full identifiers of the interlocutors can either be erased or saved and used later for simplified identification that does not require access to the identifier's storage.

It should be noted that frequent reading of the contents of a degraded sector, coupled with its erasure, can lead to its further degradation and the appearance of new "bad blocks". Therefore, after the successful identification of the chip by majority principle, newly appeared "bad blocks" should be added to the image of the identifier stored in the identifier's storage.

Thus, the conducted experiment confirmed the principal possibility of using a degraded flash memory sector as a unique identifier. Also, the procedure for identifying a network device using the degraded flash memory sector was formulated and variants of its application were proposed. The authors plan to continue research in this area and the proposed procedure is not final. It is planned to conduct research on the reliability of identification with the definition of the probabilities of correct and incorrect identification of devices. Depending on the results of future research, the identification procedure can be improved in the direction of simplification or, conversely, complication, based on the criteria of maximum reliability and speed.

Acknowledgments. The publication was financially supported by the Ministry of Education and Science of the Russian Federation (the Agreement number 02.a03.21.0008).

References

1. Kirichek, R., Koucheryavy, A.: Internet of Things laboratory test bed. In: Zeng, Q.-A. (ed.) Wireless Communications, Networking and Applications. LNEE, vol. 348, pp. 485–494. Springer, New Delhi (2016). doi:10.1007/978-81-322-2580-5_44
2. Kirichek, R., Kulik, V., Koucheryavy, A.: False clouds for Internet of Things and methods of protection. In: IEEE 18th International Conference on Advanced Communication Technology: ICACT 2016, Phoenix Park, Korea, pp. 201–205 (2016)
3. Vermesan, O., Friess, P. (eds.): Internet of Things: Converging Technologies for Smart Environments and Integrated Ecosystems. River Publishers, Aalborg (2013)
4. Russell, B., Van Duren, D.: Practical Internet of Things Security. Packt Publishing, Birmingham (2016)
5. Dhanjani, N.: Abusing the Internet of Things. Blackouts, Freakouts, and Stakeouts. O'Reilly Media, Sebastopol (2015)
6. Hu, F. (ed.): Security and Privacy in Internet of Things (IoTs): Models, Algorithms, and Implementations. CRC Press, Boca Raton (2016)
7. Leloglu, E.: A review of security concerns in Internet of Things. J. Comput. Commun. **5**, 121–136 (2017)
8. Soldatos, J., Yuming, G. (eds.): Internet of Things. EU-China Joint White Paper on Internet-of-Things Identification. European Research Cluster on the Internet of Things (2014)
9. Hegde, A.: MAC spoofing detection and prevention. Int. J. Adv. Res. Comput. Commun. Eng. **5**(1), 229–232 (2016)
10. DeJean, G., Kirovski, D.: RF-DNA: radio-frequency certificates of authenticity. In: Paillier, P., Verbauwhede, I. (eds.) CHES 2007. LNCS, vol. 4727, pp. 346–363. Springer, Heidelberg (2007). doi:10.1007/978-3-540-74735-2_24
11. Wang, Y., et al.: Flash memory for ubiquitous hardware security functions: true random number generation and device fingerprints. In: Proceedings of the 2012 IEEE Symposium on Security and Privacy. IEEE Computer Society, Washington, 20–25 May 2012. doi:10.1109/SP.2012.12
12. Jia, S., Xia, L., Wang, Z., Lin, J., Zhang, G., Ji, Y.: Extracting robust keys from NAND flash physical unclonable functions. In: Lopez, J., Mitchell, C.J. (eds.) ISC 2015. LNCS, vol. 9290, pp. 437–454. Springer, Cham (2015). doi:10.1007/978-3-319-23318-5_24
13. Jakobsson, M., Johansson, K.-A.: Unspoofable Device Identity Using NAND Flash Memory (2010). http://www.securityweek.com/unspoofable-device-identity-using-nand-flash-memory. Accessed 1 May 2017
14. Bez, R., Camerlenghi, E., Modelli, A., Visconti, A.: Introduction to flash memory. Proc. IEEE **91**(4), 489–502 (2003). doi:10.1109/JPROC.2003.811702
15. Tal, A.: Two Flash Technologies Compared: NOR vs NAND. 91-SR-012-04-8 L REV. 1.0. M-Systems Flash Disk Pioneers. Newark (2002)

DisCPAQ: Distributed Context Acquisition and Reasoning for Personalized Indoor Air Quality Monitoring in IoT-Based Systems

Tamara Belyakhina[1]($\boxtimes$), Arkady Zaslavsky[2], Karan Mitra[1], Saguna Saguna[1], and Prem Prakash Jayaraman[3]

[1] Luleå University of Technology, Skellefteå, Sweden
`tambel-6@student.ltu.se`
[2] CSIRO, Melbourne, Australia
[3] Swinburne University of Technology, Melbourne, Australia

Abstract. The rapidly emerging Internet of Things supports many diverse applications including environmental monitoring. Air quality, both indoors and outdoors, proved to be a significant comfort and health factor for people. This paper proposes a smart context-aware system for indoor air quality monitoring and prediction called DisCPAQ. The system uses data streams from air quality measurement sensors to provide real-time personalised air quality service to users through a mobile app. The proposed system is agnostic to sensor infrastructure. The paper proposes a context model based on Context Spaces Theory, presents the architecture of the system and identifies challenges in developing large scale IoT applications. DisCPAQ implementation, evaluation and lessons learned are all discussed in the paper.

Keywords: Context awareness · Indoor air quality · Internet of Things · Sensor networks

1 Introduction

The Internet of Things (IoT) connects heterogeneous devices, or "things", to the Internet. This paradigm brings new opportunities and also new challenges [1]. For example, due to the volume, velocity, and variety of the data, IoT systems require large storage, processing and communication resources. Thus, more intelligent methods and techniques for handling this amount of data are required [1].

One of the methods to overcome this challenge in is *context-aware computing* [2], an approach to system development, which involves utilization of any meaningful information (called *context*), retrieved from the devices, users, and environment. It might include the data about user's location, current time, user's activity, and users surrounding environment. Context awareness provides a possibility for integration and repurposing the IoT data across multiple systems, platforms, and applications by utilizing context information linked to sensor data [3].

© Springer International Publishing AG 2017
O. Galinina et al. (Eds.): NEW2AN/ruSMART/NsCC 2017, LNCS 10531, pp. 75–86, 2017.
DOI: 10.1007/978-3-319-67380-6_7

Studies show that air quality has a direct influence on the human health and well-being [4]. According to [5], an average person spends about 80% of his/her time indoors. A significant part of a person's life is influenced by the indoor conditions. Thus, indoor air quality is a crucial issue, that requires attention from the research community.

Indoor air quality monitoring was addressed previously such as in [6–9]. They provided users with indoor air quality monitoring services using IoT-based systems or Wireless Sensor Networks. However, some of these works [6–8] do not take into consideration person profile information. Also, all of the papers lack utilization of Humidex.

In this paper, we propose a solution for indoor air quality monitoring based on context-aware computing approach. Our system, named Distributed Context acquisition and reasoning for Personalized indoor Air Quality monitoring (DiscPAQ), was designed and implemented to provide users with real-time service for monitoring of indoor air quality. For this system we designed and developed a model, which utilized the context awareness and processing and utilization of users profile data to ensure the personalization aspect of this service. DisCPAQ also performs analysis and the prediction of air characteristics.

The rest of the paper is organized as follows. Section 2 presents the previous works. Section 3 describes the theoretical aspects of the DisCPAQ system. Description of the proposed system architecture presented in Sect. 4. Section 5 contains the DisCPAQ implementation and results analysis. Section 6 summarizes research outcomes and contribution and presents a discussion and future work.

2   Related Work

For the past few years, scientific community conducted research in the areas of context-aware computing for indoor air quality monitoring. For example, [6–8] introduce systems for indoor air quality monitoring, using Wireless Sensor Network. The system in [6] provided users with an information about air condition indoor, considering gas pollutant concentration and level of its impact, determined by Air Quality Index (AQI) model [4]. Another intelligent Smart Home system AirSense [8] was developed for monitoring of certain air characteristics (such as PM2.5, humidity, and VOCs) and forecasting. [7] focuses on the context-aware approach of personal monitoring in-house, which includes various sensor devices such as temperature, humidity, motion, and luminescence. However, none of these papers take into account user's personal (such as the existence of certain illnesses or age), preventing the provision of more personalized services.

eWall framework [9] addresses the issue of personalized services for environmental monitoring. It includes multiple sensing devices, IoT platform, and various data processing methods to handle context and utilize meaningful information, derived from it. However, as the scope of the system is only one household, this paper does not consider the distributed crowd-sensing aspect, that was used in [6,8]. In addition, another air quality indicator Humidex [10] was not included into processing in all these works, even though, according to [10], certain air temperature and humidity conditions can drastically affect human health.

Our paper tries to address the gaps in these research areas. Our system DisCPAQ deals with aspects related to distributed indoor air quality monitoring. It is built upon the developed model, one of the major aspects of which is its usage of personal user's profile data to provide an intelligent and personalized solution for data management. AQI and Humidex were used to model user-oriented air quality indicators.

3 DisCPAQ Architecture

We designed and implemented DisCPAQ to provide indoor air quality monitoring data to users using crowd-sensing approach. As shown in Fig. 1, it contains a network of sensors, which are directly connected to the smart devices, and also a base station for the data storage.

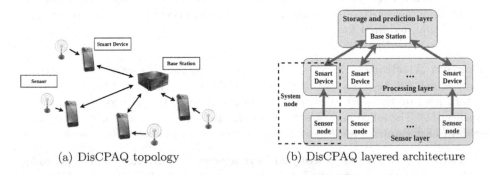

(a) DisCPAQ topology   (b) DisCPAQ layered architecture

Fig. 1. DisCPAQ system topology and architecture.

The architecture of the system, as depicted in Fig. 1, has 3 layers: sensor layer, processing layer, and storage and prediction layer. Sensor layer is responsible for collecting the measurements from sensor nodes. Sensor node can include multiple sensors for various characteristics to be measured. After obtaining the required data, communication manager of the sensor node transfers received information to the processing layer of the architecture.

Processing layer is the major part of the architecture, which performs the largest amount of data processing. The components of this layer (smart devices) operate with raw data, transforms it into context information, reasons about the current situation of the node, and then transmits it to all nodes within the system through storage and prediction layer. Each element of processing layer is coupled to a corresponding sensor node into a system node. The system node can operate as a independent unit, that is it can function even when disconnected from the network, as it provides the required processing capability. In this case, a user still can observe the air quality in his or her surrounding environment in real-time.

The storage and prediction layer works as a Publish/Subscribe Broker, that broadcasts the information, received from one node to the rest of the system. It also collects and stores historic data, which is further used for prediction of air quality. This is essential to provide users with the forecast of dust concentration, temperature, or humidity, as that can affect person's actions, especially in the case of emergency. In this research, as we operate with time-series data (values of each air characteristic in a certain place at a specific moment in time), it is reasonable to apply a State-Space model approach which is proven to be accurate for time-series prediction.

4   DisCPAQ Context Modeling

4.1   Context Model

The main approach to the context modeling in this research was based on the theory of Context Spaces [11]. This is a conceptual framework, that provides a general model for building context-aware systems. The main idea of this theory is a representation of context as a multidimensional space. This generic approach for building a context model description can be applied to any types of context information, required for the system.

Context Attributes. As the first step of context model design, the necessary data, which will be measured by the system, should be defined. This data can be determined by the set of Context Attributes, defined in the theory of Context Spaces [11]. The following list of Context Attributes was chosen for the proposed system DisCPAQ.

- **Time.** This attribute is assigned to represent the current time of the node. It in necessary to acquire time stamps in order to better understand when a certain situation occurred in the system.
- **Location.** As the system requires sharing of the context in the distributed manner and real-time air quality monitoring for the entire scale of the network, it is essential to provide an information about each node's current location.
- **Air Quality Index.** To define the level of health concern for users, the AQI [4] should be calculated from the air sensors measurements.
- **Humidex.** This attribute represents the calculated Humidex, defined in [10] value from temperature and relative humidity measurements. It is used further to determine the level of user's comfort.
- **Health Index.** This characteristic defines user personal tolerance to the air quality. In this research, we are taking into account only 2 of them: the existence of any user's illnesses of the respiratory system (and their severeness) and user's age. We introduce the health index, which is calculated from the indexes of these two characteristics with certain weights. The health index is defined the following way:

$$HealthIndex = RespiratoryToleranceIndex * 2 + AgeIndex \qquad (1)$$

where *Respiratory Tolerance Index* defines the level of user's respiratory system illnesses. The more severe is the illness, the higher is index. It can take values 0, 1, and 2. [12] presented that people at the age above 65 are more vulnerable to air pollution than the rest of population. Thus, Age Index can be 0 or 1, in case the user is at vulnerable age (under 65 years) or not. As an impact of respiratory diseases on human perception of air is much higher the impact of age, the weight for Respiratory Tolerance Index is twice greater than the one for Age Index.

- **User ID**. This feature is essential for the proper communication between nodes. It is used for identification of the nodes within the network by its elements.
- **Temperature**. This is a raw data of air temperature (in degree Celsius), that is going to be used in the further analysis and air quality prediction.
- **Humidity**. Raw data of air relative humidity (in percent) is measured for the further analysis and air quality prediction.
- **PM Concentration**. This attribute contains the data of PM pollutant concentration on the node. It is used to calculate AQI and predict possible air quality.

Situation Reasoning. The next step of DisCPAQ model is to reason about an air quality situation. An air quality situation determines the relations between the Context Attributes and their corresponding values (range). The situations were determined to address the current air quality conditions:

- **Good Air Quality.**
  This situation implies that there is no or minimum effect of air characteristics to the user.
- **Unhealthy Air Quality.**
  In this situation users experience slight discomfort and respiratory irritation.
- **Very Unhealthy Air Quality.**
  Possible problems with breathing and a stronger feeling of discomfort might occur in this situation.
- **Dangerous Air Quality.**
  The conditions of air in this situation are dangerous to the user. Urgent actions should be performed to ensure the safety of people.

Humidex and AQI values and ranges were considered as primary parameters to define possible situations. Ranges of AQI values can be divided into 6 levels of health concern [4]. At the same time, according to [10], there are 4 possibilities for human comfort, determined from Humidex value. For this research, we decided to merge these categories into 4 levels of concern (combining both comfort and health), which correspond to certain air quality situations.

According to the Health Index formula, it can take integer values in the range $[0, 6]$. Considering this, for each value of Health Index, the ranges of AQI were redefined since the impact of the same air pollutant concentration varies depending on personal respiratory issues of the users. In Table 1, the ranges for each of 4 levels of concern are presented, depending on the health index of the user.

Table 1. AQI, Humidex ranges and values of health index with corresponding air quality situations.

	0	1	2	3	4	5	Humidex
Good	[0,100]	[0,100]	[0,100]	[0,50]	[0,50]	[0,50]	[20,30)
Unhealthy	(100,250]	(100,200]	(100,150]	(50,150]	(50,100]	(50,100]	[30,40)
Very unhealthy	(250,300]	(200,300]	(150,300]	(150,250]	(100,250]	(100,200]	[40,46)
Dangerous	(300,+Inf)	(300,+Inf)	(300,+Inf)	(250,+Inf)	(250,+Inf)	(200,+Inf)	[46,+Inf)

5   DisCPAQ Implementation

5.1   Hardware Description

To implement the proposed system architecture, the appropriate hardware elements were used. For the deployment, we have chosen the following equipment to satisfy our requirements:

1. Sensor level: Raspberry Pi[1] with Sensly HAT[2], presented in the Fig. 2. It includes the temperature, relative humidity and PM10 sensors.
2. Processing level: Sony Xperia XA F3111 and M5 E5603, Asus Nexus 7 Google.
3. Base Station: Dell Inspiron laptop 5559.

Fig. 2. Raspberry Pi with a sensly HAT.

---

[1] https://www.raspberrypi.org/.
[2] http://www.instructables.com/id/Sensly-Hat-for-the-Raspberry-Pi-Air-Quality-Gas-De/.

5.2 Software Implementation

To implement the proposed system we identified tools and software solutions that are going to be used in this work. The overall architecture with more detailed information about each component implementation is presented in the Fig. 3.

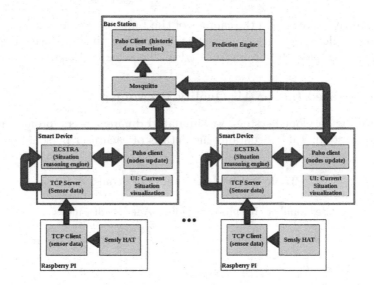

Fig. 3. DisCPAQ system's architecture.

The sensor node (in our solution Raspberry Pi and Sensly HAT) periodically (within 30 s interval) gathers data from sensors about several air characteristics (temperature, humidity, and PM10). Raspberry Pi runs a TCP client, that sends this information to the smart device. Python language was used to implement this functionality.

We developed the DisCPAQ mobile application (see Fig. 4), written in Java, that runs on the Android. It runs a TCP server in a separate thread that listens to the sensor node's queries with recently collected data about current air quality. After receiving this data, the application starts context processing and reasoning using ECSTRA engine [13]. As it has been described before, in order to define the current air quality situation, information from sensors is combined with user's personal data, and then the situation for the user is determined.

As soon as application discovers the current situation for the node, it publishes this information using MQTT[3] Paho client[4]. Other nodes, that are subscribed to this service, receive this data, and process it, considering their user's personal information. In that way, at each node, the user can observe the personalized situations for each node of the system.

---

[3] http://mqtt.org/.

[4] https://eclipse.org/paho/clients/java/.

In addition to publishing the processed information to other nodes, the smart device sends raw data. It is being received by the MQTT client on the base station, which stores it for prediction analysis.

6 Experiments and Results

The experiments were held in the campus of the Luleå University of Technology[5] in Skellefteå Building A. Sensors and smart devices were installed with DisCPAQ applications. The base station was located in the same building.

6.1 Personalized Indoor Air Quality Reasoning

To evaluate the developed system, we simulated different users' profiles in every node of the system. Table 2 contains the details of each user personal data.

Table 2. Users' personal data details.

User number	Respiratory index	Age	Health index
1	0	24	0
2	1	35	2
3	2	75	5

Table 2 shows 3 users were considered for the experiments. Each one of them has different age and respiratory index. User 1 is a young healthy person and does not suffer from any respiratory illnesses. His or her overall health index is calculated to be equal to 0 by using the formula (1), introduced in Sect. 2. User 2 has slight respiratory problems and appears to be older than user 1. Still, user's 2 age does not cross the threshold for age-vulnerability and thus, health index for user 2 is equal to 2. The last user (user 3) suffers from severe respiratory illnesses and is also more vulnerable to the air conditions due to his or her age thus, having the health index value equal to 5. Considering this, all 3 users observe the values for the same air characteristics differently for each one of them.

To test our system, we deployed the mobile application on each smart device with the corresponding user profile information. To present the situation reasoning of the system, depending on the user's personal data, the screenshots of the mobile application on different nodes are depicted in Fig. 4. It shows the screenshots of each system node's device running the application at the same time in different locations.

The UI of the application shows the following information: current location, current time, value of AQI and Humidex, and also the computed current air

---

[5] https://www.ltu.se/.

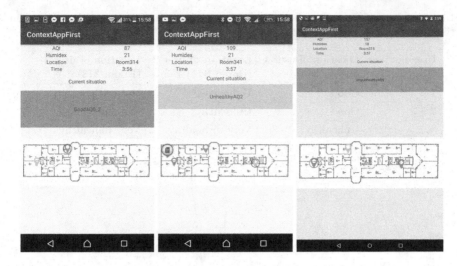

Fig. 4. Application screenshots for different user's with health index 0, 2 and 5 respectively, taken at the same point in time.

Table 3. Summary of collected results from the mobile application.

User number	Health index	Location	AQI	Humidex	Situation
1	0	Room 314	87	21	Good
2	2	Room 341	109	21	Unhealthy
3	5	Room 318	157	18	Very unhealthy

quality situation for the corresponding user. This information is also presented in the Table 3. As it can be observed from Fig. 4 and Table 3, for user 1 (with health index value 0) the current air quality in his/her location ("Room 314") is good for both user 1 and user 2. However, it is determined to be unhealthy for the user 3, who has health index equal to 5, because he or she suffers from respiratory illnesses. There are also differences between the situations in the same location for user 2 and user 3. Table 4 summarizes this information and shows the air quality situations in all 3 locations, determined for each user, considering their health index.

6.2 Indoor Air Quality Prediction

In this paper, to implement a State-Space model for air quality prediction we used MATLAB[6] software. It provides functionality to build models and analyze collected information to forecast the time-series data. We used data, collected from the sensors, for a period of 24-hours to build a model. For each of the three air characteristics, used in this study (temperature, relative humidity, and

---

[6] https://www.mathworks.com/products/matlab.html.

Table 4. Situations for each user at all locations of the system nodes at the same point in time.

User number	Location of user 1 (Room 314)	Location of user 2 (Room 341)	Location of user 3 (Room 318)
1	Good	Unhealthy	Unhealthy
2	Good	Unhealthy	Very unhealthy
3	Unhealthy	Very unhealthy	Very unhealthy

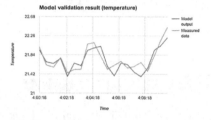   

(a) Temperature prediction model.   (b) Relative humidity prediction model.

Fig. 5. Temperature and relative humidity models validation.

PM10), we created models and validated them on the historic data. Then models were used to forecast the data and compare it with real-time measurements. The predicted data of each characteristic is displayed in the application for each user to observe. Figures 5 and 6 present the output of the models in comparison to the actual data with a 5-step time-series prediction for each air characteristic.

Figures show that overall fitness of the models output and actual measured data is quite good. More specifically, prediction model of PM10 concentration got approx. 75% fitness with the validation data. For the relative humidity calculated fitness was approx. 55%. Temperature prediction model performed worse than others with fitness equal only 48%. This result can be caused by the noisy

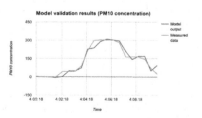   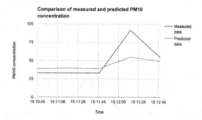

(a) PM10 concentration model valida-   (b) Comparison of 5-step predicted and
tion.   observed PM10 concentration data.

Fig. 6. PM10 concentration model validation and error in 5-step prediction.

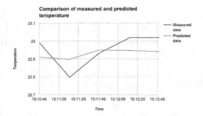
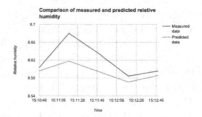

(a) Comparison of 5-step predicted and observed temperature data.

(b) Comparison of 5-step predicted and observed relative humidity data.

Fig. 7. Temperature and relative humidity errors in 5-step prediction.

sensor data. Figures 7 and 6 show the comparison of 5-step ahead predicted and observed data for each air characteristic.

Figures show that models can forecast trend of each air characteristic change for a short-step prediction. However, the accuracy of actual predicted values might be improved by implementation of more complex prediction model in case of system enhancement.

7   Conclusion and Future Work

In this paper, we proposed, developed and implemented a context-aware system DisCPAQ for indoor air quality monitoring. It provides a service for distributed monitoring of the air characteristics data indoors. The implementation of the system included real mobile sensor devices, that were carried by the users and collecting data about air quality. The developed solution provided a mobile application with a user interface, showing computed personalized air quality situation in a building for the user. The experiments were held in real-life setup, which validated our system performance.

As future work, our system can be enhanced to include more air quality features(such as gas pollutants and pressure). The system can be further integrated into more complex and sophisticated IoT solution, for instance, in areas of environmental monitoring or home automation.

Acknowledgment. The research reported here was supported and funded by the PERCCOM Erasmus Mundus Program of the European Union [14]. Part of this work has been carried out in the scope of the project bIoTope, which is co-funded by the European Commission under Horizon-2020 program, contract number H2020-ICT-2015/ 688203-bIoTope. The research has been carried out with the financial support of the Ministry of Education and Science of the Russian Federation under grant agreement RFMEFI58716X0031.

References

1. Miorandi, D., Sicari, S., De Pellegrini, F., Chlamtac, I.: Internet of Things: vision, applications and research challenges. Ad Hoc Netw. **10**(7), 1497–1516 (2012)

2. Abowd, G.D., Dey, A.K., Brown, P.J., Davies, N., Smith, M., Steggles, P.: Towards a better understanding of context and context-awareness. In: Gellersen, H.-W. (ed.) HUC 1999. LNCS, vol. 1707, pp. 304–307. Springer, Heidelberg (1999). doi:10. 1007/3-540-48157-5_29
3. Perera, C., Zaslavsky, A., Christen, P., Georgakopoulos, D.: Context aware computing for the Internet of Things: a survey. IEEE Commun. Surv. Tutorials **16**(1), 414–454 (2014)
4. Air Quality Index: A guide to air quality and your health. Washington, USEPA Air and Radiation, Environmental Protection Agency, EPA-454/K-03-002, vol. 19, pp. 11–01 (2003)
5. Brown, S.: Indoor Air Quality. Environment Australia (1997)
6. Kim, J.-Y., Chu, C.-H., Shin, S.-M.: ISSAQ: an integrated sensing systems for real-time indoor air quality monitoring. IEEE Sens. J. **14**(12), 4230–4244 (2014)
7. Kang, B., Park, S., Lee, T., Park, S.: IoT-based monitoring system using tri-level context making model for smart home services. In: 2015 IEEE International Conference on Consumer Electronics (ICCE), pp. 198–199. IEEE (2015)
8. Fang, B., Xu, Q., Park, T., Zhang, M.: AirSense: an intelligent home-based sensing system for indoor air quality analytics. In: Proceedings of the 2016 ACM International Joint Conference on Pervasive and Ubiquitous Computing, pp. 109–119. ACM (2016)
9. Kyriazakos, S., Mihaylov, M., Anggorojati, B., Mihovska, A., Craciunescu, R., Fratu, O., Prasad, R.: eWALL: an intelligent caring home environment offering personalized context-aware applications based on advanced sensing. Wirel. Pers. Commun. **87**(3), 1093–1111 (2016)
10. Canada. Service de l'environnement atmosphérique, Masterton, J., Richardson, F.: Humidex: a method of quantifying human discomfort due to excessive heat and humidity. Downsview, Ontario: Atmospheric Environment (1979)
11. Padovitz, A., Loke, S.W., Zaslavsky, A.: Towards a theory of context spaces. In: Proceedings of the Second IEEE Annual Conference on Pervasive Computing and Communications Workshops, pp. 38–42. IEEE (2004)
12. Gouveia, N., Fletcher, T.: Time series analysis of air pollution and mortality: effects by cause, age and socioeconomic status. J. Epidemiol. Commun. Health **54**(10), 750–755 (2000)
13. Boytsov, A., Zaslavsky, A.: ECSTRA – distributed context reasoning framework for pervasive computing systems. In: Balandin, S., Koucheryavy, Y., Hu, H. (eds.) NEW2AN/ruSMART-2011. LNCS, vol. 6869, pp. 1–13. Springer, Heidelberg (2011). doi:10.1007/978-3-642-22875-9_1
14. Klimova, A., Rondeau, E., Andersson, K., Porras, J., Rybin, A.V., Zaslavsky, A.: An international master's program in green ICT as a contribution to sustainable development. J. Cleaner Prod. **135**, 223–239 (2016)

Cloud Computing Solution for Investment Efficiency Measurement in Biomedicine

Petra Maresova and Vladimir Sobeslav[✉]

Faculty of Informatics and Management, University of Hradec Kralove,
Rokitanskeho 62, 50003 Hradec Králové, Czech Republic
{petra.maresova, vladimir.sobeslav}@uhk.cz

Abstract. Cloud computing is being widely adopted in various scientific areas, solutions and applications not including the biomedical research. This scientific area is characterized by the need of large amounts of data processing and these requirements are ensured by the cloud computing solutions. Cloud computing might bring many benefits, such as cost savings, elasticity and scalability of using ICT. On the other hand, the implementation might be inefficient and cost intensive. The measurement of cloud computing benefits and its efficiency is considered to be one of the crucial point for successful implementation of new systems or applications. The goal of this paper is to present specialized application for cloud computing investment efficiency with a respect to specificities of biomedical research area.

Keywords: CBA · Cloud computing · Biomedicine · Investment efficiency

1 Introduction

All kinds of scientific research are characterized by a significant growth of the ability to measure and gather data in previously unthinkable amounts. This trend results in new requirements for development and application of new methods and ICT, capable of storing, transmitting and further analyzing data, or potentially capable of using the data as input data for mathematical modelling in various contexts.

Biomedicine is a significant area of ICT utilization where huge amounts of data need to be processed. Modelled real world situations and changeable conditions are often sources of scientific calculations based on biomedical data. These issues cover a wide range of complex and highly extensive units, such as the design and implementation of automated clinical laboratories, screening facilities, multifunctional imaging centres, hospital information systems, distant monitoring, telemetry, etc. The requirements for computing power increase mainly in the evaluation of biomedical signals in real time. Various types of transformation, e.g. the wavelet, are not complicated due to the number of steps in one subscription - 2n; yet, the problem lies mainly in the use on large real biomedical data requiring fully parallel computing with high performance. Cloud computing is the technology which can be effectively used in this context [1]. The response to the previous mentioned challenges in biomedical research is the adoption of cloud computing (CC) paradigm. Cloud services create an attractive solution for the field of biomedicine, helping to better manage resources, and

© Springer International Publishing AG 2017
O. Galinina et al. (Eds.): NEW2AN/ruSMART/NsCC 2017, LNCS 10531, pp. 87–96, 2017.
DOI: 10.1007/978-3-319-67380-6_8

to provide fluid access, viewing, and sharing of medical images across organizations, departments and providers. However, the implementation of cloud computing approach in biomedical research is not a simple task, but rather a complex process that affects the whole research and creates many challenges in deploying adequate solutions.

It is highly risky and expensive to transfer from the current infrastructure to cloud computing. The main considerable issue dealing with the employment of cloud computing is its efficiency and asset recovery. This field has the subject of attention of many experts both on IT as well as on economy and management. Several studies dealing with the cost and savings related to cloud computing have been carried out recently [2–4]. Jureta et al. [5] for example evaluates the quality of cloud computing service from a customer's viewpoint. Kalepu et al. [6] proposes QoS (Quality of Service) model for cloud computing, which puts emphasis on the metrics of distributed services. Juran [7] has designed CWQM (Company Wide Quality Management) a concept for cloud computing based on three pillars: quality planning, quality control, and quality improvement; these are known as Juran's trilogy.

Thus the present paper aims to present the model and application for evaluation of investments into cloud computing in Biomedicine, which allows organizations to answer the basic question as to whether the migration from current IT infrastructure to cloud model is beneficial or not. The model reacts to given economic indicators and suitability of this solution related to the characteristics and chosen aspects of the company management, taking into consideration the financial profitability of its implementation.

2 Methods for Cloud Computing Investment Efficiency Measurement

Decision-making about new investment is crucial for the growth and prosperity of any company, especially in the rapidly changing ICT market. Companies must innovate to be successful but this invariably carries some risk and uncertainty. In recent years the issue of investment effectiveness into information technologies has been intensively solved both at private and national levels. The most common methods for evaluation can be divided into three groups:

1. Strategic and financial valuation of projects (e.g., Net Present Value (NPV), cash flow (CF), Rentability of Investment (ROI)),
2. weighting and scoring of products and product criteria (e.g., analytic hierarchy process (AHP) and conjoint analysis),
3. human decision-making (fuzzy logic, actuarial models, neural networks, technology road mapping and expert systems).

Within the first group Vernon et al. [8] described the use of NPV to determine maximum willingness to pay (WTP) using payer reimbursement signals (such as cost-effectiveness thresholds). Furthermore, Vallejo-Torres et al. [9] explained how Bayesian methods can be used to incorporate all available information at the various stages of development. However, the increased complexity of this approach as the stages of development progress may pose difficulties to its use in practice by companies. According to the surveys performed among companies, the following methods are used: discounted cash flow (DCF), unstructured peer review, scoring checklists and decision tree analysis. According to Bootman et al. [10], another way in which consequences are measured determines whether an economic evaluation is a cost effectiveness analysis (CEA), a cost-utility analysis (CUA), or a cost-benefit analysis (CBA).

The methods described above, however, face certain problems with respect to the specifics of the cloud computing deployment. In case of DCF an important business reality is ignored. Projects rarely proceed as planned and management frequently has to adapt and revise future decisions as new information becomes available or as uncertainties are resolved. The analytical hierarchy process (AHP) as one form of multi criteria decision analysis, where both quantitative and qualitative factors may be combined [11]. AHP causes another problem with potential inconsistencies in the numerical scales utilised, no theoretical foundation for their imposed link with the verbal descriptions exists [12]. Fuzzy logic concept is quite abstract, and requires expert input. An expert system is an artificial intelligence technique, which uses an input knowledge base and inference engine to solve a problem [13]. The knowledge base required by the system is time consuming, and it may be difficult for experts to express and input their knowledge in terms of specific rules and confidence factors [14].

The methods used to design the model were multi-criteria variant analysis, discussions with IT experts, and in-depth discussions with companies.

The multi-criteria variant analysis was used during the initial decision-making about which method on which the model would be based on to use. Cost Benefit Analysis method has been chosen as a basis and it has been modified to evaluate investments into cloud computing [14–16]. The initial CBA CC (Cost benefit analysis for cloud computing) model was assembled in January 2015. Discussions with IT experts have been carried out as well. The aim of these discussions was to ensure the functionality of the model related to the technical parameters of cloud computing.

Feedback on clarity and structure of entries identifying the current state of IT in the company and required level of cloud computing from the company IT experts was carried out consequently. All of that with purpose to verify the correctness of suggested entries of expenses and assets in CBA, which could be translated to numbers in the following steps. After the Cost Benefit Analysis for Cloud Computing (CBA CC) model was finished, it was possible to proceed to program an application that would automatically calculate economical return on investment and generate reports on the suitability of investing into cloud computing. This process was methodically separated into three basic steps. First of all, it was necessary to choose suitable platform, programming language, and service distribution method (full application, web pages, or mobile app).

As the most suitable method, robust JAVA-based platform and the model of primary service distribution via web pages were chosen, which would allow to expand service distribution model nearly in any way in the future, including the connection to external systems. Afterwards, the economic model had to be converted into algorithmic model, which was the first step in automating the whole process. The last step was to create a use-case and a data model. Based on them, an application was created and testing was commenced. The web application has been developing since 1 February 2016, and its functionality is currently being tested.

3   CBA CC for Biomedical Institutions

The present proposed theoretical model assessing the effectiveness of cloud computing deployment for biomedical institutions is based on the CBA method. The method proposed explores all the effects on the subjects stemming from the investment implementation. The effects are both financial and non-financial (possibly intangible). Depending on the possible perspectives, the effects are both positive (benefits), negative (costs) or neutral. The steps which should be utilized while evaluating the cloud computing investments are as follows: [15]

- CC utilization criteria specification,
- actual state and service requirements characteristics
- impact to affected subjects analysis,
- potential assets and relevant costs determination,
- all main CC implementation impacts declaration,
- all criterion's markers calculation,
- sensitivity analysis and project evaluation.

Economic methods enabling the transfer of qualitative effects into the quantitative expression as well as economic evaluation will be selected and recommended within these individual steps. Resulting assets of the CBA CC method compared to commonly used methods in IT can be summarized as:

- complexity,
- the possibility of using both static and dynamic methods,
- adjustment to biomedical institution characteristics and its activities,
- conversion of qualitative variables to quantitative ones.

The main advantage as well as difference as compared to the original Cost Benefit Analysis method lies in the following areas:

- number and structure of steps in CBA change to correspond to cloud computing deployment,
- individual content of first three steps of the method corresponds to the incentives of cloud computing deployment,
- new list of criteria queries for decision-making regarding cloud computing,
- devised scheme to specify entries of the current and desired IT state,
- subject structure that can be influenced by the technology employment has been created.

Taking into consideration biomedical infrastructure specificities, it is vital to highlight the differences in investment evaluation model in Biomedicine compared to the more general model set for commercial and industrial companies. These kinds of removed model criteria are represented by, e.g. lines of business, types of commercial relationship or employees' data access.

The most significant differences in evaluation model are technical criteria aligned with the specificities of biomedical research. The selection of cloud computing infrastructure platform highly influences the whole biomedical-medical system [17, 18]. The complete isolation of full virtualization techniques can produce negative effects due to the general binary translation techniques of the hardware resources under certain circumstances. Direct access to hardware resources of hypervisor via specialized drivers or direct instructions can be much more efficient because these drivers or instructions are programmed for the specific hardware, since biomedical research is typically compute-intensive. Three following computing types can be distinguished as follows:

- CPU intensive tasks – relatively small amount of data on compute node, complex algorithms, corresponding memory and I/O allocation.
- In memory, intensive tasks – high memory allocation on compute node, CPU intensive, corresponding I/O allocation.
- I/O intensive tasks – huge amount of stored data, storage or database intensive tasks.

The cloud computing infrastructure (IaaS) should be optimized for specific biomedical systems workloads according to different computing types in scientific research mentioned above. Cloud computing infrastructure is composed of many elements intended to cooperate together. Initially, the computing nodes, respectively hypervisors use the distribute networking model, therefore the speed of communication network, jitter, latency, delay and other networking aspects should be taken into consideration. This is vital not only for computing nodes but also for the managerial part of cloud computing infrastructure like controllers, API and network nodes. Furthermore, scientific tasks are usually memory intensive. The hypervisors handle these resources differently and virtualization memory techniques can save important amount of memory by sharing concurrent binary blocks and speed up the whole computing process. Next, the way of data distribution in cloud computing infrastructure should respect the biomedical computing types, because the extensive data operation, such as analysis, datamining, OLAP (Online Analytical Processing), distributed databases and other topics related to business intelligence in general, are highly sensitive to I/O performance issues. Data transport, management or replication can be operated on computer nodes fully, partially, locally or on a distributive basis. Finally, the balance and cooperation among each element are essential, since each part of the system can influence the rest and act as the bottle neck of the system in its entirety.

The proposed model of cloud computing efficiency evaluation is based on the setting of both quantitative and qualitative parameters. Table 1 requires specification of institution data information in the correlation with the geographical spread, data volume as well as technical requirements for availability and evaluation of the present infrastructure.

Table 2 specifies system complexity technical requirements to allow the calculation of annual costs correlating with required capacities, security and other required additional services.

The above mentioned input information enables the subsequent calculation of cloud computing efficiency. The model utilizes ROI, ČSH and TCO methods as used in the

Table 1. Specification of institution and strategic management, authors' own processing

Input information/query	Value
Geographical spread of the organization	*CR/Europe/worldwide*
What type of data work is realized? Please indicate in %, where 100% equals the total of the three below indicated possibilities.	
Digital Imaging and Picture Archiving	*%*
Teleradiology	*%*
eHealth (sharing of data from patients, doctors, health institution management, municipalities, regions)	*%*
How much data cannot be transferred to the third party? (with regards to three types of data processing)	*%*
Digital Imaging and Picture Archiving	*%*
Teleradiology	*%*
eHealth	*%*
What is the renewal cycle in years?	*Number of years*
What is the average loss per hour when the technology is unavailable	*EUR*
Estimated loss of the company caused by the infrastructure unavailability per hour	*EUR*
What is current infrastructure/SLA availability rate?	*%*
Unavailability in hours/year	*Number of hours*
Availability of cloud infrastructure	*%*
Operating expenses for current infrastructure without personal expenses per year (buildings, energies…) variables: currency	*EUR*
Amount of investment into current infrastructure	*EUR*
Amount of investment into licenses per year	*EUR*
Amount of investment into product support	*EUR*
Investment into security systems (processing, information, physical)	*EUR*

field of information technologies. The fundamental asset of CBA method is the possibility to transfer qualitative variables into quantitative formulations, which is consequently included in the aforementioned markers. The actual transfers utilize methods of shadow market prices and opportunity costs. With regards to the possible distortion, the last step of the model is sensitivity analysis which is executed for 10% and 30% of the changes in input values, which significantly contributes to the possibility of risk factor consideration.

4 Application Proposal

For a technical solution of web application capable of calculating CBA CC, it is vital to design a model which is operable and ensures that all requirements on the systems are met. The principal requirements of the whole solution are scalability, elasticity, and mobility. The design thus utilizes JAVA and WildFly 10 platforms with connection to PostgreSQL 9.x database. WildFly, previously known as JBoss, is an application server

Table 2. Technical parameters, authors' own processing

Input information/query	Value
Overall expenses for instances/servers in the cloud per year (currently according to AZURE CALCULATOR)	*Sum*
Type of instances/virtual servers	*Type*
Number of instances/virtual servers	*Number*
Expenses per instance/year	*EUR*
Overall expenses for backing up	*EUR*
Requirements for the backup system capacity in GB	*GB*
Expenses per instance	*EUR*
Number of protected instances (servers for backing up)	*Number*
Overall expenses for networks	*EUR*
Amount of data transferred from cloud/data centers in TB/year	*TB*
Expenses for data transfers from cloud/data centers	*EUR*
Number of public IP addresses for instances (not time-restricted)	*Number*
Expenses for public IP addresses	*EUR*
Overall expenses for cloud storage (not time-restricted)	*EUR*
Requirements for cloud storage capacity in GB	*GB*
Requirements for other services/administration/reporting per year	*EUR*

authored by JBoss. WildFly is written in Java, and implements the Java Platform, Enterprise Edition (Java EE) specification. WildFly is free and open-source software which supports multiple platforms. Data can be saved into other databases or formats, such as Open Source JAVA application server. Fundamental overview of application infrastructure is described on the following picture (Fig. 1).

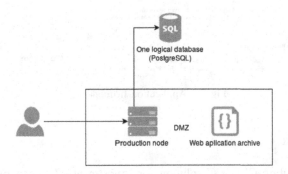

Fig. 1. Application logic.

Another, yet important issues are security and ethical aspects of biomedical research due to the data sensitivity, confidentiality and law or other policies limitations.

Transport Layer Security (TLS) utilizing certificates from trustworthy certification authorities is used to provide safety measures. Its application is accommodated to

Enterprise Edition 7 standards. The image below the diagram introduces the basic architecture of the system designed (Fig. 2). In the case of higher load when a user accesses particular node, the load is redistributed among multiple computing nodes. PostgreSQL database and application archive provides a connection to the internal logic of the production node. The CBA CC application has been developed on the basis of economic model which was described in the previous chapters. As it was stated, application is set on JAVA and web technologies, therefore is highly scalable and ready for further development. Recently, our application serves the investment efficiency reports at while is being tested by selected organizations and testers. The following picture shows the homepage of application.

Fig. 2. Application homepage.

5   Conclusion

The cloud computing CBA CC solution for investment efficiency measurement for biomedical institutions enables to provide of initial, relatively complex evaluation of technology suitability. Presently there predominantly exist calculations of price from the cloud service providers (Amazon, Azure), nevertheless, these reflect merely a narrow perspective regarding the price calculation. The advantage of the proposed model is the utilization of qualitative variables and evaluation of costs; furthermore, the solution assesses the suitability concerning the data character, legislative restrictions and geographical spread. The risk factor of the proposed method is primarily a distortion due to an inaccurate data input, which is then reflected in the calculation of the price as well as to the actual efficiency.

Acknowledgements. We would like to thank to Pavel Pscheidl, doctoral student who was helping with the application programming. This work and the contribution were also supported by project of specific science, Faculty of informatics and Management, University of Hradec Kralove, Czech Republic.

References

1. Huptych, M.: Introduction to Biomedical Informatics. Lecture at CVUT Introduction to Biomedical Informatics (2007). http://bio.felk.cvut.cz/~huptycm/Vyuka/X33BMI_prednasky/UvodDoBMI.pdf. Accessed Oct 2015
2. Kornevs, M., Minkevica, V., Holm, M.: Cloud computing evaluation based on financial metrics. Inf. Technol. Manag. Sci. **15**(1), 87–92 (2013)
3. Martens, B., Walterbusch, M., Teuteberg, F.: Costing of cloud computing services: a total cost of ownership approach. In: Proceedings of the 45th Hawaii International Conference on System Sciences, Maui, Hawaii, USA, pp. 1563–1572 (2012)
4. Freedson, P.S., Lyden, K., Kozey-Keadle, S., Staudenmayer, J.: Evaluation of artificial neural network algorithms for predicting METs and activity type from accelerometer data: validation on an independent sample. J. Appl. Physiol. **111**(6), 1804–1812 (2011)
5. Jureta, I., et al.: A comprehensive quality model for service-oriented systems. Softw. Qual. J. **17**, 65–98 (2009)
6. Kalepu, S., Krishnaswamy, S., Loke, S.: Verity: a QoS metric for selecting web services and providers. In: Proceedings Fourth International Conference on Web Information Systems Engineering Workshops, WISEW 2003, pp. 131–139 (2003)
7. Juran, J.M.: Juran's Quality Handbook, 5th edn. McGraw-Hill, New York City (1999)
8. Vernon, J.A.: Economic evaluation and cost-effectiveness thresholds: signals to firms and implications for R & D investment and innovation. Pharmacoeconomics **27**(10), 797–806 (2009). doi:10.2165/11313750-000000000-00000
9. Vallejo-Torres, L., Steuten, L., Parkinson, B., et al.: Integrating health economics into the product development cycle: a case study of absorbable pins for treating hallux valgus. Med. Decis. Making **31**, 596–610 (2011)
10. Bootman, L., Townsend, R.J., Ghan, W.F.: Principles of Pharmacoeconomics, 1996 (2008)
11. Hummel, J., Van Rossum, W., Verkerke, G., Rakhorst, G.: J. Multi-Criteria Decis. Anal. **9**(1–3), 90–98 (2000)
12. Johal, S.S., Oliver, P., Williams, H.C.: J. Med. Mark. **8**(2), 101–112 (2008)
13. Akoka, J., Leune, B.: Expert Syst. Appl. **7**(2), 291–303 (1994)
14. Marešová, P., Mohelská, H., Dolejš, J., Kuča, K.: Socio-economic Aspects of Alzheimer's Disease. Curr. Alzheimer Res. **12**(9), 903–911 (2015)
15. Marešová, P., Klímová, B.: Investment evaluation of cloud computing in the European business sector. Appl. Econ. Appl. Econ. **47**(36), 3907–3920 (2015). doi:10.1080/00036846.2015.1019041
16. Marešová, P., Hálek, V.: Deployment of cloud computing in small and medium sized enterprises in the Czech Republic. E+M. Ekonomie a Manag. **17**(4) (2014). doi:dx.doi.org/10.15240/tul/001/2014-4-012

17. Vilaplana, F., Solsona, F., Abella, R., Filgueira, R., Rius, J.: The cloud paradigm applied to e-health. BMC Med. Inform. Decis. Making **35**(13), 1–10 (2013). doi:10.1186/1472-6947-13-35
18. Fusaro, V.A., Patil, P., Gafni, E., Wall, D.P., Tonellato, P.J.: Biomedical cloud computing with Amazon web services. PLoS Comput. Biol. **7**(8), e1002147 (2011). doi:10.1371/journal.pcbi.1002147
19. Voras, I., Orlic, M., Mihaljevic, B.: An early comparison of commercial and open-source cloud platforms for scientific environments. In: Proceedings of the 6th KES International Conference on Agent and MultiAgent Systems: Technologies and Applications, pp. 164–173 (2012). http://dx.doi.org/10.1007/978-3-642-30947-2_20

Time Series Distributed Analysis in IoT with ETL and Data Mining Technologies

Ivan Kholod$^{(\boxtimes)}$, Maria Efimova, Andrey Rukavitsyn,
and Shorov Andrey

Saint Petersburg Electrotechnical University "LETI", Saint Petersburg, Russia
{iiholod,ashxz}@mail.ru, maria.efimova@hotmail.com,
rkvtsn@gmail.com

Abstract. The paper describes an approach to performing a distributed analysis on time series. The approach suggests to integrate Data Mining and ETL technologies and to perform primary analysis of time series based on a subset of data sources (primary data sources). Other data sources are only used if it is necessary to obtain additional information. This allows to reduce the number of requests to data sources and network traffic. In the result it makes it possible to use communication channels with low bandwidth (including wireless networks) for data collection.

Keywords: Data collection · Fog computing · ETL · Data Mining · Wireless networks · IoT

1 Introduction

Analyzing information collected from multiple data sources becomes more and more relevant. The humanity entered the era of Big Data which is characterized by large volumes of information, information being produced at high speed, variety of formats for data representation and storage. The majority of modern systems analyze data received not only from transactional (OLTP) systems but from various sensors and measurement tools, dash cams, CCTV, control systems, social media, financial information providers, etc. Such data sources generate time-related data forming time series. They generate data in real-time and use various communication channels for data transmission, including wireless channels (satellite, microwave, mobile networks, etc.), that have a limited bandwidth.

In order to perform data collection, processing and analysis according to the principles used for data warehouses, lambda architecture, cloud computing and Internet of Things (IoT), all the information should first be collected in a single warehouse and only then be processed. This requires large storage and computational resources (that grow with the number of data sources) and communication channels with high bandwidth. Moreover, such approach reduces the efficiency of information analysis because the information that is analyzed is collected for a certain period of time.

However, the end result of the time series analysis (and not the completeness of the data collected from all data sources), such as current state of patients, emergency warnings, changes in stock quotes, etc., is important for many practical tasks that use

© Springer International Publishing AG 2017
O. Galinina et al. (Eds.): NEW2AN/ruSMART/NsCC 2017, LNCS 10531, pp. 97–108, 2017.
DOI: 10.1007/978-3-319-67380-6_9

information from such data sources. Usually such results are based on a specific selection from all time series (for example, a selection for a specific period of time when an event occurred). Therefore practical tasks themselves do not require all the data to be transferred to the data processing centre and the data is not used after the analysis has been performed.

Reduction in volumes of information transferred can be achieved by moving the analysis closer to the data. Such concepts [1] become more and more popular: Fog computing [2], Edge computing [3], Dew computing [1] and other. However they offer an architectural solution, but not the technologies needed for practical implementation.

The paper describes an approach of integrating two technologies: ETL (Extraction, Transformation, Loading) and Data Mining for a time series analysis from distributed data sources.

2 Related Work

The problem of collecting information from multiple data sources is not new. In the end of last century and in the beginning of this century the concept of data warehouse became widespread. A data warehouse is a subject-oriented, integrated, time-variant, and nonvolatile collection of data that supports managerial decision making [4]. According to it, data from OLTP systems was collected in a data warehouse for analysis. Data integration was performed sometimes for the static, usually relational, data sources.

There are the following architecture types of data warehouses [4]:

- physical data warehouse contains consolidated data extracted from multiple operational sources [5];
- virtual data warehouse provide end-users with direct access to multiple data sources [6].

Physical data warehouse was the most functional. However unlike virtual data warehouse it did not provide access to the current information, which significantly decreased the relevance and efficiency of the results.

ETL tools were used to transfer data into physical storages [7]. Standard ETL tools extract information from data sources (Extraction), clean, merge and convert it into the format that is supported by the data warehouse (Transformation), and then load transformed information into it (Loading). Currently out of a dozen popular ETL tools the following systems are open-source: Pentaho Data Integration (PDI) [8], CloverETL [9], Talend Open Studio [10]. ETL tools can extract information in different formats from various data sources. Relevant extensions for ETL tools can be used for this.

However the variety of data source formats expanded, information started being updated in large volumes very frequently. This led to a problem of Big Data collection and processing. NoSQL technologies, stream processing [11] and systems with Lambda Architecture (LA) [12] are currently used for real-time data processing, including time series analysis.

LA was suggested for concurrent processing of real-time data and batch data. LA is a paradigm defined by Nathan Marz which provides the foundations of building real-time distributed systems in a flexible and (human/system) fault-tolerant way.

The structure of LA has three layers: batch layer, service layer and speed layer. Systems with LA integrate heterogeneous components that transfer data from one component to another (within each of the layers). Such components can include: distributed processing tools, streaming processing tools, analytics tools, etc.

It is assumed that data is collected at the service layer for further processing for both LA and ETL. Transferring all the data from data sources into a centralized storage creates a lot of network traffic and almost excludes the possibility of using wireless communication channels with a limited bandwidth. Moreover systems with LA, like ETL, do not allow scaling and dynamic use of data sources.

Development of the Internet and data sources connected to it (news resources, financial information providers, social media and it users, various devices) lead to development of Internet of Things [13] and its different variations, such as Social Internet of Things (SIoT) [14], Internet of Social Things (IoST) [15], Internet of Every things (IoE) [16] and others.

The basic architecture of IoT which is widely used to explain the approaches of IoT has three layers [13]:

- the perception layer is the bottom layer which can be regarded as the hardware or physical layer which does the data collection;
- the network layer (the middle layer), is responsible for connecting the perception layer and the application layer so that data can be passed between them;
- the application layer usually plays the role of providing services or applications that integrate or analyze the data received from the other two layers.

Cloud computing [17] technologies are used to solve the problems related to computational resources and Big Data processing on the application layer of IoT systems. A Cloud provides scalable storage, computation time and other tools to build application services.

In this case the network layer is responsible for connecting IoT devices (such as sensors, cameras, and other devices) with a cloud. It creates very large traffic through Internet. Fog computing [2] can be a solution to this problem. The fog extends the cloud to be closer to the sources that produce and act on IoT data.

Despite of Fog computing becoming popular, there are no ready solutions for it's implementation. This can be explained by the fact that such concept is every young and has a high level of abstraction.

This paper suggests an approach to integrating ETL and Data Mining technologies for practical implementation of time series analysis in Fog Computing systems.

3 Time Series Distributed Analysis

3.1 Concept of Time Series Distributed Analysis

Analysis of time series from multiple data sources assumes collecting data from data sources and applying a target model that solves the analytical problem. Let's assume there is a set S that contains n data sources:

$$S = \{s_1, s_2, \ldots, s_g, \ldots, s_n\}$$

each of them form time series:

$$s_g = \{v_g(t_0), \ldots, v_g(t_k), \ldots, v_g(t_m)\}.$$

The target model f_T at the time t_k processes a vector that contains the values received from all data sources by the time t_k:

$$r(t_k) = F_T(x(t_k))$$
$$x(t_k) = \{v_1(t_k), \ldots, v_g(t_k), \ldots, v_n(t_k)\}, \text{ where}$$

- F_T – a target model that performs a given analysis function (for example, classification or regression);
- $r(t_k)$ – result of the function for the time t_k.

Usually not all values of the vector $x(t_k)$ are equivalent for the analysis function. Some of them impact the results more than others. Such data sources can be called *primary data sources*. Therefore only the data that is the most important for the analysis should be transferred to the target model. The rest of the data should be transferred only if required.

Therefore to solve the problem either target model F_T (that is applied to the full vector $x(t_k)$ (with the values from all data sources)) or the target model F'_T, (that is applied to the reduced vector with the values from primary data sources) can be used:

$$x'(t_k) = \{v_1(t_k), \ldots, v_r(t_k)\}, \text{ where } r < n.$$

A decision about using additional data sources can be made based on application of the intermediate model F_I to the vector $x'(t_k)$ using values from the primary data sources. The target and intermediate models can be different.

Therefore the suggested approach includes the following steps:

1. choose primary data sources - the initial time series analysis is performed based on information, coming from these data sources;
2. extract data from primary data sources;
3. apply intermediate model F_I to the data to determine if it is necessary to use additional data sources;
4. if using the additional data sources is not necessary then
   (a) apply the target model F'_T to the data from the primary data sources;
   else
   (a) extract data from additional data sources;
   (b) apply the target model F_T to the full data set.

This approach is shown on the Fig. 1. The data sources s_1 and s_2 are defined as primary data sources. The data from these data sources is merged together to build a single time series on node #3. The time series is analyzed by intermediate model F_I. The analysis is performed to establish the accuracy of the time series (for example if

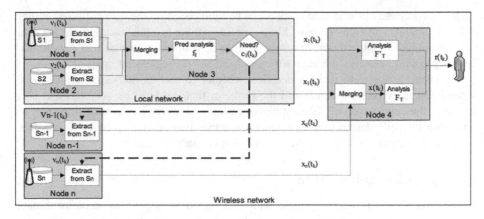

Fig. 1. Collecting data from data sources when required

there are outliers and/or spikes in the data). If the accuracy is low the request is made to the rest of the data sources.

Analyzing data only when required can be an alternative to collecting all the data prior to performing the analysis. There are two issues to implement our approach:

- selecting intermediate model F_I to make a decision about making requests to additional data sources (dynamic control of data flows).
- integrating data from the distributed data sources to the target models F_T and F'_T.

3.2   Selecting Intermediate Model

The intermediate model F_I should determine the values of $v_r(t_k)$ moments in the data from primary data sources when the result of the target model F'_T applied to the reduced vector $x'(t_k)$ is different from the target model F_T applied to the full vector $x(t_k)$:

$$F_I(x'(t_k)) = \begin{cases} true, & if \ |F'(x(t_k)) - F(x(t_k))| > \sigma \\ false, & if \ |F'(x(t_k)) - F(x(t_k))| < \sigma \end{cases}$$

One of the reasons for the differences is anomalies in the reduced vector. Those values can be either data errors or indicators of important events. Therefore it is important to check data in additional data sources when the anomalies are detected.

Regression models that perform time series approximation can be used for anomaly detection. In this case it is decided to extract data from all data sources and apply target model F_T to the full vector $x(t_k)$ if the approximated value is significantly different from the value received:

$$F_I(x'(t_k)) = \begin{cases} true, & if \ |x'(t_k) - x''(t_k)| > \sigma \\ false, & if \ |x'(t_k) - x''(t_k)| < \sigma \end{cases},$$

where $x''(t_k)$ – vector of the approximated values from primary data sources.

Regression and classification models can be built based on the data collected earlier and on situations solved by experts. Moreover when there is a feedback loop to get data from an analyst then the model can train in process, adopting to the current conditions. Recently neural networks are used for classification and regression analysis.

Choosing a model to make a decision whether to use an additional data source depends on the type of the data and the problem that needs to be solved. Next part describes the comparison of different classification and regression models that analyze data from multiple data sources, by comparing the time needed to make a decision, reduction in network traffic and the quality of the results produced.

3.3   Dynamic Data Integration

A tool that would provide dynamic data integration is needed to implement data collection, transformation and transfer to the user from distributed data sources. To implement the described approach it should meet the following requirements:

- extract information from multiple data sources;
- work with data in various formats;
- aggregate information from multiple data sources;
- be executed in a distributed environment;
- work with data that is generated in real-time;
- execute alternative branches of data processing ("if" block).

The ETL tools satisfy most of these requirements [18]. At the transformation stage information can be integrated, cleaned and transformed. Some ETL tools can perform transformations at different nodes of a distributed execution environment. Different ETL tools include different units of transformation and data processing. These units can be used to satisfy the given requirements.

As it has been said earlier, ETL tools are not used to work with flows of real time data. However adding custom modules makes it possible to implement the approach using real time data. For example, buffering with consequent reading can be used to implement processing of real-time data.

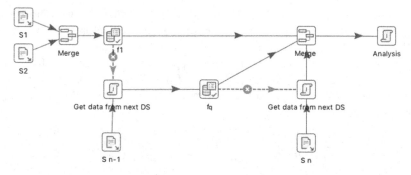

Fig. 2. Implementation of the proposed approach by Pentaho data integration

A combination of units (for example "if" units) that control the process of data transmission can be used for implementation of dynamic requests to a data source. Implementation of custom extensions for relevant algorithms can be used to perform analysis. Figure 2 shows example of a graph created using the PDI ETL tool.

Thus, ETL tools can be used as a basis to collect, convert and transfer data. It satisfies most of the requirements and can be extended by adding new units.

4 Experiments

Experimental data has been used to create regression models of appliances energy use in a low energy building [19]. The dataset contains parameter changes for energy use, temperature (T) and pressure (RH_) in various rooms in a house (kitchen, living room, laundry room, office room, bathroom, ironing room, teenager room, parents room, outside the building (north side)). The dataset contains measures taken with a 10 min interval for about 4.5 months. The house temperature and humidity conditions were monitored with a wireless sensor network. Each wireless node transmitted the temperature and humidity conditions around 3.3 min. Then, the wireless data was averaged for 10 min periods. The energy data was logged every 10 min with m-bus energy meters. Weather from the nearest airport weather station (Chievres Airport, Belgium) was downloaded from a public data set from Reliable Prognosis (rp5.ru), and merged together with the experimental data sets using the date and time column.

The target was to define the dependency between the consumed energy and the parameters measured. Generalised Boosted Regression model (GBM) [20] has been chosen for this task [19] as the target model F_T.

The analysis [19] has shown (Fig. 3) that the number of seconds from midnight for each day (NSM) and the pressure measurements from the nearest airport (Press_mm_hg) are the most important for the target GBM model. The humidity in the living room (RH_2), the kitchen (RH_1), the laundry room (RH_3), and temperature in the laundry room (T3), outside of the building (T6) have the biggest impact on classification. They have been chosen as the primary data sources.

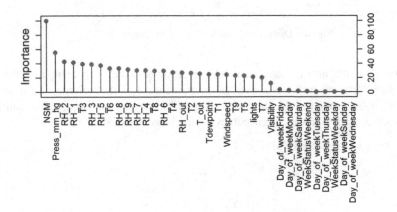

Fig. 3. Variable importance for GBM model.

To verify the approach different sets of primary data sources (sensors) have been used. In addition to that the information about the time has been used: NSM, Day of week (Monday, Tuesday, Sunday), Week status (weekend (0), or a weekday (1)). This information is not presented at the calculation node and does not load the network.

The GBM as the target model has been trained to be applied to both full vectors F_T and to reduced vectors F_T' for the listed sets of primary data sources. The following approaches have been researched to form a comparison:

- analysis of full dataset (from all data sources);
- analysis of a partial dataset (from primary data sources only);
- distributed analysis that involves applying the intermediate model F_I for the primary data sources.

The GBM and Neural network (LSTM) model [21] has been used as intermediate model F_I. Schema of time series distributed analysis for this experiment is showed on Fig. 4. Scikit-Learn Machine Learning Python library has been used for regression analysis by the GBM. Google's TensorFlow library has been used to build the neural network.

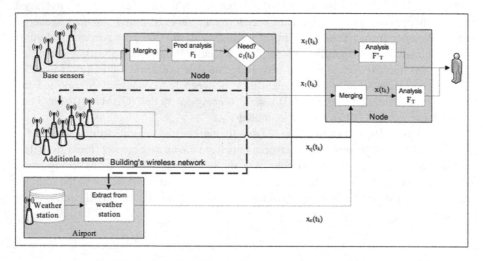

Fig. 4. Distributed analysis in a low energy building

Network traffic and the target model performance have been compared for these approaches. Network traffic is shown as the percentage of the volume of the network traffic generated when the data has been collected from all data sources. The following metrics were used to assess the accuracy of the target model:

- the mean squared error (MSE):

$$MSE = \frac{\sum\limits_{n}^{i=1}(y_i - y_i')^2}{n}$$

- the root mean squared error (RMSE):

$$RMSE = \sqrt{\frac{\sum\limits_{n}^{i=1}(y_i - y_i')^2}{n}}$$

- the coefficient of determination or R-squared/R^2:

$$R^2 = 1 - \frac{\sum\limits_{n}^{i=1}(y_i - y_i')^2}{\sum\limits_{n}^{i=1}(y_i - \bar{y})^2}$$

The results of the experiment are presented in the Table 1.

The results of the experiment showed a decrease in network traffic when using the approach of plugging in data sources dynamically based on the decisions made by the intermediate model. As expected, the network traffic and the target model performance increase with the number of primary data sources. When applying the target model to separate data sources (one, two, three and four data sources) there is a significant decrease in the volume of network traffic (4–17% from the volume of network traffic generated when the data is extracted from all data sources). However, in this case the accuracy of the target model is very low ($R^2 < 35$ when $R^2 = 54$ for all the data). When applying the suggested distributed analysis approach to three or four primary data sources an insignificant decrease in target model performance ($R^2 = 47 \ldots 51$ when $R^2 = 54$ for all data) and 25–40% decrease in network traffic is observed. This allows to conclude that the suggested approach is efficient in reducing the network traffic without a significant decrease in the quality of the analysis results.

The volume of network traffic and the target model performance do not depend much on the model and combination of primary sources used. The differences are explained by the moments when anomalies are discovered and consequently by difference in time when the additional data sources are plugged in.

It is important that when applying the approach to certain combinations of primary data sources the target model performances are the same whereas the traffic volumes are different. The following combinations can be an example: RH_2, RH_1, T3, RH_3 and RH_3, T3, T6. For all of them $R^2 = 50$, bit the first combination allows to decrease the network traffic significantly (67% against 76%), despite of using one more data source. This happens because the moments of time when the data from additional data sources is required are determined more precisely. Therefore when the accuracy of the intermediate model is increased the network traffic can be reduced even more and the accuracy of the target model can be increased. The research will be continued in this area.

Table 1. The results of the experiments

Approach	Number of primary data sources	Primary data sources	Traffic (%)	Target model performance		
				MSE	RMSE	R^2
Full data analysis	24	All	100	4736	69	0.54
Analysis on a part of the data	1	RH_2	4	11530	107	−0.11
		RH_3	4	11103	105	−0.08
	2	RH_2, RH_1	8	8399	92	0.19
		RH_3, T3	8	8683	93	0.16
	3	RH_2, RH_1, T3	12	7501	87	0.27
		RH_3, T3, T6	12	7738	88	0.25
	4	RH_2, RH_1, T3, RH_3	17	7163	85	0.30
		RH_3, T3, T6, RH_1	17	6916	83	0.33
Distributed analysis with GBM as intermediate model	1	RH_2	53	7222	85	0.30
		RH_3	53	7186	85	0.30
	2	RH_3, T3	60	5472	74	0.47
		RH_2, RH_1	57	5663	75	0.45
	3	RH_2, RH_1, T3	67	5267	72	0.49
		RH_3, T3, T6	65	5305	73	0.48
	4	RH_2, RH_1, T3, RH_3	66	5130	72	0.50
		RH_3, T3, T6, RH_1	75	5053	71	0.51
Distributed analysis with LSTM as intermediate model	1	RH_2	49	7222	85	0.30
		RH_3	49	7891	89	0.23
	2	RH_3, T3	63	5418	74	0.47
		RH_2, RH_1	60	5471	74	0.47
	3	RH_2, RH_1, T3	71	5253	72	0.49
		RH_3, T3, T6	76	5233	72	0.50
	4	RH_2, RH_1, T3, RH_3	67	5178	72	0.50
		RH_3, T3, T6, RH_1	78	5009	71	0.51

5   Conclusion

The suggested approach to dynamically plugging in data sources depending on the intermediate analysis of the data from primary data sources allows to reduce network traffic. To achieve this it is suggested to use regression function as the intermediate model built by a data mining algorithms.

The proposed approach is well implemented in the concept of Fog computing. Moving the analysis closer to the data sources and making decisions about information allows to determine whether it's necessary to transfer data from the rest of data sources and therefore avoid transferring the data that is not used for the final decision making.

As a result, it is possible to reduce the amount of information transferred several times. This allows data channels with low bandwidth (including wireless channels) to be used for data transfer.

Different intermediate models and approaches to making a decision about involving additional data sources to increase accuracy of such decisions are going to be researched in the future. This will allow to reduce the network traffic even more without reducing the target model performance.

Acknowledgments. This work was supported by the Ministry of Education and Science of the Russian Federation in the framework of the state order "Organisation of Scientific Research", task #2.6113.2017/6.7, and by grant of RFBR # 16-07-00625, supported by Russian President's fellowship.

References

1. Skala, K., Davidović, D., Afgan, E., Sović, I., Šojat, Z.: Scalable distributed computing hierarchy: cloud, fog and dew computing. Open J. Cloud Comput. (RobPub) **2**(1), 16–24 (2015). ISSN 2199-1987
2. Bonomi, F., Milito, R., Zhu, J., Addepalli, S.: Fog Computing and its role in the internet of things. In Processing of MCC (2012), 17 August 2012, Helsinki, Finland, pp. 13–16 (2012)
3. Ahmed, A., Ahmed, E.: A survey on mobile edge computing. In: 10th IEEE International Conference on Intelligent Systems and Control (ISCO 2016) (2016). doi:10.13140/RG.2.1. 3254.7925
4. Noaman, A.Y.: Distributed data warehouse architecture and design. Ph.D. thesis, University of Manitoba (2000)
5. Inmon, W.H.: Building the Data Warehouse, 2nd edn. Wiley, New York (1996)
6. Devlin, B.: Data Warehouses from Architecture to Implementation. Addison–Wesley, Boston (1997)
7. Kimball, R., Caserta, J.: The Data Warehouse ETL Toolkit. Wiley, New York (2004)
8. Pentaho Data Integration. http://www.pentaho.com/product/data-integration
9. CloverETL. http://www.cloveretl.com/products
10. Talend Open Studio. http://www.talend.com/products/talend-open-studio
11. Aggrawal, C.C.: Data Streams: Models and Algorithms. Springer, New York (2007)
12. Marz, N., Warren, J.: Big Data: Principles and Best Practices of Scalable Real-Time Data Systems. Manning Publications, Greenwich (2010)
13. Tsai, C.-W., Lai, C.-F., Vasilakos, A.V.: Future internet of things: open issues and challenges. Wirel. Netw. **20**(8), 2201–2217 (2014)
14. Atzori, L., Iera, A., Morabito, G., Nitti, M.: The social internet of things (SIoT)—when social networks meet the internet of things: concept, architecture and network characterization. Comput. Netw. **56**(16), 3594–3608 (2012)
15. Nansen, B., van Ryn, L., Vetere, F., Robertson, T., Brereton, M., Douish, P.: An internet of social things. In: OzCHI 2014, 02–05 December 2014, Sydney, NSW, Australia (2014). doi:10.1145/2686612.2686624

16. Evans, D.: The Internet of Everything. Cisco IBSG (2012). http://www.cisco.com/c/dam/en_us/about/ac79/docs/innov/IoE.pdf

17. Gubbi, J., Buyya, R., Marusic, S., Palaniswamia, M.: Internet of things (IoT): a vision, architectural elements, and future directions. Future Gener. Comput. Syst. **29**(7), 1645–1660 (2013)

18. Kholod, I.I., Efimova, M.S.: Smart collection of data for financial instruments. In: 2017 XX IEEE International Conference on Soft Computing and Measurements (SCM). pp. 705–708. IEEE Conference Publications (2017). doi:10.1109/SCM.2017.7970697

19. Candanedo, L.M., Feldheim, V., Deramaix, D.: Data driven prediction models of energy use of appliances in a low-energy house. In: Energy and Buildings, vol. 140, pp. 81–97, 1 April 2017. ISSN 0378-7788

20. Friedman, J.H.: Greedy function approximation: a gradient boosting. machine. Ann. Stat. **29**(5), 1189–1232 (2001)

21. Hochreiter, S., Schmidhuber, J.: Long short-term memory. Neural Comput. **9**(8), 1735–1780 (1997). PMID 9377276, doi:10.1162/neco.1997.9.8.1735

Exploring SDN to Deploy Flexible Sampling-Based Network Monitoring

Catarina Pires da Silva[1]([✉]), Solange Rito Lima[1], and João Marco Silva[2]

[1] Centro Algoritmi, Universidade do Minho, 4710-057 Braga, Portugal
pg27716@alunos.uminho.pt
[2] HasLab, INESC TEC, Universidade do Minho, 4710-057 Braga, Portugal

Abstract. In recent years we witnessed the arrival of new trends, such as server virtualization and cloud services, an increasing number of mobile devices and online contents, leading the networking industry to deliberate about how traditional network architectures can be adapted or even deciding if a new perspective for them should be taken. SDN (Software-Defined Networking) emerged under this framing, opening a road for new developments due to the centralized logic control and view of the network, the decoupling of data and control planes, and the abstraction of the underlying network infrastructure from the applications. Although firstly oriented to packet switching, network measurements have also emerged as one promising field for SDN, as its flexibility enables programmable measurements, allowing a SDN controller to manage measurement tasks concurrently at multiple spatial and temporal scales.

In this context, this paper is focused on exploring the SDN architecture and components for supporting the flexible selection and configuration of network monitoring tasks that rely on the use of traffic sampling. The aim is to take advantage of the integrated view of SDN controllers to apply and configure appropriate sampling techniques in network measurement points according to the requirements of specific measurement tasks. Through SDN, flexible and service-oriented configuration of network monitoring can be achieved, allowing also to improve the trade-off between accuracy and overhead of the monitoring process. In this way, this study, examining relevant SDN elements and solutions for deploying this monitoring paradigm, provides useful insights to enhance the programmability and efficiency of sampling-based network monitoring.

Keywords: SDN · Packet sampling · Network monitoring

1 Introduction

Despite the evolutionary use of traditional networks, the underneath structure of networks remain less flexible. This occurs mainly due to the integration and interconnection of many proprietary, vertically integrated devices, where vendors dictate specific methods and proprietary software [1]. It can be seen from the transition from IP (Internet Protocol) v4 to IPv6, that is taking decades to

© Springer International Publishing AG 2017
O. Galinina et al. (Eds.): NEW2AN/ruSMART/NsCC 2017, LNCS 10531, pp. 109–120, 2017.
DOI: 10.1007/978-3-319-67380-6_10

be accomplished, or the introduction of new routing protocols, that can take a decade to become fully operational, that today's networks are somehow resilient to progress or new solutions. The use of the traditional networking architecture, means that any change will affect the entire network, so it reaches a point where networks are relatively static as their operators seek to minimize the risk of disruptions.

The SDN (Software-Defined Networking) architecture proposes to structure the network in three different layers: the infrastructure layer, the control layer and application layer. This organization has the purpose of decoupling the data and control planes allowing some networking tasks such as forwarding ruling and monitoring to be held by a centralized node called *controller* [2]. Resorting to this centralized controller, the support of decision making for each device is enabled maintaining a full view of the network. Connecting the controller to the infrastructure layer requires an API (Application Programming Interface), being the OpenFlow [3] standard currently the most popular in the SDN domain. Then, the controller software, called NOS (Network Operating System), runs the data plane protocol so that the infrastructure layer and control layer can communicate with each other, enabling networking tasks to be performed [4].

Although being an encompassing innovation enabler, currently, SDN research is mostly focused on how to apply the architecture to fulfil today's needs of forwarding mechanisms, scalability and security. Taking advantage of the centralized control to enhance network monitoring and management is still an open issue, with only few works and solutions coming up in latest years. In fact, a field with large potential to evolve through SDN is the traffic monitoring based on packet sampling. Sampling provides an overview of the network dynamics by collecting only a subset of the traffic in specific nodes, at specific time or count interval, allowing to retrieve information about what happens in the network without the burden of collecting all traffic [5]. It is well documented that different sampling techniques provide distinct performance for specific measurement tasks, for instance, accounting, traffic classification, intrusion detection, among others [6]. However, most of the current network devices only support a small number of techniques, namely deterministic and probabilistic [7].

In this context, this research work explores the use of SDN concepts, architecture and elements to sustain the selection and configuration of sampling techniques in the network environment being monitored. The aim is to take advantage of an SDN controller to bring flexibility to the deployment in network devices of the most suitable sampling solutions for monitoring an SDN network, according to the requirements of each measurement task. For this purpose, this papers demystifies and clarifies the use of the SDN architecture and components to support sampling-based network monitoring, discussing the advantages and drawbacks of the possible SDNs scenarios that can be devised to deploy this monitoring paradigm. Through the provided discussion and overview, this paper contributes to foster and enlarge the domain of application of SDN and to enhance network monitoring.

This paper is organised as follows: Sect. 2 introduces relevant SDN concepts and protocols for supporting the addressed problem; Sect. 3 presents existing solutions to sustain the sampling-based monitoring process in the SDN context; Sect. 4 discusses ways to take advantage of SDN components to enhance the flexible configuration of sampling techniques in network nodes; Sect. 5 presents the conclusions of this work.

2   Software-Defined Networking Concepts

The SDN architecture proposes a change to increase flexibility and fulfil the demands and requirements of the current and next generation networks. In Fig. 1, the decoupling of data and control planes is faced with the traditional architecture scheme.

Due to the separation of planes, the neutral software and the emerging of open and free software to control and operate networks, SDN permits the introduction of new features to be more easily implemented than in most of today's environments. The capability of having a central point of control, accessing and viewing the whole system, while allowing the possibility of taking different kinds of traffic engineering decisions in different regions of the network, increases the flexibility of network management and monitoring [1].

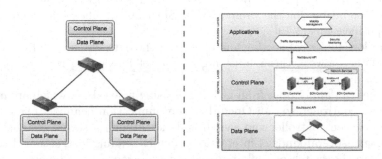

Fig. 1. Traditional architecture vs SDN architecture [2]

In more detail, the SDN architecture is divided in three different layers, as illustrated in Fig. 1, right side. The infrastructure layer is where networks devices such as switches and routers are, forming what is known as *data plane*. The middle layer, called *control layer*, is composed by one or more *controllers*, which provide several ways of centrally operating the network. The connection between these layers is achieved via an API, known as *southbound* API. If it is decided to have several controllers, their connection is accomplished via *west* and *eastbound* APIs. On the top, the application layer provides, via a *northbound* API, the possibility of communicating with the control layer, sending specific instructions through functional applications, which may have several purposes such as monitoring and controlling access for operation and management [2].

The separation of data and control planes is important as it becomes easier to address these two different functions and make each of them more flexible and manageable. It also allows data and control planes functions to be physically separated by hardware. This means that an SDN switch only has a data plane module and does not have any conventional control plane functionality, fully relying on the external controller entity to make decisions. As SDN provides a more centralized control, network operators only need to manage the controllers, enabling a possible NaaS (network-as-a-service) reality.

The vision of SDN is a key enabler for simplifying management processes leading to keen interest from both the industry and the research community. Exploring the SDN architecture in network management can solve many problems because, in this way, flexibility, programmability, simplification of tasks and application deployment can be achieved through a centralized network view [8]. Managing a network with SDN means that a single node of the network, the controller, has the power to configure, collect and store data from numerous points of the network and then analyse it.

2.1 OpenFlow

OpenFlow is a non-proprietary communication standard, which provides a way to establish connection between the control and infrastructure layers of an SDN architecture. It is supported by the ONF (Open Networking Foundation) [9], which is responsible for the promotion of SDN and publication of OpenFlow switch specifications [3, 10].

OpenFlow allows connecting and operating the data plane, enabling a direct control of the network through setting up packet forwarding rules on network devices, such as switches.

Switches can apply rules from OpenFlow using flow tables, which are manipulated by a controller, via an OpenFlow communication channel. The communication between controller and switch is done through a secure channel interface, allowing several kinds of interactions such as sending instructions from the controller to the switch or replying to a statistics request to the controller [3]. OpenFlow specifies network traffic based on what are called flows (i.e., packets that match the same entry in a flow table). These flows together with a set of headers, are combined with a set of admissible fields on the flow table. Flows can be defined by the control layer in a static or dynamic way. Actions are instructions for packet matching that allow discard, modify, queue, or forward operations to the packet. SDN architecture working together with OpenFlow, in physical or virtual networks, gives the ability to instantly respond to changes due to the granular control provided mainly by OpenFlow ability of per-flow programmability [3, 11].

2.2   Open vSwitch

Open vSwitch [12] is a virtual switch consisting on a software layer that resides in a virtual machine host. It was designed to bring flexibility and platform-free usage, meeting the needs of the open source community.

To prevent problems such as consumption of hypervisor resources, Open vSwitch implements what is called flow caching [13]. Flow caching means that traffic handling is cached on the kernel module the first time a packet from a flow, not handled previously, arrives at the switch. This occurs so subsequent packets that match the same flow entry do not have to be handled by the user space module again.

Today, Open vSwitch is very popular mainly due to its integration with OpenStack Networking service and it is also accepted as the genuine standard OpenFlow implementation. Figure 2 offers an overview of the Open vSwitch architecture and its main components.

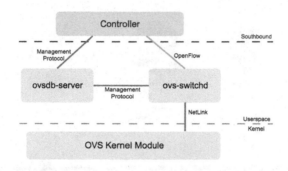

Fig. 2. Open vSwitch architecture

The Open vSwitch kernel module accesses the *ovs-switchd* daemon which implements and manages all the Open vSwitch devices. The OpenFlow protocol is used to exchange messages between the *controller* and *ovs-switchd*. The datapath (*ovs kernel* module) interacts with *ovs-switchd* daemon that implements and manages any number of ovs switches on local system, and the SDN controller interacts with *ovs-switchd* using OpenFlow protocol. The *ovsdb-server* maintains the switch table database (persistent) called *ovsdb-server*.

3   Sampling-Based Monitoring in SDN

Network monitoring based on traffic measurements is a high demanding task due to the huge amount of data currently traversing communication infrastructures. In this way, packet sampling allows retrieving information about the whole network behaviour without the need of analysing all the data, reducing the impact of monitoring operations in the network. It is widely know that different sampling strategies lead to distinct performance in diverse network tasks. However, currently, devices only offer a small set of sampling techniques, which hampers taking full advantage of sampling features. OpenFlow-based SDN are by excellence, a good way to enhance monitoring and sampling while, at the same time, providing a simplification for the introduction of new network applications [10]. Although promising, the use of traffic sampling along with SDN concepts and architecture is exploited for only few tools, namely sFlow and FleXam.

sFlow - Network operators and manufacturers are members of the sFlow.org industry consortium which focus on the dissemination and development of the sFlow standard, making its support available in OpenFlow and non-OpenFlow switches [14]. sFlow proposes that operations such as monitoring no longer be implemented on the switch, instead sampled packet headers are sent to a separate component of the control plane, called monitor, that gets packet headers, decodes them and aggregates the data through a traffic analysis application [15].

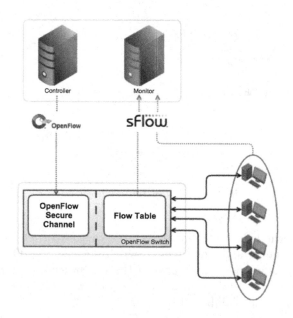

Fig. 3. sFlow OpenFlow-based SDN architecture [16]

As it can be noticed on Fig. 3, sFlow and OpenFlow work in what can be called a partnership. It is intended that the controller, using OpenFlow, configures the forwarding tables in switches and sFlow increases the visibility by providing real-time access to traffic that flows in the network. Having this type of visibility means that the network can adapt to changing demands [14,16].

One major problem regarding sFlow is that its reports do not include the entire packet, which can be a problem when more packet information is required. In addition, it only provides uniform sampling methods [17].

FleXam - FleXam is a per-flow sampling extension for the OpenFlow standard, allowing the controller to access packet-level information. Its priority is to overcome certain problems which may arise when using OpenFlow alone. One of these problems is the increase of flow-entries. From OpenFlow version 1.0 to, for instance, OpenFlow version 1.2, a flow entry went from a 12-tuple match to OXM (OpenFlow Extensible Match) based on TLV (Type-length-value) structures, that allows switches to support a wider range of header fields (for instance, OpenFlow 1.4 supports 41 different types, where TLV was also added to ports, tables and queues) [17–19].

Packets in FleXam can be sampled stochastically, meaning that a predetermined probability is set, or deterministically, which implies a pattern. This flexibility in sampling is enhanced by the possibility of the controller to define several rules on the packets, such as which should be sampled, what part of it should be selected and where they should be sent [18].

FleXam was implemented as a patch to Open vSwitch and enables the access to packet-level information at the controller where an application should run, allowing the installation of rules, processing sampled packets and collecting information. This sampling extension, in addition to presenting itself with two sampling techniques, is considered flexible for some reasons, such as providing a stochastic sampling and a generalized version of deterministic sampling. The stochastic sampling consists in the selection of packets that are included in a flow, with a probability of p. On the other hand, the generalized version of the deterministic sampling is formulated as selecting m consecutive packets from each k consecutive packets, ignoring the first δ packets.

The downside of FleXam when compared to sFlow and other sampling tools is that it offers a per-flow sampling, without the ability of performing per-packet sampling.

4   Enhancing Traffic Sampling Deployment in SDN

In this section all the previously presented elements are taken in consideration and the most suitable solution for implementing sampling-based monitoring in SDN is discussed.

To propose a solution that better explores and fits the SDN architecture several design goals were defined, namely: (i) compatibility with popular software for SDN; (ii) efficient and lightweight implementation, without compromising the SDN proposal; (iii) to consider existing solutions of data and control planes and attach monitoring to them without the need of brand new software; (iv) open and standard protocols/software support.

4.1   Interaction Between Elements

Our goal is to perform packet sampling through count or time intervals using parameters provided by the OpenFlow specification. After implementing this kind of mechanism there is still work to do in order to sampling and analysing traffic, particularly at packet-level. In this scope, tools such as FleXam may be used, as OpenFlow itself has no specific solution.

Below, two possible approaches for applying packet-level sampling are discussed: one from the controller to the switch; the other from the switch to the controller. Each strategy is characterized by where and who controls the rules applied for packet selection.

Controller to Switch - Here, the controller is responsible for making sure that sampling intervals are accomplished and that packet information is stored where it should be. This means that the controller is responsible for managing the rules of sampling, without the switch being aware that a specific selection is being made, because it is the controller who, in some way, forces that. An example consists in a controller requesting a specific subset of packets. A generic representation is shown in Fig. 4.

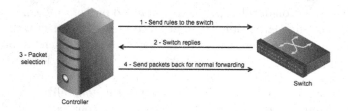

Fig. 4. Controller to switch - minimal approach

In this context, there were two solutions that appeared to be the best path to be taken:

1. Sending all packets, being them original packets or copies of the packets, matching a flow entry, to the controller [3]. In this solution, every packet is forwarded to the controller, where they are counted. When the counter reaches the intended value, the packet is collected. If the packet is not supposed to be collected, there are two options, depending if the packet sent by the switch is the original or a copy. If it is the original packet, the controller must send a *Packet-Out* message containing the packet, injecting the packet in the data plane. In the other hand, if it is a copy of the packet that is being handled, discarding the packet is the normal procedure. The problems identified with this solution include: accumulated traffic in the controller; possible bottleneck in the controller; possible delays in the packet forwarding to the switch (this is not a problem if only a copy of the packet is delivered); possible packet-loss between the controller and switch communication.
2. Request flow statistics from the switch within a pre-set time interval. When the statistics packet counter is next to the intended value, change the rule to send the packet to the controller. This solution can fit both time-based and count-based sampling. In OpenFlow specifications, statistics can be

demanded from a controller. There are several types of statistics that can be requested, such as flow and table statistics. This exchange of information starts by a request message from the controller, mentioning what statistics are wanted, and is followed by a reply from the switch containing the requested information, implying two interactions to get the statistics. The idea is having a thread which is launched from the controller, from time to time, making a flow statistics request to the switch. The flow statistics reply from the switch includes information such as the number of packets in the flow. This value is the one which the controller should pay attention to as it comes closer to the value pretended. The procedure behaves as follows: a *Flow Modify* message would be sent to the switch, followed by a *Packet-In* message sent to the controller, so that packets, matching that flow, are forwarded to the controller. When the packet arrives, it is collected. After that, another *Flow Modify* message is sent to the switch and, as the packets are forwarded normally, counters are reseted. The problems identified with this solution include: statistics requests may not match with counter values intended since the communication between requesting statistics and responding to them implies delay; to obtain higher accuracy with this solution, a previous knowledge of the network dynamic is essential and, even though it can be obtainable, it does not guarantee good levels of synchronism.

Similar solutions to the ones presented could be devised, however, to perform packet selection/collection, several messages between the controller and the switch would be involved, which ultimately leads to lack of performance.

Switch to Controller - Conversely, switch to controller interaction consists in applying rules directly on the switch. In this situation, the role of the controller is only to set the parameters of sampling required by the application, sending them to the switch, which in turn, will apply them accordingly. Since it is the switch the one to apply the proper configuration, this means that switches will be responsible for selecting and collecting packets, and responsible to send them automatically to where the controller ordered. It can be to a monitor or to the controller itself. In this way, the controller will only fill a rule with parameters to the switch and the switch will do the whole work, redirecting the outcome. In Fig. 5, a graphic scheme of this interaction is represented.

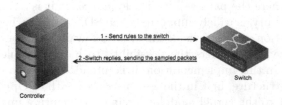

Fig. 5. Switch to controller - minimal approach

After considering the OpenFlow standard, it comes clear that the most efficient option is to implement a sampling mechanism within the OpenFlow itself by biding the operations on the switch, as presented below.

4.2 Proposed Method

The wiser approach to control sampling processes in SDN environments is to create custom actions in the OpenFlow switch to implement the sampling rules. Those customized actions would be added in the form of patches to the OpenFlow specification, resembling FleXam implementation. Opposing to FleXam approach, which unifies a reduced set of sampling techniques in a single action, the proposed solution considers that, for expandability purposes, a single action corresponds to a single sampling technique. The sampling technique implemented as example (represented in Fig. 6) is the count-based sampling [7], where packets will be counted and then sent to the controller for storage, following the configuration parameters. In Fig. 6, OFPAT_INTERVALC represents the action name, that includes two parameters: *interface* and *interval*. The *interface* parameter indicates on which switch interface will the incoming packets be sampled and *interval* represents the counting interval in which the packets will be selected as a sample.

Fig. 6. Switch to controller - minimal approach

For this patch, the OpenFlow version to be used must be at least version 1.0, since this version was considered the unified version from which all vendors start adopting the OpenFlow standard, resulting in a large software support. If the patch is implemented on a newer version, the main concern is which software should be used to provide support to work with that version. For this work, Open Flow version 1.3, the most used OpenFlow's switch version at the moment, will be used. Open vSwitch (described in Sect. 2.2) will be the OpenFlow implementation software where the path will be developed. Since it is largely used by the OpenFlow community, is widely supported from SDN software, has a solid OpenFlow implementation and there is documentation available about how it works. Therefore, Open vSwitch offers the possibility to have a usable and practical developed patch. In this implementation, it is intended to add a counter field in the packet code structure, first in the *userspace* datapath of Open vSwitch and, then, transport it to the kernel module. Having the counting implemented, it is time to implement a rule so that all packets not selected for sampling are forwarded to the right path. Selected packets should be duplicated, with one copy being sent to the controller for sampling and another copy normally processed and forwarded.

The virtual test environment will resort to Mininet [20] network emulator, which allows a user to run several networking elements, such as virtual hosts and switches, using the same Linux kernel. Mininet particularity of having switches supporting OpenFlow and SDN systems, alongside its popularity, raised the interest of using it on this work.

The ongoing work for the proposed solution involves developing a functional patch, following the recommended steps:

1. Implement the sampling action on the Open vSwitch userspace datapath;
2. Test its functionality through the DPCTL tool, which monitors and administrates OpenFlow datapaths;
3. To have it working on a real network environment, add a patch to a NOS.

5   Conclusions

The high level programmability sustained by the decoupling of data and control layers in the SDN architecture represents a promising opportunity to enhance network monitoring and management. Particularly, fostering the deployment of new packet sampling schemes using a central entity (i.e., a controller) will improve the flexibility and efficiency in traffic measurements and related tasks. Having this purpose, this work analysed SDN's architectural components aiming at identifying key aspects able to support further innovation in sampling-based monitoring activities.

As further contribution, two approaches for deploying sampling techniques in SDN environments have been discussed. The first one avoids any change in current protocols and network elements, and consists in relying all the sample process to the SDN controller, without imposing any related burden to other network elements. This approach is flexible, however may face performance issues related to latency and scalability. The second proposal consists in relying the packet selection and collection to the low level network elements. In this approach, the controller interacts with the elements by sending specific sampling rules and receiving the resulting measurement data. This solution copes with the performance issues identified in the first approach, however requires some adaptation in current protocols and devices implementation. This scenario was further explored, through the proposal of a specific method for selecting and configuring sampling techniques in network devices, using Open vSwitch, clarifying the involved challenges and benefits.

Acknowledgments. This work has been supported by COMPETE: POCI-01-0145-FEDER-007043 and FCT within the Project Scope: UID/CEC/00319/2013.

References

1. Kim, H., Feamster, N.: Improving network management with software defined networking. IEEE Commun. Mag. **51**(2), 114–119 (2013)

2. Sezer, S., Scott-Hayward, S., Chouhan, P., Fraser, B., Lake, D., Finnegan, J., Viljoen, N., Miller, M., Rao, N.: Are we ready for SDN? Implementation challenges for software-defined networks. IEEE Commun. Mag. **51**(7), 36–43 (2013)
3. McKeown, N.: OpenFlow: enabling innovation in campus networks (2008)
4. Open Networking Foundation: Software-defined networking: the new norm for networks. ONF White Paper, pp. 1–12 (2012)
5. Duffield, N.: Sampling for passive internet measurement: a review. Stat. Sci. **19**(3), 472–498 (2004)
6. Silva, J.M.C.: A modular traffic sampling architecture for flexible network measurements. Doctoral thesis, Universidade do Minho, Braga, Portugal (2015)
7. Silva, J.M.C., Paulo Carvalho, S.R.L.: Inside packet sampling techniques: exploring modularity to enhance network measurements, 29 March 2016
8. Tuncer, D., Charalambides, M., Clayman, S., Pavlou, G.: Adaptive resource management and control in software defined networks. IEEE Trans. Netw. Serv. Manag. **12**(1), 18–33 (2015)
9. ONF: Open networking foundation. https://www.opennetworking.org/
10. Kreutz, D., Ramos, F.M.V., Verissimo, P., Rothenberg, C.E., Azodolmolky, S., Uhlig, S., Kreutz, D., Ramos, F.: Software-defined networking: a comprehensive survey, pp. 1–61 (2014)
11. Blaiech, K., Hamadi, S., Valtchev, P., Cherkaoui, O., Beliveau, A.: Toward a semantic-based packet forwarding model for openflow. In: 2015 1st IEEE Conference on Network Softwarization (NetSoft), pp. 1–6. IEEE (2015)
12. OVS: Open vSwitch. http://openvswitch.org/
13. Pfaff, B., Pettit, J., Koponen, T., Jackson, E.J., Zhou, A., Rajahalme, J., Gross, J., Wang, A., Stringer, J., Shelar, P., Amidon, K., Casado, M.: The design and implementation of open vSwitch. In: Proceedings of 12th USENIX Conference on Networked Systems Design and Implementation, pp. 117–130 (2015)
14. Giotis, K., Argyropoulos, C., Androulidakis, G., Kalogeras, D., Maglaris, V.: Combining OpenFlow and sFlow for an effective and scalable anomaly detection and mitigation mechanism on SDN environments. Comput. Netw. **62**, 122–136 (2014)
15. Phaal, P.: Software defined networking (2012). http://blog.sflow.com/2012/05/software-defined-networking.html
16. Phaal, P.: OpenFlow and sFlow (2011). http://blog.sflow.com/2011/05/openflow-and-sflow.html
17. Shirali-Shahreza, S., Ganjali, Y.: Efficient implementation of security applications in OpenFlow controller with FleXam. In: 2013 IEEE 21st Annual Symposium on High-Performance Interconnects, pp. 49–54. IEEE, August 2013
18. Shirali-Shahreza, S., Ganjali, Y.: FleXam: flexible sampling extension for monitoring and security applications in OpenFlow. In: Proceedings of 2nd ACM SIGCOMM Workshop on Hot Topics in Software Defined Networking - HotSDN 2013, vol. 167 (2013)
19. Shirali-Shahreza, S., Ganjali, Y.: Traffic statistics collection with FleXam. In: Proceedings of 2014 ACM Conference on SIGCOMM - SIGCOMM 2014, pp. 117–118 (2014)
20. Lantz, B., Heller, B., McKeown, N.: A network in a laptop: rapid prototyping for software-defined networks, vol. 19, no. 6, pp. 19:1–19:6 (2010)

VNF Orchestration and Modeling with ETSI MANO Compliant Frameworks

Ales Komarek, Jakub Pavlik, Lubos Mercl, and Vladimir Sobeslav(✉)

Faculty of Informatics and Management, University of Hradec Kralove,
Rokitanskeho 62, Hradec Kralove, Czech Republic
{ales.komarek,jakub.pavlik.7,lubos.mercl,vladimir.sobeslav}@uhk.cz

Abstract. The goal of this paper is to compare existing NFV (Network Function Virtualization) management solutions and propose Salt-Stack based solution with the same goal. Modern approach to govern NFV automation in large scale IT infrastructures is done by virtualization of various network services. Hardware networking devices are gradually replaced by virtual network appliances for the lower acquisition and maintenance costs. Vendors virtualize their physical products so they can be used along or within cloud environments. The ETSI (European Telecommunications Standards Institute) standard MANO (NFV Management and Organization) aims to unify the management of physical and virtual network services and devices to single point of control for configuration and management.

The paper covers the major NFV frameworks that use MANO as their reference architecture with focus on underlying interface technologies. The main focus of the paper is NFV-MANO orchestration engine using SaltStack capabilities that can model and manage heterogeneous NFV and NFVI resources according to the ETSI NFV-MANO specification along software resources already present. This gives us unique ability to cover all data center resources, both hardware or software based in a single model.

Keywords: NFV · SDN · NFV-MANO · Cloud service · SaltStack

1 Introduction

The NVF frameworks leverage cloud technologies and network virtualization to offer services while reducing Capital and Operations expenditures and can achieve significant levels of operational automation. [7,11,14] Cloud technologies are being adopted, focusing primarily on information technology and dynamic applications. The cloud technology provides the ability to manage diverse workloads and applications dynamically [14]. The desired cloud capabilities are:

1. Real-time installation and decommission of Virtual Machines (VMs),
2. Dynamic assignment of applications and workloads to the running VMs,

© Springer International Publishing AG 2017
O. Galinina et al. (Eds.): NEW2AN/ruSMART/NsCC 2017, LNCS 10531, pp. 121–131, 2017.
DOI: 10.1007/978-3-319-67380-6_11

3. Dynamic movement of applications and dependent functions to different VMs on servers within or across data centers in different geographical locations.

Efforts have not been underway to virtualize only compute resources but also network functions that had been realized in purpose-built appliances, specialized hardware and software (e.g., routers, firewalls, switches). Network function virtualization (NFV) is focused on transforming these appliances into software [18].

Companies are implementing the metadata driven application and service catalogs by centralizing the creation of models, rules, and policies [14,18]. All orchestrator applications and controllers ingest the metadata that governs their behavior from centralized store. Implementation of the metadata-driven methodology requires an enterprise-wide paradigm change of the development and delivery process. It demands a service agreement on the overall metadata-model across the business (product development) and the software developers writing code for orchestrator, business support system and operations support system. This agreement is a behavioral contract to permit both the detailed business analysis and the construction of software to run in parallel.

The software development teams focus on building service-independent software within a common operations framework for runtime automation. The Business teams can focus on tackling business needs by defining metadata models to feed the execution environment [14]. The metadata model content itself is the glue between design time and runtime execution frameworks, and the result is that the service specific metadata models drive the common service independent software for runtime execution.

Models are created and managed centrally so that all components, both simple and complex, are easily visualized and understood together, and validated properly prior to use. Once validated and corrected for any conflicts, the models are precisely distributed to the many points of usage. This improves scalability and reduces latency of delivery [6]. Models are created by many user groups (e.g., service designers, security engineers, operations staff, etc.). Various techniques are used to validate newly created models and policies and to help identify and solve any real- world problems [15].

2   NFV Orchestration Components

The European Telecommunications Standards Institute (ETSI) developed a Reference Architecture Framework and specifications in support of NFV Management and Orchestration (MANO) [3,7]. The main components of the ETSI-NFV architecture are: Orchestrator, VNF Manager, and VI (Virtualized Infrastructure) Manager [4]. The Metadata modeling tool is the most important external part of the MANO infrastructure [2] (Fig. 1).

2.1   NFV Modeling

The modeling tools deliver product/service independent capabilities for the design, creation and lifecycle management of the target environments. Model

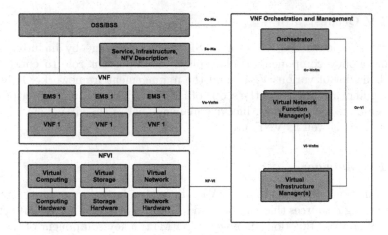

Fig. 1. ETSI MANO

design platforms are used to define and program the platform, and an execution time environment to execute the logic programmed in the design phase utilizing closed-loop, policy-driven automation [4,7]. The metadata model is managed through a design environment which guides a series of designers from 3rd party resource vendor onboarding through service creation, verification, and distribution. The modular nature of the metadata model provides a catalog of patterns that can be reused in future services. This provides approach to quickly extend resource functions to manageable services and sellable products, further realizing benefits in time to market. Orchestrator component supports companys metadata model design. Many different models of metadata exist to address different business and technology domains. Metadata modellers integrate various tools supporting multiple types of data input (e.g., YANG, HEAT, TOSCA, YAML, BPMN/BPEL, etc.). It automates the formation of an company internal metadata format to drive end-to-end runtime execution. In modeller the resource metadata is created in the description of the cloud infrastructure and the configuration data attributes to support the implementations. Subsequently, the resource metadata descriptions become a pool of building blocks that may be combined in developing services, and the combined service models to form products. In addition to the description of the object itself, modeling the management needs of an object using metadata patterns in modeller may be applied to almost any business and operations functions in company. For example, metadata can be used to describe the mapping of resource attributes to relational tables, through the definition of rules around the runtime session management of a group of related resources to the signatures of services offered by a particular type of resources.

2.2   NFV Orchestrator

Orchestrator expands the scope of ETSI MANO coverage by including Controller and Policy components. Policy plays an important role to control and manage behavior of various VNFs and the management framework. Orchestrator also significantly increases the scope of ETSI MANOs resource description to include complete meta-data for lifecycle management of the virtual environment (Infrastructure as well as VNFs).

2.3   NFV Manager

Network functions virtualization (NFV) defines standards for compute, storage, and networking resources that can be used to build virtualized network functions. The virtual network functions manager (VNFM) is a key component of the NFV management and organization (MANO) architectural framework.

The VNFM works with other MANO functional blocks, such as the virtualized infrastructure manager (VIM) and NFV orchestrator (NFVO), to help standardize the functions of virtual networking and increase the interoperability of software-defined networking elements. These standard components can help lower costs and increase new feature deployment and velocity by providing a standard framework for building NFV applications. In the virtualized network world of NFV, scalability and speed of service deployments create new challenges in the areas of network management and orchestration.

2.4   Virtualized Infrastructure Manager (VIM)

The virtualized infrastructure manager (VIM) is last key component of the MANO architectural framework. It is responsible for controlling and managing the NFV infrastructure (NFVI) compute, storage, and network resources, usually within operator's cloud based infrastructure. These functional blocks help standardize the functions of virtual networking to increase interoperability of software-defined networking elements. VIMs can also handle hardware in a multi-domain environment or may be optimized for a specific NFVI environment.

The VIM is responsible for managing the virtualized infrastructure of an NFV- based solution. VIM operations include:

1. Keeps an inventory of the allocation of virtual resources to physical resources. This allows the VIM to orchestrate the allocation, upgrade, release, and reclamation of NFVI resources and optimize their use.
2. Supports the management of VNF forwarding graphs by organizing virtual links, networks, subnets, and ports. The VIM also manages security group policies to ensure access control.
3. Manages a repository of NFVI hardware resources (compute, storage, networking) and software resources (hypervisors), along with the discovery of the capabilities and features to optimize the use of such resources.

3   Existing NFV Orchestration Platforms

Following chapter summarizes the most usual frameworks for VNF management according to ETSI MANO [7,8,12]. The ETSI NFV-MANO standard defines high-level modle, but lacks details and standards on the implementation for both managers as well as the interfaces. As a result some of the covered solutions are in fact customized with some parts based on proprietary solutions.

3.1   Open Source MANO

Open source implementation with RIFR.ware as orchestrators and OpenMANO and OpenVIM as VNF and NFVI resources [8,17]. RIFT.ware, developed by RIFT.io is the industrys first open source, NSV platform. As discussed in the previous section, network functions need to incorporate the practices of hyperscale data centers to meaningfully reduce cost and deliver agile, dynamic service provisioning. As many of these practices are common across all network functions, it follows that the data center platform needs to follow the evolutionary trend of the past decades by incorporating new, web-scale techniques. RIFT.ware redefines the networking platform by combining functions expected in legacy hardware with the technology of cloud infrastructure (Fig. 2).

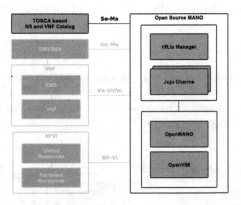

Fig. 2. OpenSource MANO

To bring hyper-scale capabilities and economics to cloud and network function builders, service providers, and enterprises, RIFT.ware simplifies the development, deployment, and management of cloud-scale network applications and VNFs. The OpenMANO is a reference implementation of an NFV-O (Network Functions Virtualization Orchestrator) [8]. It interfaces with an NFV VIM through its API and offers a northbound interface, based on REST (OpenMANO API), where NFV services are offered including the creation and deletion of VNF templates, VNF instances, network service templates and network service instances.

The OpenVIM is reference implementation of an NFV VIM (Virtualized Infrastructure Manager). It interfaces with the compute nodes in the NFV Infrastructure and an openflow controller in order to provide computing and networking capabilities and to deploy virtual machines. It offers a northbound interface, based on REST (OpenVIM API), where enhanced cloud services are offered including the creation, deletion and management of images, flavors, instances and networks. The implementation follows the recommendations in NFV- PER001.

3.2 OPEN-Orchestrator

GS-O is responsible for Global Service Orchestration to provide End-to-End orchestration of any service on any network [16]. GS-O allows the user to describe the end-to-end (global) service in a single Global Service Description. GS-O decomposes the Global Service Description to SDN Service Descriptions and NFV Service Descriptions, and then forwards these requests to the correct instances of SDN-O and NFV-O.

This NFV-O function is based on ETSI NFV MANO architecture & information model as a reference [16]. It orchestrates Network Services composed of Virtualized Network Functions and is an integral part of OPEN-O. The NFV-O function includes Network service life cycle manager, NFV resource manager, and NFV monitor. Using the south bound SBI, NFVO cooperates with several NFV SDN controllers, Multi-vendor VNFMs and different VIMs.

3.3 Cloudify

The Blueprint Composer is an editor for composing blueprint YAML files dynamically using a modern Drag and Drop Interface [5]. Cloudify's Manager comprises of Cloudify's code and a set of Open-Source applications. An elaborate explanation on these applications is provided here (Fig. 3).

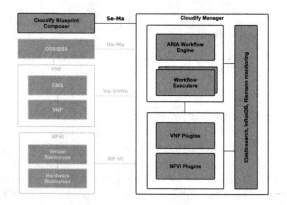

Fig. 3. Cloudify architecture

The way Cloudify users use Cloudify to deploy their application is by utilizing the power of blueprints. Blueprints are written in YAML format and their DSL is based on the TOSCA standard.

3.4   AT&T ECOMP

Capabilities are provided using two major architectural frameworks: (1) a Design Time Framework to design, define and program the platform (uniform onboarding), and (2) a Runtime Execution Framework to execute the logic programmed in the design environment (uniform delivery and lifecycle management) [1, 10]. The platform delivers an integrated information model based on the VNF package to express the characteristics and behavior of these resources in the Design Time Framework. The information model is utilized by Runtime Execution Framework to manage the lifecycle of the VNFs. The management processes are orchestrated across various modules of ECOMP to instantiate, configure, scale, monitor, and reconfigure the VNFs using a set of standard APIs provided by the VNF developers (Fig. 4).

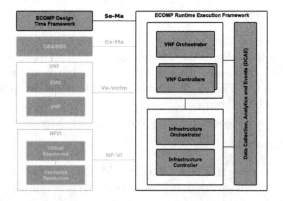

Fig. 4. AT&T ECOMP architecture

4   SaltStack Based NFV and VIF Orchestrator

In previous chapters we have covered the core components of the ETSI MANO standard and the most important implementations to the date. This chapter shows our proposed SaltStack based platform that govern resources according to the ETSI NFV-MANO model. The combination of orchestration and configuration features puts SaltStack in unique position to model, manage, and orchestrate various software stacks, network or storage devices and other infrastructure resources (Fig. 5).

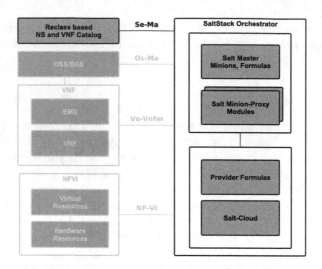

Fig. 5. SaltStack NVF-MANO orchestrator

4.1   NFV Modeller

The modeller delivers design models that are used to define and program the platform. Our modeller delivers ID centric system with simple classification using Reclass hierarchical database. All resources, both hardware and software are modeled the same way. The result model is simple data structure that can be interpreted by the Orchestrator.

4.2   NFV Orchestrator

Orchestrator is the central point of ETSI MANO infrastructure. SaltStack project began as a high-speed remote execution and communication bus designed to be used as a lightweight private cloud controller. Now it's full scale orchestration platform with configuration management capabilities. It uses modules to carry on the command execution and state layer with declarative resource description.

One of fundamental features of SaltStack is remote execution. Salt has two basic "channels" for communicating with minions. Each channel requires a client (minion) and a server (master) implementation to work within SaltStack. These pairs of channels will work together to implement the specific message passing required by the channel interface.

4.3   NFV Manager

SaltStack uses NAPALM (Network Automation and Programmability Abstraction Layer with Multivendor support) Python library that implements a set of functions to interact with different network device Operating Systems using a unified API. It supports Arista EOS, Cisco IOS, Cisco IOS-XR, Cisco NX-OS, Fortinet Fortios, IBM, Juniper JunOS, Mikrotik RouterOS, Palo Alto NOS, Pluribus, and Vyos network operating systems. YANG (RFC6020) is a data modeling language, its a way of defining how data is supposed to look like. The napalm-yang library provides a framework to use models defined with YANG in the context of network management. It provides mechanisms to transform native data/config into YANG and vice versa. We use dedicated minion to handle network and storage devices by Salt Minion Proxy service, basically wrapping each external resource by NAPALM or other service wrapper.

4.4 Virtualized Infrastructure Manager (VIM)

Salt provides powerful interfaces to interact with cloud providers. These interfaces are tightly integrated with Salt, and new virtual machines are automatically connected to your Salt master after creation. The supported cloud environments include Scaleway Rackspace, Amazon AWS, Linode, Joyent Cloud, GoGrid, OpenStack, DigitalOcean, Parallels, Proxmox and LXC. You can define wide range of infrastructure resources, networks, volumes and computers. The SaltStack uses salt-cloud capability to govern major cloud providers for compute resources using Python lib-cloud library. More complex resources are managed by individual Salt formulas.

5 Conclusion

We have presented an overview of NFV-MANO framework that has been proposed by ETSI as well as representative platforms that realize the framework architecture. Other research projects as well as industry products focused NFV MANO have been surveyed, identifying their core technologies as well as their mapping to the ETSI NFV-MANO. Based on these, we have identified and discussed opportunities to govern NFV-MANO with SaltStack configuration management platform.

We have observed that ETSI has completed the first phase of MANO standard. But the proposed MANO framework still lacks details and standards on the implementation for both managers as well as the interfaces. As a result most of the covered solutions are in fact customized and based on proprietary solutions. In addition, while some of the identified challenges such as security are being considered in some of the projects and/or industrial products, others such as resource management-and in particular automated resource management-have not received significant attention yet. Our opinion is that the success of NFV will depend on the availability of mechanisms that are able to autonomously manage network and function resources.

Our proposed SaltStack based platform can be leaveraged for complete automation of vast number of hardware and software resources from single model without any vendor-specific constraints. The Salt defined resources can be used define and manage all VNF and VI aspects of modern IT infrastructures.

Acknowledgement. This work and the contribution were also supported by project of specific science, Faculty of Informatics and Management, University of Hradec Kralove, Czech Republic.

References

1. AT&T: ECOMP (Enhanced Control, Orchestration, Management and Policy) Architecture White Paper. http://about.att.com/content/dam/snrdocs/ecomp. pdf. Accessed 21 Apr 2017
2. Arai, T., Gokurakuji, J., Oohira, M.: MANO technology supports implementation of intelligent network operations management. NEC Tech. J. **10**(3), 19–22 (2016)
3. Arai, T., Yoshikawa, N., Mibu, R.: Technology systems for SDN/NFV solutions. NEC Tech. J. **10**(3), 15–18 (2016)
4. Chayapathi, R.: Network Function Virtualization (NFV) with a Touch of SDN. Addison-Wesley Professional, Boston (2016)
5. Cloudify - Pure-Play Cloud Orchestration and Automation Based on TOSCA. http://getcloudify.org/. Accessed 20 Apr 2017
6. Coughlin, M., Keller, E., Wustrow, E.: Trusted click: overcoming security issues of NFV in the cloud. In: Proceedings of ACM International Workshop on Security in Software Defined Networks, pp. 31–36 (2017)
7. Ersue, M.: ETSI NFV Management and Orchestration - An Overview. https://www.ietf.org/proceedings/88/slides/slides-88-opsawg-6.pdf. Accessed 20 Apr 2017
8. github.com - nfvlabs/openmano: Openmano. https://github.com/nfvlabs/openmano. Accessed 02 Apr 2017
9. Goranson, P., Chuck, B.: Software Defined Networks: A Comprehensive Approach. Morgan Kaufmann, Burlington (2014)
10. Jarich, P.: AT&T, ECOMP and the Increasingly Difficult Pace of Virtualization. https://networkmatter.com/2017/02/08/att-ecomp-and-the-increasingly-difficult-pace-of-virtualization/. Accessed 21 Apr 2017
11. Kim, M.S., Do, T., Kim, Y.H.: TOSCA-based clustering service for network function virtualization. In: 2016 International Conference on Information and Communication Technology Convergence, ICTC 2016, pp. 1176–1178 (2016)
12. Kreutz, D., Ramos, F.M.V., Verissimo, P.E., Rothenberg, C.E., Azodolmolky, S., Uhlig, S.: Software-defined networking: a comprehensive survey. Proc. IEEE **103**(1), 14–76 (2016)
13. Malik, A., Ahmed, J., Qadir, J., Ilyas, M.U.: A measurement study of open source SDN layers in OpenStack under network perturbation. Comput. Commun. **102**(3), 139–149 (2016)
14. Mijumbi, R., Serrat, J., Gorricho, J.L., Latre, S., Charalambides, M., Lopez, D.: Management and orchestration challenges in network functions virtualization. IEEE Commun. Mag. **54**(1), 98–105 (2016)
15. Ming-Wei, S., Mohan, K., Taesoo, K., Gavrilovska, A.: S-NFV: securing NFV states by using SGX. In: Proceedings of 2016 ACM International Workshop on Security in Software Defined Networks, pp. 45–48 (2016)
16. Open Orchestrator. https://www.open-o.org. Accessed 19 Apr 2017
17. RIFT.io - Build Deploy Carrier-Grade Virtualized Network Services. https://riftio.com/. Accessed 21 Apr 2017
18. Sonkoly, B., Szabo, R., Jocha, D., Czentye, K., Kind, J., Westphal, F.J.: UNIFYing cloud and carrier network resources: an architectural view. In: 2015 IEEE Global Communications Conference (GLOBECOM), pp. 1–7 (2015)

19. TOSCA Simple Profile for Network Functions Virtualization (NFV) Version 1.0. http://docs.oasis-open.org/tosca/tosca-nfv/v1.0/tosca-nfv-v1.0.html. Accessed 04 Apr 2017
20. Tracker - OpenStack. https://wiki.openstack.org/wiki/Tacker. Accessed 18 Apr 2017
21. Yousaf, F.Z., Goncalves, C., Moreira-Matias, L., Perez, X.C.: RAVA - resource aware VNF agnostic NFV orchestration method for virtualized networks. In: IEEE International Symposium on Personal, Indoor and Mobile Radio Communications, PIMRC (2016)

Performance Evaluation of OpenFlow Enabled Commodity and Raspberry pi Wireless Routers

Muhammad Zeeshan Asghar$^{(\boxtimes)}$, M. Ahsan Habib, and Timo Hämäläinen

Department of Mathematical Information Technology,
University of Jyväskylä, PL 35, 40014 Jyväskylä, Finland
{muhammad.z.asghar,mahsan.habib,timo.hamalainen}@jyu.fi
http://www.jyu.fi

Abstract. Software defined network (SDN) allows the decoupling of data and control plane for dynamic and scalable network management. SDN is usually associated with OpenFlow protocol which is a standard interface that enables the network controllers to determine the path of network packets across a network of switches. In this paper, we evaluate openflow performance using commodity wireless router and raspberry pi with two different SDN controllers. Our test setup consists of wired and wireless client devices connected to openflow enabled commodity wireless router and raspberry pi. All clients used traffic generator tool to transmits data to a sink server host. The results are promising and paves the way for further research on using software defined wireless network.

Keywords: Software defined network (SDN) · Wireless SDN · Open-Flow · Performance evaluation · Raspberry-pi

1 Introduction

The modern computing environments are dynamic and require scalable computing and advanced storing needs that are not achievable by traditional network architecture. In order to address this challenge, the existing network architecture needs to be upgraded with additional features. This situation pushes operators to upgrade their network management from static to dynamic and scalable computing and opens doors for Software Defined Networking (SDN) research. The SDN is an umbrella term for various ways to use software to manage and manipulate networks. There are usually three views on SDN including, open SDN using openflow protocol, SDN via Application Programmable Interfaces (APIs) model where the functionality on networking devices is exposed using a rich API, and SDN via overlays.

The SDN has potential to offer high flexibility in network management. The key idea is to split the network forwarding function from the network control function. This can be achieved by separating the control and data planes. This allows a simpler and more flexible network control and management. SDN is usually associated with OpenFlow protocol which is a standard interface that

© Springer International Publishing AG 2017
O. Galinina et al. (Eds.): NEW2AN/ruSMART/NsCC 2017, LNCS 10531, pp. 132–141, 2017.
DOI: 10.1007/978-3-319-67380-6_12

enables the network controllers to determine the path of network packets across a network of switches. The network controller communicates with OpenFlow switches using the OpenFlow protocol through a secure channel. Using this connection, the controller is able to configure the forwarding tables of the switch. OpenFlow-based SDN technologies allow to address the high-bandwidth and dynamic nature of computer networks by adapting the network functions to different business needs. This helps to reduce the network operations and management complexity significantly. As the cost involved to replace all legacy devices by new ones is high, the possibility of using commodity wireless routers with the OpenFlow might be a smart strategy to accelerate the deployment of SDN technologies.

This work aims at performance evaluation of Open vSwitch (OVS) running in commodity wireless router and raspberry pi as a linux kernel module. The results show that Openflow router provides good performance with low cost, therefore, the adapted wireless router with openflow compatibility can be widely utilized in home networks and small organizations. The rest of this paper is organized as follows. Section 2 describes the related work. Section 3 presents the evaluation strategy and proposed testbed architecture. Section 4 shows the experimental results and analysis. We conclude by proposing future research directions.

2 Related Work

In this section we briefly describe the related work. There have been few studies about the performance evaluation of SDN architectures and openflow. Tootoonchian et al. [6] studied the SDN controller performance by focusing on the control plane only. Rotsos et al. [7] proposed a tool for evaluating performance of SDN architecture consisting of several OpenFlow implementations, and measured raw performance of OpenFlow without comparison to other SDN solutions authors presented an architecture in [4] to improve the loopup performance of the OpenFlow Linux kernel module implementation. A cost effective alternative of implementing SDN testbed with Open vSwithc (OVS) is proposed in [5]. They used raspberry-pi as an alternate choice for the SDN switch. They implemented the SDN testbed with the OpenFlow specification 1.0. The performance of net-FPGA and OpenFlow based software switched is compared in terms of maximum throughput. The results shows the similar performance with 1 Gbps net-FGPA device. The shortcoming of this work is that the testbed is implementing SDN functionalities for wired network only and therefore it is not scalable.

A comparative study of the performance evaluation of a commodity wireless router and different SDN controllers is presented in [6]. The performance evaluation is conducted using the performance metrics such as throughput, delay, jitter and packet loss. The results show bottleneck when using UDP protocol with high rates and therefore, the commodity wireless router can be only deployed in reduced size networks. Other recent research work on SDN includes [7–12]. Some recent works on SDN testbed with raspberry-pi are [5,8]. The performance of OpenFlow in commodity wireless networks are given in [6].

3　Proposed Testbed Architecture

In this section we describe our proposed testbed architecture and methodology to collect the performance measurements such as average bitrate, packet dropped, average delay and average jitter. The architecture overview of SDN testbed is shown in Fig. 1.

3.1　Network Achitecture

Our test setup consists of wired and wireless client devices and these devices connected to openflow enable commodity wireless router and raspberry pi. All the clients used traffic generator tool to transmits data to a server. The network design shown in Fig. 2 There are four clients and a server connected to the wired ports and one client connected through the wireless link. The SDN controller connected to wan port. In our testbed, we have selected TP-LINK WR1043ND ver 2.1 [13] and raspberry pi 3 [14] devices. The TP-LINK wireless router has a Qualcomm Atheros QCA9558 processor with 720 MHz clock speed, 64MB memory, which is decent amount memory for a wireless router. The router has 4 LAN and 1 WAN Gigabit Ethernet interfaces and also has a 802.11n/g/b wireless network interface. On the other hand, raspberry PI 3 has a 4 core ARM Cortex-A53, 1.2 GHz processor, 1 GB LPDDR2 memory and a Gigabit Ethernet interface. The device has also 4 USB 2 interfaces which allow for up to 4 additional Ethernet interfaces via USB-to-Ethernet adapters. We used 10/100 Mbps USB to Ethernet adapters to connected clients to the raspberry pi OpenvSwitch. All our wired and wireless clients and the SDN controller are real machines have been running the Ubuntu 14.04 operation system. The SDN controller has an Intel dual core i5-5300U with 2.30 GHz processor and 4 GB of memory.

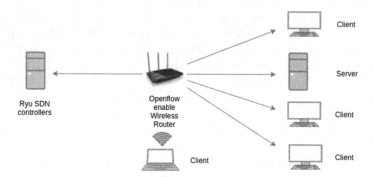

Fig. 1. Network topology used in the experiments. Wired and wireless client devices are generate traffic to a sink server. The wireless router interconnects all devices and is controlled by Ryu SDN controller

3.2 Software Design

We have used popular open source traffic generate tool called Distributed Internet Traffic generator D-ITG [15] which is a platform capable to produce IPV4 and IPV6 traffic by precisely reproducing traffic from many popular internet applications. D-ITG is also a network measurement tool able to generate most useful information from packet level such as throughput, delay, jitter, packet loss. In our experiment, Using D-ITG tools we were able restrict the packet size and control the total number of packets to be transmit and the duration of the generation experiment. In order to correctly measure the packet delay, we have implemented Network Time Protocol (NTP) [16] in our testbed environment so that all the clients and the server can be synchronization between them.

To ensure all clients and server can generate and receive rate up to 10 Gbps, we have replaced the default Libcap module in linux kernel with the PF_RING library [17]. PF_RING is a linux kernel module and user-space framework that allows us to process packets in high rates. The main idea is to bypass any and all notion of what is actually data on the wire and transmits as quickly as possible.

Raspberry-pi linux kernel is 3.7.11+ version based on debian Jessie. We installed OVS 2.3.1 to create an OpenFlow enabled switch which supports OpenFlow 1.3. The wireless router running the original equipment manufacturer (OEM) firmware The firmware OEM substitued by the OpenWRT 14.0 without OpenFlow module The OpenFlow 1.3 support in the OpenWRT firmware, using the OVS kernel module, was enabled.

3.3 SDN Controller

For this study, we have used some most popular controller in research area such as RYU [18] and Floodlight [19] to perform our experiment. Both controllers run simple switch learning application. The OpenFlow ver 1.3 protocol used to communication between OVS and SDN controller.

3.4 OpenvSwitch

In our experiment, we have installed and configured OpenvSwitch on OpenWrt [20] to showcase openflow capabilities. We complied OVS with openWrt running in linux kernel module. We have used OpenvSwitch ver 2.3.1 and OpenWrt ver 14.07 Barrier breaker image on TP-Link wireless router. The OpenFlow ver 1.3 protocol support was enabled in OVS on OpenWrt firmware. OpenVswitch is also run on Raspberry Pi as a kernel module. We have download the openvSwitch from the source code and its dependencies and complied with kernel header.

The raspberry Pi is driven by the raspbian Jessie with pixel (raspbian is a Debian based operating system) with kernel version 4.4.34. We have been running OVS version 2.7.90. OVS fail mode can be set to either standalone or secure mode.

In standalone mode, OVS will take the responsibility for forwarding the packets if the controller fails. In secure mode, only the controller responsible for forwarding the packets, and if the connect between OVS and controller lost all the packets are going to be dropped. In our test case scenarios, both OpenWrt and Raspberry Pi OVS switch set standalone as a fail mode, in order to eliminate the impact of any SDN controller in the process.

4   Experimental Results

In this section we describe experimental results with different scenarios. The performance factors for the experiments were traffic generator clients, packet size, packet sending rate and transport protocols. In the experiments we used two different controllers i.e., ryu and floodlight. We evaluated average bitrate, average packet dropped, average delay and average jitter.

Figure 2 compares average bitrate between SDN controller ryu and floodlight over openwrt and raspberry pi. The comparison shows that both sdn controllers have similar performance. Both controllers with openwrt show high average bitrate around 40–50 kpps over openwrt. However, the average bitrate in case of raspberry pi stays lower than the openwrt. Openwrt and raspberry-pi average bitrate stays at same level from 4 kpps to 6 kpps but raspberry-pi bitrate does not increase much between 7 kpps and 30 kpps, it starts to increase from 30 kpps to 50 kpps. On the other hands, openwrt shows similar pattern but openwrt bitrate increases significantly from 30 kpps and both controller shows almost similar bitrate. Raspberry-pi shows lower bitrate then openwrt because raspberry-pi has 4 (usb 2) ports which has limited data transfer rate.

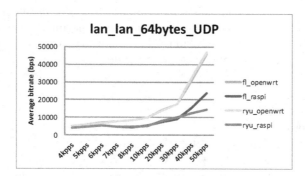

Fig. 2. Average bitrate for experiments with UDP 64-bytes packets, using different OpenFlow controllers

Figure 3 shows packet dropped for different SDN controller ryu and floodlight over openwrt and raspberry pi. The comparison shows that both sdn controllers have similar performance. There is no packet dropped in the scenario with openwrt. However in case of raspberry-pi there is significantly high. As

can be seen in the Fig. 3, the ryu controller performance is slightly better than
floodlight controller. There are not many packets drop until 40 kpps in openwrt
for both controller but the packets starts to drop around 50 kpps. If we look at
the raspberry-pi openvSwitch, we can see huge amount of packets drop for both
controller. As mentioned previously raspberry-pi has limited data transfer rate
so packets dropped increase significantly for huge number of packets specially
6 kpps.

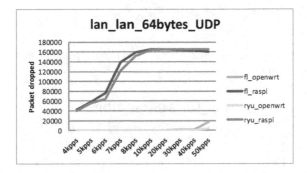

Fig. 3. Average packet dropped for experiments with UDP 64-bytes packets, using
different OpenFlow controllers

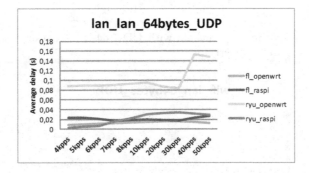

Fig. 4. Average delay for experiments with UDP 64-bytes packets, using different
OpenFlow controllers over openwrt and raspberry pi

In Fig. 4 the average delay is not much when we used OpenWrt or raspberry-
pi but average delay is high when we used openwrt with ryu controller. There
were similar kinds of delay until 30 kpps but it increase significantly around
40 kpps and 50 kpps. In Fig. 5, similar to previous figure average Jitter are similar in both openwrt and raspberry-pi but average jitter incrementally increases
in the case of ryu controller. These results show that there is a bottleneck
around 30 kpps. In Fig. 6, When wifi client is used to transmit data to LAN
client raspberry-pi bitrate does not increase because of wifi adapter data transfer rate limitation but for the openwrt it is totally opposite, bitrate increases

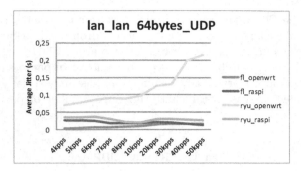

Fig. 5. Average Jitter for experiments with UDP 64-bytes packets, using ryu controller over openwrt and raspberry pi

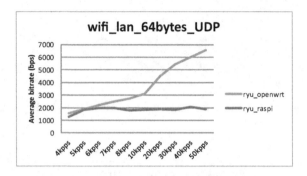

Fig. 6. Average bitrate for experiments with UDP 64-bytes packets, using ryu controller over openwrt and raspberry pi

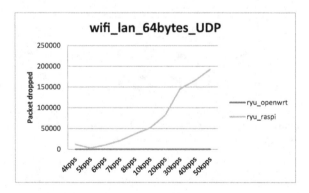

Fig. 7. Average packet dropped for experiments with UDP 64-bytes packets, using ryu controller over openwrt and raspberry pi

as more packets are transmitted from wifi client to lan client. As we can see from the Fig. 7, that openwrt shows stable results and no packet dropped at all however, raspberry-pi shows huge number of packets dropped as more packets

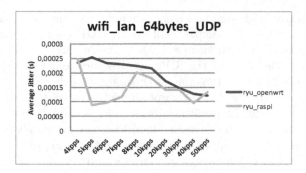

Fig. 8. Average jitter for experiments with UDP 64-bytes packets, using ryu controller over openwrt and raspberry pi

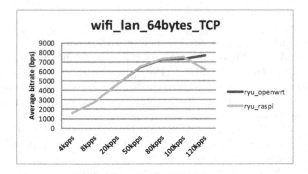

Fig. 9. Average bitrate for experiments with TCP 64-bytes packets, using ryu controller

are transmitted on from wifi client to lan client. In Fig. 8, the openwrt shows high jitter on small number of packet and it started to decrease as more packets are transmitted. In case of raspberry-pi average jitter is high at 4 kpps but it is low between 5 kpps and 7 kpps. The average jitter increases again at 8 kpps and show a decaying pattern between 10 kpps and 40 kpps. In Fig. 9, results for TCP transport protocol are shown traffic generated from wifi client to lan client. Both openwrt and raspberry-pi showed similar increasing trend. The openwrt bitrate increased around 120 kpps but average bitrate of raspberry-pi started to decrease because of external wifi adapter limitation.

5 Conclusion

In this paper we evaluated the performance of commodity wireless router and raspberry-pi used as an OVS linux kernel module. Our proposed testbed is built on latest versions of controllers. The experiments are conducted with two different OpenFlow controllers and we observed the performance variations when using with different packet size, variable packet rate generation, traffic generating clients and transport protocols. It is found out that the performance of two different controllers is similar which indicates that the performance bottleneck is at OVS module. The experiments with UDP transport protocol show that the performance bottleneck exists when traffic rate exceed 40 kpps. The results show that the usage of the SDN architecture and OpenFlow standard introduces benefits in terms of maximum throughput, delay and jitter. The performance obtained suggests that a commodity wireless router and raspberry pi can be used in scenarios of small networks. The results shows that adapted wireless router and raspberry-pi with Openflow can be widely used in small networks such as home networks, university campus, and small organizations. The proposed testbed is low-cost, less complex and scalable. The future work involves running experiments with complex scenarios using multiple controllers, thus allowing to evaluate the performance differences and bottlenecks.

References

1. Open Networking Foundation. http://opennetworking.org
2. Tootoonchian, A., Gorbunov, S.: On controller performance in software-defined networks. In: Proceedings of USENIC Hot-ICE (2012)
3. Rotsos, C., Sarrar, N., Uhlig, S., Sherwoord, R., Moore, A.W.: An open framework for OpenFlow switching evaluation: composing a complex biological workflow through web services. In: Nagel, W.E., Walter, W.V., Lehner, W. (eds.) Euro-Par 2006. LNCS, vol. 4128, pp. 1148–1158. Springer, Heidelberg (2006)
4. Foster, I., Kesselman, C.: The Grid: Blueprint for a New Computing Infrastructure. Morgan Kaufmann, San Francisco (1999)
5. Kim, H., Kim, J., Ko, Y.B.: Developing a cost-effective OpenFlow testbed for small-scale software defined networking. In: 2014 16th International Conference on Advanced Communication Technology (ICACT), pp. 758–761. IEEE (2014)
6. Lima, L., Azevedo, D., Fernandes, S.: Performance evaluation of OpenFlow in commodity wireless routers. In: Latin American Network Operations and Management Symposium (LANOMS), pp. 17–22. IEEE (2015)
7. Araniti, G., Cosmas, J., Iera, A., Molinaro, A., Morabito, R., Orsino, A.: OpenFlow over wireless networks: performance analysis. In: 2014 IEEE International Symposium on Broadband Multimedia Systems and Broadcasting (BMSB), pp. 1–5. IEEE (2014)
8. Ariman, M., Seinti, G., Erel, M., Canberk, B.: Software defined wireless network testbed using Raspberry Pi of switches with routing add-on. In: 2015 IEEE Conference on Network Function Virtualization and Software Defined Network (NFV-SDN), November 2015, pp. 20–21. IEEE (2015)
9. Brock, J.D., Bruce, R.F., Cameron, M.E.: Changing the world with a Raspberry Pi. J. Comput. Sci. Coll. **29**(2), 151–153 (2013)
10. Sezer, S., Scott-Hayward, S., Chouhan, P.K., Fraser, B., Lake, D., Finnegan, J., Viljoen, N., Miller, M., Rao, N.: Are we ready for SDN? Implementation challenges for software-defined networks. IEEE Commun. Mag. **51**(7), 36–43 (2013)
11. Shalimov, A., Zuikov, D., Zimarina, D., Pashkov, V., Smeliansky, R.: Advanced study of SDN/OpenFlow controllers. In: Proceedings of 9th Central & Eastern European Software Engineering Conference in Russia, p. 1. ACM (2013)
12. Vissicchio, S., Vanbever, L., Bonaventure, O.: Opportunities and research challenges of hybrid software defined networks. ACM SIGCOMM Comput. Commun. Rev. **44**(2), 70–75 (2014)

13. 300Mbps Wireless N Gigabit Router TL-WR1043ND. http://static.tp-link.com/resources/document/TL-WR1043ND_V2_datasheet.pdf
14. Raspberry Pi Foundation. https://www.raspberrypi.org/
15. Distributed Internet Traffic Generator (D-ITG). http://www.grid.unina.it/software/ITG/
16. The Network Time Protocol (NTP). http://www.ntp.org/
17. pf_ring. http://www.ntop.org/products/packet-capture/pf_ring/
18. Ryu. https://osrg.github.io/ryu/
19. Floodlight. http://www.projectfloodlight.org/
20. Openwrt. https://openwrt.org/

Design Issues of Information and Communication Systems for New Generation Industrial Enterprises

Valery Leventsov, Anton Radaev, and Nikolay Nikolaevskiy[✉]

Peter the Great St. Petersburg Polytechnic University, Saint Petersburg, Russia
vleventsov@spbstu.ru, TW-inc@yandex.ru,
triplenick@mail.ru

Abstract. Intensive development of technologies for information processing and transfer is the basis for creation of cutting-edge production systems in the context of "Industry 4.0". Implementation of this concept will ensure efficient development of the national production industry. The paper looks into the issues of organizational and technical engineering of such types of systems and focuses on reasoning the movement and conversion characteristics of information flows. Based on literature review a generalized structure is proposed for the info-communication systems of a production unit. The main requirements for this system and its major structural elements, which are laid down at the design stage, are formulated. The content of the design problems of the info-communication system has been proposed in the context of its structural elements. Main solution methods have been worked out to solve individual groups of problems.

Keywords: Industry 4.0 · Info-communication system · Production system engineering · Problem solving methods

1 Introduction

Given today's conditions of the national production industry, special importance is attributed to the issues concerning the process characteristics of information flow movement and conversion, which directly determine the operational consistency of the relevant technology links, when solving the problems of organizational design of industrial enterprises' production systems. This circumstance is pre-conditioned, first of all, by intensive development of information processing and transfer technologies, applied when advanced production systems are created within the framework of contemporary concepts of industrial enterprise's development. The most common concept among them is "Industry 4.0". This concept combines the principles to be used to transfer to Digital Economics as a consequence of the global digital revolution, which is largely dependent on rapid development and universal introduction of ICT. First, we should mention a considerable growth in computing capacities and network capacity parameters, as well as progressive reduction in the size of integrated circuits, which altogether have made the wide spread of ICT possible in all spheres of life.

O. Galinina et al. (Eds.): NEW2AN/ruSMART/NsCC 2017, LNCS 10531, pp. 142–150, 2017.
DOI: 10.1007/978-3-319-67380-6_13

Focusing on the production processes of industrial enterprises, it should be noted that most information generated by detectors, sensors and other recording elements is traditionally designed for an operator to control directly, analyze and make decisions. However, due to the above circumstances of the ICT distribution and development, new capabilities arise for information exchange between individual production and auxiliary systems and their elements as well as goods and environment produced in this context. Altogether they will allow forming imposing data arrays. Moreover, the results of relevant information processing will become the basis of self-organization principles and independent decision-making by active components of production systems.

2 Main Body

The above circumstances were the basis for creating the modern concept "Industry 4.0" [1, 5, 6], implying increased efficiency of industrial production facilities via integration of advanced information and production technologies to finally form prospective digital factories of the future – "Factories of the Future". In turn, operation of the "Factories of the Future" will be aimed at reducing the time characteristics for launching highly intelligent products into markets by using digital design technologies throughout the entire product lifecycle in order to increase the value added of products and expand competitive offers on the market [6]. Achievement of these goals will largely depend on the characteristics of the relevant info-communication network, ensuring movement, end-to-end acquisition and processing of information used as the most important resource throughout the entire lifecycle of a product and defining the process of industrial system functioning as a whole.

Moreover, it is important to note that, notwithstanding the high topicality of the problems related to domestic industry development and the presence of a certain number of successfully implemented projects for creation of high-tech products [8], the concept "Industry 4.0" is covered relatively modestly in scientific papers, which, in turn, allows making a conclusion about the limitedness of the existing methodological guidelines and tools for designing advanced production systems, including in terms of the relevant info-communication networks.

The above circumstance has predetermined the need for doing research aimed at developing scientific and methodological tools for justification of information flow movement and conversion processes in the context of advanced production systems of industrial enterprises.

At the initial stages, scientific papers by foreign and domestic authors were analyzed in the field of production system design of industrial enterprises within the framework of the concept "Industry 4.0" in terms of justification of the characteristics of information flow movement and conversion processes. The most important results of the above procedure are presented in Fig. 1.

According to the Figure, the main elements of the concept "Industry 4.0", which describe in most detail the processes of design and operation of production systems in terms of information flow movement and conversion processes, are as follows:

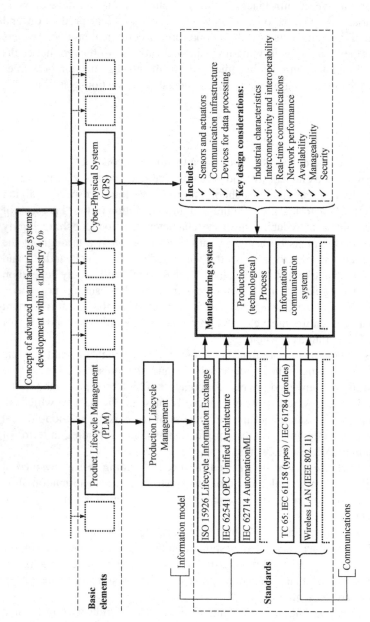

Fig. 1. Structure of production system development analysis

- the concept of Product Lifecycle Management – PLM, which combines information support methods and tools throughout the entire lifecycle (from an early stage of product design and on to its disposal) and is a particular basis for the concept of Production Lifecycle Management, described by the relevant stages and time duration; the latter concept, in particular, describes how the quality of information flow management affects the flexibility/reconfiguration ability of the production system [2, 12, 13].
- the concept of Cyber-Physical Systems – CPS, implying organization of information interaction between the production system and the external environment as well as data storage and processing with the use of advanced information technology [3, 9].

It is important to note that in some scientific papers dedicated to the subject of this research study, the term "Info-Communication System (ICS)" is used to describe the structural element of the production system ensuring movement and conversion of the relevant information flows. Hereinafter the ICS is understood as a totality of technical and program tools for acquisition, processing and transfer of information utilized in automated management of technology processes and ensuring efficient operation of the production system as a whole.

In the context of the next stage of research on the basis of the reviewed and analyzed literature sources, a generalized structure of a production unit's ICS was suggested, including the following major elements:

- a periphery subsystem – includes sensors and actuating mechanisms, which are a physical interface of a technology process and directly participate in information acquisition and interaction with the external environment;
- a commutation subsystem – includes cables and other devices ensuring movement of information flows from periphery devices to data processing and storage nodes and back again;
- a process subsystem, which includes equipment for processing, storage and interpretation of a large volume of data in order to ensure the required operational mode of the entire ICS.

At the next stage of research, on the basis of the obtained intermediate results, requirements for the ICS and its individual elements were formulated in the context of design process. The main requirements the designed ICS has to meet are the following [3, 4, 7, 10, 11]:

- stability, flexibility and reliability in the real-time operation process, which determines the need for distributed control in the process of response to constant changes in the external environment and the need for elaborating algorithms of independent decision-making by executive elements of the system;
- scalability at connecting and maintaining a large number of devices, which determines the need for a certain hierarchy to implement management and operation principles when ensuring the relevant characteristics of productivity and functional compatibility of devices;
- cybersecurity, determined by the availability of data in combination with the fail safety of system operation, including organization of access to data for reliable users

and authorized equipment exceptionally in order to prevent unwanted access, exclude a chance for accidental or deliberate change of data and other principles of operation.

The main requirements set for structural elements of ICS are as follows:

1. In respect to the periphery subsystem:

   - the use of devices for primary signal processing with enhanced functionality for fast resetting of the periphery system in accordance with the changes in the technology process;
   - application of back-up sensor devices to ensure reliability of the periphery system and high precision of data obtained from the external environment.

2. In relation to the commutation subsystem:

   - usage of modern wire and wireless technologies to ensure high rate of data transfer in real time;
   - presence of back-up communication channels (and a capability to implement an algorithm of automatic reconstruction of network topology in case of emergencies) to ensure reliable information transfer.

3. In relation to the process subsystem:

   - usage of high performance computing equipment, ensuring processing of large amounts of information and implementation of complex program algorithms for elaboration of efficient decisions in the operation process of the production system;
   - joining the elements of the system in a single digital space to ensure that the production system is flexible and adaptable to changes in the external environment, including coordination of the parameters of design products with the characteristics of implemented technology processes and parameters of processed orders;
   - application of multi-modal man-machine interfaces for enhancing functionality of management by ICS operators through process level devices.

It should also be mentioned that the formulated requirements are consistent with the existing standards in the field of design and operation of advanced production systems, including [12]:

- IEC 62541 OPC Unified Architecture – a specification, determining data transfer in production networks and interaction of devices within it, among other things M2M, including Field Device Integration, which contributes to functional consistency of the used equipment;
- ISO 15926 Lifecycle Information Exchange – standard of information representation related to design, construction and operation of a technology process within the entire lifecycle;

– IEC 62714 AutomationML – a complex of standards to ensure interrelation of tools during design and management of production processes; it embraces the issues of application of object-oriented modeling of physical and logical components of production objects and data objects, including information about topology, geometry, kinematics and logics.

At the next stage of research the content of design problems was formed for each individualized structural element of the ICS and the main solution methods for various groups of problems were identified (see Fig. 2):

– solution methods for selection problems implying determination of the most preferable option from the initially set list of alternatives by forming a system of option efficiency criteria (with the use of methods of qualimetry, expert assessment, relative preferences, etc.) and their calculation based on statistics; the content of criteria is determined by the specifics of the problem to be solved, however, in most problems, it includes various value parameters, such as cost parameters; for example, the problem of selecting a data transfer protocol in designing the commutation subsystem of the ICS implies accounting the stability criteria and possibility to reserve information transfer channels;
– analytical calculation techniques, implying evaluations of research object characteristics explicitly via consequent implementation of certain mathematical formulae;
– optimization modelling methods, implying determination of optimal values for characteristics of the research object in the implicit form via setting a mathematical optimization problem with further implementation of the relevant solution algorithms (including those based on simplex method, gradient methods, genetic methods, etc.); at this, value parameters or operation efficiency (productivity) parameters are normally considered as optimization criteria in terms of the seen structural element of the ICS; for example, when solving the problems of defining the operational process characteristics of real time data processing equipment, it is reasonable to consider the amount of delay in data processing and transfer (when providing the required productivity).
– simulation modelling techniques, based on application of modern programming languages to create algorithms (simulation models) for detailed reproduction of operation processes of research objects with changing characteristics of the internal and external environment; the stipulated specific features make it possible to provide direct accounting of time factor and stochastic properties of the researched system in the process of building the relevant simulation model (which is of high importance, for example, in solving the problems of stability analysis of the ICS commutation subsystem).

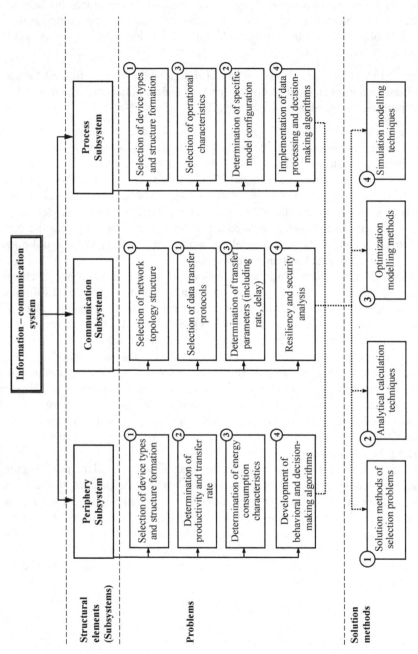

Fig. 2. Problems and solution methods for info-communication system engineering

3   Conclusion

Thus, the main findings of this paper are the following:

- scientific works have been reviewed and analyzed in the field of production system engineering of industrial enterprises within the concept "Industry 4.0" in terms of justification of information flow movement and conversion process characteristics; according to the results of the above procedure, the main elements of the concept "Industry 4.0" have been individualized, covering the issues of information communication process planning of the production system elements, and a corresponding structural element of the latter has been individualized too;
- a generalized structure of the info-communication system has been suggested, which implies marking out periphery, commutation and process subsystems;
- requirements, set for the info-communication system and its individual elements in terms of advanced production system design process, have been formulated;
- the content of the main design objectives of the info-communication system has been formed (in terms of its main structural elements) with individualization of solution methods for the relevant groups of problems.

The above results will be used at further stages of research as the basis for tool formation – analytical, optimization and simulation models – to justify the characteristics of information flow movement and conversion processes in the context of advance production systems of industrial enterprises.

References

1. Adolphs, P., Bedenbender, H., Dirzus, D., et al.: Status report: reference architecture model Industrie 4.0 (RAMI4.0). ZVEI, Düsseldorf, Germany (2015)
2. Euler-Chelpin, A.V.: Information modelling for the manufacturing system life cycle. Doctoral Thesis, KTH Royal Institute of Technology, Stockholm, Sweden (2008)
3. Geisberger, E., Broy, M., et al.: Living in networked world. Integrate research agenda Cyber-Physical Systems (agenda CPS). ACATECH (2015)
4. Gubbia, J., Buyyab, R., Marusic, S., Palaniswami, M.: Internet of things (IoT): a vision, architectural elements, and future directions. Future Gener. Comput. Syst. **29**, 1645–1660 (2013)
5. Kagermann, H., Wahlster, W., Helbig, J.: Recommendations for implementing the strategic initiative Industrie 4.0. Final report of the Industrie 4.0 Working Group. ACATECH (2013)
6. Leventsov, V., Radaev, A., Nikolaevskiy, N.: The aspects of the «Industry 4.0» concept within production process design. St. Petersburg State Polytech. Univ. J. Econ. **10**(1), 19–31 (2017)
7. Liu, Y., Candell, R., Lee, K., Moayeri, N.: A simulation framework for industrial wireless networks and process control systems. In: IEEE World Conference Factory Communication Systems (WFCS) (2016)
8. Thoben, K.-D., Wiesner, S., Wuest, T.: "Industrie 4.0" and smart manufacturing – a review of research issues and application examples. Int. J. Autom. Technol. **11** (1) (2017)
9. Wang, L., Törngren, M., Onori, M.: Current status and advancement of cyber-physical systems in manufacturing. J. Manuf. Syst. **37**, 517–527 (2015)

10. Xu, L.D., He, W., Li, S.: Internet of things in industries: a survey. IEEE Trans. Ind. Inform. **10**, 2233–2243 (2014)
11. Xu, X.: From cloud computing to cloud manufacturing. Robot. Comput.-Integr. Manuf. **28**, 75–86 (2012)
12. Lu, Y., Morris, K.C., Frechette, S.: Current standards landscape for smart manufacturing systems. Interagency/Internal report (NISTIR) – 8107, NIST (2016)
13. Wang, Z., ter Hofstede, A.H.M., Ouyang, C., et al.: How to guarantee compliance between workflows and product lifecycles? J. Inf. Syst. **42**, 195–215 (2014)

SWM-PnR: Ontology-Based Context-Driven Knowledge Representation for IoT-Enabled Waste Management

Inna Sosunova[1(✉)], Arkady Zaslavsky[1,2],
Theodoros Anagnostopoulos[1,3], Petr Fedchenkov[1], Oleg Sadov[1],
and Alexey Medvedev[4]

[1] Department of Infocommunication Technologies, ITMO University,
Kronverkskiy pr., 49, St. Petersburg, Russia
{inna_sosunova, pvfedchenkov}@corp.ifmo.ru
[2] Digital Data61, CSIRO, Box 312, Clayton South, VIC 3169, Australia
Arkady.Zaslavsky@csiro.au
[3] Research and Education, Ordnance Survey, SO16 0AS Southampton, UK
Theodoros.Anagnostopoulos@os.uk
[4] Caulfield School of Information Technology,
Monash University, Melbourne, Australia
alexey.medvedev@monash.edu

Abstract. Using knowledge-based and semantic technologies in IoT is a very active research and promising area. This paper proposes a method of ontology-based context-driven knowledge representation for IoT-enabled hard waste management as part of a wider international project that aims at building IoT ecosystems for smart cities. The paper presents the development of the waste management ontology, rules, and proposes a multistage data processing method that allows extracting knowledge about specific nontrivial situations on its basis. The paper describes implementation of the proposed system as a web application, where the content types are based on ontology, and data processing occurs according to the proposed algorithm. Benefits of the proposed knowledge-based system are discussed and demonstrated. The proposed approach will significantly improve monitoring and management of waste collection, route planning, and problem reporting.

Keywords: Waste management · Knowledge representation · Ontology · Situation awareness · Context awareness · Decision support

1 Introduction

By 2050 66% of all Earth population will live in urban environments according to estimates of scientists [1]. For this reason, the concept of Smart Cities is becoming a major area of research in developed and developing economies. A Smart City is a technology-driven paradigm that combines information technologies, communication technologies and the Internet of Things to manage the city infrastructure, data, and

© Springer International Publishing AG 2017
O. Galinina et al. (Eds.): NEW2AN/ruSMART/NsCC 2017, LNCS 10531, pp. 151–162, 2017.
DOI: 10.1007/978-3-319-67380-6_14

services. Smart Cities incorporates a large number of applications, such as intelligent energy management, intelligent transportation, smart waste management, traffic management, etc.

Efficient waste management is one of the most pressing economic and environmental problems and a significant challenge for any large city [2]. Solid Waste Management is a challenge, which requires a complex multi-criteria approach [3]. An important direction for a Smart City in this area is to provide information about the process of garbage collection. Urban services, various sensors and users of the system receive an enormous amount of heterogeneous data in different formats. Processing, presenting, obtaining information and recognizing specific situations is one of the most difficult problems in the field of Waste management in Smart Cities.

In this paper, we present the outcomes of our research which aims to develop an ontology-based adaptive context-driven knowledge representation and decision support in IoT-enabled waste management. We call this system SWM-PnR - Smart Waste Management Presentation and Recommendation. The proposed method of representing and processing knowledge using ontology makes it possible to extract knowledge from data, to display the current situation of the garbage collection on the City dashboard (map), to recognize non-trivial situations and make recommendations to dispatchers of the waste truck fleet. The paper also describes the application of the proposed method in the form of a web application, which is developed on the basis of our Waste Management Ontology.

The research described in the paper is carried out as a part of the bIoTope Project [4]. The bIoTope project builds open IoT ecosystems on the basis of innovative IoT technologies and use cases. The project is funded under the EU Horizon 2020 Program. The system described in this paper is part of St. Petersburg pilot of bIoTope project on smart waste management.

The rest of the paper is organized as follows. Section 2 includes Background and Related Work. Section 3 presents the conceptual architecture of SWM-PnR and the process of extracting knowledge from ingested data. Section 4 describes the development of the Waste Management Ontology. Section 5 provides several scenarios for the system operation and Semantic Web Rule Language (SWRL) rules for these scenarios. Section 6 presents evaluation results. Section 6 describes the implementation of the proposed method as a web application. Section 7 concludes the paper.

2 Background and Related Work

Context and Knowledge. Within the framework of our research, the context is understood as "any information that can be used to characterize situation of an entity" [5]. We define contextual information as the level of how full is the smart garbage bin (SGB), the fullness of SGB, the time of the last cleaning of SGB, type of garbage, weight of garbage, fullness of the waste truck, location of SGBs and waste trucks, etc. The quality of context depends on the quality of data and the amount of data pre-processing. Raw data is the lowest level of context. A context-aware system is a system, which "uses context to provide relevant information and/or services to the user, where relevancy depends on the user's task" [5]. We consider a situation as context abstraction and generalization. The situation-aware system is aimed at finding and highlighting situations from diverse contextual data of various formats. Context prediction provides a forecast of the future context information. Based on statistics, it is possible to suggest a possible course of events for situations that have repeatedly occurred in the past. This approach enables proactive system functionality [6].

Decision Making and Adaptive Systems. Decision making is based on logical inferences from sensor data and user input data processing, as well as statistics available in the system. Based on statistical data, a possible course of a situation that has repeatedly occurred in the past can be suggested. The results of each decision made by users of the system are entered into the knowledge base. This allows to offer users the best option for each repetition of a particular situation based on an analysis of the consequences of making different decisions in similar situations.

Ontologies. The ontology consists of a set of subject domain concepts with their attributes and relationships, limitations, and axioms. Using the ontological approach allows the most comprehensive and flexible description of the subject area. It also allows logical inferences, i.e. recognize facts/situations, not present literally, but provided by the basic semantics.

2.1   Related Work

The design of a waste management system for Smart Cities involves the use of a large amount of information from sensors, its processing, and analytics [7].

Recently, a number of decision support systems for waste management in Smart Cities have been proposed. In [8] authors propose a decision support system, which consists of a conceptual framework and the corresponding mathematical formulation, for optimum waste management solution based on a multi-objective dynamic logistics model. [9] proposes a method of Strategic Decision Support based on clusters of factors and a model with using these clusters as factors. [10] describes the development of a methodology for decision support based on analysis of appropriate metrics for a municipal solid waste management expenditure efficiency assessment using cost-effectiveness analysis.

An ontology-based approach has been attempted in the field of waste management. A model using OWL ontology to sort these smart waste items for better recycling of materials is demonstrated in [11]. A concept of intelligent decision making support in the field of waste management using knowledge-based approach is presented in [12].

3   SWM-PnR Conceptual Architecture

SWM-PnR is part of St. Petersburg pilot of bIoTope project. The Smart Waste Management system is developed in the framework of a St. Petersburg pilot project. Smart Waste Management system includes means for collecting and processing data from

SGB sensors and sensors installed on waste trucks, the model for IoT data sharing, security and privacy policies, and dynamic route optimization in real time. The functionality of SWM-PnR includes inferring knowledge from adjusted data and representing it according to the user's needs. Also, SWM-PnR is able to recognize non-trivial situations on the basis of internal relationships between data, display them with the help of a custom traffic interface and give hints to dispatchers. Figure 1 shows steps of obtaining and representing knowledge from raw data.

Data comes from the sensors installed on smart garbage bins (lid sensor, capacity sensor, gas sensor, GPS module, scales sensor, temperature sensor, battery charge

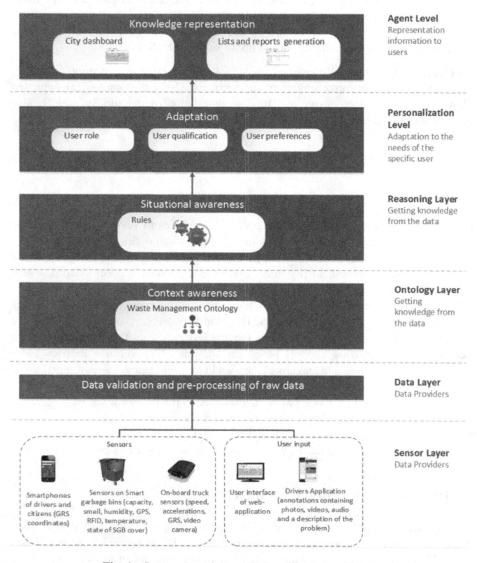

Fig. 1. Stages of receiving and representing knowledge

sensor), trucks (accelerometer, CAN bus, scales sensor, container capacity sensor, GPS module) and from user input, e.g., web application and in future mobile application for drivers [13]. The first stage is the verification and preprocessing of raw data. Then received data is transferred to the context awareness block, where the data is grouped according to the properties of ontology objects. In the Situation awareness block data is re-processed in accordance with the rules of ontology. This allows isolating specific situations and forecasts from the context. An adaptation unit allows SWM-PnR to adapt to the needs of the specific user. The information received is displayed to end users through the web application, consisting of City dashboard, means for asset management and means for generating reports.

The City dashboard is an interactive map showing SGBs, waste truck routes, and the movement of waste trucks along specified routes, waste dumps, and waste processing companies. The knowledge extracted from the data coming from sensors is displayed on the City dashboard in the form of an indication located above the objects on the map (trucks, SGBs, etc.) and in the form of tooltips for dispatchers. Full information on each of the objects on the map is available by clicking on this object. Statistics and history of each object can be viewed on the report generation page.

The proposed system can recognize the following situations and notify dispatchers about them:

- accidental or intentional damage of SGB sensor block;
- SGB sensor failure (with indicating the type of fault);
- low battery of sensor block;
- the need to change the route and schedule of the waste truck in connection with road works;
- the need to modify the route of the waste truck due to the inaccessibility of garbage containers in one of the districts. In this case corresponding city services are also informed.

4 Development of the Waste Management Ontology

We have developed the ontology-based on the analysis of business processes of garbage collection in St. Petersburg. We use the ontology language OWL [14] along with the development tool Protégé [15].

During the development the following challenges were overcome:

- the need to simultaneously analyze data from sensors on the basis of a developed ontology and visualize sensor data on the map;
- the need to provide users of the system with clear descriptions of system failures and malfunctioning.

To do this, the sensor data analysis uses data type properties of sensor classes (*battery_charge_sensor*, *capacity_sensor,* etc.). Information for users of SWM-PnR is provided using data type property *hasState* of the class *smart_garbage_bin*. Several possible values of the data type property *hasState* are shown in Fig. 2.

```
<owl:DatatypeProperty rdf:ID="hasState">
  <rdfs:range>
    <owl:DataRange>
      <owl:oneOf rdf:parseType="Resource">
        <rdf:first
rdf:datatype="http://www.w3.org/2001/XMLSchema#string"
>Missing sensor block</rdf:first>
        <rdf:first
rdf:datatype="http://www.w3.org/2001/XMLSchema#string">
Low battery</rdf:first>
        <rdf:first
rdf:datatype="http://www.w3.org/2001/XMLSchema#string">
No access</rdf:first>
        <rdf:first
rdf:datatype="http://www.w3.org/2001/XMLSchema#string">
Capacity sensor failure</rdf:first>
        <rdf:first
rdf:datatype="http://www.w3.org/2001/XMLSchema#string">
Gas sensor failure</rdf:first>
        <rdf:first
rdf:datatype="http://www.w3.org/2001/XMLSchema#string">
Damage of SGB sensors block </rdf:first>
        <rdf:first
rdf:datatype="http://www.w3.org/2001/XMLSchema#string">
Fullness 100% </rdf:first>
        <rdf:first
rdf:datatype="http://www.w3.org/2001/XMLSchema#string">
Fullness 90% </rdf:first>
...
    </owl:DataRange>
  </rdfs:range>
  <rdfs:domain rdf:resource="#smart_garbage_bin"/>
</owl:DatatypeProperty>
```

Fig. 2. Fragment of waste management ontology

Diagnostic information and the current state of garbage collection are visualized on the map using ontology data properties of SGB and waste trucks.

The structure of the Waste Management Ontology, its categories and generic items are represented in the Fig. 3. A more detailed structure with object and data type properties can be seen on our website [16].

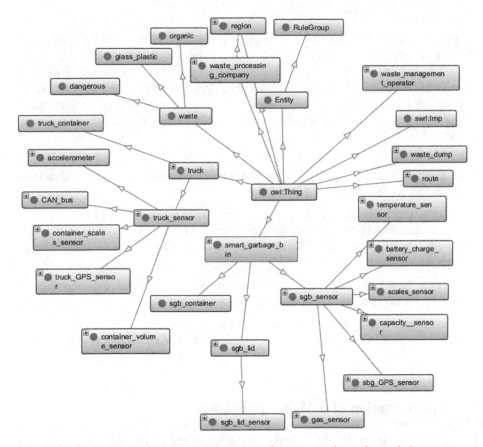

Fig. 3. Structure of waste management ontology: categories and generic items

5 Ontology Rules Composition

We investigated the process of garbage collection in St. Petersburg and identified the most common situations. In this section, we will demonstrate how the proposed method allows recognizing them. The rules for ontology were developed in the SWRL language [17]. Several scenarios and logical rules describing them are provided below. More complex scenarios have also been implemented, however due to space limitations aren't presented below.

Scenario 1: Accidental or intentional damage of SGB sensors block. The method is based on constant analysis and comparison of data from SGB sensors. If the SGB lid is open and SGB coordinates do not match the coordinates of the SGB container, the data from other sensors is analyzed. If the data does not arrive, the *hasState* datatype is assigned a value *Missing sensor block* and the notification is sent to the dispatcher.

```
smart_garbage_bin(?s) ∧ sgb_container(?c) ∧ sgb_lid(?l)
∧ is_part_of_sgb (?l, ?s)    ∧    is_part_of_sgb (?c, ?s)
∧ isOpen(?l, 1)    ∧ hasVolume(?c, 0) ∧ hasCapacity
(?c, 0)    ∧ hasCoordinates(?s, ?s_coord) ∧ hasCoordi-
nates(?c, ?c_coord)    ∧    swrlb:notEqual(?c_coord,
?s_coord)    → hasState (?s, "Damage of SGB sensors
block")
```

Scenario 2: SGB sensors failure. The method is based on constant analysis of data from SGB sensors for indicating the type of fault.

```
smart_garbage_bin(?s)    ∧    capacity_sensor(?f)    ∧
sgb_container(?c) ∧ hasVolume(?c, 0) ∧ hasCapacity (?f,
100) → hasState(?s, "Volume sensor failure ")
```

Scenario 3: Low battery. If battery of the sensor unit is discharged, it is necessary to transmit the warning to the dispatcher and display the problem on the City Dashboard.

```
smart_garbage_bin(?s) ∧ battery_charge_sensor(?b)    ∧
hasBatteryCharge(?b, 0)    → hasState(?s, "Low battery")
```

6   Implementation of a Web Application

In order to show the work of the proposed method in real conditions, we developed a web application in CMS Plone. Context types are based on ontology concepts, and the rules are implemented as external python scripts. Plone is a content management system based on the Zope application server and Zope built-in transactional object database (ZODB). To create maps in the application, we used Collective Geo add-on [18] and OpenStreetMap [19]. CMS Plone and ZODB are used for developing web applications in St. Petersburg pilot of bIoTope project on smart waste management. Thus, the web application described in this section can effectively interact with other components of the system in near real-time mode. The use of RDF based database is planned at the next stages of work on the project.

To make Plone content based on our ontology, we used the method described in [20]. After creating an ontology, we converted it into an UML class project. For making the described method functional, it was required to convert it in ZUML and then back to XML, so that it can be compatible with Plone. Next, in accordance with the method described in [20], we used ArchgenXML to add on our developed Plone Product in CMS Plone. The architecture of the developed web application is shown in Fig. 4.

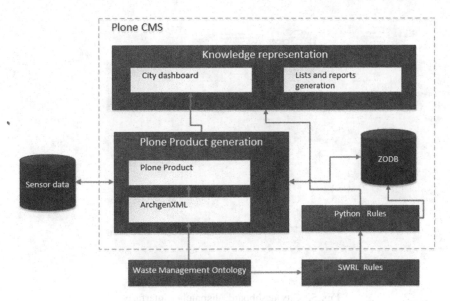

Fig. 4. Web application architecture

The developed application supports three modes of operation (3 user roles): dispatchers of waste collection companies, citizens, and city administration. For different participants in the waste collection process, different information is necessary. For citizens, the application provides information about the nearest separate waste collection points, and information about the cost of garbage disposal and rewards. Dispatchers can monitor the current state of the waste collection process, the routes of waste trucks, reports of malfunctions and problems. The city administration is able to generate reports for understanding of the big picture in the area of waste management and control pricing.

Waste collection is visualized through a City Dashboard. City Dashboard is available to all user roles. Objects on the map are "instances of the class" of the Waste Management Ontology: regions, routes, smart garbage bins, trucks, waste dumps, waste processing companies (Fig. 5).

Each SGB is equipped with the following indication: how full is an SGB, the need to immediately collect the waste, low battery, sensor failure, etc. Each waste truck has the following indication: how full is a container, the condition of the sensors (accelerometer, CAN bus, scales sensor, container volume, GPS), deviation from the route, accident, message from the driver. By clicking on the objects on the map in the pop-up, a user can see more detailed information. Transition to reports and statistics on the project is carried out by clicking on the link at the bottom of the pop-up (Fig. 6).

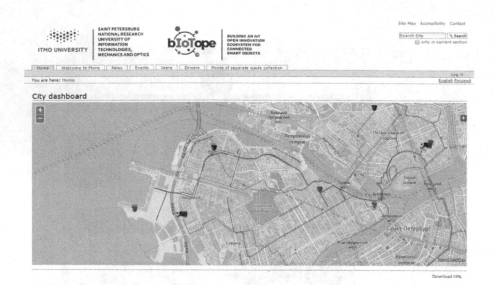

Fig. 5. City dashboard, dispatcher interface

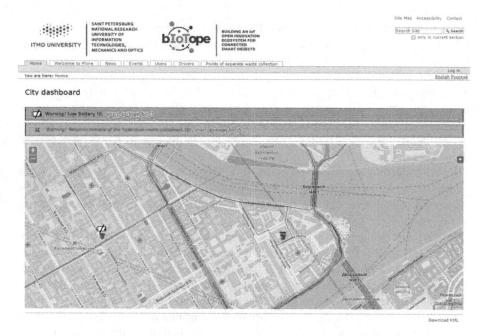

Fig. 6. City dashboard, 2 warnings for dispatcher

7 Conclusion

We investigated the waste collection process in St. Petersburg and proposed a multi-criteria approach to its visualization and decision support for multiple roles including dispatchers, citizens and city administration. The proposed approach will significantly improve various indicators: monitoring and management of waste collection, route planning, and problem reporting. The developed method allows the following:

- extract knowledge from data;
- display the current situation of the garbage collection on the City dashboard;
- recognize non-trivial situations;
- make recommendations to dispatchers of the waste truck fleet.

In future work, we will refine the developed method by including larger number of nontrivial situations and an extension of the ontology. It is also planned to integrate SWM-PnR with other components of St. Petersburg pilot of bIoTope project, such as mobile applications for drivers and citizens. This will allow SWM-PnR to use voice, photo and video data to support decision-making process [21].

Acknowledgements. Part of this work has been carried out in the scope of the project bIoTope which is co-funded by the European Commission under Horizon-2020 program, contract number H2020-ICT-2015/688203 – bIoTope. The research has been carried out with the financial support of the Ministry of Education and Science of the Russian Federation under grant agreement RFMEFI58716X0031.

References

1. United Nations Department of Economic and Social Affairs, World Urbanization Prospects Webpage. https://esa.un.org/unpd/wup/publications/files/wup2014-highlights.Pdf. Accessed 3 May 2017
2. Sakai, S., Yoshida, H., Hirai, Y., et al.: International comparative study of 3R and waste management policy developments. J. Mater. Cycles Waste Manag. **13**, 86 (2011). doi:10.1007/s10163-011-0009-x
3. Anagnostopoulos, T., Zaslavsky, A., Kolomvatsos, K., Medvedev, A., Amirian, P., Morley, J., Hadjiefthymiades, S.: Challenges and opportunities of waste management in IoT-enabled smart cities: a survey. In: IEEE Transactions on Sustainable Computing, vol. PP, no. 99, p. 1 (2017)
4. The bIoTope Project Webpage. http://www.biotope-project.eu/. Accessed 3 May 2017
5. Dey, A.K., Abowd, G.D.: Towards a better understanding of context and context-awareness. Comput. Syst. **40**(3), 304–307 (1999)
6. Anagnostopoulos, C., Mpougiouris, P., Hadjiefthymiades, S.: Prediction intelligence in context-aware applications. In: Proceedings of the 6th International Conference on Mobile Data Management, pp. 137–141 (2005)

7. Arebey, M., Hannan, M.A., Basri, H., Abdullah, H.: Solid waste monitoring and management using RFID, GIS and GSM. In: 2009 IEEE Student Conference on Research and Development (SCOReD), UPM Serdang, pp. 37–40 (2009). (Longhi, S., et al.: Solid waste management architecture using wireless sensor network technology. In: 2012 5th International Conference on New Technologies, Mobility and Security (NTMS), Istanbul, pp. 1–5 (2012))

8. Yu, H., Solvang, W.D., Yuan, S.: A multi-objective decision support system for simulation and optimization of municipal solid waste management system. In: 2012 IEEE 3rd International Conference on Cognitive Infocommunications (CogInfoCom), Kosice, Slovakia, pp. 193–199 (2012)

9. Hatwágner, M.F., Buruzs, A., Földesi, P., Kóczy, L.T.: Strategic decision support in waste management systems by state reduction in FCM models. In: Loo, C.K., Yap, K.S., Wong, K. W., Beng Jin, A.T., Huang, K. (eds.) ICONIP 2014. LNCS, vol. 8836, pp. 447–457. Springer, Cham (2014). doi:10.1007/978-3-319-12643-2_55

10. Soukopová, J., Malý, I., Hřebíček, J., Struk, M.: Decision support of waste management expenditures efficiency assessment. In: Hřebíček, J., Schimak, G., Kubásek, M., Rizzoli, Andrea E. (eds.) ISESS 2013. IAICT, vol. 413, pp. 651–660. Springer, Heidelberg (2013). doi:10.1007/978-3-642-41151-9_61

11. Sinha, A., Couderc, P.: Using OWL ontologies for selective waste sorting and recycling. In: OWLED-2012, Heraklion, Crete, Greece, May 2012

12. Kultsova, M., Rudnev, R., Anikin, A., Zhukova, I.: An ontology-based approach to intelligent support of decision making in waste management. In: 2016 7th International Conference on Information, Intelligence, Systems & Applications (IISA), Chalkidiki, pp. 1– 6 (2016)

13. Sosunova, I., Zaslavsky, A., Anagnostopoulos, T., Medvedev, A., Khoruzhnikov, S., Grudinin, V.: Ontology-based voice annotation of data streams in vehicles. In: Balandin, S., Andreev, S., Koucheryavy, Y. (eds.) ruSMART 2015. LNCS, vol. 9247, pp. 152–162. Springer, Cham (2015). doi:10.1007/978-3-319-23126-6_14

14. W3C OWL Webpage. https://www.w3.org/OWL/. Accessed 3 May 2017

15. Protege Webpage. http://protege.stanford.edu/. Accessed 3 May 2017

16. Github ITMO-SWM (St. Petersburg pilot of bIoTope project) Webpage. https://raw. githubusercontent.com/itmo-swm/Plone/master/Ontology/SWM-PnR%20Waste%20collecti on%20ontology.png. Accessed 31 May 2017

17. SWRL Webpage. https://www.w3.org/Submission/SWRL/. Accessed 3 May 2017

18. Collective Geo Webpage. https://github.com/collective/collective.geo.behaviour. Accessed 3 May 2017

19. OpenStreetMap Webpage. https://www.openstreetmap.org/#map=5/51.500/-0.100. Accessed 3 May 2017

20. Martins, C.T., Azevedo, A., Pinto, H.S., Oliveira, E.: Towards an ontology mapping process for business process composition. In: Azevedo, A. (ed.) BASYS 2008. ITIFIP, vol. 266, pp. 169–176. Springer, Boston, MA (2008). doi:10.1007/978-0-387-09492-2_18

21. Medvedev, A., Zaslavsky, A., Khoruzhnikov, S., Grudinin, V.: Reporting road problems in smart cities using OpenIoT framework. In: Podnar Žarko, I., Pripužić, K., Serrano, M. (eds.) Interoperability and Open-Source Solutions for the Internet of Things. LNCS, vol. 9001, pp. 169–182. Springer, Cham (2015). doi:10.1007/978-3-319-16546-2_13

Supporting Data Communications in IoT-Enabled Waste Management

Petr Fedchenkov[1(✉)], Arkady Zaslavsky[2], Alexey Medvedev[3],
Theodoros Anagnostopoulos[1,4], Inna Sosunova[1], and Oleg Sadov[1]

[1] Department of Infocommunication Technologies, ITMO University,
197101 Saint Petersburg, Russia
{pvfedchenkov, inna_sosunova}@corp.ifmo.ru,
theodoros.anagnostopoulos@os.uk, sadov@linux-ink.ru
[2] CSIRO Computational Informatics, CSIRO,
Box 312, Clayton South, VIC 3169, Australia
arkady.zaslavsky@csiro.au
[3] Faculty of Information Technology, Monash University, Melbourne, Australia
alexey.medvedev@monash.edu
[4] Research and Education, Ordnance Survey, Southampton SO16 0AS, UK

Abstract. The Internet of Things (IoT) enables Smart Cities with novel services. Such services demand low power, high throughput and low-cost sensor data collection technologies. The number of devices, their variety, the breadth of their distribution, and the number of standards are continuously increasing. In this paper, we explore and critically analyze emerging IoT communication technologies. LoRaWAN has been selected for deployment in the smart waste management system (SWM) for collecting data from smart garbage bins. The paper proposes an architecture for SWM data collection and delivery as part of St. Petersburg pilot of Horizon-2020 bIoTope project. Extensive experiments with actual sensors and smart garbage bins were conducted for stress-testing the LoRaWAN technology, analyzing data rates and power consumption. The paper concludes with lessons learned and LoRaWAN wider deployment feasibility and improvements discussion.

Keywords: LoRaWAN · Internet of Things (IoT) · Waste collection · Smart cities

1 Introduction

The Internet of Things (IoT) as a technology paradigm includes a wide variety of sensors and actuators. Monitoring of physical infrastructure, environment and virtual entities in real-time, processing of real-time and historical observations and actions that improve the efficiency and reliability of systems – all these tasks and many others can be solved with the help of IoT. IoT sets different requirements for devices: they should operate with less memory, less processing power and less bandwidth compare to traditional models. The long operation time of the device without additional

O. Galinina et al. (Eds.): NEW2AN/ruSMART/NsCC 2017, LNCS 10531, pp. 163–174, 2017.
DOI: 10.1007/978-3-319-67380-6_15

maintenance and charging (measured in years), the size of the device, its value, and the data transmission overhead in terms of maintenance costs are more important for IoT devices.

In Smart Waste Management (SWM) systems, IoT helps to provide interaction between components: containers with sensors, trucks, cloud infrastructure, dumps or recycling factories and user's devices. Smart Garbage Container (SGB) includes a set of sensors for determining the container's state, and, possibly, an actuator for locking or opening the lid on a command. SGB can also be considered as "a thing" which generates data. The authors in [1] consider waste management as an IoT-enabled service in Smart Cities.

The waste management system is presented in Fig. 1. All components of the system are linked through LPWAN technology between themselves and with cloud services. SWM systems helps to improve the waste collecting process with the utilization of an online decision process. Drivers of trucks should know when to collect waste from bins and which routes the collection trucks should follow. The route, for example, can also depend on route situation and prices on recycling factories and dumps. Users of the system are citizens and they should be aware of maintenance prices on local areas and service schedule.

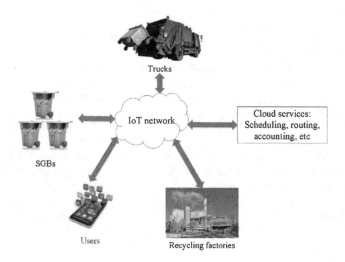

Fig. 1. Waste management system

The rest of the paper is structured as follows. Section 2 describes related work in communication technologies in the IoT. Section 3 discusses pros and cons of using LoRaWAN in Smart Waste management system. Section 4 represents methodology of testing and evaluating of LoRaWAN with several tests results, while Sect. 5 concludes the paper and discusses future work.

2 Related Work

IoT market grows, and there are many use cases for which existing cellular networks are not suitable. The provided coverage, penetration through obstacles, network capacity, battery life, and device price are insufficient to meet market demands. Existing cellular networks already offer excellent area coverage in urban areas, but require the device transmitter to operate at high power, draining the battery. A battery life of several years combined with an inexpensive device cannot be realized on existing cellular standards as they do not support mechanisms of power adaptation and saving.

Different communication technologies aimed at low power wireless IoT communication have been proposed and deployed. Below we provide a description of the most popular technologies and standards.

2.1 IEEE 802.15.4

IEEE 802.15.4 [2] is a standard specifying the physical layer and data link layer for Low-Rate Wireless Personal Area Networks (LR-WPANs). The standard focuses on low-cost, low-speed ubiquitous communication between devices. It proposes data rates up to 250 Kbit/s and transmission distance up to 1000 m, but in most real cases, the transmission distance totals only to tenths of meters. However, the provided distance is not enough for usage in Smart City applications.

2.2 Bluetooth LE

Bluetooth Low Energy (LE) is a wireless personal area network technology designed and marketed by the Bluetooth Special Interest Group. Bluetooth LE provides rapid link establishment functions (simpler pairing) and further trades off the data rate (approximately 200 kbps) for lower energy consumption, with the target to run a wireless sensor for at least one year on a single coin cell (approximately 200 mAh) [3]. Existing technology provides the possibility for Bluetooth LE devices to operate for months, even years, on coin-cell batteries.

2.3 Sigfox

Sigfox is a proprietary technology, developed and delivered by the French company Sigfox, without public specification. Sigfox operates in the 868-MHz frequency band, with the spectrum divided into 400 channels of 100 Hz [4]. Sigfox wireless systems send packages with a payload size of 12 bytes up to 140 uplink messages per day at a data rate up to 100 bps and up to 4 downlink messages per day. Each message can carry a payload of 8 Bytes. Sigfox claims that each access point can handle up to a million end-devices, with a coverage area up to 50 km in optimal conditions and up to 10 km in urban areas. Sigfox concentrates on data acquisition, but it is less useful for command-and-control scenarios.

2.4 NB-IoT

Narrowband IoT (NB-IoT) is a standards-based Low Power Wide Area technology developed to enable a broad range of new IoT devices and services. NB-IoT improves the power consumption of user devices, system capacity, and spectrum efficiency, especially in deep coverage. Battery life of more than ten years can be supported for some use cases. NB-IOT guarantees over 20 dB coverage, about 1000 connections and ten years using only 200 kHz bandwidth [5].

2.5 LoRa

In this paper, we focus on LoRa (Long Range), one of the most promising wide-area IoT technologies proposed by Semtech and further promoted by the LoRa Alliance [6]. LoRa Alliance has defined the higher layers and network architecture on top the LoRa physical layers and termed them LoRaWAN. LoRaWAN is primarily intended for IoT devices operating up to ten years on battery power alone in regional, national or global deployments. The intention of LoRaWAN is to provide secure bi-directional communication, mobility, and GPS-free localization services. LoRa targets deployments where end-devices have limited energy (for example, battery-powered), where end-devices do not need to transmit more than a few bytes at a time [7]. Data traffic can be initiated either by the end-device (when the end-device is a sensor) or by an external entity wishing to communicate with the end-device (when the end-device is an actuator).

3 LoRaWAN as the Enabling Infrastructure for SWM

The dynamic waste collection could be described as an online decision process for defining when to collect waste from bins (i.e., scheduling), and which routes the collection trucks should follow (i.e., routing). Many technologies and hardware are already used in waste management adopting different approaches in the administration of the physical infrastructure as well as the data collected in the field. Sensors enable the measurement of physical parameters and transform it to digital signals, which are transmitted wirelessly by an ad-hoc network infrastructure [8].

One of the main components of the Smart Waste Management system is an SGB. An SGB is modular and provides a different set of capabilities depending on the usage scenario. However, in any of them, data is generated from a set of sensors, and it is required to ensure the data delivery to cloud processing services. This information is used, for example, for the dynamic construction of routes, depending on SGB fullness, or garbage combustion probability and subsequent call for firefighting services.

We investigated LoRaWAN as an intermediate between end-device and cloud service. Long range of operation up to 13 km, small power consumption and confirmed packet delivery with bitrate up to 1 Kbit/s is enough for Smart City areas in which a gateway may be deployed in the district and communicate with multiple waste points.

Communication between end-devices and gateways is spread out among different frequency channels and so-called spreading factors (SF). It is the logarithmic ratio between the symbol rate (Rs) and the chip rate (Rc), i.e. SF = log2 (Rc/Rs). SFs enable simultaneous noninterfering communications between devices with the orthogonal nature of the set of codes. LoRaWAN data rates range from 0.3 Kbit/s to 27 Kbit/s for a 125 kHz bandwidth, depending on SF parameter. Data Rate (DR) is a parameter that determines the current mode of operation. The relationship between DR, SF and other parameters is presented in Table 1.

Table 1. LoRaWAN parameters [9]

DR	Configuration	Physical bit rate [bit/s]	Receiver sensitivity, dBm	Range, km
0	SF12/125 kHz	250	−137	0–2
1	SF11/125 kHz	440	−134.5	2–4
2	SF10/125 kHz	980	−132	4–6
3	SF9/125 kHz	1760	−129	6–8
4	SF8/125 kHz	3125	−126	8–10
5	SF7/125 kHz	5470	−123	10+

End-devices can transmit via any available channel at any time using any available data rate, being restricted by the necessary need to implement pseudo-random channel hopping at each transmission and to comply with the maximum transmit duty cycle. There are several modes of work for specification of LoRaWAN (for EU868), with the difference in transmission power. For a number of tx values from 1 to 5, the transmission power is respectively equals 14 dBm, 11 dBm, 8 dBm, 5 dBm and 2 dBm.

The minimum set of sensors for the proposed architecture of an SGB includes:

- Level meter, which is a set of range finders HC-SR04
- Temperature sensor, DHT-22
- RFID reader for garbage and user authentication

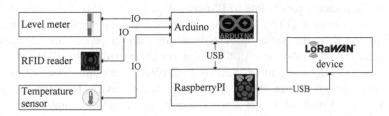

Fig. 2. Sensors connection diagram for smart container

This set of sensors is enough to determine the current state of the container and generate events about unexpected temperature increases. RFID reader can be used to authenticate the user in the system to calculate the amount of garbage ejected and determine its type.

Figure 2 presents the proposed architecture of interconnection between sensors and LoRaWAN device. We use Arduino board as a bridge between sensors and Raspberry

Pi (RPI) platform. Its main function is to convert raw sensor information to data packets that are sent for further processing and sending to RPI. This approach ensures the modularity of the device since when adding a new sensor, the Arduino firmware can be overwritten with the RPI while not disrupting the system as a whole. If we want to add pressure sensor, light sensor, or any other sensor type, we can connect it to Arduino, and flash new firmware throw RaspberryPI, to communicate with it by UART, SPI, I2C or read raw analog signal. Such an approach is useful for models, where an additional source of information can be used for refinement existing one. RaspberryPI aggregates raw data and represent a bridge to Cloud through LoRaWAN stack.

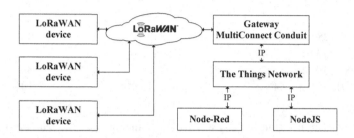

Fig. 3. Connecting diagram of SGB

Figure 3 represents the communication layer of the architecture. Each gateway serves a set of SGBs. In the proposed system, we use Multitech's MultiConnect Conduit Gateway [10]. MultiTech IoT platform includes programmable gateways, long-range RF modules, and accessory cards. Available options include a LoRaWAN mCard capable of supporting thousands of MultiConnect mDot long-range RF modules connected to remote sensors or appliances. The MultiConnect mDot is a LoRaWAN Ready, Low-Power Wide Area Network (LPWAN) RF module, capable of full duplex communication over distances more than 16 km. The main purpose of the gateway is to ensure reliable package forwarding to IP-network.

The Things Network (TTN) service is a cloud service for monitoring the queue of events from devices, decrypting packets, routing data to processing services, and sending packets back to devices. All the gateways connected to the TTN network are accessible on the map and can be used by LoRaWAN endpoint devices to communicate with the system. Packets are encrypted by endpoint devices by device key and by application key, so payload data is accessible only by the owner. Simultaneously, the openness of infrastructure is a significant step towards the spread of technology. TTN platform creators offer plug-ins for NodeRed and NodeJS for rapid prototyping of applications.

The module based on the RN2483 chip is used as an endpoint device. The RN2483 [11] is a fully-certified 433/868 MHz module based on wireless LoRa technology. The RN2483 utilizes a unique spread spectrum modulation within the Sub-GHz band to enable long range, low power, and high network capacity.

Figure 4 depicts the scheme of our prototype, which is developed in ITMO University for a pilot project of Smart Waste Management ecosystem in bIoTope project (Horizon2020).

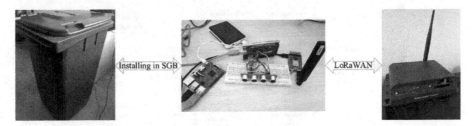

Fig. 4. Proposed architecture of smart waste bin

In Table 2 we present acceptable parameters of data flows from SGB in Smart Waste Management system. Injection of packets into the network is an event-based process. When a user interacts with an SGB, a new value of the of the garbage level parameter should be calculated. The number of packets per day per device parameter depends on the number of interactions with SGB. The upstream packet contains information about level, temperature, and RFID. The downstream direction of data flow is used by RFID system for user authentication. The lid will open only after receiving an accepting message from the cloud. The latency between the user action and response should be maintained in a sub second range.

Table 2. Acceptable parameters of data flows

	Packets count per day per device	Upstream packet size (byte)	Downstream packet size (byte)	Distance to gateway (m)	Wait for answer time (ms)
Range	20–100	40–60	20–40	1000–7000	50–2000

The next section describes the testing of the LoRaWAN technology for compliance with the specified parameters.

4 Evaluation and Testing Methodology

4.1 Testing Methodology

The main parameters determining the suitability of technology for use in the IoT-enable waste management scenarios are communication at long distances and low power consumption at a sufficient data rate.

In order to test the LoRaWAN workflow indicators, we developed an algorithm and implemented it as a module written in the NodeJS language. The algorithm includes the finite state machine represented in Fig. 5. Since the packets are sent to TTN, the connection to it is initialized. After the connection is established, it checks the availability of devices on the COM-port and initializes the connection to them.

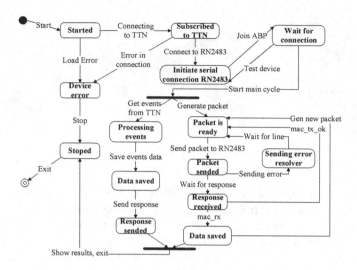

Fig. 5. LoRaWAN testing sequence diagram

The main cycle of the program with the device is as follows: a packet of the specified length (8 * N bytes) is generated, which is prepared for sending by executing the command "mac tx uncnf". The packet contains the Timestamp (the current Unix time in milliseconds) of the generation time. In the case of successful sending, the message "mac_tx_ok" comes from the device, after which the cycle proceeds to the generation of a new package. If "mac_rx" comes back, it reads the incoming data, which also contains the time of the sending, and writes them to determine the time of receiving the data packet. In the case of a send error, if there are no free channels, the data is resent. The cycle of working includes processing of incoming data and sending a response.

4.2 Packet Delivery Time

The data transfer rate, the number of packets transmitted per unit of time and the size of transmitted packets are the main parameters of any data transmission system. We investigated the dependence of the size of the payload and DR parameter. The packet size has a correlation with the data transfer rate, as we can send more information in single payload. However, such approach increases the needed time to send the packet, and affects the channel waiting time, as it is shown in Fig. 6.

Increasing the DR parameter reduces the delivery time of the packet (Fig. 7), significantly increasing the data transfer rate. However, this value affects the power consumption of the device, and the signal level decreases.

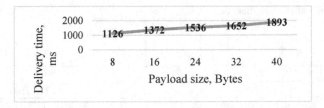

Fig. 6. Dependency between delivery time (ms) and a packet payload (byte) with DR = 1

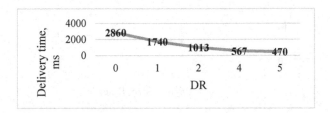

Fig. 7. Dependency between delivery time (ms) and DR with a packet size of 40 bytes

Figure 8 shows the dependence of the packet transfer rate (with 40 bytes payload) for different values of DR. As we can see in Fig. 8, the channel capacity correlates with the DR parameter.

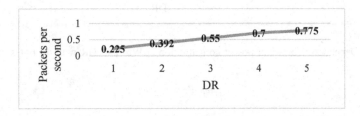

Fig. 8. The number of packets transmitted per second for DR with a packet size of 40 bytes

4.3 Power Consumption

Energy consumption is one of the determinants of the use of technology in IoT. Thus, the next series of tests were aimed at determining the power consumption of the RN2483 module when sending the data packet at various parameters. It is required to select a sufficient operation mode for providing the necessary transmission power.

In case of sending a single packet, we can see three time windows when the device is operating: the first window is used for sending a packet, and two other are used for receiving information back. The curve shown in Fig. 9 varies depending on the settings of the target device:

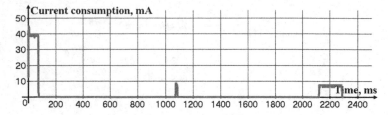

Fig. 9. Current consumption for DR = 5 and 40 bytes payload with TX power = 1

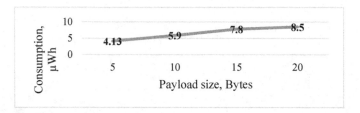

Fig. 10. Dependency between power consumption (μWh) and payload size (bytes) with DR = 5

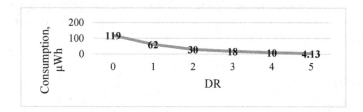

Fig. 11. Dependency between power consumption (μWh) and DR with 40 bytes payload

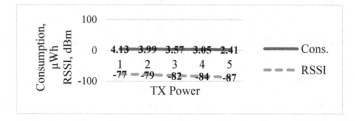

Fig. 12. Dependency between power consumption (μWh), RSSI (dBm) and TX power with 40 bytes payload and DR = 5

Figure 10 shows the dependence of power consumption on payload size for TX Power equals one. As the packet size increases, the packet sending period and the width of the first section increases. The power consumption also increases for a single packet.

As the value of DR increases, the sending period decreases: the width of the first segment decreases. The power consumption decreases for a single packet (Fig. 11).

As the value of TX Power increases, the transmission power decreases according to specification: peak current value of the first section decreases. The power consumption also decreases for a single packet (Fig. 12).

After all, the power used for delivery of a single packet depends on several parameters, and the developer is responsible for choosing the optimal values for concrete purposes. The increase in the DR value is beneficial for the system, but this parameter affects the signal level in the network. As the DR value increases, the distance of communication decreases, as we have shown in the previous section. Consequently, the proper selection of the DR parameter is the determining factor in the use of the LoRaWAN device, influencing the relationship between range, consumption, and network bandwidth.

4.4   Evaluation Results

We evaluated various operating modes of the LoRaWAN data transmission system. Taking into account the limitations on the required range of the system, the parameter DR should be chosen depending on the range equal to either 2 or 3. The "Packets per day per channel" parameter is based on information presented in Sect. 4.2 taking into account that every device operates with 3 channels. The delay of receiving an answer from the cloud system is discussed in Sect. 4.2. The power consumption is discussed in Sect. 4.3. The results of the evaluation results are presented in Table 3. We can see that LoRaWAN technology is appropriate for data collection in SWM.

Table 3. Parameters of data flows

	Packets per day per channel	Packets per day per device for fleet of 100 devices	Packet size (byte)	Distance to gateway (m)	Wait for answer time (ms)	Power per packet (µWh)	Power per day per device (mWh)
Min	11000	330	40	4000	567	18	6
Max	16000	480	60	7000	1100	30	15

5   Conclusions and Future Work

This paper proposed and discussed the challenges of architecting the communication layer in a Smart Waste Management system for St. Petersburg pilot as part of the international Horizon-2020 project bIoTope. We critically surveyed the existing IoT technologies and identified the LoRaWAN technology as the most appropriate. We discussed usability of and evaluated LoRaWAN devices in SWM systems based on the core architecture, power consumption and packet rate. We developed the architecture of an SGB with modification possibilities while maintaining the system modularity. We developed an algorithm for the evaluation of network parameters. We defined ways for reducing power consumption and increasing the data transfer speed by properly configuring LoRaWAN devices.

The future work will address the challenges of processing data in the cloud, interacting with one of the research IoT platforms, for example, OpenIoT and/or FIREWARE and optimizing the structure of packets transmitted over the LoRaWAN network. We will work on the analysis and performance evaluation of the system's backend as well as performing the evaluation of overall Smart Waste Management system with LoRaWAN devices.

Acknowledgments. Part of this work has been carried out in the scope of the project bIoTope which is co-funded by the European Commission under the Horizon-2020 program, contract number H2020-ICT-2015/688203 – bIoTope. The research has been carried out with the financial support of the Ministry of Education and Science of the Russian Federation under grant agreement RFMEFI58716X0031.

References

1. Medvedev, A., Fedchenkov, P., Zaslavsky, A., Anagnostopoulos, T., Khoruzhnikov, S.: Waste management as an IoT-enabled service in smart cities. In: Balandin, S., Andreev, S., Koucheryavy, Y. (eds.) ruSMART 2015. LNCS, vol. 9247, pp. 104–115. Springer, Cham (2015). doi:10.1007/978-3-319-23126-6_10
2. IEEE 802 Working Group and Others: IEEE Standard for Local and Metropolitan Area Networks—Part 15.4: Low-Rate Wireless Personal Area Networks (LR-WPANs); IEEE Std 802.15.4-2011, New York, NY, USA (2012)
3. Want, R., Schilit, B., Laskowski, D.: Bluetooth le finds its niche. IEEE Pervasive Comput. **12**, 12–16 (2013)
4. Margelis, G., Piechocki, R., Kaleshi, D., Thomas, P.: Low throughput networks for the IoT: lessons learned from industrial implementations. In: Proceedings of the 2015 IEEE 2nd World Forum on Internet of Things (WF-IoT), Milan, Italy, 14–16 December 2015, pp. 181–186 (2015)
5. http://www.huawei.com/minisite/iot/img/nb_iot_whitepaper_en.pdf. Assessed 17 Mar 2017
6. Sornin, N., Luis, M., Eirich, T., Kramp, T., Hersent, O.: LoRaWAN Specifications. LoRa Alliance, San Ramon (2015)
7. LoRa Alliance: White Paper: A Technical Overview of Lora and Lorawan. The LoRa Alliance, San Ramon (2015)
8. Ma, J.: Internet-of-things: technology evolution and challenges. In: IEEE MTT-S International Microwave Symposium (IMS), pp. 1–4 (2014)
9. https://www.researchgate.net/publication/309207302_Low_Power_Wide_Area_Network_Analysis_Can_LoRa_Scale. Assessed 10 Apr 2017
10. http://www.multitech.com/brands/multiconnect-conduit. Assessed 10 Apr 2017
11. http://www.microchip.com/wwwproducts/en/RN2483. Assessed 23 Apr 2017

International Workshop on Nano-Scale Computing and Communications (NsCC 2017)

Fiber-Optic Transmission System for the Testing of Active Phased Antenna Arrays in an Anechoic Chamber

Roman V. Davydov[1(✉)], Ivan K. Saveleiv[1], Vladimir A. Lenets[1],
Margarita Yu. Tarasenko[1], Tatiana R. Yalunina[2],
Vadim V. Davydov[1,2,3], and Vasily Yu. Rud'[3]

[1] Peter the Great Saint Petersburg Polytechnic University,
Saint Petersburg 195251, Russia
davydovrv@spbstu.ru
[2] The Bonch-Bruevich Saint Petersburg State University
of Telecommunications, Saint Petersburg 193232, Russia
[3] Department of Ecology, All-Russian Research Institute of Phytopathology,
Moscow Region, B.Vyazyomy, Odintsovo District 143050, Russia

Abstract. The results of the research of the developed fiber-optic transmission systems for analog high frequency signal are represented. On its basis, a new method to identify various structural defects in the active phased antenna arrays is elaborated.

Keywords: Radiophotonics · Active phased arrays antenna · Fiber-optic communication line · Anechoic chamber

1 Introduction

Now there is a continuous improvement exploited and development of new fiber-optic transmission systems (FOTS) of information, both trunk, and local assignment [1–5]. But in recent years, analog FOTS for transmission of a very high frequency of signals (radiophotonics) became an important component of development of local telecommunication area networks [2, 6–8]. Analog FOTS were given the greatest application in radar-tracking systems of different function [9–12]. Because in locations of radar-tracking systems the large number of various equipment and power stations are concentrated. In addition, the use of FOTS data can reduce the weight and dimensions of the radar systems themselves, in cases of their use on aircraft, etc.

Nowadays, many various modifications of the FOTS have been developed using direct and external modulation of laser radiation by an analog microwave signal [1, 2, 8–11]. Their designs and characteristics differ depending on the tasks to be solved with their use in radiolocation under various operating conditions [1, 2, 8–11, 13, 14]. The most difficult task is the tuning of active phased antenna arrays (APAA) at the stage of designing radar stations.

© Springer International Publishing AG 2017
O. Galinina et al. (Eds.): NEW2AN/ruSMART/NsCC 2017, LNCS 10531, pp. 177–183, 2017.
DOI: 10.1007/978-3-319-67380-6_16

One of the problem is the distortion of the signal in the transmission from the receiver to the high frequency analyzer at different APAA modes of operation during testing in an anechoic chamber (AC). These distortions are connected with the position of coaxial lines for transmitting signals to nearby high-voltage cables entering the APAA, and also location of the amplifier (without it the signal fades) in the area of direct radiation pattern of high power electromagnetic radiation (since the distance between the transmitter and receiver is smaller than 3 m) [9, 10, 13, 14]. In addition, the placement of any signal transform microwave devices (e.g. network analyzer) in the AC violates its anechoic, which makes use of the camera is not expedient.

One solution of this difficult problem is the development special construction of FOTS, where the analog signal from the receiving device placed in AC, transmits the received electromagnetic radiation of APAA in the area of instrumentation location through the region with a complex electromagnetic environment. Previously developed models of FOTS, as shown by the conducted studies, can not be used for these purposes, since their design does not consider the specific features of APAA testing in AC.

2   Fiber-Optic Transmission System and Measurement Technique

Since the distance between the active elements of APAA and the AC receiving device is not more than 3 m, it is necessary to exclude the direct effect of the pattern of the antenna radiation on the case of the optical transmission module. For this purpose, we developed a new design of a measuring probe in the form of a horn. A small design of the transmitting laser module was chosen. For this module, a small power driver has been designed and manufactured. This made it possible to place the transmitting laser module with the power driver behind the receiving horn, so that direct electromagnetic radiation from the antenna pattern did not fall on their cases. Such a construction of the FOTS was never used before.

In Fig. 1 a block diagram of the test stand for testing APAA in AC is represented. In developed by us the fiber-optic communication line the compact high-frequency laser module 1 was used ("Dilaz" company) – it is the transmitter with direct modulation with wavelength 1310 nm. The receiver is an optical module 2 ("Dilaz" company). Choosing a transmitter with direct modulation is caused by a small length of optical line (350 m long) and more stability of its parameters to temperature changes [1, 2, 6, 9, 10]. As the APAA testing with use of FOTS developed by us must be held in case of ambient temperatures from 233 to 323 K (real operating conditions of radar station) [4, 9, 10, 12]. A part of temperature conditions is provided when testing APAA in AC. Remaining temperature conditions are checked when testing APAA in the conditions of a polygon in the presence of opportunities.

The data obtained from measuring probe are coming over FOTS on vector network analyzer 4 with the subsequent processing on the personal computer 6.

Since the signals from the microwave generators are also transmitted to APAA via cables placed near the lines with high voltage for a variety of its devices (in AC there is only one entrance - it's also the output for all switching elements), so for this purpose it

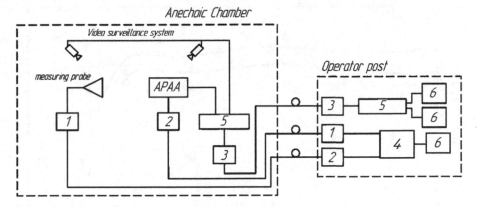

Fig. 1. Block diagram of the test stand for testing APAA in AC; 1 – laser module; 2 – receiving module, 3 – SFP module, 4 – vector network analyzer, 5 – router, 6 – personal computer

is also desirable to use FOTS. This will improve noise immunity in addition to significantly reduce the size of the connections, and unload the platform on which is placed APAA. Additionally, higher accessibility to the equipment in case of repair will be provided and to testing of APAA and the load of the mechanical elements providing movement of a platform when scanning the APAA pattern will be.

3   Results and Their Discussion

As an example, in Fig. 2 amplitude - frequency characteristic (AFC) in case of transmission of analog signals of a very high frequency on FOTS developed by us for different temperature conditions of its operation are provided.

The analysis of the received experimental result shows that losses of power during transmission of a very high frequency of a signal on FOTS developed by us in the frequency range of operation of APAA in the range of temperatures of its operation – aren't essential. It allows using FOTS developed by us for testing of APAA, as in the conditions of AC, and also at a polygon and the place of operation.

As an example, in Fig. 3 the dynamic characteristics for FOTS a developed by us at a temperature of 298.4 K are presented. This temperature regime is the main one for testing APAA in AC.

The obtained result shows, that the FOTS a developed by us can be used for testing radar station in frequency range of its exploitation [3, 9, 10, 12]. The output power is changed by less than 1 dB at changing the input power of analog microwave signal in wide limits.

As the experiments showed the use of the developed FOTS allowed, to research besides range of radiation of all APAA in different operation modes, also the ranges of radiation of its single active elements. These researches weren't possible earlier because of the existence of different noises, in particular the noise in the transferring path.

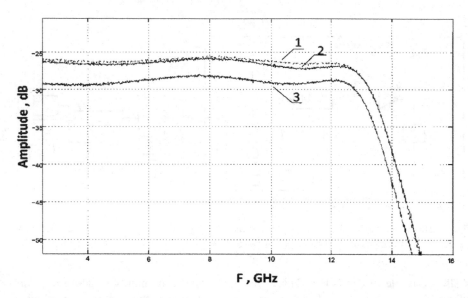

Fig. 2. The amplitude - frequency characteristic. Graphs 1, 2 and 3 corresponds to T in K: 288.1; 298.2; 328.3.

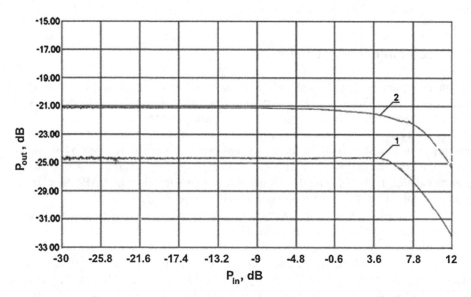

Fig. 3. The dynamic characteristics of FOTS. Graphs 1 and 2 corresponds to analog signal frequency in GHz: 9.1; 13.1.

As an example, Fig. 4 shows the emission spectra of a single active element APAA, transmitted to the control sector from the registration device of high frequency signal over coaxial cable, and developed by authors FOTS.

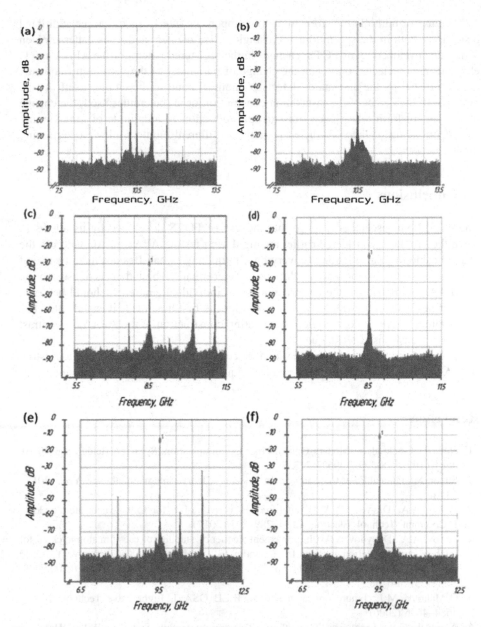

Fig. 4. The emission spectra of a single active element on 10.5, 9.5, 8.5 GHz frequency transmitted: through a coaxial cable - (a, c, e); through a FOTS - (b, d, f).

Analysis of the results, shows that the use of the developed by us FOTS allowed to completely eliminate transmission by it unpredictable RF disturbances and different induced signals, taking account of which and their further compensation (for example, subtraction) requires numerous additional measurements.

The use of multiple transformations to compensate interferences causes additional distortions, both in the spectral line of a single active element and in its pattern. When using the developed by us FOTS, changes in the registered pattern of the APAA single elements are connected only with the change of parameter of the very high frequency signal radiated by the antenna or defects of montage, and in the emitter itself. At the presence of the defect of the installation constriction or in the emitter itself, the change of the pattern form takes place (for example reduction of amplitude of a maximum, etc.). Comparing of the constructed patterns of different emitters allows determining these defects.

4   Conclusions

The carried-out researches showed that the use of FOTS developed by us allows to solve the problems considered earlier arising during testing APAA in AC with use the coaxial cable for transmission of a very high frequency signal. Besides that, the use of FOTS allowed to develop a new technique of determination, both defects of assembling of APAA, and defects in the construction of an individual emitter. It should be noted that our new technique with the use of FOTS for the first time made it expedient to test each emitter of APAA as a part of all operating construction of the APAA, in contrast to the previously used techniques.

The received results of researches confirmed feasibility of the FOTS use for testing APAA and validity of a new developed technique of its different defects determination.

References

1. Noé, R.: Essentials of Modern Optical Fiber Communication. Springer, Heidelberg (2016). doi:10.1007/978-3-662-49623-7. 337 p.
2. Silver, M.: Optical Fiber Communication Systems. Clanrye International, New York (2015). 334 p.
3. Petrov, A.A., Davydov, V.V.: Digital frequency synthesizer for 133Cs-vapor atomic clock. J. Commun. Technol. Electron. **62**(3), 289–293 (2017)
4. Petrov, A.A., Davydov, V.V.: Improvement frequency stability of caesium atomic clock for satellite communication system. In: Balandin, S., Andreev, S., Koucheryavy, Y. (eds.) ruSMART 2015. LNCS, vol. 9247, pp. 739–744. Springer, Cham (2015). doi:10.1007/978-3-319-23126-6_68
5. O'Mahony, M.J.: Future optical networks. IEEE OSA J. Light wave Technol. **24**(12), 4684–4696 (2006)
6. Agrawal, G.P.: Lightwave Technology: Telecommunication Systems. Wiley, Hoboken (2014). 480 p.
7. Marpaung, D., Roeloffzen, C., Heideman, R., Leinse, A., Sales, S., Capmany, J.: Experimental realization and application in a unidirectional ring mode-locked laser diode. Integr. Microw. Photonics (Laser Photonics Rev.) **4**(3), 506–511 (2013)
8. Davydov, V.V., Dudkin, V.I., Karseev, A.Y.: Fiber – optic communication line for the NMR signals transmission in the control systems of the ships atomic power plants work. Opt. Mem. Neural Netw. (Inf. Opt.) **23**(4), 259–264 (2014)

9. Davydov, V.V., Sharova, N.V., Fedorova, E.V., Gilshteyn, E.P., Malanin, K.Y., Fedotov, I. V., Vologdin, V.A., Karseev, A.Y.: Fiber-optics system for the radar station work control. In: Balandin, S., Andreev, S., Koucheryavy, Y. (eds.) ruSMART 2015. LNCS, vol. 9247, pp. 712–721. Springer, Cham (2015). doi:10.1007/978-3-319-23126-6_65
10. Davydov, V.V., Ermak, S.V., Karseev, A.Y., Nepomnyashchaya, E.K., Petrov, A.A., Velichko, E.N.: Fiber-optic super-high-frequency signal transmission system for sea-based radar station. In: Balandin, S., Andreev, S., Koucheryavy, Y. (eds.) NEW2AN 2014. LNCS, vol. 8638, pp. 694–702. Springer, Cham (2014). doi:10.1007/978-3-319-10353-2_65
11. Friman, R.K.: Fiber-Optic Communication Systems. Wiley, Hoboken (2012). 496 p.
12. Ermolaev, A.N., Krishpents, G.P., Davydov, V.V., Vysoczkiy, M.G.: Compensation of chromatic and polarization mode dispersion in fiber-optic communication lines in microwave signals transmittion. J. Phys: Conf. Ser. **741**(1), 012071 (2016)
13. Davydov, V.V., Dudkin, V.I., Karseev, A.U.: Nuclear magnetic flowmeter-spectrometer with fiber-optical communication line in cooling systems of atomic energy plants. Opt. Mem. Neural Netw. (Inf. Opt.) **22**(2), 112–117 (2013)
14. Davydov, V.V., Velichko, E.N., Dudkin, V.I., Karseev, A.Y.: Fiber-optic system for simulating accidents in the cooling circuits of a nuclear power plant. J. Opt. Technol. **82**(3), 132–135 (2015)

Advanced Materials for Fiber Communication Systems

Victor A. Klinkov$^{(\boxtimes)}$, Alexandr V. Semencha,
and Evgenia A. Tsimerman

Peter the Great St. Petersburg Polytechnic University,
St. Petersburg 195251, Russian Federation
klinkovvictor@yandex.ru

Abstract. Fluoroaluminate glass is a material with a number of unique properties, but at present there are no studies that relate the chemical composition and luminescence properties. A number of fluoroaluminate glasses based on the composition $(100 - x) \cdot MgCaSrBaYAl_2F_{14} - x \cdot Ba\,(PO_3)_2$, where $x = (0.5–3.0)$ mol.% were synthesized by melting technique. The tendency of concentration dependence of hydroxyl absorption peaks reduction was explained by analysis of complex IR transmittance spectra and molecular refraction. The range of composition characterized by the absence of OH-absorption peaks around 3 μm was found. The concentration dependence of molecular refraction values on barium metaphosphate composition shows the inflection points in area at 1.0 mol.% and 2.0 mol.% of $Ba(PO_3)_2$. The analysis of the data obtained by Rayleigh and Mandel'shtam-Brillouin scattering (RMBS) spectroscopy determined the composition which has the minimum value of Landau-Placzek ratio (R_{L-P}). Er^{3+} doped fluoroaluminate glasses were prepared by the same technique. The near infrared luminescence spectra of glasses corresponding to the $^4I_{13/2} \rightarrow {}^4I_{15/2}$ transition of Er^{3+} ions were observed. The mechanism of upconversion emissions Er^{3+} was discussed. Green and red emission bands at around 523, 546, and 660 nm wavelength were explained by two-photon process.

Keywords: Fluoroaluminate glass · Upconversion · Er^{3+} erbium doped · Landau-Placzek ratio

1 Introduction

Nowadays numerous fundamental and applied researches deal with synthesis of new optical materials for near infrared (NIR) and mid-infrared (mid-IR) regions of the spectrum. The NIR and mid-IR spectral windows are a useful range for chemical and biological sensing, because strong rotational-vibrational absorption lines of many molecules are situated here and the absorption strength of molecular transitions is of higher order of magnitude than those in the visible range. But existing materials are costly, which significantly limits their mass usage as optical components in this range. Therefore, investigations are being carried out to mold cheap IR optical components, for example, lenses, windows and fibers from chalcogenide materials [1–3].

© Springer International Publishing AG 2017
O. Galinina et al. (Eds.): NEW2AN/ruSMART/NsCC 2017, LNCS 10531, pp. 184–195, 2017.
DOI: 10.1007/978-3-319-67380-6_17

A lot of researches in this spectral region are focused on nanomaterials development and classical host materials modification to provide not only passive (transmission), but also active (generating, amplificating) medium. The necessity for this kind of research is due to the increasing demands of advanced science, technology solar spectra converters, communication systems and medicine. Most active media are formed by doping with rare earth elements (REE). The NIR and mid-IR luminescence and visible upconversion in crystals and glasses doped with rare-earth elements have a large number of practical applications. For example, emission of Er^{3+} ions at $\lambda \approx 1.5$ µm has already been used for eye-safe laser [4], optical temperature detectors based on fluorescent intensity ratio (FIR) of REE operated in extremely high conditions and temperature sensor based on the upconverted emission [5] have been extensively investigated in recent years [6, 7]. Moreover, many investigations have been focused on Er-doped materials because of 2.7 µm emission, which has various useful applications in military technology, room sensing, atmosphere pollution monitoring, and eye-safe laser radar.

The host materials for mid-IR are expected to have high level of transmission, a small absorption coefficient typical for OH^- groups bands and broad emission bands, low nonradiative decays and high radiative emission rates. One of the most appropriate materials with properties listed above is fluoride glasses based on heavy metals.

Fluoride glasses based on heavy metals are traditionally divided into three main groups: fluorozirconate, fluoroindate and fluoroaluminate glasses.

Among all fluoride glasses based on heavy metals fluorozirconate system, notably the ZrF_4–BaF_2–LaF_3–AlF_3–NaF (ZBLAN) glass composition, is one of the most stable against devitrification. However ZBLAN has some drawbacks such as low glass transition, fragility and high sensitivity to water and acids [8, 9].

Fluoroindate glasses (for example, system BIZYbT: $30BaF_2$-$30InF_2$-$20ZnF_2$-$10YbF_3$-$10ThF_4$) are not only more stable than the fluorozirconate ones, but also possess enhanced thermal stability and transparency that extends from the UV region to 7–8 µm. Besides this the calculated values of minimum optical losses determined by the sum of the Reyleigh scattering and multiphonon absorption are substantially smaller, for example, for the glass BIZYbT such losses reach ~ 0.002 dB km^{-1} at the wavelength of 3 µm [10], while for ZBLAN they are ~ 0.044 dB km^{-1} at the wavelength of 2.55 µm [11]. However, the synthesis technology of fluoroindate glasses is not reproducible. The least information is available about fluoroaluminate glasses, but the data obtained give evidence for their superior to other groups mentioned above. The majority of compounds of fluoroaluminate glasses are based on the mineral usovite.

In this paper we report on the spectral and luminescent properties and analysis of compositional dependence of Landau-Placzek ratio of fluoroaluminate glasses. The composition of mineral usovite was modified by the introduction of strontium and yttrium fluorides for preparing the glass-like compound.

Oxygen-free fluoroaluminate glasses have important properties: the transmission from ultraviolet up to 7 µm, selectable index of refraction, the possibility of introducing much higher than into silica concentrations of rare earth elements [12] without concentration quenching. Moreover fluoroaluminate glasses have low theoretical total losses around ~ 0.001 dB km^{-1} and low phonon energy around ~ 500 cm^{-1} (for example for Silica glasses phonon energy is around ~ 1100 cm^{-1}). It is well-known that smaller phonon energy results in smaller multiphonon transition probability thereby allowing many mid-infrared laser transitions, which are normally quenched in the glass hosts with larger phonon energy.

High crystallization ability of these compositions reduces their practical usage and that is their main drawback. Although the studies [13, 14] demonstrated the possibility of decreasing the crystallization ability of the fluoroaluminate host by introducing small additives of phosphates.

2 Object of the Study

The object of the study was a model fluoroaluminate glass of the composition: $(100 - x) \cdot MgCaSrBaYAl_2F_{14} - x \cdot Ba(PO_3)_2$, where x = (0.5–3.0) mol.%. This region of compositions was chosen to refine the result obtained earlier [15], where the researchers studying the range of compositions x = (0.0–10.0) mol.%, came to the conclusion that glass containing more than 1.7 mol.% $Ba(PO_3)_2$ is characterized by the presence of bridging -P-O-P bonds, which can modify the structure of the glass. In this paper, such research methods as Rayleigh-Mandelstam-Brillouin spectroscopy (RMBS) and spectral-luminescent analysis were used.

The composition $98MgCaSrBaYAl_2F_{14} - 2Ba(PO_3)_2$ was doped with Erbium fluoride ErF_3 with 0.1, 0.5, 1.0 mol.%. The reason to choose this composition will be clarified below. Compositions of the studied samples are given in Table 1.

Table 1. Component content $(100 - x) \cdot MgCaSrBaYAl_2F_{14} - x \cdot Ba(PO_3)_2$.

$\Sigma MgCaSrBaYAl_2F_{14}$, mol. %	x, mol. %	Sample number
99.5	0.5	11
99.0	1.0	12
98.9	1.1	13
98.7	1.3	14
98.5	1.5	15
98.3	1.7	16
98.1	1.9	17
98.0	2.0	18
97.9	2.1	19
97.7	2.3	20
97.5	2.5	21
97.3	2.7	22
97.0	3.0	23
$98MgCaSrBaYAl_2F_{14} - 2Ba(PO_3)_2$ doped with rare earth activator	ErF_3, mol. %	
98.0	2.0	0.1	24
98.0	2.0	0.5	25
98.0	2.0	1.0	26

3 Experimental Technique

The glasses were synthesized in the argon atmosphere for 1 h at the temperature of 850–950°C without mixing in glassy-carbon crucibles with the volume of 100 mL. The initial compounds were $Ba(PO_3)_2$ and stabilized usovite prepared from individual fluorides. The breakage of single crystals was used for preparing fluorides of rare earth elements. The glass was poured into a heated graphite mold and annealed in a furnace at the temperature of 400–450°C (according to the composition). Components (especially fluorides) were evaporated in the process of glass melting; however, the elaborated synthesis mode reduced the losses to 1–1.5 wt.%. Erbium was introduced in the form of fluorides (0.1, 0.5, 1.0 mol.% in excess of 100%).

The IR absorption spectra were measured using a Lumex InfraLUM FT_08 IR Fourier spectrometer in the range of 1000–5000 cm^{-1} with the resolution of 4 cm^{-1} and the accumulation time of 60 s. The optical absorption spectra in the wavelength region of 190–2000 nm were recorded at the temperature of 300 K on a Perkin Elmer LLC Lambda 900 spectrophotometer. Samples were polished on both sides. The thickness of the samples was 1 mm.

The luminescence of the samples in NIR and visible regions was excited by the radiation of a LQ 129 pulsed laser (Solar Laser System) (λpump = 975 nm). The luminescence spectra were recorded using a monochromator (Acton-300) and a receiver (model Act-441) for the NIR region, and for the visible region, a Hamamatsu photomultiplier R928 was used. The signals from the receiver were amplified and processed at a digital lock-in amplifier.

The Rayleigh and Mandel'shtam-Brillouin scattering (RMBS) spectroscopy and the Landau-Placzek ratio were used to elucidate changes in structural groups of the glass host as function of chemical composition.

RMBS spectra glass samples were taken by piezoelectric scanned Fabry-Perot interferometer. Light scattering was excited by He-Ne laser (λ = 632.8 nm, power 25 mW) and measured at the scattering angle θ = 90° [16]. The error in measuring the Landau-Placzek ratio did not exceed ±5% [17]. The measurements were made on one sample at several points (at least five), the smallest result was taken as the true one.

4 Results and Discussion

The optical absorption spectra in the wavelength region of 190–2000 nm of all undoped glasses, as expected, show high level of transparency and absence of absorption peaks. Typical absorption spectrum of undoped glasses is shown in Fig. 1b.

The IR transmittance spectra of undoped glasses, indicating the OH- content, are shown in Fig. 1a. We note the complete overlapping of the transmission spectra of some samples, for simplicity they are not given. It is clear, that the maximum level of transmittance is located in the region of 90%. The difference in the values does not exceed 2–3% in a series of studied samples, which can be related to the previously discussed problem of components losses during the melting. The losses were estimated in the

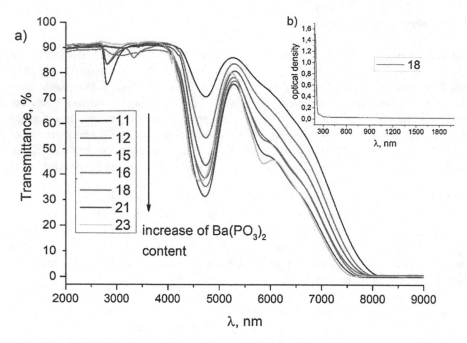

Fig. 1. (a) Optical IR transmission spectra and (b) Typical absorption spectrum of the fluoroaluminate glasses in range 190–2000 nm. Curve numbers correspond to the numbering in Table 1.

measurement of the samples density according to the standard technique and did not exceed 1.5 wt%. Another reason for the transmission losses is related to the Fresnel reflection.

The OH^- absorption peaks are located around 3 μm. The small significance of this absorption is worth noting. Considering the point of luminescent processes this is a very important circumstance. It is well known, that the content of hydroxyl groups significantly reduces the emission efficiency. These groups may take part in the energy transfer (ET) of rear-earth ions, which can lead to the energy losses of the metastable level, cause the luminescence quenching [18–20] and reduce the lifetime especially for materials with low doping concentration of rear-earth ions.

Figure 1a shows that OH^- absorption peak starts gradually disappearing with the increase of barium metaphosphate concentration from 1.5 mol.% and can't be registered at all with more than 2.0 mol.% $Ba(PO_3)_2$.

It may seem that with the increase of barium metaphosphate content the oxide content also increases, which can be joined by the hydrogen atom in the presence of unsaturated bonds, thereby forming OH^- group and hence the intensity of the absorption must grow. However, the situation is reverse. This is apparently connected with some facts: at low oxygen concentrations, oxygen cannot occupy a bridge position between the fluoride regions without significant local changes in the crystal field, thus there is a possibility of unsaturated bonds existence. With the increasing concentration of $Ba(PO_3)_2$ the structure of the glass matrix changes: besides fluoride regions the phosphate regions appear. Thus, at the concentrations of barium metaphosphate in the region of 2.0 mol.% the structure of the glass changes dramatically, as the oxygen groups participate in the formation of the glass matrix.

To verify this assumption, a number of additional measurements were made; the values of molecular refraction were calculated. It is known that the nature of the change in molecular refraction reflects structural rearrangements occurring in the matter. Molecular refraction is directly related to the polarizability of ions and molecules, that is, the ability of their electronic shells to deform in the external electric field and the local fields of other ions and molecules of the condensed matter. It is well known that the polarizability determines the dynamical response of system to external fields, and provides molecule's internal structure information. Thus, the molecular refraction R calculated from the Lorentz-Lorenz formula (1) is a measure of the average polarizability of the molecules in condensed matter:

$$R = \frac{n_D^2 - 1}{n_D^2 + 2} \frac{M}{\rho} \tag{1}$$

where M is the molecular mass of the substance, ρ is the density, n_D refractive index measured at the yellow doublet D-line of sodium, with a wavelength of 589 nm. Figure 2 shows the calculated values of molecular refraction in a series of undoped glasses.

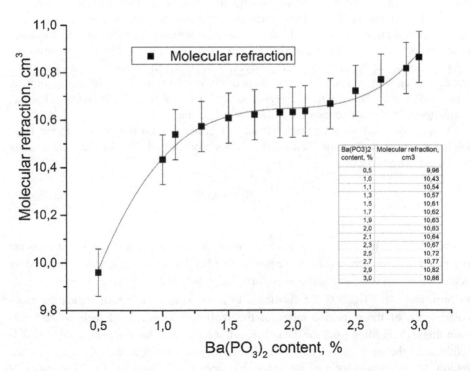

Ba(PO3)2 content, %	Molecular refraction, cm3
0,5	9,96
1,0	10,43
1,1	10,54
1,3	10,57
1,5	10,61
1,7	10,62
1,9	10,63
2,0	10,63
2,1	10,64
2,3	10,67
2,5	10,72
2,7	10,77
2,9	10,82
3,0	10,86

Fig. 2. Calculated values of molecular refraction in a series of undoped glasses. The connecting curve is drawn for clarity.

The polarizability of molecules in the electromagnetic field of radiation of the visible and near-infrared ranges, is due to the displacement of the electrons and is equal to the sum of the effects of displacements of individual electrons. The previous statement gives the molecular refraction the character of the additive constant. Considering glass the ideal solution and changing its composition, we would obtain a linear change in the indicated value. The presence of inflection points on concentration dependence in area of 1.0 mol.% and 2.0 mol.% $Ba(PO_3)_2$ may indicate changes in the local fields associated with the change in the structure and nature of the structure of the matter. This circumstance can be considered as another proof in favor of the assumption about changes in the structure of fluorophosphate host in this region of compositions [15].

In order to study the changes in structure with concentration of $Ba(PO_3)_2$ and nomenclature of structural groups the Rayleigh and Mandel'shtam-Brillouin scattering spectroscopy were used. As shown in the study [21] intensity of Mandel'shtam–Brillouin scattering (I^{MBS}) and both Rayleigh scattering losses and structural information may be extracted from the ratio of Rayleigh scattering intensity (I^{RS}) to the sum of Mandel'shtam-Brillouin scattering intensities, the so-called Landau–Placzek ratio:

$$R_{L-P} = \frac{I_\rho + I_{anis}}{2I^{MBS}} = R_\rho + R_c + R_{anis} = \frac{I^{RS}}{2I^{MBS}} \tag{2}$$

where R_ρ is the contribution of density fluctuations, R_c is the contribution of concentration fluctuations, R_{anis} is the contribution of anisotropy fluctuations, I_ρ, I_{anis}, R_{L-P} are the contributions of 'frozen-in' density and anisotropy fluctuations into Rayleigh scattering intensity and Landau–Placzek ratio, correspondingly. The Fig. 3 shows the obtained ratios for the samples under study. It is seen that the values have a sufficiently strong scatter. The presence of two regions of content of $Ba(PO_3)_2$ with a minimum of Landau–Placzek ratio is obtained: region around 1.0 mol.% and 2.0 mol.%. This fact correlates with the above data on molecular refraction.

It is known that with the predominance of concentration fluctuations in the Rayleigh scattering, the contribution of R_{anis} does not exceed 5%, so they are usually neglected. Therefore,

$$R_c \sim \frac{\left(\frac{\partial n}{\partial c}\right)^2_{PT} \langle \Delta C^2 \rangle}{2I_{MB}} \tag{3}$$

where $\langle \Delta C^2 \rangle$ is the root-mean-square value of concentration fluctuations at constant pressure P and temperature T. Consequently, the Rayleigh scattering can be reduced by achieving a high chemical homogeneity. Indeed, when $\langle \Delta C^2 \rangle$ is close to zero, Rc tends to zero, and for $\left(\frac{\partial n}{\partial c}\right)_{PT} \approx 0$, Rc also tends to zero. The latter in turn means the exact coincidence of the refractive index of the scattering centers and their nearest surroundings. Proceeding from the data obtained, we propose that the contribution of $\frac{\partial n}{\partial c}$ is minimal in the investigated ranges of compositions. Consequently, the main contribution to the scattering can be made by density fluctuations Rρ. The ratio of Landau-Placzek is the smallest for the glass composition with content of barium

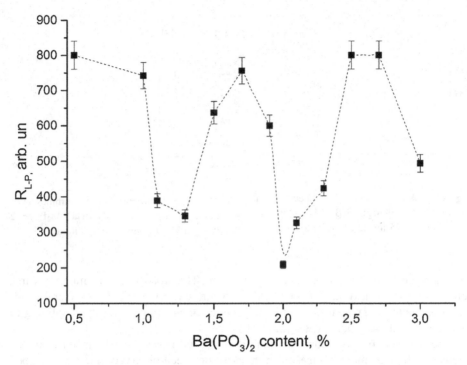

Fig. 3. Dependence of the Landau-Placzek ratio on the content of $Ba(PO_3)_2$. The curve is drawn for clarity.

metaphosphate 2.0 mol.% for which, apparently, the density fluctuations are minimal. In other compositions, the contribution of $R\rho$ remains larger.

A consistent explanation of the minimum $R\rho$ density fluctuation value for a given composition can be suggested if it is assumed that the resulting transition mixed structure is the most coordinated due to crosslinking the fluoroaluminate structure with the assist of P-O-P bridges. Another proof of this hypothesis is the presence of pyrophosphate groups providing a binding action due to the bonds PO_3F^{2-} that were found in the work [14].

Since the composition with 2.0 mol.% $Ba(PO_3)_2$ is one of the most optimal in a combination of properties, glasses doped with rare earth elements were synthesized precisely on the basis of this composition.

It is well known, the width of the luminescence spectrum of doped phosphate glasses is close to that of silicate glasses, and the width of the luminescence of fluoride glasses has record values. It was of practical interest to compare the values of the width of the luminescence with those of fluoride and phosphate because, in accordance with the above assumption, we are dealing with mixed fluoride and phosphate glass matrix.

The NIR luminescence spectrums of glasses containing 0.1, 0.5, 1.0 mol.% are shown in Fig. 4. The NIR luminescence spectra of glasses corresponding to the $^4I_{13/2} \rightarrow ^4 I_{15/2}$ transition of Er^{3+} ions are observed in the range 1400–1700 nm. The maximum of luminescence peak is located at the wavelength of 1.53 μm. Two

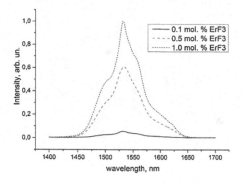

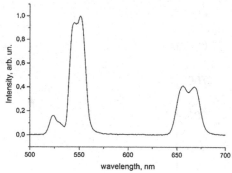

Fig. 4. NIR luminescence spectra of Er-doped 98MgCaSrBaYAl$_2$F$_{14}$ − 2Ba (PO$_3$)$_2$ upon 975 nm laser excitation.

Fig. 5. Typical upconversion luminescence spectrum Er-doped 98MgCaSrBaYAl$_2$F$_{14}$ − 2 Ba(PO$_3$)$_2$ upon 975 nm laser excitation.

shoulders are observed at 1.503 μm and 1.556 μm. The broad 1.53 μm emission with a full width at half-maximum (FWHM) of ∼65 nm was observed. This spectral band-width is due to inhomogeneous and homogeneous broadening and to Stark splitting of the excited and the ground states [22].

The increase of ErF$_3$ concentration from 0.1 to 1.0% leads to increase in the intensity. The maximum luminescence intensity has been observed in the glass doped with 1.0 mol.% of ErF$_3$.

It was found out that the values of FWHM of fluoroaluminate glasses under investigation were much higher than those of the fluoroziconate, silicate, phosphate and antimony glasses. The values of FWHM compared to other glass systems are shown in Table 2.

Table 2. Full Width at Half Maximum (FWHM, nm) for the $^4I_{13/2} \rightarrow {}^4I_{15/2}$ transition of Er^{3+} ion in different Er^{3+}-doped glasses.

Glass system	FWHM, nm
Antimony [23]	58
Phosphate [24]	34
Fluorophosphate [25]	36
Fluoride [26]	52
Silicate [27]	55
Fluoroaluminate [this work]	65

Upconversion luminescence spectrum is shown on Fig. 5. Green and red emission bands at around 523, 546, and 660 nm wavelength are attributed to the $^4H_{11/2} \rightarrow {}^4I_{15/2}, {}^4S_{3/2} \rightarrow {}^4 I_{15/2}$ and $^4F_{9/2} \rightarrow {}^4 I_{15/2}$ transitions, respectively. The most intense emission corresponds to the green emission from $^4S_{3/2}$ level.

It should be noted that both the NIR and upconversion luminescence are observed together. The particle population in the excited energy level $^2H_{11/2}$ and $^4S_{3/2}$ of Er^{3+} is attained by the two-photon process. This process consists of excited state absorption (ESA), ground state absorption (GSA) and cross relaxation (CR) [28]. The upconversion emissions process can be described as:

$$GSA :^4 I_{15/2}(Er^{3+}) + a \text{ photon } \rightarrow^4 I_{11/2}(Er^{3+});$$
$$ESA1 :^4 I_{11/2}(Er^{3+}) + a \text{ photon } \rightarrow^4 F_{7/2}(Er^{3+});$$
$$CR :^4 I_{11/2}(Er^{3+}) +^4 I_{11/2}(Er^{3+}) \rightarrow^4 I_{15/2}(Er^{3+}) +^4 F_{7/2}(Er^{3+}).$$

The Er^{3+} ions on the energy level $^4F_{7/2}$ nonradiative relax by multiphonon emission to $^2H_{11/2}$ and $^4S_{3/2}$ rapidly (due to narrow energy gap), and then transit to the ground state respectively $^4H_{11/2}/^4S_{3/2} \rightarrow^4 I_{15/2}$, emitting green light at wavelength 523 nm and 546 nm. Then electrons of $^4F_{7/2}$ level decay quickly to $^4F_{9/2}$ level by multiphonon relaxation followed by the $^4F_{9/2} \rightarrow^4 I_{15/2}$ transition. In addition, electrons in $^4I_{13/2}$ level are pumped to the $^4F_{9/2}$ level by excited state absorption process ESA2: $^4I_{13/2} +$ a photon $\rightarrow^4 F_{9/2}$ and emit red light emissions via $^4I_{9/2} \rightarrow^4 I_{15/2}$ transition.

The efficiency of the upconversion processes is determined by the probability of relaxation, absorption, transfer of electronic excitation, and electron-phonon interaction [29, 30]. These parameters are influenced by the distance between ions, the symmetry of the nearest environment, the presence of quenching impurities (including OH groups) and the transfer of energy to impurity ions, energy migration by donor ions, and the pump power density. Considering the variety of reasons described above, among all compositions researched in the article, the composition 98MgCaSr-BaYAl$_2$F$_{14}$ − 2Ba(PO$_3$)$_2$ is the most suitable one for observing luminescent processes.

5 Conclusions

The compositions of the fluoroaluminate glass with small additives of barium metaphosphate in the region of 0.5–3.0 mol.% were studied from the point of view of structural and luminescent properties. A set of methods of investigation made it possible to determine the areas of compositions characterized by formation of mixed fluorophosphate structure. One of the proofs of this can serve the inflection points on concentration dependencies of molecular refraction in the area of 1.0 mol.% and 2.0 mol.% Ba(PO$_3$)$_2$.

The composition of the glass 98MgCaSrBaYAl$_2$F$_{14}$ − 2Ba(PO$_3$)$_2$ is distinguished not only by the absence of absorption bands of OH groups on the IR transmission spectra, but also by the minimum value of Landau–Placzek ratio.

For the ErF$_3$ doped samples NIR luminescence spectra with the full width at half-maximum of $\sim$65 nm were observed. This emission corresponds to the $^4I_{13/2} \rightarrow$ $^4I_{15/2}$ transition of $Er^3{}^+$ ions. Moreover, green and red upconversion luminescence at around 523, 546, and 660 nm wavelength were observed. Upconversion emissions were explained by the two-photon process: excited state absorption (ESA), ground state absorption (GSA) and cross relaxation (CR).

Based on the evaluated results we can suggest that the composition 98MgCaSr-BaYAl$_2$F$_{14}$ − 2Ba(PO$_3$)$_2$ glass is a suitable solid state material for visible and NIR applications.

Acknowledgments. We are grateful to V. A. Aseev for the luminescence measurements and to A.V. Anan'ev for RMBS measurements of the studied samples.

References

1. Markov, V.A., et al.: Adhesive As-S-Se-I immersion lenses for enhancing radiation characteristics of mid-IR LEDs operating in wide temperature range. Infrared Phys. Technol. **78**, 167–172 (2016)
2. Lee, J.H., et al.: Thermal properties of ternary Ge–Sb–Se chalcogenide glass for use in molded lens applications. J. Non-Cryst. Solids **431**, 41–46 (2016)
3. Gibson, D., et al.: Layered chalcogenide glass structures for IR lenses. In: Proceedings SPIE 9070, Infrared Technology and Applications XL, Proceedings SPIE, vol. 9070, p. 90702I-5 (2014)
4. Simondi-Teisseire, B., Viana, B., Vivien, D., Lejus, A.M.: Yb^{3+} to Er^{3+} energy transfer and rate-equations formalism in the eye safe laser material Yb: Er:Ca$_2$Al$_2$SiO$_7$. Opt. Mater. **6**, 267–274 (1996)
5. León-Luis, S.F., et al.: Temperature sensorbased on the Er^{3+} green upconverted emission in a fluorotellurite glass. Sens. Actuators B **158**, 208–213 (2011)
6. Cai, Z.P., Xu, H.Y.: Point temperature sensor based on green upconversion emission in an Er:ZBLALiP microsphere. Sens. Actuators A: Phys. **108**, 187 (2003)
7. Manzani, D., et al.: A portable luminescent thermometer based on green up-conversion emission of Er^{3+}/Yb^{3+} co-doped tellurite glass. Sci. Rep. **7**, 41596 (2017)
8. Guery, J., et al.: Corrosion of uranium IV fluoride glasses in aqueous solutions. Phys. Chem. Glasses **29**, 30–36 (1988)
9. Fujiura, K., Hoshino, K., Kanamori, T., Nishida, Y., Ohishi, Y., Sudo, S.: Technical Digest of Optical Amplifiers and Their Applications, Davos, Switzerland, 15–17 June 1995. Optical Society of America, Washington DC (1995)
10. Bouaggad, A., Fonteneau, G.: J. Lucas Mater. Res. Bull. **22**, 685 (1987)
11. West, G.F., Hofle, W.: Non-Cryst. Solids **213–214**, 189 (1997)
12. Dmitryuk, A.V.: Tagil'tseva, N.O., Khalilev, V.D.: Terbium doped fluoroaluminate glasses. Glass Ceram. **54**(3–4), 69–72 (1997)
13. Bocharova, T.V., Karapetyan, G.O.: Tagil'tseva, N.O., Khalilev, V.D.: Investigation of the effect of barium metaphosphate additives on the structure of fluoroaluminate glasses by optical and EPR spectroscopy. Glass Phys. Chem. **27**(1), 48–53 (2001)
14. Bocharova, T.V., Karapetyan, G.O.: Tagil'tseva, N.O., Khalilev, V.D.: Optical properties of gamma irradiated Ba(PO$_3$)$_2$ containing fluoroaluminate glasses. Inorg. Mater. **37**(8), 857–862 (2001)
15. Sirotkin, S.A., et al.: Spectroscopic properties of the glass of fluoroaluminate systems with small additives of barium metaphosphate activated with the ions of rare-earth elements. Glass Phys. Chem. **41**(3), 265–271 (2015)
16. Anan'ev, A.V., et al.: Origin of rayleigh scattering and anomaly of elastic properties in vitreous and molten GeO$_2$. J. Non-Cryst. Solids **354**(26), 3049–3058 (2008)

17. Maksimov, L., et al.: Inhomogeneous structure of inorganic glasses studied by Rayleigh, Mandel'shtam-Brillouin, Raman scattering spectroscopy, and acoustic methods. In: IOP Conference Series: Materials Science and Engineering, vol. 25, p. 012010 (2011)
18. Yan, Y., Faber, A.J., Wall, H.: J. Non-Cryst. Solids **181**, 283 (1995)
19. Dai, S., et al.: Concentration quenching in erbium-doped tellurite glasses. J. Lumin. **117**, 39 (2006)
20. Yan, D., et al.: Investigation of the mechanism of upconversion luminescence in Er^{3+}/Yb^{3+} co-doped Bi2Ti2O7 inverse opal. Chin. Opt. Lett. **11**, 041602 (2013)
21. Anan'ev, A., et al.: Multicomponent glasses for electrooptical fibers. J. Non-Cryst. Solids **351**(12–13), 1046–1053 (2005)
22. Yan, Y.C., Faber, A.J., de Waal, H., Kik, P.G., Polman, A.: Appl. Phys. Lett. **71**, 2922 (1997)
23. Hamzaoui, M., Soltani, M.T., Baazouzi, M., Tioua, B., Ivanova, Z.G., Lebullenger, R., Poulain, M., Zavadil, J.: Optical properties of erbium doped antimony based glasses: promising visible and infrared amplifiers materials. Phys. Status Solidi B **249**(11), 2213–2221 (2012)
24. Linganna, K., Rathaiah, M., Vijaya, N., Basavapoornima, C., Jayasankar, C.K., Ju, S., Han, W.-T., Venkatramu, V.: 1.53 µm luminescence properties of Er^{3+}-doped K-Sr-Al phosphate glasses. Ceram. Int. **41**(4), 5765–5771 (2015)
25. Moorthy, L.R., Jayasimhadri, M., Saleem, S.A., Murthy, D.V.R.: Optical properties of Er^{3+}-doped alkali fluorophosphate glasses. J. Non-Cryst. Solids **353**(13–15), 1392–1396 (2007)
26. Ronchin, S., et al.: Erbium-activated aluminum fluoride glasses: optical and spectroscopic properties. J. Non-Cryst. Solids **284**(1–3), 243–248 (2001)
27. Zou, X., Izumitani, T.: Spectroscopic properties and mechanisms of excited state absorption and energy transfer upconversion for Er^{3+}-doped glasses. J. Non-Cryst. Solids **162**(1–2), 68 (1993)
28. Nadort, A., Zhao, J., Goldys, E.M.: Lanthanide upconversion luminescence at the nanoscale: fundamentals and optical properties. Nanoscale **8**(27), 13099–13130 (2016)
29. Lo Savio, R., et al.: J. Appl. Phys. **106**, 043512 (2009)
30. Auzel, F.: Upconversion and anti-stokes processes with f and d ions in solids. Chem. Rev. **104**(1), 139–173 (2004)

Dynamic Data Packaging Protocol for Real-Time Medical Applications of Nanonetworks

Rustam Pirmagomedov[1]([✉]), Mikhail Blinnikov[1], Ruslan Glushakov[2],
Ammar Muthanna[1], Ruslan Kirichek[1], and Andrey Koucheryavy[1]

[1] St. Petersburg State University of Telecommunication,
St. Petersburg 193232, Russian Federation
lts.pto@yandex.ru
[2] Military Medical Academy, St. Petersburg 194044, Russian Federation

Abstract. The article considers the problems of optimization of the load on communication channel, created by the traffic of a medical nanonetworks application. Previously developed operating modes were analyzed and evaluated, new protocol was suggested. This protocol allows to optimize the load on communication channel. Also simulation of traffic was carried out, result of simulation allows to evaluate the advantages of the developed protocol.

Keywords: Medical networks · Body area networks · Traffic modeling · E-health · Nanonetworks

1 Introduction

Development of the Internet of Things [1,2] and nanotechnologies allow to reach a lot of achievements in healthcare. Recent developments allow to create applications to control a lot of the physiological parameters of organism simultaneously [3,4].

Integration of healthcare and information technologies was called e-health (electronic health). E-health includes a large number of services, and telemedicine is one of them. The main purposes of telemedicine are diagnosis and healing. These processes include telemonitoring, which provides results of health indicators measurement to medical personnel [5].

Telemedicine applications work using wireless sensor and actuatory networks. These networks can be installed closely to patients, on their bodies or inside their organisms. Thus, the state of health and conditions of environment are controlled.

Medical networks [6,7] consist of following basic elements (Fig. 1):

1. sensor nodes, which measure different health indicators;
2. actuatory nodes, which effect on some processes in the organism;
3. body control unit (BCU), which collect information from sensors and transmit it to the remote server, and sends commands to actuators;

O. Galinina et al. (Eds.): NEW2AN/ruSMART/NsCC 2017, LNCS 10531, pp. 196–205, 2017.
DOI: 10.1007/978-3-319-67380-6_18

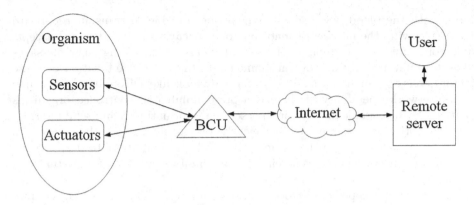

Fig. 1. The structure of the medical network application

4. remote server, analysis data and allows data access for users. In this case, the medical officer and the patient will act as a user.

Today, medical applications are realized mostly by already known technologies such as 4G, Wi-Fi, BLE, ZigBee and 6LoWPAN [8–12]. The most interesting and perspective implementation method of medical networks are nanonetworks. There are many developments in this field [13–15] but some of the issues need to be decided.

Medical apps of nanonetworks are based on usage of a huge amount of nanodevices (sensors and actuators) placed in the human body. Sensory devices are used for gathering information about the processes (to measure physical parameters) in the body, then this data (measurements) transmits to the remote server, where processing. Given the extremely limited capabilities of nanodevices that can not directly connect to existing network infrastructure (using Wi-Fi, 4G, Bluetooth, etc.), for communication with nanodevices is proposed to use an intermediate device Body Control Unit (BCU) [16]. Data transfer between the BCU and nanodevices may be carried out both directly and using intermediate nanodevices (multi-hop) if the distance to the BCU is too long for direct communication.

Model of medicine nanonetwork application is presented by project "Biodriver" [3]. Medical nanonetworks operate on the principle "master-agent", and sensors collect data according to determined rules. Traffic of such application contains packages with sensors reports. If one package contain single measurement from the one sensor, then volume of service information is more than volume of useful information [17]. The number of nanomachines in the body of one person may reach several hundred, it is necessary to improve system reliability and measurement accuracy. Such sensor nodes redundancy needs to provide required reliability. This redundancy caused by breaking down some number of devices (for example, because of troubles with power supply), some of them cannot reach the destination (implantation doesn't always allow the nanomachines to be set in the right place, therefore in the same cases devices delivery performs with the blood flow), some of them doesn't able to transmit reports to BCU (receiving the messages cannot be done, because of exceeding the acceptable communication distance to BCU). They are transmitting their messages almost simultaneously in random moment of time. The combination of these factors creates a high traffic. This traffic will look like DDoS-attack and may break network operability, if there is a large number of controlled organisms (patients). Since traffic mostly contains service information than useful information, network resources are used inefficiently. Obviously, to solve these problems it is necessary to increase the volume of useful information or to decrease the amount of sensors reports. Wherein, the information must be full, actual and reliable.

In [17] it is developed data processing mode for BCU. This processing assumes accumulation of sensors reports, combination it into a big package and further transmission it to the server. Using such mode require high performance characteristic of BCU. It is clear that such mode significantly enhance efficiency of using networks resources by decreasing volume of service information, but high delay is a disadvantage of this mode. It is caused by latency for accumulation enough amount of packages before transmission it to remote server. For certain medical applications, real-time mode is a necessary condition.

There is a contradiction in the threat of network overload on the one hand and impossibility of system operation in real-time on the another hand. This article proposes control protocol of data processing, which can solve this problem.

The article has the following structure. In Sect. 2 dynamic data packaging protocol are developed, it can decrease amount of medical application traffic, and information stays full, actual and reliable. Section 3 contains results of simulation medical application operating in preliminary data processing mode. Section 4 describes simulation of medical nanonetwork traffic using the developed protocol. Section 5 contains ideas for future works and conclusion of the paper.

2   Dynamic Data Packaging Protocol

To design remote health monitoring application it is necessary to consider individual characteristics of each organism. [3] suggested to use calibration for the first system installation. For this purpose it is required to find and fix control parameters values, which describe a state of health in normal and critical conditions. To provide full information about health indicators it needs to identify periods of time, which used to send requests by appropriate sensors. When the request is received, sensor measures the indicators value and send it to BCU.

As an improvement to the existing BCUs modes, the following algorithm is proposed. It is proposed switching of opportune mode depends on situation. If there is no danger for human health BCU uses preliminary data processing mode with low traffic and high latency. In case of critical situation device switches to real-time mode, and starts to transmit data without processing. In the first case data transmission must not be excluded fully to allow user (medical officer) observe the parameters values and to predict the critical situations. Also remote

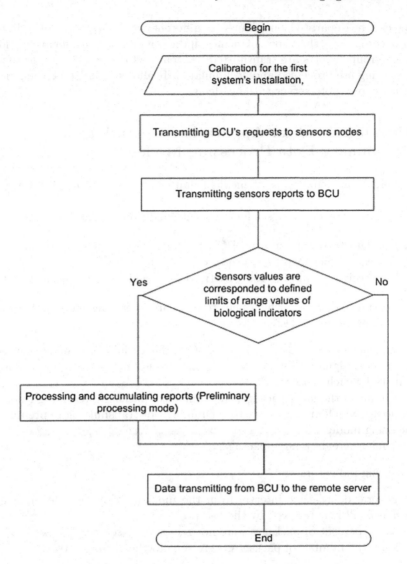

Fig. 2. Dual-mode operating algorithm

server must store all sensors reports, if previous values are requested by user (medical officer).

After finding and fixing control indicators values it is necessary to identify limits of range values, which point that situation becoming critical. For this purpose BCU compares all sensor report with critical values.

Dual-mode operating algorithm of medical application is on the Fig. 2.

Separation of sensors types on groups and developing manual method of switching the mode without processing for requesting sensors reports depending on the situation is solution, which allows to decrease the amount of packages.

Sensors can be separated on groups in different ways. For example, division occurs according to the types of health indicators or sensors location. Thus, requested group reports are transmitted in separated packages, another reports before sending are processed and accumulated. It allows to decrease the amount of messages to remote server without delays.

3 Modeling Medical Application Operating in Preliminary Data Processing Mode

For modeling and simulating of medical nanonetwork AnyLogic software was used.

Following parameters are set for constant conditions of the experiment:

- delay of data transmission from BCU to the user, $t_w = 30$ ms;
- time for each experiment, $t_{exp} = 60$ s;
- time for receiving and processing by BCU package from one sensor, $t_s = 1$ ms.

To measure indicators, which help to evaluate operating modes, there were accepted following variables:

- f - frequency of receiving packages by BCU, this variable is defined as integral parameter designated the number of sensor nodes and packages transmission intensity by each sensor node;
- V_b - volume of the large package, the number of packages, receiving from sensors nodes, which are necessary to accumulate for transmitting in preliminary processing mode;
- Δt - time to create a large package.

Measuring indicators are:

- D - general delay, time interval from the moment of reports generating by the sensor till it is received by the user;
- L - losses, percent of packages were not received by server;
- N - load, the number of packages, were sent to the remote server.

Results of experiment presented on following figures (Figs. 3 and 4) and approximating Eqs. (1–4).

$$N(V_b) = (59.8 * f_c + 20)V_b \tag{1}$$

$$D(V_b) = 0.005 * V_b + t_c p + \Delta t \tag{2}$$

During the volumes of the large package increase the delay enlarges. It happens because to transmit the message to the remote server it is necessary to accumulate more of sensors reports. Thus, calculation the general delay requires taking into account the time spent on accumulation.

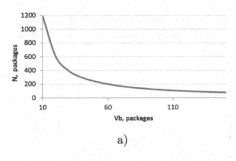
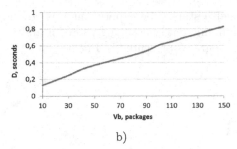

a)   b)

Fig. 3. (a) Dependence of the load on communication channel from the size of the large package; (b) dependence of the general delay from the size of the large package

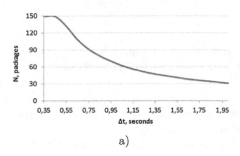
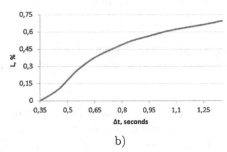

a)   b)

Fig. 4. (a) Dependence of the load on communication channel from the processing time; (b) Dependence of losses from the processing time

$$N(\Delta t) = 58.6 * \Delta t^{-1} \tag{3}$$

$$L(\Delta t) = 0.53 * ln(\Delta t) + 0.57 \tag{4}$$

During the second part of experiment, there are losses caused by insufficient capacity of the computing device BCU. To get rid of data losses it is enough to increase the amount of the buffer memory of BCU. Thereby, BCU must satisfy the condition (5) obtained by analyzing the results of the experiment:

$$V_{buff} > v_0 * (ceiling(f * D)) \tag{5}$$

V_{buff} - the buffer size required to operate the application without loss, v_0 - the package size transmitted to sensors nodes.

Provided equations can be useful for medical application design.

4   Simulation of Medical Nanonetwork Application Using the Protocol of Dynamic Data Packaging Protocol

Using the developed model the efficiency of the created protocol was evaluated. The model simulates one thousand users of application, 4% of them are in critical health state. The experiment was conducted for two modes: operated with use dynamic data packaging protocol and with preliminary processing. Traffic flow histograms presented in Fig. 5

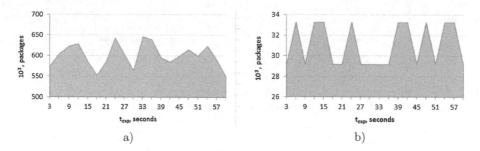

a)   b)

Fig. 5. Traffic flow of medical application of nanonetworks operating in mode (a) without preliminary processing; (b) protocol of dynamic data packaging protocol

It can be seen from Fig. 5 the number of packages in the traffic flow has decreased more than tenfold due to use of suggested protocol.

Ratio of useful information to the total volume of transmitted data is defined as the criterion for assessment efficiency of using the network resources. During the experiment, equations for this criterion for two operating BCU modes were obtained: operating without preliminary processing (6), operating with using dynamic data packaging protocol (7).

$$\frac{Q_u}{Q} = \frac{N_s * v_s}{N_s * (v_h + v_s)} = \frac{v_s}{v_h + v_s} \tag{6}$$

$$\frac{Q_u}{Q} = \frac{(1 - P) * N_b * v_b + P * N_s * v_s}{N_b * (v_h + v_b) + N_s * (v_h + v_s)} \tag{7}$$

Q - amount of transmitted information;
Q_u - amount of useful information;
P - percentage of users are in critical situation;
N_b - number of transmitted packages in a mode with preliminary processing;
N_s - number of transmitted packages in a mode without preliminary processing;
v_b - amount of useful information is contained in one package transmitted in preliminary processing mode;
v_s - amount of useful information is contained in one package transmitted in mode without preliminary processing;
v_h - header's size.

According to [17]: $v_s = 16$ bytes; $v_b = 1438$ bytes; $v_h = 62$ bytes; and accepted values: $f = 200$ packages per second; $\Delta t = 50$ ms; $V_b = 80$ packages; were obtained

Table 1. Traffics simulation results of medical nanonetworks application for three operation modes

	Without preliminary processing	With using dynamic data packaging protocol	With preliminary processing
N_s, 10^3 packages	11995	479.8	-
N_b, 10^3 packages	-	143.002	149
Q_u, GB	0,19192	0,19072	0,19072
Q, GB	0,93561	0,22933	0,19996
Q_u/Q, %	20,51	83,16	95,38

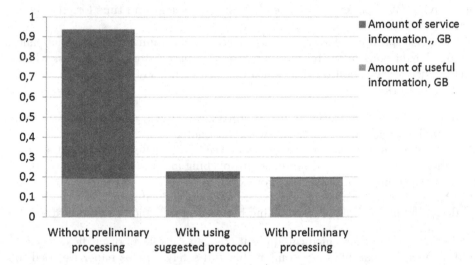

Fig. 6. Ratio of useful information to the total volume of transmitted data for three modes

following results (Table 1) and histogram of efficiency (Fig. 6). For comparison of efficiency for different modes, the simulation results for mode with preliminary processing were also presented.

During the packages transmission in preliminary processing mode and mode using dynamic data packaging protocol value of Q_u less than in mode without preliminary processing because sensor report accumulated in buffer, but quantity of such packages were not enough big for creating large package at the end of experiment.

It can be seen from the results, that developed protocol allows to decrease the load on communication channel and reduces the amount of service information (the volume of service information reaches the level of preliminary processing mode). At the same time, transmitted data stays actual, full and reliable.

5   Conclusion

Today, medical applications of IoT are developing rapidly. Such applications can help to maintain healthy lifestyle and to increase the overall vitality level. Nanonetwork's medical applications include achievements of nanotechnologies, medicine and telecommunication, thereby it attracts more and more researchers.

Except synergy issues, communications problems exist, in particular, it includes problems of efficient using of network resources. One of them was considered in this article. A new protocol was suggested, it combines advantages previously developed modes: decreases the networks load and avoids additional delay in critical situations; imitation model was designed, and three operating modes were evaluated. According to results of simulation it were obtained equations which describes networks characteristics for different variables (frequency of receiving BCU packages, volume of the large package and time for creating a large package).

Designed protocol allows decrease load on communication channel, and obtained equations allow to calculate required parameters of systems, which provide necessary condition (losses, load on channel, delay).

Future works:

1. improvement of the designed model (future model will consider probabilistic variables, which affect on the network parameters);
2. finding optimal characteristics of device (BCU), which will operate in accordance with the developed protocol; prototyping and practical testing of BCU using dynamic data packaging protocol;
3. finding criteria for creating groups of biological indicators;
4. designing manual mode of switching between groups biological indicators.

Development of new medical applications can raise level of medical service, and it makes human life more comfortable and safely. In this paper we tried to make one more step towards this aim.

This paper was supported by the RFBR grant, project No. 16-37-00215 "Biodriver".

References

1. Iera, A., Floerkemeir, C., Mitsugi, J., Morabito, G.: The internet of things. IEEE Wirel. Commun. **17**(6), 8–9 (2010)
2. Vision and challenges for realizing the internet of things. Eur. Comm. (2010)
3. Kirichek, R., Pirmagomedov, R., Glushakov, R., Koucheryavy, A.: Live substance in cyberspace biodriver system. In: Proceedings, 18th International Conference on Advanced Communication Technology (ICACT) 2016, Phoenix Park, Korea, pp. 274–278 (2016)
4. Pirmagomedov, R., Hudoev, I., Kirichek, R., Koucheyavy, A., Glushakov, R.: Analysis of delays in medical application of nanonetworks. In: 2016 8th International Congress on Ultra Modern Telecommunication and Control System and Workshops (ICUMT), pp. 80–86 (2016)

5. International Telecommunication Union: Implementing e-Health in Developing Countries: Guidance and Principles. Accessed 15 Apr 2012
6. Kwak, K.S., Ullah, S., Ullah, N.: An overview of IEEE 802.15.6 standard. In: Proceedings of 3rd International Symposium, Applied Science in Biomedical and Communication Technologies (ISABEL), 7–10 November, Rome, Italy (2010)
7. Kwak, K.S., Ameen, M.A.: Current Status of the Proposed IEEE 802.15.6 WBAN MAC Standardization (2011)
8. Culler, D.E., Hui, J.: 6LoWPAN Tutorial. IP on IEEE 802.15.4 Low-Power Wireless Networks. Arch-Rock (2007)
9. www.zigbee.com
10. www.wi-fi.org
11. IEEE 802.15.1-2005 Part 15.1: Wireless medium access control (MAC) and physical layer (PHY) specifications for wireless personal area networks (WPANs)
12. Stasiak, M., Glabowski, M., Wisniewski, A., Zwierzykowski, P.: Modelling and Dimensioning of Mobile Wireless Networks from GSM to LTE. Wiley, Hoboken (2010)
13. Jornet, J.M., Akyildiz, I.F.: Graphene-based plasmonic nano-antenna for terahertz band communication in nanonetworks. IEEE J. Sel. Areas Commun. **31**(12), 685–694 (2013). Supplement Part 2
14. Kuran, M.S., Tugcu, T., Edlis, O.: Calcium signaling overview and research directions of a molecular communication paradigm. IEEE Wirel. Commun. **19**(5), 20–27 (2012)
15. Petrov, V., Balasubramaniam, S., Koucheryavy, Y., Skurnik, M.: Forward and reverse coding for bacteria nanonetworks. In: 2013 First International Black Sea Conference on Communications and Networking (BlackSeaCom), pp. 74–78 (2013)
16. Akyildiz, I.F., Jornet, J.M.: The internet of nano-things. IEEE Wirel. Commun. **17**(6) (2010)
17. Pirmagomedov, R., Hudoev, I., Shangina, D.: Simulation of medical sensor nanonetwork applications traffic. In: Vishnevskiy, V.M., Samouylov, K.E., Kozyrev, D.V. (eds.) DCCN 2016. CCIS, vol. 678, pp. 430–441. Springer, Cham (2016). doi:10.1007/978-3-319-51917-3_38

Nano Communication Device with an Embedded Molecular Film: Electromagnetic Signals Integration with Dynamic Operation Photodetector

Dmitrii Dyubo and Oleg Yu. Tsybin[✉]

Peter the Great St. Petersburg Polytechnic University, St. Petersburg, Russia
doobinator@rambler.ru, otsybin@rphf.spbstu.ru

Abstract. Nano-devices with embedded molecular films appear as promising nano communications tools. In general, processing in such hybrid nano-device consists of initial energy of both activated electronic and atomic subsystems of an embedded molecular species, followed by intra- and inter-molecular nanoscale signals generation and reception of the transmitted signal at remote site. Actually, electromagnetic radiation can establish wireless nano communication in Visible, Infrared, and Microwave frequency ranges. In this manuscript, we estimate some properties of a microelectronics receiver in such systems, presented by novel dynamic operation pin photodetector, possesses signal integration, and required for receiving and demodulation of an electromagnetic wave. Revealed both theoretical and experimental characteristics of a photodetector have allowed designing dynamic process scenarios, as well as its possible applications inside a nano network.

Keywords: Nanotechnology · Nano device · Nano communication · Nano electromagnetic field · Molecular · Photodiode

1 Introduction

In field of emerging nano communication device technology, situated are hybrid semiconductor nano-devices, which contain embedded molecular films. Based on novel nanomaterials, such tools presented by molecular electron devices, nano-memories, biosensors, medical biochips, molecular transmitters, nano-batteries, nanoscale energy harvesting systems, and other [1–3]. Thus employed molecular mono- or multi-layer ordered films on solid-state target surface usually consist of organic molecules or bio-molecules, e.g., peptides and proteins. So-called "bottom-up" case consists of an electro conductive solid-state surface covered by a molecular layer within small operation zone ~0.1–10 nm. In contrast, a microelectronic "top-down" device contains much larger, up to or exceeding 1000 nm, solid-state surface assembled by molecules. Recently, more mechanisms of fast energetic excitation revealed in both cases of nano-devices with an embedded molecular film, resulting in nano-electromagnetic field and communication scenario [4–6]. Some agents could be applied for activation: electromagnetic field (EMF), acoustic phonons, nano-mechanics, and molecular transport. At least two

© Springer International Publishing AG 2017
O. Galinina et al. (Eds.): NEW2AN/ruSMART/NsCC 2017, LNCS 10531, pp. 206–213, 2017.
DOI: 10.1007/978-3-319-67380-6_19

of them, nano EMF and molecular transport, are directly appropriate for wireless nano communication [4]. By definition, EMF radiation appears when charge motion occurs within a molecular unit. In scientific literature, physical principles of both intra- and inter-molecular excitation, and target surface plasmons as well are under consideration widely. Surface plasmons are oscillations in the density of electrons generated by EMF excitations on electro conductive target surface. In that contest, many already existing nanoscale communication systems operate by similar ways even without molecular species [6]. Excited transport of protons, electrons, holes, and other quantum carriers could create communication channels between parts of the same molecule and/or connect a few domains inside a molecular network. Charge transfer mechanisms occur on a temporal scale that precedes notable nuclear motion. Fast dynamics of amplitude and orientation of the molecular electric dipole moment generate EMF in environmental space. From a communication perspective, the unique properties observed in molecular nanomaterials will decide on the specific bandwidths for emission of electromagnetic radiation, the time lag of the emission, or the magnitude of the emitted power for a given input energy. Control of dipole motion by external variations can actually lead to metamaterial-based nano networks [5]. Since nano communication realizes transmission and reception of electromagnetic radiation from nano components, practical nano networks can be developed with molecular metamaterials and their applications.

Nano communication receivers require special determination, investigation and design, including low-level signals acquisition inside micro- or nano-volume, denoising, and demodulation. For example, the analytical devices are widely demand, where molecular films are activated by electromagnetic waves, and reaction products are perceived by mass analyzer, spectrum analyzer, biochemical sensor transducer, surface wave phase meter, CMOS and charge-coupled device matrix, photodetectors, photomultipliers [7, 8]. Notably, comprehensive description of signal receiving phenomena for nano communication networks is still missing.

Vacuum photomultipliers have the highest performance for photon registration due to high sensitivity, wide bandwidth, low dark current and not too high noise level. However, it is available only in macroscale circuit with high voltage (kVs). Common semiconductor photodetectors possess a high noise level, demand high sensitivity broadband amplification, analog-to-digital conversion, and integration.

Recently, new high sensitivity low noise photodetector-integrator based on a hybrid PIN diode was developed, possesses signal integration properties required for receiving and demodulation of an electromagnetic wave in nano communication systems [9, 10]. The hybrid PIN dynamic operation photodetector (DPD) is gated device, having high quantum efficiency at visible light. Measuring photocurrent time delay rather than its magnitude provides a new effective light detection method. The forward current magnitude is controlling only by forward voltage and independent of the radiation intensity. Supposedly, this device has good prospects for use in visible light low signal registration systems, including activated molecular films radiation. We suggest here, that DPD has promising applications in low power radiation detectors, including nano communicators, molecular luminometers, spectral analyzers and other similar devices [9, 10]. However, not all fundamental limiting factors of DPD in nano communication network were investigated previously, which holds back the practical application of the device. Hence, the investigation of triggering time characteristics is of fundamental importance to the performance of the DPD. The objectives of that work are experimental determination of DPD temperature characteristics, as well as theoretical study of underlying physical phenomena. Here, we develop a physical model of over-barrier current that was proposed and confirmed using the computer modeling software. The obtained data have a discrepancy with the previous results demonstrated for instance in [10]. Based on obtained results, one can ascertain the potential applications of the receiver, including wireless nano network architectures operating in a wide frequency range [2, 11].

2   Methods

The experimental setup consisted of light-protected thermostat, where the thermocouple detector installed with thermal contact to the diode. The self- triggering time of the diode was measuring by a computer programmable microelectronic circuit, and controlled by digital oscilloscope Tektronix TDS684 simultaneously, to avoid the circuit temperature effects. A short pulse forward bias by switching from reverse bias creates a forward current flow after essential time delay τ_{trig} in microsecond- millisecond scale. Time delay τ_{trig} is a function of the absorbed radiation. According to the proposed model, the direct current switch on when a certain amount of electrons N_{crit} accumulates under gate in the barrier area [9, 10]. The time delay is defined as follows: $\tau = N_{crit}/(G + G_{dark})$. G is a coefficient proportional to the absorbed light power, G_{dark} is the self-generation or dark current coefficient. The dependence of the inverse triggering time $1/\tau_{trig}$ on the incident light power, according to such estimating, is linear function. The critical parameter, which is of fundamental importance, is the self-triggering temperature dependence caused by the dark current only, in the absence light. The dark current is formed by electron-hole pairs thermal generation in the substrate ($I_{thermal}$) and by cathode to anode leakage.

3   Experimental and Theoretical Results

3.1   Experimental Data

At pre-measuring stage, an anode and gate of the diode are keeping under negative bias voltage. The measuring cycle is starting when anode positive voltage applied over the negative bias. Figure 1 shows the measured dependences, as well as the theoretical dependences. Typical experimental inverse triggering time $1/\tau_{trig}$ temperature dependences are presented in the graph.

The minimum (at temperature T = 320 K) and the maximum (at 275 K) triggering time τ was approximately 150 ns and 150 ms, which corresponded to a wide dynamic range of $\tau_{trig\ max}/\tau_{trig\ min}$ up to 10^5–10^6.

Obviously, the curves slopes of the theoretical dependence [10] (dotted line) and the experimental one are significantly different. This discrepancy requires a theory revision. The solution was to apply the Richardson's low (dash line) which approximates precisely the experimental curve.

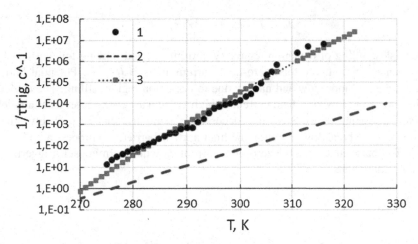

Fig. 1. Self- triggering time dependence of the DPD on temperature on a logarithmic scale. 1-experimental data, 2- theory [10] (dash line), 3- Richardson's low (dotted line). Forward anode pulsed voltage $V_{an} = 1\,V$, direct cathode and gate voltage, respectively, $V_{ct} = 0.6\,V$, $V_g = 1.2$.

In order to explain experimental facts and to identify essential physical processes, a more detailed theoretical analysis performed by computer simulation.

3.2   Simulation Results

At certain delay time after switching from reverse bias to forward, the DPD is still keeping in the closed state. The forward current strongly suppressed by means of space charge potential barriers, which formed nearby the anode due to the depletion of carriers. The triggering time τ is the neutralization time of the space charge, namely the time of electrons accumulating under the gate to full shielding. Therefore, the main process in self-switching is considered to be the electron stream formation.

When self-switching, the current consists of two components. The first one is result of electron-hole pairs thermal generation in a substrate ($I_{thermal}$), which is a common property of any solid-state photodetector.

The second component I_{leak} is due to minority carriers coming from the cathode through the junction with the p substrate, so dark current is $I_{dark} = I_{thermal} + I_{leak}$. The leakage from the cathode to the anode occurs through a low potential barrier between the p region and the substrate.

The magnitude of the applied voltage electric fields across the device quickly decrease out of embedded depletion regions. Initially, electrons and holes are trapped by electrostatic potential barriers $\Delta W(\tau)$ higher Fermi level and situated near borders of depletion regions. Due to the absence of strong DC electric field, charge carriers outside of a depletion region are quite slow. Temperature-dependent zero-field $E_{ext} = 0$ current of charge carriers flows over potential barrier (Richardson low),

$$J_0 \approx T^2 \cdot \exp(-\Delta W(\tau)/kT), \tag{1}$$

provides slow filling of depletion region by current J_{fil}, and flowing low current J_a at external anode circuit. So, total zero-field current is $J_0 \approx J_{fil} + J_a$. Potential barriers became more and more low and narrow due to depletion regions filling. The device is switching on from low zero-field current to high-current J_{high} state after a certain delay τ_{strig} (self-triggering time). Switching occurs very abruptly, that is, probably, due to tunnel avalanche breakdown through the final narrow barrier, the process amplified by positive feedback effect. So, the zero-field current J_0 will determine self-triggering, or zero-field time value,

$$\tau_{strig} \propto \frac{\Delta Q_{crit}}{J_0(\tau)} \tag{2}$$

$$\Delta Q_{crit} \propto \int_0^{\tau_{strig}} dt \cdot T^2 \cdot \exp(-\Delta W(\tau)/kT), \tag{3}$$

typically $\tau_{strig} \approx 0.1$–10 ms.

When modeling the processes using the COMSOL program, the geometry, the impurity concentration, the voltage at the electrodes of the DPD were set. The definition of the geometry and dimensions of the elements are shown in Fig. 2.

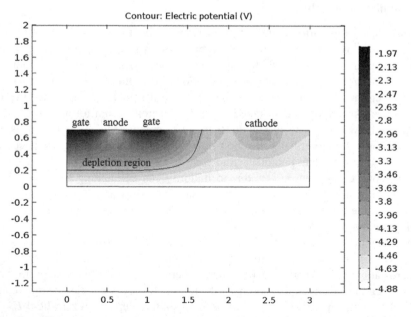

Fig. 2. Geometry and elements dimensions of the DPD in the COMSOL program. Modeling the electrostatic potential distribution in the cross section of the DPD by a direct voltage bias.

For such configuration, static spatial distributions of the potential values were obtained. The figure shows the potential distribution by forward bias, when the anode and the gate have a positive potential. Evidently, that there are electrostatic barriers trapping charges in potential wells on the trajectory of the motion of charge carriers. The charges accumulation rate determines the time of self-switching. Static potential distribution by a forward bias along the surface line of the cut (Cut Line).

Using the simulation results presented in the Figs. 2, 3 and 4 both the over-barrier and tunneling currents can be estimated. Static potential distribution has two potential barriers (Fig. 4) similar to results [9, 10]. For the first time, we conclude that discrepancy observed between time delay temperature dependence τ(T) simple theory from one side and experimental one from other side occurs due to over-barrier current mainly, approximated by Richardson low formula in form Eq. (1), and integral (3) as well. This computation gives novel insight to the further analysis of a DPD triggering mechanism.

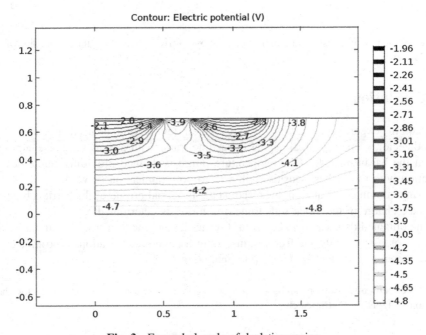

Fig. 3. Expanded scale of depletion region.

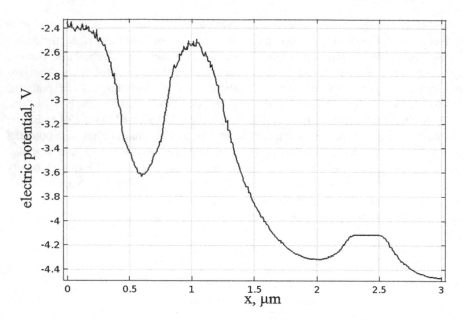

Fig. 4. Static potential distribution of the potential by a forward bias along the surface line of the cut (Cut Line).

4 Conclusion

Based on the experimental results, the investigated device could be used effectively as a photodetector, namely, a weak signal linear integrator with a dynamic noise suppression at the hardware level. The temperature range is defined providing stable device operation with the highest sensitivity in a nano communication network.

Using computer simulation, the electrostatic potential profile of the barriers holding the space charge in the diode self- triggering mode is determined.

The obtained data seems to be useful for the development of micro- and nanoscale devices possessing analytical features, including nano communications, (bio)molecular luminometers, biosensors, and spectrum analyzers.

Acknowledgements. Authors are grateful Act Light SA (Lausanne, Switzerland) for their donation of DPD for that research.

References

1. Offenhäusser, A., Rinaldi, R.: Nanobioelectronics - for Electronics, Biology, and Medicine. Nanostructure Science and Technology (2009)
2. Akyildiz, I.F., Jornet, J.M.: Electromagnetic wireless nanosensor networks. Nanocommun. Netw. **1**(1), 3–19 (2010)
3. Heath, J.R., Ratner, M.A.: Molecular electronics. Phys. Today **56**(5), 43–49 (2003)

4. Tsybin, O.: Nano-device with an embedded molecular film: mechanisms of excitation. In: Balandin, S., Andreev, S., Koucheryavy, Y. (eds.) ruSMART 2015. LNCS, vol. 9247, pp. 772–777. Springer, Cham (2015). doi:10.1007/978-3-319-23126-6_72
5. Velichko, E., Zezina, T., Cheremiskina, A., Tsybin, O.: Nano communication device with embedded molecular films: effect of electromagnetic field and dipole moment dynamics. In: Balandin, S., Andreev, S., Koucheryavy, Y. (eds.) ruSMART 2015. LNCS, vol. 9247, pp. 765–771. Springer, Cham (2015). doi:10.1007/978-3-319-23126-6_71
6. Merlo, J.M., et al.: Wireless communication system via nanoscale plasmonic antennas. Sci. Rep. **6**, 31710 (2016)
7. Awsiuk, K.: Biomolecular nanolayers on synthetic organic films: their structure and properties for biosensor applications. Ph.D. thesis, Jagiellonian University, Krakow, Poland (2013)
8. Turner, G.K.: Measurement of light from chemical or biochemical reactions. In: Bioluminescence and Chemiluminescence: Instruments and Applications, vol. I, pp. 43–78 (1985)
9. Okhonin, S., Gureev, M., Sallin, D., Appel, J., Koukab, A., Kvasov, A., Pastre, M., Polzik, E.S., Tagantsev, A.K., Uddegard, F., Kayal, M.: A dynamic operation of a PIN photodiode. Appl. Phys. Lett. **106**(3), 031115 (2015)
10. Sallin, D.: A low-voltage CMOS-compatible time-domain photodetector, device & front end electronics. Ph.D. thesis 6869, EPFL, Lausanne, Switzerland (2015)
11. Abbasi, Q.H., Nasir, A.A., Yang, K., Qaraqe, K.A., Alomainy, A.: Cooperative in-vivo nano-network communication at terahertz frequencies. IEEE **5**, 8642–8647 (2017)

A Formal Definition for Nanorobots
and Nanonetworks

Florian Büther[(⊠)], Florian-Lennert Lau, Marc Stelzner, and Sebastian Ebers

Institute of Telematics, University of Lübeck, Lübeck, Germany
{buether,lau,stelzner,ebers}@itm.uni-luebeck.de

Abstract. Nano computation and communication research examines minuscule devices like sensor nodes or robots. Over the last decade, it has attracted attention from many different perspectives, including material sciences, biomedical engineering, and algorithm design. With growing maturity and diversity, a common terminology is increasingly important.

In this paper, we analyze the state of the art of nanoscale computational devices, and infer common requirements. We combine these with definitions for macroscale machines and robots to define Nanodevices, an umbrella term that includes all minuscule artificial devices. We derive definitions for Nanomachines and Nanorobots, each with a set of mandatory and optional components. Constraints concerning artificiality and purpose distinguish Nanodevices from nanoparticles and natural life forms. Additionally, we define a Nanonetwork as a network comprised of Nanodevices, and show the specific challenges for Medical Nanorobots and Nanonetworks. We integrate our definition into the current research of Nanodevice components with a set of examples for electronic and biological implementations.

1 Introduction

Over the last 40 years, the field of nanotechnology has provided incredible new possibilities to interact with matter on the atomic and molecular level. Two important branches of nanotechnology are nano computation, concerning nanoscale computational devices, and nano communication, which investigates nanoscale information exchange. Together, these introduce the idea of machines and robots that are small enough to operate in a nanoscale environment. They may for example measure chemical environmental data, or manipulate atomic or molecular processes.

Recently, many new building parts for nanoscale devices (see Sect. 3) have been developed. With increasing variety and domain maturity, a common terminology facilitates a better comparison of nanodevice research. To this end, [1] gives a definition for nanoscale communication and [2] investigates general computational capabilities. Still, to the best of our knowledge, no common definition for nanoscale devices exists. The terms nanomachine, nanodevice, nanorobot, and nanobot are often used without differentiation, and only an intuitive description for an assumed machine model is provided.

© Springer International Publishing AG 2017
O. Galinina et al. (Eds.): NEW2AN/ruSMART/NsCC 2017, LNCS 10531, pp. 214–226, 2017.
DOI: 10.1007/978-3-319-67380-6_20

Most research agrees on a common set of components for nanomachines. These include sensors, actuators, a transceiver or antenna, a processor including memory, and a power unit [3–5]. Several approaches investigate these components in the biological domain [6,7]. Others describe nanomachines through the tasks they can accomplish, focusing on actuation, sensing and computation [8,9]. With regards to size, assumptions vary. Some papers assume an overall size of 1–100 nm [8], others consider a maximum size of a few micrometers [5]. However, neither provide a reasoning for the respective restriction. While devices of a few micrometers may be no longer nanoscale, we still feel the need to represent the corresponding literature. When talking about "nanosized" devices, we refer to the whole (vague) presented spectrum.

Next to their size, nanodevices are often described as autonomous machines or robots. Regular macroscale machines and robots are well-known research topics, and provide useful definitions for both terms [10,11].

Building upon both areas of research, we derive a precise, formal definition for nanoscale devices, machines and robots. These give way to a similar definition for nanonetworks. The definitions include the target environment of a robot or network. This in turn allows us to describe the specifics of using nanodevices in a medical application.

This paper aims for a twofold effect: First, hardware designers can precisely select and describe a machine's capabilities. Second, application and algorithm developers can identify required machine capabilities and spot possibly costly assumptions and realization difficulties.

2   Definitions

Nanoscale devices are usually characterised as small agents that chiefly interact with their physical surroundings. As such, they are less like computers as mathematical processors, but rather like controlled or autonomous robots. We thus first define normal macroscale robots, and subsequently transfer the definition to the nanoscale.

2.1   Machines and Robots

Definitions for a robot are given by Xie and the ISO Standard 8373:

"A robot is the embodiment of manipulative, locomotive, perceptive, communicative and cognitive abilities in an artificial body." [10]

"A robot [is an] actuated mechanism programmable in two or more axes with a degree of autonomy, moving within its environment, to perform intended tasks." [11]

A robot is defined by its constituting physical components, used for manipulative, locomotive, etc. aspects, as well as a set of qualities describing the design and behavior of a robot, namely that it is artificial and programmable, has a degree of autonomy, and performs intended tasks.

For the upcoming definitions, we identify two sets of relevant components, which directly correspond to the parts the respective device is made of. The first set of components covers interaction, namely *sensors S*, *actuators A*, a component for *locomotion L*, and a component for *communication C* with other devices. Second, to facilitate complex behavior, the second set includes components for *information processing I*, optionally supported by *memory M*, and a measurement of *time T*. Finally, the whole device will need a *power supply P* to power its operation. We describe these components in more detail in Sect. 2.3.

With these terms at hand, we can now give definitions for macroscale machines and robots, adopting the notions of Xie [10] and the ISO Standard [11].

Definition 1. *A machine* $\mathcal{M} = (K_{mand}, K_{opt})$ *is an artificial construct, designed to perform a predetermined actuatory task. It consists of a set of mandatory components* $K_{mand} = \{A, P\}$ *and a set of zero or more optional components* $K_{opt} \subseteq \{C, I, L, M, S, T\}$.

This definition captures the nature of machines as self-powered manipulators. For example, an excavator possesses a motor and an arm to act on its environment, thus has all mandatory components to be considered a machine. Moreover, it usually has tracks to move itself, which we consider locomotion. This neatly reflects the fact that machines may be capable of more than only manipulation, so that derived, more complex constructs are machines as well.

The qualities of machines differentiate them from natural and random phenomena: As artificial constructs, only human-made objects qualify. The requirement for a predetermined task demands intentional creation: The machine must pursue a set goal, instead of being an unintentional side effect of an unrelated process.

Definition 2. *A robot* $\mathcal{R} = (K_{mand}, K_{opt})$ *is a machine that is reprogrammable and consists of a set of mandatory components* $K_{mand} = \{A, I, M, P, S\}$ *and a set of zero or more optional components* $K_{opt} \subseteq \{C, L, T\}$.

As a subclass of machines, robots introduce additional required components and qualities. Foremost, the mandatory component for information processing enables a robot to perform complex actions. Additionally, the robot needs to be reprogrammable, so that it can switch its program logic during its deployment to perform other activities. To reprogram a robot naturally requires it to store the current programming state, thus it requires memory as well. Lastly, to adhere to the initial robot definitions given above, a robot needs to have some sort of perceptive ability, which we reflect as a mandatory sensor component.

Autonomous mobile robots are a special subclass of robots that provides useful aspects for our nanoscale definitions. They possess a mandatory component for locomotion and are autonomous, meaning that they act based on the state of their environment, and can fulfill their task without human interaction [11]. Autonomy may enable a certain degree of self-organization in networks or swarms of nanoscale devices.

2.2   Nanodevices

With the definitions of machines and robots provided, we can now define the respective nanoscale variants. In order to facilitate analysis of all kinds of nanoscale devices, we first define a suitable umbrella term.

Definition 3. *A Nanodevice $\mathcal{N}_D = (K_{mand}, K_{opt})$ is an artificial construct with an overall nanoscale size, designed to perform a predefined function in an environment Γ. It consists of mandatory components $K_{mand} = \{P\}$ and a set of zero or more optional components $K_{opt} \subseteq \{A, C, I, L, M, S, T\}$.*

Nanodevices require at least a power supply, in order to differentiate them from completely passive constructs like nanoparticles. Similar to machines, further components are optionally available, which establishes Nanodevices as the underlying term for all following definitions. Figure 1 illustrates this hierarchical approach.

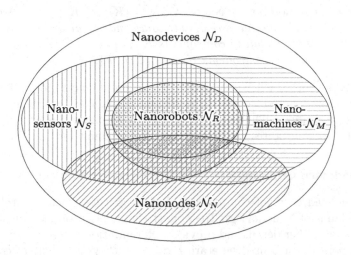

Fig. 1. The relationship between Nanodevices and the derived terms. For the formal definitions, see Definitions 3–7.

Nanodevices will operate in a possibly unusual environment, for example permanently moving along the bloodstream. In addition, the nanoscale introduces new physical effects like molecular interference and quantum effects [1], which Nanodevices need adopt to. We explicitly capture this dependency as the device's target environment Γ. It can be interpreted as a list of environmental parameters and constraints. A common constraint is the device size: As described in Sect. 2.5, the circulatory system restricts Medical Nanodevices to have a size of at most $4\,\mu\text{m}$.

Definition 4. *A Nanomachine $\mathcal{N}_M = (K_{mand}, K_{opt})$ is a Nanodevice with a set of mandatory components $K_{mand} = \{A, P\}$ and the same optional components as a macroscale machine, in an environment Γ.*

This definition interprets Nanomachines rather literally, that is, as very small machines. For example, a Nanodevice that actively assembles a specific protein constitutes a Nanomachine.

Similar to a machine, we can transfer the concept of sensor nodes, as for example employed in Wireless Sensor Networks (WSNs), to the nanoscale.

Definition 5. *A* Nanosensor $\mathcal{N}_S = (K_{mand}, K_{opt})$ *is a Nanodevice with a set of mandatory components* $K_{mand} = \{P, S\}$ *and a set of zero or more optional components* $K_{opt} \subseteq \{A, C, I, L, M, T\}$ *in an environment* Γ.

Simple Nanomachines or Nanosensors may just repeat one task, without consideration for their surroundings. However, for more complex goals they may need to adopt to environmental changes. This demands a degree of autonomy similar to autonomous mobile robots. The definition for Nanorobots includes this autonomy with respect to the environment:

Definition 6. *A* Nanorobot *or* Nanobot $\mathcal{N}_R = (K_{mand}, K_{opt})$ *is a Nanodevice that is reprogrammable, has a degree of autonomy and operates in an environment* Γ. *It consists of the mandatory components* $K_{mand} = \{A, I, M, P, S\}$ *and a set of zero or more optional components* $K_{opt} \subseteq \{C, L, T\}$.

Nanorobots adopt the lists of mandatory and optional components as well as the qualities from macroscale robots. Additionally, they extend them by autonomy as described above. The target environment Γ guides this autonomy, as it describes the possible states the Nanorobot must account for.

2.3   Nanodevice Components

Section 2.1 named a set of components that constitute machines, robots and the various kinds of Nanodevices. To support these definitions, we illustrate the components in further detail, and provide differentiation where required.

All components presented here are hardware, they are physical constructs constituting a part of the machine or robot. All software, as a kind of programming, is considered to be included in the component I, information processing.

Information Processing I. This component describes the capability of a Nanodevice to transform data. In the simplest sense, this may be a boolean operation implemented by just one transistor. Information Processing requires programmability: It must be possible to design a new Nanodevice with different behavior, given the same inputs and outputs.

Information processing relates to a robot's capability for *re*-programmability, which allows it to switch to new behavior during its lifetime. Fundamentally, reprogrammability is a kind of configuration, a bit of memory that a machine can use to guide its behavior. Reprogrammability can range from a set of bits that select behavior options up to a Turing-complete interpreter for a machine language.

Power Supply P. A mandatory component of Nanodevices is an independent energy supply. A Nanodevice may not be permanently or physically connected to an external energy supply, but has to power its components from an internal source. Two common design options are pre-charged batteries or energy harvesting mechanisms, while the latter often includes a short-term energy storage.

Communication C. As a component of a Nanodevice, communication describes the ability to send and receive environmental stimuli with the intent of exchanging information with other Nanodevices. Note that this describes only a *capability*; a Nanodevice might be able to communicate, yet never successfully exchange messages with other Nanodevices. The physical device used for communication may be similar or even equal to a common sensor or actuator component of a Nanodevice, especially in the case of biological Nanodevices using molecular communication. To differentiate actuating and sensing from communication, we examine a device's (possibly predefined) *intent*: A Nanodevice communicates if it performs an action in order to inform other devices, otherwise it acts on the environment. In the worst case, a Nanodevice may not be able to differentiate a received message from an environmental variation, for example while suffering strong molecular interference. In this situation, higher-level communication protocols must try to resolve the received signal.

Memory M. Closely linked to processing, Nanodevices may be capable of storing arbitrary data. While memory is required by many other functionalities, for example any kind of configuration or data aggregation, it is not fundamentally necessary. A Nanodevice that processes data via simple electric circuits might not require memory at all.

Actuators A and Sensors S. Actuating and sensing is a Nanodevice's capability to interact with the environment, either by measuring a physical, chemical or biological property or by manipulating it. To differentiate from communication, the *intent* to interact with the physical environment is required for sensors and actuators.

Actuators and sensors often provide continuous values like a voltage level, and will thus need an A/D converter to connect to a processing component. Even though this converter can be very complex depending on the required precision, we interpret it as part of the sensor or actuator, rather than an individual component.

Locomotion L. Nanodevice mobility can be divided into active and passive mobility. Passively mobile Nanodevices diffuse randomly in a medium, and usually do not require specific capabilities to do so. Active mobility—Locomotion—allows a Nanodevice to move deliberately, and thus requires an active component. As a simple case, a Nanodevice may move randomly within a liquid medium. Locomotion may make use of externally supplied phenomena to enable motion, for example a magnetic field as in [12].

Time T. Internal clocks are an omnipresent part of all computing devices and usually taken for granted. However, as their availability at the nanoscale is not certain yet, we need to consider devices without precise timing information. We classify three levels of timing that may be present in a Nanodevice: 1. Relative ordering, as given by the happened-before relation [13], 2. relative time, which enables a Nanodevice to measure the time difference between two events, and 3. absolute time, which provides a date-like timestamp with a given precision.

2.4   Nanonetworks

Due to the size limitations, Nanodevices will often need to collaborate to achieve a given task [3]. For example, a set of Nanosensors may detect a viral infection and then communicate the fact to surrounding Nanomachines, which in turn produce or release an antibody suitable to combat the infectious agent.

The essential basis for a Nanonetwork is the capability of its participants to communicate: It allows Nanodevices, similar to regular networked computers, to exchange information and collaborate towards a common goal.

Definition 7. *A* Nanonetwork *is a directed ad-hoc graph* $G = (V, E)$, *where* V *is a set of* Nanonodes $\mathcal{N}_N$, *operating in an environment* Γ. *Nanonodes are Nanodevices with communication as an additional mandatory component* $C \in K_{mand}$.

Like a regular network, a Nanonetwork can be modeled as a directed graph, where each vertex represents a Nanonode and the edges starting from that vertex correspond to the ability of the Nanonode to send messages to these other devices.

The environment Γ may directly influence the nature of this graph. In an unstable or mobile environment, the network structure and thereby the set of edges E may vary over time. Furthermore, due to the lack of additional network infrastructure, Nanonetworks need to form in an ad-hoc manner.

2.5   Medical Nanorobots and Nanonetworks

Medical Nanorobots exist in a more specific environment, which constraints the maximum size of Nanodevices, for example to less than 4 micrometers—the diameter of the smallest capillary in the human body [14]. For intracellular deployment, an even smaller size might be required. For applications inside the human digestive system, much bigger robots might be applicable.

Definition 8. *A* Medical Nanorobot $\mathcal{R}_M$ *is a Nanorobot whose environment* Γ *includes the environments occurring within the human body.*

These environments include the bloodstream as a distribution route, organ or interorgan tissue, and optionally the intracellular space.

The medical environment poses additional requirements to environmental compatibility: A Medical Nanorobot must not appear as a threat to the body's immune system. Furthermore, after it has performed its intended action, it must be disposed of properly to avoid accumulating waste inside the body.

It is often infeasible to position the Nanorobots manually within the body, as the target area is difficult to reach or an insertion is medically prohibitive. The body's own circulatory system provides a convenient delivery route, as it covers nearly the whole body. We thus define a Medical Nanonetwork as follows:

Definition 9. *A Medical Nanonetwork is a Nanonetwork in which the comprising Nanonodes are deployed in an environment Γ that respects the constraints occurring within the human body.*

A Medical Nanonetwork is a Nanonetwork that is explicitly designed to operate inside a human body, for example to precisely detect a pulmonary infection [15]. Like regular Nanonetworks, the Nanodevices in a Medical Nanonetwork are Nanonodes, thus mandatorily have communication capabilities.

3   Exemplary Implementations

In order to illustrate the components of Nanodevices and survey the feasibility of a practical implementation, we identify promising approaches for Nanodevice construction. These group into the two major research areas, namely electronic and biological approaches. Nevertheless, these are not intended to provide a definite separation, as hybrid solutions may be equally feasible.

3.1   Electronic Nanodevices

Nanoelectrical approaches transfer the construction principles of electronic devices to the nanoscale. While nanotechnology already investigated nanoscale construction for several decades, the discovery of Carbon Nanotubes [16] provided a vital set of new possibilities for nanoelectrical components. Examples are transistors and memory circuits M [17] and electromagnetic communication C via terahertz antennas [18]. Figure 2 shows a schematic example implementation of an artificial electronic Nanorobot.

The assembly of electronic Nanodevices will require some sort of infrastructure, to provide electric connections between components and the physical rigidness to withstand mechanical stress. This may be implicit in the construction process, for example in the case of self-assembly [19], or explicit, if the Nanodevice receives an additional casing.

Information processing I for Nanodevices can be solved in several ways. Transistors can be constructed as small as a single atom [20]. Alternate approaches employ Quantum Dots [21], which address several physical problems of nanoscale constructions. These also provide an approach to provide memory M [22].

Electromagnetic signals in the terahertz range suffer from strong path loss, rendering them viable for communications up to 2 mm [23]. Alternatively, Nanonodes may use environmental structures to communicate. In a biological deployment, Nanonodes can attach to a neural network [24], existing or custom-grown, to exchange signals.

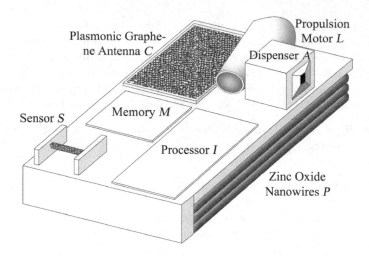

Fig. 2. An example of an electronic Nanodevice with the components $\{A, C, I, L, M, P, S\}$.

The size of batteries P is a notorious problem of miniaturization: Batteries are mass-wise the largest part of current wireless sensor devices, and much research investigates energy-saving techniques at all hardware and software levels. An approach for energy storage may involve ultra-nano-capacitors [25]. Still, they will need to be supported by energy harvesting mechanisms. A promising approach employs piezoelectric nanowires of currently about 3–4 µm in length [12].

Carbon Nanotubes can also serve as a sensing component S. Arranged between a cathode and an anode, this nanowire works as a field effect transistor. It is affected by its environment and shows fluctuations in its conductivity: A physical sensor may detect mechanical force bending the wire. A chemical sensor detects changes in gas compositions, and a biological sensor possesses hybrid receptors attached to the wire that react to specific proteins [26].

To interact purposefully with the environment, a Nanomachine has one or more actuators. For example, a droplet dispenser A releases a controlled amount of molecules into the environment [27].

With regard to locomotion L, [28] provides examples for micro- and nanosized motors. For example, bubble propulsion motors are conical objects with at least two layers. The outer layer provides a solid case for the motor, while the inner is a highly reactive sheet in combination with the surrounding liquid. As a result of the chemical reactions in the inner part, oxygen-bubbles escape through the wider end of the cone, creating propulsion.

To our knowledge, no specific work has been carried out so far to explicitly construct nanoscale clocking or timing mechanisms. It seems likely that the piezoelectric effect used for quartz crystal can be adapted alike to the above mentioned nanowires. However, verification is still required.

3.2   Biological Nanodevices

The second approach to Nanorobot construction is inspired by nature. Cells already exhibit many required characteristics, for example a mechanism for chemical energy supply P through adenosine triphosphate (ATP) fabricated in the cell's mitochondria. Similarly, the natural process of molecular communication is well suited for information exchange C. Figure 3 shows an example of a cell adopted as a biological Nanorobot including the respective components.

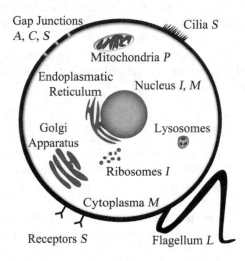

Fig. 3. Example of an eucaryotic cell as a biological Nanodevice with the components $\{A, C, I, L, M, P, S\}$.

Two methods to construct cell-like Nanodevices can be identified: Adaptation of natural cells, for example *E. Coli* bacteria [29], or construction of synthetic cells [30]. Both exploit the natural cell structure as scaffolding, thus providing stability from cell membranes, as well as component connectivity through intra-cell communication C.

Information processing I, especially with reprogrammability, is difficult to map to biological Nanorobots. While bio-chemical processes within a cell are extremely complex, it is non-trivial to adjust them to perform an intended task. Some approaches impose known concepts from circuit design onto DNA processing [31,32] located in the cell's central processing unit, the nucleus. After transcribing the DNA to RNA, the ribosomes carry out protein synthesis. Manipulating the DNA changes the cell's protein production behavior used for actuation A and communication C, and thus may serve as a kind of reprogrammability. In this way, the cell's nucleus serves as long term memory M by reading and writing DNA strands. The cytoplasma holds the current amount of proteins as a quickly changing working memory.

Some cells communicate by producing or detecting a quantity of molecules C, which they subsequently release or absorb through gap junctions in the cell's membrane [33]. Via diffusion, these molecules may reach a recipient cell at a distance of several nanometers up to more than a meter [6]. Instead of communicating with Nanodevices, the same diffusion channels can be used for sensing of or actuation on the environment A, S.

Sensing components S on the cell's outside can as well adopt natural mechanisms like antigen receptor binding [34] or cilia serving as sensing antennas. Similar to the cilia are flagella, which provide locomotion capabilites L to the cell [33].

Many of the cells organelles not yet mentioned are a vital part of the cells infrastructure, for example the endoplasmatic reticulum or golgi apparatus. They serve as intra-cell transportation, packaging and coordination organelles.

To our knowledge, no biological component to precisely estimate time exists. While it may be possible to adopt natural processes, these are of unclear precision compared to macroscale clocks.

4　Conclusion

In this paper, we present definitions for Nanodevices, Nanomachines, Nanosensors, Nanorobots, and Nanonetworks. The definitions focus on the physical components of a device along with a set of qualities and the device's environment. An incremental set of mandatory components creates an inclusion hierarchy: Nanorobots are a more specific class of Nanodevices that require components for processing, sensing and acting and which are reprogrammable and autonomous to a degree.

Our definitions build on classical machines and robots, unifying the nanoscale terminology with regular machinery. The distinction of machines, sensors and robots also arises at the nanoscale, further helping to distinguish the defined terms.

With the current trend towards increasing maturity, nanodevice research will benefit from a formal definition: A common understanding facilitates communication about nanoscale devices. Further research can avoid implicit assumptions, as the presented model illustrates expectable capabilities and the components required for them. This also helps to evaluate new nanoscale algorithms and architectures, as well as to design simulations and tests for them.

We have not yet covered the application of the Nanodevice definitions to concrete challenges posed by, for example, medicine. Future research will need to investigate the various kinds of environments $\varGamma$ Nanorobots will likely encounter.

Acknowledgements. We thank Regine Geyer and Kim Scharringhausen for much constructive discussion on biological phenomena.

References

1. Bush, S.F., Paluh, J.L., Piro, G., Rao, V., Prasad, V., Eckford, A.: Defining communication at the bottom. IEEE Trans. Mol. Biol. Multi-Scale Commun. **1**(1), 90–96 (2015)

2. Lau, F., Büther, F., Gerlach, B.: Computational requirements for nano-machines: there is limited space at the bottom. In: 4th ACM International Conference on Nanoscale Computing and Communication, ACM NanoCom 2017, Washington D.C., USA, September 2017 (in press)
3. Akyildiz, I.F., Brunetti, F., Blázquez, C.: Nanonetworks: a new communication paradigm. Comput. Netw. **52**(12), 2260–2279 (2008)
4. Stelzner, M., Dressler, F., Fischer, S.: Function centric networking: an approach for addressing in in-body nano networks. In: 3rd ACM International Conference on Nanoscale Computing and Communication, NANOCOM 2016. ACM, New York, September 2016
5. Mohrehkesh, S., Weigle, M.C., Das, S.K.: Energy harvesting in nanonetworks. In: Suzuki, J., Nakano, T., Moore, M.J. (eds.) Modeling, Methodologies and Tools for Molecular and Nano-scale Communications. MOST, vol. 9, pp. 319–347. Springer, Cham (2017). doi:10.1007/978-3-319-50688-3_14
6. Akyildiz, I.F., Pierobon, M., Balasubramaniam, S., Koucheryavy, Y.: The internet of bio-nano things. IEEE Commun. Mag. **53**(3), 32–40 (2015)
7. Nakano, T., Hosoda, K., Nakamura, Y., Ishii, K.: A biologically-inspired intrabody nanonetwork: design considerations. In: 8th International Conference on Body Area Networks, BodyNets 2013, pp. 484–487. ICST, Brussels (2013)
8. Pierobon, M., Akyildiz, I.F.: A physical end-to-end model for molecular communication in nanonetworks. IEEE J. Sel. Areas Commun. **28**(4), 602–611 (2010)
9. Dressler, F., Kargl, F.: Towards security in nano-communication: challenges and opportunities. Nano Commun. Netw. **3**(3), 151–160 (2012)
10. Xie, M.: Fundamentals of Robotics: Linking Perception to Action. Series in Machine Perception and Artificial Intelligence. World Scientific Publishing, Singapore (2003)
11. Moon, S., Lee, S.-G.: Revision of vocabulary standard for robots and robotic devices in ISO. In: 15th International Conference on Climbing and Walking Robots, pp. 849–854. World Scientific (2012)
12. Wang, X.: Piezoelectric nanogenerators—harvesting ambient mechanical energy at the nanometer scale. Nano Energy **1**(1), 13–24 (2012)
13. Lamport, L.: Time, clocks, and the ordering of events in a distributed system. Commun. ACM **21**(7), 558–565 (1978)
14. Freitas Jr., R.A.: Nanomedicine, Volume I: Basic Capabilities. Landes Bioscience, Georgetown (1999)
15. Stelzner, M., Lau, F.-L.A., Freundt, K., Büther, F., Nguyen, M.L., Stamme, C., Ebers, S.: Precise detection and treatment of human diseases based on nano networking. In: 11th International Conference on Body Area Networks (BODYNETS 2016), Turin, Italy, December 2016
16. Iijima, S.: Helical microtubules of graphitic carbon. Nature **354**(6348), 56 (1991)
17. Tans, S.J., Verschueren, A.R.M., Dekker, C.: Room-temperature transistor based on a single carbon nanotube. Nature **393**(6680), 49–52 (1998)
18. Montana, J.M.J.: Fundamentals of electromagnetic nanonetworks in the terahertz band. Ph.D. dissertation, Georgia Institute of Technology, December 2013
19. Aldaye, F.A., Palmer, A.L., Sleiman, H.F.: Assembling materials with DNA as the guide. Science **321**(5897), 1795–1799 (2008)
20. Fuechsle, M., Miwa, J.A., Mahapatra, S., Ryu, H., Lee, S., Warschkow, O., Hollenberg, L.C.L., Klimeck, G., Simmons, M.Y.: A single-atom transistor. Nat. Nanotechnol. **7**(4), 242–246 (2012)
21. Orlov, A.O., Amlani, I., Bernstein, G.H., Lent, C.S., Snider, G.L.: Realization of a functional cell for quantum-dot cellular automaton. Science **277**, 928–930 (1997)

22. Vankamamidi, V., Ottavi, M., Lombardi, F.: Tile-based design of a serial memory in QCA. In: 15th ACM Great Lakes Symposium on VLSI, GLSVLSI 2005, pp. 201–206. ACM, New York (2005)

23. Zhang, R., Yang, K., Abbasi, Q.H., Qaraqe, K.A., Alomainy, A.: Analytical modelling of the effect of noise on the terahertz in-vivo communication channel for body-centric nano-networks. Nano Commun. Netw. (2017). doi:10.1016/j.nancom.2017.04.001. ISSN 1878-7789

24. Suzuki, J., Balasubramaniam, S., Pautot, S., Perez Meza, V.D., Koucheryavy, Y.: A service-oriented architecture for body area nanonetworks with neuron-based molecular communication. Mob. Netw. Appl. **19**(6), 707–717 (2014)

25. Pech, D., Brunet, M., Durou, H., Huang, P., Mochalin, V., Gogotsi, Y., Taberna, P.-L., Simon, P.: Ultrahigh-power micrometre-sized supercapacitors based on onion-like carbon. Nat. Nanotechnol. **5**(9), 651–654 (2010)

26. Akyildiz, I.F., Jornet, J.M.: Electromagnetic wireless nanosensor networks. Nano Commun. Netw. **1**(1), 3–19 (2010)

27. Ferraro, P., Coppola, S., Grilli, S., Paturzo, M., Vespini, V.: Dispensing nano-pico droplets and liquid patterning by pyroelectrodynamic shooting. Nat. Nanotechnol. **5**(6), 429–435 (2010)

28. Wang, J., Gao, W.: Nano/microscale motors: biomedical opportunities and challenges. ACS Nano **6**(7), 5745–5751 (2012)

29. Freitas, R.A.: Current status of nanomedicine and medical nanorobotics. J. Comput. Theor. Nanosci. **2**(1), 1–25 (2005)

30. Wu, F., Tan, C.: The engineering of artificial cellular nanosystems using synthetic biology approaches. Wiley Interdisc. Rev.: Nanomed. Nanobiotechnol. **6**(4), 369–383 (2014)

31. Myers, C.J.: Engineering Genetic Circuits. CRC Press, Boca Raton (2016)

32. Nielsen, A.A.K., Der, B.S., Shin, J., Vaidyanathan, P., Paralanov, V., Strychalski, E.A., Ross, D., Densmore, D., Voigt, C.A.: Genetic circuit design automation. Science **352**(6281), aac7341 (2016). American Association for the Advancement of Science

33. Farsad, N., Yilmaz, H.B., Eckford, A., Chae, C.B., Guo, W.: A comprehensive survey of recent advancements in molecular communication. IEEE Commun. Surv. Tutor. **18**(3), 1887–1919 (2016)

34. Murphy, K., Weaver, C.: Janeway's Immunobiology. Garland Science, New York (2016)

Features of Use Direct and External Modulation in Fiber Optical Simulators of a False Target for Testing Radar Station

Margarita Yu. Tarasenko[1(✉)], Vadim V. Davydov[1,3,4],
Vladimir A. Lenets[1], Natalya V. Akulich[2], and Tatyana R. Yalunina[3]

[1] Peter the Great Saint-Petersburg Polytechnic University, St. Petersburg, Russia
margaritai.tarasenko@gmail.com
[2] AO "Zaslon", St. Petersburg, Russia
[3] The Bonch-Bruevich Saint - Petersburg State University
of Telecommunications, St. Petersburg 193232, Russia
[4] Department of Ecology, All-Russian Research Institute of Phytopathology,
B. Vyazyomy, 143050 Odintsovo District, Moscow Region, Russia

Abstract. The purpose of this paper was to develop and research the fiber optical simulators of false target with direct and external modulation for testing radar station for the X-range. The modes of testing radar station by simulator depending on external conditions were studied. Time, spectral and frequency responses have been received. The optimum modes of simulators depending on external conditions have been defined.

Keywords: Fiber optical system · Optical fiber · Analog signal of very high frequency range · Direct modulation · External modulation · Optical transmitter · Optical receiver · False target simulator

1 Introduction

One of the most perspective directions of radio photonics is transmission of analog signals of very high frequency range on communication fiber optical systems (FOS). Numerous experiments, and also operating experience showed that analog FOS can effectively be used in different paths of the multielement active phased antenna grids (APAG), switching devices of radar stations, etc. [1]. Use of analog fiber optical systems allows solving the most difficult problems connected to support of reliable operation of the send-receive antenna systems placed on moving platforms of the upper masts of the ships, different airplanes, etc. [2, 3]. But first developers need to set up and check entire antenna system for working capacity, both in the conditions of the laboratory or a polygon, and on the place of its forthcoming dislocation. And in the last two cases the flight purposes for different reasons can be absent.

The solution of this problem is the use of false target simulators, but as the most part of signals is transmitted on the fiber-optic communication line, the reliable fiber optical simulator should be checked. Fiber optical simulator has exceptional feature in comparison with other models of simulators working at other physical principles. Delay

© Springer International Publishing AG 2017
O. Galinina et al. (Eds.): NEW2AN/ruSMART/NsCC 2017, LNCS 10531, pp. 227–232, 2017.
DOI: 10.1007/978-3-319-67380-6_21

factor of an analog signal essentially doesn't depend on its frequency. It allows creating false targets even when using broadband sounding pulses. Frequency restrictions on a very high frequency of a signal are only connected to technical features of electrooptical modulators and photoreceiving modules [4]. Remaining advantages of fiber optical simulators: immunity to electromagnetic interferences, low losses (about 0.2 dB/km for external modulation and about 0.35 dB/km for direct modulation in the single-mode fiber), complete galvanic isolation, mechanical flexibility, and also small mass-dimensional characteristics compared with other simulators are well studied.

The most difficult solvable task of fiber optical simulators is determination of the optimum modes of their operation when testing the radar station. It is related to different features used in direct or external modulation in the transmitting modulex [5, 6].

We examined fiber optical simulators of false target with direct and external modulation. Prototypes for this purpose were made and the following characteristics were received: oscillograms of the reflected pulses in case of different duty cycles used in radar station; amplitude-frequency characteristics, spectrum of two types of modulators and response characteristics. Research has been done for two types of FOS: the nature of losses in frequency range from 2 to 18 GHz, response and spectral characteristics. The research looked into very high frequency signals of different forms which are used in radar stations. Research was conducted at different temperatures in the conditions of light modulation phase displacement that corresponds to real operating conditions of FOS on a polygon, etc.

The response characteristic of simulators at different temperatures formed out to be the most interesting as the purpose was to reveal an optimum operation mode depending on external conditions of radar station. Those experimental installations are shown in Fig. 1, where the developed samples of simulators are placed into a heat chamber.

2 Experimental Methods

The fiber optical simulator of false target is based on the optical fiber coil with receiving and transmitting module. The main difference between two types of simulators is in the transmitting module: the laser with wavelength of 1310 nm is used for direct modulation and modulation is performed by a pumping current, for external modulation – a 1550 nm wavelength is used and the external modulator [7]. The difference between two types of simulators [8] can be seen in Fig. 1(a, b), where the experimental installations of FOS with two types of modulation are provided.

Based on the features of the tested radar station, FOS's frequency range is 2 to 18 GHz, provided the required signal power and system performance.

The received results of measurements FOS's parameters have shown that this scheme of experimental installations (Fig. 1a, b) is the most rational from side of carrying out researches.

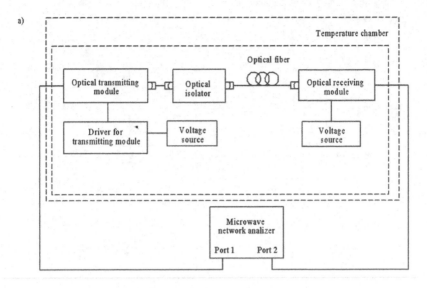

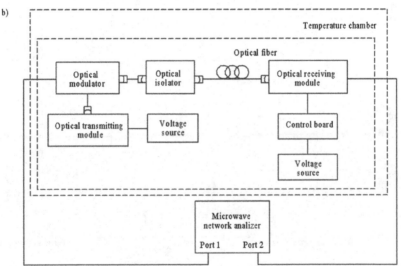

Fig. 1. Installation for measurement of parameters FODL with internal (a) and external (b) modulation.

3 Experimental Results and Discussion

The conducted researches showed that operation of simulators of false targets both types significantly doesn't change in case of different forms of signals and different duty cycles. So, time and spectral characteristics of both simulators aren't perceptible.

Frequency response of FOS with direct and external modulation at three temperatures are shown in Fig. 2(a, b).

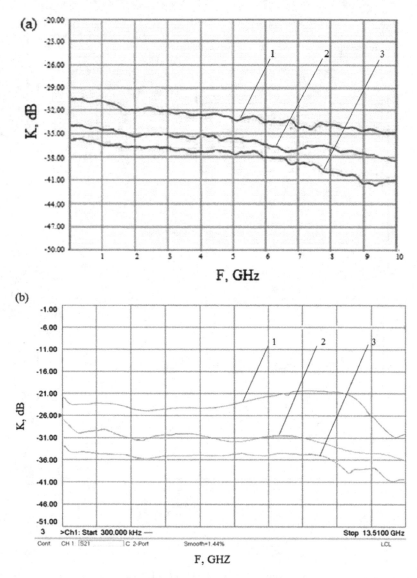

Fig. 2. Frequency response of FOS with direct (a) and external (b) modulation. Lines 1, 2, 3 corresponds to temperatures of 253 K, 293 K and 313 K.

The analysis of received frequency response for various ways of modulation shows that with increase in temperature when using external modulation the bandwidth is reduced by transfer analog by the signal microwave compared with direct modulation. It once again proves that it is necessary to use the system of thermostabilization allowing to provide a working point of the transfer characteristic at the segment with the maximum slope for the external modulator.

The response characteristic of simulators is of the greatest interest. Dependences of power output from input for three temperatures (253 K, 293 K and 313 K) are shown in Fig. 3. In view of features of a measuring equipment and assembled simulators it was succeeded to take characteristics in range from −20 dBm to 10 for direct modulation and from −20 dBm to 18 dBm for external modulation.

The analysis of the obtained experimental results allow to draw the following conclusions. The mode of external modulation is more sensitive to change of temperature that leads to increase in dynamic transmission losses of a very high frequency signal (violation of the mode of thermostabilizing on 283 K can lead to errors). In case

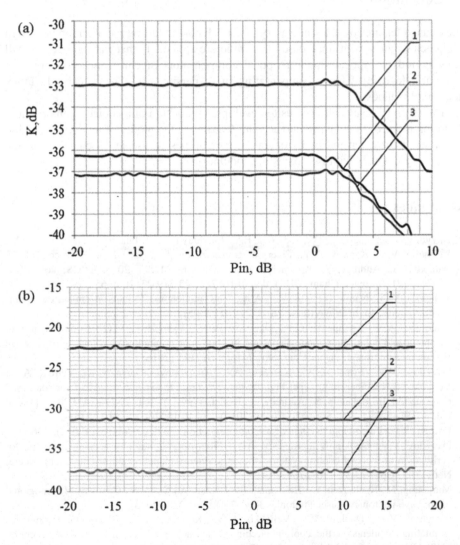

Fig. 3. Dynamic response of FOS with direct (a) and external (b) modulation at a frequency of 9 GHz. Lines 1, 2, 3 corresponds to temperatures of 253 K, 293 K and 313 K

of transmission rate more than 5 Gbit/s the mode of direct modulation enters distortions to the transmitted very high frequency signal, and also losses sharply increase, the linear range in a response characteristic is almost absent. This fact can be explained finite lifetime of carriers of charges and photons in the active layer of the laser diode. It was set that observed distortions of the form of pulses of a very high frequency of signals in case of direct modulation are connected to chromatic dispersion of the spurious frequency shift keying arising in case of distribution of radiation on an optical fiber because of existence.

4   Conclusion

The defined features by us, is necessary to consider when testing radar station very high frequencies signals with small on-off time ratio are applied. In that case in case of small delay periods of optical signal τ optical compensators of chromatic dispersion are necessary. In case of big τ compensating of dispersion needs to be carried out by electronic methods.

The received results allowed us to develop an algorithm of a choice of an optimum operation mode of FOS taking into account features of the modes of modulation and to justify need for extension of its operation capabilities to use in its construction the transferring modules with two types of modulation.

References

1. Davydov, V.V., Sharova, N.V., Fedorova, E.V., Gilshteyn, E.P., Malanin, K.Y., Fedotov, I.V., Vologdin, V.A., Karseev, A.Y.: Fiber-optics system for the radar station work control. In: Balandin, S., Andreev, S., Koucheryavy, Y. (eds.) ruSMART 2015. LNCS, vol. 9247, pp. 712–721. Springer, Cham (2015). doi:10.1007/978-3-319-23126-6_65
2. Velichko, M.A., Naniy, O.E., Susyan, A.A.: New modulation formats in optical communication systems. Light Wave Russ. Ed. **4**, 21–30 (2005)
3. Ermolaev, A.N., Krishpents, G.P., Davydov, V.V., Vysoczkiy, M.G.: Compensation of chromatic and polarization mode dispersion in fiber-optic communication lines in microwave signals transmittion. J. Phys.: Conf. Ser. **741**(1), 012171 (2016)
4. Davydov, V.V., Ermak, S.V., Karseev, A.U., Nepomnyashchaya, E.K., Petrov, A.A., Velichko, E.N.: Fiber-optic super-high-frequency signal transmission system for sea-based radar station. In: Balandin, S., Andreev, S., Koucheryavy, Y. (eds.) NEW2AN 2014. LNCS, vol. 8638, pp. 694–702. Springer, Cham (2014). doi:10.1007/978-3-319-10353-2_65
5. Friman, R.L.: Fiber–Optic Systems for Communications. Wiley, Hoboken (2012). 496 p.
6. Davydov, V.V., Dudkin, V.I., Karseev, A.Y.: Fiber-optic imitator of accident situation for verification of work of control systems of atomic energy plants on ships. Opt. Mem. Neural Netw. (Inf. Opt.) **23**(3), 170–176 (2014)
7. Marpaung, D., Roeloffzen, C., Heideman, R., Leinse, A., Sales, S., Capmany, J.: Integrated microwave photonics. Laser Photonics Rev. **7**(4), 506–538 (2013)
8. Davydov, V.V., Dudkin, V.I., Velichko, E.N., Karseev, A.Y.: Fiber-optic system for simulating accidents in the cooling circuits of a nuclear power plant. J. Opt. Technol. (A Translation of Opticheskii Zhurnal) **82**(3), 132–135 (2015)

Next Generation Wired/Wireless Advanced Networks and Systems (NEW2AN 2017)

On Detection of Network-Based Co-residence Verification Attacks in SDN-Driven Clouds

Mikhail Zolotukhin[✉], Elena Ivannikova, and Timo Hämäläinen

Faculty of Information Technology, University of Jyväskylä, Jyväskylä, Finland
{mikhail.zolotukhin,timo.hamalainen}@jyu.fi,
elena.v.ivannikova@student.jyu.fi

Abstract. Modern cloud environments allow users to consume computational and storage resources in the form of virtual machines. Even though machines running on the same cloud server are logically isolated from each other, a malicious customer can create various side channels to obtain sensitive information from co-located machines. In this study, we concentrate on timely detection of intentional co-residence attempts in cloud environments that utilize software-defined networking. SDN enables global visibility of the network state which allows the cloud provider to monitor and extract necessary information from each flow in every virtual network in online mode. We analyze the extracted statistics on different levels in order to find anomalous patterns. The detection results obtained show us that the co-residence verification attack can be detected with the methods that are usually employed for botnet analysis.

1 Introduction

Cloud environments allow users to consume computational and storage resources in an on-demand manner with low management overhead. As a rule, these resources are provided to the users in the form of virtual machines (VMs). In theory, VMs running on the same cloud server (i.e., co-resident VMs) are logically isolated from each other. However, in practice, a malicious customer can create various side channels to circumvent the logical isolation, and obtain sensitive information from co-resident VMs that may include machine workloads [1] and even cryptographic keys [2]. Unfortunately, a straightforward solution to this type of the attack that is eliminating all possible side channels [3] is not suitable for immediate deployment due to the required modifications to current cloud platforms [4]. For this reason, the problem of preventing malicious users from spawning their virtual instances on the same cloud server as the victim's VM is of greatest interest for modern cloud providers [2,4].

There are several approaches a malicious cloud customer can employ to intentionally co-locate his VMs with victim instances to run on the same physical cloud server. Probably the easiest and for this reason the most popular approach relies on networking. Co-residency verification can be carried out by measuring network round-trip time between pairs of VM instances [1], the number of network hops [5], or the network interference injected by the attacker [6].

© Springer International Publishing AG 2017
O. Galinina et al. (Eds.): NEW2AN/ruSMART/NsCC 2017, LNCS 10531, pp. 235–246, 2017.
DOI: 10.1007/978-3-319-67380-6_22

In this study, we concentrate on the problem of timely detection of such malicious customers in cloud environments that utilize software-defined networking (SDN). SDN is a hardware independent next generation networking paradigm, which breaks the vertical integration in traditional networks to provide the flexibility to program the network through centralized network control. In SDN, the control logic is separated from individual forwarding devices, such as routers and switches, and implemented in a logically centralized controller. The separation of control and data planes in SDN enables the network control to be programmable and the underlying infrastructure to be abstracted for applications and network services. Moreover, SDN enhances network security with the centralized control of network behavior, global visibility of the network state and run-time manipulation of traffic forwarding rules. Updating security policies in SDN requires updating the security applications or adding security modules to the controller platform, rather than changing the hardware or updating its firmware [7]. For these reasons, cloud computing environments can benefit from moving towards SDN technology [8].

To the best of our knowledge, there are no any studies that try to detect intentional co-residence attempts in cloud environments with the help of SDN. However, there are various approaches for detecting different sorts of cyber attacks carried out in clouds that utilize software-defined networking. Study [9] addresses man-in-the-middle and denial of service attacks caused by address resolution protocol bug in cloud centers using SDN technology. Authors propose a detection algorithm which uses Bayesian formula to calculate the probability of a virtual instance being an attacker. In [10], a new framework for DDoS detection and mitigation using sFlow and OpenFlow is proposed. The mechanism first matches an incoming flow with a legitimate sample of traffic and then installs mitigation actions if a flow found is not lying in the bounds of legitimate traffic pattern.

The rest of the paper is organized as follows. The problem in more details is formulated in Sect. 2. Section 3 describes several approaches of co-resident verification attack detection. In Sect. 4, we evaluate the performance of the techniques proposed. Section 5 draws the conclusions and outlines future work.

2   Problem Formulation

In this section, first, we specify assumptions for the cloud environment in which we detect the attack. After that, the attack vector is described in more details. Finally, we outline our solution that relies on software-defined networking.

2.1   Cloud Environment

We consider a cloud environment that consists of one controller and several compute nodes. A cloud customer can create several virtual networks and connect them to the existing public external network with the help of virtual routers. In addition, the customer is allowed to spawn several virtual instances in his own virtual networks. It is worth noting that each such instance is automatically assigned to a specific compute node according to some predefined allocation policy. For example, according to this policy, VMs can be concentrated to a number of compute nodes, in order to decrease the power consumption and maximize the utilization rate, or distributed across the whole data center, for the purpose of workload balance and higher reliability [4]. The allocation policy remains unknown to the customers.

Each customer operates inside one of the projects created by a system administrator for a particular set of user accounts. We assume that neither user or administrator accounts have been compromised. Thus, the cloud service provider guarantees that customer networks in every project are isolated from the direct intrusion of other customers. Cloud customers can only access each other's services via external public networks. Further, we assume that the networking inside the cloud is carried out with the help of SDN. SDN controller and switches communicate between each other inside the cloud's management network and are not available directly from the data center's virtual machines or external hosts. Scenarios in which either the controller or one of the switches is compromised are out of scope of this paper.

2.2 Attack Vector

In such environment, network-based co-residence verification technique such as time-to-live (TTL) probing [5] is not working. TTL values of network packets from two VMs belonging to two different cloud customers and communicating via an external public network are reduced only by routers of this network. Since virtual SDN switches usually operate on a different layer of the ISO/OSI model, they do not alter the IP payload. Furthermore, the co-location verification technique based on measuring packet round-trip times (RTT) [1] cannot be used either. Since VMs that belong to different cloud customers can communicate only over an external network, the traffic between them goes through this network's routers. For this reason, packet round-trip time to the target VM will most likely be the same from a VM located on the same server and from a VM that is on another server of the cloud. However, for machines from the same virtual network, RTT between two co-located VMs is slightly less than between VMs located on different servers. Therefore, the attacker can use this approach if one machine from the target's network has been compromised, but, in this research, we do not take this scenario into consideration.

In this study, we focus on the process of co-residence verification that relies on the fact that co-resident VMs share the same physical network interfaces. This opens an explicitly communicative channel that can be used by the malicious customer for co-residency verification. Study [6] proposes the co-resident watermarking attack which involves launching several virtual machines and performing statistical side channel tests from them. One of these machines plays role of a master host whereas the rest are flooders and sinks. The attack begins when the master initiates a web session with the target instance. Systematically, the master iterates through its list of the flooders and commands them to start injecting network activity into the outbound interface of their physical host machines by sending portions of traffic to sinks. In case one or several flooders are co-located with the target server, this activity creates delay in the legitimate server's flow initiated by the master.

The attacker can define whether one of his flooders is co-located with the target by measuring packet arrivals per interval over the length of the flow. After each measurement, the intervals are divided into two samples X_1 and X_0 based on the pre-negotiated co-resident activity respectively representing intervals when flooders were active and when they were not. In study [6], authors model packet arrivals by a Poisson distribution and employ the non-parametric Kolmogorov-Smirnov test for independence. The null hypothesis that the two samples are from the same distribution can be rejected with confidence α if it is bigger than α_{min} that is equal to:

$$\alpha_{min} = 2\exp\left(-2\frac{|X_0||X_1|}{|X_0|+|X_1|}sup(|F_1(X_1) - F_0(X_0)|)^2\right), \qquad (1)$$

where $F_i(X_i)$ for $i = 0, 1$ is empirical cumulative distribution of X_i. In other words, the less α_{min} the more confident the attacker that his flooder is co-located with the target server.

2.3 SDN Solution

There are at least two mechanisms that can help cloud data centers operating in software-defined networks in mitigation of network-based co-residence attacks. First, SDN allows the cloud provider manipulate traffic forwarding rules on fly. In particular, this means that the provider can redirect traffic of customers' VMs located in the same virtual subnet flow through different virtual switches. As a result, the traffic between VMs located on the same compute node may go via a switch that is located on another node. This would make the co-location verification technique based on measuring RTT hard for the attacker to use, especially if the forwarding rules change over time.

Second, SDN enables global visibility of the network state. Each flow from every virtual network can be monitored and analyzed in online mode. In particular, network flows can be captured on each SDN forwarding device and sent to the controller with the help of some flow collector. Once these statistics have been analyzed by the controller, it commands the SDN forwarders to either block the traffic or reroute it to security middle boxes for deeper payload-based analysis.

In this study, we rely on this second advantage of SDN for timely detection of the co-residence verification attack described in the previous subsection. For this purpose, each SDN forwarding device sends to the controller statistics of all the network flows initiated by or directed to VMs located on the corresponding compute node. These statistics may include the flow source and destination IP address and port, time stamp, duration of the flow, number of packets and bytes sent and received, presence of packets with different flags for TCP flows, and probably several others. The controller investigates the flow statistics received searching for anomalous behavior patterns of the cloud's virtual instances and external hosts. The further analysis of these patterns can help to detect potential attacks and timely mitigate them by blocking the traffic that belongs to malicious agents.

3 Co-residence Verification Detection

As can be seen from the attack vector description, the malicious cloud customer requires to spawn a large number of virtual instances on the cloud. Thus, one potential approach for co-residence verification detection would rely on the number of VMs created per a time interval by a particular customer. However, such approach cannot be scaled well on big modern cloud data centers, since there can be thousands of newly created instances that only run for a short period of time and belong to different user projects. For this reason, we try to detect malicious customers by collecting traffic in cloud's private and external public networks. The traffic is then analyzed in three different domains: flow, session and time.

3.1 Flow Domain

A flow is a group of IP packets with some common properties passing a monitoring point in a specified time interval. These common properties include transport protocol, the IP address and port of the source (client) and IP address and port of the destination (server). For each flow at each time interval, we may extract several features including the flow duration, average number of packets sent or received per second, average size of packet, and some other information. Once all relevant features have been extracted and standardized, the resulting feature vectors can be divided into several groups by applying a clustering algorithm. There are many different clustering algorithms which can be categorized based on the notation of a cluster. The most popular categories include centroid-based clustering algorithms, hierarchical clustering algorithms and density-based clustering algorithms. The flow clusters can be found during the training phase when there is only legitimate traffic in the cloud. During the detection phase, if a new feature vector does not belong to any of the clusters obtained, the corresponding flow is labeled as malicious.

3.2 Session Domain

In the next stage, we can analyze sequences of flows belonging to one user session. Such approach is often used for application-based DDoS attacks detection [11]. If packet's payload is encrypted and session ID cannot be extracted, we can group all flows which are extracted in certain time interval and have the same source IP address, destination IP address and destination port together and analyze each such group separately. We can interpret a group of such flows as a rough approximation of the user session [12].

One detection approach in the session domain is counting flows from each cluster defined by one of the clustering algorithms mentioned in the previous subsection initiated by each user during the session. If flows between a flooder and a sink are shaped to follow legitimate traffic patterns, the attacker would probably need to increase the number of flows sent by each flooder to increase delays for packets arriving at the master host, therefore, improving the co-residence verification rate. For this reason, the number of flows of at least one certain type initiated by the attacker should exceed the number of flows of this type during legitimate user sessions. This pattern can be detected with the help of one of the clustering algorithms specified above or any anomaly detection technique.

Another approach is to analyze conditional probabilities of observing a sequence of flows in a session of particular length. Given a sequence of flow labels $c_1, \ldots, c_N$ we can factorize joint probability distribution over sequences of length N as the following product:

$$P(c_1, \ldots, c_N | N) = \frac{n(c_1, N)}{\sum_{j=1}^{k} n(c_j, N)} \times \prod_{i=2}^{N} \frac{n(c_{i-1}, c_i, N)}{n(c_{i-1}, N)}, \tag{2}$$

where k is the number of flow clusters, and $n(c_i, N)$ and $n(c_{i-1}, c_i, N)$ denote respectively count of observations of label c_i and pairs (c_{i-1}, c_i) in all sequences of length N over all time intervals and sessions. Resulting joint probability values can be compared to each other in order to find anomalous sessions.

3.3   Time Domain

Finally, in order to detect the malicious customer, we can exploit the fact that the master's and flooder's network activity should be synchronized during the attack. Such approach is sometimes used for botnet detection [13]. Since virtual instances from the same project are able to communicate via the project's private network, it is very unlikely that there will be synchronized sessions between VMs over external public network.

In order to obtain feature vectors, we calculate Pearson correlation coefficient [14] between two different virtual instances:

$$\rho_{ij} = \frac{\sum_t (r_i(t) - \bar{r}_i)(r_j(t) - \bar{r}_j)}{\sqrt{\sum_t (r_i(t) - \bar{r}_i)^2} \sqrt{\sum_t (r_j(t) - \bar{r}_j)^2}}, \ \forall i \neq j, \tag{3}$$

where $r_i(t)$ is percentage of time interval t during which the i-th instance transmits some data and $\bar{r}_i$ is the average value of $r_i(t)$ over all time intervals by the moment the detection starts. Attacker's activity can be detected by searching for anomalously high values of ρ_{ij}, which correspond to the synchronized transmission of instances i and j.

4   Performance Evaluation

In order to evaluate the detection approach proposed, first, we briefly overview our virtual network environment used to generate network traffic. Then, we describe the attack implementation and analyze the attacker's strategy. Finally, we present results of the detection of the attack with different approaches overviewed in this study and discuss the results obtained.

4.1 Test Environment

We test the attack detection algorithm proposed in this study in open-source software platform for cloud computing Openstack, networks in which are carried out with the help of integrated SDN controller Opendaylight and several Open vSwitches. The cloud consists of four nodes: three compute nodes and one control node that also has compute functionality. Open vSwitch is installed on each cloud's node whereas the SDN controller is located on the cloud's control node (see Fig. 1).

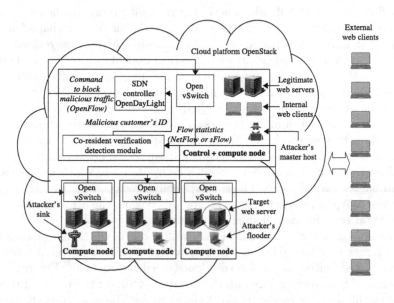

Fig. 1. Simulation environment for testing the detection algorithm.

In order to generate legitimate traffic, we spawn eight virtual web servers in the cloud and five instances that belong to different web clients. In addition, there are eight clients located outside of the cloud. Every arbitrary amount of time both internal and external clients connect to randomly selected servers and request several files from them. There are also keep-alive messages transferred time-to-time between servers and clients, and open SSH tunnels deployed in order to control VMs remotely. Statistics of all the resulting flows is recorded on switches and sent to the controller with the help of NetFlow and sFlow agents.

In addition to the normal traffic, we perform the co-residence verification attack against one of the web servers. For this reason, we spawn four virtual instances that belong to the attacker. These instances are supposed to be located on different compute nodes. In practice, this can be achieved by spawning big amount of instances and removing those that are located on the same compute node. The fact that some of the attacker's instances are co-located can be checked by measuring packet RTT as these machines operate in the same virtual network. One of the resulting instances plays role of the master host, another one is the sink, whereas the rest are flooders.

4.2   Attack Analysis

During the attack, the master host first requests a file from the target server and measures packet arrivals per time interval. Next, the master commands one of the flooders to send portions of traffic to the sink, and at the same time requests one more file from the target server. After that, the procedure is repeated using another flooder. In order flooder's activity does not arouse suspicion from the cloud service provider, the traffic generated by the flooder is shaped in such a way that it mimics the traffic sent by legitimate web servers resided in the cloud. We test two co-resident attacks: low-rate and high-rate. In the low-rate attack scenario, each flooder generates the same amount of traffic flows as any of the legitimate servers, whereas during the high-rate verification the attacker increases the number of the flows generated by the flooders up to ten times.

The lowest value of minimal level α_{min} does not necessarily mean that the corresponding flooder resides on the same compute node as the target server. The reason behind this is that there is constant outbound network activity on the external network interface of every cloud's compute node which causes various delays in packet arrival times. These delays sometimes can be even bigger than delays caused by the attacker's flooders. For this reason, the attacker is supposed to run several tests for each flooder that is under his control.

We assume that the attacker decides whether one of his flooder is co-resided with the target server based on the average value of minimal level of confidence α_{min} calculated during several co-residence verification tests. Figure 2 shows dependence of this average value on the number of the tests in case of the low-rate and the high-rate attack scenario. As one can see, the more tests the attacker conducts, the less the average value of α_{min} for the flooder resided on the same compute node as the target server and, therefore, the more confident the attacker is that the verification has been done properly. It is worth noting that, in the case of the high-rate co-residence verification, the difference between average values of α_{min} for the flooder that is co-resided with the target server and the flooder that is located on another cloud node can be clearly seen after few verification tests. Thus, an attacker that employs the high-rate attack requires less tests to guarantee that the verification is correct. However, the immediate drawback of the high-rate approach is the attack can be easily detected by analyzing the network traffic as shown in the next subsection.

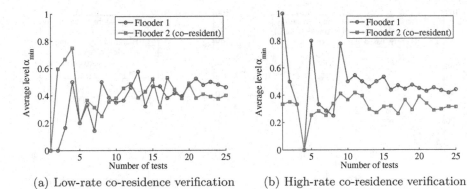

(a) Low-rate co-residence verification   (b) High-rate co-residence verification

Fig. 2. Dependence of minimal level of confidence α_{min} on the number of co-residence verification tests.

4.3   Attack Detection

First, we try to detect both versions of the attack in the flow domain. Figure 3 shows how TPR depends on FPR when detecting the attack for three different clustering approaches. We used k-means, single-linkage and dbscan as the most popular representatives of centroid-based, hierarchical and density-based clustering approaches respectively. As one can notice, all the methods fail to properly detect both the low-rate and the high-rate co-residence verification attack. In the case of low-rate verification tests, the percentage of false alarms is almost the same as the percentage of successfully detected malicious flows which results in very low accuracy values as can be seen from Table 1. The only approach that potentially can be used for the high-rate attack detection in the flow domain is density-based clustering. However, the number of false alarms when employing this approach is still high (up to 20%) making it impractical in most of the use cases.

Table 1. Detection accuracy in the flow domain

Clustering approach	Attack type	
	Low-rate	High-rate
Centroid-based	47.21–73.74%	24.56–78.72%
Hierarchical	58.79–74.30%	44.89–50.20%
Density-based	44.69–74.31%	45.30–86.11%

Next, we try to detect the attack in the session domain. Figure 4 shows how TPR depends on FPR when detecting the attack for two different approaches. As it can be noticed, the approach based on the analysis of flow distributions inside each session does not generate many false alarms, but at the same time it allows

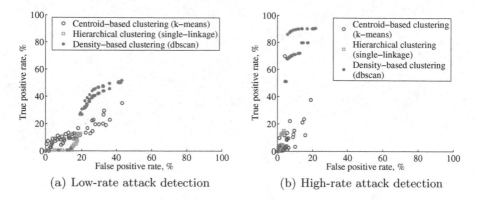

(a) Low-rate attack detection   (b) High-rate attack detection

Fig. 3. Dependence of true positive rate on false positive rate of the attack detection in the flow domain.

one to detect only 30% of attacker's sessions. On contrary, the approach based on calculating conditional probabilities generates lots of false alarms, however, most of the attacker's sessions can be detected in case of the high-rate attack. Detection accuracy in the both cases is still low as can be seen from Table 2.

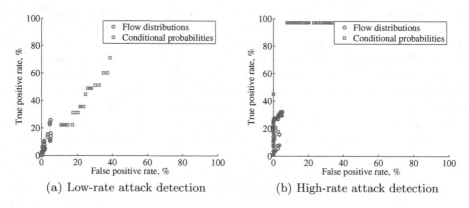

(a) Low-rate attack detection   (b) High-rate attack detection

Fig. 4. Dependence of true positive rate on false positive rate of the attack detection in the session domain.

Finally, we detect both versions of the attack in the time domain by calculating Pearson correlation coefficients. Virtual instances with extremely high coefficient values are labeled as anomalous. Dependence of true positive rate on false positive rate of the detection is shown in Fig. 5. Numbers near each point correspond to amounts of verification tests performed by the moment when the detection starts. As the number of tests increases, the detection accuracy for both versions of the attack approaches 100% (TPR = 100%, FPR = 0%). Each iteration of the high-rate attack lasts longer, as a result, more legitimate VMs transmit at the same time as flooders, which may lead to few false alarms.

Table 2. Detection accuracy in the session domain

Detection approach	Attack type	
	Low-rate	High-rate
Flow distributions	82.44–86.04%	84.44–89.09%
Conditional probabilities	48.10–79.75%	37.18–93.66%

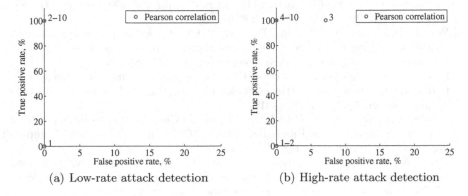

(a) Low-rate attack detection   (b) High-rate attack detection

Fig. 5. Dependence of true positive rate on false positive rate of the attack detection in the time domain.

To summarize the results obtained, the high-rate version of co-residence verification attack can be detected on both flow and session level. This can be explained by the fact that in this case the attacker initiates many look-alike flows which results in density outliers in the space of feature vectors extracted from flows and anomalously high concentration of flows of the attacker's sessions in one cluster. However, the low-rate attack can only be detected by counting for synchronized data transmissions by virtual instances from different customers' projects.

5   Conclusion

In this study, we focused on timely detection of intentional co-residence attempts in cloud environments that utilize software-defined networking. For this purpose, we analyzed statistical features extracted from network traffic flows on three different levels, and, as a result, we were able to detect both low-rate and high-rate versions of the co-residence verification attack. In the future, we are planning to build a theoretical model to estimate the time required for a malicious customer to successfully verify the co-residence with the target server, and the time for the cloud provider to detect the attack by analyzing anomalous network activity.

References

1. Ristenpart, T., Tromer, E., Shacham, H., Savage, S.: Hey, you, get off of my cloud: exploring information leakage in third-party compute clouds. In: Proceedings of the 16th ACM conference on Computer and Communications Security, pp. 199–212 (2009)
2. Zhang, Y., Li, M., Bai, K., Yu, M., Zang, W.: Incentive compatible moving target defense against VM-colocation attacks in clouds. In: Gritzalis, D., Furnell, S., Theoharidou, M. (eds.) SEC 2012. IAICT, vol. 376, pp. 388–399. Springer, Heidelberg (2012). doi:10.1007/978-3-642-30436-1_32
3. Wu, J., Ding, L., Lin, Y., Min-Allah, N., Wang, Y.: XenPump: a new method to mitigate timing channel in cloud computing. In: Proceedings of the 5th IEEE International Conference on Cloud Computing, pp. 678–685 (2012)
4. Han, Y., Chan, J., Alpcan, T., Leckie, C.: Using virtual machine allocation policies to defend against co-resident attacks in cloud computing. IEEE Trans. Dependable Secure Comput. **14**(1), 95–108 (2017)
5. Herzberg, A., Shulman, H., Ullrich, J., Weippl, E.: Cloudoscopy: services discovery and topology mapping. In: Proceedings of the ACM Workshop on Cloud Computing Security Workshop, pp. 113–122 (2013)
6. Bates, A., Mood, B., Pletcher, J., Pruse, H., Valafar, M., Butler, K.: Detecting co-residency with active traffic analysis techniques. In: Proceedings of the ACM Workshop on Cloud Computing Security Workshop, pp. 1–12 (2012)
7. Ahmad, I., Namal, S., Ylianttila, M., Gurtov, A.: Security in software defined networks: a survey. IEEE Commun. Surv. Tutor. **17**(4), 2317–2346 (2015)
8. Hao, F., Lakshman, T., Mukherjee, S., Song, H.: Secure cloud computing with a virtualized network infrastructure. In: Proceedings of the 2nd USENIX Conference on Hot Topics in Cloud Computing, pp. 16–16 (2010)
9. Ma, H., Ding, H., Yang, Y., Mi, Z., Zhang, M.: SDN-based ARP attack detection for cloud centers. In: Proceedings of IEEE 12th International Conference on Ubiquitous Intelligence and Computing, IEEE 12th International Conference on Autonomic and Trusted Computing, IEEE 15th International Conference on Scalable Computing and Communications and its Associated Workshops, pp. 1049–1054 (2015)
10. Buragohain, C., Medhi, N.: FlowTrApp: an SDN based architecture for DDoS attack detection and mitigation in data centers. In: Proceedings of the 3rd International Conference on Signal Processing and Integrated Networks, pp. 519–524 (2016)
11. Xu, C., Zhao, G., Xie, G., Yu, S.: Detection on application layer DDoS using random walk model. In: Proceedings of IEEE International Conference on Communications (ICC), pp. 707–712 (2014)
12. Zolotukhin, M., Hämäläinen, T., Kokkonen, T., Siltanen, J.: Increasing web service availability by detecting application-layer DDoS attacks in encrypted traffic. In: Proceedings of 23rd International Conference on Telecommunications (ICT), pp. 1–6 (2016)
13. Akiyama, M., Kawamoto, T., Shimamura, M., Yokoyama, T., Kadobayashi, Y., Yamaguchi, S.: A proposal of metrics for botnet detection based on its cooperative behavior. In: Proceedings of International Symposium on Applications and the Internet Workshops, pp. 82–85 (2007)
14. Herlocker, J., Konstan, J., Riedl, J.: An empirical analysis of design choices in neighborhood-based collaborative filtering algorithms. ACM Trans. Inf. Syst. **5**(4), 287–310 (2002)

Health-Care Pervasive Environments:
A CLA Based Trust Management

Omid Bushehrian[✉] and Shayeste Esmail Nejad

Department of Computer Engineering and IT,
Shiraz University of Technology, Shiraz, Iran
bushehrian@sutech.ac.ir, esmaili.shayeste@yahoo.com

Abstract. Pervasive computing environments and technologies have attracted
much attention in the health-care field recently. The application of pervasive
computing technologies in the public health and medicine caused the emergence
of pervasive healthcare as a new concept. Using pervasive healthcare, patients
are able to access to the health and medical services at anytime and anywhere.
However selecting the most trusted service provider among available ones is a
challenging task. In this paper, we have proposed a new model for pervasive
health-care systems in which the patient smart device not only uses its own past
experiences but also learns from its neighbors (friends) in a health-care social
network to update the trust of available providers using a CLA (cellular learning
automata). Comparing the proposed method with existing recommendation
based methods shows that it is more scalable, efficient and appropriate for
dynamic environments without sacrificing the precision.

Keywords: Pervasive computing environments · Pervasive health-care ·
Trust · Cellular learning automata

1 Introduction

With the growing use of online social networks in sharing a variety of professional and
personal information it is predicted that they will become an effective media for sharing
health-care information and services in the near future [1–3]. By using this media not
only the health service providers are able to advertise their services but also the patients
can find their required health service and share feedbacks about their experiences of
using those services with others in an online social network. The health-care services
encompasses a variety of domains including emergency and urgent services, consultant
and treatment services and information desk services which are delivered over a
wireless network in a pervasive and ubiquitous environment. Hence the patient smart
device is able to discover, associate and consume those services using a communication
technology such as web service technology [2, 4].

In such environments the role of intelligent applications on patient smart devices
seems to be very essential by which all the important steps of service consuming
including: looking up the advertised services in the online social network, selecting the
most trusted service from available ones, associating to the selected service, consuming
the service and finally putting feedbacks about the service in the social network is

© Springer International Publishing AG 2017
O. Galinina et al. (Eds.): NEW2AN/ruSMART/NsCC 2017, LNCS 10531, pp. 247–257, 2017.
DOI: 10.1007/978-3-319-67380-6_23

supported. Obviously designing a smart application that supports all the mentioned steps is a challenging task and in this paper we merely focus on the problem of selecting the most trusted service provider among available ones. Trust is defined as a quality of interaction between device A and service provider B based on which A believes that the service provided by B is beneficial and satisfying and is not malicious [5, 6]. Due to the importance of health-services to patients, finding a trustworthy provider is a very essential issue particularly in the emergency situations. For example an elderly patient whose vital signals are continuously monitored by a set of body sensors and his smart device is supposed to automatically invoke an urgency service (for example a web service) at the emergency conditions needs to invoke the most trusted urgency service among available ones.

Since "trust" is normally obtained based on past experiences of using a service, its computation using statistical methods, in which the log files of the devices in the environment is used to rank the providers, is prevalent [5–7]. However due to the fact that first: these methods are not scalable as the number of devices in the network increases and second: the credibility of recommendations collected from the devices are not guaranteed, in this paper a Cellular Learning Automata (CLA) [8] method is proposed to select the trustworthy provider. By using the proposed method each device merely needs to learn from its own neighbors (not the all devices) to converge to the most trusted service provider. The neighbors are the patient friends in the online social network with enough credibility.

The rest of this paper is organized as follows: In Sect. 2 the related works are explained. Section 3 presents the proposed architecture for the pervasive healthcare environments. The experimental results and discussions are presented in Sect. 4. Section 5 concludes the paper.

2   Related Works

Most previous studies in the field of trust are focused on the direct and indirect trust computation methods. In [9] a community based trust management for both direct and indirect trust computation has been proposed. In this model each node in the network acts as the center of a community of its neighboring nodes. First, trust is computed using a recommendation based method by the Trust Assistant Policy (TAP) module. To update the trust value, a non-linear function which is dependent on both the time that a node has stayed in the community and the past trust value of the node is used. Un-trusted nodes are excluded from the central node community. In [10] a probabilistic trust management model is presented where the trust is computed based on history of interactions stored in each device. This method is recommendation based and considers the possibility of false recommendation as well. The trust value is computed using the expected value of beta distribution. In [11] a multilevel trust calculation framework based on some formal definitions of trust metrics is proposed. In this framework different modules for trust evolution, trust degradation, trust filter for both direct and recommended interactions has been defined. The value of trust in this study is computed using the Bayesian formalism. In [12] a trust management model based on a combination of reputation and credentials for pervasive health-care environments has been proposed. Here with defining the community concept, communication is limited to patient neighbors. Moreover a non-linear trust computation method is selected and "gossiping" is used to frequently updating trust values. In [13] the concept of pervasive community model is proposed to reduce the number of required interactions in the recommendation based methods. Each service provider registers itself in a trusted community head by presenting its credentials. A client who has no previous history of service providers instead of gathering recommendations from all nodes may ask the community head for a trusted service provider. Moreover in this paper a global data store is used to store the trust value for each service provider.

In [6] a Bayesian-based model to evaluate the value of trust using both direct and indirect interaction history among devices has been proposed. In this paper the method is able to compute the trust both when a device has little or enough interaction history with others. The existence of malicious nodes in the environment has also been considered. The "computation method selection" sub module is used to select the appropriate module for computing the trust value. If the local observations are enough the "direct computation" sub-module is selected otherwise the "indirect computation" sub-module computes the trust based on recommendations collected from all other devices. The false recommendation is also filtered out in this method. In [14] to filter out malicious recommendations in indirect trust computation a statistical method is proposed. The method labels the recommendations which are far from the mean value of recommendation set and have low occurrence as the dishonest recommendation. It was shown that the method is resistant to different attacks.

A Cellular Learning Automata (CLA) based method is proposed in this paper to address the shortcoming of previous studies: The previous indirect trust computation methods are not scalable due to the fact that they usually collect recommendations from all devices. Moreover in the proposed method no storing of the previous interactions in each device is required and it efficiently works based on a learning algorithm of CLA. The credibility of the votes collected from adjacent cells is also guaranteed due to the fact that in the proposed method the votes are collected from friends of the patient in the online health social network with the same interests.

3   The Pervasive Health-Care Architecture

The proposed architecture for the service discovery and selection in the pervasive healthcare environment is shown in Fig. 1. Each patient device is a subscriber of a medical social network and has its own friends (neighbors) in it. This social network is periodically looked up by service discovery module on the device to find new healthcare services which are advertised by service providers. Moreover the votes of the neighbors about the last trustworthy provider they have interacted with are collected by the trust computation module once needed (the details are presented in Sect. 3.2). Since the votes come from the friends in the social network their credibility are more guaranteed opposed to the conventional recommendation based methods. Once an interaction with one service provider is required, a service provider is selected among available ones by the service selection module with a specific probability previously computed by the trust computation module. After each interaction, according to the interaction results which is either satisfactory or un-satisfactory and considering the neighbors votes, the trust value is updated by the trust computation module using the learning algorithm of the CLA.

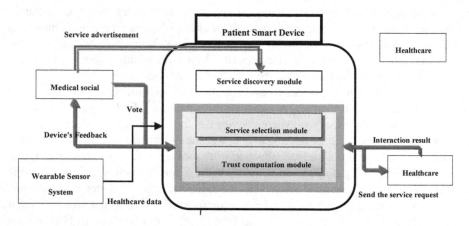

Fig. 1. Block diagram of pervasive health-care architecture

3.1 Asynchronous Cellular Learning Automata

Cellular Learning Automata [8, 15] is a combination of a learning automata and a cellular automata [16] and is very suitable for modeling environments containing distributed interactions. As shown in Fig. 2 a learning automata [17] continuously interacts with a random environment though a set of predefined actions. The emitted action is evaluated by the surrounding environment in form of positive or negative responses which are subsequently used by the automata to reward or penalize the action by increasing or decreasing its probability.

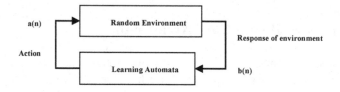

Fig. 2. The relationship between environment and learning automata

A variable structure automata [18] as a special type of learning automata is defined by quadruple {a, b, p, T}; where, a = {a_1, a_2, ..., a_r} is the automata's action set, b = {b_1, b_2, ..., b_m} is the automata's input set, p = {p_1, p_2, ..., p_r} is the action selection probability vector, and T = p(n + 1) [a(n), b(n), p(n)] is the learning algorithm. If a_i is the selected action in time n, then p is updated at time n + 1 as follows:

- For desirable response

$$P_i(n+1) = P_i(n) + \alpha[1 - P_i(n)]$$
$$P_j(n+1) = (1 - \alpha)P_j(n) \quad \forall j, i \neq j \tag{1}$$

- For undesirable response

$$P_i(n+1) = (1 - \beta)P_i(n)$$
$$P_j(n+1) = \beta/(r-1) + (1 - \beta)P_j(n) \quad \forall j, i \neq j \tag{2}$$

where α and β are reward and penalty parameters, respectively. If $\alpha = \beta$, the learning algorithm is called L_{R-P}, if $\beta \ll \alpha$, the algorithm is called $L_{R\epsilon P}$, and if $\beta = 0$ it is called L_{R-I} [18].

A cellular learning automata is in fact a cellular automata with a learning automata assigned to each cell. In cellular learning automata, the reinforcement signal after performing each action is determined based on the local rule and actions selected by neighboring cells affects the amount of reinforcement signal. One type of cellular learning automata is asynchronous cellular learning automata [15] in which only some of learning automata are autonomously activated in a specified moment.

3.2 CLA Based Trust Computing

In this section, cellular learning automata based algorithm to solve the trust issue in pervasive healthcare environment is explained. As shown in Fig. 3 each device in the medical social network is represented by a learning cell while its adjacent cells in the cellular learning automata are the device neighbors (patient friends) in the social network. The set of actions of each learning automata corresponds to the set of healthcare

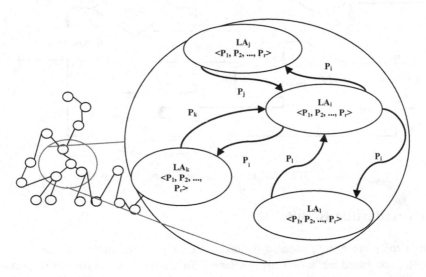

Fig. 3. CLA corresponding to each device in the social network

providers present in the environment and the selection of an action which happens propositional to the action probability means that the corresponding service provider is selected (due to its trustworthiness).

Initially, a device in the pervasive healthcare environment has no information about the trustworthiness of the available healthcare providers, therefore, the probability of selecting all of healthcare providers are equal. Thus, the initial probabilities of every learning automata's actions corresponding to a device is defined based on Eq. 3.

$$\forall i, i \leq n, P_i = 1/n \tag{3}$$

In every step of selecting a service provider, an action is selected proportional to its probability in the current probability vector p. Subsequently, the reinforcement signal is calculated according to the local rule as shown in Fig. 4. The response of the service provider which is either satisfactory or unsatisfactory and the majority vote of the neighboring cells used to compute the reinforcement signal. The learning automata updates its action probability vector depending on the reinforcement signal, in other words it rewards or penalizes its own selected action based on Eqs. 1 and 2 respectively or takes no action. The vote of an adjacent cell is its last action (last service provider) selected by that cell.

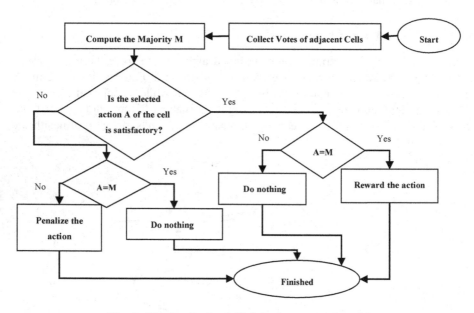

Fig. 4. The local rule of CLA in the proposed model

4   Experimental Results

In order to compare the presented method with the previous studies, a Bayesian recommendation based method in [6] is selected. In this method the history of previous interactions among all devices and the service providers are collected by a device in the environment (recommendations). Then the device calculates the trust value for each service provider using the history and selects the service provider with the highest trust value. The evaluation was performed in three dimensions:

- Precision: defined as the percentage of devices in the environment successfully were able to select the trustworthy service provider. This metric is measured after completion of each simulation round.
- Convergence time: defined as the number of rounds (successive interactions) required reaching a specific precision value.
- Elapsed time: defined as the time spent for computing the trust value for a device in the pervasive healthcare environment. This metric is used to show the scalability of the presented method.

4.1 Simulation Parameters

The medical social network is modeled by random graphs in the subsequent simulations. Each graph node represents a patient device and each graph edge represents a friendship relationship among patients. A number of service providers are also assumed in the simulated environment each is either satisfactory or malicious. Each service provider response to a device is satisfactory (response value = 1) with probability p and malicious (response value = 0) with probability (1 − p). In the following experiments the learning automata is a L_{R-P} type and the values for parameters α and β, reward and penalty respectively, were set to 0.25. The threshold value of the convergence was set to 0.9 (when the probability of action c of a device reaches this threshold it is said that the device eventual action is c). The experiment parameters are listed in Table 1. The metric to be measured in the experiment is listed in the "metric" column, the environment condition determines whether a malicious provider may convert to a satisfactory one or not. The value p in the last column is the probability P(S = "satisfactory") where S is a random variable representing a satisfactory provider response with two outcomes {"satisfactory", "malicious"}. Similarly (1 − p) is the probability P(M = "satisfactory") where M is a random variable representing a malicious provider response. Each experiment was repeated 10 times in each scenario to increase the confidence.

Table 1. The experiments parameters

	Metric	Environment	Device no.	Satisfactory provider no.	Malicious provider no.	P
Experiment 1	Precision	Fixed	30	4	2	0.8
Experiment 2	Elapsed time	Fixed	10–100	4	2	0.8
Experiment 3	Precision	Dynamic	30	Changes over time	Changes over time	Changes over time
Experiment 4	Elapsed time	Fixed	30	5	10	1

4.2   Simulation Experiments

Experiment 1: In this experiment, the precision of the proposed method is compared to the recommendation-based method. As shown in Fig. 5 all devices were able to correctly select trustworthy service provider after 90 simulation rounds in the recommendation-based method and 70 rounds in the CLA based method.

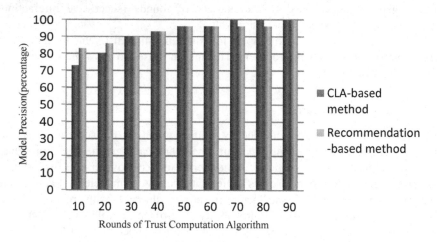

Fig. 5. Comparing between growth of models precisions

Experiment 2: This experiment compared the effect of increasing the number of service requester devices on the trust computation time spent in one round by one device. As it was expected in Fig. 6 the CLA based method is more scalable due to its more efficient trust computation model.

Experiment 3: The objective of this experiment was to study the effect of changing a malicious service provider behavior on the model precision. There are 6 providers

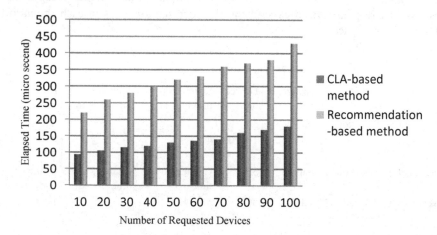

Fig. 6. Comparing the time spent by one device for computing the trust value

D1–D6 with initial p values (0.8, 0.8, 0.5, 0.5, 0.5, 0.5). Initially D1 and D2 are assumed as malicious (i.e. $P(M = \text{"satisfactory"}) = 1 - 0.8 = 0.2$), however after 20 rounds of the simulation, they are converted to satisfactory providers with p value equal to 0.9 (i.e. $P(S = \text{"satisfactory"}) = 0.9$). As shown in Fig. 7 the CLA converges correctly to the new satisfactory providers while the other method is unable to reach 100% precision even after 240 rounds of interactions.

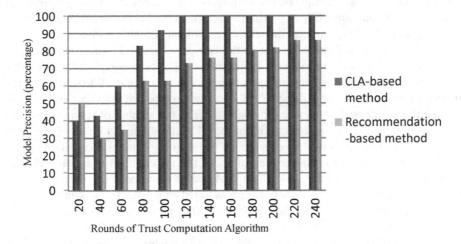

Fig. 7. Convergence in a dynamic environment

Experiment 4: In this experiment to show the scalability of the CLA method, the cumulative simulation time in successive rounds for trust computation for both methods is illustrated. As shown in Fig. 8, the CLA method outperforms the recommendation based method as the round number increases.

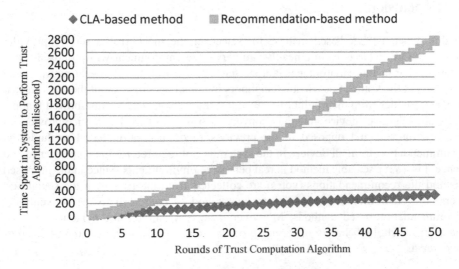

Fig. 8. The cumulative simulation time of both methods

4.3   Discussion

The previous experiments showed that when the behavior of providers are modeled by a random variables S and M (as explained previously), the CLA method not only requires fewer interactions to reach 100% precision comparing to the recommendation based method but also it is more scalable. Scalability of the method is obtained due to the voting mechanism rather than the recommendation mechanism. the former only requires collecting last actions of the neighbors (last selected service provider) to update the selection probability of probability vector of the learning automata using the local rule while the latter collects the recommendations (which are interaction logs of all providers) from all devices present in the pervasive network which is practically infeasible in real pervasive healthcare environments. The time complexity of the trust computation algorithm in recommendation-based method is $O(nvt)$, Where n shows the number of service requester devices, t shows the number of previous interactions between the device and the service providers and v is the number of service providers in the environment. The time complexity of the CLA algorithm is $O(m + v)$, where m is the number of adjacent devices of a device ($m \ll n$ in social network graphs).

The most significant characteristics of the CLA based method is its ability of adaptation to new environment conditions. As shown in the third experiment the devices were able to successfully select the new satisfactory providers, which were not initially satisfactory but overtake the previous satisfactory ones after rounds. It is due to the fact that in a CLA method even malicious providers have a little chance to be selected all the time based on the probability vector of the learning automata. The more interactions occur the lower the probability of selecting a malicious provider will be. Hence if a malicious provider is converted to a satisfactory one in the CLA method it will have the chance of selection and hence be rewarded gradually. However in a recommendation based method once a provider is found malicious, it is permanently ignored in next interactions.

5   Conclusions

In this paper a CLA based method for selecting the most trusted service provider among the available ones in a health-care pervasive environment was proposed. Each smart device in this environment is a subscriber of a health-care social network and may collect the votes of its friends (neighbors in social network) about the most trusted service provider from their point of view at last interaction. The results showed that the CLA of each smart device gradually converges to the most trusted service provider in the environment and opposed to the previous recommendation based methods the recommendation from all devices in the system is not required for the convergence and hence it is a very scalable method in real pervasive environments. Since in the proposed method the recommendations (votes) are collected from the friends in the social net-work, the credibility of votes is more guaranteed as compared to the conventional recommendation-based methods. Moreover it is very suitable for dynamic environments in which the behavior of service providers changes from trusted to un-trusted or vice versa.

As the future work we aim to consider a more general model in which (1) some adjacent cells in the CLA may abstain from voting, (2) no majority is reached, or (3) the votes are out-dated.

References

1. Smith, K.P., Christakis, N.A.: Social networks and health. Ann. Rev. Sociol. **34**, 405–429 (2008)
2. Griffiths, F., Cave, J., Boardman, F., Ren, J., Pawlikowska, T., Ball, R., Clarke, A., Cohen, A.: Social networks-the future for health care delivery. Soc. Sci. Med. **75**, 1–9 (2012). doi:10.1016/j.socscimed.2012.08.023
3. Varshney, U.: Pervasive healthcare. IEEE Comput. **36**, 138–140 (2003)
4. Lefebvre, R.C., Bornkessel, A.S.: Digital social networks and health. J. Am. Heart Assoc. **127**, 1829–1836 (2013)
5. McKnight, D.H., Chervani, N.L.: The meanings of trust. In: The MISRC Working Paper Series, Technical report 94-04, University of Minnesota (1996)
6. Denko, M.K., Sun, T., Woungang, I.: Trust management in ubiquitous computing: a Bayesian approach. Comput. Commun. **34**, 398–406 (2011). doi:10.1016/j.comcom.2010. 01.023
7. Haque, M.M., Ahamed, S.I.: An omnipresent formal trust model (FTM) for pervasive healthcare environment. In: Proceedings of the 31st Annual International Conference on Computer Software and Applications, pp. 49–56. IEEE Computer Society (2007)
8. Meybodi, M.R., Beygi, H., Taherkhani, M.: Cellular learning automata and its applications. J. Sci. Technol. **25**, 54–77 (2004)
9. Boukerche, A., Ren, Y.: A trust-based security system for ubiquitous and pervasive computing environments. Comput. Commun. **31**, 4343–4351 (2008). doi:10.1016/j. comcom.2008.05.007
10. Denko, M., Sun, T.: Probabilistic trust management in pervasive computing. In: IEEE/IFTP International Conference an Embedded and Ubiquitous Computing, Shangai, China, pp. 610–615 (2008). doi:10.1109/EUC.2008.149
11. Dong, Z., Yian, Z., Wanbao, L., Jianhua, G., Yunlan, W.: Multilevel trust management framework for pervasive computing. In: International Conference on Knowledge Discovery and Data Mining, pp. 159–162 (2010)
12. Valarmathi, J., Preethi, R., Bhuvaneshwari, N., Karthick, V.: A novel trust management scheme in pervasive healthcare. In: IEEE International Conference on Recent Trends in Information Technology, pp. 503–508 (2011)
13. Valarmathi, J., Lakshmi, K., Menaga, R.S., Abirami, K.V., Uthariaraj, V.R.: SLA for a pervasive healthcare environment. Adv. Comput. Inf. Technol. **176**, 141–149 (2012)
14. Iltaf, N., Ghafoor, A., Zia, U.: An attack resistant method for detecting dishonest recommendation in pervasive computing environment. In: IEEE, pp. 173–178 (2012). ISBN 978-1-4673-4523-1
15. Beygi, H., Meybodi, M.R.: Asynchronous cellular learning automata. Automatica **44**, 1350–1357 (2008)
16. Wolfram, S.: Cellular automata. Los Alamos Sci. **9**, 2–21 (1983)
17. Narendra, K.N., Thathachar, M.A.L.: Learning automata-a survey. IEEE Trans. Syst. Man Cybern. **SMC-4**, 323–334 (1974)
18. Thathachar, M.A.L., Sastry, P.S.: Varieties of learning automata: an overview. IEEE Trans. Syst. Man Cybern.-Part B: Cybern. **32**, 711–722 (2002)

Physical-Layer Security for DF Two-Way Dual-Hop Cooperative Wireless Networks over Nakagami-m Fading Channels

Islam M. Tanash, Mamoun F. Al-Mistarihi$^{(\boxtimes)}$,
and Amer M. Magableh

Electrical Engineering Department,
Jordan University of Science and Technology, Irbid, Jordan
islam_mohammad16@yahoo.com,
{mistarihi,ammagableh}@just.edu.jo

Abstract. Cooperative communication is a new technique that enables single antenna devices to share their antennas and help other nodes in relaying their signals. Hence, higher spatial diversity, reduced power consumption and improved reception reliability are achieved. In this paper, two way dual-hop cooperative network with source and jammer for decode and forward relaying scheme and under physical layer security is analyzed and investigated, where half-duplex mode and Nakagami-*m* fading channels are employed. The secrecy outage probability and the correct computation probability for single relay are derived to evaluate the secrecy performance of the system. Theoretical results are given for the derived expressions.

Keywords: Cooperative communication network · Physical layer security · Two-way communication · Jammer · Decode and forward · Secrecy outage probability

1 Introduction

Cooperative networks have been incorporated in the communication systems by developers and researchers due to its advantages, both the spatial diversity and the throughput will be increased while the delay, the power consumption and the interference will be decreased [1]. A two-hop one way cooperative communication system has been proposed in [2], where it consists of a single source, single destination and multiple relays with DF or AF relying scheme and the best relay was chosen based on multiple criteria such as partial relay or opportunistic relay selection [3, 4]. In partial relay selection, only the source-relay SNR is considered but in the opportunistic relay selection, the two hubs' SNR is considered leading to a better performance.

Due to the broadcast nature of wireless signals, the communication process across networks will be vulnerable to wireless attacks such as eavesdropping attacks, in which non-legitimate user tries to overhear the exchanged data between legitimate users. Hence, a big attention has been taken toward the security of the physical layer.

© Springer International Publishing AG 2017
O. Galinina et al. (Eds.): NEW2AN/ruSMART/NsCC 2017, LNCS 10531, pp. 258–269, 2017.
DOI: 10.1007/978-3-319-67380-6_24

Physical layer security was first introduced by Wyner who showed that a secure communication can be obtained when the quality of the wiretap channel is less than that of the main channel, where the wiretap channel the channel between the source and the eavesdropper [5]. In [6], a closed form expression for the secrecy capacity, which is defined as the difference between the main link capacity and the wiretap link capacity and intercept probability over Rayleigh fading channels were derived. Relay selection methods were proposed in [7] to enhance physical layer security, where optimal methods were used to select the relay that gives the best secrecy path and hence maximizing the achievable secrecy rate. On the other hand, jamming signals were exploited to provide security by degrading the quality of wiretap links [8]. A dual-hop cooperative system over Rayleigh fading channels and half-duplex mode was studied [9], where jammer and relay selection methods were investigated to provide a secure system when one or more eavesdroppers tries to overhear the transmitted data packets. Outage probability and ergodic secrecy capacity were derived to evaluate the performance of the system.

In [10], Cooperative wireless network with two communicating parties, a relay and a single eavesdropper was studied recently with half-duplex mode over Rayleigh fading channels, the operation principle took three time slots to exchange information between the two communicating parties, and the jamming signals were sent in the first and second time slots. Closed form expressions for the secrecy outage probability in case of single relay and multiple relays were derived. Also, the probability of the correct computation at the eavesdropper node was derived in order to evaluate the secrecy performance of the system.

In this paper, the given system model will be analyzed over Nakagami-m fading channels and an exact closed form expression of the outage probability in case of single relay will be derived. In addition, an exact expression of the correct computation probability of the secrecy packets x_1 and x_2 at the eavesdropper node will be derived.

The remaining of the paper is organized as follows. Sections 2 and 3 describe and analyze the proposed system model and its performance. Wiretap analysis is discussed in Sect. 4. Theoretical results are presented in Sect. 5 and finally, the paper is concluded in Sect. 6.

2   System and Channel Model

As shown in Fig. 1, the system model consists of two communicating parties A and B communicating with each other through an intermediate node (r) without a direct link and an eavesdropper tries to get the information exchanged between them. Each terminal is equipped with only one antenna and operates in half-duplex mode meaning that nodes cannot communicate simultaneously. This system model contains multiple links which are $A - r, B - r, r - E, A - E$ and $B - E$. Each link has a path loss exponent β which ranges from $2 \rightarrow 4$ and a fading channel coefficient given respectively by h_1, h_2, h_3, h_4 and h_5, all of these fading coefficients are statically independent of each other. The link distance is denoted by $d_i, i \in \{1, 2, \ldots 5\}$ which is used in calculating the mean channel power of Nakagami-m distribution. The square of the fading coefficient h_i^j is denoted by ω_i^j, where $i = \{1, 2 \ldots 5\}$ refers to the link and $j = \{1, 2, 3\}$ refers to the time slot. Since the fading coefficient h_i^j follows Nakagami-m distribution, then the square of the fading coefficient ω_i^j follows gamma distribution with average channel power $\Omega_i = \frac{m_i}{d_i^\beta}$, where m_i is the shape parameter and it is assumed to have integer values throughout this paper, whereas, β is the path loss exponent.

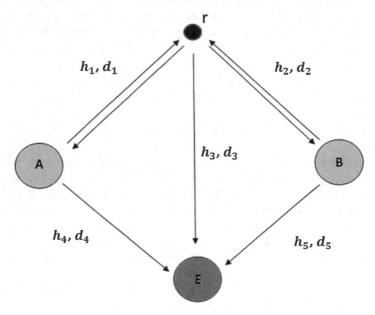

Fig. 1. System model for the spectrum sharing relay network

It is assumed that the two communicating parties and the relay know all channel state information including the CSI of the eavesdropper links through pilot sequences which are used for channel estimation. Also, the proposed system model provides secure communication by sending jamming signals which are known to the two communicating parties and to the relay and unknown to the eavesdropper and hence it is called integration of source and jammer protocol (ISJ). It should be mentioned that sending intentional jamming signals achieves more security at the expense of increasing the power consumed. ISJ protocol is based on time-division channel model, in which the operation process takes three time slots.

In the first time slot, node A transmits its data signal (x_1) to the relay and to the eavesdropper and at the same time, node B transmits its jamming signal (j_B) to the relay and to the eavesdropper. Data signal (x_1) will be decoded and stored by the relay. In the second time slot, node B transmits its data signal (x_2) to the relay and to the eavesdropper and at the same time, node A transmits its jamming signal (j_A) to the relay and to the eavesdropper. Data signal (x_2) will be decoded and stored by the relay. In the third time slot, the relay will perform XOR operation between the two received data signals x_1 and x_2 and will broadcast the resulted signal x to the two communicating parties and to the eavesdropper.

Jamming signals are used to reduce the quality of the links $A - E$ and $B - E$ but $A - r$ and $B - r$ links will not be affected since the relay knows the jamming signals j_A and j_B. Formulating the given system model by mathematical expressions, the signals received in the first time slot at the relay and at the eavesdropper are written respectively as [10]:

$$y_r^1 = \sqrt{aP}x_1h_1^1 + \sqrt{(1-b)P}j_Bh_2^1 + n_r \tag{1}$$

$$y_E^1 = \sqrt{aP}x_1h_4^1 + \sqrt{(1-b)P}j_Bh_5^1 + n_E \tag{2}$$

The received signals in the second time slot at the relay and the eavesdropper are written respectively as [10]:

$$y_r^2 = \sqrt{bP}x_2h_2^2 + \sqrt{(1-a)P}j_Ah_1^2 + n_r \tag{3}$$

$$y_E^2 = \sqrt{bP}x_2h_5^2 + \sqrt{(1-a)P}j_Ah_4^2 + n_E \tag{4}$$

where P represents the total power transmitted at each node (the two communicating parties and the relay node powers are assumed equal) and aP, bP are the power fractions assigned for transmitting the signal packets x_1 and x_2, respectively. Whereas the remaining power fractions are assigned for transmitting the jamming packets j_A and j_B and are given respectively as $(1-a)P$ and $(1-b)P$, $0 \leq a, b \leq 1$. The noise terms are assumed to be an additive white gaussian noise (AWGN) with zero mean and variance N_0, $n \sim \mathcal{N}(0, N_0)$.

The terms $\sqrt{(1-b)P}j_Bh_2^1$ and $\sqrt{(1-a)P}j_Ah_1^2$ in (1) and (3) are omitted because the jamming packets j_B and j_A are known to the relay, so, the received signal at relay r in the first and second time slots becomes [10]:

$$y_r^1 = \sqrt{aP}x_1h_1^1 + n_r \tag{5}$$

$$y_r^2 = \sqrt{bP}x_2h_2^2 + n_r \tag{6}$$

At the third time slot, the received signal at the two communicating parties and at the eavesdropper are given respectively by [10]:

$$y_A^3 = \sqrt{P}xh_1^3 + n_A \tag{7}$$

$$y_B^3 = \sqrt{P}xh_2^3 + n_B \tag{8}$$

$$y_E^3 = \sqrt{P}xh_3^3 + n_E \tag{9}$$

where $x = x_1 \oplus x_2$

Node A can extract the signal packet x_2 from the received packet x by xoring it with its packets, this can be represented as [10]:

$$x_1 \oplus x = x_1 \oplus x_1 \oplus x_2 = x_2 \qquad (10)$$

Similarly, node B can extract the signal packet x_1 from the received packet x by xoring it with its packets, this can be represented as [10]:

$$x_2 \oplus x = x_2 \oplus x_1 \oplus x_2 = x_1 \qquad (11)$$

3 Performance Analysis

The performance of the proposed system model is evaluated by finding a closed form expressions for the secrecy outage probability and the correct computation probability. In order to define the secrecy outage probability, the achievable secrecy rate (ASR) and the end-to-end ASR of a two way scheme must be introduced. The ASR of a source node is defined as the achievable data rate difference between the main link and the wiretap link, whereas the end-to-end ASR is defined as the minimum ASR across the two hops of the relay for a two way scheme.

The system is said to be in outage state when the end to end ASR is smaller than a predefined target secrecy rate (R_0). The two communicating parties receive data only in the third time slot and hence, the total received SNR at node A and node B in the third time slot are given respectively as [10]:

$$\gamma_A^3 = \frac{P|h_1^3|^2}{N_0} = \gamma \omega_1^3 \qquad (12)$$

$$\gamma_B^3 = \frac{P|h_2^3|^2}{N_0} = \gamma \omega_2^3 \qquad (13)$$

where $\gamma = \frac{P}{N_0}$

Maximal Ratio Combining (MRC) method is used at the eavesdropper node for combining the received signals. The total received SNRs at the eavesdropper of the two links $A - E$ and $B - E$ are given respectively as [10]:

$$\gamma_{E_r}^A = \frac{aP|h_4^1|^2}{(1-b)P|h_5^1|^2 + N_0} + \frac{P|h_3|^2}{N_0} = \frac{a\gamma\omega_4^1}{(1-b)\gamma\omega_5^1 + 1} + \gamma\omega_3 \qquad (14)$$

$$\gamma_{E_r}^B = \frac{bP|h_5^2|^2}{(1-a)P|h_4^2|^2 + N_0} + \frac{P|h_3|^2}{N_0} = \frac{b\gamma\omega_5^2}{(1-a)\gamma\omega_4^2 + 1} + \gamma\omega_3 \qquad (15)$$

Depending on the above signal to noise ratios, the achievable data rates of channels $B - A$, $A - B$ and eavesdropped links $A - E$ and $B - E$ are given respectively as [10]:

$$R_r^{BA} = \frac{1}{3} log_2 \left(1 + \gamma_A^3\right) \tag{16}$$

$$R_r^{AB} = \frac{1}{3} log_2 \left(1 + \gamma_B^3\right) \tag{17}$$

$$R_{E_r}^A = \frac{1}{3} log_2 \left(1 + \gamma_{E_r}^A\right) \tag{18}$$

$$R_{E_r}^B = \frac{1}{3} log_2 \left(1 + \gamma_{E_r}^B\right) \tag{19}$$

ISJ protocol takes three time slots to transmit signals that is why the factor $\left(\frac{1}{3}\right)$ is added above. The achievable data rates of channel $B - A$ and link $B - E$ are used to find the ASR of node A as follows [10]:

$$
\begin{aligned}
ASR_r^A &= \left[R_r^{BA} - R_{E_r}^B\right]^+ = \left[\frac{1}{3} log_2 \left(\frac{1 + \gamma_A^3}{1 + \gamma_{E_r}^B}\right)\right]^+ \\
&= \left[\frac{1}{3} log_2 \left(\frac{1 + \gamma\omega_1^3}{1 + \frac{b\gamma\omega_5^2}{(1-a)\gamma\omega_4^2 + 1} + \gamma\omega_3}\right)\right]^+
\end{aligned} \tag{20}
$$

Similarly, the achievable data rates of channel $A - B$ and link $A - E$ are used to find the ASR of node B as follows [10]:

$$ASR_r^B = \left[\frac{1}{3} log_2 \left(\frac{1 + \gamma_B^3}{1 + \gamma_{E_r}^A}\right)\right]^+ = \left[\frac{1}{3} log_2 \left(\frac{1 + \gamma\omega_1^3}{1 + \frac{b\gamma\omega_5^2}{(1-a)\gamma\omega_4^2 + 1} + \gamma\omega_3}\right)\right]^+ \tag{21}$$

where $[x]^+ = max\{0, x\}$.

The secrecy outage probability is expressed using mathematical formula as:

$$F_{ASR_r}(R_0) = P_{M=1}^{fixed} = Pr[ASR_r < R_0] = 1 - Pr[ASR_r \geq R_0] \tag{22}$$

where $ASR_r = min\left(ASR_r^A, ASR_r^B\right)$.

The cumulative distribution function (cdf) of ASR is found in the high SNR region as:

$$F_{ASR_r}(R_0) = 1 - e^{-p_x\left(\frac{m_1}{\Omega_1} + \frac{m_2}{\Omega_2} - \frac{B}{2\theta_x}\right)} \left(\frac{m_3}{\Omega_3}\right)^{m_3} n_x^k m_x^w \theta_x^{-m_3}$$

$$\times \sum_{i=0}^{m_1-1} \sum_{j=0}^{m_2-1} \sum_{k=0}^{i} \sum_{w=0}^{j} \frac{1}{i!j!} \left(\frac{m_1}{\Omega_1}\right)^i \left(\frac{m_2}{\Omega_2}\right)^j \binom{i}{k}\binom{j}{w} A \left(\frac{B}{\theta_x}\right)^{-\left(\frac{m_3+i+j-k-w+1}{2}\right)}$$

$$\times p_x^{\frac{m_3+i+j-k-w-1}{2}} W_{\frac{i+j-k-w-m_3+1}{2},\frac{k+w-i-j-m_3}{2}} \left(\frac{Bp_x}{\theta_x}\right)$$

$$\times G_{1,2}^{2,1}\left(\frac{n_x m_1 m_4 \Omega_5}{\Omega_1 \Omega_4 m_5} \middle| \begin{matrix} 1-k-m_5 \\ m_4-k,0 \end{matrix}\right) G_{1,2}^{2,1}\left(\frac{m_x m_2 m_5 \Omega_4}{\Omega_2 \Omega_5 m_4} \middle| \begin{matrix} 1-w-m_4 \\ m_5-w,0 \end{matrix}\right)$$

$$(23)$$

where $A = \left(\frac{m_4\Omega_5}{\Omega_4 m_5}\right)^k \left(\frac{m_5\Omega_4}{\Omega_5 m_4}\right)^w / (\Gamma^2(m_4)\Gamma^2(m_5))$, $B = \frac{m_3}{\Omega_3} + \theta_x\left(\frac{m_1}{\Omega_1} + \frac{m_2}{\Omega_2}\right)$, $\theta_x = 2^{3x}$, $G_{p,q}^{m,n}$
$\left(x \middle| \begin{matrix} a_1,\ldots,a_n \\ b_1,\ldots,b_n \end{matrix}\right)$ is the Meijer G-function and $W_{\lambda,\mu}(z)$ is the Whittaker function. It should be mentioned that in this paper, the power fractions a and b are chosen independently of successful and correct decoding conditions at the relay, and the preferred range of a and b is from 0.6 to 0.9 to obtain a good data rate and a good secrecy performance.

4 Wiretap Analysis

During the process of communication between the two communicating parties, the eavesdropper tries to overhear the data exchanged between them. In the first time slot, the eavesdropper receives the signal given by (2) which consists of the signal packet x_1 and the jamming packet j_B. The eavesdropper stores the received signal y_E^1 and tries to decode the jamming packet and stores it in the internal memory. Similarly, in the second time slot, the eavesdropper receive the signal given by (4) which consists of the signal packet x_2 and the jamming packet j_A.

The eavesdropper stores the received signal y_E^2 and tries to decode the jamming packet and stores it in the internal memory. In the third time slot, the eavesdropper uses the stored jamming packets to omit the components $\sqrt{(1-b)P}j_Bh_5^1$ and $\sqrt{(1-a)P}$ $j_Ah_4^2$ from (2) and (4) respectively (it will omit these component if the eavesdropper successfully decoded j_A and j_B), then the eavesdropper combines the remaining signals with the received signal y_E^3 given in (9) to increase the efficiency of the wiretap link. Strong and fast processing unit with large internal memory is needed at the eavesdropper to accomplish all these complicated operations.

The following condition must apply in order to successfully decode the two jamming packets j_A and j_B at the eavesdropper.

$$\begin{cases} \frac{1}{3}log_2\left(1 + \frac{(1-b)\gamma\omega_5^1}{a\gamma\omega_4^1 + 1}\right) \geq R_0 & decode\ j_B \\ \frac{1}{3}log_2\left(1 + \frac{(1-a)\gamma\omega_4^2}{b\gamma\omega_5^2 + 1}\right) \geq R_0 & decode\ j_A \end{cases} \quad (24)$$

The received SNRs of the two links $A - E$ and $B - E$ are given by:

$$
\gamma_{E_r}^A = \begin{cases} a\gamma\omega_4^1 + \gamma\omega_3 & j_B \text{ is successfully decoded} \\ \dfrac{a\gamma\omega_4^1}{(1-b)\gamma\omega_5^1} + \gamma\omega_3 & j_B \text{ is not successfully decoded} \end{cases} \tag{25}
$$

$$
\gamma_{E_r}^B = \begin{cases} a\gamma\omega_5^2 + \gamma\omega_3 & j_A \text{ is successfully decoded} \\ \dfrac{b\gamma\omega_5^2}{(1-a)\gamma\omega_4^2} + \gamma\omega_3 & j_A \text{ is not successfully decoded} \end{cases} \tag{26}
$$

From (24) and (25), the probability of correct computation of the secrecy packet x_1 is given as [10]:

$$
P_{CC}^{x_1} = Pr\left[\frac{1}{3}log_2\left(1 + \frac{(1-b)\gamma\omega_5^1}{a\gamma\omega_4^1 + 1}\right) \ge R_0, \frac{1}{3}log_2\left(1 + a\gamma\omega_4^1 + \gamma\omega_3\right) \ge R_0\right]
$$
$$
+ P_r\left[\frac{1}{3}log_2\left(1 + \frac{(1-b)\gamma\omega_5^1}{a\gamma\omega_4^1 + 1}\right) < R_0, \frac{1}{3}log_2\left(1 + \frac{a\gamma\omega_4^1}{(1-b)\gamma\omega_5^1 + 1} + \gamma\omega_3\right) \ge R_0\right] \tag{27}
$$

A closed form expression for the probability of the correct computation of the secrecy packet x_1 is given as in (28), where $B(x,y)$ is the beta function, $_1F_1(\alpha;\gamma;Z)$ is the confluent hypergeometric function and $_2F_1(\alpha,\beta;\gamma;Z)$ is the gauss hypergeometric function.

Equation (28) gives the probability that the eavesdropper will correctly overhear the secrecy packet x_1, while that of the secrecy packet x_2 is found to be the same as $P_{CC}^{x_1}$ given in (28) but with replacing a by b, m_4 by m_5 and Ω_4 by Ω_5 because of the symmetry of the system model. The detailed mathematical derivations of (23), (27), and (28) will be submitted to a journal for publication.

$$
P_{CC}^{x_1} = \left(\frac{m_4}{\Omega_4}\right)^{m_4}\frac{1}{\Gamma(m_4)}\sum_{i=0}^{m_5-1}\frac{1}{i!}\left(\frac{am_5\theta_{R_0R_0}}{(1-b)\Omega_5}\right)^i\left(\frac{m_4}{\Omega_4} + \frac{am_5\theta_{R_0R_0}}{(1-b)\Omega_5}\right)^{-m_4-i}
$$
$$
\times \Gamma\left(m_4+i, \left(\frac{m_4}{\Omega_4} + \frac{am_5\theta_{R_0R_0}}{(1-b)\Omega_5}\right)\frac{\theta_{R_0R_0}}{a\gamma}\right)
$$
$$
+ \left(\frac{m_4}{\Omega_4}\right)^{m_4}\frac{1}{\Gamma(m_4)}e^{-\frac{m_3\theta_{R_0R_0}}{\gamma\Omega_3}}\sum_{i=0}^{m_5-1}\sum_{j=0}^{m_3-1}\frac{1}{i!j!}\left(\frac{am_3}{\Omega_3}\right)^j\left(\frac{am_5\theta_{R_0R_0}}{(1-b)\Omega_5}\right)^i\left(\frac{\theta_{R_0R_0}}{a\gamma}\right)^{m_4+i+j}
$$
$$
\times B(j+1, m_4+i) \, _1F_1\left(m_4+i; m_4+i+j+1; -\left(\frac{m_4}{\Omega_4} + \frac{am_5\theta_{R_0R_0}}{(1-b)\Omega_5} - \frac{am_3}{\Omega_3}\right)\frac{\theta_{R_0R_0}}{a\gamma}\right)
$$
$$
+ \left(\frac{m_4\Omega_5}{\Omega_4 m_5}\right)^{m_4}\frac{\Gamma(m_4+m_5)max(y_0, y_1)^{-m_5}}{m_5\Gamma(m_4)\Gamma(m_5)\left(\frac{m_4\Omega_5}{\Omega_4 m_5}\right)^{m_4+m_5}} \, _2F_1\left(m_4+m_5, m_5; m_5+1; \frac{-1}{\left(\frac{m_4\Omega_5}{\Omega_4 m_5}\right)max(y_0, y_1)}\right)
$$
$$
+ \left[\left(\frac{m_4\Omega_5}{\Omega_4 m_5}\right)^{m_4}\frac{\Gamma(m_4+m_5)}{\Gamma(m_4)\Gamma(m_5)}\sum_{i=0}^{m_3-1}\sum_{k=0}^{i}\sum_{w=0}^{\infty}\frac{-1^{(2i-k)}y_1^{(i-k)}}{i!w!(m_4+w+k)}\binom{i}{k}\left(\frac{am_3}{\gamma(1-b)\Omega_3}\right)^{i+w}\right.
$$
$$
\times \left[y_1^{m_4+k+w} \, _2F_1\left(m_4+k+w, m_4+m_5; m_4+k+w+1; -\frac{m_4\Omega_5}{\Omega_4 m_5}y_1\right)\right.
$$
$$
\left.\left.-y_0^{m_4+k+w} \, _2F_1\left(m_4+k+w, m_4+m_5; m_4+k+w+1; -\frac{m_4\Omega_5}{\Omega_4 m_5}y_0\right)\right]\right]^+
$$
$$
\tag{28}
$$

5 Theoretical Results

Two closed form expressions are derived for the two way cooperation communication system model over Nakagamai-m fading channels. The derived expressions are analyzed and discussed in this section. The coordinates of the nodes in the system are distributed as follows: node A is located at the origin $(0,0)$, node B is located at $(1,0)$, the relay is located at $(x_r,0)$ and the eavesdropper is located at (x_E,x_E) where $0<x_r,x_E<1$. Hence, the distances of the wireless links which are used in finding the average channel power $\left(\Omega_i = \frac{m_i}{d_i^\beta}\right)$ where β is assumed constant and equals 3, are calculated as follows: $d_1 = x_r$, $d_2 = 1 - x_r$, $d_3 = \sqrt{(x_E - x_r)^2 + y_E^2}$, $d_4 = \sqrt{x_E^2 + y_E^2}$ and $d_5 = \sqrt{(1 - x_E)^2 + y_E^2}$.

Figure 2 shows the relationship between the secrecy outage probability of the two communicating parties and the distance from the eavesdropper to the relays (d_3), where as d_3 increases the secrecy outage probability decreases until it saturates at different levels for the different values of the parameters a and b because at that point of saturation, the eavesdropper node trivially affects the two-way communication. Hence, the performance of the system is better when the eavesdropper is further from the

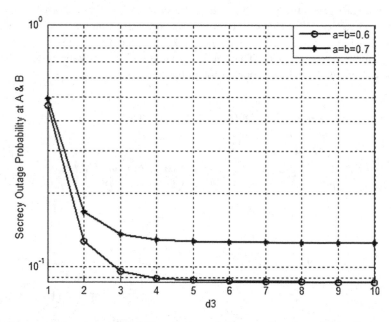

Fig. 2. The secrecy outage probability at A and B as a function of the distance of the eavesdropper node to relays (d_3) when $R_0 = 1$ bit/s/Hz, $SNR = 15$ dB, $d_1 = d_2 = 0.5$ for different parameters values (a, b) over Nakagami-m fading channel with $(m_1 = m_2 = 2, m_3 = m_4 = m_5 = 1)$

relays. It is worth mentioning that the location of the eavesdropper (x_E) is constant on the x-axis and equals 0.5, whereas the location (y_E) on the y-axis is changing from -1 to -10.

Figure 3 depicts the relationship between the sum secrecy outage probability of the two communicating parties and the location of the relays (x_r), where the best performance is achieved when the relays are located at the midpoint between the two communicating parties $(x_r = 0.5)$. Also, it can be shown that as the eavesdropper moves away from the relays on the y-axis, the secrecy outage probability decreases.

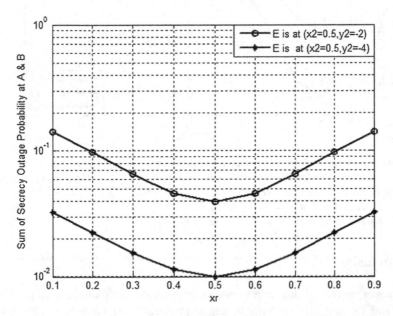

Fig. 3. Sum secrecy outage probability of the ISJ protocol as a function of location of relays x_r on x-axis when the $SNR = 15$ dB, $R_0 = 0.5$ bit/s/Hz, $a = b = 0.6$, and the location of the eavesdropper node is considered at $(0.5, -2)$ and $(0.5, -4)$ over Nakagami-m fading channel with $(m_1 = m_2 = 2, m_3 = m_4 = m_5 = 1)$

Figure 4 depicts the effect of changing the values of the power fractions a and b on the probability of correct computation, whereas the transmission power of the packets x_1 and x_2 increases, the probability of correct computation increases and hence, secrecy performance decreases. So, the performance of the eavesdropper is better when the transmission power allocated for jamming signals at the two communicating parties is small.

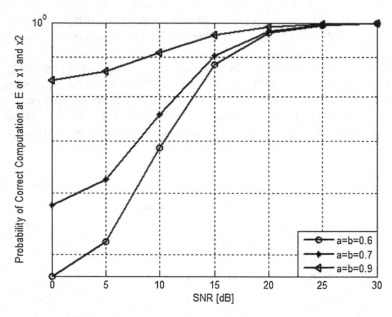

Fig. 4. The probability of the correct computation at the eavesdropper node of the secrecy packets x_1 and x_2 as a function of SNR (dB) when $R_0 = 0.5$ bit/s/Hz, the location of the eavesdropper node is considered at $(0.5, 2)$, $d_1 = d_2 = 0.5$, $d_3 = 2$, $d_4 = d_5 = 2.06$ and power fractions are considered at $(a = b = 0.6)$, $(a = b = 0.7)$ and $(a = b = 0.9)$ over Nakagami-m fading channel with $(m_1 = m_2 = 2, m_3 = m_4 = m_5 = 1)$

6 Conclusion

In this paper, two way dual-hop cooperative network with source and jammer for decode and forward relaying scheme under physical layer security is analyzed and investigated over Nakagami-m fading channels. The secrecy outage probability and the correct computation probability for single relay are derived to evaluate the secrecy performance of the system. Better secrecy performance is obtained when using jamming signals.

References

1. Nosratinia, A., Hunter, T.E., Hedayat, A.: Cooperative communication in wireless networks. IEEE Commun. Mag. **42**(10), 74–80 (2004)
2. Laneman, J.N., Tse, D.N.C., Wornell, G.W.: Cooperative diversity in wireless networks: efficient protocols and outage behavior. IEEE Trans. Inf. Theory **50**(12), 3062–3080 (2004)
3. Krikidis, I., Thompson, J., Mclaughlin, S., Goertz, N.: Amplify-and-forward with partial relay selection. IEEE Commun. Lett. **12**(4), 235–237 (2008)
4. Xia, M., Aissa, S.: Cooperative AF relaying in spectrum-sharing systems: outage probability analysis under co-channel interferences and relay selection. IEEE Trans. Commun. **60**(11), 3252–3262 (2012)

5. Wyner, A.D.: The wire-tap channel. Bell Syst. Tech. J. **54**(8), 1355–1367 (1975)
6. Yulong, Z., Xianbin, W., Weiming, S.: Optimal relay selection for physical-layer security in cooperative wireless networks. IEEE J. Sel. Areas Commun. **31**(10), 2099–2111 (2013)
7. Krikidis, I.: Opportunistic relay selection for cooperative networks with secrecy constraints. IET Commun. **4**(15), 1787–1791 (2010)
8. Krikidis, I., Thompson, J.S., McLaughlin, S.: Relay selection for secure cooperative networks with jamming. IEEE Trans. Wirel. Commun. **8**(10), 5003–5011 (2009)
9. Ibrahim, D.H., Hassan, E.S., El-Dolil, S.A.: A new relay and jammer selection schemes for secure one-way cooperative networks. Wirel. Pers. Commun. **75**, 665–685 (2014). (Springer)
10. Son, P.N., Kong, H.Y.: An integration of source and jammer for a decode-and-forward two-way scheme under physical layer security. Wirel. Pers. Commun. **79**(3), 1741–1764 (2014). (Springer)

Physical-Layer Security for DF Two-Way Full-Duplex Cooperative Wireless Networks over Rayleigh Fading Channels

Islam M. Tanash, Amer M. Magableh, and Mamoun F. Al-Mistarihi$^{(\boxtimes)}$

Electrical Engineering Department,
Jordan University of Science and Technology, Irbid, Jordan
islam_mohammad16@yahoo.com,
{ammagableh, mistarihi}@just.edu.jo

Abstract. Cooperative communication is a new method that uses multi-hop communication to send the information from source to sink. The broadcast nature of wireless signals causes the communication process to be vulnerable to wireless attacks such as eavesdropping attacks, in which a non-legitimate user tries to overhear the exchanged data between legitimate users. Physical layer security can be achieved by using jamming signals. This paper discusses and analyzes the proposed system model for a decode-and-forward two-way scheme under physical layer security where full-duplex mode and Rayleigh fading channels are employed. The secrecy outage probability in case of single relay is derived in order to evaluate the secrecy performance of the proposed system in which the transmission process takes two time slots. Theoretical results are given for the derived expression and a simple comparison with half-duplex system model is made.

Keywords: Cooperative communication network · Full-duplex · Physical layer security · Two-way communication · Jammer · Decode and forward · Secrecy outage probability

1 Introduction

Developers and researchers have showed tremendous interest in cooperative communication. In cooperative communication, when a source node sends data to the destination, multiple nodes in the network will cooperatively participate in the transmission process. Hence, the spatial diversity and the throughput will be increased. On the other hand, power consumption, delay and interference will be reduced. In [1], a one way cooperative communication system has been proposed, in which the network consists of source, destination and multiple relays. The secrecy capacity and intercept probability over Rayleigh fading channels were derived for both amplify and forward (AF) and decode and forward (DF) relaying schemes. In AF relaying scheme, the relay multiplies the received signal by a factor to amplify it and then sends the amplified signal to the destination. On the other hand, in the DF relaying scheme, the relay decodes and re-encodes the received signal and then sends it to the destination.

© Springer International Publishing AG 2017
O. Galinina et al. (Eds.): NEW2AN/ruSMART/NsCC 2017, LNCS 10531, pp. 270–279, 2017.
DOI: 10.1007/978-3-319-67380-6_25

Wireless communication has a broadcast nature, which causes the networks to be vulnerable to wireless attacks. Physical layer security has strongly emerged in wireless communication security due to the broadcast nature of wireless signals. Exploiting jamming signals is one of the effective methods that are used to provide physical layer security. Physical layer security was first introduced by Wyner [2] and jamming signals were exploited in [3] to provide security by degrading the quality of wiretap links, where the wiretap link is the channel between the source and the eavesdropper. One way cooperative wireless networks operating in full-duplex mode were studied in [4], a comparison between half-duplex and full-duplex with and without jamming was made and a closed form expressions for secrecy outage probabilities in both cases were derived. Two-way cooperative networks were proposed in [5–7], in which the information was exchanged between the two communicating parties in half-duplex mode over Rayleigh fading channels through a selected relay. In [8, 9] jamming signals were sent by one of the intermediate nodes in the first and second time phases of the operation process to provide security to the two way system. An extension to this model was proposed in [10], where the number of eavesdroppers was extended from one to multiple. Ergodic secrecy rate and secrecy outage probability were derived for this system model in case of single and multiple eavesdroppers.

In this paper, the given system model is analyzed over Rayleigh fading channels and an exact closed form expression of the outage probability is derived. The remaining of the paper is organized as follows. Sections 2 and 3 describe and analyze the proposed system model and its performance. Theoretical results are presented in Sect. 4 and finally, the paper is concluded in Sect. 5.

2 System and Channel Model

This section introduces a cooperative communication system, which consists of two communicating parties, single relay r and an eavesdropper as seen in Fig. 1. The two communicating parties and the relay are equipped with two antennas, one antenna for receiving and the other for transmitting and operate in full-duplex mode meaning that each node of them can transmit and receive simultaneously over the same channel The proposed communication process between the two communicating parties takes two time slots instead of three as used in the half-duplex system model proposed in [6]. Jamming signals are artificial noise, which are used in this system model to mislead the eavesdropper and to decrease the quality of the wiretap link and hence, enhance the secrecy system performance. Jamming signals are known to the two communicating parties and to the relay node in order not to affect the quality of the main link. Decode and forward relaying scheme is used here, in which the relay will decode the received packet and send the re-encoded packet to the end nodes. Flat Rayleigh fading channels are investigated. A closed form expression for the secrecy outage probability is derived for the proposed system model.

Node A and node B exchange data through an intermediate node without a direct link. The proposed system model has the following links: $A - r, B - r, r - E, A - E$ and $B - E$. These links have Rayleigh fading channel coefficients h_1, h_2, h_3, h_4 and h_5 which are statically independent. Also, these links have link distances, which are

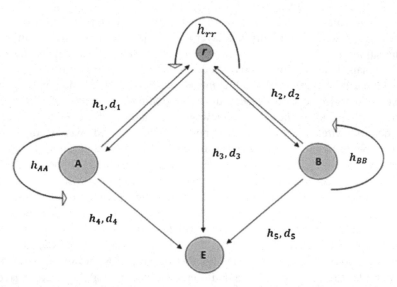

Fig. 1. Two-way system model operating in full-duplex mode under physical layer security

denoted respectively by d_1, d_2, d_3, d_4 and d_5. The square of the channel fading coefficient h_i^2 is denoted by ω_i^j where $i = \{1, 2 \ldots 5\}$ refers to the link, and $j = \{1, 2\}$ refers to the time slot. ω_i^j follows exponential distribution with rate parameter (λ), where $\lambda = d^\beta$ where β is the path loss exponent and its value ranges from 2 to 4. Full-duplex scheme suffers from self-interference at the two communicating parties and at the relay due to the simultaneous reception and transmission, hence, self-interference cancelation methods are used to suppress self-interferences at node A, node B and at the relay, which are denoted by h_{AA}, h_{BB} and h_{rr} respectively, where they are regarded as independent Rayleigh random variables. It is assumed that jamming packets that are sent by the two communicating parties are known to the two communicating parties and to the relay and unknown to the eavesdropper so that the quality of the eavesdropper links is degraded.

The proposed operation scenario is split into two time phases. In the first time phase, node B transmits its data signal x_2 to the relay and to the eavesdropper and at the same time, node A transmits its jamming signal j_A to the relay and to the eavesdropper. Relay r decodes and stores the secrecy packet x_2. Also, at the same time, relay r sends the combined signal which results from XOR operation $(v_1 = x_1 \oplus j_B)$ of the previously received signals which are the secrecy packet x_1 and the jamming packet j_B to the two communicating parties and to the eavesdropper. In the second time phase, node A transmits its data signal x_1 to the relay and to the eavesdropper and at the same time, node B transmits its jamming signal j_B to the relay and to the eavesdropper. The relay r decodes and stores the secrecy packet x_1. Also, at the same time, the relay r sends the combined signal which results from XOR operation $(v_2 = x_2 \oplus j_A)$ of the previously received signals which are the secrecy packet x_2 and the jamming packet j_A to the two communicating parties and to the eavesdropper. It is assumed that all CSI are known by the two communicating parties A and B.

Formulating the given system model, the signals received in the first time slot at the relay, the eavesdropper and the two communicating parties are written respectively as:

$$y_r^1 = \sqrt{bP}x_2h_2^1 + \sqrt{(1-a)P}j_Ah_1^1 + \sqrt{cP}v_1h_{rr}^1 + n_r \tag{1}$$

$$y_E^1 = \sqrt{bP}x_2h_5^1 + \sqrt{(1-a)P}j_Ah_4^1 + \sqrt{cP}v_1h_3^1 + n_E \tag{2}$$

$$y_A^1 = \sqrt{cP}v_1h_1^1 + \sqrt{(1-a)P}j_Ah_{AA}^1 + n_A \tag{3}$$

$$y_B^1 = \sqrt{cP}v_1h_2^1 + \sqrt{bP}x_2h_{BB}^1 + n_B \tag{4}$$

where $v_1 = x_1 \oplus j_B$, and bP, $(1-a)P$ and cP are the powers assigned for transmitting the secrecy packet x_2 at node B, jamming signal j_A at node A and the signal packet v_1 at the relay in the first time slot, respectively.

The received signals in the second time slot at the relay the eavesdropper and the two communicating parties are written respectively as:

$$y_r^2 = \sqrt{aP}x_1h_1^2 + \sqrt{(1-b)P}j_Bh_2^2 + \sqrt{(1-c)P}v_2h_{rr}^2 + n_r \tag{5}$$

$$y_E^2 = \sqrt{aP}x_1h_4^2 + \sqrt{(1-b)P}j_Bh_5^2 + \sqrt{(1-c)P}v_2h_3^2 + n_E \tag{6}$$

$$y_A^2 = \sqrt{(1-c)P}v_2h_1^2 + \sqrt{aP}x_1h_{AA}^2 + n_A \tag{7}$$

$$y_B^2 = \sqrt{(1-c)P}v_2h_2^2 + \sqrt{(1-b)P}j_Bh_{BB}^2 + n_B \tag{8}$$

where $v_2 = x_2 \oplus j_A$, and aP, $(1-b)P$ and $(1-c)P$ are the powers assigned for transmitting the secrecy packet x_1 at node A, jamming signal j_A at node B and the signal packet v_2 at the relay in the second time slot, respectively.

After node B receives the signal v_1 at the first time slot, it extracts x_1 from v_1 by using its jamming packet (j_B) as follows:

$$v_1 \oplus j_B = x_1 \oplus j_B \oplus j_B = x_1 \tag{9}$$

Similarly, after node A receives the signal v_2 at the second time slot, it extracts x_2 from v_2 by using its jamming packet (j_A) as follows:

$$v_2 \oplus j_A = x_2 \oplus j_A \oplus j_A = x_2 \tag{10}$$

Since the secrecy packet x_1 and the jamming packet j_A is sent by node A and the secrecy packet x_2 and the jamming packet j_B is sent by node B, then they are known by the two communicating parties, hence, the components $\sqrt{(1-a)P}j_Ah_{AA}^1$, $\sqrt{bP}x_2h_{BB}^1$, $\sqrt{aP}x_1h_{AA}^2$ and $\sqrt{(1-b)P}j_Bh_{BB}^2$ are omitted from (3), (4), (7) and (8) respectively, and the received signals y_A^1, y_B^1, y_A^2 and y_B^2 are rewritten respectively as:

$$y_A^1 = \sqrt{cP} v_1 h_1^1 + n_A \tag{11}$$

$$y_B^1 = \sqrt{cP} v_1 h_2^1 + n_B \tag{12}$$

$$y_A^2 = \sqrt{(1-c)P} v_2 h_1^2 + n_A \tag{13}$$

$$y_B^2 = \sqrt{(1-c)P} v_2 h_2^2 + n_B \tag{14}$$

3 Performance Analysis

Secrecy outage probability is a metric that is used to evaluate the performance of the proposed system model and is defined as the probability that the end to end ASR is smaller than a predefined target secrecy rate (R_0), where the end-to-end ASR is defined as the minimum ASR across the two hops of the relay for a two way scheme. The ASR of a source node is defined as the achievable data rate difference between the main link and the wiretap link.

The total received SNR at node A and node B are given respectively as:

$$\gamma_A^2 = \frac{(1-c)P|h_1^2|^2}{N_0} = (1-c)\gamma\omega_1^2 \tag{15}$$

$$\gamma_B^1 = \frac{cP|h_2^1|^2}{N_0} = c\gamma\omega_2^1 \tag{16}$$

where $\gamma = \frac{P}{N_0}$

The total received SNRs at the eavesdropper E of the two links $A - E$ and $B - E$ are given respectively as:

$$\gamma_{E_r}^A = \frac{aP|h_4^2|^2}{(1-b)P|h_5^2|^2 + (1-c)P|h_3^2|^2 + N_0} = \frac{a\gamma\omega_4^2}{(1-b)\gamma\omega_5^2 + (1-c)\gamma\omega_3^2 + 1} \tag{17}$$

$$\gamma_{E_r}^B = \frac{bP|h_5^1|^2}{(1-a)P|h_4^1|^2 + cP|h_3^1|^2 + N_0} = \frac{b\gamma\omega_5^1}{(1-a)\gamma\omega_4^1 + c\gamma\omega_3^1 + 1} \tag{18}$$

Depending on the above signal to noise ratios, the achievable data rates of channels $B - A, A - B$ and eavesdropped links $A - E$ and $B - E$ are given respectively as:

$$R_r^{BA} = \frac{1}{2}log_2(1 + \gamma_A^2) \tag{19}$$

$$R_r^{AB} = \frac{1}{2}log_2(1 + \gamma_B^1) \tag{20}$$

$$R_{E_r}^A = \frac{1}{2} log_2 \left(1 + \gamma_{E_r}^A\right) \tag{21}$$

$$R_{E_r}^B = \frac{1}{2} log_2 \left(1 + \gamma_{E_r}^B\right) \tag{22}$$

The proposed operation principle takes two time slots to transmit signals that is why the factor $\left(\frac{1}{2}\right)$ is added above. The achievable data rates of channel $B - A$ and link $B - E$ are used to find the ASR of node A as follows:

$$
\begin{aligned}
ASR_r^A &= \left[R_r^{BA} - R_{E_r}^B\right]^+ = \left[\frac{1}{2} log_2 \left(\frac{1 + \gamma_A^2}{1 + \gamma_{E_r}^B}\right)\right]^+ \\
&= \left[\frac{1}{2} log_2 \left(\frac{1 + (1 - c)\gamma\omega_1^2}{1 + \frac{b\gamma\omega_5^1}{(1-a)\gamma\omega_4^1 + c\gamma\omega_3^1 + 1}}\right)\right]^+
\end{aligned}
\tag{23}
$$

Similarly, the achievable data rates of channel $A - B$ and link $A - E$ are used to find the ASR of node B as follows:

$$
\begin{aligned}
ASR_r^B &= \left[R_r^{BA} - R_{E_r}^B\right]^+ = \left[\frac{1}{2} log_2 \left(\frac{1 + \gamma_B^1}{1 + \gamma_{E_r}^A}\right)\right]^+ \\
&= \left[\frac{1}{2} log_2 \left(\frac{1 + c\gamma\omega_2^1}{1 + \frac{a\gamma\omega_4^2}{(1-b)\gamma\omega_5^2 + (1-c)\gamma\omega_3^2 + 1}}\right)\right]^+
\end{aligned}
\tag{24}
$$

where $[x]^+ = max\{0, x\}$.

The secrecy outage probability is expressed using mathematical formula as:

$$F_{ASR_r}(R_0) = Pr[ASR_r < R_0] = 1 - Pr[ASR_r \geq R_0] \tag{25}$$

where $ASR_r = min\left(ASR_r^A, ASR_r^B\right)$

The cumulative distribution function (cdf) of ASR_r is found in the high SNR region as:

$$
\begin{aligned}
F_{ASR_r}(R_0) = 1 &- \frac{A_1 A_2 n_x m_x \lambda_1 \lambda_2}{\lambda_4 \lambda_5} e^{-p_1 \ddot{e}_1 - p_2 \lambda_2} \left[e^{\frac{n_x \lambda_1 \lambda_4}{(1-a)\lambda_5}} \Gamma\left(-1, \frac{\lambda_1 \lambda_4 n_x}{(1-a)\lambda_5}\right) - e^{\frac{n_x \lambda_1 \lambda_3}{c\lambda_5}} \Gamma\left(-1, \frac{\lambda_1 n_x \lambda_3}{c\lambda_5}\right)\right] \\
&\times \left[e^{\frac{m_x \lambda_2 \lambda_5}{(1-b)\lambda_4}} \Gamma\left(-1, \frac{\lambda_2 \lambda_5 m_x}{(1-b)\lambda_4}\right) - e^{\frac{m_x \lambda_2 \lambda_3}{(1-c)\lambda_4}} \Gamma\left(-1, \frac{\lambda_2 n_x \lambda_3}{(1-c)\lambda_4}\right)\right]
\end{aligned}
\tag{26}
$$

where λ is the rate parameter of the exponential distribution, $p_1 = \frac{2^{2x}-1}{(1-c)\gamma}$, $p_2 = \frac{2^{2x}-1}{c\gamma}$, $n_x = \frac{2^{2x}b}{(1-c)\gamma}$, $m_x = \frac{2^{2x}a}{c\gamma}$, $A_1 = \frac{-\lambda_3 \lambda_4}{(c\lambda_4 - (1-a)\lambda_3)}$, $A_2 = \frac{-\lambda_3 \lambda_5}{((1-c)\lambda_5 - (1-b)\lambda_3)}$, $\Gamma(a, x)$ is the upper incomplete gamma function. The detailed mathematical derivation of (26) will be submitted to a journal for publication.

The preferred range of a and b is from 0.6 to 0.9 to obtain a good data rate and a good secrecy performance, where there is a tradeoff between the amount of mutual information between the two communicating parties and the level of security. On the other hand, the power fraction c is the power allocated to transmit the received signal at the relay and usually it is set to 0.5 in order to have equal power transmission at the relay in the first and second time slots.

4   Theoretical Results

The secrecy outage probability is derived for the two way cooperation communication system model over Rayleigh fading channels. The derived expression is analyzed and discussed in this section. The coordinates of the nodes in the system are distributed as follows: node A is located at the origin $(0,0)$, node B is located at $(1,0)$, the relay is located at $(x_r, 0)$ and the eavesdropper is located at (x_E, x_E) where $0 < x_r, x_E < 1$. Hence, the distances of the wireless links which are used in finding the rate parameter $\lambda_i = d_i^\beta$ where β is assumed constant and equals 3, are calculated as follows: $d_1 = x_r$, $d_2 = 1 - x_r$, $d_3 = \sqrt{(x_E - x_r)^2 + y_E^2}$, $d_4 = \sqrt{x_E^2 + y_E^2}$ and $d_5 = \sqrt{(1 - x_E)^2 + y_E^2}$.

Figure 2 illustrates that, as the SNR increases, the secrecy outage probability at both nodes A and B decreases, hence the system secrecy performance improves. The

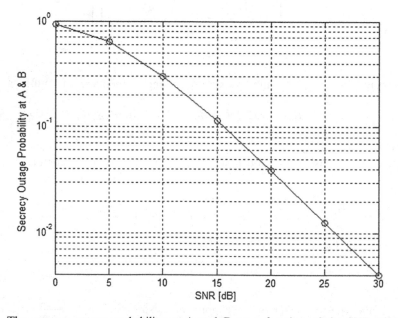

Fig. 2. The secrecy outage probability at A and B as a function of the SNR (dB) when $R_0 = 1$ bit/s/Hz, $d_1 = d_2 = 0.5$, $d_3 = 2$, $d_4 = d_5 = 2.06$, and the power fractions are considered at $(a = b = 0.6, c = 0.5)$

secrecy outage probability of node A is equal to that of node B because of the symmetry of the two way scheme.

Figure 3 shows that the secrecy outage probability of using the power fractions $(a = 0.9, b = 0.6, c = 0.5)$ is the same as that resulted of using $(a = 0.6, b = 0.9, c = 0.5)$. Moreover, when the power fractions a and b are set to 0.6, the secrecy outage probability of using the power fraction $(c = 0.1)$ at the relay is the same as that resulted of using $(c = 0.9)$ due to symmetry of the two way system.

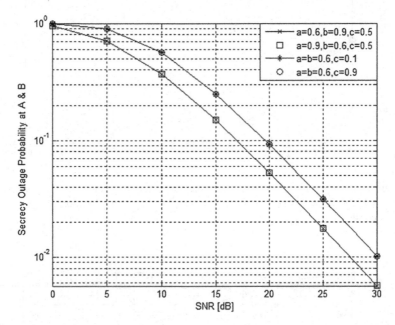

Fig. 3. The secrecy outage probability at A and B as a function of the SNR (dB) when $R_0 = 1$ bit/s/Hz, $d_1 = d_2 = 0.5$, $d_3 = 2$, $d_4 = d_5 = 2.06$, and the power fractions are considered at $(a = 0.6, b = 0.9, c = 0.5)$, $(a = 0.9, b = 0.6, c = 0.5)$, $(a = b = 0.6, c = 0.1)$ and $(a = b = 0.6, c = 0.9)$

Figure 4 illustrates the relationship between the sum secrecy outage probability of the two communicating parties and the location of the relay (x_r) on the x-axis. The best performance is achieved when the relay is located at the midpoint between the two communicating parties $(x_r = 0.5)$.

A simple comparison between the behavior of the half-duplex scheme which is handled in [6] and the proposed full-duplex scheme over Rayleigh fading channels is illustrated in Fig. 5. As expected, the secrecy performance of the full-duplex scheme is better than that of half-duplex, where the secrecy outage probability is higher for half-duplex scheme because half-duplex scenario has 66.67% loss in spectral efficiency which comes from the $\left(\frac{1}{3}\right)$ factor as three time slots are required to exchange a pair of data packets between the two communicating parties. Whereas the full-duplex scenario has 50% loss in spectral efficiency which comes from the $\left(\frac{1}{2}\right)$ factor as two time slots are needed to exchange a pair of data packets.

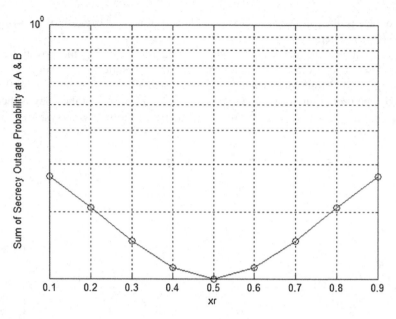

Fig. 4. Sum of secrecy outage probability as a function of the relay's location (x_r) on the x-axis when $SNR = 15$ dB, $R_0 = 1$ bit/s/Hz, $(a = b = 0.6, c = 0.5)$, and the location of the eavesdropper node is considered at $(0.5, -2)$

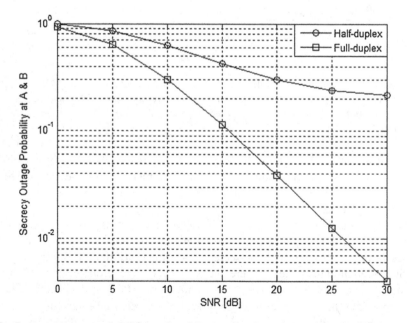

Fig. 5. Secrecy outage probability at A and B as a function of the SNR when $d_1 = d_2 = 0.5$, $d_3 = 2$, $d_4 = d_5 = 2.06$ and $(a = b = 0.6, c = 0.5)$ for half-duplex and full-duplex schemes over Rayleigh fading channel

5   Conclusion

In this paper, full duplex mode and Rayleigh fading channels were employed in a two way dual-hop cooperative network with source and jammer for decode and forward relaying scheme under physical layer security. The efficiency of the proposed system was investigated by deriving a closed form expression for the secrecy outage probability in order to evaluate the secrecy performance of the proposed system model. The theoretical results showed that full-duplex scheme performs better than half-duplex scheme and better performance is obtained at high SNRs.

References

1. Zou, Y., Wang, X., Shen, W.: Optimal relay selection for physical-layer security in cooperative wireless networks. IEEE J. Sel. Areas Commun. **31**(10), 2099–2111 (2013)
2. Wyner, A.D.: The wire-tap channel. Bell Syst. Tech. J. **54**(8), 1355–1367 (1975)
3. Stanojev, I., Yener, A.: Improving secrecy rate via spectrum leasing for friendly jamming. IEEE Trans. Wirel. Commun. **12**(1), 134–145 (2013)
4. Chen, G., Gong, Y., Xiao, P., Chambers, J.: Physical layer network security in the full-duplex relay system. IEEE Trans. Inf. Forensics Secur. **10**(3), 574–583 (2015)
5. Son, P.N., Kong, H.Y.: Exact outage probability of two-way decode-and-forward scheme with opportunistic relay selection under physical layer security. Wirel. Pers. Commun. **77**(4), 2889–2917 (2014). (Springer)
6. Son, P.N., Kong, H.Y.: An integration of source and jammer for a decode-and-forward two-way scheme under physical layer security. Wirel. Pers. Commun. **79**(3), 1741–1764 (2014). (Springer)
7. Zheng, J., Bai, B., Li, Y.: Outage-optimal opportunistic relaying for two-way amplify and forward relay channel. Electron. Lett. **46**(8), 595–597 (2010)
8. Chen, J., Zhang, R., Song, L., Han, Z., Jiao, B.: Joint relay and jammer selection for secure two-way relay networks. IEEE Trans. Inf. Forensics Secur. **7**(1), 310–320 (2011)
9. Zhang, R., Song, L., Han, Z., Jiao, B., Debbah, M.: Physical layer security for two-way relay communications with friendly jammers. In: Proceedings of IEEE Global Telecommunications Conference, GLOBECOM 2010, Miami, FL, USA, 6–10 December, pp. 1–6 (2010)
10. Ibrahim, D.H., Hassan, E.S., El-Dolil, S.A.: Improving physical layer security in two-way cooperative networks with multiple eavesdroppers. In: Proceedings of the 9th International Conference on Informatics and Systems, Cairo, Egypt, 15–17 December, pp. 8–13 (2014)

DNS Tunneling Detection Techniques – Classification, and Theoretical Comparison in Case of a Real APT Campaign

Viivi Nuojua$^{(\boxtimes)}$, Gil David, and Timo Hämäläinen

University of Jyväskylä, P.O. Box 35, FI-40014 Jyväskylä, Finland
{viivi.nuojua, gil.david, timo.t.hamalainen}@jyu.fi

Abstract. Domain Name System (DNS) plays an important role as a translation protocol in everyday use of the Internet. The purpose of DNS is to translate domain names into IP addresses and vice versa. However, its simple architecture can easily be misused for malicious activities. One huge security threat concerning DNS is tunneling, which helps attackers bypass the security systems unnoticed. A DNS tunnel can be used for three purposes: as a command and control channel, for data exfiltration or even for tunneling another protocol through it. In this paper, we surveyed different techniques for DNS tunneling detection. We classified those first based on the type of data and then within the categories based on the type of analysis. We conclude with a comparison between the various detection techniques. We introduce one real Advanced Persistent Threat campaign that utilizes DNS tunneling, and theoretically compare how well the surveyed detection techniques could detect it.

Keywords: DNS tunneling detection · Covert channels detection · APT

1 Introduction

Domain Name System (DNS) is an important protocol used when accessing the Internet. It translates domain names (e.g. www.google.com) into IP addresses and vice versa, and therefore it acts as an interpreter between the user and the Internet.

DNS is a distributed hierarchical protocol. It means that groups of servers scattered around the world are responsible for certain domains. The hundreds of authoritative name servers around the world serve the 13 named root name servers (i.e. DNS root zone), from A to M. Every new request for an IP address goes first to a root name server, which then forwards the request to the next server in the hierarchy and so on. DNS servers also cache request results for later use to avoid unnecessary search [1].

DNS is a simple and popular request and response protocol and hence can quite easily be misused. It is, indeed, widely used by the attackers for malicious activities. One huge security threat concerning DNS is tunneling. A DNS tunnel can be used for three purposes: "command and control" (C&C), data exfiltration and even for tunneling another protocol through it [1]. A typical example is to bypass a Wi-Fi access fee by tunneling the network traffic through DNS protocol. Tunnels over DNS are very powerful because DNS requests are seldom blocked by firewalls. [2] DNS tunneling

© Springer International Publishing AG 2017
O. Galinina et al. (Eds.): NEW2AN/ruSMART/NsCC 2017, LNCS 10531, pp. 280–291, 2017.
DOI: 10.1007/978-3-319-67380-6_26

has been utilized in some Advanced Persistent Threat (APT) attacks, which are stealthier and more sophisticated continuous attacks with a specific target. Unfortunately, DNS is often the least supervised protocol in many organizations.

The rest of the paper is organized as follows. Section 2 shortly reviews one real APT campaign utilizing DNS tunneling. Section 3 describes a classification for DNS tunneling detection techniques. Section 4 provides our theoretical comparison of DNS tunneling detection techniques in case of the reviewed APT campaign. Finally, Sect. 5 concludes the paper with some future work ideas.

2 Example of a Real APT Campaign Utilizing DNS Tunneling

Network security company Palo Alto Networks discovered a DNS tunneling utilizing APT attack in May 2016. The group behind this APT campaign was Wekby (aka TG-0416, APT-18, Dynamite Panda [3]), which has been quite active during the past years. It has maliciously taken advantage of the recently released exploits and targeted various organizations in the field of manufacturing, technology and healthcare. Although Wekby seems to use somewhat primitive methods, such as stolen credentials or basic malwares, it is constantly improving its toolkit [4].

Palo Alto Networks named the malware family used in the attack as "Pisloader". It was delivered via HTTP from the specific URL, and the used domains globalprint-us. com, ns1.logitech-usa.com and intranetwabcam.com were registered just before the attack. The executable file containing the Pisloader payload has similarities with the well-known Poison Ivy malware family. In addition, based on the metadata of the Pisloader sample and its malware commands, Pisloader is inferred to be a variant of the HTTPBrowser [5] malware family. In fact, HTTPBrowser was led by Wekby too [4].

Pisloader utilizes DNS tunneling not only as a C&C channel but also for data exfiltration. The malware payload can be transferred from the endpoint to the C&C server via DNS TXT transport layer as a part of a DNS request. A DNS request for TXT record consists of a randomized header and a base32-encoded data part with padding removed. The C&C server responds with a DNS TXT record with similar encoding. Only 255 bytes of data can be transported in a DNS TXT request. It is slow but won't draw attention. Additionally, Pisloader uses various obfuscation techniques, such as string and payload obfuscation, and other anti-analysis tactics to deter the detection and analysis of the sample [4].

3 Classification of DNS Tunneling Detection Techniques

Farnham and Atlasis [1] divide DNS tunneling detection techniques into two categories: traffic inspection and payload inspection. In this paper, we use the same classification into traffic and payload inspection, based on the type of data that is being analyzed. However, the grouping of techniques within the categories is implemented differently.

3.1   Traffic Inspection

According to Farnham and Atlasis [1] DNS tunneling can be detected by inspecting multiple requests or overall traffic, i.e. by traffic inspection. The following attributes are used: volume of DNS traffic per IP address, volume of DNS traffic per domain, number of hostnames per domain, geographic location of DNS server, domain history, volume of NXDomain responses, visualization, orphan DNS requests and general covert channel detection. Here, we group the detection techniques based on the type of analysis.

Statistical Analysis. A technique based on the statistical flow data analysis is introduced in [6], where flow is the sequence of packets exchanged between two hosts associated with a single service [7]. The dataset consists of unsampled flow records, each containing the following information about the flow: source IP address, destination IP address, source port, destination port, number of packets and bytes, start and end time of the flow, protocol, and the flags used (in case of TCP protocol).

They propose a module called Tunneling Attack Detector (TUNAD) for DNS tunneling detection. Its purpose is to detect anomalies in the statistics of the DNS packet sizes near real-time. TUNAD separates DNS packets to requests, responses, and unknown query types (uqr) when it is not known for sure whether those are requests or responses. For each of these packet types and each monitored circuit (receive or transmit) TUNAD calculates hourly packet size histograms from flow records. In addition, for each packet type TUNAD calculates the frequency of the packets deviated from the normal UDP/DNS packet size, that is, for responses and uqr packet types over 512 bytes and for requests over 300 bytes. The module is supposed to detect changes in the frequency of large packets. The authors managed to detect the Sinit virus in 2003 before it was reported in any external reports. No performance metrics were introduced.

Another flow-based statistical approach to detect DNS tunnels was presented in [2]. The novelty of the approach lies in combining flow information with statistical methods for anomaly detection. The four collected datasets consist of one dataset including normal DNS traffic and three datasets including tunneled traffic from three DNS tunnel misuse scenarios: C&C, data exfiltration and web browsing. These scenarios are tested by using four different DNS resolvers: a local resolver, a Norton open DNS resolver, a resolver in the network of the tunnel client and a resolver at the tunnel server. A local resolver behaves as normal DNS traffic, i.e. uses a new port number for each request. The latter three are non-local resolvers that reuse the same port number.

The analysis is based both on the raw data (unprocessed flows) and on time-binned data (preprocessed time series data). In case of normal DNS traffic, the time-of-the-day-dependency can be clearly seen. In case of tunneled traffic, the choice of resolver influences the behavior. The fact that the use of a local resolver makes tunneled traffic behave as normal traffic can easily be identified from the results. It was also observed that the number of packets sized approximately 200 bytes increase in case of a tunnel.

The authors use the following methods to detect anomalies in DNS traffic: (1) Threshold method, (2) Brodsky-Darkhovsky method [8] (BD) and (3) Distribution-based method. The anomalies under observation in each case are (1) peaks in the traffic time series, (2) changes in the average amount of traffic and (3) changes in the underlying traffic distribution. By combining different choices of input data (raw or time-binned flows), anomaly detection methods and resolvers, altogether five flow-based detectors were implemented to detect DNS tunnels. The results show that all five detectors manage to detect various tunnel misuse scenarios with high detection rates.

Machine Learning-Based Analysis. A passive DNS monitoring solution is introduced in [9]. It aims to detect anomalies in DNS traffic by using data mining, particularly a classical clustering algorithm named k-means [10]. The purpose is to detect the tunneling of IP traffic over DNS traffic.

The two datasets that were used were carefully chosen such that they differ as much as possible in geographical location, volume, duration and access networks. The k-means algorithm is used for data clustering. The clusters given by the algorithm turn out to successfully cover all different types of DNS traffic activities. The authors had no trouble in detecting DNS tunnels, when observing at the statistical profiles of different DNS record types. The number of TXT records, namely, clearly increase in case of a tunnel. No performance metrics were introduced.

Another machine learning-based approach for DNS tunneling detection is introduced in [11, 12]. The authors aim to avoid the monitoring of individual sockets and instead exploit all the information exchanged by the DNS server. Thus, it is possible to achieve real-timeness. The focus is on the exploration of simple statistical properties of DNS queries and responses, such as packet inter-arrival time and packet size statistics, with the help of supervised machine learning. The concentration isn't only on anomaly detection, but in the utilization of other statistical methods too.

The DNS databases consist of three tunnel databases based on the tunneled application and two clean databases (without tunnel), small and medium DNS, based on the size of a server. The applications tunneled through DNS protocol are a wget dump of an entire website, SSH protocol and a peer to peer (p2p) application. The DNS tunnel traffic is produced by a DNS tunneling tool called Dns2tcp [13].

In the first paper [11], the authors use the Bayes classifier [14] for classifying the DNS traffic into normal and tunneled traffic. In the second paper [12], they introduce, in addition, three other machine learning-based classifiers: k-nearest neighbor [15], neural networks [16] and support vector machine [17]. Classification error is avoided with the help of high order statistics, such as skewness and kurtosis, to detect the data positioning far from the assumed Gaussian distribution, i.e. the tunnel points. If just one of the aforementioned classifiers detects a tunnel, the tunnel is considered as detected. Classification is successful despite of the classification error of 15% in case of p2p application. The computational effort of the approach proves to be negligible. As a result, a reliable and fast DNS tunneling detection is achieved.

In the most recent paper surveyed in here [18], the authors use machine learning methods for anomaly behavior analysis of DNS protocol. The purpose is to learn the normal and abnormal behavior of DNS and develop an anomaly metric to detect DNS tunneling and other exploitations of the DNS protocol. The analysis is targeted at different levels of the domain name hierarchy and at the temporal transitions of normal DNS traffic.

Supervised machine learning methods, such as bagging classification algorithm [19], are used for observing and n-gram approach [20] for modeling the state transitions of the DNS protocol. A DNS state machine can be built based on the observed and analyzed DNS messages. In the first stage of the training phase, data structures called dnsgrams are used for capturing important properties of consecutive DNS queries and replies. The dnsgrams of DNS protocol under attack differ clearly from the normal dnsgrams, because the protocol doesn't follow the protocol state machine anymore. In the second stage of the training phase, dnsgrams are grouped periodically to dnsflows that cover a time window of t seconds, and the statistical properties of those are obtained. Finally, dnsflows are classified as normal and abnormal by comparing the extracted query indexes of a particular flow with the corresponding values of normal DNS operations. According to the authors, their general approach is able to detect both known and unknown attacks against DNS protocol. They managed to detect the executed DNS tunneling attack in every case. For known DNS attacks, such as DNS tunneling, the detection rate was high (97%) and the false positive rate low (0.014%). However, their focus was not only in DNS tunneling attacks.

3.2   Payload Inspection

In addition to traffic inspection, a DNS tunnel can be detected by inspecting a single request. This so called payload inspection uses the following attributes: size of request and response, entropy of hostnames, statistical analysis, uncommon record types, policy violation and specific signatures. [1] In this section, we introduce a different grouping based on the type of analysis. All the detection techniques introduced under the title "Statistical Analysis" can also be classified as machine learning-based analysis.

Statistical Analysis. The form of statistical analysis, character frequency analysis, is widely used for the detection of DNS tunneling. It was first introduced in the context of DNS tunneling detection in [21, 22]. In the first paper [21], the authors introduce the ability to detect DNS tunnels by analyzing the unigram (single character), bigram (two characters) and trigram (three characters) character ranks and frequencies of domain names. The approach is based on the assumptions that domain names behave similarly as natural languages, i.e. follow Zipf's law [23], whereas tunneled traffic clearly shows different, random-like, behavior.

First, the unigram, bigram and trigram character ranks and frequencies of domains are compared with the corresponding ranks and frequencies of English language. The top 1 000 000 most popular domains in 2009 (Alexa Top Sites) are used. The Top-Level Domains (TLD) and Lowest Level Domains (LLD), i.e. the furthest right and left parts of the domain names are removed. For example, in the domain name "www.example.com" TLD is "com" and LLD is "www". After removing the TLDs and LLDs, the unigram character ranks and frequencies of popular domains are compared with the unigram character ranks and frequencies of new domain names with randomly generated characters. Next, the unigram and bigram character ranks and frequencies of 1 000 000 most popular domains and 100 most popular domains are compared. On the other hand, a comparison between the unigram character ranks and frequencies of 100 randomly generated domain names and 1 000 000 popular domains is made. The purpose of these comparisons is to ensure that there won't be any problems with the character ranks and frequencies being skewed toward the outliers in the data, i.e. a smaller amount of data is sufficient to separate normal and tunneled traffic. In the detection of DNS tunneling, subdomains are the most critical ones to be analyzed. Consequently, the unigram character ranks and frequencies of popular domains are compared with subdomains and name server subdomains, with TLD and the following label removed from the domain names. The used subdomains and name server subdomains are captured by a custom website crawler visiting the aforementioned most popular sites. Finally, in order to verify the legitimacy of the introduced character frequency analysis, the unigram ranks and frequencies of subdomains are compared with the unigram ranks and frequencies of 100 Iodine [24]/Dns2tcp/TCP-over-DNS [25] packets. Thus, altogether three popular DNS tunneling tools are used.

The results support the assumptions made by the authors. The unigram character ranks and frequencies of English language and popular domain names are very much alike. The bigram and trigram character frequencies, in turn, are not that similar in ranks between English language and popular domain names, but still follow quite a similar pattern. On the other hand, the unigram character ranks and frequencies of popular domains and randomly generated domains have almost nothing in common. Since the authors presume that the randomly generated domains can represent the tunneled traffic, it seems to be quite easy to tell whether there's a tunnel or not. A comparison between the unigram and bigram character ranks and frequencies of 1 000 000 and 100 most popular domains shows that especially the unigram character ranks and frequencies are very similar. However, in case of bigrams the situation isn't bad at all either. On the contrary, there's almost no correlation between the unigram character ranks and frequencies of 100 randomly generated domain names and 1 000 000 popular domains. Consequently, a smaller amount of data is sufficient to make a clear separation between normal and tunneled traffic. The unigram character ranks and frequencies of subdomains turn out to be much alike and name server subdomains quite alike compared to popular domains. Only the ranks of the characters "n" and "s" seem to stand out in case of name server domains. Finally, subdomains have surprisingly similar unigram character ranks to the tunneling tools. Nevertheless, by inspecting the change in frequency between ranks it is easy to see the difference between normal and tunneled traffic.

As an extension to the authors' previous work, a tool called NgViz is introduced in [22]. It automates the visualization and analysis of DNS traffic. In order to detect anomalies in the n-gram frequencies, NgViz compares the input files with the fingerprint of legitimate traffic. A hash map creates a new entry for each unique n-gram by storing the number of times it is seen in the input file. The authors recommend to remove the duplicate subdomains and analyze the traffic only in case of a new subdomain.

First, the rank of each n-gram is compared with the corresponding n-gram rank in the fingerprint. After that, the similarities between the input and fingerprint files are explored by comparing the drops in frequency from one rank to the next. Significant differences may indicate to a tunnel in the DNS traffic. The detection of tunnels masked by a large amount of legitimate traffic is possible by splitting the traffic into separate files based on IP addresses, domains or specified time intervals. The results are sorted into highest and lowest matches for the fingerprint, making it easier to detect the anomaly parts of the traffic. The problem of the approach is the skewness caused by a domain consisting of similarly labeled subdomains. Therefore, the frequencies of the n-grams in that label grow so much that the domain is falsely considered as a tunnel.

The list of 1 000 000 most popular domains in 2009 (Alexa Top Sites) is utilized to form the fingerprint file of legitimate traffic. As in their previous paper [21] a custom website crawler is used to simulate a typical internet traffic by visiting thousands of popular websites, i.e. acting like a typical internet user. In the first experiment, the popular DNS tunneling tools (Iodine, TCP-over-DNS and Dns2tcp) show a poor match and the simulated traffic split every 100 unique subdomains a good match for the popular domains. In the second experiment, the 500 unique subdomains using bigrams show a great match and tunneling tools a very poor match for the popular domains. Because of this and the fact that bigrams provide significantly more data points and that way decrease the matching of tunneled and legitimate domain names, the results are even clearer. The performance evaluation is left for future research.

Character frequency analysis was used for detecting DNS tunnels in [26] too. The key idea of their approach is the score mechanism, which is able to separate normal and tunnel domain names based on bigram character frequency in real-time. Here the score of a domain is defined as the expected value of bigram character frequency. As in [21, 22], it is assumed that the normal domain names follow Zipf's law and the character frequencies of DNS tunnel domain names follow a random distribution.

Once again, the normal domain name dataset consists of the 1 000 000 most popular domains (Alexa Top Sites 2012), and the DNS tunneling domain names are generated by DNS tunneling tool called DNScat [27]. The data are divided into two parts: training and classification. The TLDs are removed from the domain names, since those don't bring that much information about the domain. The scores are calculated for both normal and tunneled domain names based on the character frequencies of the bigrams. Then, a suitable threshold is calculated according to the training data such that the classifier produces least false positives. It seems that DNS tunnel domain names have low scores that are centered within a small range. Instead, the scores of normal domain names are widely distributed. In the evaluation part, the scores calculated for the classification data are compared to the previously determined threshold. Based on that, the domain is classified either as normal or tunnel. The performance evaluation shows that the approach gets notably high accuracy of 98.74% and low false positive rate 1.24%.

Signature-Based Analysis. The use of DNS as a malicious payload distribution channel, e.g. for tunneling, was explored in [28, 29]. In the first paper [28], the malicious payload distribution channels are characterized based on the exchange behavior of the DNS request and response messages. First, all the messages with TXT resource record activities are extracted and those of a given domain name aggregated. Then, the request and response pattern analysis is applied on the domains, which are labeled as one of the four exchange pattern types: (1) Many-to-Many, (2) Many-to-Single, (3) Single-to-Many and (4) Single-to-Single relations. The naming indicates how many subdomains the client sends, and whether the server replies with one or several different TXT records.

The evaluation shows that the approach succeeds in determining different payload distribution channel patterns for malware families. Many-to-Many pattern seems to be the best choice for large datasets. Nevertheless, Single-to-Single pattern proves to be the most popular amongst malware samples. It produces the least amount of traffic, which helps the malware stay undetected. No performance metrics were introduced.

In the second paper [29], a DNS zone analysis-based approach is introduced. A near real-time detection is achieved by monitoring DNS requests and responses in passive DNS traffic. The purpose is to build DNS zone profiles out of the access counts of the resource records in order to detect payload distribution channels. The access counts indicate the number of requests for each record. The DNS request and response messages with TXT resource record activities are extracted and those of a given domain name aggregated. Finally, the messages are analyzed by the payload distribution module. A one-year malware database including malware samples that have TXT resource record activities is used for evaluation.

As a result, it is noted that in case of malware domains the number of DNS queries for TXT records is unusually high. The DNS queries received by the 500 most popular domains (Alexa Top Sites 2014) are, in turn, for different resource records. The system is able to detect payload distribution channels regardless of the payload format and malware family. The authors were actually the first ones to detect some hidden domains used by the famous Morto worm. Additionally, they emphasize that their system is supposed to detect DNS tunneling based on other resource record type than TXT too, since the detection is based on the access counts. No performance metrics were introduced.

3.3 Hybrid Inspection

The last DNS tunneling detection technique presented in this paper combines both traffic and payload inspection [30]. The purpose is to detect tunnels and other anomalies via statistical analysis in huge volume of DNS traffic near real-time. The approach doesn't focus on pattern matching, but the major interest lies in the differences between each DNS request/response.

The dataset consists of basic flow records extended with DNS fields, and collected from the monitoring probes. The analysis is targeted at the fields of DNS messages, such as TXT records. The tunneling data are produced by two DNS tunneling tools: Iodine and Dns2tcp. The module consists of two detection modes, basic and advanced. In the basic mode, all the observed network addresses are analyzed based on the statistics of flow information. The suspicious addresses are led to the advanced mode, and considered as a tunnel if a certain predetermined threshold is exceeded. Whitelisting of addresses and domain names is supported in order to avoid falsely classifying legitimate traffic as tunneled traffic.

The authors aim for a minimal resource consumption by processing smaller time windows at a time. In addition, the memory consumption is restrained by storing and analyzing only the domain names with suspicious addresses. To make the processing of huge volume of data possible, the approach utilizes efficient extended data structures. The detection module was put on a test by inputting data including both legitimate and tunneled traffic. It managed to detect all the tunnels established by the used tunneling tools. No performance metrics were introduced.

Table 1 introduces the aforementioned DNS tunneling detection techniques divided into two categories based on the type of data and then grouped based on the type of analysis. From the table, it can clearly be seen that some techniques cover more than one group. Especially, the hybrid detection technique [30] covers both traffic and payload inspection and both statistical and signature-based analysis. All the surveyed machine learning-based detection techniques can also be grouped into statistical analysis. Despite of the hybrid detection technique, the two techniques grouped into signature-based analysis, don't cover any other type of analysis.

Table 1. Classification of DNS tunneling detection techniques.

Detection approach	Traffic inspection		Payload inspection		
	Statistical analysis	Machine learning-based analysis	Statistical analysis	Signature-based analysis	Machine learning-based analysis
[6]	X				
[2]	X				
[9]	X	X			
[11]	X	X			
[12]	X	X			
[18]	X	X			
[21]			X		X
[22]			X		X
[26]			X		X
[28]				X	
[29]				X	
[30]	X		X	X	

4   Theoretical Comparison of DNS Tunneling Detection Techniques in Case of a Real APT Campaign

In this section, we theoretically compare how well the surveyed DNS tunneling detection techniques could detect the Pisloader malware attack introduced in Sect. 2.

The two, only statistical analysis-based traffic inspection techniques would succeed quite differently in the detection of Pisloader. In the first surveyed paper [6], DNS tunneling attack detector, TUNAD, regards frequent DNS requests with over 300 bytes of data as tunneled traffic. However, the Pisloader payload is transported in a DNS TXT request, which is limited to contain only 255 bytes of data. In addition, TUNAD doesn't seem to pay attention to TXT records. Thus, it wouldn't be able to detect the Pisloader attack. Pisloader utilizes DNS tunneling for C&C and data exfiltration, which are both taken into account in [2]. Unlike in case of TUNAD, the authors observed that an increase in the number of packets sized over 200 bytes indicates a tunnel. Because of that, their versatile use of different setups, analyses and methods, and high detection rates, they would manage to detect the Pisloader malware attack, in theory.

The four separate papers covering not only statistical but also machine learning-based traffic inspection techniques are all a bit uncertain in detecting the Pisloader malware. The statistical profiles of different DNS record types are explored in [9], and unlike TUNAD, the authors pay attention to TXT records. They noticed that the number of TXT records increase when a DNS tunnel is misused. However, they did not provide any threshold for packet sizes indicating a tunnel, nor performance metrics. Thus, Pisloader might stay undetected. The authors of [11, 12] focus on the simple statistical properties of DNS requests and responses, such as packet size statistics. According to them, large packet sizes refer to tunneling, but the byte numbers are different in case of each application tunneled through DNS. Thus, it is a bit difficult to evaluate whether they would be able to detect Pisloader or not. The properties of consecutive DNS requests and responses are examined in [18]. It is not clear if the authors actually utilize DNS TXT records as part of their detection method. Their focus was not only in the detection of DNS tunneling attacks but multiple different attacks. According to the authors, they managed to detect a certain executed DNS tunneling attack with high detection rate and low false positive rate. However, it is a bit difficult to evaluate if Pisloader would be detected too.

Character frequency analysis-based payload inspection techniques (covering both statistical and machine learning-based analysis) are quite tricky too when evaluating if the Pisloader attack would be detected, in theory. The authors of [21, 22] state that usually in case of a tunnel domain names seem randomly generated. The domain names used by Pisloader are not random-like, which is why Pisloader would stay undetected. Same goes for the detection technique presented in [26], although they better their detection possibilities by dividing data into training and classification parts, and by providing good performance metrics results.

A DNS tunneling detection technique that would possibly detect the Pisloader malware is introduced in [28, 29]. The authors' signature-based analysis focuses on monitoring DNS requests and responses. They noticed that in case of malware domains the number of DNS requests for TXT records is unusually high. By extracting DNS messages with TXT resource record activities, they managed to detect different payload distribution channel patterns regardless of the payload format and malware family. In fact, they were the first ones to detect some hidden domains used as malicious payload distribution channels by the famous Morto worm.

The hybrid inspection technique presented in [30] also pays attention to the differences between each DNS request/response, especially to the fields of DNS messages, such as TXT records. However, their detection module analyzes only the domain names that are classified as suspicious in advance. The domain names used by Pisloader might not be thought suspicious. Thus, Pisloader might stay undetected.

It seems that, in theory, the Pisloader malware attack would best be detected by the secondly surveyed, only statistical analysis-based traffic inspection technique [2]. In addition, the signature-based payload inspection techniques [28, 29] could theoretically be able to detect Pisloader. However, it is a bit difficult to evaluate the impact of the various obfuscation techniques and other anti-analysis tactics used by Pisloader.

5   Conclusions and Future Work

In this paper, we reviewed a real DNS tunneling utilizing APT campaign, called Pisloader. We surveyed several different DNS tunneling detection techniques and classified those, first, based on the type of data. Within those categories, we grouped the techniques based on the type of analysis. One of the problems when comparing the detection techniques was that all of those don't provide performance metrics and use different datasets. However, we made a theoretical comparison of how well the surveyed DNS tunneling detection techniques could detect the Pisloader malware attack. It came up quite clearly that the techniques with certain type of analysis approach could succeed better than others. Altogether, the surveyed techniques seem to cover the extensive range of different kind of DNS tunneling detection very well.

Next, our purpose is to actually implement, and compare in practice, the DNS tunneling detection techniques introduced in this paper, on the same dataset. That way, the actual comparison between their performance is made possible. Finally, we are going to present our own approach and compare it to the reviewed techniques.

References

1. Farnham, G., Atlasis, A.: Detecting DNS tunneling. SANS Institute InfoSec Reading Room, pp. 1–32 (2013)
2. Ellens, W., Żuraniewski, P., Sperotto, A., Schotanus, H., Mandjes, M., Meeuwissen, E.: Flow-based detection of DNS tunnels. In: Doyen, G., Waldburger, M., Čeleda, P., Sperotto, A., Stiller, B. (eds.) AIMS 2013. LNCS, vol. 7943, pp. 124–135. Springer, Heidelberg (2013). doi:10.1007/978-3-642-38998-6_16
3. Evasive Maneuvers by the Wekby group with custom ROP-packing and DNS covert channels. https://www.anomali.com/blog/evasive-maneuvers-the-wekby-group-attempts-to-evade-analysis-via-custom-rop
4. New Wekby attacks use DNS requests as command and control mechanism (2016). http://researchcenter.paloaltonetworks.com/2016/05/unit42-new-wekby-attacks-use-dns-requests-as-command-and-control-mechanism/
5. Chinese cyber espionage APT group leveraging recently leaked hacking team exploits to target a financial services firm. https://www.zscaler.com/blogs/research/chinese-cyber-espionage-apt-group-leveraging-recently-leaked-hacking-team-exploits-target-financial-services-firm
6. Karasaridis, A., Meier-Hellstern, K., Hoeflin, D.: NIS04-2: detection of DNS anomalies using flow data analysis. In: IEEE Global Telecommunications Conference, GLOBECOM 2006, pp. 1–6 (2006)
7. Copeland III, J.A.: Flow-based detection of network intrusions (2007). http://www.google.com/patents/US7185368
8. Brodsky, E., Darkhovsky, B.S.: Nonparametric Methods in Change Point Problems. Springer Science & Business Media, Heidelberg (2013)
9. Marchal, S., François, J., Wagner, C., State, R., Dulaunoy, A., Engel, T., Festor, O.: DNSSM: a large scale passive DNS security monitoring framework. In: 2012 IEEE Network Operations and Management Symposium (NOMS), pp. 988–993. IEEE (2012)
10. Hartigan, J.A., Wong, M.A.: Algorithm AS 136: a K-means clustering algorithm. J. R. Stat. Soc. Ser. C Appl. Stat. **28**, 100–108 (1979)

11. Aiello, M., Mongelli, M., Papaleo, G.: Basic classifiers for DNS tunneling detection. In: 2013 IEEE Symposium on Computers and Communications (ISCC), pp. 880–885 (2013)
12. Aiello, M., Mongelli, M., Papaleo, G.: DNS tunneling detection through statistical fingerprints of protocol messages and machine learning. Int. J. Commun. Syst. **28**, 1987–2002 (2015)
13. HSC - Tools - Dns2tcp. http://www.hsc.fr/ressources/outils/dns2tcp/
14. Moore, A.W., Zuev, D.: Internet traffic classification using Bayesian analysis techniques. In: Proceedings of the 2005 ACM SIGMETRICS International Conference on Measurement and Modeling of Computer Systems, pp. 50–60. ACM, New York (2005)
15. Cover, T., Hart, P.: Nearest neighbor pattern classification. IEEE Trans. Inf. Theory **13**, 21–27 (1967)
16. Ripley, B.D.: Pattern Recognition and Neural Networks. Cambridge University Press, Cambridge (2007)
17. Cortes, C., Vapnik, V.: Support-vector networks. Mach. Learn. **20**, 273–297 (1995)
18. Satam, P., Alipour, H., Al-Nashif, Y., Hariri, S.: Anomaly behavior analysis of DNS protocol. J. Internet Serv. Inf. Secur. JISIS **5**, 85–97 (2015)
19. Breiman, L.: Bagging predictors. Mach. Learn. **24**, 123–140 (1996)
20. Bramer, M.: Principles of Data Mining. Springer, London (2007)
21. Born, K., Gustafson, D.: Detecting DNS tunnels using character frequency analysis (2010). arXiv:1004.4358[cs]
22. Born, K., Gustafson, D.: NgViz: detecting DNS tunnels through n-gram visualization and quantitative analysis. In: Proceedings of the Sixth Annual Workshop on Cyber Security and Information Intelligence Research, pp. 47:1–47:4. ACM, New York (2010)
23. Zipf, G.K.: Selected Studies of the Principle of Relative Frequencies of Language. Harvard University, Cambridge (1932)
24. kryo.se: iodine (IP-over-DNS, IPv4 over DNS tunnel). http://code.kryo.se/iodine/
25. TCP-over-DNS tunnel software HOWTO. http://analogbit.com/2008/07/27/tcp-over-dns-tunnel-software-howto/
26. Qi, C., Chen, X., Xu, C., Shi, J., Liu, P.: A bigram based real time DNS tunnel detection approach. Procedia Comput. Sci. **17**, 852–860 (2013)
27. DNScat. http://tadek.pietraszek.org/projects/DNScat/
28. Binsalleeh, H., Kara, A.M., Youssef, A., Debbabi, M.: Characterization of covert channels in DNS. In: 2014 6th International Conference on New Technologies, Mobility and Security (NTMS), pp. 1–5 (2014)
29. Kara, A.M., Binsalleeh, H., Mannan, M., Youssef, A., Debbabi, M.: Detection of malicious payload distribution channels in DNS. In: 2014 IEEE International Conference on Communications (ICC), pp. 853–858 (2014)
30. Cejka, T., Rosa, Z., Kubatova, H.: Stream-wise detection of surreptitious traffic over DNS. In: 2014 IEEE 19th International Workshop on Computer Aided Modeling and Design of Communication Links and Networks (CAMAD), pp. 300–304 (2014)

Digital Watermarking Method Based on Image Compression Algorithms

Sergey Bezzateev[1,2]($\boxtimes$) and Natalia Voloshina[1,2]

[1] Saint-Petersburg University of Aerospace Instrumentation, St. Petersburg, Russia
bsv@aanet.ru, nataliv@yandex.ru
[2] ITMO University, St. Petersburg, Russia

Abstract. Digital watermarking is an efficient method for digital access rights management utilized in the scope of multimedia data. A possibility to combine the procedure of compression and watermarking in effective way for digital images is proposed in this manuscript. This research is focused on the compression methods considering the significance of the initial multimedia object (for example image) different elements to increase the quality of process (compressed) image. One of the most effective approaches for this task is to utilize Error Correcting Codes (ECC) allowing to maintain the number of resulting errors (distortion) as well as the value of resulting compression ratio. The application of such codes enables to distribute errors that are added during the processing procedure according to predefined significance of the initial multimedia object elements. The approach based on Weighted Hamming Metric guarantying the limitation of maximum errors (distortions) with predefined significance is represented as an example.

Keywords: Digital watermarking, DWM · Image compression · MLSB · ECC perfect in weighted Hamming metric

1 Introduction

Today, digitalization of our lives is occurring worldwide thus pushing the advancements in Cyber-Physical Systems advancement in both user [1,2], industry [3,4], and operators' directions [5]. Form the content owner's perspective, this brings a number of challenges to be solved [6,7]. Digital information may be copied, reproduced, stoled and reused during any of the transmitting and storage interaction phases [8,9]. Therefore, the need for effective watermarking methods protecting the digital data is growing.

The initial image decomposition into number of blocks with length K and changing of these blocks to ones with length less then k technique is a very useful method for digital image compression [10]. This type of image processing represents one of the lossy compression methods [11]. To execute these technique, it is necessary to define a matching rule for the initial and resulting image blocks. Apparently, when several blocks of initial image corresponds to only one

© Springer International Publishing AG 2017
O. Galinina et al. (Eds.): NEW2AN/ruSMART/NsCC 2017, LNCS 10531, pp. 292–299, 2017.
DOI: 10.1007/978-3-319-67380-6_27

in resulting image, it is so-called "surjective" mapping [12]. This fact entails the errors in blocks of resulting image that can be visually detected as distortions in the resulting image.

The manuscript is organized as following. In Sect. 2 general introduction in the problem is given. The proposed methodology of utilizing the first step of the compression procedure for digital watermarking embedding is given in Sect. 3. The last section concludes the manuscript.

2   Background

The corresponding parameters could be defined for this image compression method:

– Compression ratio η;
– Distortion ratio ρ for the resulting image compared to the initial one.

As a base assumption in this manuscript, we define the number of different values of blocks equal to N. For example, let the initial block consists of m pixels. Further, every pixel consists of n bits. Then, $N = 2^{mn}$ and initial image block is represented as binary code with length $K = mn$.

For example, the parameters for the conventional color bitmap (BMP) image could be defined as: block with size 8×8 consists $m = 64$ pixels, that for an RGB image representation, the number of bits for one block could be defined as $n = 8 \times 3 = 24$, and the value of $K = 64 \times 24$. In this case $N = 2^{1536}$. At the same time, if it is used component compression than $N = 2^{64 \times 8} = 2^{512}$.

Let us also define the overall number of different values of the resulting image blocks equal to M, i.e. after compression procedure for each block of the initial image there exists the corresponding binary vector with length $k = \log_2 M$.

In this case, the compression ratio could be defined as

$$\eta = \frac{K}{k}.$$

Generally, the mapping procedure for block a_i with length K of the initial image to compressed block b_i with length k with further decompression of this block to resulting image block c_i is schematically represented on Fig. 1, where F_1 – is the coding function, F_2 – is the coding book construction function, and $F_{2^{-1}}$ – us the decoding function.

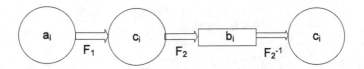

Fig. 1. Image compression and decompression processing procedure

As for compression procedure $M < N$, there could exist $\eta = \frac{N}{M}$ different
blocks of the initial image $(a_i^1, a_i^2, \ldots, a_i^\eta)$. For some the blocks, the same block
of resulting image c_i could be a match. This could occur due to compression
errors appearance and, as a result, the image quality may be decreased (Fig. 2).

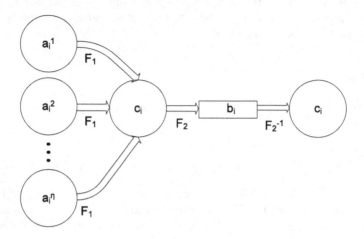

Fig. 2. Surjective mapping of initial image blocks

Image distortions occurred as a result of compression procedure could be
defined using the predefined special distance between initial and resulting blocks
c_i and a_i^j

$$d_{ij} = dist(c_i, a_i^j).$$

Wherein, if during the image compression process it is possible to optimize
(for example maximize) this distance, than the resulting image distortion could
be less perceptible. This requirement is related to image compression tasks as
well as to digital watermarking goals.

For entire image, the value of distortion could be defined as

$$\rho = \frac{R}{\sum\limits_{i,j} d_{ij}},$$

where R – is some coefficient that define property of the entire image.

In order to calculate ρ, the well-known for image quality measuring parameter, peak signal-to-noise ratio (PSNR), could be applied. It is calculated as:

$$PSNR = 20 \log_{10} \frac{\max\limits_i |P_i|}{RMSE},$$

where $RMSE = \sqrt{MSE}, MSE = \frac{1}{L} \sum\limits_{i=1}^{L} (P_i - Q_i)^2$ L – is the overall number of
pixels of initial image, P_i, Q_i – are values of i-th pixel of initial and resulting
decompressed image consequently.

Thus, the PSNR value for the image is calculated as

$$R = 20 \log_{10} \max_i |P_i|.$$

Trivially, for n-bit pixel image the value is calculated as

$$\max_i |P_i| = 2^n - 1.$$

This metric is well concerted with human visual system properties and it's maximization allows to obtain minimum visibility of the image distortion. It is considered that the perceptible quality is reached if $RMSE \geq 30\,\text{dB}$.

The possible distance metric to be utilized is Hamming distance between two vectors: (i) one vector corresponds to the initial image block and the other corresponds to (ii) the resulting image block

$$d_H(a_i^j, c_i) = \#_{l=1,\ldots,K}(a_{il}^j \neq c_{il}),$$

where $c_i = (c_{i1}, c_{i2}, \ldots, c_{iK})$ – is the resulting image block decompressed with errors, $a_i^j = (a_{i1}^j, a_{i2}^j, \ldots, a_{iK}^j)$ - j - is the initial image block. Then,

$$\rho = \frac{R}{\displaystyle\sum_{i,j} d_H(a_i^j, c_i)}. \tag{1}$$

This metric is well concerted with the error bits number occurred during the image compression procedure. Maximizing of ρ means that minimal number of error bits occurred while compression procedure.

By applying the Hamming metric, it fairly easy to implement the error correcting coding procedure to realize coding function F_1, i.e. the initial image block a_i^j is looked at as some vector of length K that was distorted during the transmission through noisy channel. Therefore, it is necessary to find closed codeword in Hamming metric c_i for predefined code C. In this case, the value of ρ is defined by covering radius of code C.

Covering radius $\mathcal{R}(C)$ of the linear code C with the length n could be represented as

$$\mathcal{R}(C) = max\left\{min\left\{d_H(\mathbf{x}, \mathbf{c}), \mathbf{c} \in C\right\}, \mathbf{x} \in \mathrm{F}_2^n\right\}.$$

If the compression ratio is defined as η than the length of code word λ an number of information symbols μ of predefined binary ECC are connected to each other by the ratio $\eta = \frac{\lambda}{\mu}$. By the proposed metric (1) the optimal code should has minimal covering radius. Therefore the perfect code is an optimal code but the set of perfect codes is exhausted by Hamming and Golay codes [13]. The use of other classes of error-correcting codes, such as Reed-Solomon, BCH, Goppa codes is not optimal and require implementation of covering radius calculation procedure for each code with defined parameters λ and μ to find the appropriate code. Unfortunately, for most classes of ECC codes, only the lower and upper boundaries of the covering radius are known.

It should be noted that the considered version of the mapping function F_1 does not take into account different significance of elements $a_{il}^j, l = 1, \ldots, K$ of the initial image block. Evidently, the distortion of components related to the high-order bits of the pixel has more impact on the recovered quality than the distortion of the low-order bits.

To create the mapping procedure that is agreed with different significance of image zones, it was proposed to assign to these components various reliability levels, in the LDPC encoding definition, followed by soft decoding [11]. Also in [14] a method considering different significance by dedicating corresponding weight to different parts of code words was proposed. As a result, construction of optimal ECC code in Weighted Hamming metric with corresponding parameters became feasible. A similar approach considering the significance of DWT embedding was proposed in [15].

The main advantage of the first approach is the adaptive assignment of the components reliability levels for each block of the initial image. It allows to correct the embedded errors in the decoding process by taking into account the previously occurred errors.

On the other hand, it has also a strong disadvantage since it requires a very labor-intensive preprocessing of initial image blocks. Another disadvantage of this procedure is a very rough estimate of the total number of distortions of the initial image being added to the block when executing coding function F_1.

For the estimation of this type, it is possible to utilize the covering radius boundaries for the ECC class for those ones where the pre defined code belongs, or to calculate its exact value for a particular code generating matrix, which is generally a very laborious task.

The main advantage of the second method is the ability to guarantee the total weighted number of distortions occurred in the block of the initial image not exceeding a certain predetermined value, which is equal to the code's covering radius. Moreover, for the perfect codes, the coverage radius coincides with the radius of the spherical packing and is equal to the minimum distance of the code and accordingly does not require any additional computations.

3   Using the First Step of the Compression Procedure for Digital Watermarking Embedding

The variant of the first step (the F_1 transformation) of the compression procedure usage for digital watermarking embedding is discussed in [16]. Wherein, it was proposed to utilize the class of Goppa codes in the usual Hamming metric. An obvious generalization of this approach considering the results obtained earlier is the use of perfect codes in the weighted Hamming metric. As an example, we consider the set of perfect in the weighted Hamming metric Goppa codes for the block of the original image consisting of 8 pixels, each represented by 8 bits, i.e., $m = 8, n = 8, K = mn = 64$. To compress this block, we apply two various Goppa codes C_1 C_2 with parameters $\lambda_1 = 35, \mu_1 = 29, \lambda_{11} = 8, \lambda_{12} = 27$ and $\deg G_1(x) = 2$ and $\lambda_2 = 29, \mu_2 = 23, \lambda_{21} = 4, \lambda_{22} = 6, \lambda_{23} = 19$ respectively.

$\deg G_2(x) = 3$. $G_1(x)$, $G_2(x)$ – are irreducible polynomials with coefficients ranging from $GF(2^3)$ to $GF(2^2)$. The total number of information symbols for such a pair of codes is $k = \mu_1 + \mu_2 = 29 + 23 = 51$. In this case, the block of the original image is divided into three sub-block with lengths $K_1 = \lambda_{11} + \lambda_{21} = 12$, $K_2 = \lambda_{12} + \lambda_{22} = 33$, $K_3 = \lambda_{23} = 19$. And weights (values of significance from the point of view of influence on distortion of the original image) are respectively $v_1 = 1, v_2 = 2, v_3 = 3$. Both codes are perfect in the weighted Hamming metric and the covering radius for these codes is defined as follows:

$$\mathcal{R}(C_1) = \max_{v_1 t_1 + v_2 t_2 \leq \deg G_1(x)} t_1 + t_2 = 2,$$

$$\mathcal{R}(C_2) = \max_{v_1 t_1 + v_2 t_2 + v_3 t_3 \leq \deg G_2(x)} t_1 + t_2 + t_3 = 3.$$

One of the possible variants of the decomposition block of the original image into codewords of the first and second Goppa code is depicted in Fig. 3.

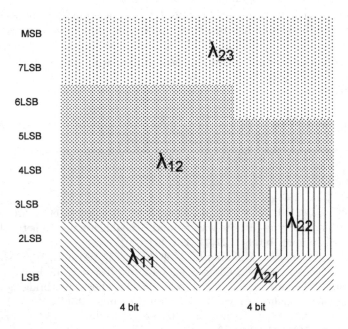

Fig. 3. Splitting the block of the source image into codewords of the first and second code. LSB – is the least significant bit, iLSB – is the significant bit of ith-level, MSB – is the most significant bit

Let the number S_1, S_2 of possible different polynomials $G_1(x), G_2(x)$ that define different Goppa codes with parameters

$$S_1 = 28, S_2 = 20.$$

Trivially, each block of the original image can be mapped to one of 560 pairs of codewords belonging to Goppa codes with polynomials $G_1(x), G_2(x)$ as a result of the procedure F_1. A set of such pairs can be used as digital watermark. As noted earlier in the manuscript [16], such system provides an additional protection of the digital label from random and intentional distortions.

4   Conclusion

In this paper we have presented a possible to embed digital watermark during the first stem of the compression algorithm for coding quantization compression approach. It was shown that it is possible to mark an image by one or set of Goppa codes with equal parameters but different polynomials $G_i(x)$ so that, if the resulting sequence of $G_i(x)$ at the decompression step corresponds to one on the compression step with good accuracy, the watermark is detected. Main goal of this approach is to minimize distortions by minimizing number of changed bites and considering the significance of image zones.

Resulting efficiency could be estimated by value penalty function, i.e. the smaller is the value of penalty function the smaller is visible distortion of the resulting image. The proposed approach of multilevel significant bit embedding is based on error-correcting codes constructed on weighted Hamming metrics allows to pre-estimate the maximum of penalty function to guaranty resulting image quality.

Acknowledgment. This work was partly financially supported by Russian Foundation for Basic Research in 2017 (grant 17-07-00849-A).

References

1. Yahya, A.N., Jalab, H.A., Wahid, A., Noor, R.M.: Robust watermarking algorithm for digital images using discrete wavelet and probabilistic neural network. J. King Saud Univ.-Comput. Inf. Sci. **27**(4), 393–401 (2015)
2. Ometov, A., Orsino, A., Militano, L., Moltchanov, D., Araniti, G., Olshannikova, E., Fodor, G., Andreev, S., Olsson, T., Iera, A., Torsner, J., Koucheryavy, Y., Mikkonen, T.: Toward trusted, social-aware D2D connectivity: bridging across the technology and sociality realms. IEEE Wirel. Commun. **23**(4), 103–111 (2016)
3. Mosterman, P.J., Zander, J.: Industry 4.0 as a cyber-physical system study. Softw. Syst. Model. **15**(1), 17–29 (2016)
4. Yan, X., Zhang, L., Wu, Y., Luo, Y., Zhang, X.: Secure smart grid communications and information integration based on digital watermarking in wireless sensor networks. Enterp. Inf. Syst. **11**(2), 223–249 (2017)
5. Orsino, A., Ometov, A.: Validating information security framework for offloading from LTE onto D2D links. In: Proceedings of the 18th Conference of Open Innovations Association FRUCT, pp. 241–247 (2016)
6. Chandrakar, N., Bagga, J., et al.: Performance comparison of digital image watermarking techniques: a survey. Int. J. Comput. Appl. Technol. Res. **2**(2), 126–130 (2013)

7. Ometov, A., Masek, P., Malina, L., Florea, R., Hosek, J., Andreev, S., Hajny, J., Niutanen, J., Koucheryavy, Y.: Feasibility characterization of cryptographic primitives for constrained (wearable) IoT devices. In: Proceedings of International Conference on Pervasive Computing and Communication Workshops (PerCom Workshops), pp. 1–6. IEEE (2016)
8. Andalibi, M., Chandler, D.M.: Digital image watermarking via adaptive logo texturization. IEEE Trans. Image Process. **24**(12), 5060–5073 (2015)
9. Bajracharya, S., Koju, R.: An improved DWT-SVD based robust digital image watermarking for color image. Int. J. Eng. Manuf. (IJEM) **7**(1), 49 (2017)
10. Lai, C.C., Tsai, C.C.: Digital image watermarking using discrete wavelet transform and singular value decomposition. IEEE Trans. Instrum. Meas. **59**(11), 3060–3063 (2010)
11. Belogolovyi, A.: Image compression based on LDPC codes. In: Proceedings of International Conference GraphiCon (2004)
12. Guyeux, C., Bahi, J.M.: Topological chaos and chaotic iterations application to hash functions. In: Proceedings of the 2010 International Joint Conference on Neural Networks (IJCNN), pp. 1–7. IEEE (2010)
13. Bezzateev, S., Shekhunova, N.: Class of generalized Goppa codes perfect in weighted Hamming metric. Des. Codes Crypt. **66**(1–3), 391–399 (2013)
14. Bezzateev, S., Voloshina, N., Zhidanov, K.: Steganographic method on weighted container. In: Proceedings of XIII International Symposium on Problems of Redundancy in Information and Control Systems (RED), pp. 10–12. IEEE (2012)
15. Dariti, R., Souidi, E.M.: An application of linear error-block codes in steganography. Int. J. Digit. Inf. Wirel. Commun. (IJDIWC) **1**(2), 426–433 (2011)
16. Bezzateev, S., Voloshina, N., Minchenkov, V.: Special class of (L, G) codes for watermark protection in DRM. In: Proceedings of Eighth International Conference on Computer Science and Information Technologies, pp. 225–228 (2011)

Development of the Credit Risk Assessment Mechanism of Investment Projects in Telecommunications

Sergei Grishunin[(✉)] and Svetlana Suloeva[(✉)]

St. Petersburg State Polytechnical University, St. Petersburg, Russia
sergei.v.grishunin@gmail.com, emm@spbstu.ru

Abstract. We developed a mechanism of modelling of internal credit ratings (ICRs). It is applied in investment controlling to assess the credit quality of projects of telecommunication companies. Its advantages over the conventional credit risk modelling approaches are higher robustness and incorporation of all modelling operations in one mechanism. The mechanism gives the possibility to compare of modelled ICRs to the public credit ratings assigned by reputable international credit agencies. To achieve higher accuracy, the mechanism converts the input financial data, presented in different accounting standards, to the common basis. The explanatory variables in the mechanism are closely aligned with the credit risk assessment factors listed in the methodologies of international credit rating agencies. The testing of the mechanism shows that ICRs modelled with application of our mechanism had the accuracy ratio of 55% for testing sample and 65% for the design sample. This exceeds the accuracy ratios of ICRs modelled with application of conventional approaches (37%–42%).

Keywords: Project management · Investment controlling · Credit risk · Internal credit ratings · Artificial neural networks

1 Introduction

In recent years, executives of telecommunication companies (TCs) and investors in development projects of TCs have demanded the enhancement of projects' risk assessment tools of investment controlling (IC). This is underpinned by the growing instability and uncertainty of TC's business environment. The complexity of risks is increasing; the speed of risks onset is growing while losses from their realization become more severe. The risk identification and assessment mechanisms, which proved to be efficient in the past, currently do not ensure the required accuracy and robustness [9].

The mechanism of credit risks (CRs) assessment of telecommunication projects (TPs) is one of the most important tools in IC. Its significance is underpinned by the high capital intensity of TPs which, in turn, requires TCs to increase the share of debt in their capital structures. The accurate assessment of CRs is important for investors to make decision of funding TPs as well as for TPs' owners and executives to diagnose, control and monitor of projects' creditworthiness.

© Springer International Publishing AG 2017
O. Galinina et al. (Eds.): NEW2AN/ruSMART/NsCC 2017, LNCS 10531, pp. 300–314, 2017.
DOI: 10.1007/978-3-319-67380-6_28

The analysis of literature shows that the conventional mechanisms of assessing CRs of TPs need development. In mechanisms based on manual scoring systems, the CRs are assessed with fundamental analysis of TPs' business plans and project documentations [8, 11]. The disadvantages of this approach are its high cost, time and labor intensity. Another approach assumes reliance on public credit ratings (PCR) assigned to TPs by reputable international rating agencies (IRAs, such as Moody's, Standard and Poor or Fitch Ratings). However, many TPs lack PCRs as they are assigned only to few very large projects. Yet, there is the third type of mechanisms in which CRs are assessed by modelling of internal credit ratings (ICRs). The ICRs emulate the PCRs and are built with econometric models on the large array of financial, operational, market and macroeconomic data [11, 13]. We consider such mechanisms as the most promising because they are easy to use, low cost and require limited involvement of experts [10, 11, 13]: However, they still have some limitations. The ICRs have lower than required accuracy (37%–42%) as shown in [11] and/or unproven robustness. The latter is caused by utilizing insufficient data for models' training or methodological framework with unproven ability to predict defaults [10, 13].

2 The Scope of the Paper

We will develop the mechanism of ICRs' modelling which overcomes the limitations of conventional approaches mentioned above. The special attention is paid to the modelling infrastructure (processing of the data, selection of explanatory variables, selection of the models' typology and specification). We then test the mechanism by comparing the modelled ICRs to existing PCRs and evaluating ICRs' accuracy ratio.

3 Telecommunication Industry Outlook

The global telecommunication industry is in its way of transformation to the info-communication industry from one just providing core voice and data services [7]. The new opportunities in the next 3–5 years, apart from growing penetration of smart phones and mobile Internet, come from Internet of Things, 5G network technology, autonomous vehicles, biometrics and machine learning [1]. However, these opportunities will result in raising competition among industry players which, in turn, increases the capital intensity in the industry. The areas for investments include development of new products and services, core connectivity and network infrastructure, operational efficiency and customers care [1]. The share of capital expenditures in TCs' sales will remain between 15% and 20% [8]. Other challenges come from viability of new technologies and adoption of new products by customers, global economic uncertainties, regulatory requirements and regulatory restrictions on mergers and acquisitions [1].

The industry outlook for the next 3–5 years depends on geographies [1]. In Europe and Asia, the telecommunication industry will grow in line with that of gross domestic product (GDP) in the countries of TCs' operations (1%–2% in Europe and 3%–4% in Asia). The TCs' profitability will remain strong and supported by operators' efforts to improve efficiency with deploying less labor-intensive technologies. However, there is a risk of declining margins due to intensifying competition, higher costs for providing data services and investments in margin-dilutive digital businesses. In USA, the business profile of TCs will weaken due to growing competition resulting in price pressure, especially at the wireless market while the event risks in the industry remain high. In Russia, TCs' revenue and profitability will be under pressure. This is underpinned by (1) reduced purchasing power of consumers resulting from weak country's economic environment; and (2) limited ability to pass rising costs to consumers as the domestic market is a highly competitive and saturated. However, financial profiles of domestic TCs remain strong due to moderate leverage and decent cost management.

There is common feature in the outlook for all TCs regardless the region. Large capital spending will pressure TCs free cash flow thus challenging their debt service capacity. In such an environment, owners and creditors need a reliable project management system such as investment controlling to manage risks with higher precision.

4   Credit Risk Management in Investment Controlling

Investment controlling (IC) is an application of methods of controlling to project management, a combination of processes, methods, skills and tools which ensuring the achievement of projects' goals in an uncertain business environment [8, 9]. In TCs, it is applied for managing strategically important projects: development and implementation of new technologies and/or services or construction of critical infrastructure [8]. The functions of IC are presented in Fig. 1.

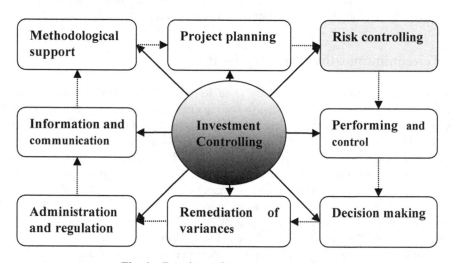

Fig. 1. Functions of investment controlling

In turn, risk-controlling is IC's critical function responsible for risk management at all stages of projects' lifecycle [9]. It is responsible for (1) identification and assessment of projects' risks; (2) regular tracking the performance of TPs against plans to prevent realization of risks at the earliest stages; and (3) timely performing remediation of variances caused by the realization of risks [8].

Consequently, a project's credit risk is a probability of TC's inability or unwillingness to repay its obligations on time. One way of assessing the credit risk is to emulate its PCR, if it absent, by modelling the ICRs. ICRs provide the relative ranking of TP's creditworthiness for the next 12–18 months expressed by the rating symbols and related probabilities of default [12]. In the next chapters, we will develop a mechanism of building such rating systems.

5 The Specifics of Projects in Telecommunications

TPs possess several special features which distinguish them from investment projects performed in other capital intensive industries (such as power generation or infrastructure) [4, 6] (Table 1).

Table 1 shows that CRs for TP are usually higher than for projects in other industries due to weaker creditors protection and absence of projects' ring-fencing. Therefore, CR for TP is equal to CR of the entire TC with adjustments on lenders protection measures:

$$ICR_i \cong M(CR) \pm \Delta_i \tag{1}$$

where ICR_i – internal credit rating of project "i"; $M(CR)$ – econometric model; Δ_i – adjustment to modelled ICR to account for lenders protection measures in i-th project.

6 Developing the Mechanism of ICRs Modelling for TPs

6.1 Selection of Methodological Framework

We selected rating methodologies of Moody's as a framework to build the ICRs. This is underpinned by (1) proved ability of PCRs of Moody's to predict default with high accuracy [2]. The application of these methodologies will also ensure the comparability of ICRs and PCRs from Moody's. We applied Moody's rating scale for ICR system with alpha scores from Ca to Aaa (Table 2). This scale consists of 20 grades. Each grade is also mapped to a numerical scale from 1 to 20 [14] and also to idealized default probabilities [6].

6.2 Selection of Explanatory Variables for ICR Modelling

The starting point for selection of explanatory variables (EVs) was Moody's methodology for telecommunication service providers [14]. According to it, the creditworthiness of TC is assessed by analysis of its six components: (1) size and scope; (2) business profile and characteristics of external environment; (3) quality of corporate governance; (4) profitability and operational efficiency; (5) debt leverage and coverage; and (6) liquidity. Out of these six components, four of them are directly inferred from TC's financial reporting: the size, the profitability; the leverage and coverage and liquidity.

Table 1. TP's features in comparison to projects in other capital intensive industries

Features	Projects in telecommunications	Project in other industries
Contracts	No single contract is fundamental to the business of the TC. The projects bear elevated price and volume risks as well as supply risks	There is a fundamental contract which gives the right to issuer to operate and generate revenues for a given period. Other contracts include a supply, an operating, a construction contract and financing contracts. They allocate risks to the parties who can best manage them
Number of projects	The large number of not ring-fenced projects at different stages of life cycle. The TC and/or project company has few restrictions on the scope of business	One single project, ring-fenced in a special purpose entity with limited ability to change its scope of business as defined in the contracts
Projects' length	Short (<2 years) or medium (3–5 years)	High (15–20 years or more)
Limitation on cash flow movement	No or weak restrictions on cash flow movement among projects	Strong restrictions on resources and cash movement among projects
Creditors protection	Low level of protection (usually some fixed assets pledge and covenants)	Many structural/contractual protection measures available to creditors
Creditors involvement in management	Management discretion	Creditor control and oversight over project's performance through budgets, reliance on specialist due diligence consultants. In stress scenario, control passes from equity to debt to achieve timely rectification
Debt service	Reliance on TC's cash flows or value of assets to repay debt. Weak limitations on cash flow usage. Recourse to owners	Reliance on specific asset/reserve future cash flows to repay debt. Non-recourse to owners
Capital structure	Stable capital structure, large share of equity capital. Few restrictions on ability to incur additional debt. Assumption is that the company will keep refinancing its loans	Usually, the project is fully financed by debt capital. All or most of the debt must be repaid by end of concession/license/useful life of asset

The remaining two are evaluated by subjective analysis of TC's environment. For these components, we selected EVs which are the best match to the rating factors mentioned in the methodology. To ensure robustness, these EVs obtained from the reputable international organizations – World Bank, World Economic Forum, International Telecommunication Union. The list of EVs is presented in the Tables 3 and 4.

Table 2. Rating scale of ICRs proposed in the mechanism

Rating grade	Aaa	Aa1	Aa2	Aa3	A1	A2	A3	Baa1	Baa2	Baa3
Numerical scale	1	2	3	4	5	6	7	8	9	10
Default rate, %	0.00	0.02	0.05	0.10	0.19	0.35	0.54	0.83	1.20	2.38
Rating grade	Ba1	Ba2	Ba3	B1	B2	B3	Caa1	Caa2	Caa3	Ca
Numerical scale	11	12	13	14	15	16	17	18	19	20
Default rate, %	4.20	6.80	9.79	13.9	18.1	24.0	32.5	43.9	66.2	80

Table 3. Explanatory variables from TC's financial statements

Factors of creditworthiness[a]	Explanation variable	Explanation
Company's scale and size	**Revenue (R_t). $ million**	Scale influences attributes which drive TC's competitive strengths: the breadth of business, access to customers, economies of scale, operational and financial flexibility, pricing power and access to financial markets. For service providers, scale enhances TC's ability to bundle products, increasingly a competitive advantage. Large size often means broader diversification which reduces exposure to risks
Profitability and operating efficiency	**EBITDA margin (EM_t), %** $EM_t = \frac{EBITDA_t}{R_t}$ (2) EBITDA - earnings before interest, tax, depreciation and amortization **Revenue trajectory (DR_t), %** $DR_t = \frac{R_t}{R_{t-2}} * 100\%$ (3)	The higher the margin, the lower the risk that the TC will not have sufficient funds to pay debt after it performed reinvestment in marketing, research, facilities, and human capital Revenue trajectory in past two years shows the company's lifecycle. If revenue declines there is a risk of company's inability to cut costs thus putting profitability under pressure
Leverage and coverage	**Debt loading ratio (DLR_t)** $DLR_t = \frac{Debt_t}{EBITDA_t}$ (4) **Debt coverage by operating cash flows ($DCCF_t$), %** $DCCF_t = \frac{FFO_t}{Debt_t} * 100\%$ (5) FFO_t – funds flow from operations; equals cash flow from operations before changes in working capital	Leverage and coverage metrics are key indicators for a TC's financial flexibility and long term viability. Financial flexibility is critical to respond to changing consumer preferences, regulatory changes or competitive challenges

(continued)

Table 3. (*continued*)

Factors of creditworthiness[a]	Explanation variable	Explanation
	Interest coverage ratio (ICR$_t$) $$ICR_t = \frac{(EBITDA_t - CE_t)}{Int_t} \ (6)$$ CEt – capital expenditures Int$_t$ – cash interest expenses	DLR$_t$ – is an indicator of debt leverage, comparative financial strength and TC's ability to service debt DCCF$_t$ – characterizes TC's ability to service the debt from cash flows from operating activities ICR$_t$ – characterizes TC's ability to cover ongoing cash interest payments. The presence of capital expenditures in the ratio reflects the capital intensity of the industry
Liquidity	**Cash asset liquidity ratio (QL$_t$)** $$QL_t = \frac{Cash_t + MS_t}{CL_t} \ (7)$$ MS$_t$ – marketable securities and short term deposits CL$_t$ – current liabilities **Cash flow ratio (OCF$_t$)** $$OCF_t = \frac{CFO_t - Div_t}{CL_t} \ (8)$$ CFO$_t$ – operating cash flow Div$_t$ – cash dividends	Liquidity indicates the sufficiency of company's cash to quickly repay current obligations. It is particularly important for TC in highly seasonal operating environment

[a]t denotes the reporting period

Table 4. Explanatory variables describing business environment and the governance

Factors of creditworthiness[a]	Explanation variable	Explanation
Macroeconomic environment[b]	**Growth in real gross domestic product (gGDP$_t$) in TC's domicile, %** $$gGDP_t = \frac{GDP_t}{GDP_{t-1}} * 100\% \ (9)$$ **GDP per capita in TC's domicile, cGDP$_t$, \$ per person a year** **Inflation in TC's domicile (IN$_t$) (%)** Measured by consumer price index, CPI$_t$	Volume of services of telecommunication company depends on economic cycle (as reflected in growth or in reduction of GDP) Demand for telecommunication services depends on population's income reflected in GDP per capita High inflation is a sign of economic and political instability and weaknesses in country's financials

(*continued*)

Table 4. (*continued*)

Factors of creditworthiness[a]	Explanation variable	Explanation
Competition[c]	**Global competitiveness index (GCI$_t$) of TC's domicile** Measured on scale from 1 (the lowest) to 7 (the highest competitiveness)	The high level of domicile's competitiveness can reflect the high level of competitiveness of domestic telecommunication industry
Penetration of telecommunication services[d]	**IUI$_t$ – the share of individuals using Internet, (% of total population) MS$_t$ – the number of mobile phone subscriptions per 100 of population)**	The higher the % of new technology users the higher the demand for telecommunication services
Domicile institutional strength[e]	**Institutional framework and effectiveness in domicile (IFEt)** $IFE_t = \frac{1}{2}W_1 + \frac{1}{4}W_2 + \frac{1}{4}W_3$ (10) W_1 – government effectiveness index W_2 – rule of law index W_3 – control of corruption index Each index is measured with the scale from -2.5 (the lowest) to 2.5 (the highest)	The higher is the government effectiveness index the higher is the probability that the government will implement the policy fostering economic growth and supporting competition The higher the rule of law index, the more predictable company's financial results and the higher the level of investors' protection is The higher is the control of corruption index the higher is the level of protection of investors rights from unlawful actions of the government
Corporate governance[c]	**IBD$_t$ – index of efficacy of corporate' boards in the domicile** Measured by the poll of diversified group of respondents. The possible responses ranges from 1 = management has little accountability to investors and boards to 7 = management is highly accountable to boards	The higher the quality of corporate governance in the companies the more interests of all investors are protected and the less probability that management will intentionally lead TC to default

[a]t denotes the reporting period
[b]Source: World Bank, http://data.worldbank.org/
[c]Source: World Economic Forum, https://www.weforum.org/reports/
[d]Source: International Telecommunication Union, http://www.itu.int/
[e]Source: World Bank, http://info.worldbank.org/governance/wgi/index.aspx#home

6.3   Selection the Core Model

The dependence of the ICR from explanatory variables is nonlinear; and the dependent variable (ICR) is categorical. Consequently, we propose the following options for a core model M(CR): (1) ordered multinominal logistic regression (logit): (2) ordered multinominal probit regression; (3) artificial neural networks (ANN); (4) genetic algorithms; (5) support vector machine; (6) committee machine learning; (7) decision tree algorithm or other models which estimate the probability that an observation with values of EV falls into a specific rating grade. To increase the model accuracy we suggest the following options of model's specification:

1. Specification 1AY assuming the time lag between ICR and EV of 1 year:

$$ICR_t = M(EV_{t-1}) \pm \Delta \tag{11}$$

   It assumes that the credit analyst at the year t uses data from available financial statement of TC for the year t − 1 to assign the PCR to TC. This corresponds with the conclusions in [11].
2. Specification 3AY using average values of EV for the three years t, t − 1 and t − 2:

$$ICR_t = M\left(\overline{EV}_{jt} = \frac{EV_{j,t-2} + EV_{j,t-1} + EV_{j,t}}{3}\right)^K_{j=1} \tag{12}$$

   K-number of EVs in the model. It assumes that the credit analyst at the year t uses TC's financial statements for years t − 1 and t − 2 and have a good forecast of TC's financial for the year t. The advantage of this specification is that it allows analyzing company's credit history and checking adequacy of the forecast for year t.

6.4   Collection and Processing the Input Data

PCRs from Moody's of telecommunication companies at the year-end are obtained from Bloomberg. IFRS or GAAP financial statements of TCs are obtained from EDGAR (for US companies), FactSet, Capital IQ and Thomson One. Financial statements should be translated into the single currency (e.g. US dollars). Macroeconomic and market data are obtained from public databases of World Bank, World Economic Forum and International Telecommunication Union.

   However, data from TCs' financial statements cannot be directly used in M(CR). This is underpinned by (1) differences in accounting treatment of certain transactions between IFRS and GAAP; and (2) the fact that some transactions are not reflected in accounting by their economical substance. Consequently, certain items in financial statements are adjusted to overcome these limitations [3]. The information for adjustments can be found in relative disclosures in financial statements. Examples of such adjustments are listed in Table 5.

Table 5. Financial statement adjustments to improve presentation of financial data

Accounting objects	Recommended analytical adjustments
Operating leases	Operating lease is capitalized and recognized as a debt obligation. Rent expenses are recognized in profit and loss statements as interest payment and depreciation
Capitalized interest	Expensed in the current year, in cash flow statement capitalized interest is reclassified to an operating cash flow
Capitalized development costs	Expensed in the current year, in cash flow statement these costs are reclassified to an operating cash flow
Defined benefit pension plans	Recognize underfunded or unfunded pension obligations as debt
Securitizations and factoring arrangements	To classify off balance sheet securitization and factoring arrangements as collateralized borrowings
Interest expense related to discounted long-term liabilities other than debt	To adjust interest expense to reclassify the accretion of discounted long-term liabilities other than debt as an operating expense
Inventory reported on a LIFO cost basis	To adjust inventory recorded on a LIFO cost basis to FIFO value
Unusual items	Exclude from calculations of ratios

6.5   Selection the Core Model with the Highest Predictive Ability

The goal of this step is to select the best core model among the alternative which ensures the highest predictive power. The accuracy criterion is an error (δ) calculated as a difference between the modelled ICRs and actual PCRs. The best model provides the lowest number of discrepancies. The errors δ are additionally analyzed for classifying them into type I errors ($\delta < 0$, indicating that ICR overestimates the actual PCR) and type II errors ($\delta > 0$, indicating that ICR underestimates the actual PCR). The losses from errors type I are more material.

To estimate the parameters of the model and to assess its prediction power (1) the input data is divided into the training and testing sample; (2) the model is trained on the training samples and model's parameters are estimated; (3) the model with estimated parameters is tested by using the testing sample; and (4) the errors δ are calculated and analyzed and the accuracy ratios are evaluated. The accuracy ratio for the given threshold X can be written as:

$$AR_{\delta \leq X} = \frac{N_{\delta \leq X}}{N} \qquad (13)$$

$AR_{\delta \leq X}$ – the accuracy ratio, the ability of ICRs to predict PCRs with the error not exceeding the threshold X

$N_{\delta \leq X}$ – the number of modelled ICRs with the error, not exceeding the threshold X
N – the total number of modelled ICSs.

The ICR system built in this mechanism will be acceptable if its $AR_{\delta=0}$ is above that of ICRs modelled with application of conventional mechanisms which provided an accuracy ratio of 37%–43%. Consequently, the k-th core model among l-alternatives ($l \in [2, L]$ will be the best if (1) the quantity of error I type is minimal; and (2) $AR_{k,\delta=0} > AR_{l,\delta=0} \ \forall l, l \neq k$.

6.6   Adjusting ICRs for Creditor Protection Measures and TP Features

Adjustments in ICRs for creditor protection measures (Table 1) and TP's special features are done manually by the evaluators (e.g. bank's credit committee or TC's risk committee). These adjustments (Δ) can increase or decrease the ICR by one or two grades [4]. Examples of TPs' features which can require adjustments in ICR are listed in Table 6.

Table 6. TP's features which can require adjustments to modelled ICR

Features which decrease TP's credit risks	Features which increase TP's credit risks
Existence of liquidity reserves (committed credit lines, reserve funds, etc.) for more than 6 months (for projects with duration less than 1 year) or for more than 12 months (for projects with duration of more than 1 year)	Application of new, untested or unique technology without a track record of successful implementation. New product without track record of customers' acceptance
TP is ring-fenced in dedicated special purpose entity with limited recourse to other creditors	Past track record of similar project/technology failures both in the company and/or in industry
Structural protection to creditors (permitted distributions to shareholders subject to pre-agreed tests, cash waterfall, controlled accounts, restriction on new business, new debt, on sale of assets and acquisitions etc.)	Customers and/or supplier concentration
Specific security for TP's debt on material contracts, accounts, revenues and account receivables, shares, all assets, etc.	The necessity to refinance significant share of indebtedness after the project's completion
Support from the state and/or TP's sponsor with a higher creditworthiness. Debt is guaranteed by the sponsor with higher creditworthiness	Structural and/or contractual subordination of project indebtedness to other obligations of TP and/or TC
Project management system (e.g. IC) with proved ability of early detection of variances vs. plans and rectification of these variances	Project management system with a track record of projects' failures and significant cost overruns. Project managers with not sufficient qualification and experience
Guaranteed offtake of TP's products or services and/or guaranteed supply of critical components	TP is subject to numerous event risks

6.7   Periodic Calibration of ICR Model

Periodic calibration of M(CR) captures cumulative changes in Moody's methodologies (they are revised by the agency every 3–5 years) as well as changes in macroeconomic environment, technological progress, etc. In calibration process the new input data for the past 10 years is obtained and processed, the model is trained on new data and updated model's parameters are estimated. We recommend revisiting the models ones in five years or more frequently in case of material and/or fast changes in TCs' business environment.

7   Testing of the Mechanism

For testing we used (1) macroeconomic and telecommunication market data for 2006–2016 from World Bank, World Economic Forum and International Telecommunication Union; and (2) financials from annual reports of 112 TCs from 40 countries for the same period. The number of observations was 899. More than half of TCs in the sample are representatives of 9 countries: USA (41), Canada (7), UK (6), Germany (4), the Netherlands (4), Mexico (4), Luxemburg (3), Indonesia (3) и Russia (2). Figure 2 shows distribution of PCRs of TCs in the sample.

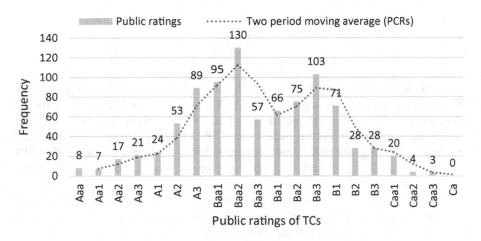

Fig. 2. Input data rating distribution by rating categories

This data was processed in accordance to requirements of model specification 3AY and split into the training sample ($n_1 = 201$) and testing sample ($n_2 = 519$). For M(CR) we selected feedforward artificial neural network with Levenberg-Marquardt training algorithm (LMA) [5]. In this algorithm, for k-ordered ICRs (presented in set of ordered integers from 1 to 20), the probability that the creditworthiness (Y) of m-th TC with the set of explanatory variables Y_m will be categorized to one of the classes k equals:

$$
\begin{cases}
P(Y_m = 1) = F(c_1 - Y_m\beta) \\
P(Y_m = 2) = F(c_2 - Y_m\beta) - F(c_1 - Y_m\beta) \\
\qquad \dots\dots\dots\dots\dots\dots\dots \\
P(Y_m = k - 1) = F(c_{k-1} - Y_m\beta) - F(c_{k-2} - Y_m\beta) \\
P(Y_m = k) = 1 - F(c_{k-1} - Y_m\beta)
\end{cases}
\tag{14}
$$

where, F-function of probability distribution, β – vector of model's parameters and $c = (c_1, c_2\dots\dots c_{k-1})$ – vector of threshold values. LMA implies successive approximation of initial values of parameters to the desired local optimum c. The method is based on steepest descent which, in case of long linear approximation, requires the calculation of second derivatives for the defined target function [5]. Testing outcome is shown in the Table 7.

Table 7. Accuracy ratios of modelled ICRs

Accuracy ratio (AR)	Training set (n_1 = 201)	Testing set (n_2 = 519)	Design set (n = 720)
$\delta \leq -3$	50.2%	30.8%	36.2%
$\delta \leq -2$	74.7%	45.8%	53.9%
$\delta \leq -1$	85.5%	52.4%	61.6%
$\delta = 0$	**90.3%**	**55.3%**	**65.1%**
$\delta \leq 1$	88.2%	54.1%	63.6%
$\delta \leq 2$	73.5%	45.1%	53.0%
$\delta \leq 3$	54.8%	33.6%	39.5%
$|\delta| \leq 1$	**96.4%**	**85.0%**	**88.2%**
$|\delta| \leq 2$	**98.6%**	**89.3%**	**91.9%**

Testing shows that ICRs, modelled with the application of our mechanism with ANN as a core model, have a precision of 55% for training sample and 65% for the design sample. These accuracy exceed that of ICRs built with application of conventional mechanisms which had the precision of 37%–43% [11]. The accuracy ratio with an error not exceeding one rating grade was around 85% for testing set and around 89% for design set. This is comparable to those of conventional mechanisms (90%–92%) [10, 11].

However, errors type I and errors type II in outcome are distributed almost evenly. Therefore, the task of reducing the number of errors type I remains actual. We also observe a material difference between accuracy ratios obtained in the model training stage and accuracy ratios obtained in running the model with estimated parameters on the testing set (90.3% vs. 55.3%). Additional research is necessary to investigate the reasons for this difference.

8   Conclusion

We developed a mechanism of modelling of internal credit ratings (ICRs) of investment projects in telecommunication companies. This tool is applied in TCs investment controlling systems to assess the creditworthiness of projects if PCRs from IRAs do not exist. Alternatively, the mechanism can be applied by investors (banks or investment funds) as part of broader process of providing financing to TPs. The necessity of development of the mechanism was underpinned by unsatisfactory accuracy and robustness of ICRs built with the application of conventional methods and approaches.

The advantage of developed mechanism is that it combines all procedures and operations of ICRs' modelling in one system, thus improving managers and/or investors understanding of ICR-building process and leading to reduction of cost and time spend on modelling. Furthermore, the application of Moody's rating system and methodologies as a framework for the mechanism ensures the comparability of ICRs to the existing PCRs and robustness of the ICRs. To increase the accuracy of modelling, the mechanism provides (1) conversion of input data, presented in accordance to different accounting standards, to the common basis; (2) recording of certain transactions in accordance to their economic substance rather than the form; and (3) taking in account both past and future performance of TCs.

To test the mechanism, we used artificial neural network as a core model and financial data of 112 telecommunication companies from 40 countries for the period from 2006–2016. The testing shows that ICRs modelled with application of our mechanism had the accuracy ratio of 55% for testing sample and 65% for the design sample. This exceeds the accuracy ratios of ICRs modelled with the application of the conventional approaches (37%–42%).

References

1. Telecommunications Industry Outlook. https://www2.deloitte.com/content/dam/Deloitte/us/Documents/technology-media-telecommunications/us-tmt-2017-telecommunications-industry-outlook.pdf
2. Corporate Debt Ratings Performance Insights. https://www.moodys.com/researchdocument contentpage.aspx?docid=PBC_1057582
3. Financial Statement Adjustments in the Analysis of Non-financial Corporations. https://www.moodys.com/researchdocumentcontentpage.aspx?docid=PBC_1055812
4. Gatti, S.: Project Finance in Theory and Practice. Elsevier, Amsterdam (2013)
5. Gavin, H.P.: The Levenberg-Marquardt method for non-linear least squares curve-fitting problems. Environ. Eng. **4**, 5–18 (2016)
6. Generic Project Finance Methodology. https://www.moodys.com/researchdocumentcont entpage.aspx?docid=PBC_127446
7. Glukhov, V., Balashova, E.: Economics and Management in Info-communnication: Tutorial. Piter, Saint Petersburg (2012)
8. Grishunin, S., Suloeva, S.: Development of project risk rating for telecommunication company. In: Galinina, O., Balandin, S., Koucheryavy, Y. (eds.) NEW2AN/ruSMART - 2016. LNCS, vol. 9870, pp. 752–765. Springer, Cham (2016). doi:10.1007/978-3-319-46301-8_66
9. Grishunin, S., Suloeva, S.: Project controlling in telecommunication industry. In: Balandin, S., Andreev, S., Koucheryavy, Y. (eds.) ruSMART 2015. LNCS, vol. 9247, pp. 573–584. Springer, Cham (2015). doi:10.1007/978-3-319-23126-6_51

10. Huang, Z., Chen, H., Hsu, C.-J., Chen, W.-H., Wu, S.: Credit rating analysis with support vector machines and neural networks: a market comparative study. Decis. Support Syst. **37** (4), 543–558 (2004)
11. Karminsky, A.M., Polozov, A.A.: Handbook of Ratings: Approaches to Ratings in the Economy, Sports, and Society. Springer, Cham (2016)
12. Mishenko, A., Chizhova, A.: Methodology of credit risk management and optimization of credit portfolio formation. Financ. Manag. **1**, 91–105 (2008)
13. Morgunov, A.: Modeling the probability of default of investment project. J. Corp. Financ. Res. **1**(37), 23–45 (2016)
14. Rating Methodology. Telecommunication Service Providers. https://www.moodys.com/researchdocumentcontentpage.aspx?docid=PBC_1055812

High-Tech Sector in the Conditions of Institutionalization of the Smart Economy (on the Example of the Telecommunication Industry)

N.V. Vasilenko[1(✉)], A.J. Linkov[1], and V.V. Glukhov[2]

[1] Russian State Herzen Pedagogical University, Saint Petersburg, Russia
nvasilenko@mail.ru, al308-linkov@yandex.ru
[2] Peter the Great St. Petersburg Polytechnic University, Saint Petersburg, Russia
vicerector.me@spbstu.ru

Abstract. The article contains a theoretical analysis of the institutionalization of the Smart Economy as a new type economy on the materials of communications industry. The analysis is based on the research of interrelations between the concepts of "cluster", "network transactions" and "collaboration". The perspectives of participation of the telecommunications industry enterprises in the clusters formation are shown. The conjunction of the merits of the hierarchical and market structures in the clusters is due to the network nature of clusters, the possibility of combining of the cooperation and the competition between participants. The proliferation of clusters in the high-tech sectors of economy is associated with the advantages of network transactions. Specific characteristics of network type transactions and their advantages to the interacting parties are identified. The awareness of the levels of institutionalization of the Smart Economy and the consistent movement on the chain "clusters - network transactions - collaboration" when building relations between the cluster participants will allow a qualitative solution of a number of problems facing the Russian communications industry.

Keywords: Telecommunication industry · Telecommunications services market · The smart economy, institutionalization · Cluster · Network transactions · Collaboration

1 Introduction

In the conditions of institutionalization of the Smart Economy, the opportunities for economic growth are determined by the level of development of the high-tech sector. An integral part of the high-tech sector of the economy is the modern telecommunications industry, which includes a variety of data transfer activities. According to "TMT Consulting", the volume of Russian telecommunications services in value terms increased by 11.3 times in the period from 2000 to 2015 and reached 1674 billion rubles. In the structure of the Russian telecommunications market, 67% account for cellular communications services, 26% for wired telephony, 6% for fixed broadband access, 1% for wireless broadband access [27].

© Springer International Publishing AG 2017
O. Galinina et al. (Eds.): NEW2AN/ruSMART/NsCC 2017, LNCS 10531, pp. 315–325, 2017.
DOI: 10.1007/978-3-319-67380-6_29

The main trends in the development of telecommunications are currently [13, 22, 32]:

- convergence of telecommunication, information and computer technologies, which contributes to the intensive computerization of telecommunications equipment;
- creation of a worldwide network based on national personal communication networks;
- provision of services in accordance with the individual needs of the consumer, regardless of time and space;
- liberalization and internationalization of the market, free movement of new developments across borders of states, which contributes to the further innovative development of the telecommunications industry.

The special importance of the telecommunications industry in the context of the formation of the Smart Economy is determined by the following circumstances:

- application of "high" technologies;
- the fulfillment by the industry of the infrastructure function for the innovative development of other industries;
- the social significance of the industry, which is currently actualized by the problem of "digital divide".

High-tech nature of the telecommunications industry is determined by the implementation of research and development and the application of knowledge-intensive technologies. Scientific developments in the field of microelectronics, computer technology, space research, material technologies, etc. are used to design new and improve existing technical devices and communication networks [42]. The status of the telecommunications industry as a high-tech sector (along with the aviation and rocket and space industry, shipbuilding, nuclear energy complex, power engineering and radio electronic industry) is enshrined in the "Concept of Long-Term Social and Economic Development of the Russian Federation for the Period to 2020".

The infrastructural function of telecommunications is realized through ensuring the coordinated work of all economic structures, as well as government bodies. The opportunities for innovative development of other industries, created by the telecommunications industry, are determined by the essence of the technologies used, providing opportunities for rapid exchange of information, creating favorable conditions for investments attracting, support of the population employment, etc.

The social importance of telecommunications services is due to their positive impact on the quality of life of the population. This influence is expressed in the possibility of providing access to different types of goods and, on this basis, increasing the level of consumption, safety and qualification [14]. The State Program of the Russian Federation "Information Society (2011–2020)" provides for the implementation of a subprogram aimed at "providing high-quality and affordable communication services, as well as services to provide access to the information and telecommunications infrastructure".

It follows that the telecommunications industry is one of the main factors for the sustainable growth of other sectors of the economy and the basis for increasing national competitiveness.

At the same time, the further development of the telecommunications industry as a high-tech sector of the Russian economy is constrained by a number of factors [29, 39].

The main factors among them are:

- imperfection of the regulatory mechanisms;
- uneven development of various segments of the telecommunications market;
- resource limitations and risks of innovation activity.

Imperfection of regulatory mechanisms is manifested in several ways. Firstly, the legislation on communication has a vague terminology and ambiguous wording, making it difficult to make decisions. Secondly, the legal consolidation of norms and rules lags behind the practice of developing telecommunications. An example is the problem of legalizing the technological neutrality of the frequency spectrum, which hinders the transition of telecommunications enterprises to modern standards and contributes to reducing the level of competition and increasing the technological gap.

The uneven development of various segments of the telecommunications market is manifested in the underdevelopment of telecommunications in rural and sparsely populated areas due to the unprofitable provision of telecommunications services in these areas due to low population density. As a result, the restriction of access to modern means of communication for certain population groups is fixed. The emerging problem of "digital inequality" is exacerbated by the vastness of the territory of Russia, the complex natural and climatic conditions, etc.

Resource limitations consist, first of all, in insufficient volume of investments for conducting full-scale telecommunication activity and dependence on imported equipment.

2 Research Problem

The formation of the Smart Economy is accompanied by a system of interconnected transformations. These transformations at a certain stage begin to affect not only the set of basic technologies, but also the institutional environment, including at the level of the prevailing organizational structures. At the same time, some of the changes are observable, and some can be detected only on the basis of special studies that presume a rather high level of abstraction. This is due to the fact that the formation of the Smart Economy is accompanied by institutional transformations taking place at various levels.

The current state of the telecommunications market in Russia and in the world as a whole is characterized by profound structural changes. The formation of new elements and characteristics of the infrastructure of telecommunications services creates the prerequisites for the inclusion of enterprises in the telecommunications industry in various cluster alliances [4, 23].

The authors believe that the proliferation of clusters as an effective organizational form providing for the creation of new technologies, products and services in the high-tech sector is associated with the advantages of network transactions, the effectiveness of which in the innovative environment is conditioned by the formation of a new, higher form of interaction of network partners - collaboration.

The transition through the chain "cluster - network transactions - collaboration" assumes an increasingly deeper level of analysis of changes that determine the institutionalization of the Smart Economy.

The process of research used general scientific methods of generalization, analysis and synthesis, comparative analysis, quantitative methods.

3   Methodology and Results

The process of clustering, according to the research results of foreign and Russian authors by 2010, covers about 50% of the economies of the leading countries of the world, forming "structures more flexible than the hierarchy model, and at the same time more integrated than the traditional market model" [34, 35]. Currently, clustering is characteristic of many countries, for example, the winemaking cluster in Chile and California, the automobile cluster in Austria, Hungary and Japan, the chemical and machine building cluster in German, the food and perfume cluster in France, the engineering and marine cluster in Norwegian, the steel cluster in Sweden. In the US there are more than 40 large clusters that create 61% of GDP and provide jobs for more than half of the employable population [18].

Porter understood by clusters geographically concentrated groups of interrelated companies in certain areas that are competing but at the same time working together [28]. The similar concept of industrial clusters is fixed in Article 20 of the Federal Law of the Russian Federation "On Industrial Policy in the Russian Federation" of December 31, 2014 N 488-FZ. In Resolution of the Government of the Russian Federation No. 779 of July 31, 2015 (Edited on September 26, 2016) "On Industrial Clusters and Specialized Organizations of Industrial Clusters," it is stated that "the purpose of creating an industrial cluster is to create a certain set of stakeholders associated with relations due to territorial proximity and Functional dependence".

The emphasis in the definitions given above is made on the geographical and technological features of the cluster. The geographical feature makes it possible to obtain a win from the distribution between companies of the common resources costs of maintaining and developing from reducing the time of the necessary raw materials and services relocation between firms and from the possibility of implicit knowledge spreading from firm to firm [36, 41].

The technological sign was initially interpreted as the use of similar technologies by firms belonging to a cluster. In modern conditions, the technological sign in the modern sense focuses on the innovative capabilities of the cluster organization. Industrial clusters can be considered as innovative if they have specialization in knowledge-intensive industries [3, 7].

The main document setting the framework for cluster policy in the Russian Federation is the Concept of Long-Term Social and Economic Development of the Russian Federation for the period up to 2020. In it, the creation of a network of territorial clusters is a condition for modernizing the economy and realizing the competitive potential of the regions. In 2012, simultaneously with the launch of the Cluster Initiatives Support Program in Russia, a list of such clusters was approved.

According to the map of clusters (http://map.cluster.hse.ru) developed by specialists of the Russian cluster observatory of the Institute of Statistical Studies and Economics of Knowledge at the Higher School of Economics, taking into account the experience of the projects of the European cluster observatory and US Cluster Mapping, in Russia currently are 54 clusters maintained by the Center of Cluster Development within the framework of the program of the Ministry of Economic Development of Russia on support of small and medium-sized enterprises, 28 pilot innovative territorial clusters and 7 industrial clusters of the Ministry of Industry and Trade of Russia.

The cluster's effectiveness depends on the combined effect of the infrastructure supporting processes, as well as on the innovative activity of resident enterprises. An analysis of the territorial location of clusters allows to conclude that the regions that concentrate a powerful production, social and infrastructure potential have priority [7, 12].

In the later works of other authors, the geographic-technological conditionality of the cluster is supplemented by the following its interpretations [16, 19, 40]:

– organizational form of achieving a high level of competitiveness;
– market form of development of cooperative interaction;
– a network of suppliers, producers, consumers, elements of industrial infrastructure, research institutes, interrelated in the process of creating added value.

In these approaches, emphasis is placed on the contractual feature, reflecting the specifics of the interaction of firms within the cluster. The contractual properties of the cluster are manifested in the form of a special institutional environment, including a set of common norms, values and knowledge that connects firms to each other and creates an effective balance between cooperation and competition. Geographical localization and technological interconnection determine the content of contracting, and the cluster is its organizational form.

The interaction between the cluster participants is based on the balance between competition and cooperation [1, 30, 31]. The competition of enterprises within the cluster creates prerequisites for increasing the effectiveness of its members. Cooperation is beneficial for the participants of the cluster for the maintenance and development of the necessary infrastructure, expanding sales markets, but most importantly - for sharing the accumulated knowledge and experience among themselves.

Hence, the cluster unites enterprises and firms of different profile and size, primarily interested in mutually beneficial cooperation and development, which corresponds to the conditions of networked mutual assistance [24]. Clusters with the participation of enterprises in the telecommunications industry are networks of independent manufacturing and service firms, bringing together their suppliers, developers of technologies and know-how, financial and consulting institutions and consumers interacting with each other within the single value chain [8]. Interaction between the enterprise of key and related specializations ensures the introduction of innovations both in the creation of the final product and in the line of technological equipment and components. The main task of the cluster - ensuring the efficiency and competitiveness of the products produced by the cluster - is a special case of solving the problem of "survival together", that is characteristic for the networked communities.

The main advantage of the cluster as an institutional form in the Smart Economy is that the successes of one cluster enterprise, primarily in the creation of the results of intellectual activity, quickly become the property of the other. This leads to a reduction in costs at all stages of the innovation process, and the benefits are spread in all directions [21]. Therefore, clusters, as a rule, are formed where there is or expected a "breakthrough" in the field of technology or technology with the subsequent release of an innovative product to the market [19]. This explains the participation in the clustering of enterprises in the telecommunications industry.

The growth of firms' competitiveness in the cluster is based on network effects and the division of labor. Some activities are disadvantageous or not profitable for large companies, so in a cluster the large companies can delegate the production of intermediate products and services to small firms [37].

Insufficient attention to the contractual sign of the cluster limits efforts on innovative development by creating only an organizational infrastructure without the formation of real mutually beneficial network type relationships.

The generalization of existing approaches makes it possible to identify the following institutional features of the cluster as a network form of organization of economic activity:

- implementation by enterprises of joint activities in a limited territory for the production of a certain type of products and services;
- cooperative relationships between cluster participants, mediated by "team culture";
- formal and informal mutual influence, complementarity and cooperation of enterprises;
- the rapid spread of new knowledge and experience from one enterprise to another.

The reliance on network interactions among firms inside the cluster necessitates of the research and identification of specific characteristics of network type transactions.

Distribution of networks as an organizational form of economic activity in modern conditions is connected with the strengthening of the interactive mode of contacts, as well as with the low efficiency of systems with a single control center in an environment with increasing information flows and interactive nature of innovations that generate continuous updates [5, 9, 26]. Networks form a special space with open borders and a mobile structure, the bases of development of which are laid in a positive resonant interaction of internal elements. This flexibility gives the economy the ability to continuously update [34, 35].

From an institutional point of view, relationships within the network are based on network transactions. In the modern economy, the transition to network transactions is carried out from two sides - both from market (for obtaining network effects), and from managerial (for increasing flexibility) types of interaction of economic entities.

Transition to network transactions gives to the interacting parties the advantages that associated primarily with the following properties:

- as opposed to market transactions - the prevalence of interdependencies in the relations of participants in the network, which perform the functions of monitoring the implementation of their interests; the development of the activity and creativity of the human capital of an individual economic entity and its integration action in

the associated networked human capital; reliance on coherent strategic behavior of participants in network interactions;
- in contrast to management transactions- the preponderance of mobility of the main processes over their scale, achieved when network services, on the one hand, penetrate into the sphere of operational activity, on the other - they lead a number of types of operating activities to the external network; diversity of links as the basis for choosing the most successful partnership option; the predominance of informal professional communication and mechanisms of mutual agreement, through which linear hierarchical links and traditional schemes of formalized behavior are eroded.

The study of the advantages of network transactions allows us to determine their specific properties, among which an important place is occupied by:

- integration of economic actors in the course of their interaction;
- the ability to achieve synchronization of changes in the needs of network customers with the formation of new values of products of network manufacturers, which creates the prerequisites for the formation of network equilibrium;
- orientation of network transactions to specific mega-initiatives and megaprojects in conditions of prevalence of non-price equilibrium factors.

Network transactions have a number of shortcomings, which probably will not allow them in the near future to oust other forms of organization of interaction of economic entities. Among such shortcomings:

- in comparison with market transactions - the duration of the process of networking, which can be accompanied by high enough costs; the possible emergence of opportunistic sentiments in the inter-company network formed around a large company; a high probability of a preponderance of the desire for stability over the search for better opportunities; the threat of absorption for small enterprises - network participants;
- in comparison with management transactions - the lack of material support for the interaction of network participants due to the replacement of classical long-term contractual forms and labor relations with network ones; high dependence on market conditions and the need to cross the interests of participants.

A more detailed analysis of the institutional network norms leads to the conclusion that the advantages of network transactions are related to the specific type of communication between economic entities, which creates for them new opportunities and new competitive advantages. At the heart of these new opportunities lies a collaboration that reveals the nature of the connection in network transactions [2, 11, 15, 20, 35, 43].

As a form of cooperation for obtaining a new intellectual product, collaborations can be formed in any field related to mental work [38]. Most widely they are distributed in the scientific sphere [4], where they are connected with innovative processes. There are five areas of intensive international cooperation with the participation of Russian enterprises in the telecommunications industry: regional telephone networks, mobile communications, satellite communications; Internet and e-commerce; Production of telecommunication equipment [17].

The systematization of existing approaches makes it possible to distinguish the following types of association of enterprises and organizations in collaboration on different grounds:

- the areas of activity: interdisciplinary and intradisciplinary;
- a unifying basis: territorial and virtual [33];
- terms of audience coverage: regional, national and international;
- subjects of collaborative interaction [6];
- on the predominant type of interaction: formal and informal [25].

Continuous interactive exchange of explicit and implicit knowledge among the participants of the collaboration allows to create the prerequisites for making effective decisions in the face of uncertainty and changes in the external environment. This increases the flexibility of the collaboration as an organizational structure and gives it incentives for further development.

One of the important characteristics of the collaboration is the absence of borders both between branches and spheres of economic activity, and between countries in the process of interaction of collaborators. The latter is especially important, as in the course of the collaboration a new unified institutional environment for the creative activity of participants is formed [1, 2]. As a result of the implementation of projects based on international collaboration, specific collective ownership and use objects are created that generate commodity objects and bring commercial effects at the global level of the economic system.

The need for an effective collaboration in the high-tech sector and, in particular, in the field of telecommunications services, is determined by the continued dependence of the Russian industry on the transfer of technology from abroad. This dependence remains quite high. According to Rosstat data for 2014, about 60% of Russian imports are represented by high-tech goods and services, while exports of high-tech products from Russia do not exceed 8%.

The recognition of the three levels of institutionalization of the Smart Economy and the consistent movement of the chain of "clusters - network transactions - collaboration" when building relations between the cluster participants will allow a more rapid and qualitative solution of a number of pressing problems facing the Russian communications industry. The main ones are:

- modernization of obsolete telecommunication services networks, taking into account the trends of their long-term development with the aim of reducing the digital divide;
- use of transitional technologies to erase the boundaries between fixed and mobile operators in terms of potential consumers of network telecommunications services [10];
- search and introduction of new effective types of monetization of telecommunication services networks;
- attracting more investment, including through a different mix of stakeholders.

4   Conclusions

1. The most important features of the telecommunications industry are the use of high technologies, the performance of infrastructural functions for other sectors of the economy and the social significance that is manifested in creating conditions for improving the quality of life.
2. Currently the telecommunications services market is approaching the saturation point. Its further development is associated with a qualitative improvement of the services provided on the basis of integration with the IT sector.
3. Due to the trend towards mobile wireless technologies, the participation of enterprises in the telecommunications industry in the process of clustering is due, first of all, to the contractual sign of the cluster as an organizational form.
4. Transition from the level of analysis of the organizational form to the level of analysis of the type of interaction allows, when implementing structural changes in the telecommunications industry, to rely on the advantages of network transactions in comparison with competitive market relations and intra-company hierarchy.
5. The preconditions for cooperation in the field of telecommunications services are associated with a combination of product features and a service component in them and the need to develop effective technological solutions to ensure national interests in the market liberalization.
6. Taking into account the levels of institutionalization of the Smart Economy as a new type economy creates opportunities for the formation of effective strategies for the development of enterprises in the telecommunications industry, as well as for the development of public policy priorities in relation to the high-tech sector.

5   Directions for Further Research

The authors see the directions of further research in the development of an institutional-cluster approach to the organization of innovative activities in the telecommunications sector, including the identification of conditions and factors for increasing the competitiveness of telecommunications enterprises in the Russian and international markets in the conditions of the Smart Economy.

References

1. Al-Hakim, L., Lu, W.: The role of collaboration and technology diffusion on business performance. Int. J. Prod. Perform. Manag. **66**(1), 22–50 (2017)
2. Antipina, E.A.: Regional'naja politika novogo pokolenija: postanovka zadachi v sfere chelovecheskogo kapitala. Chelovecheskij kapital **4**(64), 41–45 (2014). (Russian)
3. Arthurs, D., Cassidy, E., Davis, C.H., Wolfe, D.: Indicators to support innovation cluster policy. Int. J. Technol. Manag. **46**(3–4), 263–279 (2009)
4. Bogatov, V.V., Syroezhkina, D.S.: Kollaboracii nauchnyh organizacij kak jelement infrastruktury nauki. Nauka. Innovacii. Obrazovanie **4**(22), 30–44 (2016). (Russian)

5. Choj, M.K.: Funkcionirovanie setevyh struktur v sovremennom mire. Modeli, sistemy, seti v jekonomike, tehnike, prirode i obshhestve **3**(11), 95–99 (2014). (Russian)

6. Diane, H.: Sonnenwald Scientific Collaboration: A Synthesis of Challenges and Strategies, Swedish School of Library and Information Science Göteborg University and University College of Borås Sweden. http://datafedwiki.wustl.edu/images/f/fc/SonnenwaldScientific CollabOverview.pdf. Accessed 7 July 2016

7. Gagarina, G.J., Arhipova, L.S.: Innovacionnye territorial'nye klastery kak instrument povyshenija konkurentosposobnosti rossijskoj jekonomiki. Vestnik Novgorodskogo gosudarstvennogo universiteta im. Jaroslava Mudrogo **82**, 28–33 (2014). (Russian)

8. Golubev, A.A.: Sfera vysokih tehnologij: sovremennye rossijskie realii. Bjulleten' nauki i praktiki **9**(10), 95–98 (2016). (Russian)

9. Gordenko, G.: Kontraktnyj podhod k upravleniju setevymi organizacijami. Vestnik Instituta jekonomiki Rossijskoj akademii nauk **4**, 98–101 (2013). (Russian)

10. Hatunceva, E.A., Hatuncev, A.B.: Analiz osnovnyh tendencij razvitija setej svjazi na telekommunikacionnom rynke Rossii. T-Comm: Telekommunikacii i transport **10**(7), 71–74 (2016). (Russian)

11. Inshakov, O.V.: Kollaboracija kak global'naja forma organizacii jekonomiki znanij. Jekonomika regiona **3**(35), 38–44 (2013). (Russian)

12. Ketels, C.: Recent research on competitiveness and clusters: what are the implications for regional policy? Cambridge J. Reg. Econ. Soc. **6**(2), 269–284 (2013)

13. Kim, G.: The shape and implications of Korea's telecommunication industry: crisis, opportunity and challenge. Aust. J. Telecommun. Digit. Econ. **4**(4), 214–233 (2016)

14. Kiselev, S.V., Shakirov, M.M.: Soderzhanie, harakteristika i klassifikacija telekommunikacionnyh uslug kak ob#ekta issledovanija. Vestnik Kazanskogo tehnologicheskogo universiteta **16**(10), 311–316 (2013). (Russian)

15. Klasternaja jekonomika i promyshlennaja politika: teorija i instrumentarij: kollektivnaja monografija/Pod red. d-ra jekon. nauk, prof. A.V. Babkina. SPb.: Izd-vo SPbPU, pp. 113–130 (2015). (Russian)

16. Korobov, S.A., Mosejko, V.O., Fomina, S.I.: Primenenie klasternoj koncepcii v razvitii regional'nogo predprinimatel'stva. Biznes. Obrazovanie. Pravo **2**(27), 108–114 (2014). (Russian)

17. Larichkina, D.A.: Osobennosti razvitija otrasli telekommunikacionnyh uslug v Rossii. Mezhdunarodnyj zhurnal prikladnyh i fundamental'nyh issledovanij **5–3**, 473–477 (2015). (Russian)

18. Lenchuk, E.B., Vlaskin, G.A.: Klasternyj podhod v strategii innovacionnogo razvitija zarubezhnyh stran. Problemy prognozirovanija **5**, 38–51 (2010). (Russian)

19. Lisovskaja, N.V.: Klasternyj podhod k razvitiju jekonomiki regiona. Biznes. Obrazovanie. Pravo **3**, 135–139 (2012). (Russian)

20. Markowski, P., Dabhilkar, M.: Collaboration for continuous innovation: routines for knowledge integration in healthcare. Int. J. Technol. Manag. **71**(3–4), 212–231 (2016)

21. Mishura, N.A.: Formirovanie regional'nyh territorial'no-proizvodstvennyh klasterov: samoorganizacija i organizacija. Biznes. Obrazovanie. Pravo **3**, 168–172 (2012). (Russian)

22. Mumuni, A.G., Luqmani, M., Quraeshi, Z.A.: Telecom market liberalization and service performance outcomes of an incumbent monopoly. Int. Bus. Rev. **26**(2), 214–224 (2017)

23. Muntjaev, S.S.: Sovremennoe sostojanie i perspektivy razvitija kachestva uslug svjazi i telekommunikacij Rossii. Aktual'nye problemy jekonomiki i menedzhmenta **4**(12), 29–34 (2016). (Russian)

24. Navickas, V., Malakauskaite, A.: The impact of clusterization on the development of small and medium-sized enterprise (SMA) sector. J. Bus. Econ. Manag. **10**(3), 255–259 (2009)

25. Oldham, G.: International scientific collaboration: a quick guide. Bringing science and development together through original news and analysis. http://www.scidev.net/global/policy-brief/internationalscientific-collaboration-a-quick-gui.html. Accessed 7 July 2016
26. Pickernell, D., Rowe, P.A., Christie, M.J., Brooksbank, D.: Developing a framework for network and cluster identification for use in economic development policy-making. Entrep. Reg. Dev. **19**(4), 339–358 (2007)
27. Popova, T.N.: Tendencii razvitija mirovogo rynka telekommunikacionnyh uslug. Vestnik Dal'rybvtuza **3**, 94–98 (2014). (Russian)
28. Porter, M., Konkurencija, M.: Vil'jams, 608 p. (2005). (Russian)
29. Proskura, N.V., Efremenko, D.V.: Perspektivy razvitija telekommunikacionnogo sektora v Rossii. Upravlenie jekonomicheskimi sistemami: jelektronnyj nauchnyj zhurnal **12**(60), 39 (2013). (Russian)
30. Quandt, C.O., De Castilho, M.F.: Relationship between collaboration and innovativeness: a case study in an innovative organisation. Int. J. Innov. Learn. **21**(3), 257–273 (2017)
31. Rivera, L., Sheffi, Y., Knoppen, D.: Logistics clusters: the impact of further agglomeration, training and firm size on collaboration and value added services. Int. J. Prod. Econ. **179**, 285–294 (2016)
32. Rudnik, I.V.: Mirovoj rynok telekommunikacionnogo oborudovanija: drajvery kon#junkturnyh izmenenij. Rossijskij vneshnejekonomicheskij vestnik **2016**(2), 113–117 (2016). (Russian)
33. Salmela, E., Juvonen, P.: Differences in motivational mindset affecting collaboration of virtual research organisation. Int. J. Netw. Virtual Organ. **16**, 298–313 (2016)
34. Smorodinskaja, N.V.: Smena paradigmy mirovogo razvitija i stanovlenie setevoj jekonomiki. Jekonomicheskaja sociologija **13**(4), 95–115 (2012). (Russian)
35. Smorodinskaja, N.V.: Trojnaja spiral' kak novaja matrica jekonomicheskih sistem. Innovacii (4), 66–78 2011. (Russian)
36. Swords, J.: Michael Porter's cluster theory as a local and regional development tool: The rise and fall of cluster policy in the UK. Local Econ. **28**(4), 369–383 (2013)
37. Tatarkin, A.I., Lavrikova, J.G.: Klasternaja politika regiona. Promyshlennaja politika v Rossijskoj Federacii **8**, 11–19 (2008). (Russian)
38. To, C.K.M., Ko, K.K.B.: Problematizing the collaboration process in a knowledge-development context. J. Bus. Res. **69**(5), 1604–1609 (2016)
39. Trapeznikova, E.V., Shelud'ko, E.F.: Perspektivy razvitija rynka telekommunikacionnyh uslug v Rossii. Trudy Bratskogo gosudarstvennogo universiteta. Serija: Jekonomika i upravlenie **1**(1), 160–167 (2014). (Russian)
40. Tret'jak, V.P., Klastery predprijatij, M.: Avgust Borg, 132 p. (2006). (Russian)
41. Wolman, H., Hincapie, D.: Clusters and cluster-based development policy. Econ. Dev. Q. **29** (2), 135–149 (2015)
42. Zavivaev, N.S., Proskura, N.V.: Podhody k formirovaniju jetapov razvitija telekommunikacionnyh uslug. Vestnik NGIJeI **12**(67), 91–95 (2016). (Russian)
43. Glukhov, V.V., Balashova, E.: Operations strategies in info-communication companies. In: Balandin, S., Andreev, S., Koucheryavy, Y. (eds.) ruSMART 2015. LNCS, vol. 9247, pp. 554–558. Springer, Cham (2015). doi:10.1007/978-3-319-23126-6_48

Virtual Telecommunication Enterprises and Their Risk Assessment

N.V. Apatova[1]([✉]), O.V. Boychenko[1], Tatyana P. Nekrasova[2], and S.V. Malkov[3]

[1] V.I. Vernadsky Taurida National University, Simferopol, Russia
apatova@list.ru, oleg_boychenko@cfuv.ru
[2] Peter the Great St. Petersburg Polytechnic University, Saint Petersburg, Russia
dean@fem.spbstu.ru
[3] TAIR Enterprise, Simferopol, Russia
stas_malkov@tair.in.ua

Abstract. The article describe virtual telecommunication enterprises risks, their possible causes and ways of elimination, using the economic-mathematical model of risks assessment.

Keywords: Virtual telecommunication enterprise · Risk assessment · Mathematical model

1 Introduction

Internet economics is a new scientific and practice area which develops very quickly and the theory is not only ahead of reality but does not exist in enough volume. Many virtual enterprises start in the world every day; each organization has its website and realizes the activity in Internet in different forms. A lot of virtual enterprises sale of telecommunication goods and provide telecommunication services which include the production of information, telecommunication equipment and it's future support, also mobile connection, mobile phones and Internet providing. There are many books and articles about transforming the principles, forms, types and structures of traditional firms and organizations when they transfer their activities into Internet (see, for example, Davidow and Malone 1992; Filos and Banahan 2001; William 2006).

Moving to the Internet many economic activities served as the basis for the creation of virtual enterprises, producing primary network product - the information and carrying out trade and service, including financial functions (Beddle 2010). Internet has changed the demand for goods and services, expanded the proposal; user-consumer has involved in the production process, has allowed to personalize consumers, to take into account their wishes and meet the diverse needs (Strauss and Frost 2011). This business is largely corresponds to the traditional in terms of content, but has a number of features related to the virtual environment that imposes certain requirements and restrictions on its success (Fellenshtein 2000).

Despite the positive effect of the Internet on social and economic processes, new types of risks for both entrepreneurs and consumers (Kumar and Harding 2011). These

O. Galinina et al. (Eds.): NEW2AN/ruSMART/NsCC 2017, LNCS 10531, pp. 326–336, 2017.
DOI: 10.1007/978-3-319-67380-6_30

risks examined in the literature and there are many recommendations of management on the business risks of virtual enterprises (Ponis 2010). However, currently there is no common point of view on the virtual enterprise as economic actors on the Internet, the mechanisms of their functioning and risk management because cyber-attack and other risks change very quickly. Necessary to solve several problems as to systematize the virtual enterprises risks, to find the causes and roots of each group of risks, to define the characters of virtual enterprises and their risks, methods and instruments of risk's evaluation.

2 Risk's Factors of Virtual Enterprises

A virtual enterprise is exposed to both traditional commercial risks and risks of the virtual environment, therefore for virtual enterprises there are special principles of sustainability that help to successfully manage it (Apatova and Malkov 2013).

1. Whole system, which is an enterprise, may be in a state in which the strategic goals of development realized, regardless of the impact of internal and external factors.
2. The internal factors need to include those risk factors that reduce the sale, including technical failures of computers and networks, databases and violations of software errors, violation of terms of purchases and sales, failures in service delivery.
3. The impact of external factors shaping introduced by external risks manifested in the actions of competitors to win more of the market, changes in legislation or tariffs delivery.
4. It is necessary to introduce a system of assessments, indicators for destabilizing factors, as well as some integral index - the signal of economic stability violations of the enterprise.
5. The functions of management of virtual trade enterprise includes: assessment of economic sustainability, simulation and analysis of the indicators, the formation of the optimal range.

The basic principles of commercial policy are the integration of the management of all business units, focus of the competitive environment, the consistency of all kinds of resources, flexibility, control and timely prioritization.

Organization of commodity flows associated with the formation and use of inventory, supply chain organization. The advantage of virtual stores is that they usually do not have large warehouses, logistics chains formed in the course of execution of orders of customers shopping, speed of delivery the product to the buyer often depends on the availability of his stock and by the coordinated work of all participants in the chain, from the manufacturer required product. The goods flow and working capital change their parameters, in the virtual stores they are minimized and become more dynamic the risk of excess inventories is minimizing and do not need a large number of parameters to optimize the flow of goods. Working capital is reduced in the process of minimizing the assets of virtual trading enterprise because to start an Internet store in some cases, the necessary equipment is a computer and a modem to connect to the Internet (in extreme cases may be sufficient only the mobile phone).

Internet trade and telecommunication service were very dependent on providing a network connection, a dedicated lines and modems until the advent of mobile phone. The benefits of trade using mobile to connect to the Internet are: universal access, localization, personalization, flexibility and ease of use, data portability and speed product payments.

In modern enterprises in telecommunication service an important task of management objective is to manage information flows for save time by achieved by combining in a single flow of goods and information throughout the supply chain from producer to consumer. The special management system of the informational flows is necessary for the control the time factor as a new production factor and the information when is separated from the physical flows can be itself a product or service. This is particularly evident in Internet trading when the buyer has the opportunity to assess the quality of the goods only according information provided him. In this case, the buyer gets primarily an information service that provided free for him, and only after the decision about purchase, the process of transaction starts.

For a virtual enterprise the nature-geographical and environmental barrier factors are insignificant, the space action of virtual enterprise has no physical boundaries and virtual enterprises do not cause damage to the environment in terms of environmental barriers.

From market's risks for the virtual enterprise are commercial risk as the risk of trading activity and the impact of the risks of competition. Virtual enterprises have good competitive potential, which consists of an innovative, social and economic potentials and the potential of information technology. Effect of scale, which is working for Internet trade, makes it possible not to take into account the impact of factor of competitive environment.

One of the factors affecting the stability of the virtual trading enterprise is its advertising, which is a showcase of online store, as well as brief banners on other websites.

Marketing in virtual trade enterprises including logistics and communications functions. Marketing software regards goods, prices, sales promotion, marketing; logistical support covers issues of needed goods, the required quantity and quality, minimize costs, required time, place and the necessary consumers. The use of Internet technologies in marketing also allows to provide the communicative functions which are of production and transmission to the target audience the content of the information. There are three groups of Internet technology's functions.

1. Customer service (commercial communications, electronic consumer survey, monitoring visits to the site server or virtual enterprise, sent questionnaires).
2. Communication (advertising and promotion products, advertising campaigns, support for consumer testing of new products to the organization of feedback from customers).
3. Sales and logistics functions (sales via the Internet shops, Internet sales sites of manufacturers, sales through Internet portals).

For the implementation of these two actions is necessary to realize several steps: to conclude an agreement with the buyer for the sale and payment of goods or services; to provide the customer with the necessary information with fixing it on the server of the virtual trade enterprise; to track the process of delivering the goods; to support the payment through online payment systems and delivery of the goods. At each stage of the transaction with the buyer exist mutual risks: for the seller there are the risks of supply disruptions, non-receipt of payment from the buyer and for the buyer - receiving the product which nonequivalent his image, removal of extra money from the account when paying goods and other types of risks. Analyze of business activity in the computer network and features the work of virtual trade enterprise leads to the conclusion of the main sources of risks inherent in the economic processes and data objects. The following main causes of each kind of business risks allow their means of control, including the prevention and minimization.

Causes of Information Risks. Information on the electronic site of the seller may not be accurate. License, copyrights, trade secrets can be stolen because of unauthorized access to the site of the seller. Information contained on the website of the seller, may violate laws of other countries. Information, which has been introduced by customers with plastic cards, can be intercepted and used for personal gain. Intellectual property leakage may occur due to the transition of the virtual enterprise workers to competitors.

Causes of Technological Risks. Negligence and errors in the design of website of the seller. Unauthorized access to the site of the seller. Infecting virus the software of as a result of cyber-attacks. Crash the server of Internet service provider. Interception by a third party, such as a plastic card buyer introduced by the client code and the card number when purchasing online seller. Inadequate of speed of seller server to receive and data processing. Outdated hardware and software and as the result is the inability of seller's server to process incoming applications of clients. Wrong e-commerce integration with the internal base of the company and the internal management processes of the virtual enterprise.

Causes of Business Risks. Legal risks associated with the use of information that may violate the laws of other actors both traditional and virtual marketplace. Risks associated with the site developers to pay the seller. Lack of financial resources to support the site of the seller. Negative impact on the business because of the transition of personnel to competitors, changes in relationships with suppliers. Lack of mass of commodities due to poor interaction with suppliers. High costs for the delivery of goods and the distribution of the mass of commodities. The inconvenience of returning the goods by the client as a result of lack of coordination with customers. Inability to control the temporary cyclical commodity mass due to incorrect miscalculations turnover ratio. Risks of the site name of the seller, as others can already use this name. Improper integration of e-commerce with the internal management processes within the company. Lack of integration of e-commerce channels of delivery of goods from suppliers.

Many of the risks of the two above-stated groups are interlinked and they tend to repeat. A characteristic feature of these risks is their immediate importance for the Internet trade, but these risks also apply to business is e-commerce, but because at the present stage of development is traced a clear relationship between business and e-commerce, then these risks are common.

3   Methods for Assessing Business Risks of Virtual Enterprises

The main problem of enterprises in the face of uncertainty is risk management, which requires a risk assessment, development of methods and models of management.

Various methods are used to assess risks, including the construction of a decision tree with predictable risk factors, the construction of econometric risk factors and obtain an integrated numerical evaluation, expert risk assessment, options for its occurrence and management, theoretical-game approach.

In order to more fully characterize the risks and their level (degree), it is necessary to build a system of hypotheses with alternative solutions, which in ideal can be turned into an expert system for the entrepreneur. The system of hypotheses requires the testing on truth, it will be the process of alternatives development to the use of economic and mathematical methods and models. As suggested (Sigal 2014), risk management carried out according to the following conceptual scheme: the qualitative analysis - quantitative assessment - accounting and risk management decision-making. Quantitative analysis uses techniques such as analogies method, sensitivity analysis, risk analysis of losses, the method of simulation, statistical methods, expert estimates. Game-theoretic modeling of economic risk more accurately reflects the economic and market situation where the interests of the parties are not the same, and there is no need for undue risk.

The main feature of Internet trading and telecommunication online service is that they are performed in real time at any time and in any geographic location. In this regard, the corresponding software must maintain a dialogue with the user automatically, creating the appearance of an open shop with traditional retailers.

Commercial communication effectiveness can be evaluate on the following criteria (Lebedenko 2008):

1. Profitability (profitability of promotion and return on assets).
2. Sales (growth rates of the average monthly turnover volume of purchases in monetary terms, the frequency and amount of purchases) income (revenue dynamics and the level of sales).
3. Market share (market share in the number of sales of goods and the proportion of regular customers).
4. Earnings (the size of the gross income).

Overall indicators of communicative efficiency of an electronic resource may be as follows. On the technological norms the indicators mean the concordance the name of the site its thematic focus, the ease of finding resources on the network, the speed of access to the resource, the price of hour job site, the speed and qualities of pages load and its open ability. The content indicators of the information provided in the site are accordance it to the intended audience and purpose of the site; lack of grammatical and stylistic errors; completeness and accuracy of the information filed; timely periodic updating of data, their relevance and more information is available. On the criterion of navigation: the logical structure of the site organization, correct references and correspond to the contents, usability navigation. By design criteria: the quality of graphics and its load; accordance for marketing objectives the line style and color; ease of page layout, efficient use of page space; lightness perception interface and text. Interactivity criteria: the presence of contact information, a variety of communication options; the availability of additional services; virtual assistance to users; responsiveness to customer requests; various forms of payment for the goods and their delivery. These criteria are universal for all virtual trading companies and can use for qualitative and quantitative assessment of its work, planning and adjustment activities at any stage.

In the activities of the virtual trade enterprise also important the marketing based on relationships, which allows you to build long-term relationships with customers. The added value of goods or services is created in the course of the relationship; the buyer becomes the follower of the virtual trading company, its brands and, as a result, participates in advertising of goods or services, but the e-shop.

Agents of advertising services market are the advertisers, advertising producers, distributors and consumers of advertising, but in the case of Internet advertising a software package is also included and facilitates the work of the first three groups of agents. Advertising is classified on the following signs: the basis of the breadth of coverage of the target audience; the type of presentation of the advertising information and the area occupied by it; a method of influence on the consumer; the effect of the duration of the viewing of advertising; the supplying technology; the position on the website and the method of contacting the consumer advertising.

When managing Internet advertising and interactive communications implementation must take into account the specifics of Internet users, their knowledge of other goods and on similar products of other commercial enterprises, greater openness of new products, preference the value of the cost of goods or services. Databases of Internet store contains information about the acquired specific consumer goods, time and frequency of their purchases, money spent, as well as personal data that ask many sellers, registering buyers play a role in predicting consumer behavior. The stored information is about each buyer, the history of his purchases, information about age, sex.

Creation of a virtual enterprise is an innovation and investment project, for manage risks they can be divided into three groups of risks. The first group of risks are into account in the discount rate, the project's revenues and expenses, it evaluated by the expert or probabilistic methods. This group of risk will affect in different ways: to assess the loss of income is necessary to consider the likelihood of breach of contract on the part of partners and the share of the budget, which is lost in the event of a risk. The second group - market risks, they can be taken into account in the project scenario analysis, which is based on the use of probability theory and mathematical statistics methods, involves a numerical definition of the values of individual risks and the risk of the project as a whole. There are various investment processes management concepts, one of which is a project-based approach, considering the movement of investment cash flows, taking into account risk.

There are four stages of analysis investment risks (Glebova, 165).

1. Forecasting sales to formalize uncertainty.
2. To assess the investment risk, taking into account the current activities of the individual investor's tolerance to determine the optimal pricing strategy of the

enterprise for specific conditions. The optimization criteria are to minimize the risk of claims and the maximization of revenue.

3. The calculation of the risks of investing in the current activities of the company to introduce new production technologies using the methods of scenario analysis and simulation.

4. To takes into account the risk of the formation of the optimal structure of portfolio sales. The variance of sales for each product can use also as a risk indicator. This technique is used and convenient for virtual enterprises, because the using information technology can automatically records sales in any given period of time, allows us to calculate the risk index to represent him in visualized form of, that contributes to the rapid and accurate management decisions.

4 Economic Risks Evaluation

The economic risks of virtual enterprises can be simulation with main following elements of model.

1. Economic and statistical model of the risks of external nature: uncertainty of the macroeconomic business development strategies, unforeseen changes in legislation, changes in the foreign policy of the state, changes in demand for goods and services.

2. Modeling the impact of emergencies on the activities of the company, made on indicators of economic stability, the simulation of market and financial risks, based on multivariate analysis. These models allow us to evaluate internal and external risks of virtual enterprises.

We have analyzed the activities of five commercial enterprises operating in the computer network of the Internet as a virtual trade and telecommunication enterprise. The main factors of their activity were determined: Y - income, x_1 - cost price of products, x_2 - working capital, x_3 - trading margin, x_4 - advertising costs, x_5 - the number of employees. For each factor the group of experts from among business leaders was define the risk as a parameter in the range from 1 to 10 (10 is the highest risk). Risk of income in the virtual enterprise has been estimated at 6 points, the risk of cost price of products has 9 points (the highest), working capital - 8, trading margin - 5, advertising costs - 7 and the number of employees - 5. To calculate the significance of the data factors averaged and normalized (Table 1).

The calculated correlation coefficients between pairs of factors and the value of the function (Table 2) have shown a high correlation factor, which makes it impossible to build a single linear regression equation for the function of the income of all factors.

Therefore, we have constructed a regression equation for each pair of factors (see Table 3), t-statistics calculated for each factor and coefficient of determination R^2 for each equation. In the results of calculations, the significant factors were the cost price of products and working capital (Table 4).

Table 1. Normalized values of factors

Years/factors	Y	x_1	x_2	x_3	x_4	x_5
2006	0.69396	0.83036	0.79406	0.6	0.58294	0.72727
2007	0.7353	0.9053	0.86006	0.6	0.90225	0.80909
2008	1	1	1	1	1	1
2009	0	0	0	0	0	0.14242
2010	0.38002	0.26118	0.29281	0.6	0.20306	0.18182
2011	0.40763	0.31124	0.33689	0.6	0.26789	0
2012	0.48767	0.45634	0.46468	0.6	0.31804	0.03333
2013	0.24262	0.01207	0.07342	0.6	0.49542	0.03333

Table 2. Correlations between factors

	Y	x_1	x_2	x_3	x_4	x_5
Y	1	0.95859	0.97479	0.8414	0.8863	0.8389
x_1	0.95859	1	0.99797	0.6569	0.84297	0.8978
x_2	0.97479	0.99797	1	0.7029	0.85876	0.8911
x_3	0.84147	0.65694	0.70297	1	0.75268	0.4999
x_4	0.8863	0.84297	0.85876	0.7526	1	0.8480
x_5	0.83894	0.89785	0.89116	0.4999	0.84809	1

Table 3. Regression models of commercial factors, taking into account the risks

Regression equation	Coefficient of determination R^2
$Y = 0.562x_1 + 0.429x_3 - 0.018$	0.998881
$Y = 0.637x_2 + 0.355x_3 - 0.015$	0.999228
$Y = 0.527x_4 + 0.463x_3 - 0.021$	0.925022
$Y = 0.426x_5 + 0.648x_3 - 0.035$	0.970214

Table 4. The importance of commercial factors in the regression models

Factor	Point of risk	t-statistics	Importance of factor
Income (Y)	6		
Cost price of products (x_1)	9	3.782491	Yes
Working capital (x_2)	8	4.420603	Yes
Trading margin (x_3)	5	0.849437	No
Advertising costs (x_4)	7	1.091264	No
Number of employees (x_5)	5	1.378232	No

5 Price Risks Evaluation

Virtual enterprise begins with the development of the investment project and there are first risks at this stage. In the analysis of the risk of the investment project and its implementation are invited to consider the loss of certain types of risks, the loss of one of the elements of the investment portfolio, as well as the probable maximum loss. To determine the level of risk using statistical techniques, methods appropriateness of costs, expert assessments, analogies and analytical method. In many economic and mathematical models as a risk indicator, be taken of the standard deviation of the random variable characterizing the results, but for the market environment, this indicator does not reflect the real situation. In our opinion, if the investors of trading virtual enterprise are the customers - Internet users, the investment risks at each stage of the sales (or each sale) it is advisable to measure as the index of volatility (financial statistical measure of the variability of prices), acting on the Forex financial market.

Volatility measured in order to keep track of price fluctuations as one of the components of risk.

Historical volatility calculated as the standard deviation of the price of the average price of this product (share). Historical volatility calculated for the year, it makes it possible to obtain the expected volatility of the index.

Expected volatility is dependent on:

1. The historical volatility;
2. Political and economic situation;
3. Supply and demand on the market.

Expected volatility calculated based on the current value of a financial instrument on the assumption that the market value reflects the expected risks.

Volatility characterized by certain features: cyclical, consistency and desire for the middle and because it is more predictable than price, since is in a predetermined range, this quality of volatility can predict the behavior of an entrepreneur in the market.

Given that the Internet works as an auction, price fluctuations on the same goods during the day can be quite high. Prices are easily tracked by the buyer and compared among different vendors.

With falling the sales, the investments reduced and the risk of non-repayment of the core investor (the seller) will increase. To determine the indicator of volatility, you must have special software to monitor the actions of competitors in the virtual market or collection of such data must engaged the persons in the marketing service of the virtual enterprise.

Volatility reflects the rate of price movements, for the virtual enterprise, it allows you to adjust pricing and reflects the saturation of the market this type of product or group of products.

The higher the volatility of this product, the higher the supply, and therefore, the employer must either reduce the price, or to update the assortment.

Classical calculation of historical volatility suggests the comparing prices over time and calculation the ratio of two prices of two adjacent periods. We will calculate the price pairs for virtual online shopping market for the same goods, acting on the following algorithm.

1. The construction of a triangular matrix of pairs of relations eponymous commodity prices (the dimension of the matrix $(n - 1) \times (n - 1)$, where n - the number of stores that sell this product).
2. The calculation the matrix of the natural logarithm of pairs of relations prices.

3. Calculation of the mean value logarithms of pairs of relations prices.
4. Calculation of volatility as the standard deviation (dispersion) logarithms of pairs of relations prices.

We analyzed the prices of household appliances in Internet shops and telecommunication services in Russia and Ukraine, 7 stores investigated and 5 types of telecommunication products and services (Satellite receiver, Satellite antenna, Tablet PC, Laptop, Mobile phone) each product of the same brand.

After the calculation and the analysis of volatilities, we see that the highest volatility, which is equal to 0,011, it is for the prices of Satellite antenna. It means that consumers bay this type of products seldom and they prefer to use the service unite the Internet and TV from provider, so is necessary to reduce the price for sales success.

Low volatility, which is equal to 0.001, is the Mobile phone price, which means that further sales of this product will be success at current prices. The possibilities of modern Mobile phone is the same of Tablet PC, so the prices of Tablet PC is less than good Mobile phone and them popularity fell. This conclusion confirmed by the actual sales. In the proposed approach, the volatility serves, as a measure of the price change is not in time, as it understood in the traditional sense, but as a gradient index of prices and price risk of virtual telecommunication enterprise in the virtual space of Internet.

6 Conclusion

Use of the Internet in economic activity has led to a constant increase in the number of virtual enterprises and new risks for entrepreneurs. These enterprises have special principles of works and stable, their advantages are the possibility of permanent use of information technology, monitoring the behavior of competitors and consumers and methods of mathematical modeling directly in the operation of the Internet.

Risks of virtual telecommunication enterprises united in three groups: information, technological and commercial risks.

The more important is the first group:

1. Misinformation of buyers and sellers;
2. Cyber-attacks which violated the integrity of the information of the seller and her abduction;
3. Interception and use for personal gain information from the cards and leaked knowledge of the virtual enterprise and its intellectual property as a result of human error.

Virtual trade and telecommunication enterprises are the economic activity similar to the traditional enterprise, but carried out in a computer on the Internet, so the indicators of its economic activities coincide with the traditional, but each of these indicators also reflects the factor of risk. Regression models for commercial factors, taking into account the risks showed that represent the greatest risk are the prime cost and capital turnover.

The Internet affects the choice of the consumer, who can analyze the prices of various virtual enterprises at a given time. This causes the need to track virtual market prices and sellers change the prices on specific products, depending on the saturation of the market segment.

Volatility index, adapted to the space of the Internet virtual market, helped to solve the problem and to determine thereby, the goods with the highest offer and, accordingly, the goods with the lowest offer that allows the seller to adjust the own range of products and prices.

Summary, the classification of virtual enterprises risk gave the possibility to define three main groups of risks and they causes; to build the mathematical model of traditional economic indicators with risk account; to use the volatility index to Internet market for assessment of virtual enterprises risks and identify the demand, supply and the price on the commodity.

References

Apatova, N.V., Malkov, C.V.: Riscology of virtual entrepreneurship, 353 p. Simferopol (2013)

Beddle, D.A.: Wide open world. In: Global Finance, pp. 23–25, January 2010

Davidow, W.H., Malone, M.S.: The virtual corporation. In: Structuring and Revitalizing the Corporation for the 21th Century, p. 294. Harper Collins (1992)

Fellenshtein, C.: Exploring E-Commerce, Global E-Business and E-Societies, p. 269. Prentice-Hall, Upper Saddle River (2000)

Filos, E., Banahan, E.P.: Why the organisation disappear? The challenges of the new economy and future perspectives. In: Camarinha-Matos, L.M., Afsarmanesh, H., Rabelo, R.J. (eds.) PRO-VE 2000. ITIFIP, vol. 56, pp. 3–20. Springer, Boston (2001). doi:10.1007/978-0-387-35399-9_1

https://books.google.ru/books?id=Kb_cBwAAQBAJ&pg=PA192&lpg=PA192&dq=virtual+ent erprises&source=bl&ots=_N5fLvLNAa&sig=ZUkApK3Wh04-IcR8pNhuBo3_vWI&hl=ru& sa=X&ved=0ahUKEwidgZT8nOPKAhUFdXIKHSWTCO04ChDoAQgaMAA#v=onepage& q=virtual%20enterprises&f=false

Kumar, S.K., Harding, J.: Risk assessment in the formation of virtual enterprises. In: Camarinha-Matos, L.M., Pereira-Klen, A., Afsarmanesh, H. (eds.) PRO-VE 2011. IAICT, vol. 362, pp. 450–455. Springer, Heidelberg (2011). doi:10.1007/978-3-642-23330-2_49

Lebedenko, M.S.: Marketing communication in Internet. Econ. Bull. Kiev Polytech. Inst. **5**, 332–338 (2008)

Ponis, S.: Managing Risk in Virtual Enterprise Networks: Implementing Supply Chain Principles, 384 p. Business science reference, New York (2010). https://books.google.ru/ books?id=ShSgAyhKbQC&pg=PA25&lpg=PA25&dq=risk+of+virtual+enterprises+pdf&so urce=bl&ots=UoVDxHYqOB&sig=2tPR7B2O1Qv0PmyYg7KCFvKQ9Q&hl=ru&sa=X&ve d=0ahUKEwj6nsXM2PKAhWCJXIKHeSoAMUQ6AEIGzAA#v=onepage&q=risk%20of% 20virtual%20enterprises%20pdf&f=false

Sigal, A.V.: Game Theory in Economy Decision Making. Taurida National V.I. Vernadsky University, Simferopol (2014)

Strauss, J., Frost, R.: E-marketing, 6th edn. Pearson Education International, New Jersey (2011). 519 p.

William, L.: Evolution of the new economy business model. In: Brousseau, E., Curine, N. (eds.) Internet and Digital Economics, pp. 59–113. Cambridge University Press, Cambridge (2006)

Peculiarities of Creation of Information System at the Enterprises of Telecommunication Branch

Ye. Yu. Vinogradova[1(✉)], A.I. Galimova[1], Natalya V. Mukhanova[2], and S.L. Andreeva[1]

[1] Ural State University of Economics, Yekaterinburg, Russia
{katerina, svetlana}@usue.ru, anna.baibuz8@gmail.com
[2] Peter the Great St. Petersburg Polytechnic University, Saint Petersburg, Russia
nmukhanova@spbstu.ru

Abstract. Nowadays it is hard not to notice a priority of development of hi-tech production in different countries including the Russian Federation. The fact that unique (high) technologies are a subject of interest of our country is confirmed by existence of plans of implementation of the programs developed by various ministries and departments of the Russian Federation. Their financing from the state budget is measured by trillions of rubles. The special place among the knowledge-intensive and high-tech industries is occupied by telecommunication branch. Many enterprises of this sphere think of introduction of corporate information systems for the purpose of optimization of management accounting and improvement of quality of planning. Nevertheless, the existing information systems are unsuitable for the accounting of this type of hi-tech production owing to its considerable uniqueness. The article considers major factors of unfitness of the information systems functioning at the moment for management accounting of production of telecommunication branch. The authors analyse necessary set of characteristics of an information system for optimization of its use in telecommunication branch. The authors define necessary modules of information systems at the enterprise, which introduction will allow automating the accounting of the resources necessary for development of scientific and technical innovations. That will increase the level of innovative development of the Russian Federation.

Keywords: Hi-tech production · Telecommunication branch · Information system · Management accounting · Consumer value

1 Introduction

For many years Russia has the status of "the raw power" that is shown by highly production of natural resources and a low share of their high-quality processing. Despite existence in the territory of our country of almost all table of Mendeleyev, the extracted raw materials generally are delivered to other countries. Then we buy production made from our resources. In such situation its prime cost is manifold above expenses, which would be incurred at independent production. Dependence of the

© Springer International Publishing AG 2017
O. Galinina et al. (Eds.): NEW2AN/ruSMART/NsCC 2017, LNCS 10531, pp. 337–350, 2017.
DOI: 10.1007/978-3-319-67380-6_31

country on export of raw materials does it vulnerable to changes of the world prices for raw materials, to world economic crises. Since 2007 a numerous talk on need of diversification of the Russian economy and plans of its disposal of raw dependence are carried on. But, despite positive shifts in recent years, in the analysis of indicators in the long-term period is observed the falling of a share of the processing productions in GDP from 15,2% in 2005 to 13,0% in 2015 against the background of growth of production of natural resources. One of solutions of these problems is increase in innovative development of the Russian economy by development of the unique equipment and advanced technologies in various branches: aircraft, telecommunication, automotive industry, space technologies, shipbuilding, power, agricultural machinery and others [7, 9]. Development and production of hi-tech production becomes harder and harder process, which demands the detailed accounting of used means and high quality planning in connection with attraction of considerable volumes of financial resources of the state budget. Especially this problem is particularly acute concerning telecommunication branch, which develops the advancing rates now and successfully overcomes lag from the international standards.

2 The Problems of Application of the Existing Information Systems for Management Accounting of the Enterprises of Telecommunication Branch

High-tech production represents the advanced technology product created on the basis of application of unique productions or production, which realizes the consumer functions by using of the latest physics and technology effects [1, p. 134]. As for telecommunication as one of the types of hi-tech production, it represents the communication by means of a certain electronic equipment (for example, phone, the optical fiber cable, the satellite) intended for information transfer on long distances.

The existing at the enterprises automated systems, which primary activity is creation of hi-tech samples, are the unified information systems of "box" type with in advance defined modules. They aren't capable to consider specifics of the organizations therefore divisions of companies often create their own automated systems. But they also have the disadvantages: similar systems are oriented on the decision of local tasks of divisions, repeated duplicating of data entry, absence of automation of information flows of data between systems, the analysis of the obtained data requires a lot of time and doesn't allow achieving desirable results for improving management of a company. Information support of management accounting represents dynamically distributed databases, which characteristic is paper current documentation and the statistical reporting [6, p. 118]. There is not always information of coordination of these operations, of creation of the integrated information systems, structures of their functional part.

The existing standards of information systems, to which belong concepts of MRP (Material Requirements Planning), MRP II (Manufacturing Resource Planning) and ERP (Enterprise Resource Planning) aren't able to consider specifics of projects of production of telecommunications in view of a difficult assessment of its cost when forming plan documents. It is caused by the following factors:

- wide range of the types of production created at various stages of life cycle during realization of projects;
- using the basic data possessing various degree of reliability and specification for estimation of cost of a project;
- existence of a significant amount of the factors influencing the cost of creation of production;
- duration of realization of a project for production can reach more than ten years;
- lack of the developed competitive environment;
- high probability of change of cost of production during the project.

Projects represent research and development on creation of new product samples, modernizations of the samples, which are in operation and also their research, purchase of serial samples and work on ensuring their operation and repair.

As for the systems functioning in Russia, the complexity of decisions, which is present at them, causes redundancy on functionality and insufficient flexibility of system in the course of its application [4, p. 82]. The high cost of the acquired system in combination with duration of its introduction and discrepancy to specifics of production are the cause of considerable unproductive costs of acquisition and maintenance of an information system.

3 Development of an Information System for the Enterprises of Telecommunication Branch on the Basis of Decomposition of Life Cycle of Hi-Tech Production

Creation of perspective samples of hi-tech production, including in the sphere of telecommunications, represents the sequence of the following stages:

- plan generation;
- placing orders for creation of production;
- implementation of contracts by the enterprises for its creation.

As a result of performance of these stages the efficiency of using of budgetary funds is controlled [8, 10].

Within this research it is necessary to make decomposition of life cycle of hi-tech production with the subsequent accounting of each stage in a corporate information system. Change of a condition of hi-tech production from the beginning of a research and justification of development before removal from operation and utilization is presented in Fig. 1 in the form of Gant's chart.

Development of hi-tech production begins with carrying out research and development. That includes creation of working and technological documentation, technical documentation on a prototype, manufacture, conducting preliminary and acceptance tests of a hi-tech sample, adjustment and the approval of the listed documentation. Within this stage the status of a prototype changes from the specifications on execution

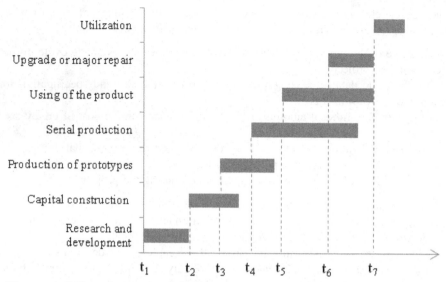

where t_1 – start time of research and development, it is a basis for creation of hi-tech samples; t_2 – start time of a stage of capital construction under mounting; t_3 – start time of production of prototypes; t_4 – start time of serial production; t_5 – start time of using of the product; t_6 – start time of upgrade or major repair; t_7 – start time of a utilization of the product.

Fig. 1. Gant's chart of stages of life cycle of a hi-tech sample

of research and development set by requirements before setting of a sample of industrial production.

The government makes the decision on creation of new samples of hi-tech production or the upgrade existing based on studies of technical and economic character, performing at the same time function of the customer.

After the inference of the contract with the head performer for creation of a hi-tech sample operation of design offices and research institutes begins. They develop technical sentences on creation of a certain prototype by means of conducting theoretical and pilot studies. Generally development of a hi-tech sample is carried out by a number of the organizations, one of which plays a role of the head performer [2, p. 20]. In this regard process of carrying out research and development is divided into the significant amount of interdependent operations, which coordination is carried out by means of monitoring of the course of their performance. The total of a stage of advanced development is obtaining working and technological documentation, approved for the organization of production. The stage comes to the end with manufacture and carrying out tests of a prototype.

In case of the accounting of volume of necessary resources on development of production during creation of an enterprise information system it is necessary to consider possible options of creation of a new prototype: full-scale development or upgrade of the existing prototype, because the size and the characteristic of the required resources will differ considerably.

Full-scale development is possible in a revolutionary or evolutionary type. The revolutionary type implies changes in design-layout solutions of a prototype of new generation in comparison with the same prototype.

As a result scientific and technical innovations are realized by giving to a perspective prototype essentially new properties. This type is high-cost and it is rather high risk of unsuccessful development of a prototype. Nevertheless, the possible result of creation of a prototype with essentially new properties and increases in innovative development of the Russian economy superimposes all possible expenditures.

The evolutionary type of development of a perspective prototype differs from revolutionary in the fact that it doesn't acquire essentially new properties by means of change of design-layout decisions, but there is considerably an improving of all its qualitative characteristics.

As for upgrade of the existing prototype, it can be presented in the following forms: deep, average and insignificant.

In case of deep upgrade is made a replacement of the main part of subsystems of a prototype by subsystems of new generation, which include set of the scientific and technical innovations leading to significant improvement of characteristics.

Average upgrade is a changeover of separate subsystems of a prototype by subsystems of new generation, which includes set of the scientific and technical innovations leading to improving of characteristics.

Insignificant upgrade represents marginal improvement of characteristics of a prototype by finishing of separate subsystems.

After development a prototype and carrying out all necessary tests begins a production of hi-tech product, which includes the following stages:

- setting of a prototype on production;
- production mastering;
- mass/batch production;
- construction, mounting, assembly and adjustment;
- delivery;
- removal from production.

Production of telecommunication often demands attraction of considerable number of suppliers of materials and accessories. Quality of production directly depends on the level of the production and technological potential, which is saved up at the assembly enterprise [3, p. 87]. At a stage of production control of the set scientific and technical level and quality of a prototype is carried out.

The following stage of life cycle of hi-tech product is its operation. It covers process of transfer of finished goods by manufacturer or repair organization to the consumer, the subsequent commissioning. Further reduction in the established degree to use to destination, maintenance of readiness for operation, storage and transportation, maintenance and repair follows. Operation of each prototype is carried out by qualified specialists according to the established plans, annual norms of application of resources of the equipment, the general standard and individual operational documentation. The source of means for elimination of possible malfunctions (the supplier or the operating company) depends on the reasons and the period of their emergence. At a stage of operation of a prototype control of quality of prototypes and of implementation of their operation to destination continues. Important aspect is combination of providing the set requirements for efficiency with profitability of operation.

The stage of capital repairs includes development of repair documentation and technological repair documentation, production of means of technological equipment of own production for ensuring repair and carrying out tests of prototypes, inclusion in repair, the conducting preliminary and acceptance tests of skilled repair prototypes, maintaining set like repair production and delivery of the repaired prototypes to the customer.

Process of capital repairs can include modernization of product. This stage allows carrying out technically feasible and economically expedient restoration of parameters and characteristics of product, which are capable to change in use.

At a stage of modernization the complex of works on the samples of product, which are in operation or laid off on bringing characteristics of samples to compliance to new requirements, is carried out. Implementation of modernization allows to increase terms of an obsolescence of samples and to prolong terms of their stay in operation at rather small amounts of financing. Essential economy arises because on modernization much less means, than on creation of new generations of samples are required. Nevertheless, modernization doesn't exclude need of development of new samples but only postpones the date of implementation of innovations. Definition of the direction of modernization is a task both the customer within performance of the corresponding research and development, and the developer and the manufacturer of samples of hi-tech production.

Utilization is a final stage of life cycle of hi-tech production and represents a complex of the organizational and technical, economic, scientific, ecological actions providing processing of the taken out of service morally and physically outdated equipment, which has developed the resource.

4   Algorithm of Determination of Consumer Value of Hi-Tech Production

The samples made by one enterprise and each of the listed stages of life cycle demand unique approach to an assessment of an economic component and determination of effect of use of resources. The comprehensive characteristic requires studying of cost and temporary factors. Cost factors characterize planned or actual costs of financial resources on realization of life cycle of a sample of production and its components. Control of the temporary factors characterizing the planned or actual duration of life cycle of a sample and its components is applied to distribution of financial resources by years or quarters of planning period. From efficiency of the tools used for the account and formation of plans of creation of high-tech production in the sphere of telecommunications, quality of the methodical providing applied to definition of cost and temporary factors feasibility of planned documents and efficiency of use of the funds allocated for this purpose significantly depends.

By means of implementation of an information system the problem of estimation of cost of telecommunications, which has dual character, can be solved. On the one hand, cost is formed under the influence of expenses, on the other hand, according to its consumer value, which represents the benefit brought to the buyer [5, p. 157]. Specifics of determination of consumer value of such production consist in the accounting of social consumer value for ensuring economic security of the country. Within this research for an assessment of consumer value use of the following groups of indicators is offered:

- definition of characteristics of a sample of hi-tech product in telecommunication branch;
- cost factors;
- effect of application of a sample.

According to characteristics of a sample it is possible to judge the level of its perfection by comparing with alternative samples. The perfection level is higher, so the consumer value is higher.

The cost of realization of separate works and all life cycle of a sample, specific cost of unit of the main characteristic of a sample, effect of use belong to number of cost factors, which are a basis for determination of consumer value of hi-tech production. Determination of consumer value is based on comparison of the listed indicators of the considered project and its analog.

Characteristics of hi-tech production have to be considered without separation from the external environment, in which use of a sample is carried out. The consumer value of hi-tech production directly depends on the effect reached as a result of using production.

The analysis of the consumer value determined proceeding from the characteristic of a sample and effect of using can be made on the basis of the following calculations:

1. Analytical expressions on condition of existence of one indicator in each group:

- absolute characteristic of consumer value of hi-tech production:

$$\Lambda_{tcv} = R_{C\,tcv} - R_{A\,tcv} \tag{1}$$

where $R_{C\,tcv}$ – value of the indicator used for a technical (functional) consumer value of the considered hi-tech production; $R_{A\,tcv}$ – value of the indicator used for a technical (functional) consumer value of alternative hi-tech production;
- relative characteristic of consumer value of hi-tech production:

$$\varepsilon_{tcv} = \frac{R_{C\,tcv}}{R_{A\,tcv}} \tag{2}$$

2. Analytical expressions on condition of existence of several indicators in each group:

- absolute integrated characteristic of consumer value of hi-tech production:

$$\Delta_{tcv}^{i} = R_{C\,tcv}^{i} - R_{A\,tcv}^{i} \tag{3}$$

where $R_{C\,tcv}^{i}$ – value of the integrated indicator used for an assessment of consumer value of the considered hi-tech production,

$$R^i_{C\,tcv} = \sum_{j=1}^{N} r_{Pj}\rho_{Pj} \tag{4}$$

where N - total number of the indicators used for an assessment of technical or functional consumer value of alternative hi-tech production; r_{Pj}- the indicator used for an assessment of technical or functional consumer value of alternative hi-tech production; ρ_{Pj}- the indicator characterizing importance of value for an assessment of technical or functional consumer value of alternative hi-tech production,

$$\sum_{j=1}^{N} \rho_{Pj} = 1, \quad 0 \le \rho_{Pj} \le 1 \tag{5}$$

$R^i_{A\,tcv}$ – value of the integrated indicator used for a technical (functional) consumer value of alternative hi-tech production,

$$R^i_{A\,tcv} = \sum_{j=1}^{N} r_{Aj}\rho_{Aj} \tag{6}$$

where r_{Aj} – the indicator used for a technical (functional) consumer value of alternative hi-tech production; ρ_{Aj} - the indicator characterizing importance of value for an assessment of technical or functional consumer value of alternative hi-tech production,

$$\sum_{j=1}^{N} \rho_{Aj} = 1, \quad 0 \le \rho_{Aj} \le 1 \tag{7}$$

- relative integrated characteristic of consumer value of hi-tech production:

$$\varepsilon^i_{tcv} = \frac{R^i_{C\,tcv}}{R^i_{A\,tcv}} \tag{8}$$

if $\Delta_{tcv} > 0$ or $\varepsilon_{tcv} > 1$ ($\Delta^i_{tcv} > 0$ or $\varepsilon^i_{tcv} > 1$), then it means that the considered sample represents for the customer great technical (functional) consumer value, than an alternative sample. If $\Delta_{tcv} = 0$ or $\varepsilon_{tcv} = 1$ ($\Delta^i_{tcv} = 0$ or $\varepsilon^i_{tcv} = 1$) – the considered and alternative samples are equivalent for the consumer. At the same time it is supposed that with an increase in values of indicators of $R_{C\,tcv}$ and $R^i_{C\,tcv}$ the technical (functional) consumer value of hi-tech production also increases.

It is expedient to determine cost consumer value on the basis of $R_{C\,tcv}$, $R_{A\,tcv}$, $R^i_{C\,tcv}$ and $R^i_{A\,tcv}$, which are used for calculation of technical and functional value.

For determination of cost consumer value we will enter the following designations: $R^{TV}_{C\,tcv}$, $R^{TV}_{A\,tcv}$ – the indicators characterizing the technical consumer value of the considered and alternative hi-tech production, respectively; $R^{FV}_{C\,tcv}$, $R^{FV}_{A\,tcv}$ - the indicators characterizing the functional consumer value of the considered and alternative hi-tech production.

For an assessment of cost consumer value of one project as $R_{C\,tcv}$ и $R_{A\,tcv}$ in formulas (1) and (2) can use expected cost indexes of the considered hi-tech production (C_C) and alternative (C_A), such as:

$$R_{C\,tcv} = C_c, \quad R_{A\,tcv} = C_A \tag{9}$$

The cost consumer value of hi-tech production in this case is defined provided that characteristics and effects of use of the considered and alternative samples coincide or differ slightly. Analytical record of the specified condition has the following appearance:

- in case for an assessment of technical consumer value of hi-tech production is used one indicator:

$$R_{C\,tcv}^{TC} \cong R_{A\,tcv}^{TC} \tag{10}$$

$$R_{C\,tcv}^{FC} \cong R_{A\,tcv}^{FC} \tag{11}$$

- in case for an assessment of technical consumer value of hi-tech production are used several indicators:

$$R_{C\,tcv}^{i\,TC} \cong R_{A\,tcv}^{i\,TC} \tag{12}$$

$$R_{C\,tcv}^{i\,FC} \cong R_{A\,tcv}^{i\,FC} \tag{13}$$

If performing formulas (10), (11), (12), (13) are carried out, then cost consumer value is defined from expressions:

- in case for an assessment of technical and functional consumer value of hi-tech production is used one indicator:

$$\Delta_{tcv} = C_C\left(R_{C\,tcv}^{TC}, R_{C\,tcv}^{FC}\right) - C_A\left(R_{A\,tcv}^{TC}, R_{A\,tcv}^{FC}\right) \tag{14}$$

- in case for an assessment of technical and functional consumer value of hi-tech production are used several indicators:

$$\Delta_{tcv}^{i} = C_C\left(R_{C\,tcv}^{i\,TC}, R_{C\,tcv}^{i\,FC}\right) - C_A\left(R_{A\,tcv}^{i\,TC}, R_{A\,tcv}^{i\,FC}\right) \tag{15}$$

For an assessment of cost consumer value of life cycle of hi-tech production as factors $R_{C\,tcv}$ и $R_{A\,tcv}$ can also be used expected costs of realization of life cycle of the considered sample of hi-tech production $C_{C\,tcv}^{LC}$ and alternative sample of hi-tech production $C_{A\,tcv}^{LC}$:

$$R_{C\,tcv} = C_{C\,tcv}^{LC}, \cdots R_{A\,tcv} = C_{A\,tcv}^{LC} \tag{16}$$

If performing formulas are carried out, then the cost consumer value of life cycle of a sample of hi-tech production is defined according to the following expressions:

1. If for an assessment of technical and functional consumer value of hi-tech production is used one indicator,

$$\Delta_{tcv} = C_C^{LC}\left(R_{C\,tcv}^{TC}, R_{C\,tcv}^{FC}\right) - C_A^{LC}\left(R_{A\,tcv}^{TC}, R_{A\,tcv}^{FC}\right) \tag{17}$$

2. If for an assessment of technical and functional consumer value of hi-tech production are used several indicators:

$$\Delta_{tcv}^{i} = C_C^{LC}\left(R_{C\,tcv}^{i\,TC}, R_{C\,tcv}^{i\,FC}\right) - C_A^{LC}\left(R_{A\,tcv}^{i\,TC}, R_{A\,tcv}^{i\,FC}\right) \tag{18}$$

By means of studying of consumer value it is possible to define the prospects of development of hi-tech production and to analyse feasibility of the created plans for the long-term, medium-term and short-term periods. Therefore such analysis is a significant component of one of the modules of the accounting of specifics of telecommunication production within an information system, which is presented further.

5 The Main Modules of the Accounting of Specifics of Hi-Tech Production Within an Information System

The information system has to contain the following modules, which represent six consecutive stages.

The first module includes identification of the purpose of realization of a project and tasks, which are planned to be solved during its implementation. The tasks can be such as:

- creation of a sample of new generation with a certain set of characteristics;
- modernization of the existing sample by replacement of separate components on newer;
- creation new or improvement of the existing methodology in any subject domain;
- formation of new types of the materials, which have defined characteristics.

Among the examples of such tasks can be: improvement of methodical ensuring forecasting of temporary or cost indexes of creation of high-tech production, justification of the direction of development of telecommunication complexes, creation of the terminal device with different set of user interfaces and others. Results of the analysis of a goal of a project and necessary to the decision tasks represent a basis for formation of the second module.

The second module represents forecasting of the expected consumer value of a project with application of economic-mathematical methods. Within this module the analysis of an analog project is carried out with reduction of cost of realization by settlement year. As an analog project is chosen a project, which performing the problems belonging to one subject domain with the tasks of the planned project and also characterized by creation of production of identical or close functional purpose. Process of the choice of an analog project is one of the key moments, which allow defining basic values of cost, duration and dynamics of an expenditure of financial resources. Besides, it is important to pay attention to reduction of cost of an analog project to a comparable form. It is connected with a variety of reasons:

- the actual expenses can't coincide with planned because of distinction of planned annual indexes deflators and their actual values;
- various sources of financing on a covering of costs of realization of an analog project: own sources of financing, state budget and others;
- in the actual expenditures can be the expenses connected with inappropriate use of financial resources and also expenses, which were a consequence of violation by the performer of safety measures, manufacturing techniques of production or quality control of production of a sample, its components. It leads to increase in number of prototypes and duration of tests.

The purpose of reduction of cost of an analog project to a comparable form is the accounting of all money regardless of a source of financing or conditions of implementation, and also the accounting of different time of an investment of financial resources on realization of an analog project and the planned one. Within this module it is necessary to allocate the following stages:

1. Summation of all used financial resources is made for every year of implementation of a project. This step will allow avoiding a situation of other share distribution of financial resources to the budgetary and own funds for the planned project and for an analog project.
2. Financial expenses, which are spent in an inappropriate way are excluded from calculation or are connected with implementation of work of force majeure character.
3. A result of the first two stages is recalculated in the prices of settlement year for the purpose of determination of total volume of necessary money in real terms.

By means of performance of the listed stages the cost of an analog project is given to basic conditions. It allows comparing consumer properties of production created and planned to creation: characteristics of a sample, quantity of tasks, which need to be solved for ensuring economic security, and achievement of the required effect. Studying of information on consumer properties will allow carrying out the analysis of change of cost of the perspective hi-tech production, which rather exists.

One more feature of telecommunication production is existence of probability that the customer won't receive production provided by the contract after the termination of planned terms of carrying out research and development. It is connected with unique character and complexity of solvable tasks, which demand involvement of highly qualified specialists, using of advanced scientific and technical base that rather often is absent in practice. Also difficult technologies, element base and modern materials, strict observance of technical regulations are necessary for development and production of samples. Their violation can lead to accidents and destruction of expensive samples. Besides, absence of the highly qualified specialists and headed by them scientific, production and engineering schools of modern scientific and technical base doesn't allow to produce competitive telecommunication product with necessary consumer properties in the plan terms.

As a result it is possible to note existence of the risks exerting considerable impact on formation of cost of a project, which extent of influence directly depends on complexity and novelty of the production planned to creation.

Therefore planning of implementation of research and development, and also production of skilled and serial samples has to be carried out by means of application of the automated information system with a possibility of timely adjustment.

The purpose of formation of the third module is calculation of the top limit price and comparison with the price of a planned project. Excess of the expected top limit price leads to realization of a project is inexpedient from the technical and economic point of view. That demands a project exception of planned documents or its analysis and adjustment. The main objective of the module is determination of effect of using of a sample. In classical understanding effect is the indicator characterizing result of activity regardless. Such definition is used for carrying out the technical and economic analysis. If the effect is applied to an assessment of expediency of implementation of investment projects in the civil sphere of economy, then it means the net discounted income (the sum of annual current effects in the project). Nevertheless, use of one of the given approaches to an assessment of the actions connected with creation of hi-tech production isn't possible owing to the following reasons:

- the circle of consumers is rather limited as the main part of production can't be sold to the population in domestic market;
- use of financial, temporary, human and material resources is directed not to maximizing profit, but to ensuring economic security of the state;
- implementation of commercial realization of hi-tech production under contracts with foreign customers.

In that case, determination of effect as the indicator characterizing effectiveness of use of hi-tech production will be optimum. The concrete type of an indicator depends on specifics of a project and the available basic data. For the enterprises, which effect of realization is difficult to express in number, the top limit price isn't defined and the assessment of technical and economic expediency of a project isn't made.

The necessary amount of financing often depends on many factors, such as a political and financial and economic situation in the country and beyond its limits, the required characteristics of a sample, prices for materials and semi-finished products. These factors can't be predicted absolutely precisely therefore the top limit price of hi-tech production represents a random variable, and its calculated value – an assessment of population mean, around which possible values are grouped. It is connected with the fact that owing to casual character of factors it is impossible to define in advance authentically extent of their influence on the price.

The fourth module is intended for conducting check of technical and economic expediency of realization of a project that is connected with need of the accounting of consumer properties of production planned to creation. The indicator of effect of use is for this purpose entered and the analysis of probable effectiveness of operation of a sample taking into account influence of the external environment is carried out.

Within the fifth module indicators of the risk accompanying realization of a project are defined. It is necessary to consider that the basic data used for estimation of cost of a project aren't reliable as changes of internal and external factors can exert impact on indicators of production not only at approach by the beginning of realization of a project, but also in the course of its implementation.

The sixth module includes an assessment of accuracy and verification of expected value of cost of a project. This module is necessary for definition of concrete actions for decrease in expenses following the results of the analysis of range of their possible values. It is possible to refer application of economic-mathematical models for the purpose of optimization of expenses to such projects.

6 Conclusion

The analysis of the existing regulatory legal and instructive-methodological base showed that planning and estimation of cost of hi-tech production of telecommunication branch is the least regulated. Using already developed information systems, which seem optimum for using by any enterprises, during the planning and production management of telecommunications it is possible to face low efficiency of an information system in view of uniqueness of such type of production.

Thus, introduction of a corporate information system at the enterprises of telecommunication branch demands inclusion in the standard set of subsystems unique modules. Their application will allow to consider specifics of production of branch, and also to carry out the analysis of effect of development and production of hi-tech samples.

References

1. Batkovskiy, A.M., Kalachanov, V.D., Kravchuk, P.V.: Tehnologicheskie riski razrabotki vyisokotehnologichnoy produktsii [Technological risks of development of hi-tech production]. Voprosyi radioelektroniki, seriya RLT **1**, 129–139 (2014)
2. Georgieva, T.N.: Corporate telecommunication network with integrated ERP system. Sci. Eur. **6-2**(6), 19–22 (2016)
3. Koch, S., Mitteregger, K.: Linking customization of ERP systems to support effort. Enterp. Inf. Syst. **10**(1), 81–107 (2016)
4. Vinogradova, Y.Y.: Intellektual'nye informacionnye tekhnologii – teoriya i metodologiya postroeniya informacionnyh system [Intellectual information technologies – the theory and methodology of creation of information systems]. Ekaterinburg: Izd. Ural. gos. ekon. un-ta, 263 p. (2011)
5. Vinogradova, Y.Y., Galimova, A.I.: Formirovanie kompleksnoy sistemyi ekonomicheskogo planirovaniya i upravleniya kak instrument povyisheniya privlekatelnosti ERP-sistem dlya rossiyskih organizatsiy [Formation of complex system of economic planning and management as instrument of increase in appeal of ERP systems to the Russian organizations]. BI-tehnologii i korporativnyie informatsionnyie sistemyi v optimizatsii biznes-protsessov: materialyi IV Mezhdunar. nauch.-prakt. ochno-zaoch. konf./ Ekaterinburg: Izd. Ural. gos. ekon. un-ta, pp. 156–159 (2016)
6. Volkova, V.N., Golub, Y.A.: Informatsionnaya sistema: k voprosu opredeleniya ponyatiya [Information system: to a question of definition of a concept]. Nauchno-praktich. zhurnal: Prikladnaya informatika **5**(23), 112–120 (2009)

7. Babkin, A.V., Kudryavtseva, T.J.: Identification and analysis of instrument industry cluster on the territory of the russian federation. Mod. Appl. Sci. **9**(1), 109–118 (2015)
8. Glukhov, V.V., Ilin, I.V., Levina, A.I.: Project management team structure for internet providing companies. In: Balandin, S., Andreev, S., Koucheryavy, Y. (eds.) ruSMART 2015. LNCS, vol. 9247, pp. 543–553. Springer, Cham (2015). doi:10.1007/978-3-319-23126-6_47
9. Babkin, A.V., Karlina, E.P., Epifanova, N.S.: Neural networks as a tool of forecasting of socioeconomic systems strategic development. In: Proceedings of the 28th International Business Information Management Association Conference - Vision 2020: Innovation Management, Development Sustainability, and Competitive Economic Growth, pp. 11–17 (2016)
10. Vlasova, M., Pimenova, A., Kuzmina, S., Morozova, N.: Tools for company's sustainable economic growth. Procedia Eng. **165**, 1118–1124 (2016)

A Game-Theoretic Model for Investments in the Telecommunications Industry

Sergey A. Chernogorskiy[✉] and K.V. Shvetsov

Peter the Great St. Petersburg Polytechnic University, St. Petersburg, Russia
Chernog_sa@spbstu.ru, shvetsov@inbox.ru

Abstract. The rent-seeking behavior of domestic and foreign investors of the telecommunications sector was considered and analyzed. A game-theoretic model determining the behavior of investors was developed. Comparative results for three or more investors were obtained. The model of rent-seeking behavior, when each player is not indifferent about who will get the prize if he does not receive it himself, was proposed. The model is analyzed in cases of identical and different estimates of the rent by the players. The formulas for total equilibrium costs of the rent-seeking behavior as well as the individual equilibrium costs for each of the domestic and foreign firms under identical and different estimates of the rent were obtained. A situation with two Russian and one foreign investor of the telecom sector of the Russian Federation was deeply examined.

Keywords: Game theory · A game-theoretic model · Search for rent · Investment · Telecommunication industry

1 Relevance of the Research

Rent-seeking behavior is an attempt to get an economic rent i.e. a part of income from a factor of production in excess of an average market income by means of manipulating the social or political environment in which the economic activity is being performed but not by means of creating new wealth.

Such a behavior of companies and organizations turns out to inhibit the economic growth by leading to a high level of the received rent at a very low level of economic efficiency. In this case the rent-seeking behavior hurts the economy more than other negative factors because it results in failure of innovation [1].

Ultimately the rent-seeking behavior is considered as a negative phenomenon entailing considerable losses in public welfare. However in many free market economies a large part of rent-oriented behavior is legal regardless of the losses it could cause to the economy [2].

An example of rent-seeking in modern economy is lobbying of firms engaged in tenders and auctions for development and production of telecom products in order to get preferential treatment in contrast to the competitors and/or to increase its market share [3].

© Springer International Publishing AG 2017
O. Galinina et al. (Eds.): NEW2AN/ruSMART/NsCC 2017, LNCS 10531, pp. 351–364, 2017.
DOI: 10.1007/978-3-319-67380-6_32

The purpose of this article is to find the conditions that minimize the loss in public welfare from the rent-oriented behavior of Russian and foreign investors involved in the development of the Russian telecom industry.

2 Introduction

The Government can support or protect a monopolistic position of certain producers, creating or increasing the rent at the expense of the buyers and customers. The specified rent turns out to be some kind of prize, stimulating the efforts to obtain it. The activity aimed at getting the rent created by the Government is called a rent-oriented behavior. The term "rent-seeking" was proposed by Krueger [4]. The origin of the term is associated mainly with getting a control over land or other natural resources.

The efforts to obtain the rent correspond to certain costs. Tullock has shown that considerable resources from the public viewpoint can be squandered on obtaining of the economic rent [5].

According to Buchanan et al. [6–8], there are three types of costs associated with rent-oriented behavior:

1. The costs of lobbyists trying to motivate officials, which are within the competence of the decision on the rent creation.
2. The costs of the Government related to the proceeds of the relevant activities and response to incentives, that create a rent-seeking behavior (because the revenues of the officials that manage the privileges increase by the volume of bribes, there exists some "excessive" competition for appropriate positions; those who wish to take them, spend more time and effort) [9].
3. The cost of third parties resulting from the activities of firms or officials involved in rent-seeking behavior (e.g., competition from other interested groups to receive subsidies, tax exemptions, etc.).

2.1 The Costs of Rent-Oriented Behavior

On the basis of the Tallock model, one can see how the costs of rent-oriented behavior correspond to the rent that stipulates this activity. In particular, we will obtain the conditions when the costs and the rent are equal.

Suppose that "competitors" are participating in the competition for the rent of size ; we assume that the probability to get the rent for the participant i is the amount [10]

$$p_i(I_1, \ldots, I_n) = \frac{I_i^r}{I_i^r + \sum_{j \neq i} I_j^r}$$

where I_i - investments of the competitor i in the rent-seeking behavior, r - parameter, which characterizes the investment efficiency. Assuming that all the participants are risk-neutral, the gain of the participant i equals

$$\left(\frac{I_i^r}{I_i^r + \sum_{j \neq i} I_j^r}\right) R - I_i$$

Provided that the Nash equilibrium of the game is an internal (each participant invests in rent-seeking behavior) and symmetrical one ($I_i = I_j$ for all $i, j = 1.2, \ldots, n$), the value $I_i = I$ turns out to be equal

$$I = \frac{n-1}{n^2} rR$$

In a symmetric equilibrium, each participant invests in the rent-oriented activity a value of $\frac{n-1}{n^2} rR$, if

$$\frac{I^r}{nI^r} R - I \geq 0 \tag{1}$$

and does not invest anything ($I = 0$) otherwise.

When $r \leq 1$ the condition (1) is always true. In other words, if the rent-oriented activity is characterized by non-growing return $r \leq 1$, there is a symmetric equilibrium with positive investments in the rent-seeking behavior.

The total investment and share of the investment in the rent equal $nI = \frac{n-1}{n} rR$ and $\gamma = \frac{nI}{R} = \frac{n-1}{n} r$, respectively.

When return of the rent-oriented activity is constant $r = 1$, the share of the rent offsetting the investments with $n = 2$ equals to ½. If the number of the participants increases, the share of rent offsetting the investment approaches to 1.

Let us consider the case of increasing return $r > 1$. If $r > 2$ the condition (1) is not true at any n, because there is no symmetric equilibrium in pure strategies. If $r = 2$ the condition (1) is true only when $n = 2$. In this case, each of the two parties invests exactly the half of potential rent in the rent-oriented activity, i.e. that the total investment equals exactly the rent [11].

If $r < 2$ the equilibrium can exist when n is greater than two. For example, if $r = 1.5$ the condition (1) holds true when $n = 3$. If $n = 2$ the value of total investment amounts to two thirds of the rent, and when $n = 3$ it exactly equals the rent. We should note that until the expected gain from rent-oriented activity proves to be positive, there is an incentive attracting additional participants in the game. Thus, free entrance and constant return make a well-predicted "rent-dissipation" as a result of competition for the rent.

3   Incomplete Information in the Model of Rent-Competition

We have examined the models of rent-oriented behavior, assuming that the decisions were made by the participants with all the information available [12]. Now we abandon the assumption of full awareness of the participants competing for rent. We will consider the situation when the number of the competing participants equals two, whereas the first participant doesn't know the prize evaluation of the second participant. Therefore, according to the terminology of the game theory, player 1 doesn't know the type of player 2. Let's also assume that the competition for rent goes as in the Stackelberg model, where the player 1 (leader) takes the decision first, and player 2 decides, knowing about the decision of the leader. In order to make the assumption of information scarcity true, we'll suppose that estimates of the prize are different with different players. Finally, we assume that the technology of the rent-oriented activity is characterized by constant return $r = 1$.

Let us assume that the prize estimation of the player 2 can take two values: either v_1, or v_h ($v_2 \in \{v_1, v_h\}$), where $v_h > v_1$. In the terms of game theory, it means that player 2 can be of two types. Suppose that player 2 can be type of player v_h with probability q, and, accordingly, type of player v_1 with probability $(1 - q)$. It is easy to see that the equilibrium strategy in the game goes as follows:

$$x_2^*(x_1) = \begin{cases} \max\{0, (v_h x_1)^{1/2} - x_1\}, \text{ если } v_2 = v_h \\ \max\{0, (v_1 x_1)^{1/2} - x_1\}, \text{ если } v_2 = v_1 \end{cases} \qquad (2)$$

The player 1 must choose x_1 in a way to maximize the expected payoff, i.e. to maximize the following expression:

$$E[U_1(x_1)] = q v_1 \Pr(npu\ v_2 = v_h) + (1 - q) v_1 \Pr(npu\ v_2 = v_1) - x_1$$

In the beginning, we suppose that the costs of the player 2 are positive. We substitute in the Eq. (2) the expression for $x_2(x_1)$. Then we will see that

$$E[U_1(x_1)] = q \frac{v_1 x_1}{(v_h x_1)^{1/2}} + (1 - q) \frac{v_1 x_1}{(v_1 x_1)^{1/2}} - x_1$$

The decision of the player 1 is characterized by the first-order condition:

$$\frac{v_1 x_1^{*-1/2}}{2} \left[q v_h^{-1/2} + (1 - q) v_1^{-1/2} \right] = 1$$

This expression determines the costs of the player 1 unambiguously as:

$$x_1^* = \frac{v_1^2}{4} \left[q v_h^{-1/2} + (1 - q) v_1^{-1/2} \right]^2$$

Besides, we know that if $x_1^* = v_1^2/4 \left[q v_h^{-1/2} + (1 - q) v_1^{-1/2} \right]^2 \geq v_1$, then in case the player 2 does not appreciate the prize, he chooses not to participate in the competition. This happens when $v_1 \geq 2 v_1^{1/2} \left[q v_h^{-1/2} + (1 - q) v_1^{-1/2} \right]^{-1}$. In this case the best answer of the player 2 in the equilibrium will be abandoning the investment. Then player 1 would maximize the following expression:

$$E[U_1(x_1)] = q\frac{v_1 x_1}{(v_h x_1)^{1/2}} + (1-q)v_1 - x_1$$

provided that $x_1 \geq v_1$.

When the constraint is not important, the maximum of the first order condition is written as $v_1 q(v_h x_1)^{-1/2}/2 = 1$. Here we get the value of optimal costs for the player 1 as $x_1^* = \max\{v_1, q^2 v_1^2/(4v_h)\}$.

If $x_1^* = q^2 v_1^2/(4v_h) \geq v_h$, then the best answer of the player 2, regardless of its type, will be zero cost. This condition is fulfilled when $v_1 \geq 2v_h/q$, in this situation the player 1 will choose the amount of the costs as v_h.

Let us put all the results together:

$$x_1^* = \begin{cases} v_1^2/4\left[qv_h^{-1/2} + (1-q)v_1^{-1/2}\right]^2, \ if \ v_1 < 2v_1^{1/2}\left[qv_h^{-1/2} + (1-q)v_1^{-1/2}\right]^{-1} \\ \max\{v_1, q^2 v_1^2/(4v_h)\}, \ if \ 2v_1^{1/2}\left[qv_h^{-1/2} + (1-q)v_1^{-1/2}\right]^{-1} \leq v_1 < 2v_h/q \\ v_h, \ if \ v_1 \geq 2v_h/q \end{cases}$$

That is, if the probability q is low, then the equilibrium choice of the player 1 is close to the solution with full information. The same is true if q is close to 1. Moreover, x_1^* is a non-decreasing function of v_1 and q and non-increasing function of v_h and v_1.

4 A Competition-for-Rent Model, When Each Player Cares About Who Will Get the Prize If He Does Not Receive It Himself

In seeking-for-rent situations, the participants are often not indifferent for who will get the rent, even when it is not himself [13]. In other words, the rent has a well-known characteristic of a public good, i.e. is a distributed public good [14]. In order to implement the idea of such an assessment, we'll consider the rent evaluations as a vector. The rent estimation of each player is a vector $\mathbf{v_i} = (v_{i1}, v_{i2}, \ldots, v_{in})$, where v_{ij} is the value for the player i if the player j gets the rent. The probability that a player will get the rent equals to the share costs in the total expenditure, i.e. the probability that a player i will receive the rent, $p_i(\mathbf{x})$ equals to x_i/s, where $s = \sum_{j=1}^{n} x_j$, $\mathbf{x} = (x_1, x_2, \ldots, x_n)$ and x_i denotes the costs of player i.

Let $\mathbf{p}(\mathbf{x})$ be the vector of corresponding probabilities $\mathbf{p}(\mathbf{x}) = (p_1(\mathbf{x}), p_2(\mathbf{x}), \ldots, p_n(\mathbf{x}))$. The expected profit of each player depends not only on how he assesses the gain for himself (in case he wins), but also on how he evaluates the gain of the other players. If we look at the game of, say, three players, then the expected gain of the first player is the sum of the rent value in case he wins multiplied by the probability of winning the game, plus the gains of the other players for the cases of their winning multiplied by the corresponding probabilities, minus the costs of the first player. In mathematical symbols, it looks as follows $U_1(x_1, x_2, x_3) = v_{11}\frac{x_1}{x_1 + x_2 + x_3} +$

$v_{12} \frac{x_2}{x_1+x_2+x_3} + v_{13} \frac{x_3}{x_1+x_2+x_3} - x_1$. We can express the expected profit of player i in the competition of n players more concisely using matrix notation $U_i(\mathbf{x}) = \mathbf{v}_i^T \mathbf{p}(\mathbf{x}) - x_i$. These utility functions assume that the players are risk-neutral [15].

We will define an $n \times n$ matrix composed of vectors of assessments $\mathbf{v}_1^T, \mathbf{v}_2^T \ldots \mathbf{v}_n^T$. The Nash equilibrium can be obtained by finding the solutions of n equations that are the first-order terms, which can be written in matrix form as follows:

$$
\begin{pmatrix}
v_{11} & v_{12} & \cdots & v_{1n} \\
v_{21} & v_{22} & \cdots & v_{2n} \\
\vdots & \vdots & \ddots & \vdots \\
v_{n1} & v_{n2} & \cdots & v_{nn}
\end{pmatrix}
\begin{pmatrix}
\frac{x_1}{s} \\
\frac{x_2}{s} \\
\vdots \\
\frac{x_n}{s}
\end{pmatrix}
-
\begin{pmatrix}
x_1 \\
x_1 \\
\vdots \\
x_n
\end{pmatrix}
=
\begin{pmatrix}
U_1 \\
U_1 \\
\vdots \\
U_n
\end{pmatrix}
$$

If we assume that $\mathbf{U}(\mathbf{x}) = (U_1, U_2, \ldots, U_n)^T$ is the vector of gains of the players, we can summarize the above expression, using matrix notation $\mathbf{U}(\mathbf{x}) = \mathbf{V}\mathbf{p}(\mathbf{x}) - \mathbf{x}$.

Let us define the condition of the maximum of the first order for the expected profit of the first player using the mathematical analysis, then:

$$
\frac{\partial U(x_1, x_2, \ldots, x_n)}{\partial x_1} = \frac{v_{11}s - v_{11}x_1}{s^2} - \frac{v_{12}x_2}{s^2} - \cdots - \frac{v_{1n}x_n}{s^2} - 1 = 0
$$

or

$$
\frac{(v_{11} - v_{12})x_2}{s^2} + \ldots + \frac{(v_{11} - v_{1n})x_n}{s^2} = 1
$$

In the same way, we can find the first order condition of maximized expected utility of each player and summarize the results, getting the following expression:

$$
\begin{pmatrix}
0 & v_{11} - v_{12} & \cdots & v_{11} - v_{1n} \\
v_{22} - v_{21} & 0 & \cdots & v_{22} - v_{2n} \\
\vdots & \vdots & \ddots & \vdots \\
v_{nn} - v_{n1} & v_{nn} - v_{n2} & \cdots & 0
\end{pmatrix}
\begin{pmatrix}
\frac{x_1}{s^2} \\
\frac{x_2}{s^2} \\
\vdots \\
\frac{x_n}{s^2}
\end{pmatrix}
=
\begin{pmatrix}
1 \\
1 \\
\vdots \\
1
\end{pmatrix}
$$

To make it a bit more concise we can simplify the expression above to matrix notation $(\mathbf{J}_{n \times n} - \mathbf{V}) \frac{\mathbf{x}}{s^2} = \mathbf{1}_n$, where $\mathbf{J}_{n \times n}$ is an $n \times n$ matrix, with a row i consisting of v_{ii} and $\mathbf{1}_n$ is an $n \times 1$ vector.

We can analyze how the results will change as the number of player's increases. To do this, we add some players with interests similar either to the interests of players 1 and 2 or to the interests of player 3 from the previous example. We will consider mobile phone manufacturers, which are divided into two groups: domestic and foreign ones. We suppose that there are n manufacturers who have common interests with players 1 and 2 and m manufacturers that have the same interests as player 3. We can analyze a lot of players, consisting of two parts. Domestic manufacturers are 1, 2, ..., n and foreign ones are n + 1, n + 2, ..., n + m.

Many players that share common interests and the value they attribute to the winning company of the group can characterize payments to the firms in these two groups. For domestic manufacturers each player estimates the rent as 1 in the case of winning and γ if some other domestic producer wins. For a foreign manufacturer the rent is estimated as 1 in the case of winning and δ if another foreign manufacturer wins. The expected utility, which players want to maximize, can be written as follows [16]:

$$
\begin{pmatrix}
1 & \gamma & \cdots & \gamma & 0 & \cdots & 0 & 0 \\
\gamma & 1 & \cdots & \gamma & 0 & \cdots & 0 & 0 \\
\vdots & \vdots & \ddots & \vdots & \vdots & \ddots & \vdots & \vdots \\
\gamma & \gamma & \cdots & 1 & 0 & \cdots & 0 & 0 \\
0 & 0 & \cdots & 0 & 1 & \delta & \cdots & \delta \\
0 & 0 & \cdots & 0 & \delta & 1 & \cdots & \delta \\
\vdots & \vdots & \ddots & \vdots & \vdots & \vdots & \ddots & \vdots \\
0 & 0 & \cdots & 0 & \delta & \delta & \cdots & 1
\end{pmatrix}
\begin{pmatrix}
\frac{x_1}{s} \\
\vdots \\
\frac{x_n}{s} \\
\frac{x_{n+1}}{s} \\
\vdots \\
\frac{x_{n+m}}{s}
\end{pmatrix}
-
\begin{pmatrix}
x_1 \\
\vdots \\
x_n \\
x_{n+1} \\
\vdots \\
x_{n+m}
\end{pmatrix}
=
\begin{pmatrix}
U_1 \\
\vdots \\
U_n \\
U_{n+1} \\
\vdots \\
U_{n+m}
\end{pmatrix}
$$

The first-order condition is determined in the same manner as in the previous example. The matrix $(\mathbf{J}_{(n+n)\times(n+m)} - \mathbf{V})$ can be inverted for any γ, $\delta \neq 1$ and $n, m \geq 1$. Using the notation $\mathbf{I}_{k \times k}$ for the $k \times k$ identity matrix, we can get a solution for vector of equilibrium costs $\mathbf{x}^*$.

$$
\mathbf{x}^* = (s^*)^2 (\mathbf{J}_{(n+n)\times(n+m)} - \mathbf{V})^{-1} \mathbf{1}_{(n+m)}
$$

For the above expression we have

$$
(\mathbf{J}_{(n+n)\times(n+m)} - \mathbf{V})^{-1} = \begin{pmatrix} W_{11} & W_{12} \\ W_{21} & W_{22} \end{pmatrix}
$$

with W_{ij} defined below.

$$
W_{11} = -\frac{1}{(1-\gamma)} \left[\mathbf{I}_n - \frac{(1-\gamma)(1-\delta)(m-1) - m}{(n-1)(m-1)(1-\gamma)(1-\delta) - nm} \mathbf{J}_{n \times n} \right]
$$

$$
W_{12} = \frac{-1}{(n-1)(m-1)(1-\gamma)(1-\delta) - nm} \mathbf{1}_{n \times m}
$$

$$
W_{21} = \frac{-1}{(n-1)(m-1)(1-\gamma)(1-\delta) - nm} \mathbf{1}_{m \times n}
$$

$$
W_{22} = -\frac{1}{(1-\delta)} \left[\mathbf{I}_m - \frac{(1-\gamma)(1-\delta)(n-1) - n}{(n-1)(m-1)(1-\gamma)(1-\delta) - nm} \mathbf{J}_{m \times m} \right]
$$

Now we can deduce an expression both for total equilibrium costs s^* and individual equilibrium costs. We will use x_d^* to denote the equilibrium costs for each of the domestic firms (players 1, 2, ... n) and x_f^* for each of the foreign manufacturers (players $n + 1, n + 2, \dots n + m$). Then,

$$s^* = \frac{nm - (1 - \gamma)(1 - \delta)(n - 1)(m - 1)}{n[m - (1 - \delta)(m - 1)] + m[n - (1 - \gamma)(n - 1)]}$$

$$x_d^* = (s^*)^2 \left[\frac{m - (1 - \delta)(m - 1)}{nm - (1 - \gamma)(1 - \delta)(n - 1)(m - 1)} \right]$$

$$x_f^* = (s^*)^2 \left[\frac{n - (1 - \gamma)(n - 1)}{nm - (1 - \gamma)(1 - \delta)(n - 1)(m - 1)} \right]$$

These expressions can be solved for x_d^* and x_f^* in terms of game properties. It should be noted that each firm will make contribution to the equilibrium.

Based on the results obtained for the equilibrium costs, we can deduce some simple comparative results. Firstly, if the members of each group increase the assessment of the gain of the other members within the group (γ or δ increases) then the total costs will decrease i.e. $\partial s^*/\partial \gamma, \partial s^*/\partial \delta < 0$. In addition, the contribution of a group member with common interests will decrease as the gain assessment attributed to another group member increases, i.e. $\partial x_d^*/\partial \gamma, \partial x_f^*/\partial \delta < 0$. One of the ways to see what happens if we change the parameters is to review the relationship $nx_d^*/mx_f^* = (nm - (1 - \delta)(m - 1)n)/(nm - (1 - \gamma)(n - 1)m)$, which is the ratio of the winning probability of the domestic manufacturer to the winning probability of the foreign one. It is easy to see that this ratio will increase if foreign manufacturers estimate the rent for each of them higher (so that δ increases) or if domestic manufacturers give less value to the rent winning by each other (γ decreases). Moreover, as the number of players in a group increases, simultaneously increases the likelihood that the group will win. However, an increasing number of players with common interests mean a greater likelihood for "free riders" to pop up, when there are more players with interests that are the same as yours. This increases your chance of winning, as long as the interests of players are not identical [17].

4.1   The Difference in the Rent Estimates

Now let us analyze the game, where players have their own estimates of the prize. We'll consider three players-investors of the Russian telecom industry. Let the first player be the concern "Almaz-Antey" (Almaz-Antey), the second one - JSC "RTI-Group" (RTI) and third player - Intel. We denote the gain estimates of the players as follows: v_{11} for Almaz-Antey, v_{22} for RTI and v_{33} for Intel. However, the gain estimates of each other for Almaz-Antey and RTI make up a share of their own estimate. We assume that the share is equal for the two. Then RTI estimates the gain of Almaz-Antey as γv_{22} and while Almaz-Antey attributes the value of γv_{11} to the gain of RTI. In this case, Intel estimates the gain of Almaz-Antey or gain of RTI as zero.

We can take these asymmetric rents as the result of various circumstances of the players. For example, if the stock price of RTI is more sensitive to foreign competition, then the company may evaluate the protectionist legislation higher than Almaz-Antey. For simplicity, we will analyze only this competition of three players. The matrix of estimates is presented in the following form:

$$\mathbf{V} = \begin{pmatrix} v_{11} & \gamma v_{11} & 0 \\ \gamma v_{22} & v_{22} & 0 \\ 0 & 0 & v_{33} \end{pmatrix}$$

Determining of the conditions of the first order in this task is somewhat more difficult than in the previous examples, because we assumed that $v_{ii} \neq v_{jj}$. However, the same logic still holds, and the solution, provided that all players participate, gives the following equilibrium costs:

$$x_1^* = \frac{2v_{11}v_{22}v_{33}[v_{11}v_{33} - v_{22}v_{33} + v_{11}v_{22}(1-\gamma)]}{[v_{11}v_{22} + v_{22}v_{33} + v_{11}v_{33} + v_{11}v_{22}\gamma]^2(1-\gamma)} \tag{3}$$

$$x_2^* = \frac{2v_{11}v_{22}v_{33}[v_{22}v_{33} - v_{11}v_{33} + v_{11}v_{22}(1-\gamma)]}{[v_{11}v_{22} + v_{22}v_{33} + v_{11}v_{33} + v_{11}v_{22}\gamma]^2(1-\gamma)} \tag{4}$$

$$x_3^* = \frac{2v_{11}v_{22}v_{33}\left[v_{22}v_{33}(1-\gamma) + v_{11}v_{33}(1-\gamma) - v_{11}v_{22}(1-\gamma)^2\right]}{[v_{11}v_{22} + v_{22}v_{33} + v_{11}v_{33} + v_{11}v_{22}\gamma]^2(1-\gamma)} \tag{5}$$

$$s^* = \frac{2v_{11}v_{22}v_{33}}{[v_{11}v_{22} + v_{22}v_{33} + v_{11}v_{33} + v_{11}v_{22}\gamma]^2(1-\gamma)} \tag{6}$$

It is important to note that the above costs form equilibrium only with the involvement of all players. Of course, if the numerator of the Eqs. (3)–(6) is negative, the players will decide not to play. It turns out that the same conditions that stipulate strictly positive contributions in the Eqs. (3)–(6) have a very beautiful interpretation. These conditions answer the following question: "If two players are already in the game, then when the third player chooses to carry the cost of participation in the game?" We also would like to know whether the order of the players entering the game influence the final set of players, and under what conditions all three players will participate in this game.

We suppose first that Almaz-Antey and RTI are playing against each other whereas Intel has to decide whether to join the competition for the rent. Of course, Intel will only participate if its marginal expected profit is positive. That is, Intel wants to participate in the competition, if $\partial U_3(x_1, x_2, x_3)/\partial x_3 > 0$. In this example, the condition of Intel participation goes as following:

$$\frac{\partial U_3(x_1, x_2, x_3)}{\partial x_3} = \frac{(x_1 + x_2)v_{33}}{s^2} - 1 = \frac{(s - x_3)v_{33}}{s^2} - 1 > 0$$

Since we evaluated the derivative with $x_3 = 0$, then s is the sum of the equilibrium cost of both Almaz-Antey and RTI, if they play with each other $s = v_{11}v_{22}(1 - \gamma)/(v_{11} + v_{22})$. Therefore, we obtain that Intel wants to participate if:

$$v_{33} > v_{11}v_{22}(1 - \gamma)/(v_{11} + v_{22})$$

We should note that the condition that gives a positive contribution for Intel in Eq. (5) is the same that gives the company a positive marginal expected utility. It is also interesting to note that as the mutual evaluation of the gain of Almaz-Antey and RTI increases (γ increases), the minimum value of the prize required by Intel to participate is reduced. In the case when the prize turns out to be a pure public good for Almaz-Antey and RTI (γ approaches 1), Intel chooses to participate in the game. The degree of competition of Almaz-Antey and RTI will decline as the mutual evaluation of the gain of each other increases (γ increases) until ultimately the companies will participate in the game as a single player. In this case, Intel joins the competition with a lower gain evaluation, and if the prize is a pure public good for two other players, Intel will always participate.

If we consider the conditions of entry for RTI or Almaz-Antey, the analysis is only slightly more complicated. We assume that the Almaz-Antey and Intel are leading the game. When RTI wants to participate? Using the same arguments as before, we get that RTI will decide to contribute if it increases its expected utility or if $\partial U_2(x_1, x_2, x_3)/\partial x_2 > 0$. This requirement means that the first-order condition goes as follows:

$$\frac{\partial U_2(x_1, x_2, x_3)}{\partial x_2} = \frac{v_{22}(1 - \gamma)x_1}{s^2} + \frac{v_{22}x_3}{s^2} - 1 > 0$$

This expression can be simplified by multiplying it by s and regrouping the summands. This will give $\frac{v_{22}(s - x_2)}{s} + \frac{\gamma v_{22}x_1}{s} > s$. Almaz-Antey and Intel will implement their equilibrium strategies in the game of two players, where Almaz-Antey evaluates the prize as v_{11} and Intel evaluates the prize as v_{33}. The equilibrium contributions in the game of two players are $x_1 = v_{11}^2 v_{33}/(v_{11} + v_{33})^2$, $x_3 = v_{11}v_{33}^2/(v_{11} + v_{33})^2$ respectively, and $s = v_{11}v_{33}/(v_{11} + v_{33})$. If we make the substitution and evaluate the derivative with $x_2 = 0$, then we can deduce a condition for participation of RTI in the game:

$$v_{22} > \frac{v_{11}v_{33}}{v_{11}(1 - \gamma) + v_{33}} \tag{7}$$

It is again the same condition, which provides positive costs in Eq. (4). If RTI evaluates the gain of Almaz-Antey as zero ($\gamma = 0$), then it will be again the same result obtained earlier. While considering the condition (7) we see that in order to participate RTI have to evaluate the gain higher as the estimates of Almaz-Antey or Intel (v_{11} or v_{33}) are increasing or if both Almaz-Antey and RTI will evaluate the gains of each other higher (γ increases). It is intuitively clear that Almaz-Antey will make a greater contribution to the game of two players, if its own gain estimate increases (v_{11}), while RTI will be able to take a "free ride". If RTI evaluates the gain of Almaz-Antey higher (γ increases), then RTI would make a smaller contribution, since it gets a higher expected utility from the contribution of Almaz-Antey. Moreover, the increase in gain estimates by Intel (v_{33}) will make the contribution of RTI less profitable, so it must assess the gain higher prior to choosing for taking part in the game. Let us examine what would happen if the gain estimates of each other at RTI and Almaz-Antey were getting close to their own estimates (γ approaches 1). In this case, RTI chooses to participate in the competition only if its estimate exceeds the one of Almaz-Antey.

We should note that at least two players are always active in the game, with no dependence on the order in which players enter the competition. The conditions for switching the players from non-participation to participation are identical to the conditions determining that their expenditure is greater than zero, if they are already in game [18].

If we assume that all players are active, and this situation will stay the same with small changes of the parameters, then we can draw some conclusions regarding the game. It is not surprising, that if Almaz-Antey and RTI attributed higher importance to the gain of each other (γ increases), then their contributions and overall costs (x_1^*, x_2^* and s^*) would reduce while the cost of Intel x_3^* would increase. It will be clear if we consider the probability of winning of Almaz-Antey or RTI divided by the probability of winning of Intel:

$$\frac{x_1^* + x_2^*}{x_3^*} \frac{2v_{11}v_{22}}{v_{33}(v_{11} + v_{22}) - v_{11}v_{22}(1 - \gamma)}$$

It is obvious, that the probability of winning of Almaz-Antey or RTI decreases as the values, which they attribute to the gain of each other increase (γ increases) or as the prize assessment of Intel increases (v_{33} increases). An increase in "self-assessment" (v_{11} or v_{22}) will increase also the ratio, indicating that win of Intel is less probable.

4.2 An Example

Now we will evaluate the behavior of the three players assuming that Almaz-Antey estimates its proposal as 1. However, the company prefers that the proposal of RTI were accepted rather than the proposal of Intel [19]. Almaz-Antey evaluates the win of RTI as $\gamma < 1$. RTI also evaluates its proposal as 1, and the company prefers that the proposal of Almaz-Antey were accepted rather the one of Intel. Intel also evaluates its proposal as 1 and the company wishes that only its proposal were accepted, while the other two have no value for Intel. Let us assume that Almaz-Antey and RTI evaluate the proposals of each other equivalently, so the expected profit of these firms can be expressed as follows:

$$\begin{pmatrix} 1 & \gamma & 0 \\ \gamma & 1 & 0 \\ 0 & 0 & 1 \end{pmatrix} \begin{pmatrix} \dfrac{x_1}{s} \\ \dfrac{x_2}{s} \\ \dfrac{x_3}{s} \end{pmatrix} - \begin{pmatrix} x_1 \\ x_2 \\ x_3 \end{pmatrix} = \begin{pmatrix} U_1 \\ U_2 \\ U_3 \end{pmatrix}$$

Based on first-order conditions we obtain:

$$\begin{pmatrix} x_1^* \\ x_2^* \\ x_3^* \end{pmatrix} = \frac{(s^*)^2}{2(1-\gamma)} \begin{pmatrix} -1 & 1 & 1-\gamma \\ 1 & -1 & 1-\gamma \\ 1-\gamma & 1-\gamma & -(1-\gamma)^2 \end{pmatrix} \begin{pmatrix} 1 \\ 1 \\ 1 \end{pmatrix}$$

We can find the values of s^*, x_1^*, x_2^* and x_3^* from the above equation:

$$s^* = 2/(3+\gamma)$$
$$x_1^* = x_2^* = 2/(3+\gamma)^2$$
$$x_3^* = 2(1+\gamma)/(3+\gamma)^2.$$

Thus, the result is that if the Almaz-Antey and RTI give a greater value to the gain of each other in the competition, then there will be a smaller overall activity in the struggle for rent ($\partial s^*/\partial\gamma < 0$) and lower costs for both companies ($\partial x_1^*/\partial\gamma$, $\partial x_2^*/\partial\delta < 0$). However, Intel pays more as γ increases ($\partial x_3^*/\partial\gamma > 0$). In other words, if Almaz-Antey and RTI were in greater accordance on the legislation, then their individual contributions in the equilibrium would decrease, whereas the contribution of Intel would increase. The increase of the contribution of Intel is smaller than the overall reduction in the contributions of both Almaz-Antey and RTI. This shows that as RTI and Almaz-Antey are becoming closer in the estimates, the competition for Intel is getting more expensive. The fact that rent is an imperfect public good means lower costs for rent competition. In addition, the player, whose winning is socially undesirable, will spend more and the probability of his winning in the equilibrium increases.

5   Conclusion

1. The game-theoretic models of rent-seeking behavior under complete and incomplete information in the struggle for rent were elaborated.
2. The model of rent-seeking behavior, when each player is not indifferent about who will get the prize if he does not receive it himself, was proposed. The model is analyzed in cases of identical and different estimates of the rent by the players.
3. The formulas for total equilibrium costs of the rent-seeking behavior as well as the individual equilibrium costs for each of the domestic and foreign firms under identical and different estimates of the rent were obtained.
4. Some conclusions as a result of the model implementation were drawn:

   - if the members of each group improved the assessment of the gain of the other members within the group, the total costs of the group will decrease;
   - the contribution of a member of a group with common interests will decrease if the value attributed to the gain of the other group member increases;
   - as the number of players in a group increases the likelihood that the group will win also rises. On the other hand, a greater number of players with common interests mean a greater likelihood for so-called "free riders" to emerge.

5. An example with three players-investors within the telecommunications industry of the Russian Federation was considered. The first and the second player (Almaz-Antey and RTI) are Russian companies, and the third player - Intel - is a foreign firm.

The result is that if the Almaz-Antey and RTI give a greater value to the gain of each other, then their costs in the struggle for rent will be lower. This concerns both the total cost and the individual cost of the Russian companies. In other words, if Almaz-Antey and RTI were in greater accordance on the legislation, then their individual contributions in the equilibrium would decrease, whereas the contribution of Intel would increase. This shows that as RTI and Almaz-Antey are becoming closer in the estimates for the best proposal, the competition for Intel is getting more expensive.

In this case, the probability of winning an auction for creating of new telecommunications equipment for domestic companies is higher than that for foreign ones.

References

1. Murphy, K.M., Shleifer, A., Vishny, R.W.: Why is rent-seeking so costly to growth? Am. Econ. Rev. **83**(2), 409–414 (1993)
2. Levine, M., Satarov, G.: Rent-oriented Russia. Voprosy ekonomiki, no. 01, pp. 61–77. Publishing House Editorial Office of the Journal "Questions of Economics", Moscow (2014)
3. Feenstra, R., Taylor, A.: International Economics. Worth Publishers, New York (2008)
4. Krueger, A.: The political economy of the rent-seeking society. Am. Econ. Rev. **64**(3), 291–303 (1974)
5. Tullock, G.: the welfare costs of tariffs, monopolies, and theft. W. Econ. J. **5**(3), 224–232 (1967)
6. Buchanan, J., Tullock, G., Tollison, R.: Toward a Theory of the Rent-Seeking Society. Texas A&M University Economics Series, p. 367. Texas A&M University, Texas (1980)
7. Tullock, G.: Efficient rent-seeking. In: Buchanan, J.M., Tullock, G., Tollison, R.D. (eds.) Toward a Theory of the Rent-Seeking Society. Texas A&M University Press (1980)
8. Corcoran, W.J.: Long-run equilibrium and total expenditures in rent-seeking. Public Choice **43**(1), 89–94 (1984)
9. Leeson, P.T.: The Invisible Hook: The Hidden Economics of Pirates, p. 191. Princeton University Press, Princeton (2009)
10. Busygin, V.P., Zhelobod'ko, E.V., Tsyplakov, A.A.: Microeconomics - third level, Novosibirsk: SB RAS, 704 p. (2003). (in Russian)
11. Thijssen, J.J.J., Huisman, K.J.M., Kort, P.M.: Symmetric equilibrium strategies in game theoretic real option models. J. Math. Econ. **48**(4), 219–225 (2012)
12. Nitzan, S.: Collective rent dissipation. Econ. J. **101**(402), 1522–1534 (1991)
13. Mokyr, J., Nye, J.V.C.: Distributional coalitions, the industrial revolution, and the origins of economic growth in Britain. South. Econ. J. **74**(1), 50–70 (2007)
14. Baik, K., Lee, S.: Collective rent seeking with endogenous group sizes. Eur. J. Polit. Econ. **13**(1), 121–130 (1997)
15. Cavusoglu, H., Raghunathan, S., Yue, W.T.: Decision-theoretic and game-theoretic approaches to IT security investment. J. Manag. Inf. Syst. **25**(2), 281–304 (2008)

16. Chernogorskiy, S.A.: Rent-seeking behavior of russian and foreign car manufacturers. In: Integration in Economy of the World Economic System: Collection of Scientific Works of the 10th International Conference, pp. 181–185. Publishing House of St. Petersburg State Technical University, St. Petersburg (2005). (in Russian)
17. Pasour, E.C.: Rent seeking: some conceptual problems and implications. Rev. Austrian Econ. 1, 123–145 (1987)
18. Shastitko, A.E.: Institutional Economic Theory. Theis Publishing House, Moscow (1997). (in Russian)
19. Schenk, R.: Rent Seeking. In: CyberEconomics, Archived from the original on January 3 (2006). Accessed 02 Nov 2007

Business Perception Based on Sentiment Analysis Through Deep Neuronal Networks for Natural Language Processing

Mónica Pineda Vargas[1], Octavio José Salcedo Parra[1,2(✉)], and Miguel José Espitia Rico[2]

[1] Departamento de Ingeniería de Sistemas e Industrial,
Universidad Nacional de Colombia, Bogotá D.C., Colombia
{mppinedav, ojsalcedop}@unal.edu.co
[2] Facultad de Ingeniería, Universidad Distrital "Francisco José de Caldas",
Bogotá D.C., Colombia
{osalcedo, mespitiar}@udistrital.edu.co

Abstract. In recent years, the machine-learning field, deep neural networks has been an important topic of research, used in several disciplines such as pattern recognition, information retrieval, classification and natural language processing. Is in the last that this paper it's going to be our principal topic, in this branch exist an specific task that in literature is called Sentiment Analysis were the principal function is to detect if an opinion is positive or negative.

In the paper we show how use this subset of the machine learning knowledge and use it for give us an insight in the question: what is the perception in a business or a product by means of the opinion of the consumers in social networks?

Keywords: Deep learning · Natural language processing · Neural networks · Machine learning

1 Introduction

In a globalized world the market competition has been increased and become stronger, the technology sector it is one with greater growing behavior in the last years, by that this economy sector has been gaining a lot of dynamism; thereby the companies became forced to launch new products in a shorter period. Therefore, nowadays it's a necessity make a profile of the market, analyzing the social perception of the company and their products with the objective of improve the business policies and by that way gain a bigger segment of the market.

The social networks have an incommensurate potential, because could bring us a psychological profile of the users; that information can help or even replace classical market analysis such as ethnography, statistics, polling etc. That potential remains undeveloped, impeding the access of this knowledge to the CEOs to overcome weak commercial strategies an enhance the positive ones, this kind of knowledge could be give an important competitive advantage over another companies.

© Springer International Publishing AG 2017
O. Galinina et al. (Eds.): NEW2AN/ruSMART/NsCC 2017, LNCS 10531, pp. 365–374, 2017.
DOI: 10.1007/978-3-319-67380-6_33

A common application of Natural Language Processing is a Sentiment Analysis, whose goal is to extract the sentiments and emotional content in text. This have many useful such as business intelligence, where we know about consumer reactions to a specific product detecting in a online comments. Extract this kind of knowledge is possible with the use of machine learning techniques, such as presented by [1–3], these techniques are based in the Bag-of-Words paradigm, this paradigm works rating every word in a language; this rating varies between positive and negative but have a drawback that don't take in account the context of the phrase in which the word appears. For example employing a Naive Bayes classifier such as the implementation shown in [19] using the word "bad" in a different context we would obtain the same negative result i.e. "it's bad" return a probability of 0.8 to be a negative sentiment, but if we change the context given the phrase "it's not bad" remains with the same probability although we know that it's a positive sentence, if we use a highly positive context like "it's not bad at all" the probability of be a negative sentence remains. This behavior can be explained because the word "bad" have a high negative weight in the English language with comparison with the rest.

Another similar perspective that attempt to preserve the semantically information of the phrases is the Bag-of-n-grams this technique treat several words as only one, then for the combination of words only have a unique score; but this improvement have the drawback that only preserve an small amount of the semantically information, we can try longer and longer combinations let is say 5 words (Bag-of-5-g), but then we need to store near to the 5 permutation scores of the entire English language, and this perspective don't improve in a substantial way the results.

This paper is organized as follows, in the Sect. 2, we show the related work in this topic, in the Sect. 3 the mathematical background behind the Word2Vec, in the Sect. 4 we explain the experimentation framework that we used and the results derived from the experimentation and in the Sect. 5, we discuss the conclusions and the future work.

2 Related Work

In recent years, a lot of works has been done in the field of Sentiment analysis. This field started since the beginning of the century and it was intended for binary classification, which assigns opinions or reviews to bipolar classes such as positive or negative, i.e. the opinions was assigned into two classes such as positive or negative. Nowadays, one of the most used models are neural networks, as a work presented 1 by Zweig and Mikolov in [4] using Recurrent Neural Network language model (RNNLMs). Other works implements a classical feedforward neural network language models such as [5–9], that unlike (RNNLMs) this not maintain a hidden-layer of neurons with recurrent connections to their own previous values.

A characteristic of neural network language models is their representation of words as high dimensional real valued vectors, i.e. the words are converted via a learned lookuptable into real valued vectors which are used as the inputs to a neural network. The first proposals of distributed word representations, was including by [11, 12] in the 90's and most recently, studied in the context of feed-forward networks by Bengio et al. in [5] though a linear projection layer and a non linear hidden layer, used to learn jointly the word vector representation and a statistical language model. Later in RNNLMs by Mikolov, demonstrated outstanding performance in terms of word-prediction.

Today, Le and Mikolov proposed in [13] a Paragraph Vector, an unsupervised algorithm that learns fixed-length feature representations such as paragraphs, sentences and documents, achieving a new state-of-the-art results on sentiment analysis tasks. We apply these methods to perform an sentimental analysis based on some comments from twitter.

3 Prior Knowledge

3.1 Feed-Forward Neural Network

This was proposed by Bengio et al. in [5] as a language model. It consist in a of an input layer, a hidden layer with an output layer interconnected by modifiable weights, represented by links between layers as in the Fig. 1. Furthermore exists, a single bias unit that is connected to each unit other than the input units. These networks compute a series of transformations between their input and their output and the activities of the neurons in each layer are a nonlinear function of the activities in the layer below. The input units represent feature vector components and signals emitted by output units will be discriminant functions used for classification. In [5], consist in a N previous words that are encoded through 1-of-V coding (V is the size of vocabulary).

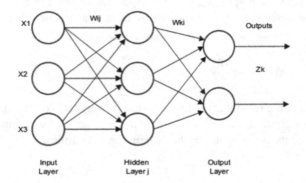

Fig. 1. Feed forward neural network. Source: Authors.

Each hidden unit realize the weighted sum of its inputs to form its (scalar) net activation, i.e. the net transference is the inner product of the inputs with the weights at the hidden unit, as can be seen in Eq. 1.

$$z = b + \sum_{i=1} x_i w_{ji} = b + \mathbf{W}^T \mathbf{x} \qquad (1)$$

Where w_{ji} denotes the input to hidden layer weights at the hidden unit j and b is the bias.

3.1.1 Activation Function

The activation function of a neuron defines the output of that neuron given an input or set of inputs. The most common activation function used is the logistic function, also known as the sigmoid function.

$$\varphi(z) = \frac{1}{1+e^{-z}} \tag{2}$$

3.2 Backpropagation

Is one of the most general methods for supervised learning of multilayer neural networks, proposed by Rumelhart et al. in [14]. This algorithm is based on delta rule and it looks for the minimum of the error function in weight space using the method of gradient descent. The method calculates the gradient of a loss function with respects to all the weights in the network. The gradient is fed to the optimization method, which in turn uses it to update the weights, thus be able to minimize the loss function.

The error is a scalar function of the weights such that the network outputs match the desired outputs, it is minimized. To reduce this measure of error, the weights are adjusted. The training error is defined by the Eq. 3:

$$E(w) = \frac{1}{2}\sum_{n \,\in training} (t^n - y^n)^2 \tag{3}$$

Where y is the network output and t is the target output. As the backpropagation is based on gradient descendent, the weights are updated in a direction that will reduce the error:

$$\Delta w = -\varepsilon\frac{\partial E}{\partial \omega} = -\varepsilon\sum_n x_i^n y^n (1 - y^n)(t^n - y^n) \tag{4}$$

Where ε is the learning rate indicates the variation in weights. To actualize the weights used a recursive algorithm, starting with the output neurons and working backwards to the input layer, adjusting weights for the m iteration it as follow:

$$w(m+1) = w(m) + \Delta w(m) \tag{5}$$

Finally, the error regarding weights is:

$$\frac{\partial E}{\partial z_j} = \frac{dy_i}{dz_j}\frac{\partial E}{\partial y_j} = y_i(1 - y_j)\frac{\partial E}{\partial y_j} \tag{6}$$

$$\frac{\partial E}{\partial y_i} = \sum_j \frac{dz_j}{dy_i}\frac{\partial E}{\partial z_j} = \sum_j w_{ij}\frac{\partial E}{\partial z_j} \tag{7}$$

$$\frac{\partial E}{\partial w_{ij}} = \frac{\partial z_i}{\partial w_{ij}}\frac{\partial E}{\partial z_j} = y_i\frac{\partial E}{\partial z_j}. \tag{8}$$

3.3   The Skip-Gram and Continuous Bag-of-Words (CBOW) Models

Distributed representations of words was proposed by Rumelhart et al. in [16] and have become extreme successful. The advantage of this model is that the representations of similar words are close in the vector space. They showed that in distributed representation of words are many types of similarities among words that can be expressed as linear translations. For example, vector operations "king" − "man" + "woman" = " queen" [17].

Two models for learning word representation proposed in 2013 by Mikolov et al. [15] was the Continuous Bag-of-words (CBOW) and Skip-gram models. In CBOW the goal is to predict one target word given the surrounding words. Similarly, the training objective of the Skip-gram model is predict a context, a window of word given a single word. This two models are based on artificial neural networks [5, 7].

3.4   Cbow

The Fig. 2, shows the architecture of this models, the input of CBOW are vectors with one-hot encoded vector, i.e., V is the vocabulary vector and the input is a vector with size V; the values of input represents the times to appears each word of the context in the vocabulary.

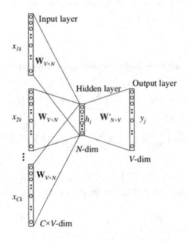

Fig. 2. Continuous Bag-of-words architecture. Source: Authors.

The weights between the input and the output layer are represented by a $V \times N$ matrix W. Each row of W is a vector v_w with dimension N, that represents the associated word of the input layer. Given a context, then:

$$h = \frac{1}{C}W(x_1 + x_2 + \ldots + x_C) \tag{9}$$

$$h = \frac{1}{C}W(v_{w1} + v_{w2} + \ldots + v_{wc}) \tag{10}$$

Where $w_1 + w_2 + \ldots + w_C$ are the words in the context and v_w is the input vector of a word w.

The hidden layer to the output layer have the weight matrix $W' = \{w'_{ij}\}$ with size $N \times V$. Using this weights, the score u_j can be calculated for each word in the vocabulary.

$$u_j = V'^T_{w_j}.h \tag{11}$$

Where V'_{w_j} is the $j - th$ column for the weights matrix W'. Using soft-max, we can be obtain the posterior distribution of words.

$$p(w_j|w_I) = y_j = \frac{\exp\left(V'^T_{wO}V_{wI}\right)}{\sum_{j'=1}^{V}\exp\left(V'^T_{wj}V_{wI}\right)} \tag{12}$$

Where y_j is the output of the $j - th$ node in the output layer. V_w is the input vector and V'_w is the output vector of the word w.

The loss function is E.

$$E = -\log p(w_o|w_{I,1}, \ldots, w_{I,C}) \tag{13}$$

$$= -u_{j*} + \log \sum_{j'=1}^{V} \exp(u_{j'}) \tag{14}$$

$$= -V'^T_{wo}.h + \log \sum_{j'=1}^{V} \exp\left(V'^T_{wo}.h\right) \tag{15}$$

The goal is minimize $E.j_*$ is the index of the actual output word in the output layer. The prediction error e_j of the output layer is given by the derivative of E :

$$e_j = \frac{\partial E}{\partial u_j} = y_j - t_j \tag{16}$$

Where t_j is 1 only when the $j - th$ node is the actual output word, in other case is 0. To obtain the gradient on the hidden to output layer weights is necessary to take the derivative of loss function on w'_{ij} :

$$\frac{\partial E}{\partial w'_{ij}} = \frac{\partial E}{\partial u_j} \cdot \frac{u_j}{\partial w'_{ij}} = e_j.h_i \tag{17}$$

The weight updating equation for hidden to output weights is calculated using stochastic gradient descendent:

$$w_{ij}^{\prime(new)} = w_{ij}^{(old)} - \in .e_j.h_i \tag{18}$$

$$V_{wj}^{\prime(new)} = V_{wj}^{(old)} - \in .e_j.h, \text{ for } j = [1, \ldots, V] \tag{19}$$

Where $\in$ is the learning rate and h_i is the $i - th$ node in the hidden layer.

We obtain the update of the input to hidden layer weights as:

$$V_{w_{I,c}}^{(new)} = V_{w_{I,c}}^{(old)} - \frac{1}{C} . \in .EH, \text{ for } c = [1, \ldots, C] \tag{20}$$

Where:

$$EH = \frac{\partial E}{\partial h_i} = \sum_{j=1}^{V} e_j.w_{ij}^{\prime} \tag{21}$$

Where h_i is the output of the $i - th$ node of the hidden layer and EH is the sum of the output vectors for all words in the vocabulary, weighted by the prediction error e_j.

3.5 Hierarchical Soft-Max

It was first introduced by Morin and Bengio [18] in the context of neural network language models. It uses a Huffman tree representation of the output layer based on word frequencies with the V words as it leaves and for each node, represents the relative probabilities of its child nodes. Each step in the search process from root to the target word is a normalization. The final probability for finding the target word is the continuous multiplication of probabilities in each search step.

Each word w can be reached by an appropriate path from the root of the tree. Then the hierarchical softmax defines the probability of a word being the output word $p(w_O|w_I)$ as follows:

$$p(w|w_I) = \prod_{j=1}^{L(w)-1} \sigma [\![n(w, j+1) = ch(n(w,j))]\!] . v_{n(w,j)}^{\prime}{}^{T} h \tag{22}$$

Where $n(w, j)$ is the $j - th$ node on the path from the root to word w. $L(w)$ is the length of the path. So, $n(w, 1)$ is the root and $n(w, L(w))$ is $w.ch(n)$ a left child of n and $[\![x]\!]$ is 1 if x is *True* or -1 in otherwise. $\sigma(x) = \frac{1}{(1+\exp(-x))}, v_{n(w,j)}^{\prime}$. is the vector representation for the inner node $n(w, j)$ and h is $\frac{1}{c}\sum_{c=1}^{C} V_{w_c}$ for the CBOW model.

3.6 Word2vec

Word2vec is an efficient implementation of the CBOW and the Skip Gram model for computing vector representations of words. It was developed by Google in 2013. This tool takes a text corpus as input and produces the word vectors as output. It first constructs a vocabulary from the training text data and then learns vector representation of words. The resulting word vector file can be used as features in many natural language processing and machine learning applications. This means that the model learns to map each word into a low dimensional continuous vector space from their distributional properties observed in some raw text corpus.

4 Experiments

The Fig. 3 shows the implemented process to carried out the experiment. The text corpus was a Google News Dataset with 100 billion words in representation vectorial. This is our vocabulary.

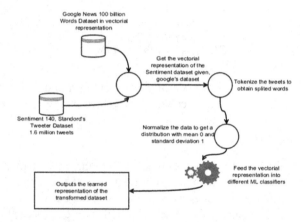

Fig. 3. Experiment process. Sources: Authors

To train the model, we use Sentiment 140, a Stanford's Tweeter Dataset with 1'600.000 tweets, 800.000 positive and 800.000 negatives, that they have been classified by humans. The format of this database consists of 6 fields:

- Polarity of the tweet (0 = negative, 2 = neutral, 4 = positive)
- ID of the tweet
- Date of the tweet
- The query. If there is no query, then this value is NO QUERY
- User that tweeted
- Text of the tweet

As the tweets have a symbols, links, etc., is necessary tokenize each tweet and then, obtain the vector representation of each one of them. As we mentioned above, the vector representation is for a word, for this we have to obtain the representation for a single word and then calculate the average for all words in the tweet. In this way, we obtain the vector representation for the full tweet.

The next step, as can be seen in the Fig. 3, is to normalize where we move our dataset into a gaussian distribution with a mean of zero and standard deviation 1, meaning that values above the mean will be positive, and those below the mean will be negative.

Finally as shown in Fig. 3, a classifier is used. For this experiment we used Logistic Regression, Support Vector Machine and Random Forest (Table 1).

Table 1. Partial results

Algorithm	Accuracy
Logistic Regression	0.746 (±0.009)
Support Vector Machine	0.748 (±0.010)
Random Forest	0.672 (±0.005)

5 Conclusions

The results of this naive experimental framework offers a performance far from the state of the art algorithms, but taking in account the lack of preprocessing over the twitter database such as removing the misspelling, removing urls, treat in a special way the emoticons that clearly express sentiments, and another techniques that can remove noise from the data we can introduce all of this enhancements to achieve better results in future work.

Also provide another sources of knowledge such as topic related information, with the aim to offer a better semantical meaning, moreover the syntactical information isn't taking in account this kind of information can give to the model another perspective that helps to define more accurately the meaning not only of the window of the neural network but also about the entire context of the tweet phrase.

References

1. Gautam, G., Yadav, D.: Sentiment Analysis of Twitter Data Using Machine Learning Approaches and Semantic Analysis, Department of Computer Science and Engineering Jaypee Institute of Information technology Noida, India (2014)
2. Maharani, W.: Microblogging Sentiment Analysis with Lexical Based and Machine Learning Approaches, Faculty of Informatics Telkom Institute of Technology, Bandung Indonesia (2013)
3. Waila, P., Marisha, R., Singh, V.K., Singh, M.K.: Evaluating Machine Learning and Unsupervised Semantic Orientation Approaches for Sentiment Analysis of Textual Reviews, Banaras Hindu University, Varanasi India (2012)
4. Zweig, G., Mikolov, T.: Context Dependent Recurrent Neural Network Language Model. Microsoft, Redmond (2012)
5. Bengio, Y., Ducharme, R., Vincent, P., Jauvin, C.: A neural probabilistic language model. J. Mach. Learn. Res. 3(6), 1137–1155 (2003)
6. Schwenk, H.: Continuous space language models. Comput. Speech Lang. 21(3), 492–518 (2007)

7. Mnih, A., Hinton, G.: Three new graphical models for statistical language modelling. In: Proceedings of the 24th International Conference on Machine Learning (2007)
8. Le, H.S., Oparin, I., Allauzen, A., Gauvain, J.-L., Yvon, F.: Structured output layer neural network language model. In: ICASSP (2011)
9. Hai Son, L., Allauzen, A., Yvon, F.: Measuring the influence of long range dependencies with neural network language models. In: Proceedings of the Workshop on the Future of Language Modeling for HLT (NAACL/HLT 2012) (2012)
10. Mikolov, T., Karafiat, M., Cernocky, J., Khudanpur, S.: Recurrent neural network based language model. In: Proceedings of Interspeech 2010 (2010)
11. Hinton, G.E.: Learning distributed representations of concepts. In: Proceedings of the Eighth Annual Conference of the Cognitive Science Society, Amherst, MA (1986)
12. Elman, J.L.: Distributed representations, simple recurrent networks, and grammatical structure. Mach. Learn. **7**(2), 195–225 (1991)
13. Le, Q., Mikolov, T.: 31st International Conference on Machine Learning, Beijing, China (2014)
14. Rumelhart, D.E., Hinton, G.E., Williams, R.J.: Learning internal representations by error propagation. In: Rumelhart, D.E., McClelland, J.L., CORPORATE PDP Research Group (eds.) Parallel Distributed Processing: Explorations in the Microstructure of Cognition, vol. 1, pp. 318–362. MIT Press, Cambridge (1986)
15. Mikolov, T., Chen, K., Corrado, G., Dean, J.: Efficient estimation of word representations in vector space (2013). arXiv preprint arXiv:1301.3781
16. Rumelhart, D.E., Hinton, G.E., Williams, R.J.: Learning representations by back-propagating errors. Nature **323**(6088), 533–536 (1986)
17. Mikolov, T., Yih, S.W., Zweig, G.: Linguistic regularities in continuous space word representations. In: NAACL HLT (2013)
18. Morin, F., Bengio, Y.: Hierarchical probabilistic neural network language model. In: Proceedings of the International Workshop on Artificial Intelligence and Statistics, pp. 246–252 (2005)
19. Natural Language Processing APIs and Python NLTK Demos, 23 March 2015. http://text-processing.com/demo/sentiment/5

Cognitive Models for Access Network Management

Vladimir Akishin$^{(\boxtimes)}$, Alex Goldstein, and Boris Goldstein

Federal Bonch-Bruevich Telecommunication University St. Petersburg,
Saint Petersburg, Russia
v.akishin@argustelecom.ru, {agold, bgold}@niits.ru

Abstract. Research has highlighted the necessity of a cognitive nature of the access network (AN) management model intended as a cognitive representation describing a new NGN/IMS and post-NGN business processes. The AN management problems are solved more or less for existing digital fixed and mobile telecommunications networks. But applying the same principles for management new services in new post-NGN telecommunication networks with IoT, Cloud computing, Big Data analysis, CEM, etc. seems not to be perspective. Usage of cognitive approach with fuzzy cognitive maps (FCMs) might be very useful in this new telecommunication paradigm. The goal of this paper is to provide modern cognitive AN management approaches with FCMs survey.

Keywords: Business intelligence · Cognitive maps · Customer Experience Management · OSS/BSS

1 Introduction

The traditional OSS/BSS approach suited for traditional networks. The shift of a telecommunication paradigm happening now caused the need for the new ideas and methods of management of considerably become complicated service of calls/sessions for network elements of NGN/IMS which isn't satisfied by only one gain of local intelligence of these network elements [1]. Problems of routing and QoS of a multimedia traffic at the IP/MPLS transport plane can be resolved by a research of probabilistic and time characteristics and traditional methods of the teletraffic theory [2], but the session plane and also the service plane require already attraction of cognitive mechanisms and models for control of this plane. The reason of it that the telecommunication industry already begins to face, and in the nearest future will even more seriously face tasks which complexity considerably exceeds possibilities of the OSS/BSS systems existing today. In particular, modern telecom operator also faces the issues of linking client experience and aspects of network management, including in the context of OSS/BSS.

Figure 1 shows two approaches to control of info-communications which we with the known level of conditionality can define as the hard planning and as cognitive models of control. Shifting to the right - towards cognitive models of control - just corresponds to above-mentioned shift of a paradigm of the modern info-communications.

© Springer International Publishing AG 2017
O. Galinina et al. (Eds.): NEW2AN/ruSMART/NsCC 2017, LNCS 10531, pp. 375–381, 2017.
DOI: 10.1007/978-3-319-67380-6_34

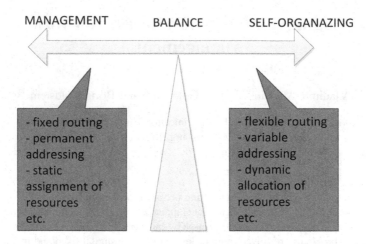

Fig. 1. The balance between two approaches

An exponentially growing number of infocommunication technologies used in the modern access network, as well as the complication of logical interfaces between NGOSS and the functional subsystems of the business support system BSS, subsystems of the life cycle management system for services and resources, subsystems for the production of services on the network, the definition of basic business entities in the NGOSS information model based on the SID (Shared Information and Data) and the dynamics of their changes, the more complex interconnection of these business entities with Functional areas of Frameworx - all this and causes the involvement of new models and methods of cognitive management of info-communications.

2 Related Works

The above-mentioned paradigm shift in telecommunications network management is described in number of articles in [1, 2]. Standards and modern aspects of telecommunication process management are considered in TMForum [3–5, 7, 8] latest developments. The evolution of telecommunications network management approaches is shown in [13].

Works by Michael Glykas [12] propose the fundamental methods which are based on fuzzy cognitive maps and allow to solve the problem of interconnection between different factors of operation management.

Problems of interconnections between aspects of network management and customer experience management are discussed in [6, 9–11].

3 Applicability of Fuzzy Cognitive Maps to Control AN

Fuzzy Cognitive Maps (FCMs) consist of concept nodes (factors) and weighted arcs, which are graphically illustrated as a signed weighted graph with feed back. Signed weighed arcs, connecting the concept nodes (factors), represent the causal relationship that exists among concepts. In general, concepts of a FCM, represent key-factors and characteristics of the modeled complex system and stand for: events, goals, inputs, outputs, states, variables and trends of the complex system been modeled. This graphic display shows clearly which concepts influences with other concepts and what this degree of influence is (Fig. 2).

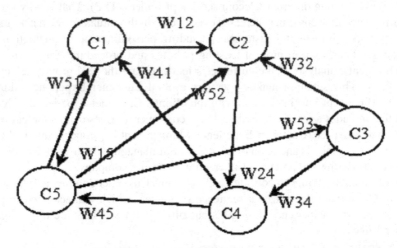

Fig. 2. Example of cognitive model

A Fuzzy Cognitive Map is a graph shows the degree of causal relationship - among concepts of the map knowledge expressions and the causal relationships are expressed by and fuzzy weights. Existing knowledge on the behavior of the system is stored in the structure of nodes and interconnections of the map. Each one of the key-factors of the system. Relationships between concepts have three possible types; (a) either express positive causality between two concepts ($W_{ij} > 0$) (b) negative causality ($W_{ij} < 0$) and (c) no relationship ($W_{ij} = 0$). The value of W_{ij} indicates how strongly concept C_i influences concept C_j. The sign of W_{ij} indicates whether the relationship between concepts C_i and C_j is direct or inverse. The direction of causality indicates whether concept C_i causes concept C_j, or vice versa. These parameters have to be considered when a value is assigned to weight W_{ij}. A new formulation for calculating the values of concepts at each time step, of a Fuzzy Cognitive Map, is proposed:

$$A_i^t = f(k_1 \sum_{\substack{j=1 \\ j \neq i}}^{n} A_j^{t-1} W_{ji} + k_2 A_j^{t-1})$$

The k1 expresses the influence of the interconnected concepts in the configuration of the new value of the concept Ai and k2 represents the proportion of the contribution of the previous value of the concept in the computation of the new value and the. This new formulation assumes that a concept links to itself with a weight Wii = k2 [12].

4 Aims and Research

Today customer loyalty and churn rate are the most important issues for companies in telecommunication market they have to deal with. Realities of the telecommunications market force companies to operate in a situation when the majority of customers have already divided among different telecom service providers (TSP). That is why the cost of acquiring each new customer increases every year. In this situation, such activities as competent work with existing customers, including personification of interaction with each specific client, optimization of existing policies and procedures for communications with clients, analysis of the customer's feedback are the real growth drivers for the company. This situation makes operators embed Customer Experience Management (CEM) concept into interaction with the clients. Customer experience (CX) is a concept that includes a set of impressions that a customer receives from interaction with telecom operator. Then, Customer Experience Management is a concept that includes a set of processes, methods and technologies aimed at managing the impressions that the client receives during the process of interaction with the company.

First of all authors offer to consider such terms as Customer Journey and Customer Lifecycle. Customer Lifecycle is a sequence of stages, which characterized different aspects of customer experience during his customer journey. There are nine stages of Customer Lifecycle.

- Be Aware (Observe, Learn, React)
- Interact (Inquire, Request Detail, Reserve)
- Choose (Select, Place, Receive)
- Consume (Use, Review Use, Evaluate Value)
- Manage (Manage Profile/Service, Get Help, Request to Resolve)
- Pay (Receive Notification, Verify or dispute, Pay)
- Renew (Enhance Selection, Re-contract)
- Recommend (Refer Product/Service, Gain Loyalty)
- Leave (Feedback, Discontinue)

There are a lot of different indicators, which can determine customer experience during his relations with telecom operator. TM Forum offers over 450 metrics for measuring customer experience that extend across the entire customer lifecycle []. This metrics are presented in the projection on stages of customer Lifecycle and types of operators KPIs.

Nowadays, there isn't any formal model describing how the customer experience is formed in the specifics of the operator's activity, in particular how the network characteristics, the quality of the provided service, the level of development of the OSS landscape and other aspects of the telecommunications environment can influence it.

One of the main aims of investigation is to offer a variant of a functional model, which describes how customer experience can be influenced by different operations KPI and network characteristic. The model is based on a fuzzy cognitive map.

It is supposed, that the functional model which connects customer experience and network aspects of operator's management will be able to:

- To establish a correlation between numerical operator's KPIs and an integral indicator of customer experience;
- Take into account the experience accumulated by the client at various stages of his lifecycle;
- Take into account standards (TM Forum recommendations) and current experience of operator's management.

5 Investigation

Determination of factors which can be a foundation of network management functional model became the first step of our investigation.

TM Forum Lifecycle metrics were selected as regulatory factors for our functional model [4]. It has been selected as the foundation of functional model for two reasons:

- TM Forum model of identifies all the key stages in the relationship between the customer and telecom operator – it gives an opportunity to get integrated assessment of Customer Experience throughout all Customer Journey.
- It is widely recognized that Customer Experience cannot be assessed from single metrics in isolation, but instead should be assessed in the context of the customer's unique journey, taking into account the customer's history, preferences, mood and expectations.

For our research we took 42 metrics from 4 Lifecycle Stages: Choose, Consume, Manage, Pay. The choice was justified by the fact that this metrics can be received from traditional OSS/BSS of modern telecom operator.

Examples of metrics are presented below:

- Orders Successful – Used to evaluate the popularity of Product and Service Plan IDs and sales performance across all Channels (Choose)$
- Service Activations Successful – Used to measure efficiency of Service Activation process (Choose);
- % TCP Connect Success – Used to measure media streaming experience (Consume);
- Service Interruptions - Used to measure service availability (Consume);
- First Contact Resolutions – Used to measure the performance of support channels (Manage);
- % Service Configurations Failed - Used to measure efficiency of Service Configuration process (Manage);
- % Top-Up Success – Used to measure the Top-Up success rate, and identify channels with faults (Pay);
- % Bills Incorrect – Used to measure experience of billing function (Pay).

As part of the investigation, the vectors of the relationship between the factors are represented as a mutual interaction matrix. Some cases for example are presented for stage Manage (Table 1).

Table 1. Mutual interaction matrix

Factors	Seconds manage profile service	First contact resolutions	Incidents resolved	...	Minutes per service configuration
Seconds manage profile service		−	0	...	−
First contact resolutions	0		0	...	+
Incidents resolved	0	+		...	+
...	...	...	...		...
Minutes per service configuration	+	−	0	...	

As a result of analysis there was formed a problem field. The problem field is a combination of factors and interrelations between them with influence vector. Cognitive mapping of network's and customer's aspects is carried out in the form of a cognitive map – a weighted-oriented graph.

$$G = <V, E>,$$

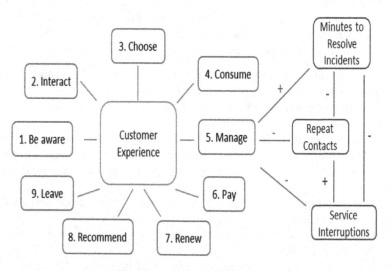

Fig. 3. A bit of cognitive map

where
V   – vertex set; Vi ∈ V, i = 1, 2, …, k;
E   – set of arcs.

Arc eij ∈ E, i, j = 1, 2, …, n connects vertices of the graph. Vertices of graph are significant factors (TM Forum metrics) in operator's network management. A few significant factors with their interrelations are presented for example on cognitive Map (Fig. 3).

6   Conclusion

New aspects of access network management were considered in the paper. Particularly there were investigated actual problems of interconnection between network management and customer experience. One of the main achievements of the investigation is functional model based on fuzzy cognitive map which shows relation between quantitative aspects of network management and customer experience on the different lifecycle stages. The next step of the investigation is quantitative estimation of the weights of the graph.

References

1. Goldstein, B., Kucheryavy, A.: Post-NGN Communication Networks. BHV, Sankt-Petersburg (2013)
2. Goldstein, A., Yanovsky, G.: Traffic engineering in MPLS tunnels. In: International Conference on "Next Generation Teletraffic and Wired/Wireless Advanced Networking (NEW2AN 2004)", 02–06 February 2004
3. Samuylov, K., Chukarin, A., Yarkina, N.: Business processes and information technologies in the management of a modern infocommunication company. Alpina Pablisherz (2015)
4. TMForum. GB962A_Lifecycle_Metrics R15.0.1. TMForum, December 2015
5. Enhanced Telecom Operations Map (eTOM) GB921: Concepts and Principles Release 8.0, TM Forum (2009)
6. Lankevich, K., Habaev, N., Skorinov, M.: OSS complex as a tool to control customer loyalty. T-Comm – Telecommun. Transp. 36–38 (2015)
7. TM FORUM. GB921CP. Business Process Framework (eTOM). Concepts and Principles. Release 13.0. TM Forum, August 2013
8. «TMF_GuideBk_Big_Data_Analytics_Addendum_A_v1.0.1», TeleManagement Forum (2013)
9. Goldstein, A., Kislyakov, S., Skorinov, M.: Telecom Aikido: NGOSS. Mob. Telecommun. (6-7), 10 (2015)
10. Goldstein, A., Skorinov, M., Fenomenov, M.: Big data - releasing the genie out of the bottle? Technol. Commun. (5), 34–38 (2015)
11. Butcher, S.A.: Customer Loyalty Programmes and Clubs. Vil'yams, Moscow (2004)
12. Glykas, M.: Fuzzy Cognitive Maps. Springer, Heidelberg (2010)
13. Pavlou, G.: On the evolution of management approaches, frameworks and protocols: a historical perspective. J. Netw. Syst. Manag. 15(4), 425–445 (2007)

Bargaining in a Dual Radar and Communication System Using Radar-Prioritized OFDM Waveforms

Andrey Garnaev[1,2]($\boxtimes$), Wade Trappe[2], and Athina Petropulu[3]

[1] Saint Petersburg State University, St. Petersburg, Russia
garnaev@yahoo.com
[2] WINLAB, Rutgers University, North Brunswick, USA
trappe@winlab.rutgers.edu
[3] Department of Electrical and Computer Engineering, Rutgers University,
Piscataway, USA
athinap@rutgers.edu

Abstract. This paper examines a dual radar/communication system employing OFDM style waveforms with two objectives: a radar task involving target tracking, and a communication task involving communicating with a distant receiver. One challenge with such two objective problems is that the optimal solution for one of objective could be non-optimal for the other. We examine the problem of resolving dual radar/communication objectives and solve the associated two objective problem in two steps: *First*, for each fixed level of performance for the radar's objective, the optimal solution is found in a water-filling form which includes the classical OFDM waterfilling solution as a boundary case. The formulation returns a continuum of optimal solutions (a solution per radar performance level). *Second*, from this continuum of solutions, we find the Nash bargaining solution, which returns the best tradeoff between the two objectives, and an algorithm to find the Nash bargaining solution is presented.

Keywords: Radar · Communication · OFDM · Waveforms · Bargaining

1 Introduction

Wireless communication and radar systems are two distinctly different systems that utilize radio frequency (RF) spectrum. While these systems have traditionally been designed separately from each other, they nonetheless share many properties that can facilitate a new generation of dual communication and radar systems that can handle the emerging trend towards less exclusionary spectrum policies. It is anticipated that radar and communication functions, and their associated systems, will be asked to share spectrum bands, in large part due to government policies that allow for more freedom in spectrum licensing, and due to opportunities that might arise for these different systems to support

© Springer International Publishing AG 2017
O. Galinina et al. (Eds.): NEW2AN/ruSMART/NsCC 2017, LNCS 10531, pp. 382–394, 2017.
DOI: 10.1007/978-3-319-67380-6_35

each other's functions. A further reason arises because radar systems typically use backbone communication infrastructure for relaying communications data to the operators and users of the data, which can be costly in many realistic operational scenarios. Examples of this include scenarios where the operators are not close to the radar, or when the radar is in sparsely populated areas with little existing communication infrastructure, or for a networked radar system where many communication backplanes would be needed. Using the radar transceivers for both target tracking and communication could offer a large life cycle cost savings to the operation of these systems, eliminating the cost and maintenance for communications services.

This paper examines a scenario involving a dual radar-communication system that faces two objectives: a radar task involving target tracking, and a communication task that involves communicating with a distant receiver. While there is active research into designing systems that can benefit from using the same spectral band for both radar and communication functions, most of the research has been focused on designing new waveforms with jointly desirable communications and radar functions. In this work, though, we take a complementary view of the problem by recognizing that the joint radar-communication system will face choices between better serving the radar function or better serving the communication function, and thus there are fundamental *fairness* challenges associated with waveform design and resource allocation in a joint radar-communication system. In order to explore the problem of fairness in such a multi-function system, we explore the specific case where an OFDM-style waveform is used for simultaneously supporting a radar tracking and communication throughput objective.

We have chosen to focus our attention on multicarrier waveform design since multicarrier modulation is an effective approach to building communication systems that address fading while also supporting high throughput demands associated with modern wireless applications. OFDM has become the basis for many communication standards, such as LTE, WiFi and WiMax, due to its several advantages like robustness against multi-path fading, easy synchronization and equalization, and a high flexibility in system design. While multicarrier methods have been very popular for communication system design, their applicability to the radar scenario is more nascent and there is strong evidence that multicarrier waveforms can have desirable radar properties, such as reported in [25, 26]. Also, in [20], suitability of OFDM-like signals for radar applications was demonstrated, and implementation of radar networks with integrated communications functions based on OFDM signals was studied in [18].

A review of the state of the art on radar and communication system coexistence and bandwidth sharing can be found [15]. In [14], a design of integrated radar and communication systems that utilize weighted pulse trains with the elements of Oppermann sequences serving as complex-valued weights is suggested. An integrated radar-communication system based on multiple input and multiple output (MIMO) that shares hardware and spectrum was investigated in [27]. An approach was introduce in [11] that recognizes the primary radar function of the system and employs beam sidelobes to transmit the digital communication signals. In [19], it is shown that radar waveform design for multiple target estimation can be accomplished using a linear combination of mutual information between each target signal and the related received beam. A two-stage resource allocation optimization method was developed [10] to assign bandwidth to the UEs of cellular communications networks operating in the vicinity of a radar such that the radar spectrum was shared with certain sectors of the communications infrastructure.

One of challenge facing two objective problems is that the optimal solution for one objective could be non-optimal for the other, while another challenge is that the objectives are in different metrics. We deal with these challenges in two steps. *First*, for each fixed level of performance for radar's objective, the optimal solution is found in water-filling form, which includes classical OFDM solution as a boundary case. This formulation allows one to combine radar and communication metrics in one optimization problem and returns a continuum of optimal solutions (a solution per level of radar performance). *Second*, from this continuum of solutions we find the solution that corresponds to the Nash bargaining solution. Numerical modeling illustrates the efficiency of the associated solutions.

The problem of bargaining arises when two agents must determine how to cooperate in choosing between different equilibria with different payoffs for each agent. The foundation of modern bargaining theory was presented by Nash in [22]. A survey of different bargaining concepts used in wireless communication is given in [28]. Here we just mention a few works. In [21], cooperative bargaining solutions, in the context of LTE-A networks, for resource allocation over the available component carriers was investigated. In [9], a bargaining solution for fair performing dual radar and communication task was designed. In [3,4], α-fair channel sharing strategy for joint access by Wi-Fi and LTE-U networks with equal priority was derived. In [17], an axiomatic approach for fairness measures in resource allocation was developed. In [8], a problem of bargaining over the fair trade-off between secrecy and throughput in OFDM communications was studied.

The organization of this paper is as follows: in Sect. 2, we present the basic system model and formulate the corresponding optimization problems. In Sect. 3, auxiliary notations are introduced and auxiliary results are obtained. In Sect. 4, the optimal strategy is found for each fixed level of performance for radar's objective. In Sect. 5, the bargaining problem over a continuum of optimal strategies is formulated and solved. In Sect. 6, conclusions are given, and in Appendix the proofs of the obtained results are supplied.

2   Formulation of the Problem

We begin our formulation by considering an operational scenario involving a single RF transceiver that is attempting to support two different objectives: communicate with a communication receiver that is distant and separate from the transmitter (the *communication objective*), while also supporting the tracking of a radar target through the reflections witnessed at the RF transmitter (the *radar's objective*). To support these two different objectives, the transmitter uses a spectrum band that consists of n adjacent subchannels, which may be associated with n different subcarriers in a transmission scheme like OFDM. Associated with these n different subcarriers are two different collections of (fading) channel gains. Specifically, we let h_i^R correspond to the i-th radar channel gain associated with the *round-trip* effect of the transmitted signal, reflected off the radar target, and received at the RF transceiver. Similarly, we denote by h_i^C the i-th channel gain associated with the i-th communication subcarrier between the transmitter and the communication recipient.

In this work, we assume that the dual radar-communication system has knowledge of the channel states for both tasks. The assumption that the radar task has awareness of the channel states is consistent with radar tracking applications where the tracker maintains state information associated with past estimates and performs the prediction of future channel states, which are typically used by modern waveform schedulers in radar systems. Likewise, knowledge of the channel state in a communication system is often achieved through channel state feedback mechanisms.

Although there are two different objectives, *there is nonetheless only a single transmit controller responsible for deciding how to allocate power across the different subcarriers to best meet radar and communication tradeoffs.* We consider a power-allocation strategy for the transmitter to be the power vector $\boldsymbol{P} = (P_1, \ldots, P_n)$, where P_i is the power assigned for transmitting on subcarrier i, and $\sum_{i=1}^{n} P_i = \overline{P}$ where $\overline{P}$ is the total power budget allocated for transmission. Let $\Pi = \{\boldsymbol{P} = (P_1, \ldots, P_n) : \sum_{i=1}^{n} P_i = \overline{P}, P_i \geq 0, i = 1, \ldots, n\}$ be the set of all strategies (Fig. 1).

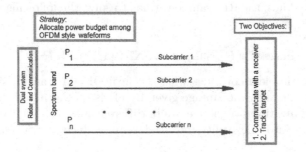

Fig. 1. Model for two different objectives.

To unify the examination of radar and communication metrics, we note that radar detection/tracking and communication throughput are both closely related to the associated signal-to-noise ratios (SNR) as witnessed at the appropriate recipient. For the radar system, the SNR for the i-th subcarrier corresponds to the monostatic, round-trip (squared) magnitude impact upon the corresponding power P_i, which we shall denote by $h_i^R P_i$. We may then sum over all the sub-carriers to obtain the total radar SNR, which is given by: $\mathrm{SNR}^R(\boldsymbol{P}) = \sum_{i=1}^{n} h_i^R P_i$. Thus, $\mathrm{SNR}^R(\boldsymbol{P})$ can be considered as a utility function for the RF transceiver when considering the fulfillment of the radar task. See, for example, [2,5–7,23] about detection of illegal signals in bandwidth by SNR. Let $\boldsymbol{P}^R$ be the strategy that maximizes the radar's objective, i.e., $\boldsymbol{P}^R = \arg\max\{\mathrm{SNR}^R(\boldsymbol{P}); \boldsymbol{P} \in \Pi\}$. We call such a mode when the RF transceiver system applies strategy $\boldsymbol{P}^R$ as the *radar mode*.

For fulfilling the communication task, the total (Shannon) communication capacity (which we use as a surrogate for throughput) for the RF transceiver's transmission to its intended communication recipient is considered, i.e., $\mathrm{T}^C(\boldsymbol{P}) = \sum_{i=1}^{n} \ln\left(1 + h_i^C P_i\right)$, where h_i^C is the communication fading sub-carriers gains. Let $\boldsymbol{P}^C$ be the strategy that maximizes the communication's objective, i.e., $\boldsymbol{P}^C = \arg\max\{\mathrm{T}^C(\boldsymbol{P}); \boldsymbol{P} \in \Pi\}$. We call such a mode when the RF transceiver applies strategy $\boldsymbol{P}^C$ as *communication mode*. Note that $\boldsymbol{P}^C$ is a classical OFDM solution. A survey on OFDM solutions can be found in [13]. Here we mention just a few works: closed form solution of the water-filling equation was obtained in [1], its generalization for individual peak power constraints was studied in [12], power increment water-filling and power decrement water-filling algorithms were suggested in [24].

In this paper, we consider the situation where the radar's objective is considered a *primary task* for the RF transceiver system, which means that a fixed *level of performance* for the radar's objective has to be maintained with certainty. In other words, the communication object should not disrupt the performance of the radar's objective, but beyond this restriction communication can be maximized. To formulate the problem, denote the level of performance for the radar's objective by ϵ^R. Thus, the RF transceiver has to solve the following optimization problem:

$$\text{maximize } \mathrm{T}^C(\boldsymbol{P}), \text{ subject to } \mathrm{SNR}^R(\boldsymbol{P}) \geq \epsilon^R, \boldsymbol{P} \in \Pi. \tag{1}$$

For $\epsilon^R = 0$ this problem coincides with classical OFDM transmission problem. Also, note that one of the advantage given by (1) is that although radar's objective and communication's objective are in different metrics, due to they are not combined in (1) into one utility, no need to introduce metric transformation coefficients.

Of course, for correct formulation of problem (1), the set of feasible strategies Π^C has to be non-empty, where $\Pi^C := \left\{\boldsymbol{P} \in \Pi : \sum_{i=1}^{n} h_i^R P_i \geq \epsilon^R\right\}$. It is clear that this set is not empty if and only if $\max_{\boldsymbol{P} \in \Pi} \sum_{i=1}^{n} h_i^R P_i \geq \epsilon^R$. This condition is equivalent to the following:

$$\epsilon^R \leq h_{\max}^R \overline{P} \text{ with } h_{\max}^R = \max_i h_i^R. \tag{2}$$

3   Auxiliary Results

In this section we introduce an auxiliary notation and give auxiliary results. Let

$$p_i = p_i(\omega, \nu) := \left[\frac{1}{\omega - \nu h_i^R} - \frac{1}{h_i^C} \right]_+ \quad \text{for } i = 1, ..., n, \tag{3}$$

$$S(\omega, \nu) := \sum_{i=1}^n p_i(\omega, \nu) = \sum_{i=1}^n \left[\frac{1}{\omega - \nu h_i^R} - \frac{1}{h_i^C} \right]_+, \tag{4}$$

$$S^R(\omega, \nu) := \sum_{i=1}^n h_i^R p_i(\omega, \nu) = \sum_{i=1}^n h_i^R \left[\frac{1}{\omega - \nu h_i^R} - \frac{1}{h_i^C} \right]_+ \quad \text{with } 0 \le h_{\max}^R \nu < \omega. \tag{5}$$

Theorem 1

(a) *For a fixed $\nu \ge 0$ both functions $S(\omega, \nu)$ and $S^R(\omega, \nu)$ are continuous and decreasing from infinity for $\omega \downarrow h_{\max}^R \nu$ to zero for $\omega = \max_i(h_i^R \nu + h_i^C)$.*

(b) *For a fixed $\omega > 0$ both functions $S(\omega, \nu)$ and $S^R(\omega, \nu)$ are continuous and increasing from $S(\omega, 0)$ and $S^R(\omega, 0)$ for $\nu = 0$ to infinity for $\nu \uparrow \omega / h_{\max}^R$.*

(c) *There exists the unique $\omega^C \in (0, h_{\max}^C)$ with $h_{\max}^C = \max_i h_i^C$ such that*

$$S(\omega^C, 0) = \overline{P}. \tag{6}$$

(d) *There is a uniquely defined continuous function $\Omega(\nu)$ increasing from ω^C for $\nu = 0$ to infinity as ν tends to infinity such that*

$$S(\Omega(\nu), \nu) = \overline{P}. \tag{7}$$

(e) *There is a unique $\omega^R \in (0, h_{\max}^C)$ such that*

$$S^R(\omega^R, 0) = \epsilon^R. \tag{8}$$

(f) *There is a uniquely defined continuous function $N(\omega)$ increasing from zero for $\omega = \omega^R$ to infinity as ω tends to infinity such that*

$$S^R(\omega, N(\omega)) = \epsilon^R. \tag{9}$$

Due to the monotonicity properties of $S(\omega, \nu)$ and $S^R(\omega, \nu)$ given by (a) and (b), ω^C, ω^R, $\Omega(\nu)$ and $N(\omega)$ can be found by the bisection method.

Note that function $p_i(\omega, \nu)$ has a water-filling structure and for $\nu = 0$ (so, when the radar's objection of the RF transceiver system is not taken into account) it coincides with a classical water-filling solution maximizing throughput in OFDM transmission. Also, ω^C is the Lagrangian multiplier for such classical water-filling solution.

4   Optimal Strategy

In this Section the optimal strategy is found in water-filling form.

Theorem 2. *For a fixed ϵ^R the optimization problem (1) has the unique optimal strategy $\boldsymbol{P}_{\epsilon^R}$.*
   (a) Let

$$\sum_{i=1}^{n} h_i^R p_i(\omega^C, 0) \geq \epsilon^R. \qquad (10)$$

Then
$$\boldsymbol{P}_{\epsilon^R} = \boldsymbol{p}(\omega^C, 0).$$

(b) Let

$$\sum_{i=1}^{n} h_i^R p_i(\omega^C, 0) < \epsilon^R. \qquad (11)$$

Then

$$P_{\epsilon^R, i} = p_i(\Omega(\nu), \nu) \, for \, i = 1, \ldots, n, \qquad (12)$$

where $\nu > 0$ is the unique root of the equation:

$$S^R(\Omega(\nu), \nu) = \epsilon^R. \qquad (13)$$

Thus, (10) reflects that the power budget is large enough to keep the classical water-filling solution for OFDM as the solution for dual radar/communication system with radar priority; while (11) means that the classical water-filling solution has to be adjusted by an auxiliary boundary condition in such a way to maintain radar's priority.

The theorem has the following Corollary giving an algorithm to find the optimal solution.

Corollary 1. *The optimal strategy $\boldsymbol{P}_{\epsilon^R}$ is given as follows:*
   (a) Let

$$\omega^C \leq \omega^R. \qquad (14)$$

Then, $\boldsymbol{P}_{\epsilon^R} = \boldsymbol{p}(\omega^C, 0)$.
   (b) Let

$$\omega^C > \omega^R. \qquad (15)$$

Then, $\boldsymbol{P}_{\epsilon^R} = \boldsymbol{p}(\omega, \nu)$ with ω and ν are the unique roots of the equations:

$$\omega = \Omega(\nu) \ and \ \nu = N(\omega). \qquad (16)$$

These ω and ν can be found by Algorithm 1 to calculate a fixed point of superposition of functions Ω and N, i.e., $\omega = \Omega(N(\omega))$.

Algorithm 1. Iterative algorithm, where θ is the tolerance of the algorithm

let $\omega_0 = \omega^C$, $\nu_0 = N(\omega_0)$, $i = 0$
repeat
   let $i = i + 1$
   let $\omega_i = \Omega(\nu_{i-1})$, $\nu_i = N(\omega_i)$
until $\max \{|\omega_i - \omega_{i-1}|, |\nu_i - \nu_{i-1}|\} > \theta$

5 Bargaining Problem

The bargaining solution is based on the concept of negotiation between *agents* to get a solution that is appropriate for both of them [22].

First, to formulate the bargaining problem, we have to define *agents* who are bargaining. We consider the radar's objective and the communication's objective as such agents.

Second, we have to define a pair $(\Gamma, \boldsymbol{g}_0)$. Here Γ is the so-called *feasibility* set, which is a subset of $\mathbb{R}^2$. Its elements reflect all possible payoffs that these two agents could get if they achieve an agreement. $\boldsymbol{g}_0 = (g_{01}, g_{02})$ is a *disagreement* point where g_{01} and g_{02} are payoffs to the agents if they do not achieve an agreement.

In the considered case, since the payoff for the radar's objective is SNR^R, and payoff for the communication's objective is T^C, Γ (all possible payoffs) can be defined as a parameterized curve with respect to ϵ^R in the plane $(\text{SNR}^R, \text{T}^C)$ as follows: $\Gamma = \{(\text{SNR}^R(\boldsymbol{P}_{\epsilon^R}), \text{T}^C(\boldsymbol{P}_{\epsilon^R})) : \epsilon^R \in [E_{\min}, E_{\max}]\}$ with $E_{\min} = \sum_{i=1}^{n} h_i^R p_i(\omega^C, 0)$ and $E_{\max} = h_{\max}^R \overline{P}$. As the disagreement point, we can consider $(\text{SNR}_d^R, \text{T}_d^C) = \left(E_{\min}, \text{T}^C(\boldsymbol{P}^R)\right)$.

Third, the Nash bargaining solution (NBS) is defined as a solution of following optimization problem:

$$\text{maximize NP}(g_1, g_2) := (g_1 - g_{01})(g_2 - g_{02})$$
$$\text{subject to } (g_1, g_2) \in \Gamma, \tag{17}$$

with $\text{NP}(g_1, g_2)$ is called *the Nash product*.

NBS is the *Pareto optimal* point of set Γ. Recall that the point $(g_1, g_2) \in \Gamma$ is said to be Pareto optimal if and only if there is no $(g_1', g_2') \in \Gamma$ such that $g_i' \geq g_i$ for $i = 1, 2$ and $g_i' > g_i$ for at least one i.

In the considered case, Γ is a concave curve in the plane $(\text{SNR}^R, \text{T}^C)$, since $\boldsymbol{P}_{\epsilon^R}$ is the unique solution of the concave optimization problem (1), and $\text{SNR}^R(\boldsymbol{P}_{\epsilon^R}) = \epsilon^R$ for $\epsilon^R \in [E_{\min}, E_{\max}]$. Thus, each point of this Γ is Pareto optimal, and the following result holds.

Theorem 3. *NBS is* $(SNR^R(\boldsymbol{P}_{\epsilon^R}), T^C(\boldsymbol{P}_{\epsilon^R}))$, *where*

$$\epsilon^R = \arg \max_{\epsilon^R \in [E_{\min}, E_{\max}]} NP_{\epsilon^R} \tag{18}$$

Algorithm 2. Nash bargaining solution, where θ is the tolerance of the algorithm

> let $\epsilon_a = E_{\min}$, $\epsilon_b = E_{\max}$
> **repeat**
>   let $\epsilon_c = (\epsilon_a + \epsilon_b)/2$, $\epsilon_{a+} = (\epsilon_a + \epsilon_c)/2$, $\epsilon_{b-} = (\epsilon_c + \epsilon_b)/2$
>   **if** $\mathrm{NP}_{\epsilon_{a+}} \leq \mathrm{NP}_{\epsilon_c} \leq \mathrm{NP}_{\epsilon_{b-}}$ **then**
> $\epsilon_a = \epsilon_{a+}$
>   **end if**
>   **if** $\mathrm{NP}_{\epsilon_{a+}} \geq \mathrm{NP}_{\epsilon_c} \geq \mathrm{NP}_{\epsilon_{b-}}$ **then**
> $\epsilon_b = \epsilon_{b-}$
>   **end if**
>   **if** $\mathrm{NP}_{\epsilon_c} \geq \max\{\mathrm{NP}_{\epsilon_{a+}}, \mathrm{NP}_{\epsilon_{b-}}\}$ **then**
> $\epsilon_a = \epsilon_{a+}, \epsilon_b = \epsilon_{b-}$
>   **end if**
> **until** $\alpha_b - \alpha_a > \theta$

with $NP_{\epsilon^R} := \left(\epsilon^R - E_{\min}\right)\left(T^C(\boldsymbol{P}_{\epsilon^R}) - T^C(\boldsymbol{P}^R)\right).$

Since the maximum in (18) is unique, it can be found by the Nelder-Mead simplex algorithm [16] which is described in Algorithm 2.

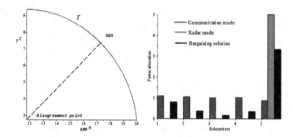

Fig. 2. Feasibility set Γ and NBS in the plane $(\mathrm{SNR}^R, \mathrm{T}^C)$ (left) and optimal strategies for communication mode and radar mode, and the bargaining solution (right) when $\overline{P} = 5$, $n = 5$, $h^C = (9, 6.5, 5, 5.5, 3)$, $h^R = (3, 2, 1, 2, 4)$.

Figure 2 illustrates the feasibility set Γ and the NBS in the plane $(\mathrm{SNR}^R, \mathrm{T}^C)$, where the Nash bargaining solution which is (17.298, 7.369). Also shown are the optimal strategies for communication mode, for radar mode and the bargaining solution. Since h_i^R is maximum for $i = 4$, the optimal strategy in the radar's mode is focused on subcarrier 4. The optimal strategy for communication mode has a waterfilling form and uses all of the subcarriers. The strategy corresponding to NBS inherits these properties for the single-mode solution, namely, it spreads power among all the sub-carriers like the communication mode solution and at the same time focuses most of the power on fulfilment of the radar objective.

Figure 3 illustrates the parameterized curve of Nash bargaining solutions versus total power $\overline{P}$, which is concave. Thus, for small $\overline{P}$, the communication objective gains more than the radar objective from an increase in $\overline{P}$, while for large

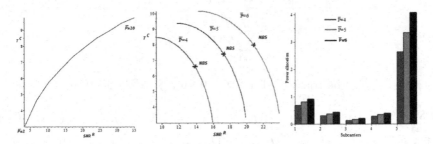

Fig. 3. Parameterized curve of Nash bargaining solutions versus total power $\overline{P}$ (left), feasibility sets Γ and NBS's (middle) and the optimal strategies (right) for $\overline{P} \in \{4, 5, 6\}$.

$\overline{P}$, they gain in vice-versa style. Also, the figure illustrates how the feasibility set Γ and strategy corresponding to NBS varies with an increase in total power $\overline{P}$.

6   Conclusions

In this paper we examined a dual-function OFDM-style wireless system that can support radar and communication functions. We have suggested two step methods to find the bargaining solution for maintaining both of these objectives. *First*, for each fixed level of performance for the radar objective, the optimal solution is found in a water-filling form which includes the classical OFDM waterfilling solution as a boundary case. The formulation returns a continuum of optimal solutions (a solution per radar performance level). *Second*, from this continuum of solutions, an algorithm is developed for effectively finding the Nash bargaining solution, which returns the best tradeoff between these two objectives. The behavior of the Nash bargaining solution is illustrated through a numerical example.

A   Appendix

Proof of Theorem 1. Since $p_i(\omega, \nu)$ is continuous and decreasing in ω and increasing on ν for $\omega > h_i^R \nu$, it tends to infinity for ω tending to $h_i^R \nu$, (a) and (b) follow. Then, (a) implies (c) and (e). Finally, (a), (b) and (c) imply (d), while (a), (b) and (e) imply (f). ∎

Proof of Theorem 2. Since $\mathrm{T}^C(\boldsymbol{P})$ is strictly concave and $\mathrm{SNR}^R(\boldsymbol{P})$ is linear on $\boldsymbol{P}$, there is a unique optimal $\boldsymbol{P}$. To find it we define the Lagrangian as follows:

$L_{\omega, \nu}(\boldsymbol{P}) = \sum_{i=1}^{n} \ln(1 + h_i^C P_i) + \nu(\sum_{i=1}^{n} h_i^R P_i - \epsilon^R) + \omega(\overline{P} - \sum_{i=1}^{n} P_i)$ where $\nu \geq 0$ and ω are Lagrange multipliers. Thus, by the Karush-Kuhn-Tucker Theorem, the strategy $\boldsymbol{P}$ is optimal if and only if the following conditions hold:

$$\frac{h_i^C}{1 + h_i^C P_i} + h_i^R \nu - \omega \begin{cases} = 0, & P_i > 0, \\ \leq 0, & P_i = 0, \end{cases} \tag{19}$$

$$0 = \nu(\sum_{i=1}^{n} h_i^R P_i - \epsilon^R). \tag{20}$$

Two cases have to be considered: (i) $\nu = 0$ and $\sum_{i=1}^{n} h_i^R P_i > \epsilon^R$, and (ii) $\nu > 0$ and $\sum_{i=1}^{n} h_i^R P_i = \epsilon^R$. Let (i) hold. Then, since $\nu = 0$, by (19), $P_i = p_i(\omega, 0)$. Thus, by Theorem 1, $\omega = \omega^C$ since $\boldsymbol{P} \in \Pi$. This and (i) imply that (10) also must hold, and (a) follows. Let (ii) hold. Then, since $\nu > 0$, by (19), $P_i = p_i(\omega, \nu)$. Then, since $\boldsymbol{P} \in \Pi$ and $\sum_{i=1}^{n} h_i^R P_i = \epsilon^R$, this and Theorem 1 imply (13), and the result follows.   ∎

Proof of Corollary 1

(a) By Theorem 1(a), the condition (10) is equivalent to (14). This, by Theorem 2(a), implies (a).

(b) By Theorem 2 the optimal strategy is unique, it is given by (3) where $\omega = \Omega(\nu)$ and ν is the unique root of the Eq. (13). By Theorem 1(d) and (f), this is equivalent to the fact that ω and ν are the unique solutions of the equations $\omega = \Omega(\nu)$ and $\nu = N(\omega)$. Thus, ω is a fixed point for the problem $\omega = \Omega(N(\omega))$. By Theorem 1(d) and (f), both functions $\Omega(\nu)$ and $N(\omega)$ are increasing. Thus, (fixed point) Algorithm 1 converges, and the result follows.   ∎

References

1. Altman, E., Avrachenkov, K., Garnaev, A.: Closed form solutions for water-filling problem in optimization and game frameworks. Telecommun. Syst. J. **47**, 153–164 (2011)
2. Garnaev, A., Garnaeva, G., Goutal, P.: On the infiltration game. Int. J. Game Theory **26**, 215–221 (1997)
3. Garnaev, A., Sagari, S., Trappe, W.: Fair channel sharing by Wi-Fi and LTE-U networks with equal priority. In: Noguet, D., Moessner, K., Palicot, J. (eds.) CrownCom 2016. LNICSSITE, vol. 172, pp. 91–103. Springer, Cham (2016). doi:10.1007/978-3-319-40352-6_8
4. Garnaev, A., Sagari, S., Trappe, W.: Bargaining over fair channel sharing between Wi-Fi and LTE-U networks. In: Koucheryavy, Y., Mamatas, L., Matta, I., Ometov, A., Papadimitriou, P. (eds.) WWIC 2017. LNCS, vol. 10372, pp. 3–15. Springer, Cham (2017). doi:10.1007/978-3-319-61382-6_1
5. Garnaev, A., Trappe, W.: Stationary equilibrium strategies for bandwidth scanning. In: Jonsson, M., Vinel, A., Bellalta, B., Marina, N., Dimitrova, D., Fiems, D. (eds.) MACOM 2013. LNCS, vol. 8310, pp. 168–183. Springer, Cham (2013). doi:10.1007/978-3-319-03871-1_15
6. Garnaev, A., Trappe, W.: A bandwidth monitoring strategy under uncertainty of the adversary's activity. IEEE Trans. Inf. Forensics Secur. **11**, 837–849 (2016)

7. Garnaev, A., Trappe, W.: Bandwidth scanning when facing interference attacks aimed at reducing spectrum opportunities. IEEE Trans. Inf. Forensics Secur. **12**, 1916–1930 (2017)
8. Garnaev, A., Trappe, W.: Bargaining over the fair trade-off between secrecy and throughput in OFDM communications. IEEE Trans. Inf. Forensics Secur. **12**, 242–251 (2017)
9. Garnaev, A., Trappe, W., Petropulu, A.: Bargaining over fair performing dual radar and communication task. In: 50th Asilomar Conference on Signals Systems, and Computers, Pacific Grove, CA, pp. 47–51, November 2016
10. Ghorbanzadeh, M., Abdelhadi, A., Clancy, C.: A utility proportional fairness resource allocation in spectrally radar-coexistent cellular networks. In: IEEE Military Communications Conference (MILCOM), pp. 1498–1513 (2014)
11. Hassanien, A., Amin, M.G., Zhang, Y.D., Ahmad, F.: Dual-function radar-communications: information embedding using sidelobe control and waveform diversity. IEEE Trans. Sig. Process. **64**, 2168–2181 (2016)
12. He, P., Zhao, L., Zhou, S., Niu, Z.: Water-filling: a geometric approach and its application to solve generalized radio resource allocation problems. IEEE Trans. Wireless Commun. **12**, 3637–3647 (2013)
13. He, P., Zhao, L., Zhou, S., Niu, Z.: Radio Resource Management Using Geometric Water-Filling. Springer, Heidelberg (2014)
14. Jamil, M., Zepernick, H.-J., Pettersson, M.I.: On integrated radar and communication systems using Oppermann sequences. In: IEEE Military Communications Conference (MILCOM), pp. 1–6 (2008)
15. Khawar, A., Abdelhadi, A., Clancy, T.C.: MIMO Radar Waveform Design for Spectrum Sharing with Cellular Systems: a MATLAB Based Approach. Springer, Heidelberg (2016)
16. Lagarias, J.C., Reeds, J.A., Wright, M.H., Wright, P.E.: Convergence properties of the Nelder-Mead simplex method in low dimensions. SIAM J. Optim. **9**, 112–147 (1998)
17. Lan, T., Kao, D., Chiang, M., Sabharwal, A.: An axiomatic theory of fairness in network resource allocation. In: IEEE INFOCOM, pp. 1–9 (2010)
18. Lelloucha, G., Nikookar, H.: On the capability of a radar network to support communications. In: 14th IEEE Symposium on Communications and Vehicular Technology in the Benelux (2007)
19. Leshem, A., Naparstek, O., Nehorai, A.: Dual-function radar-communications: information embedding using sidelobe control and waveform diversity. IEEE J. Select. Topics Sig. Process. **1**, 42–55 (2007)
20. Levanon, N.: Multifrequency complementary phase-coded radar signal. IEE Proc. Radar, Sonar Navig. **147**(6), 276–284 (2000)
21. Militano, L., Niyato, D., Condoluci, M., Araniti, G., Iera, A., Bisci, G.M.: Radio resource management for group-oriented services in LTE-A. IEEE Trans. Veh. Technol. **64**, 3725–3739 (2015)
22. Nash, J.: The bargaining problem. Econometrica **18**, 155–162 (1950)
23. Poor, H.V.: An Introduction to Signal Detection and Estimation. Springer, New York (1994)
24. Qi, Q., Minturn, A., Yang, Y.: An efficient water-filling algorithm for power allocation in OFDM-based cognitive radio systems. In: International Conference on Systems and Informatics (ICSAI), pp. 2069–2073 (2012)
25. Sen, S., Nehorai, A.: OFDM radar waveform design for sparsity-based multi-target tracking. In: 2010 International Waveform Diversity and Design Conference (2010)

26. Sen, S., Tang, G., Nehorai, A.: Multiobjective optimization of OFDM radar wave-form for target detection. IEEE Trans. Sig. Process. **59**(2), 639–652 (2011)
27. Xu, R., Peng, L., Zhao, W., Mi, Z.: Radar mutual information and communication channel capacity of integrated radar-communication system using MIMO. ICT Express **1**, 102–105 (2015)
28. Yang, C., Li, J., Anpalagan, A.: Strategic bargaining in wireless networks: basics, opportunities and challenges. IET Commun. **8**, 3435–3450 (2014)

Discrete Time Bulk Service Queue for Analyzing LTE Packet Scheduling for V2X Communications

Vitalii Beschastnyi[1]([⊠]), Valeriy Naumov[2], Pasquale Scopelliti[4],
Irina Gudkova[1,3], Claudia Campolo[4], Giuseppe Araniti[4], Iliya Dzantiev[1],
and Konstantin Samouylov[1,3]

[1] Peoples' Friendship University of Russia (RUDN University), 6 Miklukho-Maklaya
St., Moscow 117198, Russian Federation
{beschastnyy_va,gudkova_ia,samuylov_ke}@rudn.university
[2] Service Innovations Research Institute, Annankatu str. 8, 00120 Helsinki, Finland
valeriy.naumov@pfu.fi
[3] Institute of Informatics Problems, FRC CSC RAS, 44-2 Vavilov St., Moscow
119333, Russian Federation
[4] Mediterranea University of Reggio Calabria, Via Graziella, Loc. Feo di Vito, 89124
Reggio Calabria, Italy
{pasquale.scopelliti,claudia.campolo,araniti}@unirc.it

Abstract. Cellular technologies are promoted by different stakeholders as a key enabler to support vehicle-to-everything communications. In this paper, we shed light into the performance of the LTE (Long Term Evolution) technology in delivering messages generated by vehicles under different network configurations and workload settings. To this purpose, by employing methods of queuing theory, we derive closed-form expressions and provide numerical results for reliability and latency metrics.

1 Introduction

Vehicle-to-everything (V2X) communications will allow vehicles to communicate with nearby vehicles, pedestrians and the roadside infrastructure with the main aim to improve road safety, by supporting the delivery of messages informing drivers of potentially dangerous objects/conditions on their way, who could react by taking actions to avoid accidents [5,11,16]. Rapid deployment of V2X networks is expected in the early 2020s, pushed by governments, telco vendors, automotive industries and standardization bodies, when both the legislative regulation basis will be adopted and road infrastructure will be enabled with proper capabilities.

The publication was financially supported by the Ministry of Education and Science of the Russian Federation (the Agreement no. 02.a03.21.0008) and RFBR according to the research projects No. 16-37-00421 mol a and No. 16-07-00766 a.

© Springer International Publishing AG 2017
O. Galinina et al. (Eds.): NEW2AN/ruSMART/NsCC 2017, LNCS 10531, pp. 395–407, 2017.
DOI: 10.1007/978-3-319-67380-6_36

IEEE 802.11p[1] is considered as the *de facto* standard to enable V2X communications [9]. However, it requires costly on-board equipment and the deployment from scratch of a road-side infrastructure made of road side units (RSUs).

Another solution for V2X is to leverage the existing cellular infrastructure, based on the Long Term Evolution (LTE) specifications and its upcoming evolutions. LTE exhibits all the advantages of broadband wireless technologies like high reliability, ubiquity and low cost due to the already deployed infrastructure [4]. Moreover, compared to 802.11p, LTE is expected to achieve a higher market penetration. Indeed, the LTE network interface is integrated in common user devices like smart phones, and passengers are accustomed to being connected to the Internet through these devices while on the road. The usage of cellular technologies to support V2X raise a lot of concerns including proper resource allocation techniques under high mobility conditions, strict quality assurance to support the stringent latency and reliability requirements of V2X applications and the effect of the support of vehicular services on traditionally served LTE network users [7]. Moreover, the majority of V2X interactions are expected to occur among vehicles. The infrastructure-based architecture does not natively support vehicle-to-vehicle (V2V) communications (crucial for safety-critical services); end-to-end latency could increase due to message passing through infrastructure nodes before being redistributed back to destination vehicles. This issue cannot be overstepped, unless resorting to device-to-device (D2D) communications [13,15] included in 3GPP specifications, since Release 12, and properly customized to meet the demands of V2X in Release 14 [1].

In this work, we theoretically analyze the performance of messages exchanged between vehicles and the cellular infrastructure under different vehicle density when varying LTE [2] and application settings. We apply the methods from queuing theory [6] to develop a mathematical framework that is based on discrete modeling [3,10] of data transmission process and allows for obtaining close upper and lower bound estimates of the considered metrics, i.e., the packet loss probability and mean sojourn time in the queue, representative of reliability and latency performance, respectively.

The rest of the paper is organized as follows. In Sect. 2, we describe the system model and introduce our main assumptions. In Sect. 3, we develop the analytical model and derive expressions for estimations of packet loss probability and mean sojourn time. In Sect. 4 we provide some numerical results to illustrate the system performance under different input parameters. Conclusions are drawn in the last Section.

2   System Model

Consider a circularly shaped service area of a single LTE base station (BS), the eNodeB (eNB), with a fixed radius R and a road crossing this area (Fig. 1).

---

[1] The amendment p is superseded and it is now part of the 802.11 standard. Nonetheless, the customization of 802.11 to the vehicular environment is still referred to as 802.11p.

Assume that exactly K vehicles are uniformly distributed over the area of interest. Each vehicle is equipped with a cellular interface that allows it to periodically send packets (e.g., carrying its current state, speed, position, etc.) to the BS. The size of packets is assumed to be constant for each vehicle and equals to s bytes.

The BS receives packets from individual vehicles. Once received, such packets can: *(i)* be sent in broadcast in the downlink direction to all the other vehicles within the service area, i.e., when the V2V connectivity is sparse, *(ii)* disseminated to farther areas in the downlink direction, *(iii)* sent to a remote/edge server for further processing. However, at this stage of research, the model focuses only on capturing the performance on the uplink direction. The intervals between the generation times of packets by individual vehicles are constant and equal to τ ms. Packet transmission is considered to be successful if only it is transmitted to the BS before the next packet (likely conveying fresher information) is generated by the same vehicle. If a packet is not transmitted within τ, it is considered outdated and dropped out of the system not to waste the systems resources. In our baseline scenario, we assume that as the vehicles arrive at the service area, they follow the conventional service procedure as defined in the recent 3GPP specifications. Accordingly, once the first packet to be transmitted is generated, the vehicle follows a random access procedure to inform the BS about its transmission needs. The BS scheduler may accept the transmission request and place it into a virtual queue. There is no need to pass this procedure for the following packets, as the scheduler will automatically attempt to provide the vehicle with RBs till it departs from the service area, i.e., a semi-persistent scheduling policy is assumed. The BS allocated a bandwidth to be exclusively used for V2X communications, to avoid interference with other cellular users.

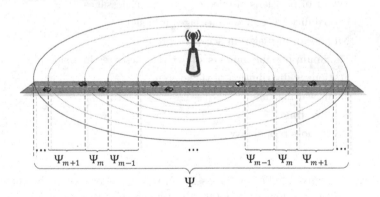

Fig. 1. CQI-based division of the service area

Depending on the quality of channel between a vehicle and BS, the BS defines an appropriate CQI (Channel Quality Information) with respective modulation and coding scheme (MCS), which determines available guaranteed minimum

Table 1. Notation used in this paper.

Parameter	Definition
K	Number of vehicles in service area
c	Number of RBs within one TTI
r	Number of RBs in virtual buffer
τ	Interval between moments of packet generation by a vehicle (i.e. packet lifetime)
l_τ	Number of slots within τ
Δ	Length of a slot
s	Packet size
v	Minimum guaranteed transmission rate
Ψ	Service area
$\mathcal{M}$	Set of CQI groups
M	Cardinality of the $\mathcal{M}$ set
Ψ_m	Subarea of m-th CQI group
χ_m	Number of messages in system in slot n
a_n	Number of arrived messages
d_n	Number of served messages
γ_k	Probability to have k arrived groups of messages in a slot
h_m	Size of a group of messages corresponding to m-th CQI group
β_m	Probability to have a groups of h_m messages
$\mathcal{N}$	Set of all possible sequences of groups of messages
$A_j(k, n)$	Vector of positions of accepted groups of messages
$B_j(k, n)$	Vector of positions of rejected groups of messages
$\alpha_i(s)$	Probability to receive s messages in a slot
q_α	Number of messages in accepted groups
N	Maximum possible number of arrived messages in a slot
π	Stationary distribution of the system
W	Mean waiting time
Q	Sojourn time
B	Packet loss probability

transmission rate v provided by single assigned RB [12]. Consequently, the poorer the channel quality (the lower the CQI) the bigger the number of RBs that are needed to transmit a fixed size packet. The number of needed RBs to accommodate a packet transmission at a given channel quality can be easily found as $\lceil \frac{s}{v\Delta} \rceil$. It may happen that numbers of needed RBs per packet are the same for several CQI values, so for the sake of simplicity we join such CQIs into groups $\mathcal{L}_m \in \mathcal{L}$ with the same demand for RBs, where $m \in \mathcal{M}$, $|\mathcal{M}| = M$. An example of mapping between such groups and number of needed RBs for

$s = \{100, 300, 500, 1000\}$ bytes is presented in Table 2. Since the channel quality goes in hand with the distance between the communicating devices, it is possible to represent the service area as a set of adjacent subareas Ψ_m as shown in Fig. 1.

We assume that the average number of vehicles inside each of the subareas is stationary, i.e., it is not changed significantly over time. This means that any type of vehicles mobility may be applied provided by satisfaction of the above-mentioned constraint. The virtual queue is introduced to avoid excessive delays, as the communicated information is delay-sensitive. Main notations used in this paper are listed in Table 1.

Table 2. CQI-RB mapping.

CQI, l	v_l, [bit/ms]	h_m, [RB] $s = 100$ B	h_m, [RB] $s = 300$ B	h_m, [RB] $s = 500$ B	h_m, [RB] $s = 1000$ B
1	25.59	32	94	157	313
2	39.38	22	62	102	204
3	63.34	14	38	64	127
4	101.07	8	24	40	80
5	147.34	6	18	28	55
6	197.53	6	14	21	41
7	248.07	4	10	17	33
8	321.57	4	8	13	25
9	404.26	2	6	10	20
10	458.72	2	6	9	18
11	558.72	2	6	8	15
12	655.59	2	4	7	13
13	759.93	2	4	6	11
14	859.35	2	4	5	10
15	933.19	2	4	5	9

3 Analytical Model

To analyze the proposed model, we first represent the service process of vehicle-to-BS data traffic as a queuing system. In this model a typical LTE bandwidth F is dedicated to V2V communications (e.g., 10 MHz). This means, that there are $c = \frac{F}{(180+20) \cdot 10^3}$ RBs available in a time slot of length Δ ms, which is presented in Fig. 2. The number of slots available for packet transmission is defined as $l_\tau = \lfloor \frac{\tau}{\Delta} \rfloor$. For the last $l_\tau - 1$ ms of packet's time to live a virtual buffer of size $r = c \cdot (l_\tau - 1)$ is organized. The requested RBs for a packet transmission hereafter is designated in terms of queuing theory as a batch of group messages. Each group is composed of messages that are assigned to a number of waiting positions in the virtual buffer and represents a packet. Therefore, one of the modeling choices is to represent the service process as the discrete-time queuing system [14] with bulk arrival and service processes, whereas the service time of each message is constant. Note that a packet is not necessarily served within a single slot, and accordingly the messages belonging to the same batch may occupy waiting positions in multiple slots.

The system time diagram is displayed in Fig. 3. The system time is slotted with the frame duration of τ ms determined by the interval between the generation times of packets from an individual vehicle. According to our model, groups of messages arrive either before the ends (LAM, Late Arrival Model) or after beginnings (EAM, Early Arrival Model) of slots, depending on the selected observation points ξ_n [8]. It is easy to see that the observation points form an embedded one-dimensional Markov chain $\{\xi_n, n = 0, 1, ...\}$, describing the number of messages in the system for both EAM and LAM. Hereafter we denote the number of messages arrived in a slot as a_n. The number of served messages at n-th slot equals to $d_n = \min(\xi_n, c)$, as the BS cannot provide more than c RBs.

In LAM ξ_n depends on the number of messages in the previous observation point ξ_{n-1} in the following way: $\xi_n = \min(a_n + \xi_{n-1}, c + r) - \min(\xi_{n-1}, c)$, where the minuend is the number of messages after batch arrival with respect to queue capacity, while the subtrahend is the number of served messages at previous slot.

On the other hand, in EAM $\xi_n = \min(a_n + \max(\xi_{n-1} - c, 0), c + r)$, where the first argument is the number of messages in arrived groups and the messages that remained after previous slot, while the second argument provides limit for ξ_n to be compliant with the queue capacity. The queue size in EAM is always lower than in LAM, since in EAM we observing the queue size before possible arrival point in a slot meaning that in EAM waiting time will be also lower. Messages depart from the system in bathes of groups as well, at the slot boundaries. The

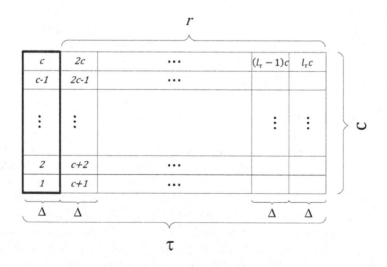

Fig. 2. Virtual queue

messages are not allowed to occupy the server immediately and the service of any arrived messages may start no sooner than the beginning of the slot next to the arriving point.

The sojourn time estimates are obtained as the number of slots spent by the first and the last messages of groups in the system. We assume batch acceptance service strategy, that is, whenever a group of h_m messages arrives and there are ξ arrivals in the system, such that $\xi + h_m > c + r$, the entire group is rejected.

Generally, a vehicle starts transmission as soon as it is provided with channel through RACH (Random Access Channel) procedure. Thus, we assume that the number of packets generated within τ is uniformly distributed in the time domain as the successful execution of RACH procedure does not depend on current TTI, and a vehicle may start transmission at any slot with the same probability. Therefore, we can derive the probability γ_k that exactly k packets are generated within one slot, which follow the principals of Bernoulli trials. Let a packet generation in the considered slot be a success, then a failure corresponds to generation of the packet in one of the rest $l_\tau - 1$ slot. Thus the probability of success is $\frac{1}{l_\tau}$ and probability of failure is $\frac{l_\tau - 1}{l_\tau}$. Consequently, the probability γ_k can be found as

$$\gamma_k = \binom{K}{k} \left(\frac{1}{l_\tau}\right)^k \left(\frac{l_\tau - 1}{l_\tau}\right)^{K-k} = \frac{K!(l_\tau - 1)^{K-k}}{k!\,(K-k)!\,l_\tau^{\,K}}, \tag{1}$$

where $k = 0, ..., K$ and the number of trials K is the number of vehicles that generate packets within τ.

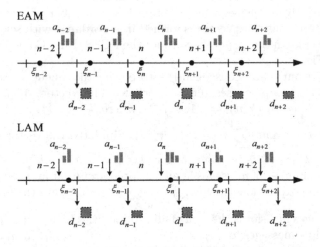

Fig. 3. System time

Knowing the correspondence between the m-th CQI group and the area $|\Psi_m|$ to which this CQI belongs, and the whole area of the cell $|\Psi_m|$, it is possible to find the probability that a vehicle reports CQI of m-th group

$\beta_m := \frac{|\Psi_m|}{|\Psi|}$, $m \in \mathcal{M}$. Such probabilities form the group size distribution of the arrival process. In each slot we have r arriving messages which form a sequence $(n(1), ..., n(k))$, where h_i is the size of the i-th group, $i = 1, ..., k$. Then, the set of all possible sequences that may occur in a slot is $\mathcal{N} = \{(k, (n(1), ..., n(k))), k = 1, ..., K, n(i) \in \{h_1, ..., h_m\}\}$ with cardinality $|\mathcal{N}| = \sum_{k=1}^{K} M^k$, which is the number of possible variations with repetitions in respect to all possible numbers of arrived groups of messages.

Since the number of packets generated in a slot and their sizes are mutually independent, the probability to have a sequence at an arbitrary slot is simply the product of probabilities of the following events: we have $(k, \boldsymbol{n})$ packets in a slot, each of the packets has its own size h_m, and the packets are ordered in one of M^k ways. Summarizing the above-mentioned statement the probability to find a sequence $(k, \boldsymbol{n})$ is defined as

$$P\{k, (n(1), ..., n(k))\} = \frac{\gamma_k}{M^k} \prod_{m \in \mathcal{M}} (\beta_m{}^g), \qquad (2)$$

where $g = \sum_{j=1}^{k} \mathbb{1}\{n(j) = h_m\}$ counts the number of groups of the same length h_m in a sequence $\boldsymbol{n}$, $\mathbb{1}\{x = a\} = 1$ if $x = a$ and $\mathbb{1}\{x = a\} = 0$ otherwise. In this way the arrival process in the model can be classified as composition on batched Bernoulli processes.

An arriving sequence of groups in a slot can be naturally divided into two vectors $\boldsymbol{A}_j(k, \boldsymbol{n})$ and $\boldsymbol{B}_j(k, \boldsymbol{n})$, vectors of positions of accepted and rejected groups respectively in the sequence where j is the number of messages in the system in the previous slot. Vector is formed in accordance with selected packet scheduling scheme. In this paper we consider the FCFS (First Come First Served) scheme (see Fig. 4).

In FCFS groups are consequently stored in the queue until the next one, $(k^* + 1)$-th group, causes queue overflow. The vector $\boldsymbol{A}_j(k, \boldsymbol{n})$ in this case is defined as $\boldsymbol{A}_j(k, \boldsymbol{n}) = (1, ..., k^*)$, where $k^* \in \{1, ..., k\}$, such that $\sum_{i=1}^{k^*} n(i) \leq c + r - j$ and $\sum_{i=1}^{k^*+1} n(i) > c + r - j$. Similarly, the vector $\boldsymbol{B}_j(k, \boldsymbol{n}) = (k^* + 1, ..., k)$, where k^* is the same as for $\boldsymbol{A}_j(k, \boldsymbol{n}) = (1, ..., k^*)$.

To find the number of messages in all accepted groups at an arbitrary slot we use the following expression: $q_\alpha(k, \boldsymbol{n}, j) = \sum_{i \in \boldsymbol{A}_j(k, \boldsymbol{n})} n(i)$. Thus the probability to receive s messages in a slot is the sum of all probabilities to accept sequences that provide exactly s messages: $\alpha_i(s) = \sum_{k, \boldsymbol{n} \in \mathcal{N}: q_a(k, \boldsymbol{n}, i) = s} P\{k, \boldsymbol{n}\}$. In the same time, the maximum possible number of arrived messages in a slot N are generated in case when all K vehicles belong to the first CQI group and generate packets in this slot, therefore $N = K \cdot h_1$. Then it is possible to compose transition matrix $\boldsymbol{P}_E = [p_{i,j}^E]$, $i, j = 0, ..., c + r$, for EAM with the elements defined by

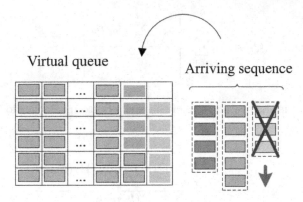

Fig. 4. FCFS packet scheduling

$$
p_{i,j}^E = \begin{cases}
\alpha_i\left(j\right), i = 0, .., c+1, j = 0, ..., c+r-1, \\
\sum\limits_{k=c+r-i}^{N} \alpha_i\left(k\right), i = 0, ..., c+1, j = c+r, \\
\alpha_i\left(j-i+c\right), i = c+1, ..., c+r, j = i-c, ..., c+r-1, \\
\sum\limits_{k=c+r-i}^{N} \alpha_i\left(k\right), i = c+1, ..., c+r, j = c+r, \\
0, i = c+1, ..., c+r, j = 0, ..., i-c-1,
\end{cases} \tag{3}
$$

while the elements of transition matrix $\boldsymbol{P}_L = \left[p_{i,j}^E\right]$, $i, j = 0, ..., c+r$, for LAM are defined by

$$
p_{i,j}^L = \begin{cases}
\alpha_i\left(j\right), i = 0, .., c+1, j = 0, ..., c+r-1-i, \\
\sum\limits_{k=c+r-i}^{N} \alpha_i\left(k\right), i = 0, ..., c+1, j = c+r-i, \\
\alpha_i\left(j-i+c\right), i = c+1, ..., c+r, j = i-c, ..., r-1, \\
\sum\limits_{k=c+r-i}^{N} \alpha_i\left(k\right), i = c+1, ..., c+r, j = r, \\
0, i = 1, ..., c+r, j = \max\left(c+r-i, r+1\right), ..., c+r, \\
0, i = c+1, ..., c+r, j = 0, ..., i-c-1.
\end{cases} \tag{4}
$$

The two models provide us with upper and lower bound estimates of state distribution $\boldsymbol{\pi}_t$, respectively, by solving the equation $\boldsymbol{\pi}_t = \boldsymbol{\pi}_t \boldsymbol{P}_t$, where $t \in \{E, L\}$.

As we obtain the $\boldsymbol{\pi}_t$ distribution it is possible to estimate mean waiting time W_t for a packet as

$$
W_t = \Delta \sum_{z=c+1}^{r} \left\lceil \frac{z}{c} \right\rceil \pi_{t,z}, \tag{5}
$$

where $\left\lceil \frac{z}{c} \right\rceil$ quantifies how many slots it will take to serve all the z messages before a newly arrived group. It is also possible to estimate mean sojourn time Q_t for a packet as

$$Q_t = \Delta \sum_{z=c+1}^{r} \left\lceil \frac{z + \sum_{m=1}^{M} \beta_m h_m}{c} \right\rceil \pi_{t,z}, \tag{6}$$

where $\sum_{m=1}^{M} \beta_m h_m$ is the average size of the arrived group. Packet loss probability B_t can be found as

$$B_t = \frac{\sum_{z=0}^{c+r} \pi_{t,z} \sum_{(k,\boldsymbol{n})\in\mathcal{N}} P(k,\boldsymbol{n}) \dim(\mathbf{B}_i(k,\boldsymbol{n}))}{\sum_{(k,\boldsymbol{n})\in\mathcal{N}} P(k,\boldsymbol{n}) k}, \tag{7}$$

where the numerator is the average number of rejected groups and the denominator is the average number of arrived groups in a slot, $t \in \{E, L\}$.

4   Performance Evaluation

In this section we present the results for a numerical experiment conducted under varying application and LTE settings: $c = \{25, 50, 75\}$ RBs, $r = (\tau - \Delta) \cdot c$ RBs, $50 \leq K \leq 300$ vehicles, $s = \{100, 300, 500, 1000\}$ bytes, $\tau = 100$ ms, $\Delta = 1$ ms. In Fig. 5 a comparison between EAM and LAM is shown. The difference between the two models under the proposed input parameters is almost unnoticeable in terms of the considered metrics. For this reason we further confine the presentation to the results of LAM as it provides the upper estimates of the desired parameters, and therefore seems to be more interesting.

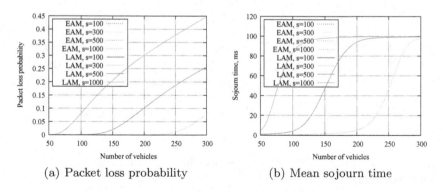

(a) Packet loss probability   (b) Mean sojourn time

Fig. 5. Early and late arrival models comparison

As can be noticed from Figs. 6(a) and 7(a) there is almost linear dependency between the data traffic load (the number of vehicles) and packet loss probability, while Figs. 6(b) and 7(b), in contrast, exhibit rapid growth in case the system is close to be 'saturated', i.e. the number of accepted messages in a slot is relatively equal to the number of V2X-dedicated RBs per one slot.

As expected, performance gets worsened as the number of available resources, c, decreases and the packet size, s, gets larger. For instance, in Fig. 7(a) it is shown that nearly half of 1000 bytes-long packets are lost for a common LTE system bandwidth (i.e., 10 MHz bandwidth) for a road with 300 transmitting vehicles.

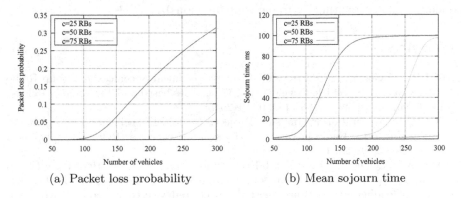

(a) Packet loss probability   (b) Mean sojourn time

Fig. 6. Parameters estimation by varying c, ($s = 300$ bytes)

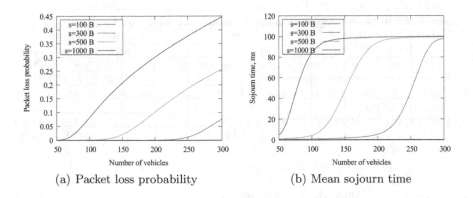

(a) Packet loss probability   (b) Mean sojourn time

Fig. 7. Parameters estimation by varying s, ($c = 50$ RBs)

5 Conclusions and Future Work

In this paper, we analyzed key performance metrics of vehicular communications over LTE by deriving a theoretical framework based on discrete modeling and queuing theory. The developed analytical framework allows for close estimation of packet loss probability and average experienced latency analytically without the need for time-consuming simulations. Results clearly indicate the need to explore alternative packet scheduling schemes compared to the simplistic FCFS considered scheme, in order to accommodate the transmission of large packets from more vehicles. Indeed, with the increasing level of cooperation, sensing and automation of vehicles information-rich messages will be likely delivered. Thus, as the subject of our ongoing work we consider extension of the proposed model with support of different packet scheduling schemes, closer LTE behavior modeling (e.g., accounting for frame configurations).

References

1. 3GPP. 3rd Generation Partnership Project; Technical Specification Group Radio Access Networks; Vehicle to Vehicle (V2V) services based on LTE sidelink. Release 14. Technical report 36.785 v14.0.0, 10–2016
2. 3GPP. 3rd Generation Partnership Project; Technical Specification Group Radio Access Networks; Radio Frequency (RF) system scenarios (Release 9). Technical report 25.942 v9.0.0, 12–2009
3. Alfa, A.S.: Queueing Theory for Telecommunications. Springer, Heidelberg (2010)
4. Andreev, S., Pyattaev, A., Johnsson, K., Galinina, O., Koucheryavy, Y.: Cellular traffic offloading onto network-assisted device-to-device connections. IEEE Commun. Mag. **52**(4), 20–31 (2014)
5. Araniti, G., Campolo, C., Condoluci, M., Iera, A., Molinaro, A.: LTE for vehicular networking: a survey. IEEE Commun. Mag. **51**(5), 148–157 (2013)
6. Basharin, G.P., Gaidamaka, Y.V., Samouylov, K.E.: Mathematical theory of teletraffic and its application to the analysis of multiservice communication of next generation networks. Autom. Control Comput. Sci. **47**(2), 62–69 (2013)
7. Bazzi, A., Masini, B.M., Zanella, A., Thibault, l.: Beaconing from connected vehicles: IEEE 802.11p vs. LTE-V2V. In: IEEE 27th Annual International Symposium on Personal, Indoor, and Mobile Radio Communications (PIMRC), pp. 1–6 (2016)
8. Bose, S.K.: An Introduction to Queuing Systems. Springer, Heidelberg (2002)
9. Campolo, C., Molinaro, A., Scopigno, R.: Vehicular ad hoc networks. Stand. Solutions Res. (2015)
10. Efimushkina, T., Gabbouj, M., Samuylov, K.: Analytical model in discrete time for cross-layer video communication over LTE. Autom. Control Comput. Sci. **48**(6), 345–357 (2014)
11. ETSI EN 102 962. Intelligent Transport Systems (ITS); Framework Public Mobile Network (2012)
12. Militano, L., Condoluci, M., Araniti, G., Molinaro, A., Iera, A., Muntean, G.-M.: Single frequency-based device-to-device-enhanced video delivery for evolved multimedia broadcast and multicast services. IEEE Trans. Broadcast. **61**(2), 263–278 (2015)
13. Mumtaz, S., Huq, K.M.S., Ashraf, M.I., Rodriguez, J., Monteiro, V., Politis, C.: Cognitive vehicular communication for 5G. IEEE Commun. Mag. **53**(7), 109–117 (2015)
14. Petrov, V., Samuylov, A., Begishev, V., Moltchanov, D., Andreev, S., Samouylov, K., Koucheryavy, Y.: Vehicle-based relay assistance for opportunistic crowdsensing over narrowband IoT (NB-IoT). IEEE Internet Things J. (2017)

15. Piro, G., Orsino, A., Campolo, C., Araniti, G., Boggia, G., Molinaro, A.: D2D in LTE vehicular networking: system model and upper bound performance. In: Ultra Modern Telecommunications and Control Systems and Workshops (ICUMT), pp. 281–286 (2015)
16. Vinel, A.: 3GPP LTE Versus IEEE 802.11p/WAVE: which technology is able to support cooperative vehicular safety applications? IEEE Wireless Commun. Lett. 1(2), 125–128 (2012)

Structure Analysis of an Explanatory Dictionary Ontological Graph

Yu. N. Orlov[1,2(✉)] and Yu. A. Parfenova[1]

[1] Keldysh Institute of Applied Mathematics of Russian Academy of Sciences,
125047 Miusskaya Pl., 4, Moscow, Russia
yuno@kiam.ru
[2] Peoples' Friendship University of Russia (RUDN University), 117198 6
Miklukho-Maklaya St., Moscow, Russia

Abstract. When working with large connected systems, it is important to present the data in a form that is easy to analyze and process. For that purpose an ontological graph is often used. In this paper, developed tools for constructing and analyzing an ontological graph of the connections in the Russian explanatory dictionary are described. The concepts of specificity of oriented graphs with a large number of loops, vertices and edges are introduced: the vertex hierarchy, word basin, basin shell and basin volume are determined. The statistical properties of ontological graph are investigated.

Keywords: Ontological graph · Cycle statistics · Vertex hierarchy · Word basin · Basin basis

1 Introduction

Traditionally, ontological graphs are used for modelling different types of social networks, for example, enterprise management schemes, links to the dictionary articles, in particular for describing structures of links between different Internet resources, as well as communication schemes between recipients of various kinds of messages.

The purpose of this research was to develop a software for text processing and analyzing the connections inside the explanatory dictionary. In addition to structuring and analyzing relations in a system with large number of objects, the problem of determining the distance between sets of dictionary objects was also investigated, i.e. it was necessary to develop a metric that would help to determine the semantic proximity between the vocabulary definitions.

Consider the following example. In one of the dictionaries the following definition is given: "An ontology is a detailed structure specification of a particular problem area". At the same time, in another article we read: "An ontology is a category system that belongs to a certain area of knowledge." These definitions

The publication was financially supported by the Ministry of Education and Science of the Russian Federation (the Agreement number 02.a03.21.0008).

O. Galinina et al. (Eds.): NEW2AN/ruSMART/NsCC 2017, LNCS 10531, pp. 408–419, 2017.
DOI: 10.1007/978-3-319-67380-6_37

have only one common word, however, it would be useful to be able to determine that these two statements define the same word, and also to evaluate the closeness of such definitions to find words with similar meaning without relying on expert opinion.

For that purpose, it may be useful to choose a set of basis words from the dictionary, into which the component words of the statement will be decomposed if the adjacency relation between them is given.

It may also be useful to find clusters of the dictionary ontological graph according to cycles in connections between words. The closeness between statements in this case can be estimated from the degree of intersection of the unions of clusters corresponding to the component words of these statements.

The ontological graph of a dictionary of nouns is constructed as follows: its vertices are nouns, that are defined in a dictionary, and there is a directed edge from vertex x to vertex y if the word y or its other form with the same root occurs in the dictionary article that defines x.

Such graph has following features:

- each vertex has at least one edge,
- there are vertices which do not have incoming edges, i.e. each path in such a graph closes into a cycle,
- the adjacency matrix of this graph is very sparse (for 20–30 thousand cells in each row there are an average of only 10 nonzero ones, and also there are many zero columns, because each word is determined in the dictionary with about 10 words after the processing, and there are no incoming links for most of the words).

To simplify the analysis of the connections in this graph, to find clusters and shortest paths, to count the statistics of its cycles and numbers of outgoing and incoming edges etc. a program was created for pre-processing the dictionary text, visualize the structure of the resulting graph as well as count various statistics using built-in algorithms for graph analysis.

2   Basic Concepts and Problems of the Analysis of the Ontological Graph of a Dictionary

The process of explaining the meaning of a word with the explanatory dictionary articles is as follows: starting with some word (zero level of the word explanation hierarchy), we read the statement from the dictionary that determines it (the first level of the hierarchy), then for each word in this statement we look up its determination in the same dictionary (the second level of the hierarchy) and so on. The explanation procedure ends when each hierarchy branch is looped, i.e. when at a certain level of the hierarchy there are no words that were not referenced from previous levels.

When analyzing the ontological graph of an explanatory dictionary, the following definitions and related questions naturally arise.

Let $n(A)$ be the hierarchy depth of the word A in which all paths are looped and $N(A)$ – the number of vertices in the resulting subgraph. We will call this subgraph the word basin and denote as $B(A)$. We will call A the generative word for the basin $B(A)$. Other words A_i from $B(A)$ also have their basins $B(A_i)$ that belong to $B(A)$. $\bigcup_i B(A_i)$ belongs to $B(A)$. If the number of edges entering A is greater than zero, then

$$B(A) = \bigcup_i B(A_i) \tag{1}$$

Each vertex A_i of the basin $B(A)$ is characterized by n – the number of outgoing edges and m – the number of incoming edges. We denote the vertex with its degree as $A_i(n, m)$. In this study, we are interested in the main statistics characterizing word basin $B(A)$, for example, the distribution of the basins contained in it according to the number of vertices $N(A_i)$, the distribution of word basins according to the depths of hierarchies $n(A_i)$, the distribution of the basins $B(A_i)$ according to the number $\nu(A_i)$ of its cycles and the correlation between these values. Corresponding distribution functions will be denoted as $F_A(N), f_A(n), \rho_A(N, n)$.

One of the main questions we plan to study is the distribution of the basin $B(A_i)$ according to the number of different cycles it contains. The cycle starting (and ending) in the vertex $A_i' \in B(A_i)$ will be called $C(A_i, A_i')$. To uniquely identify the cycle, we consider its starting point the vertex which is the closest in the hierarchy level to the basin generative vertex. The number of edges in a simple path (without self-intersections), first vertex of which is also the end of that path, will be called the length $L(A_i, A_i')$ of a cycle $C(A_i, A_i')$. How close is the distribution function $U_{A_i}(L)$ of basin cycle lengths to the to the distribution function $U_A(L)$ of the basin, which includes all the basins $B(A_i)$? Is there a correlation between the volume of the basin, the number of cycles in it, and the maximum cycle length?

The presumed basin basis, as well as the basis of the entire dictionary, is based on the topology of this basin subgraphs. Let's consider all the basins $B(A_i)$ inside $B(A)$. Do they have a non-empty intersection? How different is $B(\Omega)$ from $B(A)$, where $\Omega = \bigcap_i B(A_i)$? Is it possible to select basis words that are not connected by a path? If there is a reasonable definition of a dictionary basis, is it possible then to associate each word to a real function that characterizes the meaning of this word? This paper is dedicated to the study of the above questions.

3   Theoretical Estimation of the Dictionary Graph Parameters

Let's estimate the average width of the ontological graph for some word from the dictionary. The absolute ontology width is the volume of the basin $N(A)$. At each hierarchy level of defining words there are k_i words, the number of levels in the hierarchy equals $i = 1, 2, \ldots, n(A)$ with $\sum_{i=1}^{n(A)} k_i = N(A) - 1$. Let each word correspond to an average of v words on the next hierarchy level. If there were no restriction on the total number of words in the dictionary, then on a new level of hierarchy the number of new words would increase by v times, i.e. $k_{t+1} = \nu k_t$ with $k_0 = 1$. However, not all the words on a next level of the hierarchy are new, because words from the previous levels begin repeating. The deviation of a real dependence from the ideal one ($k_t = \nu^t$) depends on the fraction of already used words from the whole basin. On a t level of the hierarchy this fraction is

$$\mu(t) = \frac{1}{N(A)} \sum_{i=0}^{t} k_i. \tag{2}$$

We will assume that the number of vertices on a given level of a hierarchy basin is defined by the formula

$$k_{t+1} = \nu k_t \left(V(\mu(t)) \right) \tag{3}$$

where $V(\mu)$ is some function monotonously decreasing from 1 for $\mu(0) = \frac{1}{N}$ to 0 for $\mu(n) = 1$ at some maximum depth of the hierarchy n. Strictly speaking, since the number of words is a nonnegative integer, integer part of the expression 3 should be used. The average width of a graph for a given word is the average number of vertices per hierarchy level:

$$s(A) = \frac{N(A)}{n(A)} \tag{4}$$

$n(A)$ – the number of hierarchy levels – is determined using the condition that the expression 3 is positive. Let's estimate $n(A)$ from the differential equation, which we get from 3 in the case of a continuous index (further in this section $n(A) = n$, $N(A) = N$). Thus we have the following expression:

$$k(t + 1) - k(t) = \frac{dk(t)}{dt} = k(t) \cdot (\nu V(\mu(t)) - 1),$$
$$\mu(t) = \frac{1}{N} \left(1 + \int_0^t k(\tau)d\tau \right), \int_0^n k(\tau)d\tau = N - 1 \tag{5}$$

From 5 it follows that

$$\frac{d\mu(t)}{dt} = \frac{k(t)}{N} \tag{6}$$

The Eq. 5 for $\frac{dk}{dt}$ in terms of a variable μ (a fraction of accumulated basin words) is the following:

$$\mu'' = \mu' \cdot (\nu V(\mu) - 1) \tag{7}$$

Since the index of the hierarchy level is not included explicitly in the Eq. 7, the degree of the equation can be reduced by introducing a function $g(\mu) = m u'$. For a basin, $g(\mu) > 0$. Thus, after substituting $\mu'' = g g''$ into 7, we get

$$\frac{dg}{d\mu} = \nu(\mu) - 1 \qquad (8)$$

and the solution is reduced to a simple quadrature. The maximum occupancy of the level is determined by the condition $g' = 0$. From 8 we get the accumulated fraction $\tilde{\mu}$ at the maximum of the graph width

$$V(\tilde{\mu}) = \frac{1}{\nu} \qquad (9)$$

The choice of the function can significantly affect the estimates of the graph parameters. As we will see later, one of the approximations of V is logarithmic:

$$V(\mu) = \frac{\ln(\mu)}{\ln(N)} \qquad (10)$$

For this case using 8 we obtain the following solution:

$$g(\mu) = \frac{d\mu}{dt} = C - \left(1 - \frac{\nu}{\ln(N)}\right)\mu - \frac{\nu}{\ln(N)}\mu \ln(\mu), \qquad (11)$$

where the integration constant can be found using the conditions $\mu(0) = \frac{1}{N}$, $\mu'(0) = \frac{1}{N}$. The last condition follows from 6 with $k(0) = 1$. As a result, we get

$$\frac{d\mu}{dt} = \frac{1}{N}\left(2 - \nu - \frac{\nu}{\ln(N)}\right) - \mu(1 - \lambda) - \lambda\mu \ln(\mu), \lambda = \frac{\nu}{\ln(N)} \qquad (12)$$

The general solution of 12 cannot be obtained in elementary functions. However, we can consider the approximation $N \gg 1$, which can be observed in practice. Then the free term in 12 can be neglected and the solution of this equation can be found analytically:

$$\mu(t) = \exp\left(\left(\frac{1}{\lambda} - 1 - ln(N)\right)e^{(-\lambda t)} - \frac{1}{\lambda} + 1\right) \qquad (13)$$

This model corresponds well to the actual distribution of words according to hierarchical levels.

4 Algorithms for the Ontology Graph Statistical Analysis

To solve the tasks described above, a program was developed to convert an explanatory dictionary in the most effective way to a format suitable for computer use.

For dictionary text processing and for construction and analysis of the resulting ontological graph lemmatization and Porter stemming algorithms were used

[1–3], as well as the Kosaraju's algorithm for finding the strongly connected components and the Floyd-Warshall algorithm [4–6] for finding the shortest paths in a graph. The radius of the graph was calculated, the dimensions of the strongly connected components and the number of short cycles in it. These algorithms were implemented in Python.

Using javascript, an editor was developed for manual processing of the dictionary articles. The editor page was made in a form of a word definition from the dictionary: the original statement and the corresponding word forms – results of the automatic processing, which can be edited if necessary. The program displays possible editing options (previously processed nouns), and the result of the editing is generalized to the entire dictionary, which makes the manual processing process more efficient. The current state of the whole system at any moment can be saved in JSON format.

Then, a program was created for visualizing the connections in this ontological graph in a convenient way, to an arbitrary depth for any selected word. For any displayed word in the resulting figure, the connections and statistics over them, such as number of edges and vertices, can also be displayed.

The running time of the algorithm for different system and browser parameters may differ, average statistics (Google Chrome, Win8, Intel i3) are displayed in Table 1.

Table 1. Algorithm running time

Hierarchy depth	Number of edges	Time, sec
<2	<3500	<1
3	3500–4000	1–2
4–5	4000–6000	3–5

Both of these programs are optimized for Google Chrome 56 and above, Windows 8/10. Due to the use of javascript, they are cross-platform and do not require installation.

5   Study of the Ontological Graph Using an Example of a Word "предмет"

It is worth mentioning that before creating the ontological graph the dictionary [7] was pre-processed. All union words in it were removed, and adjectives, adverbs and verbs were replaced by same-root nouns. In the resulting ontological graph there are no loops, because a word can not be determined by itself. Every word occurs in the ontology only once.

After that, the numeric research of parameters of the obtained graph was carried out. In this chapter, we will present results of the study for the basin of russian word "предмет" in this ontological graph. Overall, the basin of the word "предмет" has 19 hierarchical levels with 3836 different words. The total in-degree of vertices is 24628 i.e. the average in-degree of the vertex in this obtained graph approximately equals 6.

Figure 1 shows the distribution of words by levels. In the legend, the "number of words" is the total number of outgoing edges from the given level hierarchy, "number of different words" is the number of vertices which are connected to the outgoing edges of a given level, and "number of vertices" is the number of new words on a given level, i.e. number of words not used before.

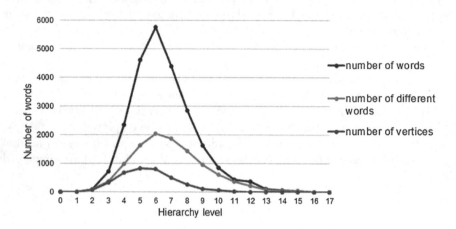

Fig. 1. The distribution of words by levels

The largest number of words, as well as the largest number of edges, is found at the sixth hierarchy level, the number of vertices appearing at a certain level of the hierarchy, has its maximum at the fifth level. The number of vertices is close to the number of edges for the first three levels, then the number of edges sharply increases. It means that there are almost no new words explaining the original word on higher hierarchy levels.

Figure 2 shows the fraction of new words at a certain level of the hierarchy out of the total number of outgoing edges of this level. With high accuracy, this dependence is polynomial:

$$K_{new}(t) \approx at^\gamma, \gamma \approx -\frac{3}{2} \tag{14}$$

Figure 3 shows how function $V(\mu)$ introduced in 14 depends on the fraction of new words. The polynomial approximation has a significantly lower determination ($R^2 = 0.88$) than the logarithmic one, which equals $R^2 = 0.97$.

The fraction of edges connected to the most frequent words from each level of the hierarchy is approximately constant for levels from the 3rd to the 10th and does not have a significant correlation with the level number. For other levels with small number of words, this fraction is negligible. Hence, the model, in which the function was considered independent of the level number, is acceptable.

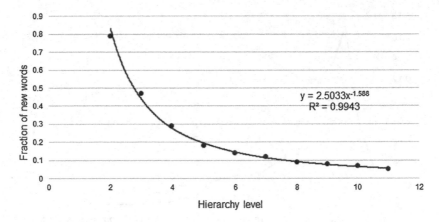

Fig. 2. New words at hierarchy levels

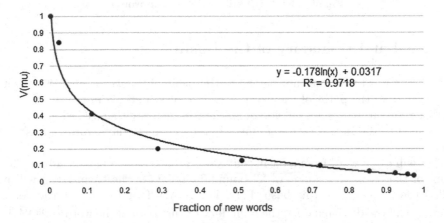

Fig. 3. The correction factor for the word "предмет" hierarchy

The distribution of the basin words according to the number of outgoing edges (i.e. the distribution of the parameter with mean value equal to ν from 14) is shown in Fig. 4. It can be approximately described by a function $n^{\alpha}exp(-\beta n)$, which is similar to the approximation of distances between identical letters in the literary text distribution (see [8]).

It turned out that for almost any pair of words in the basin there is a cycle passing through that pair. The exception is the cycles formed by two (or three) words, determined one by one another. Thus, all words, except for such cycles, are generative for the basin, and form the kernel of the dictionary.

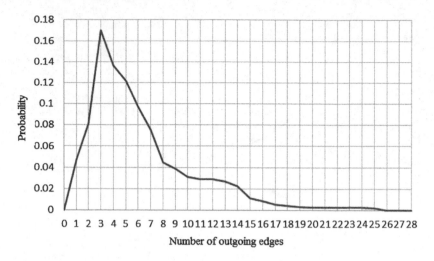

Fig. 4. The distribution of the basin words

6   Shell and Boundary of the Basin

Let's add a word that does not belong to the basin of the word "предмет" to its ontology. If all the paths of the word A' hierarchies continue in the basin $B(A)$ so that there are no cycles outside of it, then we will say that A' belongs to the basin shell $O(A)$. In this case, the union of all such dictionary words forms that shell.

We will call a word from a basin that has only one incoming edge a boundary point of the basin. The union of all boundary points of the basin $B(A)$ will be called the boundary of this basin $\Gamma(B(A))$. It was found out that the kernel boundary of the dictionary, i.e. the boundary of the basin of the word "предмет", consists of 1360 words.

It is not adequate to use the kernel words as the basis "vectors", since the dimension of such basis is too large for determining the semantic nuances of some sentences. So the basic element of the dictionary is not a word, but a cycle of a certain set of words. The number of such cycles is $N!$, where N is the kernel volume, i.e. from a practical point of view it is infinite.

On the other hand, the elements of the kernel ultimately determine the entire dictionary. Words can be considered vectors in some linear space, with all the elements of the kernel connected by cycles. Such vector systems are called over-flowed [9, 10]. For them, there is a representation of so-called coherent states, originally developed to describe quantum-optical phenomena, but it has applications in other fields. We will describe it in terms of our problem.

So, let there be a finite basis of meanings (cycles or their linear combinations) which is the structure of the kernel. Let's denote nth element of this basis as $|n\rangle$ and assume that these vectors are orthonormal $\langle k \mid n \rangle = \delta_{kn}$. We will consider an individual word as an analog of the "coherent state" and denote it as $|z\rangle$, where

z is a complex number. It is known [10] that there is the following representation for the coherent state [10]:

$$|z\rangle = \frac{1}{\sqrt{G(|z|^2)}} \sum_{n=0}^{\infty} \frac{z^n}{\sqrt{Q_n}} |n\rangle,$$

$$z \in M : \left\{ |z|^2 < R = \lim_{n \to +\infty} \frac{Q_{n+1}}{Q_n} \right\},\qquad(15)$$

$$G(|z|^2) = \sum_{n=0}^{\infty} \frac{|z|^2}{Q_n}$$

Here Q_N are some real numbers that have to be determined, and R is the radius of convergence of the normalizing sum in 15. From 15 it follows that

$$\langle n \,|\, z\rangle = \frac{z^n}{\sqrt{Q_n G(|z|^2)}} \qquad(16)$$

Formula 16 means that $\langle n \,|\, z\rangle^2$ is a probability of a word $|z\rangle$ being in a cycle of a length n. In a circle $M : |z|^2 < R$ a measure can be chosen $d\mu_z = \rho(|z|^2)dzd\bar{z}$ such that

$$|n\rangle = \frac{1}{\pi\sqrt{Q_n}} \int_M \bar{z}^n |z\rangle \, d\mu_z, \quad Q_n = \frac{1}{\pi} \int_M \frac{|z|^{2n}}{\sqrt{G(|z|^2)}} d\mu_z \qquad(17)$$

The measure $\frac{\rho(x)}{\sqrt{G(x)}}$, moments of which are calculated in 17, can be in our case found empirically: this is the probability density function of the kernel words according to the frequency of its occurrence, in descending order (Fig. 5).

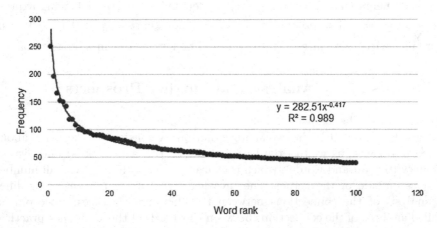

Fig. 5. The distribution of words in the basin of the word "предмет" by the frequency of their occurrence

The frequency decreases approximately as $\frac{1}{\sqrt{(n)}}$, where n is the index of the word. The radius R of the circle M in 17 is equal to the square of the maximum word number in the above distribution, i.e. the index of the rarest word in the kernel.

Thus, the "measure of the dictionary" is defined in the ring $1 \leq R^2 \leq N$. Q_N in 17 are calculated, and the function $G(x)$ in 15. As a result, we can calculate the measure $\rho(x)$.

Now we can represent non-orthogonal decomposition of a unity, related to the described coherent states, as follows:

$$\hat{I} = \sum_{n=0}^{\infty} |n\rangle \langle n| = \frac{1}{\pi} |z\rangle \langle z| \sqrt{G(|z|^2)} \rho(|z|^2) dz d\bar{z} \tag{18}$$

It follows from 15 that if we have two words, i.e. two coherent states $|z\rangle$ and $|w\rangle$, then the projection of one of them onto the other, understood as a scalar product, is defined by the formula

$$
\begin{aligned}
\langle z \,|\, w \rangle &= \frac{1}{G(|z|^2)G(|w|^2)} \sum_{k,n=0}^{\infty} \frac{\bar{z}^k w^n}{\sqrt{Q_k Q_n}} \langle k \,|\, n \rangle \\
&= \frac{1}{G(|z|^2)G(|w|^2)} \sum_{k,n=0}^{\infty} \frac{(\bar{z}w)^n}{Q_n} = \frac{G(\bar{z}w)}{G(|z|^2)G(|w|^2)}
\end{aligned}
\tag{19}
$$

This expression can be considered a quantitative measure of semantic closeness. In practice, the corresponding value of $G(zw)$ is calculated on the real line. Similarly, the angle between statements can be determined. They are determined as follows: $|\psi\rangle = \sum_k a_k |z_k\rangle$, $|\varphi\rangle = \sum_l b_l |w_l\rangle$. The closeness between these statements (sentences, texts, etc.) is determined by the following expression: $\frac{\langle \varphi \,|\, \psi \rangle}{\sqrt{\langle \varphi \,|\, \varphi \rangle \langle \psi \,|\, \psi \rangle}}$ The proximity evaluation that guarantees understanding in accordance with 19 requires a study that will be carried out in a separate paper.

7   Results of the Analysis and Further Prospects

In this research, the following results were obtained.

It was found out that the explanatory dictionary consists of a kernel of about 4 thousand words, as well as the kernel shell (25 thousand words) and distant periphery (6 thousand words), which is characterized by having a small number of cycles outside of the kernel. The revealed structure significantly simplifies the analysis of the connections between the dictionary articles, since only a detailed analysis of the connections between the words of the kernel has practical importance.

Also, in the analyzed dictionary, logical errors were found, such as words determined by themselves (loops in the graph), definition of different words with the same article, but not linked to each other, incorrectly defined words with several meanings and so on.

One of the questions that require special study is how can the dictionary definitions be corrected to avoid direct logical errors, and what is the kernel sensitivity to the choice of defining words.

Of course, some remarks about the correctness of the dictionary can be considered insignificant or easily fixable. But they lead to the problem of the stability of the ontology depth depending on the formulations choice. That question, however, is beyond the scope of this research.

References

1. Segalovich, I.: A fast morphological algorithm with unknown word guessing induced by a dictionary for a web search engine. In: Proceedings of the International Conference on Machine Learning; Models, Technologies and Applications. MLMTA 2003, Las Vegas, pp. 273–280 (2003)
2. Porter, M.: An algorithm for suffix stripping. Program **14**(3), 130–137 (1980)
3. Larsen, B., Chinatsu, A.: Fast and effective text inning using linear-time document clustering. In: Proceedings of KDD, pp. 16–22 (1999)
4. Kukich, K.: Techniques for automatically correcting words in text. ACM Comput. Surv. **24**(4), 377–439 (1992)
5. Cormen, T.H., Leiserson, C.E., Rivest, R.L., Stein, C.: Introduction to Algorithms, 3rd edn. The MIT Press, Cambridge (2009)
6. Manning, C.D., Raghavan, P., Schütze, H.: Introductionto Information Retrieval. Cambridge University Press, Cambridge (2008)
7. Ozhegov, S.I., Shvedova, N.: Explanatory dictionary of the Russian language: Russian language (1989)
8. Orlov, Y.N., Osminin, K.P.: Methods of statistical analysis of literary texts. Editorial URSS. Book House "LIBROKOM" (2012)
9. Berezin, F.A.: Second Quantization Method. Nauka, Moscow (1986)
10. Orlov, Y.N.: Fundamentals of Quantization of Degenerate Dynamical Systems. MIPT, Moscow (2004)

Performance Optimization of a Clustering Adaptive Gravitational Search Scheme for Wireless Sensor Networks

Elham Pourabdollah[1], Reza Mohammadi Asl[2], and Theodore Tsiligiridis[3(✉)]

[1] Information Technology Engineering Department, Faculty of Electrical and Computer Engineering, University of Tabriz, Tabriz, Iran
[2] Control Engineering Department, Faculty of Electrical and Computer Engineering, University of Tabriz, Tabriz, Iran
[3] Informatics Laboratory, Department of Economics and Rural Development, Agricultural University of Athens, Athens, Greece
tsili@aua.gr

Abstract. In this research we propose a new clustering scheme based on a combination of a well known stochastic, population-based Gravitational Search Algorithm (GSA) and the k-means algorithm to select optimal reference nodes in a Wireless Sensor Networks (WSN). In the proposed scheme the process of grouping sensors into clusters reference nodes is based on a K-means clustering algorithm to divide the initial population and select the best position in the neighbourhood to exchange information between clusters. In cases when sensor nodes receive multiple synchronization messages from more than one reference node a weighted average method is used. In this paper we limit our research on a number of benchmark functions which are used to compare the performance of the proposed algorithm with other important meta-heuristic algorithms to show its superiority.

1 Introduction

Wireless sensor networks (WSNs), are a set of connected tiny sensors, which collaborate with each other in order to reach a specific goal. In a WSN, sensors capture information from the environment and do some processes on them, then send results to the sink or base station. WSNs are being in use in a wide range of applications such as agro-environmental to monitor temperature, soil moisture, light, humidity and other micro-climatic parameters, military to track the path military vehicles, traffic control, health, as well as home applications. Characteristics and limitations of WSNs, especially their limited energy source, encounter them with some challenges, one of which is the clock or time synchronization, used to determine the timing of the event in many applications. For example, in the application of forest fire monitoring wireless sensor nodes at different positions report about fire when fire enters into their range. Sensor readings about fire and times are reported to sink node where data is combined to get a result. A further example could be the target tracking. In this application, the wireless

© Springer International Publishing AG 2017
O. Galinina et al. (Eds.): NEW2AN/ruSMART/NsCC 2017, LNCS 10531, pp. 420–431, 2017.
DOI: 10.1007/978-3-319-67380-6_38

sensor nodes are deployed in fixed positions and report position and time to sink node. At sink node position and time data from different nodes is combined to find position of target. In both cases clock synchronization of different sensor nodes is essential. Clock synchronization is also crucial for medium access control (MAC) protocol such as TDMA. Sensors use battery, however, because of their small size they have limited power. To save energy all sensor nodes can sleep and wake-up at the same time. Thus, synchronization of sensor nodes in network is essential [26], however, low-cost sensor nodes may cause offset and skew drifts, which cause the sensors be asynchronous with each other. For this reason, sensors should have different method of synchronization for different requirements.

The clock synchronization problem, which is used to synchronize the local clocks of sensor nodes in the WSN, have been extensively studied over the last two decades and it is still an important issue to study [1–3,6,8,9,20,23]. There are lots of strategies and protocols which addressed the problem; however, low-cost sensor nodes may cause offset and skew drifts, which cause the sensors be asynchronous with each other. Some other protocols are based on distributed consensus-based approaches for global clock synchronization [15,17,22,24]. The limitations of these kind of protocols is that they do not consider the propagation delay, which is an important issue and it cannot be neglected in real clock synchronization scenarios, and also they have slow convergence. A Group Neighbourhood Average (GNA) protocol is proposed in [16], which is a distributed, network-wide clock synchronization scheme.

In previous protocols, reference nodes are selected randomly. Therefore, some of the synchronized nodes may be located near each other, while some other parts of the network may be not synchronized at all. In the present work a new method to select the correct reference node is proposed, by means of selecting the one which should have higher accuracy and also higher remaining energy. Note that selecting an incorrect reference node, increases the message exchanging, which decreases remaining energy of sensor nodes. Therefore, the goal is to use an intelligence algorithm to select reference nodes in a way that decreases exchanging messages, while increases the synchronization accuracy. In other words, a node should selected as a reference node which has large number of neighbour nodes that it can synchronize them, and also it has higher remaining energy.

There are many heuristic evolutionary optimization schemes which have been developed so far and can be used as a basis to solve the synchronization problem in WSNs. These include Genetic Algorithm (GA), Deferential Evolution (DE), Particle Swarm Optimization (PSO), Ant Colony (AC), and Gravitational Search Algorithm (GSA). Among different evolutionary algorithms, Gravitational Search Algorithm (GSA) [21] is a relatively new and strong algorithm, which can be applied. There are some studies in the literature which have been done to synthesize PSO with the above algorithms such as hybrid PSOGSA [19], PSOGA [7,12], PSODE [14], and PSOACO [10]. These hybrid algorithms are aimed at reducing the probability of trapping in local optimum. Using above concepts, we propose a novel protocol for clock synchronization in WSNs, which is a modified version of the Clustered Adaptive GSA (CA-GSA). The modification is based on a weighted group neighbourhood average approach combining the GSA and PSO schemes in order to select reference nodes and to increase the convergence speed. This work contributes specifically to the validation of the performance of the proposed CA-GSA scheme. We use 10 benchmark functions [25] to compare the performance of new CA-GSA to the performance of other meta-heuristic algorithms including Genetic Algorithm (GA), differential evolution (DE), Particle Swarm Optimization (PSO), and Gravitational Search Algorithm (GSA) and to show its superiority.

The rest of this paper is organized as follows: Sect. 2 presents the background information used, by means the GSA and data clustering schemes. Section 3 presents the methodology of the CA-GSA scheme and explains its steps. The experimental results obtained by the comparison are reported in Sect. 4. Finally Sect. 5 concludes the paper.

2   Background

2.1   Gravitational Search Algorithm

The standard GSA introduced in [21] is inspired by the Newtonian laws of gravity and motion. This swarm optimization technique provides an iterative method that simulates object interactions, and moves through a multi-dimensional search space under the influence of gravitation. The effectiveness of the GSA in solving a set of nonlinear benchmark functions has been proven also in [21].

In the basic model of the GSA, which has been originally designed to solve continuous optimization problems, a set of agents is introduced in the D-dimensional solution space of the problem to find the optimum solution. The position of each agent in GSA demonstrates a candidate solution to the problem; hence each agent is represented by the vector X_i in the solution space of the problem. Agents with a higher performance get a greater mass value, because a heavy agent has a large effective attraction radius and hence a great intensity of the attraction. During the lifetime of GSA, each agent successively adjusts its position X_i toward the positions of best agents of the population.

To describe GSA in more detail, consider a D-dimensional space with N agents, say A_i, in which its position X_i is defined as:

$$X_i = (x_i^1, x_i^2, \ldots, x_i^d, \ldots, x_i^D); \; i = 1, 2, \ldots, N \tag{1}$$

where x_i^d presents the position of the agent A_i in the $d - th$ dimension. The number N of agents is assumed fixed, whereas the gravitational mass value $M_i(t)$ of the agent A_i in the iteration t is calculated after computing the current fitness value (solution) $fit_i(t)$ as follows:

$$q_i(t) = \frac{fit_i(t) - worst(t)}{best(t) - worst(t)} \tag{2}$$

$$M_i(t) = \frac{q_i(t)}{\sum_{j=1}^{s} q_j(t)} \tag{3}$$

In the above, $best(t)$ and $worst(t)$ is the best and worst solution in the iteration t defined as follows:

$$best(t) = min_{j\in\{1,...,N\}}fit_j(t) \text{ and} \tag{4}$$

$$worst(t) = max_{j\in\{1,...,N\}}fit_j(t) \tag{5}$$

By using law of motion the acceleration of the $i - th$ agent is calculated as:

$$\alpha_i^d(t) = G(t) \sum_{j\in N_{best},j\neq i} r_j \frac{M_j(t)}{R_{ij}(t) + \epsilon}(x_j^d(t) - x_i^d(t)) \tag{6}$$

where:

- $r_j \in U[0,1]$ is a uniformly distributed random number.
- ϵ is a very small value (usually $\epsilon = 0.1$) used to escape from division by zero error whenever the Euclidean distance between two agents i and j is zero.
- $R_{ij}(t) = ||X_i(t), X_j(t)||_2$ is the Euclidean distance between agents A_i and A_j.
- *Exploration control*: From the total of N agents, only N_{best} will attract the others. Thus N_{best} is the set of first N_{first} agents with the best fitness value and biggest gravitational mass. N_{first} is a function of time, with the initial value N_0 at the beginning and decreasing with time, and
- $G(t)$ is the gravitational constant at time t that will take an initial value, G_0, at the beginning of the search and it will be reduced with time toward a final value, G_{end}, to control the search accuracy as follows:

$$G(t) = G(G_0; G_{end}; t) = G_0 e^{-\gamma \frac{t}{T}} \tag{7}$$

In the above T is the total number of iterations (also presents the maximum number of iterations allowed). G_0 is the gravitational constant of the initial time t_0 and γ is a parameter. These two parameters are set by simulation to 100 and 20 respectively [21].

Then, the next velocity of the agent A_i is calculated as a fraction of its current velocity added to its acceleration:

$$v_i^d(t + 1) = r_i v_i^d(t) + \alpha_i^d(t) \tag{8}$$

$$= r_i v_i^d(t) + G(t) \sum_{j\in N_{best},j\neq i} r_j \frac{M_j(t)}{R_{ij}(t) + \epsilon}(x_j^d(t) - x_i^d(t)) \tag{9}$$

whereas, its next position can be calculated as:

$$x_i^d(t + 1) = x_i^d(t) + v_i^d(t + 1) \tag{10}$$

or in a more compact form

$$V_i(t + 1) = r_i V_i(t) + \alpha_i(t), \tag{11}$$

$$X_i(t + 1) = X_i(t) + V_i(t + 1) \tag{12}$$

where $V_i = (v_i^1, v_i^2, \ldots, v_i^d, \ldots, v_i^P)$ is the velocity of agent A_i to move in the direction of the optimal position. Algorithm 1 presents the pseudo code of the original GSA:

Algorithm 1. Template of Gravitational Search Algorithm

---

1 **begin**
2 | Initialization
3 | Generate initial population
4 | Evaluate the fitness for each agent

5 **while** *While stopping criteria are not satisfied* **do**
6 | Update G, N_{first}, and N_{best}
7 | Calculate the acceleration of each agent by Eqn (6)
8 | Calculate the velocity of each agent by by Eqn (9)
9 | Update the position of each agent by by Eqn (10)
10 | Evaluate the fitness for each agent

  Output: Best solution found

In the GSA, at first, all agents are initialized with random values. Over the course of iteration, the velocity and position is defined by Eqs. (9) and (10) respectively. Finally, the GSA algorithm is terminated by satisfying a stop criterion. Rashedi et al. [21] have been compared GSA with some well known heuristic search methods. The high performance of GSA has been confirmed in solving various non-linear functions [25].

2.2   Data Clustering Scheme

Clustering is a method of partitioning a given set of N agents $A = \{A_1, A_2, \ldots A_N\}$, each of them explained by d components or features into a finite number of k clusters $C = \{C_1, C_2, \ldots, C_k\}$, so that the agents in the same clusters are similar and related and agents in different clusters are dissimilar and unrelated. The clusters should contain at least one agent i.e. $C_i \neq \emptyset$, $\forall i \in \{1, 2, \ldots, k\}$, two different clusters should not have common agents i.e., $C_i \cap C_j = \emptyset$, $\forall i, j \in \{1, 2, \ldots, k\}$ with $i \neq $ J, and finally, each agent should be assigned to a cluster i.e. $\cup_{i=1}^k C_i$.

Since there are different ways to cluster a given set of agents, a fitness function for measuring the goodness of clustering should be defined. A widely used function for this purpose is the total Mean-Square quantization Error (MSE) [11] which is defined as follows:

$$f(A, C) = \sum_{l=1}^k \sum_{A_i \in C_l} d(A_i, \mu_l)^2 \quad \text{where} \quad d(A_i, A_j) = \sqrt{\sum_{p=1}^d (A_i^p - A_j^p)^2} \quad (13)$$

In the above $d(A_i, \mu_l)$ is the distance between agent A_i and the cluster centroid μ_l, whereas to measure the dissimilarity between two agents A_i and A_j the most known metric, by means of the distance Euclidean distance is used. Note that the agent A_i, $i = 1, 2, \ldots, N$ can be represented by the d–dimensional array as $A_i = (s_i^1, s_i^2, \ldots, s_i^d)$, where $s_i^j \in A_i$, $j = 1, 2, \ldots, d$ is a scalar denoting the $j - th$ component of the corresponding agent. The main steps of the K-means Algorithm 2 [11] are summarised as follows:

Algorithm 2. Template of K-means Algorithm

Input: No of clusters, say k

Output: Clustering solution found

1 Initialization:
2 Randomly choose k agents as the initial centroids
   $\mu_i = $ some value, $i = 1, \ldots, k$
3 Assign each agent to the closest centroid
   $c_i = \{j : d(x_j, \mu_i) \leq d(x_j, \mu_l),\ l \neq i,\ j = 1, \ldots, N\}$
4 Set the position of each cluster to the mean of all data points belonging to that cluster $\mu_i = \frac{1}{|c_i|} \sum_{j \in c_i} x_j,\ \forall i$
5 Update the locations of each centroid by calculating the mean value of the agents assigned to it
6 Repeat Steps 2 and 5 until a termination criterion is met (the maximum number of iterations is reached or the means are fixed)

The algorithm takes the number of the clusters as input parameter and organizes the agents into exactly k partitions where each partition represents a cluster. This method generates initial centroids and assigns each data object in the dataset to the nearest centroid and evaluates the objective function. It then gradually moves to a new centroid and evaluates the objective function again and compares to the previous objective value. If the new centroid provides a lower objective value, then the new centroid is kept and movement continues along the same direction. If the objective value is higher, the new centroid is ignored and movement continues in the opposite direction. The move will be stopped when the maximum number of iterations is reached or the quality of solution is acceptable. Details of the method explaining the generation and relocation of the centroids can be seen in [11, 18].

K-mean is a fast and simple algorithm, however it is very sensitive to the selection of the initial centroids, meaning, that the different centroids may produce completely different results. In addition, an incorrect choice of the number of clusters will invalidate the whole process. An empirical way to find the best number of clusters is to try K-means clustering with different number of clusters and measure the resulting sum of squares. A benchmarking of some procedures for estimating the number of clusters may be seen in [18], although some approaches to solve this problem can be found in [4,5]. Finally, the algorithm eventually converges to a point, although it is not necessarily the minimum of the sum of squares. That is because the problem is non-convex and the algorithm is just a heuristic, converging to a local minimum. The algorithm stops when the assignments do not change from one iteration to the next.

3   Clustered Adaptive GSA: Methodology

The purpose of incorporating clustering methodology into the process of evolutionary algorithm is to split a population of agents based on the property
of objects. This leads to form multiple sub-populations in a dynamic structure which will improve the grouping behaviour of agents. Thus, a new velocity
updating is provided to adjust the movement of an object in which the effects
from best ones in the dynamic sub-populations are considered. After creating
sub-populations, some agents should explore new areas and some others should
exploit local areas. Therefore, an adaptive parameter adjustment method is used
to evaluate the positions of the sub-populations after clustering. The proposed
algorithm has two main parts: First, using K-means algorithm as a clustering method to split initial population into sub-population. Then, applying ring
topology to exchange information between clusters. Second, adaptive parameters
to adjust the behaviour of agents based on states of the sub-populations. The
main procedures of the proposed algorithm are as follows.

- N agents with random positions X_i and velocities v_i are generated.
- For each agent calculate the fitness f_i, $i = 1, 2, \ldots, N$.
- For each agent initialize the weights w_{max}, w_{min} and w_i.
- Clustering
    - Randomly select k agents and set them as centres of the clusters.
    - Use a clustering method (in this case the k-means algorithm) to obtain
    sub-populations C_j, $j = 1, 2, \ldots, k$.
    - Select the agent with highest mass, in cluster C_j. Set this as the best
    agent Cb_j.
    - Select best position in the neighbourhood, say $nbest_i$, by comparison
    among Cb_{j-1}, Cb_j and Cb_{j+1}.

Note that the best position $nbest_i$ of the best agent A_i, indicated by Cb_j,
in class C_j has an important effect on the process of movement of the agents
towards $nbest_i$ which is selected by comparison among Cb_{j-1}, Cb_j and Cb_{j+1}.
Therefore the velocity of each particle will be updated by using a modification
of the Eq. (10) as follows:

$$v_i(t + 1) = \omega v_i(t) + \alpha_i(t) + c(nb_i - x_i) \tag{14}$$

Obviously, each agent should update its position based on previous best and
neighbourhood best provided by the Eq. (10). In the above we used the following
definitions:

- The average fitness of the class C_j is:

$$\alpha_j = \sum_{i=1}^{|C_j|} \frac{f_{ij}}{|C_j|} \tag{15}$$

where $|C_j|$ states for the population of cluster C_j and f_{ij} represents the fitness
of the agent A_i in cluster C_j.

- The average fitness of swarm will be:

$$A = \sum_{i=1}^{N} \frac{f_i}{N} \tag{16}$$

where N is the population of agents and f_i is the fitness of agent A_i.
- ω is the inertia weight coefficient used in velocity update formula instead of a random number, in order to bring adaptation for GSA. It actually controls the impact of the previous velocity of a particle on its current one. Therefore, ω_{min} and ω_{max} show initial minimum and maximum inertia weights of agents, respectively. Inertia weight of agent A_i in generation t is denoted by $\omega_i(t+1)$. Based on its state and using ω_{min}, ω_{max} and $\omega(t)$ it can be adjusted.
- c is a parameter controlling how far an agent will move in one generation.
- nb_i shows the best position in the neighbourhood

Note that, for the adaptation strategy, if an agent is not in a good position, it should speed up, whereas, in case its position is appropriate, it should slow down its movement to find better positions. Therefore, using clustering method, the positions of the agents will be improved. The average fitness of clusters is different with each other and we use it to adjust inertia weight of an agent. For the maximization problem, the adjustment for two mentioned situations will be as follows:

case-1: $\alpha_j \geq A$. This means that sub-population has a better position and therefore, it is good to reduce the velocity in order to improve local search ability and find optimal solution. This can be done by reducing ω as follows:

$$\omega_i(t+1) = \omega_i(t) - min\{|\omega_{max} - \omega_i(t)|, |\omega_{min} - \omega_i(t)|\} \tag{17}$$

case-2: $\alpha_j < A$. This means that current position is far from best solution. Thus, the velocity should increase to improve global search ability and escape from local search. This can be done by increasing ω as follows:

$$\omega_i(t+1) = \omega_i(t) + max\{|\omega_{max} - \omega_i(t)|, |\omega_{min} - \omega_i(t)|\} \tag{18}$$

The Algorithm 3 describes implementation of proposed CA-GSA for a maximization problem.

4   Results

This section compares the performance of CA-GSA to the performance of other meta-heuristic algorithms including Genetic Algorithm (GA), differential evolution (DE), Particle Swarm Optimization (PSO), and Gravitational Search Algorithm (GSA) on 10 benchmark functions [25]. The first five functions are two-dimensional ($D = 2$), whereas the rest are thirty-dimensional ($D = 30$). All functions may be separated into the type categories of multi-modal/uni-modal and separable/non-separable. Table 1 provides benchmark function details. In all runs and in all algorithms we kept the number of agents equal to N = 50. For PSO we used $\omega = 0.85, \alpha = \beta \approx 2$.

Algorithm 3. CA-GSA algorithm

---

1 **begin**
2 | Initialization
3 | Initialize constant parameters
4 | Create N agents and make randomized initialization of their $D-$dimensional positions X_i and velocities V_i
5 | Initialize first iteration $t = 1$

6 **for** $i \leftarrow 1$ **to** N **do**
7 | Calculate fit_i^t
8 | Initialize ω_i, ω_{min}, **and** ω_{max}

9 End Initialization
10 **repeat**
11 | Clustering: Select randomly k agents as the centres of clusters. Use clustering method (i.e., k-means), to divide population and obtain sub-populations C_j, where j is a member of N_k.
12 | Calculate best fitness in each cluster, $bestC_j^t$: agent with highest mass.
13 | Set it as the best position of cluster: Cb_j^t.
14 | Calculate worst fitness in each cluster, $worst_{C_j}^t$: agent with minimum mass.
15 | Choose the best position in the neighbourhood as nb_i for each agent by comparison among Cb_{j-1}, Cb_j and Cb_{j+1}.
16 **until** *End Clustering*
17 **repeat**
18 | **for** $j \leftarrow 1$ **to** N_k **do**
19 | | **for** $i \leftarrow 1$ **to** N_j **do**
20 | | | Calculate the fit_{ij}
21 | | | Calculate mass M_{ij}^t using Eqn (3)
22 | | | Calculate gravitational constant using Eqn (7)
23 | | | Calculate Euclidean distance and force between agent i and nb_i
24 | | Calculate average fitness α_j of cluster C_j, using Eqn (15)
25 | Calculate average fitness of swarm, A, using Eqn (16)
26 | **if** $\alpha_j \geq A$ **then**
27 | | Adjust inertia weight using Eqn (17)
28 | **else**
29 | | Adjust inertia weight using Eqn (18)
30 | **for** $j \leftarrow 1$ **to** N_k **do**
31 | | **for** $i \leftarrow 1$ **to** N_j **do**
32 | | | Calculate acceleration of agent i using Eqn (6)
33 | | | Update the velocity using Eqn (8)
34 | | | Update the position using Eqn (10)
35 | $t = t + 1$
36 **until** *termination criterion is satisfied*
   Result: Write the solution found
37 END Adaptive GSA in Clusters
38 END

Table 1. The benchmark functions

Function	Formula	Min
Beale	$f(x) =$ $(1.5-x_1+x_1x_2)^2+(2.25-x_1+x_1x_2)^2+(2.625-x_1+x_1x_2)^2$	0
Bohachevsky 1	$f(x) = x_1^2 + 2x_2^2 - 0.3\cos(3\pi x_1) - 0.4\cos(\pi x_2) + 0.7$	0
Bohachevsky 2	$f(x) = x_1^2 + 2x_2^2 - 0.3\cos(3\pi x_1)(4\pi x_2) + 0.3$	0
Bohachevsky 3	$f(x) = x_1^2 + 2x_2^2 - 0.3\cos(3\pi x_1 + 4\pi x_2) + 0.3$	0
Booth	$f(x) = (x_1^2 + 2x_2 - 7)^2 + (2x_1 + x_2 - 5)^2$	0
Achley	$f(x) = -20\exp\left(-0.2\sqrt{\frac{1}{n}\sum_{i=1}^{D}x_i^2}\right) -$ $\exp\left(\sqrt{\frac{1}{n}\sum_{i=1}^{D}cos(2\pi x_i)}\right) + 20 + e$	0
Schwefel 2.22	$f(x) = \sum_{i=1}^{D}|x_i| + \Pi_{i=1}^{D}|x_i|$	0
Rosenbrock	$f(x) = \sum_{i=1}^{D-1}100(x_{i+1} - x_i^2)^2 + (x_i - 1)^2$	0
Rastrigin	$f(x) = \sum_{i=1}^{D}(x_i^2 - 10\cos(\pi x_i) + 10)$	0
Griewank	$f(x) = \frac{1}{4000}\left(\sum_{i=1}^{D}(x_i - 100)^2\right) - \left(\Pi_{i=1}^{D}\cos\left(\frac{x_i-100}{\sqrt{(i)}}\right)\right)$	0

Table 2. Comparative results of CA-GSA with GA, DE, PSO and GSA

Function	GA	DE	PSO	GSA	CA-GSA
Beale	0	0	0	0	0
Bohachevsky-1	0	0	0	0	0
Bohachevsky-2	0.0527	0	0	0.0638	0
Bohachevsky-3	0	0	0	0	0
Booth	0	0	0	2.41E−05	0
Achley	0.0778	5.5E−06	0.0665		0
Schwefel 2.22	0	0	0	0	0
Rosenbrock	2.1E−07	4.1E−09	4.8E−07	7.1E−08	1.1E−09
Rastrigin	8.7E−08	7.7E−08	1.3E−06	2.4E−07	7.9E−08
Griewank	0	7E−07	5.7E−07	0.0753	0

DE is similar to GA except that the candidate solutions are not considered as binary strings (chromosome) but, usually, as real vectors. It is an optimisation technique which iteratively modifies a population of candidate solutions to make it converge to an optimum of the function we are studying. One key aspect of DE is that the mutation step size is dynamic, that is, it adapts to the configuration of the population and will tend to zero when it converges. This makes DE less vulnerable to genetic drift than GA. For the DE algorithm the crossover parameter and the mutation factor are taken equal to 0.95 and 0.6, respectively. The stop criteria in all of the algorithms is 10000 maximum number of function evaluations. Any value less than 1E−10 is noted as 0.

Table 2 delineates the respective performance of CA-GSA and other algorithms in solving benchmark functions. All of tests show better or at least comparable efficiency of the proposed algorithm in comparison with the basic algorithm.

5   Conclusion

In this research, a new clustering algorithm is introduced utilizing strengths of PSO and GSA. The main idea is to integrate the abilities of PSO in exploitation and GSA in exploration. Ten benchmark functions are used to validate the performance of the CA-GSA compared to standard GA, PSO and GSA and DE. The results show that CA-GSA performs better in almost all function minimization. The results are also proved that the convergence speed of CA-GSA is faster that PSO and GSA.

References

1. Bagul, D., Kurumbanshi, S., Verma, U.: Survey on clock synchronization in WSN. Int. J. Eng. Sci. Invention **2**(12), 24–31 (2013)
2. Bholane, S., Thakore, D.: Time synchronization in wireless sensor networks. Int. J. Sci. Eng. Res. **3**(7), 1–6 (2012)
3. Cena, G., Scanzio, S., Valenzano, A., Zunino, C.: Evaluation of the reference broadcast infrastructure synchronization protocol. IEEE Trans. Ind. Inf. **11**, 801–811 (2015). IEEE Press
4. Das, S., Abraham, A., Konar, A.: Automatic clustering using an improved differential evolution algorithm. IEEE Trans. Syst. Man Cyber. Part A Syst. Hum. **38**(1), 218–237 (2008)
5. Das, S., Abraham, A., Konar, A.: Automatic hard clustering using improved differential evolution algorithm. Stud. Comput. Intell. 137–174 (2009)
6. Elson, J., Girod, L., Estrin, D.: Fine-grained network time synchronization using reference broadcasts. ACM SIGOPS Operating Syst. Rev. **36**(SI), 147–163 (2002)
7. Esmin, A., Lambert-Torres, G., Alvarenga, G.: Hybrid evolutionary algorithm based on PSO and GA mutation. In: Proceeding of the Sixth International Conference on Hybrid Intelligent Systems (HIS 2006), p. 57 (2007)
8. Ganeriwal, S., Kumar, R., Srivastava, M.: Timing-sync protocol for sensor networks. In: Proceedings of the 1st International Conference on Embedded Networked Sensor Systems, SENSYS 2003, pp. 138–149. ACM (2003)
9. Garone, E., Gasparri, A., Lamonaca, F.: Clock synchronization protocol for wireless sensor networks with bounded communication delays. Automatica **59**, 60–72 (2015)
10. Holden, N., Freitas, A.: A hybrid PSO/ACO algorithm for discover in classification rules in data mining. J. Artif. Evol. Appl. (JAEA) **2008** (2008). Article ID 316145, Hindawi Publishing Corporation
11. Jain, A.K.: Data clustering: 50 years beyond K-means. Pattern Recogn. Lett. **31**, 651–666 (2010)
12. Lai, X., Zhang, M.: An efficient ensemble of GA and PSO for real function optimization. In: 2nd IEEE International Conference on Computer Science and Information Technology, pp. 651–655 (2009)

13. Lasassmeh, S., Conrad, J.: Time synchronization in wireless sensor networks: a survey. In: Proceedings of the IEEE SoutheastCon 2010 (SoutheastCon), pp. 242–245 (2010)
14. Niu, B., Li, L.: A novel PSO-DE-based hybrid algorithm for global optimization. In: Huang, D.-S., Wunsch, D.C., Levine, D.S., Jo, K.-H. (eds.) ICIC 2008. LNCS, vol. 5227, pp. 156–163. Springer, Heidelberg (2008). doi:10.1007/978-3-540-85984-0_20
15. Li, Q., Rus, D.: Global clock synchronization in sensor networks. IEEE Trans. Comput. **55**(2), 214–226 (2006)
16. Lin, L., Ma, S., Ma, M.: A group neighborhood average clock synchronization protocol for wireless sensor networks. Sensors **2014**(14), 14744–14764 (2014)
17. Maggs, M., O'Keefe, S., Thiel, D.: Consensus clock synchronization for wireless sensor networks. IEEE Sens. J. **12**(6), 2269–2277 (2012)
18. Milligan, G., Cooper, M.: An examination of procedures for determining the number of clusters in a data set. Psychometrika **50**(2), 159–179 (1985)
19. Mirjalili, S., Hashim, S.Z.M.: A new hybrid PSOGSA algorithm for function optimization. In: International Conference on Computer and Information Application, ICCIA 2010, pp. 374–377. IEEE (2010)
20. Ranganathan, P., Nygard, K.: Time synchronization in wireless sensor networks: a survey. Int. J. UbiComp **1**(2), 92–102 (2010)
21. Rashedi, E., Nezamabadi-Pour, H., Saryazdi, S.: GSA: a gravitational search algorithm. Inf. Sci. **179**(13), 2232–2248 (2009)
22. Schenato, L., Fiorentin, F.: Average timesynch: a consensus-based protocol for clock synchronization in wireless sensor networks. Automatica **47**(9), 1878–1886 (2011)
23. Wu, Y.-C., Chaudhari, Q., Serpedin, E.: Clock synchronization of wireless sensor networks. IEEE Sig. Process. Mag. **28**(1), 124–138 (2011)
24. Wu, J., Zhang, L., Bai, Y., Sun, Y.: Cluster-based consensus time synchronization for wireless sensor networks. IEEE Sens. J. **15**(3), 124–138 (2015)
25. Yao, X., Liu, Y., Lin, G.: Evolutionary programming made faster. IEEE Trans. Evol. Comput. **3**(2), 82–102 (1999)
26. Yadav, K.-G., Kumar, A., Raghuvanshi, R.: Analysis of time synchronization protocols for wireless sensor networks: a survey. Int. J. Comput. Sci. Mob. Comput. **4**(5), 1062–1068 (2015)

A Retrial Queueing System with Preemptive Priority and Randomized Push-Out Mechanism

Alexander Ilyashenko[✉], Oleg Zayats, Maria Korenevskaya, and Vladimir Muliukha

Peter the Great St. Petersburg Polytechnic University, St. Petersburg, Russia
ilyashenko.alex@gmail.com, zay.oleg@gmail.com,
masha_kor95@mail.ru, vladimir@mail.neva.ru

Abstract. A single-server finite-buffer retrial system with a Poisson stream of arrivals and an exponential service time distribution is analyzed. If an arriving customer finds the buffer filled up, this customer joins a special retrial waiting group (orbit) in order to seek service again in an exponentially distributed period of time. The customers seeking service for the first time have the preemptive priority over retrial ones, thus, one can think of the original system as of the priority one with retrial customers being low-priority. We also introduce a randomized push-out buffer management mechanism, which makes it possible to control the loss probability of the high-priority and low-priority customers efficiently.

It is shown that such queueing model can be reduced to a similar one but without retrial. Using the generating function technique, main probabilistic characteristics (e.g., loss probabilities) for both types of the customers are obtained. The dependency of the loss probabilities on the model parameters, i.e., the push-out and retrial probabilities, is investigated. Regions of the load parameters are found such that the system is locked for the low-priority customers, or such that there is a linear dependence of the loss probabilities on the push-out probability.

Keywords: Priority queueing system · Retrial queueing system · Randomized push-out mechanism · Poisson arrivals · Exponential service time · Markov process · Steady-state distribution · Finite buffer · Preemptive priority

1 Introduction

The main analytic approach for the telematics devices investigation is to consider them as queuing systems of a specific type [1]. In order to describe real network interactions correctly one has to use rather complicated queuing system models. Firstly, the actual structure of information streams demands to use the queueing systems with multiple incoming flows. Secondly, the particular streams should be prioritized in order to make the system more efficient. The priority means that the specific types of customers are given some service privilege relative to other types. Such the models were successfully used in a series of space experiments "Kontur" and allowed one to change the

O. Galinina et al. (Eds.): NEW2AN/ruSMART/NsCC 2017, LNCS 10531, pp. 432–440, 2017.
DOI: 10.1007/978-3-319-67380-6_39

characteristics of telematics devices in a wide range, to improve the quality of the remote control with telepresence, and to strengthen feedback on a control device [2–4]. It this article will be considered retrial model and it will allow to make remote control system configuration more precise and thin-tuned. As a result remote control system will provide better quality and more accurate results for predicting network channel characteristics behavior.

The recent research has demonstrated that besides the service priority there should be a queueing priority as well, that is, some push-out mechanism. The push-out mechanism concept was initially proposed and developed by Basharin in 1960–1970s [5]. However, in the early stages only the deterministic push-out mechanism was considered. In that case the high-priority customers always push the low-priority ones out of the buffer when it is filled up.

Such a push-out mechanism is able to solve the packet switching control task only partially, as it causes the prevalence of the high-priority customers in the buffer. In [6–11] the randomized push-out mechanism, so that the pushing out occurs randomly with some probability α, was studied. The parameter α plays the role of the control parameter and is used to adapt the telematics devices to the network environment.

In practice it is interesting to the investigate retrial priority systems [12, 13]. Firstly for these systems the length of the low-priority customers queue was finite, while the length of the high-priority customers queue could be infinite [12]. Two-stream queueing systems with retrial customers and a randomized push-out mechanism are considered in [14].

We should especially mention the work of Bocharov et al. in which the priority is granted to the primary customers, but when a customer is reapplied it becomes of a low priority [15–17]. Such problems are extremely interesting in practice, but they are clearly not sufficiently investigated. As it was noted above, the presence of the retrial customers can radically change the behavior of a queueing system, even if the priorities are used. This is due to a large number of secondary customers trying to get into the system again. In many cases it is reasonable and natural to grant a priority to the customers entered into the system for the first time. This is the case, for example, when modeling of switching nodes in the information systems [17].

Earlier in [2–4] was shown that randomized push-out mechanism allowed one to change characteristics of a queueing model in a wide range. In this article the combination of the above mentioned extensions for queueing models similar to [15–17] in combination with a randomized push-out mechanism will be considered. It makes the fine tuning of the primary and secondary customers balance possible, because all customers will be splitted into three groups: primary, primary after retrial and secondary. This balance is important, for example, for the tasks of control of robotic complexes in space experiments, which is described in detail in [4].

2 Queueing System Model

Let us consider the queueing system with a single incoming elementary stream of intensity $\lambda_{1,0}$. We denote this flow of customers as the primary one. These customers are granted with the highest priority. When a system has a limited buffer size, there is a chance for these primary customers to be lost due to a lack of available storage in the buffer. Usually in these cases such customers are just lost irretrievably. However, in retrial queueing systems the lost customers have a possibility to return to the system again. Such customers form one more, secondary flow with lower priority than first one. This stream actually turns the model into the one with two incoming streams. In more detail, the behavior of retrial customers in the system can be described as follows.

The primary customers that are lost due to the lack of available storage in the buffer, fall into the orbit of the repeated customers with the probability equal to one. In the classical formulation of the problem of repeated customers, the probability of returning them back into the system from the orbit is equal to one for all types of customers. To prevent the orbit from clogging with the pushed-out customers, let us introduce one more parameter into the model. It is a probability q to fall to the orbit for the pushed out secondary customers (which have already returned to the system from the orbit before). So, if a secondary customer is pushed out then it comes back to the orbit again with a probability q and is irretrievably lost with the probability $(1 - q)$. The described queueing system is presented on the Fig. 1.

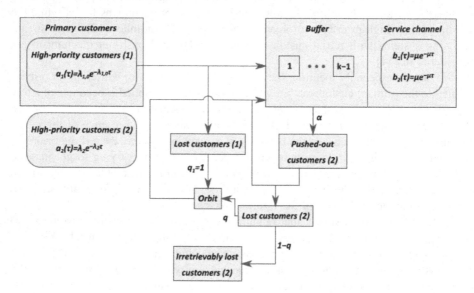

Fig. 1. The scheme of the retrial queueing system $\overrightarrow{M_2}/M/1/k/f_2^1$ with a single primary stream.

The investigation of such a system can be reduced to the case of an ordinary two-stream system with randomized pushing-out and without retrial customers. However, for this purpose it is necessary to use the intensities of the primary and secondary incoming streams derived from the considered system. Let us consider the process of obtaining these intensities in details.

In this case the intensities of the incoming streams can be derived as follows:

$$
\begin{cases}
\lambda_1 = \lambda_{1,0}; \\
\lambda_2 = \lambda_{1,rep} + \lambda_{2,rep} = \dfrac{\lambda_{1,0}P_{loss}^{(1)}}{1-P_{loss}^{(1)}} + \dfrac{\lambda_2 q P_{loss}^{(2)}}{1-q P_{loss}^{(2)}},
\end{cases}
\tag{1}
$$

where $\lambda_{1,rep}$, $\lambda_{2,rep}$ are the intensities of primary and secondary streams, respectively; $P_{loss}^{(1)}$ and $P_{loss}^{(2)}$ are the loss probabilities for the primary and secondary customers, respectively.

It can be concluded from the results of [7, 8] that the loss probabilities depend on four parameters: ρ_1, ρ_2, k and α, the first two of which are the load factors, respectively, for high- and low-priority traffic, and the third one is the total system capacity. Then, in order to determine the load factors, we can use the system of equations obtained from the Eq. (1) by dividing both of them by the service intensity μ in the service channel:

$$
\begin{cases}
\rho_1 = \rho_{1,0}; \\
\rho_2 = \dfrac{\rho_{1,0}\varphi_1(\rho_1,\rho_2,k,\alpha)}{1-\varphi_1(\rho_1,\rho_2,k,\alpha)} \cdot \dfrac{1-q\varphi_2(\rho_1,\rho_2,k,\alpha)}{1-2q\varphi_2(\rho_1,\rho_2,k,\alpha)},
\end{cases}
\tag{2}
$$

The functions of the four arguments mentioned above state the expressions for the loss probabilities.

Using the above expressions for the load factors (2) and for the loss probabilities obtained by the methods of [7, 8], it is easy to found a numerical value of the intensity of the secondary stream, and hence the load factor for this stream. After that one can find values of the probabilities of losing the customers, given all the system parameters and the probability q of repeated service.

3 Generating Functions Method Application

Let us consider a system with a randomized push-out mechanism, preemptive priority and retrial customers in a steady state. The process in this system is Markovian. It is also ergodic, which ensures that the final probabilities exist and do not depend on the initial state of the system. They satisfy the stationary system of Kolmogorov equations.

Because initial model was reduced to model considered in [6–11], we will not provide detailed algorithm for obtaining main characteristics and mention only main results.

We first introduce the phase space and state probabilities of this model

$$
\Omega = \{(i,j) : i = \overline{0,k}, j = \overline{0,k-i}\}, \ P(i,j;t) = P\{N_1(t) = i, N_2(t) = j\}
\tag{3}
$$

where $N_q(t)$ denotes the number of customers of q-th type in the system at moment t.

Generating function of the final probabilities $P_{i,j}$ of the phase space (3) is defined as

$$G(u,v) = \sum_{i=0}^{k} \sum_{j=0}^{k-i} P_{i,j} u^i v^j \tag{4}$$

Using the state graph same as in [6–11], we obtain following equation for generating function (4)

$$[\lambda_1 u(1-u) + \lambda_2 u(1-v) + \mu(u-1)]vG(u,v)$$

$$= \mu(u-v)G(0,v) + (1-\alpha)\lambda_1 P_{0,k} v^k u(u-v)$$

$$+ \mu u(v-1)G(0,0) + \alpha\lambda_1 u^{k+1}(v-u)P_{k,0} \tag{5}$$

$$+ [\alpha\lambda_1(u-v) + \lambda_1(1-u)v + \lambda_2(1-v)v]u \sum_{i=0}^{k} P_{i,k-i} u^i v^{k-i}$$

In a way as it was described in [8, 9] expand it into the series in powers of u and v. For this purpose let's use the expressions of the system characteristics well-known from the investigation of conventional single-stream system of the class $M/M/1/k$. Specifically, the distribution of the total number of customers in the system is given by

$$r_n = \sum_{i=0}^{n} P_{i,n-i} = \sum_{i=0}^{n} P_{n-i,i} \tag{6}$$

Substituting the expression for the final probabilities from [8] into (6), we obtain the system of equations with respect to the "diagonal" probabilities $p_i = P_{k-i,i}$:

$$r_z = \sum_{j=0}^{z} P_{z-j,j} = p_0 \rho_1^{-1} \zeta_z - \alpha\varphi_z p_{z+1}$$

$$+ \sum_{j=1}^{z} p_j \xi_{z,j} + p_k(\alpha-1)\delta_{z,k-1}, (0 \le z \le k-1), \tag{7}$$

where

$$\zeta_z = \sum_{j=0}^{z} \rho_1^{(1-k+z-j)/2} \beta^j \left(C_{k-z-1}^{j+1} - \rho_1^{1/2} C_{k-z-2}^{j+1} \right);$$

$$\varphi_z = \rho_1^{(z+1-k)/2} C_{k-z-1}^1;$$

$$\beta = -\frac{\rho_2}{\sqrt{\rho_1}}; \tag{8}$$

$$\xi_{z,j} = \sum_{s=j}^{z} \rho_1^{(z-k-s+j)/2} \left(\rho_1^{-1/2} C_{k-z-1}^{s-j+1} - C_{k-z-2}^{s-j+1} \right) \beta^{s-j}$$

$$- \alpha\rho_1^{(j-k)/2} \beta^{z+1-j} C_{k-z-1}^{z-j+2}.$$

This system should be complemented by the Eq. (9) to make it of full rank.

$$r_k = \sum_{i=0}^{k} P_{k-i,i} = \sum_{i=0}^{k} p_i. \tag{9}$$

4 Computational Results

One of the most important characteristics of any telematics device model is a customer loss probability, which depends on its type and can be obtained for both types of customers from the above Eqs. (7)–(9) as:

$$P_{loss}^{(1)} = p_0 + (1 - \alpha) \sum_{i=1}^{k-1} p_i; \quad P_{loss}^{(2)} = r_k + \alpha \frac{\rho_1}{\rho_2} \sum_{i=1}^{k-1} p_i + \frac{\rho_1}{\rho_2} p_k. \quad (10)$$

The numerical computations of the dependencies of $P_{loss}^{(i)}$ on the probability α revealed interesting results presented on the figures below. Let's consider the dependency of the loss probability for each type of customers on parameters and q for different values of load factor $\rho = \rho_1$. The results for high- and low-priority customers are shown on Figs. 2, 3, 4 and Figs. 5, 6 and 7, respectively. All these results correspond to the values of $\rho_1 = 1.0; 1.6; 2.2$.

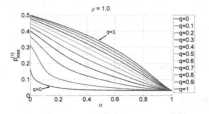

Fig. 2. Dependency of loss probability of high-priority packets on α and $q(\rho_1 = 1.0)$.

Fig. 3. Dependency of loss probability of high-priority packets on α and $q(\rho_1 = 1.6)$.

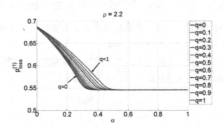

Fig. 4. Dependency of loss probability of high-priority packets on α and $q(\rho_1 = 2.2)$.

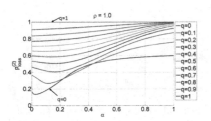

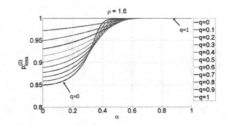

Fig. 5. Dependency of loss probability of low-priority packets on α and $q(\rho_1 = 1.0)$.

Fig. 6. Dependency of loss probability of low-priority packets on α and $q(\rho_1 = 1.6)$.

Figures 2, 3, 4, 5, 6 and 7 demonstrate that when the system load factor increases, the separate curves corresponding to the different values of q tend to be closer to each other. This means that the dependence of the loss probability on q gets weaker. And since the curves themselves are getting higher, then we can conclude that the greater the load factor is, the greater the loss probability will be.

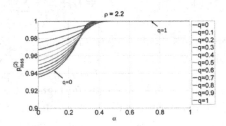

Fig. 7. Dependency of loss probability of high-priority packets on α and $q(\rho_1 = 2.2)$.

The papers [8–11] introduce the notion of areas of the "linear behavior", where the dependence of loss probability on the pushing out probability α is approximately linear. In our case, such a behavior of loss probability can be observed for only quite small values of load factor.

Figures 5, 6 and 7 demonstrate that some values of α lead to the situation when the probability of the low-priority customers is close to one, thus the low-priority customers have an extremely low chance to get into the system. This effect is referred to as "closing" effect in the papers [8–11], where the areas of this effect are found.

For the present system we can conclude that when the load factor is large enough, the variation of the probability q does not affect significantly on the closing effect. But if the load factor is close to the system throughput then $P_{loss}^{(2)}$ can be varied in a wide range and the closing effect can be achieved.

A one more interesting result we obtained is the dependency of a ratio $\frac{\rho_2}{\rho_1}$ on the system load factor ρ_1 shown on Figs. 8, 9, 10 and 11.

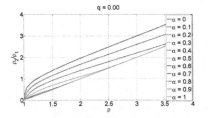

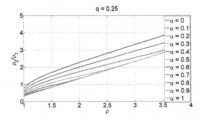

Fig. 8. Dependency of $\frac{\rho_2}{\rho_1}$ on load factor of system $q = 0$.

Fig. 9. Dependency of $\frac{\rho_2}{\rho_1}$ on load factor of system $q = 0.25$.

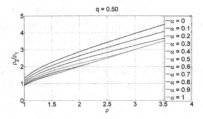
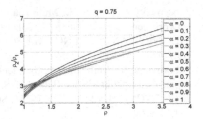

Fig. 10. Dependency of $\frac{\rho_2}{\rho_1}$ on load factor of system $q = 0.5$

Fig. 11. Dependency of $\frac{\rho_2}{\rho_1}$ on load factor of system $q = 0.75$

As it can be seen from Figs. 8, 9, 10 and 11, when pushing out probability α increases, the ratio of the load factors decreases, so the primary flow starts dominating over the secondary one. However, this ratio increases when the probability q increases. In all cases the maximum value of this ratio is reached when $\alpha = 0$ and $q = 1$.

5   Conclusion

The single-flow queueing system with retrial customers and randomized pushing out mechanism was investigated. The analytical expressions for the load factors of the system with the retrial customers are obtained. The way of reducing such a system to the two-flow model without retrial customers is demonstrated. The generating functions method is applied to find the final probabilities of system states. The numerical dependencies of loss probabilities on the model parameters are obtained as well.

The most significant results are obtained areas of parameters corresponding to the "closing" effect and to the "linear behavior" of the loss probabilities. These results may be used to reduce the computational load on the telematics devices operating in the real time mode. The computations of such areas performed in advance make it possible to choose the necessary parameter values immediately instead of computing them in runtime. This will lead to increase in the systems performance and improvement of controlling them.

Acknowledgement. This work was financially supported by the Ministry of Education and Science of the Russian Federation in the framework of the Federal Targeted Programme for Research and Development in Priority Areas of Advancement of the Russian Scientific and Technological Complex for 2014–2020 (No. 14.578.21.0211, ID RFMEFI57816X0211).

References

1. Vishnevsky, V.M.: Theoretical Bases if Designing Computer Networks, 512 p. Technosphere, Moscow (2003)
2. Muliukha, V., Ilyashenko, A., Laboshin, L.: Network-centric supervisory control system for mobile robotic groups. Procedia Comput. Sci. **103**, 505–510 (2017)

3. Zaborovsky, V., Guk, M., Muliukha, V., Ilyashenko, A.: Cyber-physical approach to the network-centric robot control problems. In: Balandin, S., Andreev, S., Koucheryavy, Y. (eds.) NEW2AN 2014. LNCS, vol. 8638, pp. 619–629. Springer, Heidelberg (2014). doi:10.1007/978-3-319-10353-2_57

4. Zaborovsky, V., Muliukha, V., Ilyashenko, A.: Cyber-physical approach in a series of space experiments "kontur". In: Balandin, S., Andreev, S., Koucheryavy, Y. (eds.) ruSMART 2015. LNCS, vol. 9247, pp. 745–758. Springer, Heidelberg (2015). doi:10.1007/978-3-319-23126-6_69

5. Basharin, G.P.: Some results for priority systems, queueing theory in data transfer systems, pp. 39–53 (1969)

6. Avrachenkov, K.E., Shevlyakov, G.L., Vilchevsky, N.O.: Randomized push-out disciplines in priority queueing. J. Math. Sci. 22(4), 3336–3342 (2004)

7. Avrachenkov, K.E., Vilchevsky, N.O., Shevlyakov, G.L.: Priority queueing with finite buffer size and randomized push-out mechanism. Perform. Eval. 61(1), 1–16 (2005)

8. Ilyashenko, A., Zayats, O., Muliukha, V., Laboshin, L.: Further investigations of the priority queueing system with preemptive priority and randomized push–out mechanism. In: Balandin, S., Andreev, S., Koucheryavy, Y. (eds.) NEW2AN 2014. LNCS, vol. 8636, pp. 433–443. Springer, Heidelberg (2014). doi:10.1007/978-3-319-10353-2_38

9. Muliukha, V., Ilyashenko, A., Zayats, O., Zaborovsky, V.: Preemptive queueing system with randomized push-out mechanism. Commun. Nonlinear Sci. Numer. Simul. 1(3), 147–158 (2015)

10. Ilyashenko, A., Zayats, O., Muliukha, V., Lukashin, A.: Alternating priorities queueing system with randomized push-out mechanism. In: Balandin, S., Andreev, S., Koucheryavy, Y. (eds.) ruSMART 2015. LNCS, vol. 9247, pp. 436–445. Springer, Heidelberg (2015). doi:10.1007/978-3-319-23126-6_38

11. Ilyashenko, A., Zayats, O., Muliukha, V., Lukashin, A.: Randomized priorities in queuing system with randomized push-out mechanism. In: Galinina, O., Balandin, S., Koucheryavy, Y. (eds.) ruSMART 2016. LNCS, vol. 9870, pp. 230–237. Springer, Heidelberg (2016). doi:10.1007/978-3-319-46301-8_19

12. Falin, G.I., Templeton, J.G.: Retrial Queues, 328 p. Chapman and Hall, Boca Raton (1997)

13. Falin, G.I.: A survey of retrial queue. Queueing Syst. 7, 127–168 (1990)

14. Zayats, O., Korenevskaya, M., Ilyashenko, A., Lukashin, A.: Retrial queueing systems in series of space experiments "kontur". Procedia Comput. Sci. 103, 562–568 (2017)

15. Bocharov, P.P., Pavlova, O.I., Puzikova, D.A.: A M/G/1/r retrial queueing system with priority of primary customers. Math. Comput. Model. 30(1), 89–98 (1999)

16. Bocharov, P.P., Pechinkin, A.V., Fong, N.: Stationary probabilities of the states of the retrial system MAP/G/1/r with priority servicing of primary customers. Autom. Remote Control 8(61), 68–78 (2000)

17. Agalarov, Y.M.: On a numerical method for calculating stationary characteristics of the switching node with retrial transmissions. Autom. Remote Control 1, 95–106 (2011)

NGN/IMS and post-NGN Management Model

Alex Goldstein[(⊠)]

Federal Bonch-Bruevich Telecommunication University, St. Petersburg, Russia
agold@niits.ru

Abstract. The goal of this paper is to provide a new telecommunications network management model that will be adequate to the modern and future NGN/IMS and post-NGN telecommunications networks. Instead of the traditional OSS/BSS approach in the article we propose a new comprehensive approach for the communication networks management, which is adequate to the paradigm shift in telecommunications that takes place currently, as well as a comprehensive approach to building mathematical models of the management organization and self-organization in the new post-NGN era.

Keywords: Network management · post-NGN · OSS/BSS · Business Intelligence · Queuing network · G-networks · IP/MPLS · Multi-agent system

1 Introduction

Already today, NGN (Next Generation Network) subscribers are provided with an ever-expanding class of various information and communication services and applications, which presuppose a wide range of various protocols, technologies, transmission rates, QoS parameters and cost characteristics [1, 2]. In addition, new information and communication services require from a telecommunications Operator revolutionary changes in OSS/BSS (Operation Support Systems/Business Support Systems) too, which are determined, in particular, by the need to manage servicing customer requests in the real time [11].

Complexities of building these OSS/BSS of the next-generation are also conditioned by the fact that modern NGN networks [3, 4] are used not so much for voice transmission as for providing the mobility of subscribers, accessing knowledge bases on-line, viewing video, listening music, organizing a multimedia conference call, machine-to-machine communications (M2M) and Internet of Things [4], organizing network games, IPTV [13] and other entertainment industry applications on a real-time basis, as well as many other services of modern multi-service networks. All this drastically changes the architecture of modern NGN information and communication networks and principles of their management.

The article deals with a new comprehensive approach to building the management of communication networks, which is adequate to the paradigm shift in telecommunications that takes place currently, as well as a comprehensive approach to building mathematical models of the management organization and self-organization in new information and communication networks.

© Springer International Publishing AG 2017
O. Galinina et al. (Eds.): NEW2AN/ruSMART/NsCC 2017, LNCS 10531, pp. 441–454, 2017.
DOI: 10.1007/978-3-319-67380-6_40

2 Related Works

The above-mentioned paradigm shift in telecommunications is described in monograph [1] and in number of articles in [2]. Telecommunications network management aspects are considered in monograph [3], which is based on TMForum [4] latest developments. The evolution of telecommunications network management approaches is shown in [11].

Problems of transport IP/MPLS network management are discussed in [5, 6, 13]. Works by Le Gall [8] and Boxma [19] propose the fundamental methods to solve this problem in Sect. 4 of the article for transport plane control. The very beginning of this investigation was published in [7].

These teletraffic models and algorithms [15, 16], sought to solve the requirements of new self-organization processes in call/session control in new post-NGN networks. However such approaches are not sufficient enough. The ideas of multi-agent systems [12, 17] were added in order to investigate the control/switching and service applications planes on Fig. 1.

The multi-agent systems methods that solve a relatively similar problem, which is solved in Sects. 5 and 6 of the article, are discussed in [9, 14, 18].

Another fruitful idea by Gelenbe [10] – idea of G-networks as a model for queuing systems with specific control functions, such as traffic re-routing or traffic destruction - is very useful in investigation in Sect. 6.

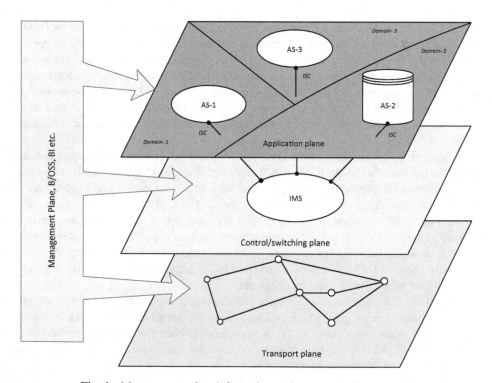

Fig. 1. Management of an information and communication network

3   Complex Network Management Model

The above-mentioned integrated approach, in its turn, can be divided into three target groups - conceptual models that are capable to adequately display three telecommunication network control planes, as it is shown in Fig. 1.

The first group (the lower plane) includes transport network management problems, the analysis of the probability-time characteristics of the mechanism of tunnels management in the IP/MPLS network, ensuring the specified quality of the multimedia traffic service, development of the algorithm for the effective organization of tunnels in the IP/MPLS network based on this analysis.

Research of multi-agent self-organizing architectures and processes of the organization of switching and management of calls/sessions service, distribution and redistribution of the multimedia traffic in the NGN/IMS next generation network of a modern telecommunications Operator, construction of the IT-landscape of such network, advanced systems for technical record-keeping and databases management of telecommunications network elements constitute the second group. This group corresponds to the second plane in Fig. 1 - the switching and control plane, which consists of network elements responsible for establishing a connection/organizing a session, service management of calls/sessions, breaking connection/session, authorization, authentication of a subscriber, the storage of a user information. Such functions, as collection and processing of statistics for billing, modification of subscriptions of users (provisioning), operational support of the network and the network management (O&M), are worked out with the management system means at this level.

The third group comprises problems of the management of new information and communication services, machine-machine M2M communications, social networks, etc., the most promising methods of their solution are based on multi-agent systems and cognitive network architectures. In accordance with the IMS principles, specialized server platforms of applications are arranged on this plane in Fig. 1, which function as standard servers of applications in the IMS environment with the already implemented ISC interface to IMS and open API (Application Program Interfaces) to developers of third-party applications. The determinant factor here was that the voice turned into just one of many services along with the Video Sharing, Instant Messaging, Presence, Gaming and all possible variants of Push to Talk + Push to X, that predetermined the need for innovative approaches to the telecommunications management at this level.

4   Tunneling Model for the IP/MPLS Transport Plane Complex Network Management Model

This section of the article is devoted to the creation of the IP/MPLS transport network management model [5]. The range of problems of the unified management of transport networks is actively discussed in the industry in the course of the last decade, as result of that the decision was made to move evolutionarily, developing a sufficiently successful multi-protocol MPLS label switching [6]. In this regard, GMPLS was developed as the set of MPLS management protocols with a gradual expansion of them to the specific character of the operation of circuit switching networks. The confidence in the correctness of such approach was supported by arguments that use of the same protocols can lead to the smart, automated and unified level of the management for various networks with different principles of the data transfer: packets, time slots, wavelengths, etc.

An extremely important advantage is the possibility, within the framework of the MPLS architecture, along with the packet to transmit not one label, but a whole stack of labels. Operations of adding/removing a label are defined as stack operations (push/pop). The result of switching sets only the top label of the stack, while the lower ones are transmitted transparently before the operation of removing the top label. Such approach allows creating the hierarchy of flows in the MPLS network and organizing tunneling transmissions. The question is the possibility to manage the entire packet transmission path in MPLS without specifying intermediate routers in an explicit form. This is achieved by the creation of tunnels through intermediate routers, which can cover several network segments.

The mathematical model of the MPLS tunneling mechanism represents the queuing network with sequential queues. Probabilistic-temporal characteristics of the waiting time in such systems with sequential queues were studied in [7] for the tunneling mechanism in MPLS, where the Poisson packet stream with the intensity λ and the average service time of $1/\mu$ comes to the edge node 1 input [7].

The complexity of packet behavior in the transport network is conditioned by two phenomena: chaining of bursts, originating from the first node, and fragmentation of the same bursts.

The first chaining phenomenon refers not only to the second, but also to any not the first node n (n $\neq$ 1) and is connected with the fact that the first packet of the k-th burst overtakes the last packet of the (k − 1)-th burst at this node, and both bursts - k-th and (k − 1)-th are appropriately chained.

The second phenomenon of fragmentation is not so obvious and takes place only in the second node, but it is also quite clearly evident. Let the packet number j from the k burst is serviced in the first node and at this moment the next packet number j + 1, whose service time exceeds the service time of the packet j, arrives at the same first node. Let there is no queue at the next second node at this time, and packet j is serviced as soon as it arrives at the node 2, packets j + 1 and j begin to be served simultaneously at the nodes 1 and 2, respectively. When packet j next leaves the node 2, the packet j + 1 still continues to be processed at the node 1, since its service time is longer. Any packet belonging to the burst number k at the output of the node n at n $\geq$ 2 has the service time, which is less than or equal to the service time of the first packet of this burst. Any burst at the output of the node n = 2 and all subsequent nodes is retained, i.e. all packets in it remain hard tied to each other.

Chaining bursts k and k + 1 of the node n − 1 at the node n is possible if and only if the first packet of the burst k + 1 leaves the node n − 1 before the burst k finish to be serviced by the node n in the course of the interval, which does not exceed the service time of the first packet of the previous burst t_{n-1}^{k} [3, 4]. The average length of the burst k of the node n (n > 2) expressed in the number of packets K_n is determined by the average length of the burst in the node n − 1 in the form $K_n = K_{n-1} + \frac{\rho}{1-\rho}$. The average length of the burst k in the node n (n > 2), expressed in the number of packets K_n, is equal to $K_n = K_2 + (n - 2)\frac{\rho}{1-\rho}$. It was noted above that the complex situation at the node 2 does not allow obtaining the exact analytic formula for n = 2. Therefore, the following approximation for this node 2, and also for an arbitrary node n is proposed. The average length of the burst k in an arbitrary node n, expressed in a number of packets K_n, is equal approximately $1 + \frac{\rho}{1-\rho}$, at n = 1 and $1 + (n - 1)\frac{\rho}{1-\rho}$, at n ≥ 2. Above expressions actually determine the tunnel busy period of the IP/MPLS network. On the basis of them under the stationary conditions at ρ < 1 for the distribution function of the tunnel total stay time of N nodes and for greater N values, the following approximation is proposed: $V(t;N) \cong \left(\frac{1-\rho}{1-\rho F(t)}\right)^{N+1} F(t)$, where F(t) – function of service time distribution in each node.

In graphs, shown in Fig. 2, dependences are represented of the number of packets on the load in the range from 0.2 up to 0.8.

Graphs are built for four values of the load ρ = 0.2, 0.4, 0.6 and 0.8. Regardless of the load, the average burst length grows with the node number, beginning from the node 1. This growth is linear, but only starting from the node 3. It should be noted that the growth of the average burst length is insignificant between nodes 1 and 3. Above in this chapter, the explanation of this behavior is given: at the node 2 the burst-chaining phenomenon (from the node 1) is supplemented by the phenomenon of the burst fragmentation. Consequently, the fragmentation phenomenon, related to this particular node, cannot be neglected. This approach is recommended to be included in the composition of engineering methods of managing the IP/MPLS network as a part of the BI (Business Intelligence) system. Passing on from mathematical models to the study of engineering technologies of the transport network, their important feature should be emphasized - the longer-term process of their modernization in comparison with other

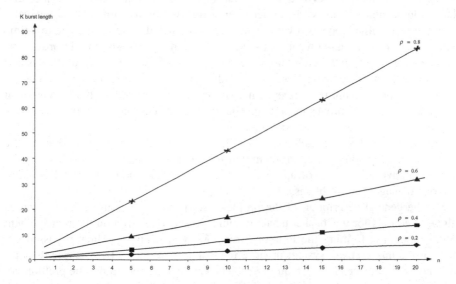

Fig. 2. Dependences of the burst K average length on the node number n and its load ρ

two planes in Fig. 1. Operators with a much greater degree of the incredulity consider proposals to improve the existing transport infrastructure than, for example, updating the operating access network or service management equipment.

5   Calls/Sessions Control Plane Management Model

The second class of models for next two planes in Fig. 1 is based on multi-agent systems - more new area of analysis and modeling of telecommunications management processes, in comparison with the previous class, which become increasingly important, as far as the modern info-communications are drastically complicating, and the number of parameters in management models increases.

Models of such multi-agent systems, compiled by teams of interacting independent agents, are considered, in particular, in [8, 9].

Here, the comparative description of approaches to the mathematical modeling of these two classes, involved into the thesis, is proposed too. And here this description and the comparison itself will be carried out to study extremely simplified telecommunications management scheme, and therefore the material of this section can be considered as the introduction to a deeper study of more complex telecommunication network management schemes, presented in the next section.

For this extremely simplified management scheme, only one question is studied: how results of servicing requests for telecommunication services depend on strategies of managing resources of the telecommunications network. Probabilistic-temporal characteristics of servicing requests for various telecommunications services are calculated taking into account the probabilistic nature of various telecommunication resources, provided by the network at each moment of time t. The output data of models will be the numerical values of some generalized service quality index QoS (Quality of Service) at specified PTC requests for telecommunication services and telecommunication resources for services being released by the time of their arrival.

In these simplified models, two scenarios are considered: with immediate real-time recruiting of newly released resources to service incoming requests (that improves the resource usage factor and thereby improves the quality of service, but complicates and slows down the management process itself and thereby deteriorates the service quality) and without such prompt recruitment (that simplifies and accelerates the management process and thereby improves the service quality, but reduces the resource usage factor and thereby deteriorates the service quality).

We remind that we are talking about the ultimate simplification of this, in the general form, very complicated problem, but formulated in this section exclusively for the comparative description of approaches to mathematical modeling of telecommunications management problems.

So, the general description of the problem, set in this way, can be formulated as follows. Requests for this or that telecommunication service arrive in the management system randomly, forming the Poisson stream with intensity λ_1. Having passed the queue at the management system input, a request gets in one of two service devices. One (the lower, device A) - for requests, which service does not require the preliminary analysis of resources (and making appropriate managerial decisions). Examples of such service can be request replies about some information of the type that is contained in the personal account of the subscriber, or for the call to his personal voice mail, etc. Of course, even for these requests, the memory resource or throughput of the service platform can be insufficient at the moment, but the simplicity of scenarios, as a rule, does not require any deep research. Then these arrive to device C, wherein initiating of corresponding business processes 1, 2, ..., N to service the request begins.

Another device B receives requests, which require the analysis and the resource mobilization (and making appropriate managerial decisions). These are, for example, requests for expanding the Internet access band, IPTV services [13]. They also arrive to the same device C, but also in the corresponding queue to the device D, to where released resources with the intensity λ_2 arrive too. The device D, which actually performs the functions BI, directs results of its work to corresponding business processes with the period τ (for example, the planning and rescheduling mode) or immediately, i.e. with $\tau = 0$ (on-line control).

Such description in terms of the theory of queuing networks can be represented by the scheme in Fig. 3, and formulas for calculating PTC of such systems are built on the basis of these theory results, which were described, in particular, in [7, 13].

In the multi-agent system (MAS), the model consists of descriptions of agents, their behavior and their initial states. The behavior of an agent can be described in a rather simple form, for example, in the form of the finite state automation. But taking into account the context of the dissertation research, the main emphasis here is made on the research of the PTC, and in this connection the dynamics of the agent's behavior also has the probabilistic character.

Although such representation lacks the conciseness of Fig. 3, but it contains much more informal information related to the interaction of request agents for telecommunications services and resource agents. More formalized interaction between agents in such MAS will be described on the basis of the mathematical theory of relations. In it, the classical n-ary relation is defined as a subset of the Cartesian product of arbitrary n sets:

$$R \subseteq X_1 \times X_2 \times \ldots \times X_n$$

Below we consider basically binary relations on the set of agents A, including ordinary, and also, perhaps, and fuzzy relations. So the aRb record means that the agent a is in relation R with the agent b, and is understood as the degree of intensity of this relationship manifestation.

Then E is the unit (diagonal) ratio that is, $\forall a, b \in A$

$$\mu_E(a, b) = \left\{ \begin{array}{l} 1, \ if \ a = b \\ 0 \ in \ otherwise, \end{array} \right\}$$

and the inverse relation R^{-1}

$$R^{-1}(a, b) = R(b, a), \forall a, b \in A.$$

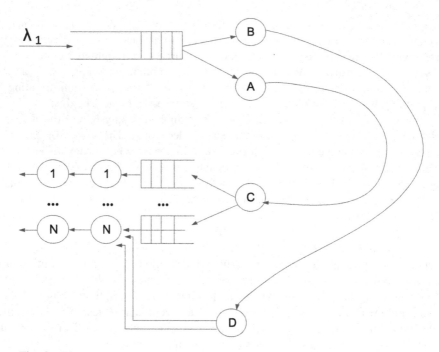

Fig. 3. The approach to research with the help of the theory of queuing networks

The operation of composition of relations can be set in the form

$$\mu_R(a,c) = \vee_y (\mu_R(a,b) \times \mu_R(b,c))$$

for $\forall a, b, c \in A$.

The resulting mathematical model of the multi-agent communication network management system can be included in the advanced Business Intelligence BI tools based on NRI (Network Resources Inventory) systems, next-generation billing systems and other OSS/BSS systems [11]. Let us designate the total number of them with N (Fig. 2). The same value of N determines the number of smart agents {1,2, …, N}, each of which performs the appropriate corresponding jobs. Jobs arrive in the multi-agent system at different moments of time and at different nodes. At each moment of time t, the state of the agent i {i = 1, …, n} is described by the corresponding probabilistic-temporal characteristics (PTC).

6   Applications/Service Plane Management Model

Returning to the paradigm shift, mentioned at the beginning of this article, we note that thanks to this shift today there are two approaches to the modern info-communication services management, which can be defined, with some degree of conventionality, as the centralized fixed planning and the decentralized dynamic self-organization. And if problems of the IP-traffic multimedia routing on the transport plane level can be resolved by the research of probabilistic-temporal characteristics with traditional methods of the teletraffic theory (as it was done, in particular, in the Sect. 4 of this article), then the service management plane demands to a greater extent already the attraction of self-organization mechanisms and multi-agent models for managing this plane.

And in BI, two types of multi-agent models are used. Multi-agent models for the second control/switching plane in Fig. 1 are formed from a small number (essentially, from 5 to 20) of intellectual agents as it is shown on Fig. 3. A separate article is devoted to the study of this class of models on Fig. 3 more deeply.

For tasks of the upper plane management in Fig. 1 is proposed another MAS model on Fig. 4, which is, on the contrary, composed of the extremely large number (essentially, thousands) of reactive agents. Let us note that, as is usually the case in the real world around us too, in the MAS models the intellectuality of each individual agent is inversely related with a number of agents, which form the multi-agent system [16, 18]. The small number of cognitive agents of the second plane is compensated by the fact that each such cognitive agent, as a rule, represents itself a complex program system and can contain an expert system or a whole learning neural network, which is just discussed in the separate publication.

Here, also, for modeling cognitive processes for management of the most complex upper plane in Fig. 1 (Application Plane), we will build MAS with a large number of weak agents, characterized by the reactive, stochastic behavior and the extremely low intelligence, which, nevertheless, despite the simplicity of their components, are able to solve the important class of problems in the cognitive management of info-communications. Although the behavior of each agent is determined by simple reactive rules, their combined reaction has a complex probabilistic character which simulates the complex behavior of management processes in the Application Plane. It is noted in [10, 14], that MAS composed of a plurality of such agents, in general, can manifest the so-called swarm intelligence. The result is achieved here exactly due to the large number of agents and the probabilistic nature of their behavior.

The MAS model on the Fig. 4 built on the basis of two counteracting agent streams - agents of requests for differentiated services (S-agents) and agents of resources, needed to provide differentiated services (R-agents). In this case, S-agents interact with R-agents in an obvious way: The S-agent captures the R-agent, i.e. the R-agent ceases to exist (that takes place with a high probability, if the resource is very limited, and the request takes it in a considerable volume) or the S-agent is satisfied without the significant damage to the R-agent, i.e. the S-agent ceases to exist (that happens with a high probability if the resource is unlimited, and/or the request takes it in a small volume). Thus, it is possible to speak of R-agents of two types: P_1-agents, when the resource is almost unlimited for service requests, and R_2-agents, when the resource is relatively limited for these requests. In the interaction of the S-agent with the R_1-agent S-agent is, and the P_1-agent remains in the system. When the interaction of the S-agent with the R_2-agent, the S-agent is serviced with the inconsiderable possibility and the R_2-agent with the high possibility goes out of the system.

The state of such system is described by the equation $N(t) = (N_y(t), N_{P1}(t), N_{P2}(t))$, where elements of the vector N(t) are the total number of agents in the system

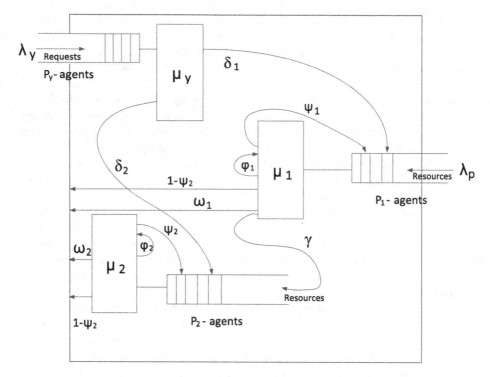

Fig. 4. Multi-agent model of the service plane

at the moment t, and $N_y(t), N_{P1}(t), N_{P2}(t)$ of the number of agents of types S, R_1 and P_2, correspondingly [10].

In the model in Fig. 3, the system receives the Poisson stream of S-agents (requests for differentiated services) arrives in the system with the intensity of λ_y and also the Poisson stream, independent from it, of R-agents (resources) with the intensity of λ_P. Let us note that this refers to R_1-agents, since only they enter the system as they appear/release. R_2-agents result from R_1-agents under the effect of S-agents as the telecommunication resources starvation.

When S-agents enter the system, they are serviced with the intensity of μ_y. The service in this case is the interaction with the R-agent, if there is one in the system (in the absence of any type of R-agents in the system, the S-agent leaves the system being unserviced). At that, the S-agent does not know, with which type of P-agent he interacts: with the P_1-agent or with the P_2-agent, when a service request is received, it can not be known whether the required resource is in excess or in short supply. Let us consider, that the S-agent interacts with the R_1-agent with the possibility of δ_1, and with the P_2-agent with the possibility of δ_2. If the S-agent interacted with the R_1-agent, he removed him from the system with the possibility of ω_1, and if he interacted with the R_2-agent, then he removed him from the system with the probability of ω_2.

Now let us turn to R_1-agents. They are processed with the intensity μ_1, and after processing the service, expressed by the S-agent, the R_1-agent remains for satisfaction of following S-agents, or it turns into the R_2-agent, when the resource for S-agents is assumed to be limited. The first event occurs with the probability of γ, and the second event, accordingly, - with the probability of $(1 - \gamma)$. When the R_1-agent with the probability of δ_1 provides the resource for the S-agent service (the request for the service), and the serviced S-agent leaves the system (the request is satisfied), and this R_1-agent remains in the system and returns to its stack of R_1-agents with the probability of ψ_1 (the resource is inexhaustible and remains for other S-agents) or also leaves the system with the probability of $(1 - \psi_1)$, when the resource is exhausted. The situation is possible, of course, when the R_1-agent failed to organize the service for the S-agent, whether due to his insufficiency or excessive requests from the S-agent, or for some other reasons, then the R_1-agent leaves the system after an unsuccessful attempt. The probability of such event is $(1-\varphi_1)$. Let us note, that this last probability is, as a rule, small, because, the most often, as the resource is exhausted, S-agents beforehand turn R_1-agents into the swarm of R_2-agents with the limited resource. But, of course, the scarcity of a resource can come instantly, that happens sometimes - with the probability of $(1-\varphi_1)$.

Finally, let us consider R_2-agents. They are processed with the intensity of μ_2. With the probability of φ_2 the agent R_2 is enabled to organize the service, expressed by the S-agent, and successfully fulfills this task. then this R_2-agent either returns to his swarm of R_2-agents with the probability of ψ_2 or leaves the system with the probability of $(1 - \psi_2)$, when the resource of the R_2-agent is completely exhausted. It is possible that already for this S-agent the resource was not enough, the service could not be organized, and the R_2-agent leaves the system without result. This event comes with the probability of $(1-\varphi_2)$.

As for IP/MPLS transport network models in the previous section, we will keep assumptions about Poisson streams of requests for differentiated information and communication and self-releasing resources, as well as also about the exponential character of a service. Then it is possible to describe the model, presented in Fig. 3, by the uniform circuit of Markov [15], for which states for N(t): t > 0 can be obtained with the help of Kolmogorov-Chapman equations for p (n, t) = P[N(t) = n], where n = (n_y, n_1, n_2), and n_y, n_1, n_2 - non-negative integers, corresponding to the numbers of S-agents, R_1 – agents and R_2 – agents in the system at the time moment t. Found out in this way the analytical expression for these probabilities and the mathematical expectations of a number of agents of 1 and 2 types in the system, we will leave beyond this article, but instead focus on the physical interpretation of the above probabilities.

R_1-agents and R_2-agents interact with S-agents of requests for information and communication services with probabilities of ω_1 and ω_2, accordingly. Results of this interaction are successful (the S-agent is serviced = necessary resources are provided for the request for the service) with probabilities of φ_1 for P_1-agents and φ_2 for P_2-agents, accordingly. The R-agent itself, after allocation of the necessary resource for the S-agent, retains the ability to remain in the system: R_1-agent with the probability of ψ_1 and R_2-agent with the probability of ψ_2.

Difference between R_1-agents and P_2-agents is also interpreted in a natural way. Resources, which are in excess, can be provided to S-agents with much less doubts and hesitations, therefore $\mu_1 > \mu_2$, as well as $\omega_2 > \omega_1$. For the same reason, after interaction with the U-agent in the system R_1-agents remain and return to the swarm with much higher probability, than R_2-agents, i.e. $\psi_1 > \psi_2$.

Now, let us consider a number of numerical examples of such system behavior in Fig. 4. Let us suppose for clarity $\mu_y = \mu_1 = \mu_2 = 1$, and λ_y, λ_1 and λ_2 to be such, that the system load would be limited $0 < \rho_1 < 1$, $0 < \rho_2 < 1$ and $0 < \rho_y < 1$ and take on values closer to 0.99. Let us investigate the influence of the probability q (the probability of that the R_1-agent completely satisfies the request for the service, expressed by the S-agent and this R_1-agent retains enough resources, and its transformation into the R_2-agent with the probability of $(1 - \gamma)$ is not the issue. It is natural to assume that probabilities δ_1 and δ_2 are equal, i.e. $\delta_1 = \delta_2 = 0.5$, since S-agent is absolutely not aware of the state of the required resource: whether it is in excess as the R_1-agent or in deficiency as the R_2-agent. Accordingly, it selects them equally probable.

Let us assume also, that probabilities $\psi_1 = \psi_2 = 0$, i.e. after servicing a request for the service, provided by the S-agent, involved in it resource already does not remain in the system. Let us note that if for R_2-agents this happens most often, but for R_1-agents this is a rather rare phenomenon. Nevertheless, this regime is the most interesting for us to study, since in conditions of super-sufficient resources and modest S-agents in the system everything will be fine with the service anyway.

Under these conditions, let us turn to the problem of the exhaustion of resources, or more precisely - to two probabilities, to the probability γ of turn P_1-agent in the R_2-

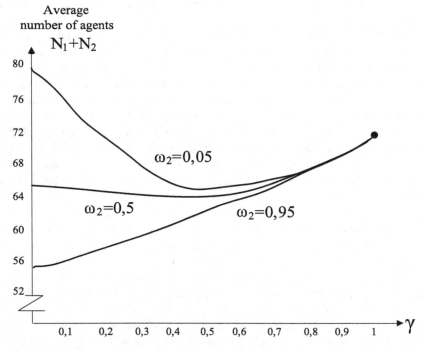

Fig. 5. The dependence of the number of R-agents on the probability γ of the turn of R_1-agent in R_2-agent

agent when servicing S-agent and to the probability ω_2 of the withdrawal of R_2-agents from the system. Let us calculate the dependence of the number of R-agents in the system (R_1-agents and R_2-agents in aggregate) on the value of γ at different values of ω_2. Let us choose three values $\omega_2 = 0.05$, $\omega_2 = 0.5$ and $\omega_2 = 0.95$.

If the probability ω_2 of the withdrawal from the system of the R_2-agent is close to one ($\omega_2 = 0.95$ in Fig. 5) the total number of R-agents in the system almost linearly increases with increasing probability γ of the R_1-agent turn in the R_2-agent when servicing S-agent. When the low probability of ω_2 ($\omega_2 = 0.05$ in Fig. 4), the dependence of the total number of R-agents in the system on the probability γ has a pronounced minimum in the area of $\gamma = 0.3 - 0.6$.

In general, specified dependencies create a good base for man oeuvres in the field of management in the telecommunications operator BI at different target functions: minimization of the number of idle S-agents or, on the contrary, maximization of their number, if it is necessary to reserve resources for the expected upsurge in arrival of S-agents. Setting the probability value γ, and the rest the multi-agent system in Fig. 4 will make itself in fact, carries out this management.

7   Conclusion

In this paper the new integrated approach and three different models for NGN/IMS and post-NGN management were considered. It is shown; how these three models successfully complement each other in the research process of all the time more complicating processes of the modern and future telecommunications network management.

References

1. Janevski, T.: NGN Architectures, Protocols and Services, 1st edn. Wiley, London (2014)
2. Balandin, S., Andreev, S., Koucheryavy, Y. (eds.): Internet of Things, Smart Spaces, and Next Generation Networking. Springer, Heidelberg (2010). 13th International Conference, NEW2AN 2013, and 6th Conference, ruSMART 2013, St. Petersburg, Russia, 28–30 August 2013
3. Rahul, W.: OSS/BSS for Converged Communication Networks. ICT Series, 2nd edn. Elixir Panacea Consultancy Services, Pune (2016)
4. Enhanced Telecom Operations Map (eTOM) GB921: Concepts and Principles Release 8.0, TM Forum (2009)
5. Morrow, M.J., Sayeed, A.: MPLS and Next-Generation Networks: Foundations for NGN and Enterprise Virtualization. Cisco Press, Indianapolis (2006)
6. Liu, C.J.: MPLS TE queue creation as a mechanism for QoS routing. In: Proceedings of Applied Telecommunication Symposium/ASTC, Orlando, Florida, pp. 23–27 (2003)
7. Goldstein, A., Yanovsky, G.: Traffic engineering in MPLS tunnels. In: International Conference on NExt Generation Teletraffic and Wired/Wireless Advanced Networking, NEW2AN 2004 (2004)
8. Le Gall, P.: Single server queuing networks with varying service times and renewal input. J. Appl. Math. Stochast. Anal. 13, 4 (2000)

9. Gardelli, L., Viroli, M., Casadei, M., Omicini, A.: Designing self-organising MAS environments: the collective sort case. In: Weyns, D., Parunak, H.V.D., Michel, F. (eds.) E4MAS 2006. LNCS, vol. 4389, pp. 254–271. Springer, Heidelberg (2007). doi:10.1007/978-3-540-71103-2_15

10. Gelenbe, E., Fourneau, J.N.: G-networks with resets. Perform. Eval. **49**, 179–192 (2002)

11. Pavlou, G.: On the evolution of management approaches, frameworks and protocols: a historical perspective. J. Netw. Syst. Manag. **15**(4), 425–445 (2007)

12. Wooldridge, M.: An Introduction to Multi-agent Systems, p. 366. Wiley, London (2002)

13. Goldstein, B., Mansour, G.: Improving IPTV on-demand transmission scheme over WiMax. In: Balandin, S., Koucheryavy, Y., Hu, H. (eds.) NEW2AN/ruSMART -2011. LNCS, vol. 6869, pp. 541–549. Springer, Heidelberg (2011). doi:10.1007/978-3-642-22875-9_49

14. Sahan, A., Yakhno, T.: A multi-agent traffic controller with distributed fuzzy intelligence. In: Proceedings of the 5th Annual International Conference on Mobile and Ubiquitous Systems: Computing, Networking, and Services, Mobiquitous 2008, Article no. 6, Dublin, Ireland (2008)

15. Feller, W.: An Introduction to Probability Theory and Its Applications, vol. I. Wiley, London (1968)

16. Frost, V., Melamed, D.: Traffic modeling for telecommunications networks. IEEE Commun. Mag. **32**(3), 70–81 (1994)

17. Drener, K., Stone, P.: Multi-agent traffic management: an improved intersection control mechanism. In: Proceedings of the 7th International Joint Conference on Autonomous Agents and Multi-agent Systems, AAMAS 2008, vol. 3 (2008)

18. Wang, Y., Garcia, E., Casbeer, E., Zhang, D.: Cooperative Control of Multi-agent Systems: Theory and Application, 1st edn. Wiley, London (2017)

19. Boxma, O.J.: On a tandem queuing model with identical service times at both counters. Adv. Appl. Prob. **11**, 616–659 (1979)

The Principles of Antennas Constructive Synthesis in Dissipative Media

Roman U. Borodulin[1(✉)], Boris V. Sosunov[1],
and Sergey B. Makarov[2]

[1] Military Academy of Telecommunications named
after Marshal of the SU S.M. Budenny, St. Petersburg, Russia
borodulroman@yandex.ru, bsosunov@gmail.com
[2] Peter the Great St. Petersburg Polytechnic University, St. Petersburg, Russia
makarov@cee.spbstu.ru

Abstract. The results of tasks solving of the antennas constructive synthesis (CS) in the dissipative media are presented in this article. The function schemes of the conceptual models immersed in the antennas dissipative media are presented, which are suitable for the analysis by means of finite element method (FEM) and finite difference time domain method (FDTD). The special models parameters which allow us to take into account the influence of the infinite boundary and dissipative media on the electric characteristics of the immersed antennas are discovered.

The general approach to cost functions forming up is presented in the article. The cost functions allow us to conduct the synthesis of the immersed antennas. The presented cost functions are used for the optimization of the immersed antennas constructive parameters by means of multicriteria optimization of the genetic algorithm (GA). The results of the constructive synthesis of the immersed laminate kind antennas by the FEM-GA methods link are presented. The strong and weak points of the latter are shown in the article. The practical usage recommendations of the developed antennas are given.

Keywords: Constructive synthesis · Conceptual models of the immersed antennas · Finite elements method · Finite difference time domain method · Genetic algorithm

1 Introduction

The antennas synthesis is the electrodynamics inverse problem consisting in the definition of the sources system which generates electromagnetic field having set structure [1].

The classical application tasks of the radiating systems synthesis presume the known in advance construction and the antenna kind, the demands to the direction diagram and the limitations setting for the amplitude-phase field distribution operability in the antenna aperture. It is necessary to choose such implementing distribution, which makes a normalized diagram satisfying the set demands [2–7].

© Springer International Publishing AG 2017
O. Galinina et al. (Eds.): NEW2AN/ruSMART/NsCC 2017, LNCS 10531, pp. 455–465, 2017.
DOI: 10.1007/978-3-319-67380-6_41

The antennas **constructive synthesis** (CS) compared to the antennas synthesis in the traditional understanding refers to the **internal** inverse tasks of electrodynamics, i.e. finding antenna geometric shape which dimensions are less than the wave-length and the electric characteristics are in the given boundaries. The antennas dimensions limitation by the wave-length caused by the fact that in case of high size the received antennas can be referred to the continuous or discrete antenna arrays for which synthesis methods are developed. The constructive synthesis is the analogy of the structural synthesis in other scientific knowledge fields [8].

Currently, the constructive synthesis primarily is applied for the finding of the current density distribution on the surface of the small electric size bandwidth antennas which are in free space. The given natural coordination characteristic (coordination with the transmission line without special coordinating device) in the service band with required covering within the order of several units is in the first place for them.

It is necessary to take into account the influence of the boundary on the electric characteristics for the antennas which are in the dissipative media especially near the infinite boundary with the medium which parameters defer greatly.

The interest to the antennas immersed on some depth in the dissipative medium is rather high. These antennas are known for a long time. Comprehensive monographs and specialists' technical reports were dedicated to these antennas research [9–12].

In practice the application of the immersed antennas (IA) is connected with some peculiarities of the radio waves distribution in the dissipative media. There were attempts of the immersed antennas using for the radio lines noise immunity increasing, for radio communication in mines at great depth, in the instrumental landing systems, radio geology, radio communication with sunk and immersed objects and all that [9]. However, the gain of these antennas is rather low as compared with antennas in air because of the loss connected with the medium characteristics influence.

It shaped historically that the majority of the electric characteristics research and immersed antennas parameters research were carried out experimentally. However, there are examples of such antennas calculation with transmission lines theory usage with losses. This theory evolved in the latter half of the twentieth century. For example, thanks to professor Sosunov and Filippov the influence of electrically closely-adjacent boundary between two media (air-medium with losses) is analytically taken into account at the expense of the original Somerfield's task [13]. The received expressions became the basis for calculating of the isolated and nonisolated immersed antennas of cylindrical section. However, the problem of the electrodynamics analysis of the immersed antenna form, differing from cylindrical, has not been solved until recently.

With the appearance of the modern numerical electro dynamics methods, algorithms and programs development based on them, it is possible to calculate the immersed antennas by means of these methods which are well-reputed while calculating antennas in free space (air). The numerical methods of electrodynamics allow us to simulate the antennas of arbitrary shape. The tasks solutions of the immersed antennas simulating by means of finite element method [14] and finite difference time domain (FDTD) [1] method using artificially given absorbing layers were found lately. These layers are known as perfectly matched layers (PML) [15]. The opportunity of implementing modern numerical methods for the immersed antenna calculation pursued lately, allowed us to begin the work of investigating the ways of increasing their gain without carrying out painstaking and expensive experiments [10]. It found out that it is possible to increase the immersed antenna gain at the expense of constructive synthesis using.

2 Functional Schemes of the Immersed Antennas Conceptual Models Calculating by Means of the Full Field Methods

If we calculate the electromagnetic field (EMF) which is made at all the points of the space by the current of the antenna immersed in the dissipative medium which works for the transmission, it is necessary to take into account the presence of the reflections from the media boundaries. These are the internal reflection and the wave distribution inside the dissipative medium in parallel with the boundary.

In case of this situation modeling by the moment method, it is necessary to process the initial equations for the automation of the infinitely extended media boundary account that is rather difficult mathematical task.

The side wave and the wave went through the wave boundary reflect from the calculation field walls fully for the methods based on differential equations solution. It is necessary to set the absorbing boundary conditions (ABC) or simulate the perfectly matched layers (PML). It allows us to develop the conceptual models of the immersed antennas simulating the infinite extension of the half-spaces surrounded by them.

As the absorbing layers (AL) refer to the artificially given anisotropic materials unparalleled in nature, their parameters (ε_{AL}, σ_{AL}) are chosen on the basis of the media parameters contacting with them for the reflections excluding.

The variant of the absorbing boundary conditions for the air half-space and perfectly matched layers for the ground half-space (Fig. 1) or the variant of two perfectly matched layers with different parameters using is appropriate for the conceptual FEM-models (Fig. 2). But two half-spaces should be put in each other, the m parameter defining the distance between the surfaces with the boundary conditions set on them in

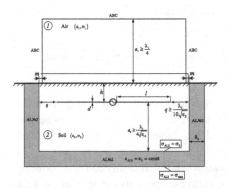

Fig. 1. The first variant functional scheme of the conceptual FEM-model of the immersed vibrator

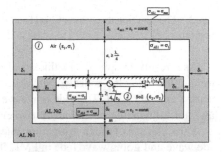

Fig. 2. The second variant functional scheme of the conceptual FEM-model of the immersed vibrator

both variants should be small to exclude the introduced distortions but not equal to 0 otherwise the accuracy of the boundary conditions setting will not be pursued.

The parameters of the model placing of the investigating antenna relatively half-infinite media limited by the perfectly matched layer are chosen on the basis of the electrodynamics peculiarities of the models and the limitations of the numerical method used. The parameters a_1, a_2 define the distance of the computational region walls limited by the absorbing boundary conditions or PML from the radiating antenna model (Figs. 1 and 2). These values are chosen on the basis of non-admission of the waves at an acute angle falling on the computational regions boundaries, at which the reflection coefficient becomes more than 0 in FEM. As a rule, it is sufficient to have the distance of quarter wave length inclusive of the reduction in the medium to fulfill this condition. The q parameter is chosen on the basis of maximum size of the near zone of the antenna in the analyzed frequency bands.

It is sufficient that the dielectric capacity of the absorbing layer is chosen as the least from the dielectric capacities of the limited PML media for the methods based on the finite difference time domain method. In this case this is dielectric capacity of air воздуха ε_1. The original conductivity of the internal PML surface implemented by FDTD can be set by the least equal from all the limited by them modeling media i.e. in this case $\sigma_{AL} = \sigma_1 = 0$ (Fig. 3). The value of the maximal PML σ_{max} conductivity on their external boundary is chosen based on PML δ depth and necessary degree of the distributing electromagnetic wave weakening.

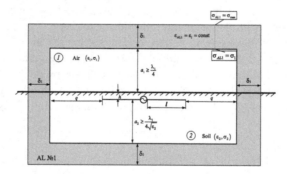

Fig. 3. The functional scheme of FDTD antenna model immersed in the semi-infinite dissipative medium

As a result, in spite of the thicker dissipative medium parameters differing from air the amplitude of the reflected wave becomes so small that it does not influence the nearest antenna field. The side wave went inside PML fades in it by the exponential law, not influencing on the nearest antenna zone field.

Thus, in case of the investigating antenna model work for transmission, the forming side wave and going wave fade in PML. The numerical experiments carried out showed that the parameters of the conceptual FEM- and FDTD-models q and a_2, for the antennas placed in the homogeneous isotropic media influence the nearest antennas field weakly. Their change within reason does not affect the received current values distributed on the antenna surface and the currents on the entrance of the antenna.

3   The General Approach to the Object Characteristics of the CS Task Forming

As an investigating object let us take a plane antenna (Fig. 4) immersed in the dissipative medium, for example, with tight coil parameters. If we use FEM than the mathematical model of the antenna can be constructed as a system of linear algebraic equations which are the discrete transformation of the inhomogeneous equation of Helmholtz for one of the EMF with posed peculiar to the immersed antennas boundary conditions. The solution of this system on the discrete frequency samples ω_i, $i = 1 \ldots n$, $n \in N$ by computer leads to finding vectors distribution $\vec{E}(x,y,z)$, $\vec{H}(x,y,z)$ in the enclosed calculating area V including infinitely expanded air space model, infinitely expanded dissipative medium model and the plane antenna model immersed on $h = 0.5$ м depth from the boundary, which curtains lie in parallel to the boundary.

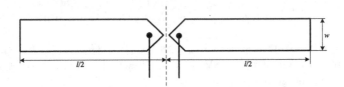

Fig. 4. The general appearance of vibrator

For FDTD method IA mathematical model is Maxwell's equations set in finite difference form with coefficients corresponding to the electric parameters of the model materials. All the EMF vectors components as time dependence their instantaneous amplitudes in the course of task solution. There are separate spectral EMF vectors components in the FDTD cells of the calculating area after the Inverse Fourier transform.

Thus, the received after solution by FEM or FDTD EMF vectors values in every point (cell) of the calculating area allow us to calculate many values which are used for receiving electrical antennas characteristics. The example is the distribution of surface currents densities vectors and the current on the entrance of the antenna model which is the response from the applied to the antenna model tension. They allow us to find the gain ratio (gain, G) and voltage standing wave ratio (WSVR).

It is appropriate to describe the quality evaluation of the essential IA properties by its electric characteristics in the limited lower ω^- and upper ω^+ by the working area frequencies $\omega \in [\omega^-, \omega^+]$.

The external parameters $\mathbf{q} = (q_1, q_2, \ldots, q_{nn})$, characterizing the external medium in relation to the object (medium 1–air, medium 2–soil) is recorded as

$$\mathbf{q} = (\varepsilon_1, \varepsilon_2, \sigma_1, \sigma_2, h, \omega) = (1, 5, 0, 10^{-3}, 0, 5, \omega). \tag{1}$$

They do not change. The joint complex of the external and internal parameters is traditionally called the set of the synthesis input parameters.

Gain ratio G and WSVR can be taken as the input parameters (improving or deteriorating characteristics) characterizing the object quality indicators

$$\gamma = (G(\theta = 0, \varphi, \omega), \text{WSVR}(\omega)). \tag{2}$$

The managing **x variables** limited by the task conditions belong to the range of $\mathbf{x} \in D$ – the search area being permissible solution. It is necessary to choose the best permissible solution of $\mathbf{x}^*$.

In this case, it is impossible to extract the most important output parameter as improving of some electric characteristics (gain) can lead to other characteristics deteriorating (WSVR) that is why two optimality criteria are formed on the basis of the imposed limitations:

$$Q_1(\mathbf{x}, \omega) = \begin{cases} 0, & \text{if } \gamma_1(\mathbf{x}) \geq \gamma_1^+ \text{ or } G(\mathbf{x}, \omega) \geq G_{\text{HWSV}}; \\ \gamma_1^+ - \gamma_1(\mathbf{x}) = G_{\text{HWSV}} - G(\mathbf{x}, \omega), & \text{if no,} \end{cases} \tag{3}$$

where $\gamma_1(\mathbf{x}) = G(\mathbf{x}, \omega)$- the current value gain IA $\gamma_1^+ = G_{\text{HWSV}} = 1.64$ – gain of the half-wave symmetric vibrator (HWSV) without losses in free space as the maximum limit gain of the underground antenna

$$Q_2(\mathbf{x}, \omega) = \begin{cases} 0, & \text{if } \gamma_2(\mathbf{x}) \leq \gamma_2^+ \text{ or } \text{WSVR}(\mathbf{x}, \omega) \leq 2; \\ \gamma_2(\mathbf{x}) - \gamma_2^+ = \text{WSVR}(\mathbf{x}, \omega) - 2, & \text{if no,} \end{cases} \tag{4}$$

where $\gamma_2(\mathbf{x}) = \text{WSVR}(\mathbf{x}, \omega)$ is the current value WSVR IA; $\gamma_2^+ = 2$ is the set maximum limit WSVR.

The expression (3) shows that the chosen optimality criterion maximizes G IA unlike the expression (4) where WSVR maximizing is performed.

As a result we receive multicriterion setting of the optimizing task.

$$\begin{cases} Q_1(\mathbf{x}*, \omega) = \min\limits_{\mathbf{x} \in D} Q_1(\mathbf{x}, \omega); \\ Q_2(\mathbf{x}*, \omega) = \min\limits_{\mathbf{x} \in D} Q_2(\mathbf{x}, \omega). \end{cases} \tag{5}$$

Let us make the scalar cost function $F(\mathbf{x}^*, \omega^*)$, corresponding to our task. We receive:

$$F(\mathbf{x}*, \omega*) = \min\limits_{\mathbf{x} \in \tilde{D}} [f(\mathbf{Q})]$$

$$= \min\limits_{\mathbf{x} \in \tilde{D}} \left[\left(\frac{1}{2} \left(\max\limits_{\omega^- \leq \omega \leq \omega^+} [Q_1(\mathbf{x}, \omega)] \right) + \frac{1}{2} \left(\max\limits_{\omega^- \leq \omega \leq \omega^+} [Q_2(\mathbf{x}, \omega)] \right) \right)^{1/2} \right], \tag{6}$$

where $\tilde{D} = D \cap D'$; $D' = \{\mathbf{x} | Q_k(\mathbf{x}) \leq Q_k^0, \ k = 1, 2\}$.

The evolutionary methods, for example, genetic algorithm (GA) [16] approved themselves for finding the minimum cost function (CF) of (5) kind. We can minimize the cost functions of (6) kind by the search methods by the pattern search, for example, by Hooke-Jeeves method [17]. Let us conduct CS on the basis of both cost function FEM-GA.

4 The Numerical Experiment

Let us choose the plated antenna «ungeometrical» rhombic form as the output for conducting CS with CF (5) (Fig. 5). Let us demand that the antenna should work in SW-band of the frequencies 3…30 MHz. The dissipative medium where it is placed has the parameters weakly conductive soil ($\varepsilon = 5$, $\sigma = 0,001$ S/m).

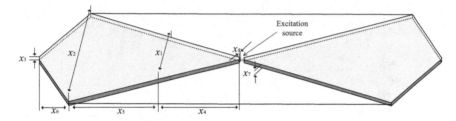

Fig. 5. The original antenna

As the synthesized plated antenna has rather complicated form, it is appropriate to use 7 controllable variables x_1, x_2... x_7. It is obvious that $x_1 - x_3$ values of the antenna width coordinates $x_4 - x_6$ – the values of the antenna coordinates length. The indicated variables in the course of optimization will change greatly. The controlled x_7 variable characterizes the thickness of the plated antenna and changes within narrow limits.

The x_8 variable characterizes the antenna width in the connection point to the excitation source. It is defined experimentally that the parameter can be taken approximately equal to 0.01 m, that is why it is set as constant originally.

The lower limit of the controlled variables x_1, x_2, x_3, x_4, x_5, and x_6 corresponds to 0.8 from the corresponding constructional parameters of the original antenna 2.5 m. It is connected with the fact that the coordination minimum frequency shift to the higher frequency area happens at the proportionally antenna size decreasing. And the task is to provide a good level in WSVR on the lower frequencies (Table 1).

The maximum length is limited by the reiterated increasing of the controlling variables x_4, x_5, и x_6 size (more increasing leads to reducing gain ratio in the upper frequency area and poses problems with antenna array constructing). It is necessary to reach the goal set to cover the whole frequencies diapason. The choice of the limits of x_1, x_2, и x_3 values upper boundary is implemented on the basis of the conducted parameter analysis and limitations impose on the antenna size.

As a result of the calculations carried out using FEM-GA band, the minimum value of the cost function received the value equal to 4.3 on 304 step (Fig. 6). The

Table 1. The x variables changing diapasons

Variable name	Original value, m	Minimum value, m	Maximum value, m
x_1	2	1	4
x_2	4	2	8
x_3	0.1	0.05	5
x_4	1	0.8	2
x_5	1	0.8	2
x_6	0.5	0.4	2
x_7	0.005	0.002	0.01

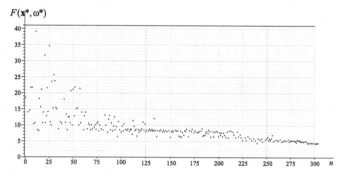

Fig. 6. The diagram of CF convergence (6) in the process of CF minimization (5) with gain using

calculations were made in ANSYS HFSS. The program stopped the calculation according to the set stop criterion «slow convergence».

The cost function convergence to the minimum is seen on the diagram. As a result of the CS conducted was received the plated structure of the symmetric antenna (Fig. 7). Bypasses on the antenna allowed us to expand the diapason to the lower frequencies.

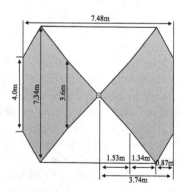

Fig. 7. The symmetrical bandwidth IA consisting of six segments received in result of CS

The results of synthesis of plated antenna model construction are presented in the Tables 2, herewith the rounding-up of all the values of the controlled variables.

Table 2. FEM-GA optimization results

The synthesizing antenna constructive parameter	The size before optimization, m	The size after optimization, m
x_1	1	3.6
x_2	2	7.34
x_3	0.001	4
x_4	1	1.53
x_5	1	1.34
x_6	0.5	0.87
x_7	0.005	0.007

The measurements presented show that the limitations for the optimizing structure are performed fully. Because of the necessity of the diapason low frequencies covering, the antenna became bigger than the original one.

The results of the G calculation showed that on the frequencies less than $f = 3.7$ MHz G goes lower than -15 dB (Fig. 8 line α),. It does not satisfy the set cost function. On the frequencies lower than $f = 26.3$ MHz, G turned out to be more than at the original antenna (it is connected with big size of the antenna received in consequence of CS). Nevertheless,

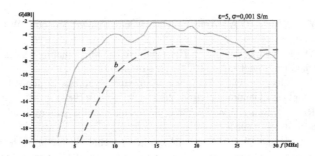

Fig. 8. The diagrams of the dependence of G from symmetric plated antennas frequency: a – received by CS; b – of the original one

The diagram of WSVR calculated (Fig. 9) showed that the received antenna has better coordination in all the diapason of the considering frequencies and covers it practically in full excluding the frequencies lower than 3.1 MHz. However, the diapason 3...30 MHz in CS process was considered, small spectral region left beyond $WSVR^{opt} \leq 2$ limit set consequently. From the 3.8 picture it is seen that the frequency equal to 3 MHz corresponds to $WSVR = 2.05$ value, which also can be called appropriate in view of the subtle deviation from the value demanded.

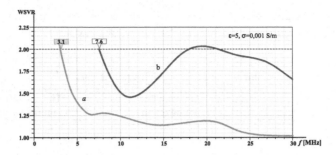

Fig. 9. The diagram of WSVR dependence from the symmetric plated antennas frequency a – received by CS, b – the original one

5 Conclusion

Thus, CS of the antennas in the dissipative media is possible. It has specific peculiarities caused by the medium influence character of the antennas immersing. If it is required the maximal uniformity G in the wide diapason, synthesized antennas will be enormously wide. It is explained by fuller metallization which influences positively on the uniformity of the electrical characteristics. In case of possibility to implement several IA typical sizes, it is appropriate to use in phase system of the plated vibrators. In some cases such antennas can substitute mast antennas.

References

1. Stutzman, W.L., Thiele, G.A.: Antenna Theory and Design, 3rd edn. Wiley, New York (2012)
2. Bakhrak, L.D., Kremenetskiy, S.D.: The synthesis of the radiating systems (theory and method calculation). Soviet radio (1974). (in Russian)
3. Bakhrakh, L.D., Litvinov, O.S.: The place of aperture synthesis in general antenna theory (review). Radiophys. Quantum Electron. **26**(11), 953–962 (1983)
4. Michelson, R.A., Schomer, J.W.: A three parameter antenna pattern synthesis technique. Microw. J. **8**(9), 88–94 (1965)
5. Smith, W.T., Stutzman, W.L.: A pattern synthesis technique for array feeds to improve radiation performance of large distorted reflector antennas. IEEE Trans. Antennas Propag. **40**, 57–62 (1992)
6. Safaai-Jazi, A.: A new formulation of the design of Chebyshev arrays. IEEE Trans. Antennas Propag. **42**, 439–443 (1994)
7. Chaplin, A.F.: The analysis and the antennas arrays synthesis. IRE. AS USSR (1984). (in Russian)
8. Borodulin, R.U., Sosunov, B.V., Makarov, S.B.: Principles of constructive synthesis of ‚electromagnetic wave radiators. In: Galinina, O., Balandin, S., Koucheryavy, Y. (eds.) NEW2AN/ruSMART -2016. LNCS, vol. 9870, pp. 717–730. Springer, Cham (2016). doi:10.1007/978-3-319-46301-8_63
9. Lavrov, G.A, Knyazev, A.S.: Ground antennas and underground antennas. The antennas theory and practice, placed near the ground surface. Soviet radio (1965). (in Russian)

10. King, R.W.P.: Dipoles in dissipative media. Cruft Laboratory Scientific Report 336, Harward University (1961)
11. Iizuka, K., King, R.W.P.: An experimental study of the properties of antennas immersed in conducting media. Cruft Laboratory Scientific Report 2, Harward University (1961)
12. Galleys, J.: Antennas in Inhomogeneous Media. Pergamon Press Inc., Oxford (1969)
13. Sosunov, B.V., Filippov, V.V.: The principles of the underground antennas calculation, 82 p. MTA (1990). (in Russian)
14. Jin, J.: The Finite Element Method in Electromagnetics, 2nd edn. Willey, New York (2002)
15. Berenger, J.P.: A perfectly matched layer for the absorption of electromagnetic waves. J. Comput. Phys. **114**(2), 185–200 (1994)
16. Johnson, J.M., Rahmat-Samii, Y.: Genetic algorithms and method of moments (GA/MOM) for the design of integrated antennas. IEEE Trans. Antennas Propag. **47**(10), 1606–1614 (1999)
17. Bandi, B.: The optimization methods. Radio and Communication, 128 p. (1988). (in Russian)

Amplitude and Phase Stability Measurements of Multistage Microwave Receiver

Yuriy V. Vekshin[1] and Alexander P. Lavrov[2(✉)] (iD)

[1] Institute of Applied Astronomy of Russian Academy of Sciences,
St. Petersburg, Russian Federation
yuryvekshin@yandex.ru
[2] Peter the Great St. Petersburg Polytechnic University,
St. Petersburg, Russian Federation
lavrov@ice.spbstu.ru

Abstract. The measurement technique and results of amplitude and phase stability measurements of microwave receiver and its separate stages are presented in this paper. Investigations were performed for prototype of new wideband receiving system for 13-m radio telescope RT-13 of the Institute of Applied Astronomy of the Russian Academy of Sciences. The receiver has two parallel channels for receiving linear orthogonal polarization and each of them is heterodyne type ones. The main sources of fluctuations and their types on different time intervals (from milliseconds to hundreds of seconds) are identified using Allan variance, power spectral density and correlation functions calculations. Phase stability measurements of unit with frequency conversion were performed by using 4-port R&S vector network analyzer. The contribution of the fluctuations of receiver separate stages to the overall receiver stability is determined taking into account the correlation between their fluctuations also.

Keywords: Heterodyne receiver · Multistage receiver · Gain stability · Phase stability · Allan variance

1 Introduction

Amplitude and phase stability is one of the main receiver characteristics that determine receiver performance in different applications. For example, in radio astronomy amplitude stability limits receiver sensitivity – ability to detect weak signals, phase stability is important for coherent receiver in radio interferometers. Receiver gain fluctuations are noise-type ones with different power spectral density – white noise, $1/f^\alpha$ noise and drifts. The fluctuation analysis using Allan variance [1] allows to distinguish noise types dominating on different time intervals and to determine time stability interval, at which minimum fluctuation variance is achieved [2]. Radio astronomy receiver (radiometer) is complex microwave heterodyne type receiver and it consists of several serially connected stages – input low-noise microwave (RF) amplifier, mixer-frequency converter, IF amplifier, registration and processing systems at the end. It is of great importance to identify main sources of amplitude and phase fluctuations and their contribution to total output signal. The measurement technique resolving this task is introduced in this paper.

© Springer International Publishing AG 2017
O. Galinina et al. (Eds.): NEW2AN/ruSMART/NsCC 2017, LNCS 10531, pp. 466–472, 2017.
DOI: 10.1007/978-3-319-67380-6_42

2   The Measurement Technique and Measurement Setup

Main investigations were performed using Allan variance analysis. The Allan variance $\sigma_A^2(\tau)$ – is the variance of the averaged over the time interval τ first differences of the processed data y values, sampled with a period T, and time interval τ is increased successively as $\tau = n \cdot T$, $n = 1,2,...N$:

$$\sigma_A^2(\tau) = \frac{1}{2(N-1)} \cdot \sum_{j=1}^{N-1} (\bar{y}_{j+1}(\tau) - \bar{y}_j(\tau))^2. \tag{1}$$

The Allan variance $\sigma_A^2(\tau)$ is associated with the noise power spectral density (PSD) $S(f) = h_\alpha/f^\alpha$: $\sigma_A^2(\tau) = K_\alpha \tau^{\alpha-1}$, where $K_0 = h_0/2\tau$, $K_1 = h_1 2 \cdot \ln 2$, $K_2 = h_2 \pi^2 \cdot 2\tau/3$ [3]. For linear drift with time dependency $y = pt$ $\sigma_A^2(\tau) = p^2\tau^2/2$.

For characterization receiver limiting sensitivity one can consider case with thermal noise source connected to its input, and measure receiver output fluctuations ΔP_{OUT}. Proceeding from the relation $P_{OUT} = k \cdot T_{SYS} \cdot G \cdot \Delta f$, relative variance of receiver output power fluctuations when registration system noise is negligible can be represented as

$$\left(\frac{\Delta P_{OUT}}{P_{OUT}}\right)^2 = \left(\frac{\Delta T}{T_{SYS}}\right)^2 = \frac{1}{\Delta f \cdot \tau} + \left(\frac{\Delta G}{G}\right)^2, \tag{2}$$

where T_{SYS} – the system noise temperature (referred to receiver input), Δf – receiver bandwidth, τ – output averaging time, G – total receiver gain, G is the product of receiver single stages gains $G = G_1 \cdot G_2 \cdot G_3$, ΔG – gain fluctuations. The first term in (2) is due to the so-called "radiometric noise" at radiometer output with no gain fluctuations. Radiometric noise is sensitivity for ideal radiometer.

The relative total gain fluctuation is the sum of relative gain fluctuations of individual stages [4]:

$$\left(\frac{\Delta G}{G}\right) = \left(\frac{\Delta G_1}{G_1}\right) + \left(\frac{\Delta G_2}{G_2}\right) + \left(\frac{\Delta G_3}{G_3}\right). \tag{3}$$

The total phase fluctuation $\Delta\varphi$ is the sum of phase fluctuations $\Delta\varphi_i$ of individual stages:

$$\Delta\varphi = \Delta\varphi_1 + \Delta\varphi_2 + \Delta\varphi_3. \tag{4}$$

The total gain (or phase) variance σ^2 is the sum of variances of single stage relative gain (or phase) fluctuations and mutual covariances K of their fluctuations:

$$\sigma^2 = \sigma_1^2 + \sigma_2^2 + \sigma_3^2 + 2K_{12} + 2K_{23} + 2K_{13}. \tag{5}$$

The covariance of two samples of the random processes X and Y is determined as follows:

$$K_{12} = \frac{1}{N} \sum_{i=1}^{N} (X_i - \overline{X}) \cdot (Y_i - \overline{Y}), \qquad (6)$$

where N – samples' length, $\overline{X}$ и $\overline{Y}$ denotes the expectation of X and Y samples. It is convenient to represent the results for each term in (5) as the Allan variance versus averaging time plot, while the covariance K of two signals is calculated for the same series of pairwise differences (as well as in Allan variance calculation).

Investigations were performed for the prototype of new wideband receiving system being developed for 13-m radio telescope RT-13 [5] of the "Quasar" Russian VLBI network. The prototype of wideband receiving system operates in 3–16 GHz band at two linear orthogonal polarization simultaneously [6], receiving system block-diagram is presented at Fig. 1. Cryoelectronic focal receiving unit (cryo unit, cooled to the temperature below 20 K) contains quad-ridged flared horn feed, low-noise amplifiers and directional couplers for calibration signals injection from noise generator unit (NGU). Signals of each polarization are divided into 4 ways by the splitter units. The splitter unit also includes preamplifiers under room temperature. Four dual-channel frequency converter units (FCU) use up-down conversion to select from input total range 3–16 GHz only 1 GHz bandwidth section and then to pass it to broadband acquisition system (BRAS) with working frequencies 1–2 GHz. Power, control and monitoring equipment (PCM) includes following parts: frontend power unit (FEPU) which provides primary power suppliers for cryo unit and splitter units, and cryogenic temperature measurement also; main power unit that provides supply voltages for another receiver units; Ethernet switch that connects all controllable units to common network.

One pair of two polarization channels of the prototype of the wideband receiving system was investigated. Amplitude gain fluctuations measurements of receiving system stages were performed by measurement setup, its scheme is presented at Fig. 2.

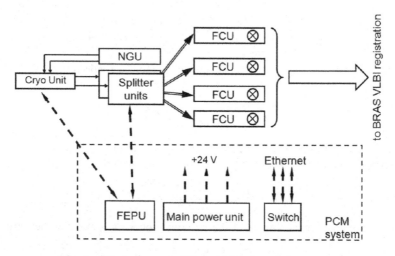

Fig. 1. Block diagram of the wideband receiving system

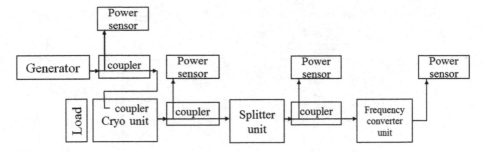

Fig. 2. Measurement setup scheme

Harmonic signal from microwave generator is passed to low-noise amplifiers in cryo unit through directional coupler, wherein special wide-aperture match load was mounted on the cryo unit input (on horn feed). Input and output powers of all receiver units (stages) were sampled by power sensors connected to power meter. Power meter readouts (samples) were long term recorded into PC using LabVIEW software. So, this measurement setup allows simultaneously measuring gain of all stages, and it is important when studying correlations between stage's readouts.

Phase stability measurements were performed by using vector network analyzer (VNA) R&S ZVA on fixed frequency in time domain mode. Phase φ is common measured as $\arg[S_{21}]$. Three ports of VNA were used to measure phase of transmission coefficient (gain) of frequency converter unit. Port 1 was exploited as generator of RF frequency, port 2 – as receiver of IF frequency after frequency converter and port 3 – as generator of the same IF frequency. Phase difference φ of ports 2 and 3 was measured at long time. It is important, that all VNA channels were synchronized between themselves and with the reference signal for frequency converter unit; reference signals were taken from laboratory hydrogen standard.

3 Investigation Results

Fluctuation of the measurement (registration) system should be less than fluctuation of the device under test. So, own stability of two available test setups was investigated before gain fluctuation measurements. The first setup is CW generator Agilent 8257D and Agilent N1914A power meter with 8487D power sensors, the second setup is vector network analyzer R&S ZVA. Amplitude gain fluctuations for these setups were measured and relative Allan deviation for these fluctuations is presented at Fig. 3, left: 1 – for generator + power meter setup, 2 – for vector network analyzer. Noise floor for the first setup is less, so this setup was used for amplitude gain fluctuation analysis. Noise floor for phase φ measurements with vector network analyzer is presented at Fig. 3, right.

Results of amplitude stability measurements of the wideband receiving system stages are presented at Fig. 4, left as relative Allan deviation $\sigma_{A\ REL}(\tau)$ plots. The relative Allan deviation is the Allan deviation for relative gain fluctuations $\Delta G/G$. At Fig. 4, left: 1 – $\sigma_{A\ REL}(\tau)$ for cryo unit gain, 2 – $\sigma_{A\ REL}(\tau)$ for splitter unit gain, 3 –

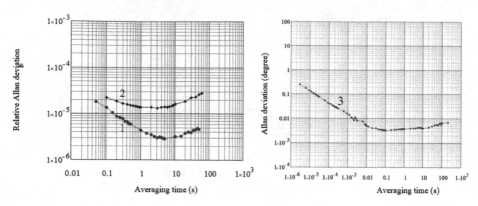

Fig. 3. Measurement system noise floor: *1* – generator + power meter setup, amplitude, *2* – vector network analyzer, amplitude, *3* – vector network analyzer, phase.

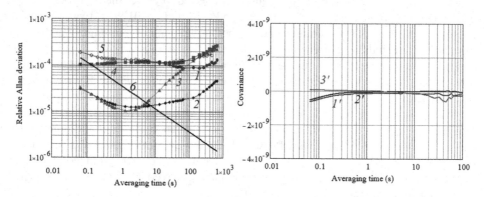

Fig. 4. Amplitude fluctuations of the wideband receiving system: *1* – cryo unit gain, *2* – splitter unit gain, *3* – frequency converter unit gain, *4* – total receiving system gain, *5* – total power, *6* – radiometric white noise for 1 GHz bandwidth, *1'* – K_{12}, *2'* – K_{23}, *3'* – K_{13} – covariances.

$\sigma_{A\ REL}(\tau)$ for frequency converter unit gain, *4* – $\sigma_{A\ REL}(\tau)$ for total receiving system gain, *5* – $\sigma_{A\ REL}(\tau)$ for total power $\Delta P/P$ fluctuations, *6* – $\sigma_{A\ REL}(\tau)$ radiometric "white" noise for $\Delta f = 1$ GHz bandwidth (theoretical plot for no gain fluctuations). Covariances of these signals are presented at Fig. 4, right: *1'* – K_{12}, *2'* – K_{23}, *3'* – K_{13}. Covariances of fluctuations are negligible. Experimentally was approved, that $\sigma_{A\ REL}(\tau)$ plot *4* for total receiving system gain coincides with the sum of gain fluctuations of stages *1, 2, 3* and their covariances K_{12}, K_{23}, K_{13}, see (5). Also, measured total power $\Delta P/P$ fluctuations *5* is the sum of total gain $\Delta G/G$ fluctuations *4* and radiometric white noise for $\Delta f = 1$ GHz bandwidth *6* (formula (2)). It is obvious, that cryo unit gain fluctuation *1* makes main contribution to overall amplitude system stability. This fluctuation is flicker-noise (Allan deviation is constant). The contribution of splitter unit gain fluctuations *2* and frequency converter unit gain fluctuations *3* is lower, but at intervals more than 100 s increase on plot *5* is due to $1/f^2$ noise of *3*. Since

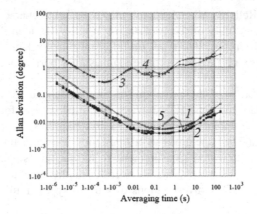

Fig. 5. Phase stability measurements of the wideband receiving system prototype: *1* – cryo unit phase, *2* – splitter unit phase, *3* – frequency converter unit phase, *4* – total receiving system phase, *5* – phase difference between 2 channels of frequency converter unit.

the receiving system noise temperature is about 30 K, the minimum radiometric sensitivity is 30 K·2·10^{-4} = 6 mK, and it can not be reduced by additional averaging (see Allan deviation plot *5*).

Results of phase stability measurements of the wideband receiving system stages are presented at Fig. 5 as Allan deviation σ_A (τ) phase plots. At Fig. 5 such marks are used: *1* – σ_A (τ) for cryo unit phase, *2* – σ_A (τ) for splitter unit phase, *3* – σ_A (τ) for frequency converter unit phase, *4* – σ_A (τ) for total receiving system phase. It is obvious, that total receiving system *4* phase instability is determined by frequency converter unit *3*. Phase fluctuations of cryo unit *1* and splitter unit *2* are small and slightly exceeds phase measuring floor of VNA (Fig. 3, right). But after 10 s averaging time phase drift appears (Allan deviation increases). Since frequency converter unit has two channels with one common heterodyne, its instability can be excluded by transferring one input signal to both channels and measuring phase difference between channels. Allan deviation for this case is presented at Fig. 5, plot *5*. There is a local maximum at 1 s, that corresponding to harmonic component 0.5 Hz. Allan deviation *5* is much smaller, than Allan deviation for one channel *3*. So, the main source of phase instability in the prototype of wideband receiving system is heterodyne in frequency converter unit. Standard deviation of receiver phase fluctuation *4* is about 6°, which leads to 1% signal-to-noise ratio decrease of correlation response when two such receivers will be used in radio interferometer, according [7].

4 Conclusion

The stability measurement technique of multistage receiver was developed and realized. Relative instability of receiver total gain is the sum of relative variances of receiver separate stages (cascades) and their covariances. This consideration was experimentally approved and used in precision measurements on developed measurement setup. Relative Allan deviation plot is useful to characterize and compare amplitude and phase stability of receiver stages. It allows to identify main sources of fluctuations and to determine their types on different time scales. Phase measurement technique for devices with frequency conversion using 4-port vector network analyzer was developed.

Stability investigation of the prototype of new wideband (3–16 GHz) receiving system for 13-m radio telescope RT-13 was performed. It was found, that the main source of amplitude instability is input low-noise amplifier of cryo unit, the main source of phase instability is heterodyne of frequency converter unit.

We consider that presented measurement technique gives valuable information about contribution of stages instability to overall stability and it can be applied for amplitude and phase stability investigations of various types' multi-stage receivers.

References

1. Rutman, J.: Characterization of phase and frequency instabilities in precision frequency sources: fifteen years of progress. Proc. IEEE **66**, 1048–1075 (1978). doi:10.1109/PROC.1978.11080
2. Vekshin, Y.V., Lavrov, A.P.: The Allan variance usage for stability characterization of weak signal receivers. In: Galinina, O., Balandin, S., Koucheryavy, Y. (eds.) NEW2AN/ruSMART-2016. LNCS, vol. 9870, pp. 648–657. Springer, Cham (2016). doi:10.1007/978-3-319-46301-8_56
3. Gonneau, E., Escotte, L.: Low-frequency noise sources and gain stability in microwave amplifiers for radiometry. IEEE Trans. MTT **60**(8), 2616–2621 (2012). doi:10.1109/TMTT.2012.2202681
4. Shan, W., Li., Z., Shi, S., Yang, J.: Gain stability analysis of a millimeter wave superconducting heterodyne receiver for radio astronomy. In: Proceedings of Asia-Pacific Microwave Conference, pp. 477–480 (2010)
5. Ipatov, A.V.: A new-generation interferometer for fundamental and applied research. Phys.-Usp. **56**(7), 729–737 (2013). doi:10.3367/UFNe.0183.201307i.0769
6. Evstigneev, A.A., Evstigneeva, O.G., Lavrov, A.S., Mardyshkin, V.V., Pozdnyakov, I.A., Khvostov, E.Y.: Development results of ultra wideband receiving system of radio telescope RT-13. Trudy IPA RAN [Trans. IAA RAS] **35**, 98–103 (2015). (in Russian)
7. Thompson, A.R., Moran, J.M., Swenson, G.W.: Interferometry and Synthesis in Radio Astronomy, 2nd edn. Wiley-VCH Verlag GmbH & Co. KGaA, Weinheim (2004). doi:10.1007/978-3-319-44431-4. 710 p.

Evolving Toward Virtualized Mobile Access Platform for Service Flexibility

Seung-Que Lee[✉] and Jinup Kim

Electronics and Telecommunications Research Institute (ETRI), Daejeon, Korea
{sqlee,jukim}@etri.re.kr

Abstract. This paper deals with the application of virtualization technology, which is regarded as important technology for securing service flexibility in 5G mobile communication system, to base stations. 5G mobile communication should provide users with personalized, localized and intelligent new concept services. Therefore, it is necessary to secure design and implementation technology that guarantees flexibility of its structure in addition to communication technology such as high speed, low delay, and concatenation. The 5G mobile communication infrastructure with virtualization can adapt and reconfigure communication functions adaptively to the environment, so that it can quickly reflect the needs of users and markets. The services of the 5G can be accommodated by sharing resources in one equipment, rather than requiring different devices for execution. In this paper, we design a virtualized access platform with Cloud, Virtualization, and SDN concepts. This includes the virtualization design of wireless transport modem, wireless access protocol, and radio resource management functions. The design is implemented as Service Virtualization subset where SON and RRM are virtualized and confirmed that LTE basic service and SON Network Monitoring functions operate in the standard including commercial user terminal and core network.

Keywords: 5G · Cloud · Virtualization · Software Defined Network · Mobile access platform · Service virtualization

1 Introduction

5G mobile communication is a future mobile communication infrastructure that reflects innovative communication technologies such as high speed, low delay, and hyper connectivity. Based on this, it is expected that user services will be provided in a very progressive form featuring personalization, localization, and intelligence. 5G services will be very diverse, and the so-called service life cycle, which includes planning, implementation, testing, launching and destruction of services, is expected to be shortened. In order to cope with such a service environment, the infrastructure of 5G mobile communication has to be as flexible as possible in terms of structure, and virtualization is becoming an important technology for this [1].

Virtualization is a new design and development paradigm that separates SW and HW by allowing only virtualized resources to be used through the abstraction layer rather than directly using actual hardware resources. With the concept of virtualization,

© Springer International Publishing AG 2017
O. Galinina et al. (Eds.): NEW2AN/ruSMART/NsCC 2017, LNCS 10531, pp. 473–481, 2017.
DOI: 10.1007/978-3-319-67380-6_43

the SW can be ported anywhere regardless of the underlying HW, dramatically increasing system flexibility. With the development of virtualization technology, it is applied to cloud server and edge server existing in wired network, and now it is extended to wireless communication service providing nodes such as eNB (enhanced Node B), MEC (Mobile Edge Computing) Server, and EPC (Evolved Packet Core).

The 5G mobile communication infrastructure with virtualization can adapt and reconfigure communication functions adaptively to the environment, so that it can quickly reflect the needs of users and markets. The services of the 5G can be accommodated by sharing resources in one equipment, rather than requiring different equipment for launching. In addition, the capacity of the system can be scaled up/down adaptively to the environment. For example, if traffic surges unexpectedly in an area, it is possible to reconfigure virtual resources and virtual functions to cope with it quickly and flexibly without adding or replacing hardware.

Since the mobile communication base station is mostly composed of a modem and a protocol for wireless connection, it includes many elements requiring high computing such as wireless signal processing, channel coding, encryption and compression. Therefore, the virtualization of base stations for mobile communication is regarded as the most difficult area for application of virtualization of communication nodes. The concept of Network Slicing, which can provide any service easily, quickly, and flexibly, is completed when the virtualization of the base station is completed.

In this paper, we propose a virtualization design for a base station for mobile communication that is responsible for the access function of mobile communication users. We present the design concept of access platform to support virtualization, and offer virtualization design of wireless transmission modem, wireless access protocol, and resource management function which applied presented concept. Also, we show implemented service virtualization subset in which SON (Self-Organizing Network) and RRM (Radio Resource Manager) parts are virtualized, and confirm that LTE attach, video streaming, and SON network monitoring functions correctly under commercial UE (User Equipment) and EPC environment. Finally, we summarize and conclude the research to be done in the future.

2 Design

2.1 Cloud, NFV and SDN

The virtualization access platform to be presented in this paper includes several important concepts, namely Cloud, NFV and SDN. Figure 1 shows the design concept of a virtualized access platform [2–4].

The concept of Cloud is characterized by a centralized processing unit and a distributed subscriber interface unit. This leads to an increase in system utilization due to the integrated processing, an increase in gain due to the integrated control, and a convenient configuration due to the integrated management. This paper applies this concept to divide the entire device into two units, a centralized CU (Central Unit) and a distributed RU (Remote Unit) [5].

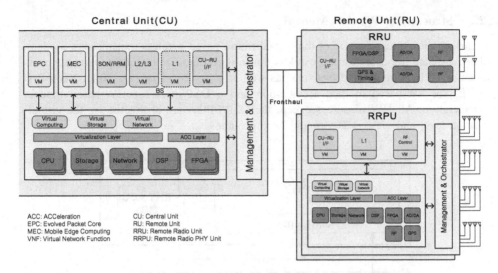

Fig. 1. Concept of virtualized access platform

NFV (Network Function Virtualization) concept that seeks to separate SW (Software) and HW (HardWare) by abstracting actual resources by virtualization and providing them as virtual resources. Under this concept, all programs will use only virtualized resources, leaving porting issues that are caused by hardware dependencies. The virtualization access platform has a virtualization layer between the hardware and the software so that all hardware can be abstracted to the software.

SDN (Software Defined Network) is a concept to separate control and data. This allows the virtualized program to directly participate only in the control part, and the data part does not need to be up to the area of the virtualized program, thus speeding up the data processing speed. This paper overcomes the performance degradation in virtualization by accelerating the 3GPP data plane using the SDN concept.

In the figure, the CU accommodates all protocols beyond the L1 layer except RF (Radio Frequency). For some high performance, it is possible to use the so-called Acceleration function which uses acceleration processing devices such as DPDK, DSP and FPGA. RUs can exist in various forms, such as a RRU (Remote Radio Unit) type for processing only RF functions and a RRPU (Remote Radio PHY Unit) type for processing functions of a modem together with RF. In case of RRU, most of it is constituted by passive HW, so it is composed of non-virtualization system. In case of RRPU, virtualization system is required because it requires execution environment of physical layer.

The CU and the RU are connected by a front-haul. The medium of the front-haul can be connected to various forms such as 1G/10G Ethernet, optical, and USB 3.0, and there can be various communication methods such as RoF, CPRI, and nFAPI. This is selected according to the split type of the base station, the bandwidth of the required front-haul and the latency [6].

2.2 Structure of Wireless Modem

Figure 2 shows the structure of the virtualization based wireless transmission modem. The hardware resources for the wireless transmission modem include general elements such as CPU, Storage, and Network, as well as a co-processor (e.g., Intel Xeon PHI KNL) for SW-based modem acceleration and a DSP/PGA for HW-based modem acceleration. All HW resources except the DSP/FPGA are abstracted by the virtualization layer and viewed as virtualization resources.

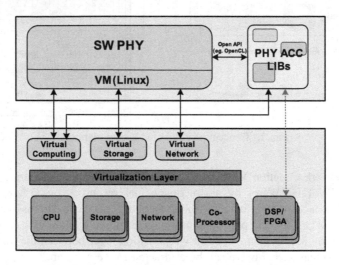

Fig. 2. Structure of wireless transmission modem

The SW PHY uses the virtualization resources to perform the LTE radio transmission modem function. It uses Linux (Redhat or Ubuntu) as a virtual machine (VM). SW PHY codes do not use hardware resources directly, so they have a feature that is independent of platform HW.

PHY ACC LIBs is a module that processes some functions of a modem with high complexity in the form of a library in case that the SW PHY needs to be accelerated due to the speed limit during operation. There are two ways in which LIBs can handle acceleration. The first way is a so-called SW-based acceleration method that performs vectorization/parallelization using a plurality of virtual computing resources based on Co-Processor resources, and the second method is a so-called HW acceleration method using DSP/FPGA. We will proceed with SW-based acceleration as early as possible and proceed with HW-based acceleration as a lane. Both methods increase the portability of the acceleration module by allowing access to the SW PHY via standardized/open APIs such as OpenCL.

2.3 Structure of Wireless Protocol

Figure 3 shows the structure of the wireless access protocol based on virtualization. As with the virtualization-based wireless transmission modem, the hardware resources for the wireless connection protocol include the co-processor (e.g., Intel Xeon PHI KNL) for SW-based modem acceleration in addition to general elements such as CPU, Storage and Network. However, it does not use HW-based acceleration like DSP/FPGA. On the other hand, DPDK (Data Plane Development Kit) of Intel is used for data plane acceleration. The DPDK allows SDN based routing and switching by allowing protocol SW to handle data and control separately.

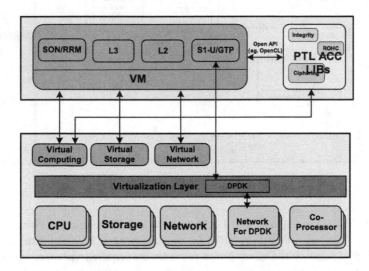

Fig. 3. Structure of wireless access protocol

The wireless access protocols such as SON, RRM, L3, L2, S1-U and GTP performs the LTE radio access protocol function using virtualization resources. As in the case of modems, all protocol codes are made up of hardware-independent code based on virtualization.

PTL ACC LIBs can be used to process some functions (e.g. Integrity, ROHC, Ciphering, etc.) of a wireless access protocol with high performance complexity is needed due to the speed limit of the protocol. The way PTL ACC LIBs uses the SW-based acceleration method by vectorization/parallelization. To increase portability, the module is connected to the main code through a standardized and open interface.

2.4 Structure of Resource Management

Figure 4 shows the resource management structure of the virtualization platform. The CRM (Central Resource Manager) is responsible for network resource management. Resource Integration (RI), Resource Optimization (RO), Resource Capacity

(RC), etc., of the virtual base station while interfacing with SON and RRMs in a plurality of virtual base station (VNF-BS). The CRM is seen as VNF of the same kind as VNF-BS in view of NFVI (NFV Infrastructure).

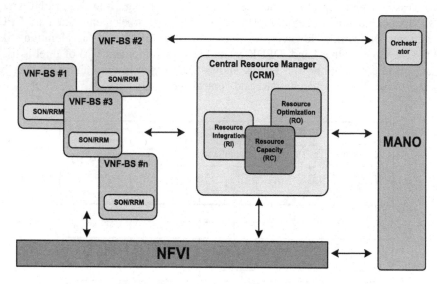

Fig. 4. Structure of resource management

The MANO (Maintenance And Orchestration) is responsible for virtual resource management functions. It controls NFVI for management of HW resources and VNFs for operation of network functions. It also provides the required virtual resources to the platform statically and dynamically according to service contents and service capacity.

For the integrated resource management of the virtualization platform, interaction between the CRM and the MANO is needed. The virtualized base stations (VNF-BS) can be installed and executed by the MANO. After the execution, the virtualized base stations are controlled by the CRM in an integrated manner.

3   Implementation and Testing

In this paper, we implement the service virtualization part based on the above design and verify its function. Service virtualization refers to the virtualization of the part responsible for resource management such as SON and RRM among various functions of the virtualization base station. From a layered viewpoint, it is also referred to as L3+ layer virtualization in the sense of virtualization for higher ones than layer 3.

Figure 5 shows the functional arrangement of CU and RUs according to service virtualization. The L3+ elements, SON and RRM are placed in the CU, and the other protocols such as RRC, GTP, L2, L3, modem, and RF are placed in the RU. The CU is implemented in VNF on the virtual infrastructure and the RU is implemented in PNF in the non-virtualized infrastructure.

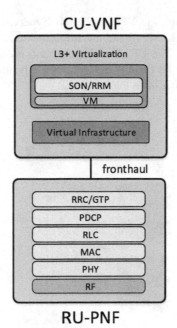

Fig. 5. Service virtualization

Figure 6 shows the interfaces between CU-VNF and RU-PNF. The nFAPI 3.0 standard interface defined by SCF is applied between SON and PNF. Through this, SON sends and receives various control commands to the PHY in the PNF. On the other hand, the interface between RRM and PNF is via ASN.1. Currently, this section does not have an international standard and will be changed if future standards are enacted. Through the ASN.1 interface, the RRM and the L2/L3 protocols in the PNF exchange various radio resource management commands.

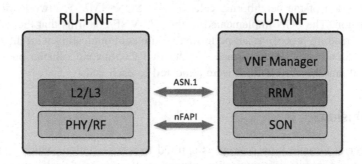

Fig. 6. Interfaces

Figure 7 shows the test environment for implementing and verifying Service Virtualization. The CU-VNF with virtualization SON and RRM was virtualized on the

Intel Xeon-based HW through the VMware ESXi hypervisor. The L1/L2/L3 and RF functions are located in the RU-PNF and are implemented on a Freescale 9131 based small cell base EVM board. The UE (User Equipment) uses a commercially available commercial LTE terminal, and the EPC is connected behind the CU-VNF and configured to be able to interwork with the Internet.

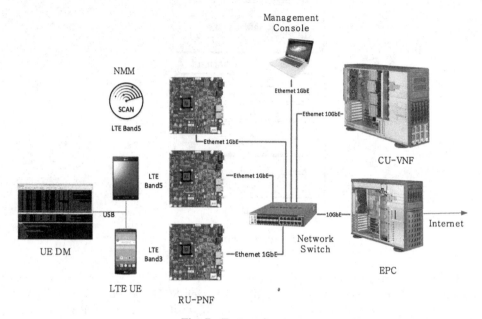

Fig. 7. Test environment

Commercial terminals operated LTE Band3 and Band5, and two RUs for wireless connection were operated for each band. It confirmed that there is no problem making LTE attach and playing YouTube video stream. In addition, one RU is responsible for the scan function of the neighboring cells for the SON NMM (Network Monitoring Mode) function. The SON implemented in the CU-VNF confirms that the cell search and the SIB reading procedure are performed while communicating with the RU-PNF via nFAPI. As a result, we confirmed that the PCI Confusion/Collision problem was resolved and that the correct PCIs were allocated to each cell.

4   Conclusions

In this paper, a virtualization concept is applied to a base station, which is a user 's access point in a mobile communication system, and a service virtualization prototype is implemented by virtualizing the L3+ layer to verify its operation. It is confirmed that LTE basic services and SON NMM functions according to the specification under the test environment including commercial UE and EPC.

Virtualization is a technology that makes communication functions very flexible. Virtualization enables fast service introduction, adaptive network configuration/reconfiguration, and economical HW usage. So virtualization technology is becoming an important component of 5G. In the wired network, virtualization has become an essential implementation paradigm, and base stations for mobile communication are no longer an exception to virtualization.

Base stations for mobile include many difficulties in realizing virtualization because they involve highly complex communication elements such as wireless transmission modems and wireless access protocols. Therefore, a so-called step-by-step virtualization methodology is needed that increases the weight of virtualization after confirming the feasibility of virtualization through easy networking. In this study, based on this methodology, virtualization of L3 layer, L2 layer, and physical layer will be gradually progressed. In addition, various acceleration techniques such as vectorization, parallelization, and data plane switch are required to overcome the performance degradation caused by virtualization.

Acknowledgment. This work was supported by ICT R&D program of MSIP/IITP. [2016-0-00183, Development of 5G mobile access platform technology based on virtualization converged with computing]

References

1. Lee, S.-Q., Kim, J.U.: Local breakout of mobile access network traffic by mobile edge computing. In: ICTC 2016, October 2016
2. NGMN: NGMN 5G White Paper, February 2015
3. ETSI: Mobile Edge Computing Technical White Paper, September 2014
4. ETSI: Network Function Virtualization (NFV): Architectural Framework. ETSI GS NFV 002 v1.1.1, October 2013
5. Small Cell Forum, Release Document, Functional Splits and Use Cases for SC Virtualization, June 2015
6. Lee, S.-Q., et al.: The Design Specification for the Development of Virtualized Access Platform, ETRI TDP FLEXELL-SYS-DES-120A, August 2016

Model of Photonic Beamformer for Microwave Phased Array Antenna

Sergey I. Ivanov⬤, Alexander P. Lavrov$^{(\boxtimes)}$ ⬤, and Igor I. Saenko⬤

Peter the Great St. Petersburg Polytechnic University, St-Petersburg, Russia
{s.ivanov,lavrov,saenko}@ice.spbstu.ru

Abstract. Photonic beamformer model for linear phased array antenna in the receive mode has been developed, investigated and experimentally tested. The proposed beamformer scheme based on true-time-delay technique, DWDM technology and fiber chromatic dispersion has been designed. The beamforming arrangement for ultra wideband antenna array can currently provide the required true time delay capabilities by using modern fiber-optical telecom components available at the market. The attention was paid to measurements of intrinsic time delays in used fiber-optic components which are significant for beamformer design. Experimental demonstration of the designed model beamforming features included actual measurement of 5-element receiving microwave linear phase array antenna far-field patterns in 6–18 GHz frequency range for antenna pattern shifts up to 40° which show no squint effect. The influence of inaccuracy of photonic beamformer channels time delays setting and amplitude aligning on the antenna beamforming has been also analysed.

Keywords: Phased array antenna · Photonic beamforming · True time delay · Far-field pattern · Microwave fiber optic link · Squint effect

1 Introduction

Microwave phased array antennas (PAA) beamforming by using photonic technology have been a subject of intense researches for last two decades due to their well known advantages compared to electronic beamforming systems [1–3]. Especially ultrawideband large antenna arrays could benefit from development of photonic beamformers based on true-time-delay (TTD) technique instead of traditional phase-control technique. So, in the last two decades it has been shown that applying the microwave photonic devices for PAA beamforming could afford the means to overcome conventional radio electronic beamformer limitations. Therefore, many beamforming architectures based on microwave photonics for PAA have been proposed and evaluated, both in transmitting and receiving modes. Most of the proposed architectures using fiber optic components to provide control of one- or multibeam antenna array pattern require specially developed units, such as low-noise, fast tunable lasers with narrow spectral linewidths, chirped fiber Bragg gratings, reflection fiber segments with precise lengths, optical filter arrays and others. Numerous research over the past ten years have yielded significant improvements in the key performance parameters of fiber optic telecom components, so one can use it in development beamformer (BF)

© Springer International Publishing AG 2017
O. Galinina et al. (Eds.): NEW2AN/ruSMART/NsCC 2017, LNCS 10531, pp. 482–489, 2017.
DOI: 10.1007/978-3-319-67380-6_44

arrangements [4–6]. Reported here photonic beamformer model for PAA in receive mode is based on dense wavelength division multiplexing (DWDM) components and chromatic dispersion of a single mode optical fiber and implemented with units and components available at the market of modern fiber-optic telecom systems [7, 8]. We consider the results of a type of photonic BF experimental model assembling and adjustment and its performance investigation including the microwave receiving linear (1-D) PAA far-field pattern measurement in the 6–18 GHz frequency range.

2 Photonic Beamformer Model Design and Setup

In order to provide required time delays in photonic BF based on DWDM technique one can use optical comb and DWDM multiplexing and choose some different approaches to realize wavelength depending interchannel time delays. We interested in approaches for interchannel delays setting in Time Delay Unit (TDU) by exploiting of chromatic dispersion in optical fiber segments or in chirped fiber grating, as shown at Fig. 1. However, for the present we have developed the BF model comprising ready-made components and not investigated yet TDU with chirped fiber grating which has been specially manufactured. The scheme in Fig. 1 is designed for the N element's linear PAA and uses optical comb – a set of N lasers with different but uniformly spaced (step $\Delta\lambda$) wavelengths, the total wavelength band is $(N - 1)\cdot\Delta\lambda$. The microwave (RF) signals from the antenna elements A_1 ... A_N modulate a set of laser diodes. Electrooptic conversion is achieved either by direct modulation of lasers or by using an external modulators (Mach-Zander modulators) as represented in Fig. 1. Further, the intensity-modulated optical carriers are combined into a single fiber by a multiplexer (MUX $N \times 1$ unit) and fed into a time delay unit (TDU). As all the optical carriers

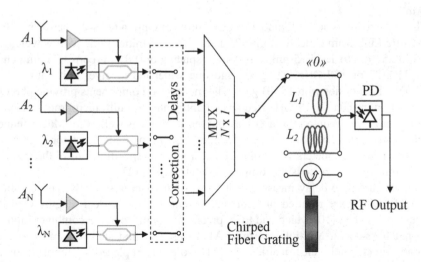

Fig. 1. Architecture of photonic beamformer based on DWDM and wavelength depending interchannel delays.

share with the same light paths, the time delay differences between adjacent channels are produced by the fiber chromatic dispersion D measured in ps/(nm km). Time delay $\tau = \Delta\lambda \cdot L \cdot D$ introduced between adjacent channels results in the respective tilting of PAA beam. Photodiode (PD) at TDU output converts sum of delayed intensity-modulated optical carriers back to microwave domain. In Fig. 1 one can see 2 fiber segments (single mode fibers) with lengths L_1 and L_2 successively inserted between the multiplexer and the photodiode thus giving (together with "0" - length position) 3 interchannel delay values: 0, τ_1 and τ_2 corresponding to 0, θ_1 and θ_2 beam direction angles according to relation $d \cdot \sin \theta / c = \tau$, where d is antenna elements spacing, and c – speed of EM-wave. To align the electrical delays from RF sources (antennas $A_1 \ldots A_N$) to the MUX output the "Correction Delays" unit comprises fiber-cords with strictly sized up lengths Hence, it corresponds to a flat phase law tilted 0° to PAA base line – the PAA beam directed normal to PAA base. The RF signals modulating the comb of optical wavelengths are coherently summed at the photodetector output while optical carriers are summed incoherently. So due to chosen fiber segment lengths and thus specified interchannel delays RF signals received from the corresponding direction (angle) are combined in phase.

In the considered architecture of 1-D photonic BF the number of switching beams can be increased that only requires increased number of dispersive fiber segments. Also a multibeam mode could be realized (for example as in [4]) by using of a higher splitting ratio after MUX, a set of fiber segment with photodiode at the end of each segment.

The beamformer model developed for initial demonstration of linear PAA beam steering is based mainly on microwave photonic components commercially available on market of fiber optic telecom links (radio-over-fiber links). Among the photonic key components depicted in Fig. 1 the units converting radio signal into optical one (laser jointly with modulator) and backward (photodiode) are crucial for BF correct performance.

The designed photonic BF model includes units of Optiva OTS-2 Microwave Band Fiber Optic Link from Emcore Corp. [9] namely 5 transmitter units with wavelengths from 1552.52 nm to 1555.75 nm, wavelength spacing of 0.8 nm (corresponding channels # 31 .. 27 of 100 GHz ITU grid), and one wideband receiver unit with in-built additional RF amplifier with 15 dB gain. The mentioned components provide photonic link operation in 0.05 … 18 GHz frequency bandwidth. In-built microprocessor-based control (in transmitter units) for laser bias and temperature as well as for Mach-Zehnder modulator bias provides more stable performance operation and allows for appreciably reduce BF model parameters variations. The performance specifications of the main BF model components were considered in more detail in [10, 11].

The accurate time delay measurements of used transmitters jointly with combiner instead of multiplexer resulted in a number of channel's electrical lengths with differences up to 0.4 m [10] which had to be precisely equalized at the combiner input by additional fiber-cords. Multiplexer (from AFW) used at first step research had great differences in channel's electrical lengths [11], so it was replaced by simple photonic light circuit 8 × 1 combiner.

Further 5 corrective patch cords with the calculated lengths: 200, 226, 254, 550, 624 mm, have been fabricated with deviations in range –0.02 … +2.43 mm from required values and inserted in BF model channels at the combiner inputs. The measurements have been made to check time delays alignment over 5 BF channels at TDU output for 3 beam positions at RF frequencies near 4 GHz. The results of these measurements show their good conjunction with calculated values (maximal difference less than 2 ps [11]). Figure 2 shows calculated dispersion delays for two lengths of standard SMF 28 type fiber with dispersion parameter 17 ps/nm·km at λ = 1550 nm as well as measured relative channels delays to provide two PAA beam positions. Time delay versus channel number line approximation shows, that inserted interchannel delays mean values were 21.6 and 41.8 ps. Figure 2 shows also differences in delays between measured and line approximated values for two fiber lengths. RMS value for difference was approx. 0.5 ps. These error values will be used for estimation of errors influence on PAA beamforming, see Sect. 4.

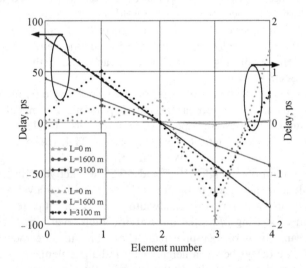

Fig. 2. Calculated and measured channels delays for two beam point angles.

3   BF with PAA Testing and Results

The photonic BF model assembled with the 1-D antenna array is characterized through far-field pattern measurements in anechoic chamber. For experimental research we used 5 radiating elements of available multielement wideband linear PAA with 10.2 mm element spacing, but PAA elements were connected to the BF model through one, giving resulting spacing d = 20.4 mm. PAA radiating elements are Vivaldi patch antennas. For providing 3 beam point angles we used 2 SMF 28 fiber coils with lengths 1500 m and 1600 m (2 coils in Rack-mount Fiber Lab 250 from M2 Optics) inserted individually or in cascade with short patch-cord. Selected fiber lengths correspond to calculated antenna array beam point angles considered earlier in [11]. For PAA with $d = 20.4$ mm realized interchannel delays mean values 21.6 and 41.8 ps give estimated beam point angles 18.6° and 37.9°, respectively. The BF model was connected to PAA by 5 RF cables with equal lengths 40 cm and the whole arrangement was located on an azimuth scan drive.

In our pattern measurements a microwave signal from a vector network analyzer (with additional microwave power amplifier) fed a wideband reference irradiant horn. An azimuth scan drive holds the PAA under test located in front of the reference horn. The distance between the horn and array is about 6 m to fulfill far-field condition requirement. Microwave signal received by PAA elements and processed in the photonic BF is returned to vector network analyzer and compared with initial signal feeding reference horn during azimuth scanning. A network analyzer scans wide frequency range from 6 to 18 GHz for each azimuth position varying with 1° step. These measurements were done for 3 different fiber lengths L pointed above.

The PAA response in power $|H(\theta, f)|^2$ can be calculated in accordance with $|H(\theta, f)|^2 = |H_E(\theta, f) \cdot H_N(\theta, f)|^2$, where $H_E(\theta, f)$ is the field radiation pattern of single antenna element surrounded by other ones in array, and $H_N(\theta, f)$ – frequency-dependent array factor. By measuring response $|H_E(\theta, f)|^2$ and considering theoretical $H_N(\theta, f)$ for array with $N = 5$, and elements spacing $d = 20.4$ mm we can calculate the response $|H(\theta, f)|^2$ of our PAA, see Sect. 4 down.

Quasi third-dimensional representations of measured angle-frequency responses $|H(\theta, f)|^2$ for 5-element PAA in conjunction with photonic BF model are shown in Fig. 3 These system responses corresponding to three sets of delays following from insertion of 0, 1600 or 3100 m optical fibers lengths into TDU (see Fig. 1) are depicted in figures (a), (b), (c), respectively. We note, that main lobes in figures (b) and (c) corresponding beam point angles 18.8° and 36.8°, respectively illustrate no squint effect in the whole frequency band from 6 to 18 GHz. Also, high level of the side lobes at high frequencies are caused by wide bandwidth of one element pattern $|H_E(\theta, f)|^2$ and rather small element spacing ($d = 20.4$ mm) giving high level lobes in array factor | $H_N(\theta, f)|$. The beam points of measured pattern exhibit fair agreement with the calculated ones for time delays between neighboring radiating elements 0, 21.6, 41.8 ps. Some additional experimental results on the BF with PAA performance investigation one can find in [10, 11].

We designed BF for PAA in receive mode, so we had paid attention also to BF noise characteristics such as the output signal-to-noise ratio and dynamic range. The simulation technique that we have developed allows performing calculations of transfer function and signal-to-noise ratio at the photonic BF output. The mathematical model underlying this consideration takes into account all linear and nonlinear signal conversions in different radio and photonic members of the specific BF structure in accordance with its performance. Some results of the simulation are published in [8].

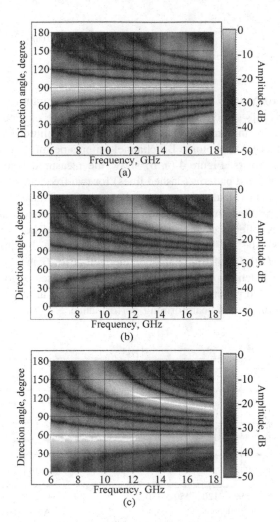

Fig. 3. Quasi 3-D representation of measured PAA angle-frequency response for dispersion optical fiber length in TDU: 0 m – (a); 1600 m – (b); 3100 m – (c).

4 Beamforming Errors Estimation

To estimate the influence of BF channels amplitude and time delay fluctuations on total system performance we have developed a calculation method based on known results for phased antenna array performance investigations but taking account of photonic nature of BF components in which the errors concerned are arisen and using the results of BF channels amplitude and phase errors measurements. Here we consider only some resulting ultimate estimations from that method.

The normalized pattern of a linear array of isotropic antenna elements with uniform distribution can be expressed as [13].

$$|H|^2 = e^{-\sigma_\varphi^2}|H_0|^2 + (1 + \sigma_a^2 - e^{-\sigma_\varphi^2})^{\eta_0} / |H_{0max}|^2$$
$$= e^{-\sigma_\varphi^2}|H_0|^2 + (1 + \sigma_a^2 - e^{-\sigma_\varphi^2})^1 / N \cdot S_0$$

where H_0 is no-error pattern, σ_a and σ_φ are the relative amplitude and RMS phase errors, respectively, η_0 – efficiency factor, S_0 is the aperture efficiency, and N is the number of elements of the array. Our measurements of amplitude and phase fluctuations at the 5-element PAA photonic BF channel outputs permit to get the estimations of expected pattern errors. Figure 4 (a, b) show the measured beam patterns (solid lines) as well as calculated patterns (dashed lines) for two frequencies: 6 GHz – (a) and 14 GHz – (b), wherein calculated patterns take account of measured amplitude and phase fluctuations resulted from correction fiber-cords lengths deviations, channel's frequency responses differences and temperature variations of lasers wavelengths and fiber characteristics. The dot lines also depicted in Fig. 4 (a, b) show the measured radiation patterns $|H_E(\theta, f)|^2$ of single antenna element at $f = 6$ and 14 GHz, which are used for 5-element antenna patterns calculation. One can see a good agreement of measured and calculated patterns which justifies insignificant influence of photonic BF parameters fluctuations on its beamforming performance.

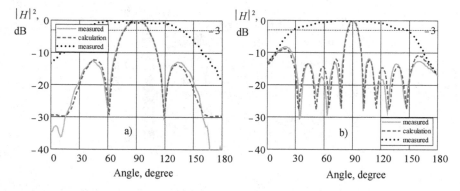

Fig. 4. 5-element PAA in conjunction with BF model far-field pattern at frequencies 6 GHz – (a) and 14 GHz – (b), measured (solid line) and calculated (dash line), and measured single element pattern (dot line).

5   Conclusions

The model of photonic beamformer for ultrawideband microwave phased array antenna in the receive mode has been developed and tested. The beamformer realizes true-time delay principle and contains the components available on the market of fiber-optic analog and DWDM telecom links. 5-element microwave linear array antenna far-field patterns have been measured in 6–18 GHz frequency range, which demonstrated the pattern control up to 40° with no squint effect as desired. The results of measured antenna patterns comparison with the calculated ones in view of channels amplitude and phase fluctuations show their good agreement. Future work will include more detail study of time and temperature stability of developed BF model and also its updating with chirped fiber grating. The number of PAA beams in the considered architecture of 1-D beamformer can be increased that only requires to split signals after combiner and to use another set of dispersive fibers and photodiodes.

References

1. Riza, N.A. (ed.): Selected papers on photonic control systems for phased array antennas. In: SPIE Proceedings, vol. MS136. SPIE Optical Engineering Press (1997)
2. Yao, J.P.: A tutorial on microwave photonics – part II. IEEE Photonics Soc. Newslett. **26**(3), 5–12 (2012)
3. Urick, V.J., Mckinney, J.D., Williams, K.J.: Fundamentals of Microwave Photonics. Wiley, Hoboken (2015)
4. Hunter, D.B., Parker, M.E., Dexter, J.L.: Demonstration of a continuously variable true-time delay beamformer using a multichannel chirped fiber grating. IEEE Trans. Microw. Theor. Tech. **54**(2), 861–867 (2006). doi:10.1109/TMTT.2005.863056
5. Meijerink, A., Roeloffzen, C.G.H., Meijerink, R., Zhuang, L., Marpaung, D.A.I., Bentum, M.J., Burla, M., Verpoorte, J., Jorna, P., Huizinga, A., van Etten, W.: Novel ring resonator-based integrated photonic beamformer for broadband phased array receive antennas – Part I: design and performance analysis. J. Lightwave Technol. **28**(1), 3–18 (2010). doi:10.1109/JLT.2009.2029705
6. Blanc, S., Alouini, M., Garenaux, R., Queguiner, M., Merlet, T.: Optical multibeamforming network based on WDM and dispersion fiber in receive mode. IEEE Microw. Theor. Tech. Trans. **54**(1), 402–411 (2006). doi:10.1109/JLT.2009.2029705
7. Lavrov, A.P., Ivanov, S.I., Saenko, I.I.: Investigation of analog photonics based broadband beamforming system for receiving antenna array. In: Balandin, S., Andreev, S., Koucheryavy, Y. (eds.) NEW2AN 2014. LNCS, vol. 8638, pp. 647–655. Springer, Cham (2014). doi:10.1007/978-3-319-10353-2_60
8. Ivanov, S.I., Lavrov, A.P., Saenko, I.I.: Optical-fiber system for forming the directional diagram of a broad-band phased-array receiving antenna, using wave-multiplexing technology and the chromatic dispersion of the fiber. J. Opt. Technol. **82**(3), 139–146 (2015). doi:10.1364/JOT.82.000139
9. Optiva OTS-2 18 GHz Amplified Microwave Band Fiber Optic Links. http://emcore.com/wp-content/uploads/2016/03/Optiva-OTS-2-18GHz-Unamplified.pdf
10. Ivanov, S.I., Lavrov, A.P., Saenko, I.I.: Application of the fiber-optic communication system components for ultrawideband antenna array beamforming. In: Proceedings of 10th International Conference on Antenna Theory and Techniques, Kharkiv, Ukraine, pp. 366–368 (2015). doi:10.1109/ICATT.2015.7136886
11. Ivanov, S.I., Lavrov, A.P., Saenko, I.I.: Application of microwave photonics components for ultrawideband antenna array beamforming. In: Galinina, O., Balandin, S., Koucheryavy, Y. (eds.) NEW2AN/ruSMART 2016. LNCS, vol. 9870, pp. 670–679. Springer, Cham (2016). doi:10.1007/978-3-319-46301-8_58
12. Yang, Y., Dong, Y., Liu, D., He, H., Jin, Y., Hu, W.: A 7-bit photonic true-time-delay system based on an 8 × 8 MOEMS optical switch. Chin. Opt. Lett. **7**, 118–120 (2009)
13. Skolnik, M.I.: Introduction to Radar Systems, 3rd edn. McGraw-Hill, New York City (2001)

Nanosecond Miniature Transmitters for Pulsed Optical Radars

Alexey V. Filimonov[1]([✉]), Valery E. Zemlyakov[2],
Vladimir I. Egorkin[2], Andrey V. Maslevtsov[1],
Marc Christopher Wurz[3], and Sergey N. Vainshtein[4]

[1] Peter the Great St. Petersburg Polytechnic University,
195251 Polytechnicheskaya 29, St. Petersburg, Russia
filimonov@rphf.spbstu.ru, avm@spbstu.ru
[2] National Research University of Electronic Technology – MIET,
124498 Bld. 1, Shokin Square, Zelenograd, Moscow, Russia
vzml@rambler.ru, egorkin@qdn.miee.ru
[3] Institute for Microproduction Technology,
Leibniz University of Hanover, Garbsen, Germany
wurz@impt.uni-hannover.de
[4] CAS Group, Department of Electrical Engineering,
University of Oulu, 4500, 90014 Oulu, Finland
vais@ee.oulu.fi

Abstract. The-state-of-the-art in long-distance near-infrared optical radars is utilization the laser-diode-based miniature pulsed transmitters producing optical pulses of 3–10 ns in duration and peak power typically below 40 W. The bandwidth of the receiving channel nowadays exceeds 300 MHz, and thus the duration of the optical pulses exceeding 3 ns is a bottleneck in the task of high practical importance, namely increase in the radar ranging precision. Nowadays the speed of the high-current drivers is limited by the speed of a semiconductor switch that is typically field-effect transistor or an avalanche switch. The last one provides faster switching, and development of new avalanche switches is very challenging and important task, but this is not the only factor limiting the transmitter speed. Here we show that not only the switch, but also parasitic inductance in the miniature assembly and type of the capacitor play very important role in solving the problem of long-distance decimeter-precision radar.

Keywords: Optical radars · High-speed switching · Avalanche drivers · Miniature assembly · Peak power

1 Introduction

There is a need to generate current pulses of a few nanoseconds in length with an amplitude of $\sim 10\text{--}10^2$ A across a low-ohmic load for a number of commercial applications. This concerns particularly the pumping of high-power broad-stripe laser diodes for pulsed LIDARS and other systems [1–3]. Large variety of commercial generators providing nanosecond and sub-nanosecond current pulses approaching and

© Springer International Publishing AG 2017
O. Galinina et al. (Eds.): NEW2AN/ruSMART/NsCC 2017, LNCS 10531, pp. 490–497, 2017.
DOI: 10.1007/978-3-319-67380-6_45

even exceeding 100 A are available (see, e.g., http://www.fidtechnology.com/), but large size and prices make their application problematic. Much cheaper and compact is making use of avalanche transistor-based Marx circuits [4] providing reliable operation after simple modifications [5] in nanosecond and even picosecond switching at dozens amperes [6], or several hundred amperes for a pulses approaching a dozen nanoseconds. In all those examples, however, the transmitter cannot be considered as low cost or miniature (comparable in price and size with encapsulated laser diode), which are the criteria of principal importance for most of applications. The state-of-the-art competing solutions satisfying those criteria are pulsed switches based on latest developments of GaN field-effect transistors (FET), or properly optimized avalanche transistors. Despite GaN FETs can operate at higher repetition rate (>100 kHz), they obviously lose to avalanche transistors in switching speed, when nanosecond and sub-nanosecond high-current switching is required together with the transmitter miniaturization. Thus simplest, cheapest and most effective ways is to make use of high-voltage ($\sim$150–300 V) avalanche transistors [7, 8], which is the state-of-the art in optical radars (lidars) utilizing 3–10 ns in duration optical pulses from the laser diodes in 10–100 W power range. This duration of the optical pulse becomes a bottleneck of the lidars aiming at longest possible ranging (preferably several kilometres) with high (around a decimetre) ranging precision. Indeed, the bandwidth of the receiving channels based on avalanche detectors exceeds nowadays 300 MHz, and thus even 1 ns in duration optical pulses can be easily detected without noticeable reduction in the detector sensitivity and critical growth in the noise level. Thus development of a laser diode transmitter emitting optical pulses around 1 ns in duration with as high as possible (at least several dozen W) peak power is the challenging task for long-ranging high-precision radars, and the main problem consists in high-current nanosecond drivers. It is additionally worth noting that a reduction in optical pulse duration well below 1 ns is possible when pumping current pulse of the laser diode becomes comparable with lasing delay, and various gain-and Q-switching picosecond modes [9–11] can be in principle realized, and even sub-micron precision of the distance measurement was demonstrated for the distance of 700 m using femtosecond laser [12]. Those modes are used in laboratory practice or for low-power laser transmitters, but they are not very reliable when broad laser diode chips with intrinsic structure inhomogeneity make the stability and reproducibility of the laser transmitter problematic in serial production quantities. Thus the challenging practical task for miniature low-cost laser transmitter for large-range and high-precision optical radars can be specified as optical pulse duration around 1 ns with as high as possible peak power (aiming at $\sim$ 100 W). For the current driver this means current pulses of several dozen amperes with the pulse duration around 1 ns. Obligatory condition are miniature assembly, simplest construction, and low-cost components.

The goal of this work is to analyze main problems to be solved for development miniature low-cost, high-power nanosecond transmitter optimized for long-distance decimeter-precision lidar.

2 Driving Circuit and a Role of Each Component

It is possible to show that for a given electrical switch the simplest circuit shown in Fig. 1 provides maximum in amplitude and shortest in duration current pulse, which can be used also in the transmitter when replacing the load resistor by a laser diode.

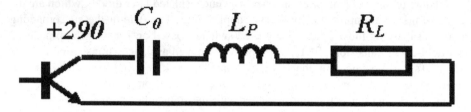

Fig. 1. Simplest, most effective and miniature current driver to be realized in miniature assembly. The switch is presented by a bipolar junction transistor (BJT) operating in so called "second breakdown" avalanche mode; capacitor C_0 accumulates the energy between the pulses (charges) and discharges across the switch across the load resistor R_L and entire parasitic inductance of the circuit loop L_P. L_P limits the current ramp and should be diminished as well as R_L, which is needed only for measuring the current waveform (in our case $R_L = 1\ \Omega$). In case of optical transmitter the load resistor is replaced with the laser diode.

Very useful for simple but very instructive estimates are the following relations for peak current amplitude I_m and FWHM of the current pulse t_w valid for an "ideal" switch with infinitely short switching time between the levels U_0 (maximum voltage across the switch) and U_R (residual voltage across the switch):

$$t_w = 2.2 \times \sqrt{L_P \times C_0} \tag{1}$$

and

$$I_m = \frac{U_0 - U_R}{\sqrt{\frac{L_P}{C_0}} + R_L} \tag{2}$$

Given the switch is ideal ($U_R = 0$, switching time is infinitely short) and load resistor is replaced with a laser diode with zero impedance we obtain the following "quality criterion" for our transmitter: $I_m/t_w = \frac{U_0}{2.2 \times L_P}$. It is obvious that for a long-range high-precision radar the quality criterion I_m/t_w has to be as large as possible, and thus preferable is highest possible biasing U_0 and lowest possible inductance. The biasing voltage is determined by the compromise between mentioned criterion, switching speed of the particular transistor and the maximum voltage that a particular user allows in his system. The parasitic inductance L_P should always be diminished, and its fundamental limit is determined by the size of the components of circuit shown in Fig. 1. (It is obvious that the assembly has to be realized as a 3-D construction of minimal size, and that all components in the construction have to be represented as semiconductor chips).

3   Results and Discussions

Among commercial avalanche transistors optimal in our case is apparently FMMT415 (ZETEX Semiconductors Inc.) as it combines fairly high biasing ($U_0 \sim 300$ V) with reasonably short switching time (~ 2 ns). Shown in Fig. 2 are our measurement results of the voltage and current waveforms across this switch at different C_0 values and relatively large parasitic inductance of 5 nH, which cannot be reduced when multilayer ceramic capacitors NP0 type (well operating in nanosecond pulses) have been used with surface mounted transistor package, (but not a chip transistor).

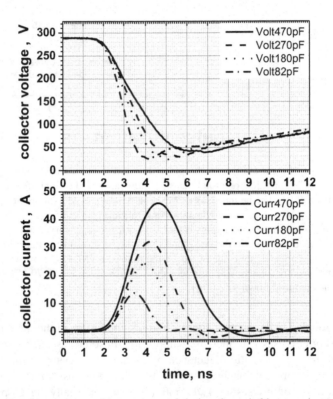

Fig. 2. Voltage and current waveforms measured during the switching transient of commercial avalanche transistor FMMT415 (Zetex semicond. Inc.) switched in the circuit shown in Fig. 1 with different multilayer ceramic capacitors of NP0 dielectric type, total inductance of the circuit of 5 nH and load resistor 1 Ω. Shown current pulses are recovered from the measured voltage waveforms across the load by a numerical procedure on account of 0.8 nH parasitic inductance of the load resistor.

Most instructive for analyzing ways of reduction in the current pulse duration and increase in the pulse amplitude is the presentation of the experimental results in the graph shown in Fig. 3. Comparison of the experimental points with the predictions given by simplest relations (1) and (2) allows to arrive to important practical recommendations.

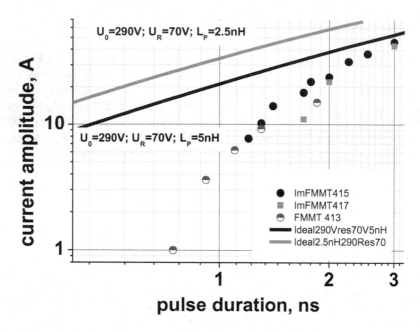

Fig. 3. Dependence of peak current on the pulse duration measured using multilayer ceramic NP0 capacitors of different value, one total parasitic inductance of the circuit 5 nH, and different commercially available avalanche transistors FMMT 415, 413 and 417 (Zetex Semiconductors). Black and gray solid curves present calculations using formulas (1) and (2). The value of residual voltage U_R = 70 V used in the calculations is taken from the voltage curves in Fig. 2 in a timing point corresponding to the peak in the corresponding current waveform. Black curve has to be compared to the experimental points measured for the transistor FMMT415, while the gray curve provides expectations for a switch with infinitely short switching time with 70 V residual voltage but the inductance reduced down to 2.5 nH.

This very rough, but descriptive approach gives good ideas of the main bottle-necks in achieving shorter pulses with increased amplitude. In some cases it require first of all a reduction in the total parasitic inductance by assembling means, in other case main limitation origins from the switching speed of avalanche transistor, and in third case most effective way of increasing I_m/t_w ratio is using specially developed surface-mounted capacitors, which will not only reduce parasitic inductance in the assembly, but will provide releasing the charge accumulated in the dielectric in sub-nanosecond range at large-signal mode. (According to our experience the dielectric losses tangent measured in small-signal mode may contradict the results obtained in high-current nanosecond pulse measurements, and optimization of the dielectric require additional labor-consuming tests.) The third factor will be considered elsewhere, while here we concentrate on first two factors.

One can see from Fig. 3 that the best of all available from Zetex avalanche transistors from point of view of maximizing quality criterion I_m/t_w is FMMT415. Then, for an ideal switch (infinitely short switching time) the calculated for L_P = 5 nH curve is presented by black solid line. It gives a bit higher I_m/t_w values, but comparable with the experiment for

FMMT415 and large C_0 values (pulse duration t_w exceeding 2–3 ns). This means that the switching time of the transistor is comparable with 2 ns, and the current pulse amplitude and duration estimated using formulas (1) and (2) are not far from the experiment, while for shorter pulse durations the difference increases drastically.

We can conclude from Fig. 3 that (i) when current pulses longer than 2 ns are required, commercial avalanche transistor FMMT415 is suitable, and current amplitude can be efficiently increased by reduction in the parasitic inductance: for example if inductance were reduced from 5 nH to 2.5 nH, the pulse amplitude can be increased by a factor of 1.5–2 (see gray curve in Fig. 3); (ii) when current pulse below 2 ns in duration were required, no significant win can be obtained from reduction in the inductance, but special avalanche switch has to be designed with shorter switching time, and only then reduction in the inductance value can provide significant win in the current. Actually we have clear vision of how higher-speed avalanche transistors can be designed even using standard Si technology, and the work is already underway. If a current (optical) pulse of 1 ns in duration is needed, even FMMT415 or 413 can be used, but the current pulse will not exceed 4–5 A, and the optical power even with advanced laser diode will be around 10 W at best. Thus significant increment in optical power for 1 ns pulses requires simultaneously development of faster avalanche transistor, low-inductance assembly of properly configured components, and, as will be

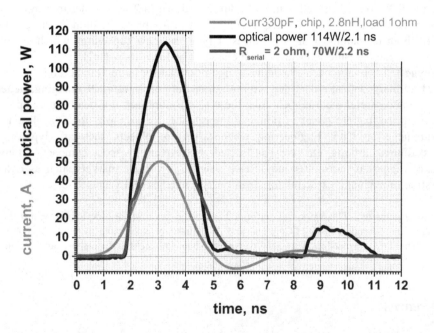

Fig. 4. Pumping current pulse using chip of FMMT415 transistor, NP0 capacitor 330 pF 1 Ω load resistor for current measurement and laser diode; optical response or 114 W/2.1 ns in presence, however, of a second pulse caused by a current relaxation oscillation. The oscillation is dumped using additionally a serial resistor of 2 Ω, which has allowed the second optical pulse to be suppressed, but this reduced the peak power down to 70 W.

seen from another paper, proper development of the capacitor. Optimistically we may expect an increase in the peak current up to ~ 20 A, which may provide with an appropriate laser diode peak power of ~ 40 W.

Shown in Fig. 4 is optical pulse obtained with a driver based on FMMT415 transistor and NP0 capacitor. The emitter is a broad-stripe (100 µm) three active layer near-infrared (900 nm) laser diode with efficiency of 2.3 W/A. After pulsing was observed in optical response due to relaxation current oscillation, which was successfully suppressed by a serial resistor of 2 Ω, but the peak power of optical pulse in this case reduced from 114 W down to 70 W. More technologically advanced way would be reducing the parasitic inductance still further below 2.8 nH, which apparently will dump the relaxation oscillation below lasing threshold. This requires, however, replacing NP0 capacitor by specially designed surface-mounted capacitor, which is a subject of another work [13].

4 Conclusion

Commercial avalanche transistors allow current/optical pulse generation up to 2 ns in duration, while at shorter pulses the efficiency of the switch reduces drastically, and special development of optimized Si avalanche transistor for ns/sub-ns range is needed. Using commercial avalanche transistor chip allows a peak power in a single optical pulse of 70 W to be achieved with pulse duration down to 2 ns. Further improvement in the peak power and very challenging reduction in the pulse duration down to 1 ns requires both development of new transistor type and new capacitors, which should also reduce the parasitic inductance of the assembly down to ~ 1 nH.

Suggested here approach should have high potential for large-volume civil market. Indeed, military optical radars that require larger time-of-flight (TOF) ranging, better precision and shorter measurement time loose to our approach drastically in compactness and price that are issues of principal importance, for example, for TOF distance measurements in Civil Engineering when large-size objects such as Hypermarkets, stadiums, airports, etc. are constructed. Moreover, low-cost compact systems utilizing suggested transmitter may create new large-volume markets such as radar utilization in navigation systems for small boats, launches and yachts.

Acknowledgements. This work was performed under the government order of the Ministry of Education and Science of RF. M.C. Wurz. and A. Filimonov would like to thank the DAAD Grant "Scientific Partnership with St. Petersburg State Polytechnic University and Leibniz Universität Hannover".

References

1. Biernat, A., Kompa, G.: Powerful picosecond laser pulses enabling high-resolution pulsed laser radar. J. Opt. **29**, 225–228 (1998)
2. Kilpela, A., Kostamovaara, J.: Laser pulser for a time-of-flight laser radar. Rev. Sci. Instrum. **68**, 2253–2258 (1997)

3. Myllylä, R., Marszalec, J., Kostamovaara, J., Mäntyniemi, A., Ulbrich, G.-J.: Imaging distance measurements using TOF lidar. J. Opt. **29**, 188–193 (1998)
4. Li, J., Zhong, X., Li, J., Liang, Z., Chen, W., Li, Z., Li, T.: Theoretical analysis and experimental study on an avalanche transistor-based Marx generator. IEEE Trans. Plasma Sci. **43**(10), 3399–3405 (2015). doi:10.1109/TPS.2015.2436373
5. Duan, G., Vainshtein, S.N., Kostamovaara, J.T.: Modified high-power nanosecond Marx generator prevents destructive current filamentation. IEEE Trans. Power Electron.: Manuscript ID TPEL-Reg-2016-10-2004 (in Press)
6. Vainshtein, S.N., Duan, G., Filimonov, A.V., Kostamovaara, J.T.: Switching mechanisms triggered by a collector voltage ramp in avalanche transistors with short-connected base and emitter. IEEE Trans. Electron Devices **63**(8), 3044–3048 (2016). doi:10.1109/TED.2016. 2581320
7. Herden, W.B.: Application of avalanche transistors to circuits with a long mean time to failure. IEEE Trans. Instrum. Meas. **IM-25**, 152–160 (1976)
8. Streetman, B.Q.: Solid State Electronic Devices, pp. 344–346. Englewood Cliffs, Inc., New York (1972)
9. Vainshtein, S.N., Simin, G.S., Kostamovaara, J.T.: Deriving of single intensive picosecond optical pulses from a high-power gain-switched laser diode by spectral filtering. J. Appl. Phys. **84**(8), 4109–4113 (1998)
10. Vainshtein, S.N., Kostamovaara, J.T.: Spectral filtering for time isolation of intensive picosecond optical pulses from a Q-switched laser diode. J. Appl. Phys. **84**(4), 1843–1847 (1998)
11. Vainshtein, S.N., Kostamovaara, J.T., Myllyla, R.A., Kilpela, A.J., Maatta, K.E.A.: Automatic switching synchronization of serial and parallel avalanche transistor connections. Electron. Lett. **32**(11), 950–952 (1996)
12. Lee, J., Kim, Y.J., Lee, K., Lee, S., Kim, S.W.: Time-of-flight measurement with femtosecond light pulses. Nat. Photonics **4**, 716–720 (2010)
13. Zemlyakov, V.E., Egorkin, V.I., Vainshtein, S.N., Maslevtsov, A.V., Filimonov, A.: Investigation of electro-physical and transient parameters of energy accumulating capacitors applied in nanosecond and sub-nanosecond high-current avalanche switches. In: Galinina, O., Balandin, S., Koucheryavy, Y. (eds.) NEW2AN/ruSMART -2016. LNCS, vol. 9870, pp. 731–737. Springer, Cham (2016). doi:10.1007/978-3-319-46301-8_64

Fog Computing for Telemetry Gathering from Moving Objects

Ivan Kholod$^{(\boxtimes)}$, Nikolai Plokhoi, and Andrey Shorov

Saint Petersburg Electrotechnical University "LETI",
Professora Popova Str. 5, Saint Petersburg, Russia
{iiholod,ashxz}@mail.ru, elefus@yandex.ru

Abstract. The paper describes two approaches to gathering measurement data about moving objects in wireless networks. The use of Fog computing technology makes it possible to relocate a part of calculations closer to measuring devices. The first approach suggests an estimation of telemetry quality into measuring points. The second approach uses prediction of telemetry quality by mining models. As a result, it became possible not only to redistribute the computational load, but also to significantly reduce the network traffic, which in turn brings the possibility to decrease the requirements for communication channels bandwidth and to use wireless networks for gathering telemetry.

Keywords: Gathering telemetry · Fog computing · Data mining · Moving object

1 Introduction

Traditionally, telemetry (from Greek roots tele = remote, and metron = measure) has been used to fulfill people's needs for remote data gathering and control. The task of telemetry gathering is very relevant. Following examples illustrate this:

- transportation industry - telemetry provides meaningful information about driver's performance;
- space science - telemetry are used by spacecraft for data transmission;
- rocketry telemetry equipment forms an integral part of the rocket range assets used to monitor the position and health of a vehicle launch;
- motor racing - telemetry allows race engineers to interpret data collected during a test or a race and use it to properly tune a car for optimum performance.

These examples illustrate gathering telemetry from moving objects (MO). This task has become even more relevant with development of the unmanned vehicles industry. Measurements (speed, vibration, temperature, etc.) transferred from the vehicle to the control point using telemetry channels are used for unmanned vehicle management.

More recently, wireless networks, cell phone networks, satellite communication networks, and terrestrial microwave networks have increasingly been used for data gathering, including telemetry [1]. This is explained by convenience of their use: no need for cable installation, ease of connection, etc. However, one of the principal disadvantages of networks of this type is limited bandwidth.

O. Galinina et al. (Eds.): NEW2AN/ruSMART/NsCC 2017, LNCS 10531, pp. 498–509, 2017.
DOI: 10.1007/978-3-319-67380-6_46

This paper proposes to use Fog computing and dynamic management of telemetry data in order to decrease network traffic by transferring only necessary measurements.

2   Related Works

The problem of telemetry gathering is not new. First data-transmission circuits go live between the Russian Tsar's Winter Palace and army headquarters in 1845. In the end of last century and in the beginning of this century the concept of Machine to Machine communications (M2M) became widespread. M2M technology continues where telemetry left off: wireless telemetry systems today cost substantially less compared to what they did 10 to 15 years ago. Applications that were previously not economically feasible are now cost effective. M2M brings discipline to telemetry through open standards and protocols [2].

At present time we can observe rapid growth in the number of the Internet of Things (IoT) devices. The basic idea of IoT is to connect all things (devices) in the world to the Internet. According to Gartner, Inc. (a technology research and advisory corporation), there will be nearly 26 billion devices on the IoT by 2020 [3].

The basic architecture of IoT which is widely used to explain the approaches of IoT has three layers [4]:

- the perception layer is the bottom layer which can be regarded as the hardware or physical layer which does the data collection;
- the network layer, is responsible for connecting the perception layer and the application layer so that data can be passed between them;
- the application layer usually plays the role of providing services or applications that integrate or analyse the data received from the other two layers.

Cloud computing [5] technologies are used for solving the problems related to computational resources and Big Data processing on the application layer of IoT systems. A Cloud provides scalable storage, computation time and other tools to build application services. It can be also used to process telemetry gathering from objects.

For example, space situational awareness program in European space agency and the Egyptian space program use Cloud computing for telemetry processing [6]. It enables software and hardware decoupling and makes flexible telemetry data analysis possible. Similar approach was also used in the Cloud-Based Ground System for space telemetry processing [7].

In this case the network is responsible for connecting the sensors/actuators and a Cloud. All telemetry data are transferred into it. It creates very large traffic in the network. Fog computing [8] can be a solution to this problem. The fog extends the Cloud to be closer to the devices that produce and act on data gathering systems.

Despite of Fog computing becoming popular, there are no ready solutions for it's implementation. This can be explained by the fact that such concept is every young and has a high level of abstraction. The paper describes approaches to telemetry gathering based on Fog computing technology. These approaches suggest to move part of computations closer to telemetry gathering points. It allows to redistribute the computational load and reduce network traffic.

3   Generic Approach of Telemetry Gathering

3.1   Typical System of Telemetry Gathering

Gathering telemetry from MO involves the use of measuring points (MP) located along the route of the movement of the MO. Each of them measures various parameters of the objects' performance. Measurements can be obtained by different measuring systems (MS): radar, lidar, optics, telemetry, navigation and etc.

During measurements, all MSs work in the measurement mode, regardless of the presence of the object in their detection zones. It is necessary that measurements of an object are taken as soon as it enters the MS detection zone. Typically, all measurement data that has been received is continuously transferred from MS to the measurements processing center (MPC).

At the MPC the data is filtered and combined into a main stream of measurements. This stream can be used to identify potential hazardous situations with MO and make decisions to manage them. In addition, tuning of MS can be performed based on the results of measurement processing (e.g. aiming of radar antennas, camera lenses, etc.) using current and predicted parameters of MO (its location, speed, etc.). A MPC can be created base on Cloud computing technology.

The MO detection zones are different for each MP, but they may have significant intersections. As a consequence, the MPC receives duplicate data of different quality (Q) from all MPs throughout the period of measurement. Figure 1 shows the described problem. Thus, the collected data have a significant percent of duplicated measurements and noise.

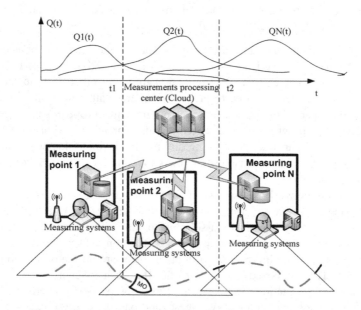

Fig. 1. Gathering of measurement data from moving objects.

3.2   Formal Representation of a Telemetry Gathering System

The entire measurements data gathering system can be represented formally as a set:

$$CS = <mo, mpc, MP>$$

where *mo* is a controlled moving object; *mpc* is a MPC; *MP* is a set of measuring points.

The state of any MO is characterized by a set of parameters:

$$S_{mo} = S_{mo}^k \cup S_{mo}^t = \{x, y, z, vx, vy, vz\} \cup \{p_1, p_2, \ldots, p_k, \ldots, p_u\}$$

Among them we can distinguish the following kinematic parameters S_{mo}^k: coordinates *(x, y, z)* and the velocity vector *(vx, vy, vz)*; parameters transferred through the telemetry channel S_{mo}^t: temperature, vibration, fuel level, etc.

Typically, the volume of telemetry data is much higher than the volume of kinematic parameters of a moving object. Therefore, parameters transferred through telemetry channels result in the biggest traffic load.

The *MP* is a set of MPs:

$$MP = \{mp_1, mp_2, \ldots, mp_r, \ldots, mp_m\}$$

Each measuring point mp_r includes: MSs that measure kinematic parameters of MO and MSs receiving telemetry data from MO.

Therefore, a vector of MO parameters is formed at each measuring point mp_r at each time instant t_j:

$$s_r(t) = s_r^k(t) \cup s_r^t(t) = \{x_r(t), y_r(t), z_r(t), vx_r(t), vy_r(t), vz_r(t)\} \cup \{p_{r.1}(t), \ldots, p_{r.k}(t), \ldots, p_{r.u}(t)\},$$

where $s_r^k(t_j)$ is a set of kinematic parameters obtained at time t_j at the point mp_r; $s_r^t(t_j)$ is a telemetry frame containing a set of parameters of the object MO obtained at time t_j by telemetry at the point mp_r.

For each telemetry frame it is possible to measure its quality $q(s_r^t(t_j))$, characterising the integrity and reliability of the received data. The quality of telemetry frame can be estimated in many ways and is not considered in this paper (verification of built-in markers (e.g. Reed-Solomon codes) for each frame, can serve as an example of estimation of telemetry data quality).

The main purpose of a measuring system is to produce a main stream containing the best measurements of all parameters of a MO at each time instant of the measurement:

$$s_{mo} = \{s_{mo}(t_1), s_{mo}(t_2), \ldots, s_{mo}(t_j), \ldots, s_{mo}(t_w)\},$$

where $s_{mo}(t_j)$ is the set of values of MO parameters at time t_j, produced from all telemetry frames obtained from the measuring points:

$$s_{mo}(t_j) = best(s_1(t_j), \ldots, s_r(t_j), \ldots, s_m(t_j))$$

3.3   Typical Procedure of Telemetry Gathering

The generic approach of telemetry gathering (Fig. 1) involves the following procedure performed at each time instant t_j:

1. obtaining a telemetry frame $s_r(t_j)$ on each MP mp_r:
   ```
   sr(tj) = mpr.collect();
   ```
2. transferring measurements $s_r(t_j)$ from each MP mp_r, $r = 1..m$ into the processing center MPC:
   ```
   forall mpr from MP
   sr(tj) = mpr.transf(mpc, sr(tj))
   ```
3. assessing the quality of the stream received from each MP:
   ```
   forall mpr from MP
   q(sr(tj)) = mpc.quality(sr(tj));
   ```
4. selecting the best measurements:
   ```
   sb(tj) = mpc.best(q(s1(tj)), ..., q(sr(tj)), ..., q(sm(tj)))
   ```
5. producing the main stream:
   ```
   mpc.aggregate(smo, sb(tj))
   ```

4   Suggested Approaches to Telemetry Gathering

4.1   Approach of Telemetry Gathering with a Distributed Quality Assessment

In order to reduce network traffic during measurements data gathering, it is proposed to transfer only the snapshots that will eventually be included into the main stream. To achieve this, assessment of a telemetry frame quality must be assessed at the MP and transferred to the MPC. The MPC collects assessments of quality and selects the best one. The best frame is requested from the corresponding MP and sent to the MPC to be included into the main stream.

Thus, the following procedure is performed at each time instant t_j (Fig. 3):

1. obtaining a telemetry frame $s_r(t_j)$ from each MP mp_r, $r = 1..m$:
   ```
   sr(tj) = mpr.collect();
   ```
2. assessing the quality of the frame at each MP mp_r, $r = 1..m$:
   ```
   q(sr(tj)) = mpr.quality(sr(tj));
   ```
3. transferring the quality assessment $q_r(t_j)$ from each MP mp_r, $r = 1..m$ into the MPC:
   ```
   forall mpr from MP
   q(sr(tj)) = mpr.transf(mpc, q(sr(tj)))
   ```
4. selection of the best telemetry frame:
   ```
   sb(tj) = mpc.best(q(s1(tj)), ..., q(sr(tj)), ..., q(sm(tj)))
   ```
5. sending the command's' to transfer the frame to the selected MP mp_b and the quality command'q' to the all MPs mp_r, $r = 1..m$:
   ```
   mpc.transf(mpr, 's')
   forall mpr from MP
   mpc.transf(mpr,'q')
   ```

6. transferring the snapshot from MP mp_b to the processing center MPC:
   ```
   sb(tj) = mpb.transf(mpc, sb(tj));
   ```
7. producing the main stream:
   ```
   mpc.aggregate(smo, sb(tj))
   ```

4.2 Gathering with Quality Assessment Prediction

In order to reduce the main stream formation time, is suggested to predict quality of telemetry frame at each time instant $q(s_r^t(t_j))$. By predicting the change in the quality of the telemetry, it is possible to preemptively determine the time instants when it is necessary to transfer a frame to be included in the main stream.

The main problem is to calculate $q(s_r^t(t_j + n))$ for a period of time n sufficient for MS switching. Such prediction can be achieved by applying mining models, obtained through the use of data mining algorithms, including those used for time series analysis (logical regression, etc.).

In this case, quality value $q(t_j + n)$ will be approximated for the following period. However, to improve the accuracy of the prediction not only the existing values, but also additional attributes that affect the quality of the telemetry can be used. In this case, it is possible to produce the following vector that would characterise the quality of telemetry frame $s_r^t(t_j)$ at time instant t_j from MP mp_r:

$$x_r(t_j) = \{x(t_j), y(t_j), z(t_j), vx(t_j), vy(t_j), vz(t_j), q(s_r^t(t_j)), q(s_r^t(t_j + n))\}.$$

Thus training set is the time-ordered sequence of all vectors from the start time instant of measurement t_0 till the finish t_w on each MP mp_r:

$$X_r = \{x_r(t_0), x_r(t_1), \ldots, x_r(t_j), \ldots, x_r(t_j + n), \ldots, x_r(t_w)\}$$

A mining model can be constructed by data mining algorithms using such vector sets obtained from previous measurements of this MP's objects. In this case, the quality assessment $q(s_r^t(t_j + n))$ is known for each $x_r(t_j)$. In addition, the mining model can be corrected in the course of measurements, using the measurements that have already been performed.

Using this approach, the formation of the main stream $s_{mo}(t)$ at each time instant t_j is performed concurrently on each MP mp_r and at the MPC (Fig. 4):

• The following actions are performed at each measuring point MP mp_r r = 1..m:

1. obtaining a telemetry frame $s_r(t_j)$ from each MP mp_r r = 1..m:
   ```
   sr(tj) = mpr.collect();
   ```
2. assessing the stream quality at each MP mp_r r = 1..m:
   ```
   q(sr(tj)) = mpr.quality(sr(tj));
   ```
3. transfer of the quality assessment $q(s_r(t_j))$ and kinematic parameters $s_r^k(t_j)$ from each MP mp_r, r = 1..m into the MPC:
   ```
   forall mpr from MP
   q(sr(tj)) = mpr.transf(mpc, q(sr(tj)))
   ```

$$s_r^k(t_j) = mp_r.transf(mpc, s_r^k(t_j))$$

4. transmission of a telemetry frame $s_b(t_j)$ from the selected MP mp_b to the MPC:

```
sb(tj) = mpb.transf(mpb, sb(tj));
```

- The following actions are performed at MPC:

1. producing the main stream:
```
mpc.aggregate(smo, sb(tj));
```
2. predicting the stream quality for each MP mp_r $r = 1..m$ at time $t_j + 1$:
```
forall mpr from MP
q(sr(tj + 1)) = mpc.predicte(sr^k(tj) ∪ q(sr(tj)))
```
3. selection of the best measurements:
```
b = mpc.best(q(s1(tj + 1)),...,q(sr(tj + 1)),...,q(sm(tj + 1)));
```
4. sending's' command for transferring the telemetry frame to the selected MP mp_b and quality command 'q' to the all MP mp_r $r = 1..m$:
```
mpc.transf(mpr, 's')
forall mpr from MP
mpc.transf(mpr,'q')
```

4.3   The Suggested Approaches Versus the Generic Approach

Let compare the generic approach (Sect. 3) and suggest approaches to telemetry gathering with a distributed quality assessment and a quality assessment prediction. Figure 2 illustrates the distribution of the computing operation of telemetry gathering procedure for these approaches. Time diagrams for the main stream formation at each time instant is shown in Fig. 3.

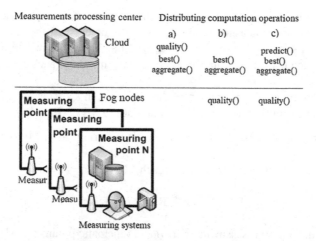

Fig. 2. Distributing computation operations for: (a) the generic approach; (b) approach with a distributed quality assessment; (c) approach with a predicted quality assessment.

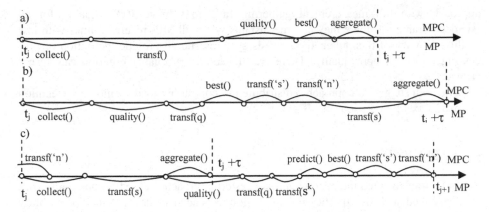

Fig. 3. A time diagram for the main stream formation by: (a) the generic approach; (b) approach with a distributed quality assessment; (c) approach with a predicted quality assessment.

The advantage of the generic approach is the highest possible quality of telemetry because it gathers telemetry from all MPs and selects the best frames from them. The disadvantages of this approach are:

- high network traffic between MP and the MPC due to the need to transfer all measurements from MP, even those that will not eventually be included in the main measurement stream;
- a large number of calculations in the MPC associated with the quality evaluation quality() of the incoming streams from each $mp_r \in MP$, selection of the best stream best() and their aggregation into the main stream aggregate() (Fig. 2).

Thus the following problems of telemetry gathering and processing arise during measurement:

- geographical isolation of MPs from each other and from the MPC;
- duplication of data received from different MPs, as their detection zones intersect;
- use of wireless networks with limited bandwidth (radio relay, cellular, etc.);
- need to control both MSs and MO itself, requiring real-time analysis of the collected data.

The situation becomes even more difficult if there are several MOs and measurement streams multiply depending on the number of MOs.

The advantage of the approach with a distributed quality assessment is low network traffic because only the best telemetry frame is transferred. The disadvantage of this approach is a possible increase in the time of formation of the main stream after the moment of measurement, because additional functions are executed (Fig. 3).

The approach with a predicted quality assessment has minimal network traffic and the time of formation of the main stream τ (Fig. 3). The disadvantage of this approach is a potential decrease in quality of main stream. The extent of the decrease depends on the quality of the developed predictive model. However, it should be noted that even an approach with full measurement gathering fails to achieve 100% quality for all telemetry frames, as there may be time instants when all MPs receive frames with low quality. In view of this, telemetry analysis systems should provide the option of processing frames of poor quality. Therefore, the telemetry quality criterion can be not critical.

The both suggested approach redistribute the computing load by moving execution of `quality()` operation on MPs (Fig. 2).

5 Experiments

We have implemented the generic and suggested approaches based on our framework for distributed analysis [9]. The implementation uses actor model [10] and Akka library [11] to implement units that are deployed on the MPs and the MPC. The actors execute functions which are described above.

Experiments were performed with the distributed system where each of the actors was located on an individual machine. Four physical computers with two virtual machines deployed on each of them were used. Ubuntu Server 16.04 was installed on virtual machines.

For experiments we used data sets of really telemetry from flight object. Each set contains 2 700 telemetry frames. Following measurement parameters of the MO were selected: kinematic parameters: coordinates *(x, y, z)* and the velocity vector *(vx, vy, vz)*; measurement time from object; measurement time from MPs.

The first parameter used to assess different implementations was the frame delay, i.e. the time interval from the moment of the telemetry frame formation to the moment of its aggregation into the main data stream. This characteristic demonstrates minimal possible speed of online decision-making by the calculation centre operators or third-party decision-making systems using main stream as input data.

From retrieved data, it follows that the average delay value shall be:

- for generic approach – about 0.2629 s
- for approach with distributed quality assessment – 0.2921 s
- for approach with quality assessment prediction – 0.2645 s

Delay in the prediction implementation is close to the result received from the generic approach. The value is increased insignificantly due to necessity to switch between data sources when swapping the best MP.

Another criterion for implementation comparison is network traffic volume analysis. Measurements were made with Wireshark utility [12], a traffic analysing software for Ethernet network.

Traffic measurement was made with different system configurations, in particular with varying PC quantity used in test emulations. In this kind of test, quality of the formed main stream was not taken into account. Thus, for each implementation variant, 4 system configurations were prepared: with 1, 2, 4 and 6 PCs. Test were performed three times for each configuration, and results were averaged and generated traffic volume was assessed based on those values.

Diagrams (Fig. 4, 5) represent results of the implementation testing. The size of one source data file is ~339 Mb. It can be seen that the dependence of the traffic volume from quantity of MPs is linear, wherein actual data confirm theoretical assumptions about linear dependence of the traffic volume from source data volume and, at the same time, number of MPs in the system.

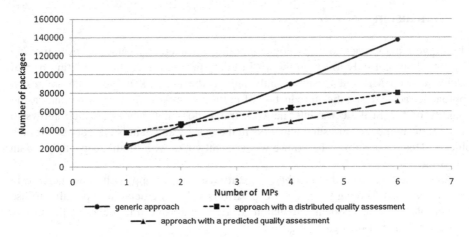

Fig. 4. Summary diagram of packet quantity dependence from quantity of PCs

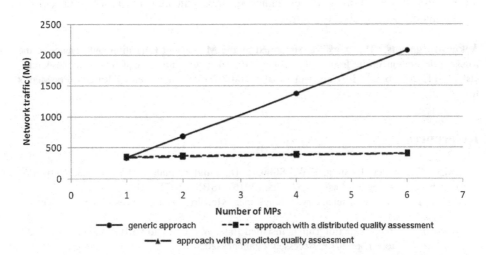

Fig. 5. Summary diagram of traffic volume dependence from quantity of PCs

Analysis of the above data allows to conclude that the implementations of proposed approaches use much greater number of TCP-packets for data transfer (37241 compared to 21752 with a single MP). Yet, specific size of these packets is significantly smaller, as the general traffic volume for one used MP is 355.31 Mb compared to

342.715 Mb in the implementation of generic approach. This can be explained by smaller size of the packets formed for transferring. At the same time, traffic volume for implementation of proposed approaches is not directly dependant on quantity of MPs in the system. This significantly reduces total volume with their greater amount (411.98 Mb compared to 2077.23 Mb with 6 active MPs).

6 Conclusion

The suggested approach to using Fog computing for telemetry gathering from moving objects allows not only to lessen the load to data processing centre by putting part of calculations on MP, but to significantly reduce the network traffic. Both suggested approaches achieve significant reduction of in the network load. Also, when increasing quantity of MP in these variants, network traffic does not increase, which allows to increase the number of measurement tools in the system and gain better control over the objects. This makes it possible to use wireless networks with limited capacity for data gathering.

The disadvantage of the distributed quality assessment approach is an increase in time delay when forming the main stream. If this value is crucial, it is possible to use the approach with assessment prediction. It gives the same time delay as the implementation with full data gathering.

Quality of the formed main stream with this approach is fully dependent on prediction model implementation. Prediction can be improved also by model learning in the process of measurement values gathering, thus adjusting the model with current conditions of moving objects control.

Acknowledgments. This work was supported by the Ministry of Education and Science of the Russian Federation in the framework of the state order "Organisation of Scientific Research", task#2.6113.2017/6.7, and by grant of RFBR# 16-07-00625, supported by Russian President's fellowship.

References

1. Miao, G., Zander, J., Sung, K.W., Slimane, B.: Fundamentals of Mobile Data Networks. Cambridge University Press, Cambridge (2016). ISBN 1107143217
2. M2M: The new age of telemetry. White Paper. Metrilog Data Services GmbH, Schwechat, Austria (2005)
3. Gartner Says the Internet of Things Installed Base Will Grow to 26 Billion Units By 2020. Gartner. 12 December 2013. Accessed 2 January 2014
4. Tsai, C.-W., Lai, C.-F., Vasilakos, A.V.: Future internet of things: open issues and challenges. Wirel. Netw. 20(8), 2201–2217 (2014)
5. Gubbi, J., Buyya, R., Marusic, S., Palaniswamia, M.: Internet of things (IoT): a vision, architectural elements, and future directions. Future Gener. Comput. Syst. 29(7), 1645–1660 (2013)

6. El-Sharkawi, A., Shouman, A., Lasheen, S.: Service oriented architecture for remote sensing satellite telemetry data implemented on cloud computing. Int. J. Inf. Technol. Comput. Sci. **5**(7), 12–26 (2013)
7. Mirchandani, C.: Cloud-based ground system for telemetry processing. Procedia Comput. Sci. **61**, 183–190 (2015)
8. Bonomi, F., Milito, R., Zhu, J., Addepalli, S.: Fog computing and its role in the internet of things. In: Processing of MCC, Helsinki, Finland, 17 August 2012, pp. 13–16 (2012)
9. Kholod, I., Petuhov, I., Kapustin, N.: Creation of data mining cloud service on the actor model. In: Balandin, S., Andreev, S., Koucheryavy, Y. (eds.) ruSMART 2015. LNCS, vol. 9247, pp. 585–598. Springer, Cham (2015). doi:10.1007/978-3-319-23126-6_52
10. Hewitt, C., Bishop, P., Steiger, R.: A universal modular ACTOR formalism for artificial intelligence. In: IJCAI, pp. 235–245 (1973)
11. Akka Documentation. http://akka.io/docs/
12. Wireshark Developer's Guide. https://www.wireshark.org/docs/wsdg_html_chunked/

Stability and Delay of Algorithms of Random Access with Successive Interference Cancellation

Nikolay Apanasenko, Nikolay Matveev$^{(\boxtimes)}$, and Andrey Turlikov

State University of Aerospace Instrumentation, Saint-Petersburg, Russia
{n.apanasenko,n.matveev,turlikov}@vu.spb.ru

Abstract. The article is devoted to the work of class of Random Access algorithms with Successive Interference Cancellation. This class of algorithms includes Irregular Repetition Slotted ALOHA, Contention Resolution Diversity Slotted ALOHA and other algorithms in which the Random Access is combined with successive interference cancellation. We propose method of stabilization and lower bound of average delay for this class of algorithms.

Keywords: Random access · ALOHA · SIC · IRSA · CRA · Stability · mMTC

1 Introduction

Currently new programs for the development of new mobile communication standard 5G are maintained. An idea of the 5G standard is tightly connected with a paradigm of massive Machine Type Communication (mMTC). The paradigm is characterized by fully automatic data generation, exchange, processing and actuation among intelligent machines, without or with low interactions with of humans. It is expected that in mMTC scenarios number of supported devices will be $10^6/\text{km}^2$ [1]. Today, a variety of access technologies are used in mMTC scenarios: long range-coverage, such as Heterogeneous Networking (LTE), and short-range, such as WiFi, BLE, ZigBee, etc. [2–4].

The uncoordinated access channel is efficient for mMTC scenario, that is shown in METIS I [5]. Uncoordinated Medium Access Control (MAC) protocol, such as Slotted ALOHA (SA) jointly with Successive Interference Cancellation (SIC) technique of Physical layer was proposed in [5,6]. Presented analysis results [6,7] show the possibility to achieve the throughput $T \simeq 0.8$ [users/slot] instead of $T \simeq 0.37$ [users/slot]. However, [8,9] have shown that the direct use of this approach is unstable in random flow. It leads to unlimited increasing of active devices number. It is well known for Random Multiple Access algorithm such as ALOHA. We propose the method of stabilization which is based on approach proposed in [6,7] and show that our algorithm is stable.

O. Galinina et al. (Eds.): NEW2AN/ruSMART/NsCC 2017, LNCS 10531, pp. 510–518, 2017.
DOI: 10.1007/978-3-319-67380-6_47

2   System Model

2.1   Assumptions

The system model is based on the following assumptions:

1. Wireless system contains one Base Station (BS) and a large number of user equipments (UEs). UEs use random access channel to transmit data to BS. BS uses a common broadcast channel to transmit a feedback data to UEs.
2. Time is divided into equal intervals – frames. Each frame contains n slots. Number of slots in each frame is constant. Slot length is equal to a time of message transmission. The boundaries of frames and slots are precisely known by all UEs. An UE can start transmitting a message only at the beginning of the slot.
3. New users arrive in slot according to a homogeneous Poisson process with the arrival rate λ. Number of active users at the current frame is a random value and it is equal to a sum of newly arrived users at the previous frame and number of users awaiting retransmission.
4. Each UE receives feedback data which contains information about number of slots for retransmission to BS.
5. In each slot one of three events can occur:
   (a) Successful/received – one slot is used by one user.
   (b) Empty – slot is not used.
   (c) Interfered/collision – one slot is used more than by two users.
6. Each message from UE contains index of slots for retransmission. The channel impulse response between each UE and BS is known. SIC procedure is performed in BS.

2.2   SIC Algorithm

Figure 1 represents the concept of SIC algorithm. Two messages of UE 3 are received successfully since there are not interfered in first frame. But two messages of UE 1 and UE 2 are interfered, hence these users will transmit message by next frame. Second replica of UE 2 is received successfully, first replica of the UE 2 and second replica of the UE 1 are interfered in the second frame so as if each message from UE contains index of slots for retransmission the second replica of UE 1 can be decoded. This process is then iteratively repeated for decoding the message of UE 5.

2.3   Algorithms of Base Station and User Equipment

When current frame is received, BS calculates number of empty slots – N_{empty}, number of slots with only one message – N_{only}, number of collisions before SIC procedure – N_a and N_b – number of collisions after SIC procedure. BS assigns transmission probability of j copies in one frame based on the calculated values:

$$p_j = AL(N_{empty}, N_{only}, N_a, N_b), \tag{1}$$

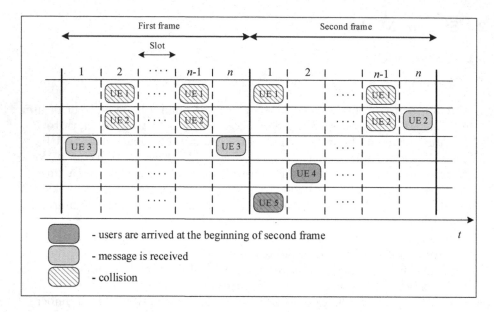

Fig. 1. Concept of SIC algorithm.

wherein $\sum_j p_j = 1$, where j denotes the number of slot for retransmission. For example, if $\mathbf{P} = \{p_0 = 0, p_1 = 0, p_2 = 0.5, p_3 = 0.28, p_4 = 0, p_5 = 0, p_6 = 0, p_7 = 0, p_8 = 0.22\}$ so, each UE transmits two copies of message with probability $p_2 = 0.5$ or three copies with probability $p_3 = 0.28$ etc. [6]. Then, the list of users from which the message is received and probability of transmission of j copies are transmitted to all users by feedback channel. UEs, which are not in the list, will send message in the next available slot according to the probabilities contained in the feedback data.

3   Stability and Optimization of CSA

The model represented above is a generalization of the model proposed in [6,7] with the following two differences. The first difference is that the number of active users at the current frame is a random value which depends on characteristics of average arrival rate and operation algorithm of UE and BS (see assumption Sect. 2.1). The second is that behavior of users depends on a feedback data. We show unstable operation of approaches proposed in [6,7] and we represent a method of ensuring stability.

Not centralized algorithms of Coded Random Access (CRA) are analyzed in [6,7]. We consider two of them. The first algorithm is Stability of Contention Resolution Diversity Slotted ALOHA (CRDSA), which is easiest of CRA algorithms. CRDSA can increase throughput ($T \simeq 0.55$) compared with SA. All UEs transmit two copies of message in this algorithm. The second algorithm is

Irregular Repetition Slotted ALOHA (IRSA) which provides a throughput equal to $T \simeq 0.80$. Number of copies transmitted by UEs is chosen randomly from a fixed set [6,7].

Let us demonstrate instability of IRSA at short-term increase in intensity of an entrance stream. Figure 2 shows number of users in system for ISRA and stable of SA for 40 frames. The effect of instability is observed after appearance of 100 users at the beginning of the twentieth frame for IRSA algorithm. One can see a number of active users at the beginning of each frame is increased.

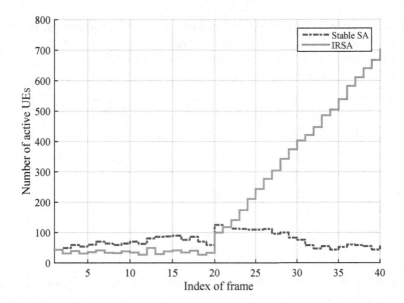

Fig. 2. Example of unstable operation of IRSA and stable operation of ALOHA with average arrival rate $\lambda = 0.34$ and $n = 100$

In case when the number of active users at the beginning of frame $t - L_t$ is more than number of slots in each frame $- n$, stability can be reached if transmission probability for SA without SIC is determined by following ratio:

$$p_t = \frac{n}{L_t}. \tag{2}$$

It is shown in [10]. However, the strategy is stable when average arrival rate is less than value e^{-1}. It is known that SIC algorithm increases a value of average arrival rate, but it is needed to change the strategy proposed in [10]. It means that average number of active users (arrive rate) have to correspond to number of users such that value of throughput must be maximum.

We introduce an auxiliary model for defining the conditions of stability. Let us consider the case, where there are L users at the beginning frame. L is a Poisson random variable and $E[L] = g$. All UEs use single strategy (for instance

ALOHA, IRSA and etc.). Let us denote this strategy by A. SIC is operate in BS. Let M is a random value, which equals a number of successful transmitted messages in the current frame. Let us introduce the following function:

$$f_A(g) = \frac{E[M]}{n}. \tag{3}$$

Then, for algorithm of ALOHA the function can be rewritten as follows:

$$f_{ALOHA}(g) = ge^{-g}. \tag{4}$$

For strategy with using SIC procedure, value of function $f_A(g)$ can be obtained by method of Monte Carlo with any precision for any finite value of g. Figure 3 represents view of functions for different strategies with $n = 100$.

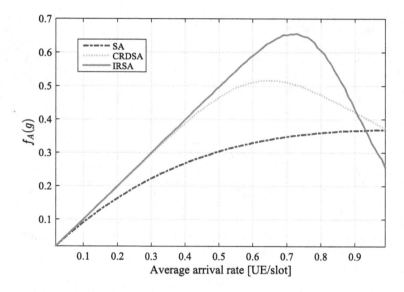

Fig. 3. Functions $f_A(g)$ for different strategies with $n = 100$

Let us form a theorem for introduced model.

Theorem 1. *Let g_0 be fixed a value, then stability of the system can be reached by following algorithm:*

(1) If there are L_t active users at the beginning of frame t and $L_t < ng_0$, thus every user transmits a message according to strategy A with probability 1.

(2) If $L_t \geq ng_0$, thus every user transmits a message with probability $\frac{g_0 n}{L_t}$ according to strategy A and do not transmits with probability $1 - \frac{g_0 n}{L_t}$.

In this case, average delay is finite for $\lambda < f_A(g)$.

The proof is based on the fact, that the sequence of random variables forms aperiodic homogeneous Markov chain. Chain is ergodic for $\lambda < f_A(g)$, as follows from criterion of Foster [11]. It means, that if A is fixed, we need to choose g_0, such that $f_A(g)$ is maximized.

One can see that $f_A(g)$ is maximal when average arrival rate is $g_0 = 0.7$ UEs/slot for IRSA algorithm (Fig. 3). Then BS assigns transmission probability for stabilization of algorithm by following expression:

$$p_t = \begin{cases} 1, & \text{for } \frac{L_t}{n} < g_0; \\ g_0 \frac{n}{L_t}, & \text{otherwise.} \end{cases} \tag{5}$$

Therefore, BS will assign transmission probability of j copies in one frame according to the expression (5) given above. Then, the example represented above can be written: $\mathbf{P} = \{p_0 = 1 - p_t,\ p_1 = 0,\ p_2 = 0.5p_t,\ p_3 = 0.28p_t,\ p_4 = 0,\ p_5 = 0,\ p_6 = 0,\ p_7 = 0,\ p_8 = 0.22p_t\}$.

Figure 4 represents an operation of IRSA with the proposed modification. One can see, that when 100 users appear at the beginning of the twentieth frame, RSA algorithm stays stable. Increasing of number of active UEs is not observed (in contrast with Fig. 2).

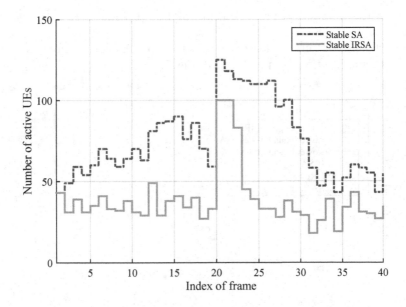

Fig. 4. Example stable operation of IRSA with proposed modification and ALOHA with average arrival rate $\lambda = 0.34$ and $n = 100$

4   Analysis of Delay for CRA with Stabilization

The lower bound of average delay can be obtained for a class of CRA algorithms defined in section II. We assume that number of active users at the beginning of each frame is known and proposed modification of stabilization IRSA is used. In practice, number of active UEs is not known, so estimation of number of active UEs at the beginning of each slot is needed (assumption Sect. 2.1).

Let us formulate the following theorem for this model.

Theorem 2. *If strategy A and g_0 are fixed and BS and UEs use algorithm of access with stabilization, then average number of users in frame for all values $0 < \lambda = f_a(g_0)$ can be obtained solution of the following inequality:*

$$\lambda < f_A(g). \tag{6}$$

Second theorem is a generalization of statements and evidences in [12], which considers algorithm of SA and $g_0 = 1$. Proof of second theorem is similar to proof in [12].

According to the Littles law average delay can be estimated as follows:

$$d = \frac{\bar{g}}{\lambda} + 0.5, \tag{7}$$

where $\bar{g}$ is the minimal positive solution of Eq. (6), 0.5-is average delay of beginning of new frame.

Lower bound and curves for delay obtained by simulations for three algorithms with stabilization are presented in Fig. 5.

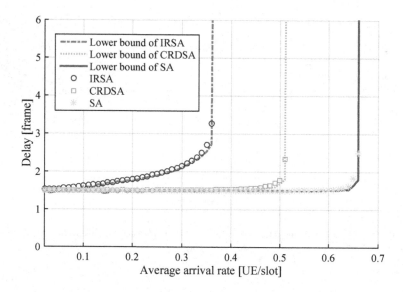

Fig. 5. Delay vs. average arrival rate $n = 100$

One can see in Fig. 5, IRSA with proposed modification has considerably smaller delay than SA.

5   Conclusion

In this paper, unstable algorithms of CRA are shown. Approach which provides stability of algorithms CRA by added feedback between BS and USs is proposed. General approach for the construction of the lower bound of average delay of algorithms with stabilization is shown. Theoretical and practical curves of average delay are given.

Future work will aim to generalize the result of this work on case when SIC procedure does not always out puts correct results.

Acknowledgment. The author Turlikov is supported by research project RFBR No. 17-07-00142.

References

1. ITU: Recommendation ITU-R M.2083-0 (09/2015), IMT Vision Framework and overall objectives of the future development of IMT for 2020 and beyond
2. Ometov, A.: Short-range communications within emerging wireless networks and architectures: a survey. In: 2013 14th Conference of Open Innovations Association (FRUCT) (2013)
3. Orsino, A., Ometov, A., Fodor, G., Moltchanov, D., Militano, L., Andreev, S., Yilmaz, O.N., Tirronen, T., Torsner, J., Araniti, G., et al.: Effects of heterogeneous mobility on D2D-and drone-assisted mission-critical MTC in 5G. IEEE Commun. Mag. **55**, 79–87 (2017)
4. Stusek, M., Masek, P., Kovac, D., Ometov, A., Hosek, J., Kröpfl, F., Andreev, S.: Remote management of intelligent devices: using TR-069 protocol in IOT. In: 2016 39th International Conference on Telecommunications and Signal Processing (TSP) (2016)
5. METIS: ICT-317669 METIS project. Proposed solutions for new radio access. Deliverable D2.4, February 2015
6. Liva, G.: Graph-based analysis and optimization of contention resolution diversity slotted ALOHA. IEEE Trans. Commun. **59**, 477–487 (2011). ISSN 0090-6778
7. Paolini, E., Stefanovic, C., Liva, G., Popovski, P.: Coded random access: applying codes on graphs to design random access protocols. IEEE Commun. Mag. **53**, 144–150 (2015). ISSN 0163-6804
8. Meloni, A., Murroni, M.: CRDSA, CRDSA++ and IRSA: stability and performance evaluation. In: 2012 6th Advanced Satellite Multimedia Systems Conference (ASMS) and 12th Signal Processing for Space Communications Workshop (SPSC) (2012)
9. Kissling, C.: On the stability of contention resolution diversity slotted ALOHA (CRDSA). In: 2011 IEEE Global Telecommunications Conference (GLOBECOM 2011) (2011)
10. Galinina, O., Turlikov, A., Andreev, S., Koucheryavy, Y.: Stabilizing multi-channel slotted aloha for machine-type communications. In: 2013 IEEE International Symposium on Information Theory (2013)

11. Foster, F.G.: On the stochastic matrices associated with certain queuing processes. Ann. Math. Stat. **24**, 355–360 (1953)
12. Hajek, B., Likhanov, N., Tsybakov, B.: On the delay in a multiple-access system with large propagation delay. IEEE Trans. Inf. Theory **40**, 1158–1166 (1994). ISSN 0018-9448

Throughput Analysis of Adaptive ALOHA Algorithm Using Hybrid-ARQ with Chase Combining in AWGN Channel

Artem Burkov, Nikolay Kuropatkin, and Nikolay Matveev(✉)

State University of Aerospace Instrumentation, Saint-Petersburg, Russia
{a.burkov,n.matveev}@vu.spb.ru, nickolay.kuropatkin@gmail.com

Abstract. This article is devoted to the class of algorithms of random multiple access, such as adaptive Aloha, in which throughput is limited by e^{-1}. This class can be expanded by using Hybrid-ARQ with Chase combining. For this expansion, an upper bound was obtained, which turned out to be 0.523.

Keywords: Random multiple access · Aloha · Chase combining · Hybrid-ARQ · Throughput

1 Introduction

Currently, the development of new mobile communication standard 5G is maintained [1,2]. An idea of the 5G standard is tightly connected with using of massive Machine Type Communication (mMTC) [3,4]. This type of communication is characterized by fully automatic generation, exchange and processing of data with low interactions with humans [5]. It is expected that within the mMTC the number of supported devices will be large [6]. The basic algorithm of the adaptive Aloha has a finite delay at the arrival rate below e^{-1} [7]. In mMTC scenarios, arrival rate can top this value [8–10]. In this regard, the most acute issue is the analysis of the communication channel throughput, as well as the development of methods to improve the efficiency and reliability of information transfer in mobile networks. There are a number of manuscripts [11], [12] in which the modernization algorithm of ALOHA for increasing throughput is examined, including articles devoted to the analysis of hybrid automatic repeat request [13], [14] with different models of channel.

In one of such works [15], the system with hybrid automatic repeat request does not correspond to the scenarios of new standards, due to the fact that the number of users in the model proposed in this article is determined as constant, and the base station already knows which user in which window attempts to transmit a message.

In the current work, we examine the modernization of the algorithm of the adaptive ALOHA based on the use of HARQ methods with Chase combining. We offer a number of assumptions for this system. For the considered algorithm, the throughput is determined in terms of this assumptions.

© Springer International Publishing AG 2017
O. Galinina et al. (Eds.): NEW2AN/ruSMART/NsCC 2017, LNCS 10531, pp. 519–525, 2017.
DOI: 10.1007/978-3-319-67380-6_48

2   System Model

2.1   Assumptions

First, let us describe the basic model of the system with random multiple access. In this paper, we consider a channel with Gaussian noise without Rayleigh fading. The channel output is available to all users.

In this article, we will consider a model with the Galagher's system of assumptions [16]:

1. All messages of all users have the same length and the transmission time of one message is taken as a unit of time (slot). All system operating time is divided into windows and users can start the transfer only at the beginning of the window (no user is attached to window).
2. In each slot there can be one of 3 events:
   (a) in the slot transmits only 1 user, it is informed that the message was delivered to the receiver (event SUCCESSFUL);
   (b) Two or more users transmit in the slot, then the receiver gets no information about the contents of the transmitted packets (event CONFLICT);
   (c) no one transmits in the slot (EMPTY).
3. At the end of the slot, all subscribers can reliably know which event occurred in the channel.

In this work, Assumptions 2 and 3 are modified as follows:

Assumption 2. If two or more users transmit in the window (slot), then the conflict that has arisen can be successfully resolved. Users who transmitted simultaneously with the subscriber affect the probability of successful decoding of his message, in which case the SUCCESSFUL event occurs in the channel. The method of handling the message on the receiving side is described in more detail below.

Assumption 3. We add checksum to all transmitted messages on the transmitting side, and then the obtained word go through error-correcting coding and transmitted to the channel where noise is added to the signal. At the receiving end, noise error-correcting decoding is first performed, and then we validate the checksum. With the correct checksum, an ACK is sent, in case of an unsuccessful one – NACK. When user receive an ACK, the transmitter assumes that SUCCESSFUL event occurs in the channel, when user receive a NACK – Conflict. It is assumed that the ACK is reliably received by the transmitter in the same slot in which the message was transmitted.

We also introduce additional assumptions.

Assumption 4. Each user always has a message ready to be sent, which user transmit with a probability of $1/M$, M is the number of active transmitters, which is known at the base station. The same message will be repeatedly transmitted for as long as SUCCESSFUL event does not occur in the channel.

Assumption 5. Users appear in the channel according to Poisson distribution.

Assumption 6. We use the upper bound to determine the probability of successful decoding for the Chase combining. The procedure used for this described in the next subsection.

2.2   The Upper Bound for the Probability of Successful Decoding of a Message Using Chase Combining

Using the approach from [15], the probability of successful decoding can be estimated by upper bound using the following step-by-step procedure:

Step 1. Calculating Signal Interference + Noise Ratio (SINR) for user k in slot s assuming that decoder for user k has a perfect knowledge of the path loss:

$$\text{SINR}_{k,s} = \frac{\alpha_{k,s} E}{\sum_{j \in K(s): j \neq k} \alpha_{j,s} E + N_0}, \tag{1}$$

where N_0 denotes energy of the noise, $\alpha_{k,s}$ denotes the path loss of user k over slot s, E – user signal energy (all user signals have the same energy), $K(s)$ denotes the set of active users over slot s.

We assume that $\alpha_{k,s}$ is equal 1 for the channel without Rayliegh fading. With this in mind, the formula for SINR takes the following form:

$$\text{SINR}_{k,s} = \frac{E}{NE + N_0}. \tag{2}$$

Step 2. Using SINR obtained at the first step mutual information [17] takes the following form:

$$I_{k,s} = \sum_{s \in S_{k,m}} \log_2 \left(1 + \text{SINR}_{k,s} \right), \tag{3}$$

where $\text{SINR}_{k,s}$ - SINR of user k over slot s, $S_{k,s}$ denotes the sequence of slots, and user k transmits the same message and m is a number of attempts.

Step 3. We determine if an error has occurred, using comparison between mutual information and channel spectral efficiency. The probability of decoding error is arbitrarily small if $R < I$ and large if $R > I$. In this paper, we consider following particular cases of Hybrid-ARQ mechanism:

1. Classical Aloha:
$$\log_2 \left(1 + \text{SINR}_{1,1} \right) \leq R, \tag{4}$$

   triggers an error event.
   In this case we assume that decoder considers only the last received signal.
2. Chase combining:

$$\log_2 \left(1 + \sum_{s \in S_{k,m}} \text{SINR}_{1,s} \right) \leq R, \tag{5}$$

   triggers an error event.

In the case of successful decoding the ACK signal is sent over the feedback channel. In the case of a decoders error, the NACK signal is sent to the transmitter, and transmitter repeats the transmission. In the case of Chase combining receiver may combine the first received packet with the following ones [18].

3   Throughput Analysis

Based on the article [19], we define the throughput of an algorithm random multiple access as

$$T = \sup\{\lambda : D < \infty\}, \tag{6}$$

i.e., as the maximum (more precisely, the supremum) intensity of the input traffic that can be transmitted by the algorithm of random multiple access with finite delay.

The throughput for the classical ALOHA algorithm is equal to $e^{-1} \approx 0.36$ [15]. We show how it is possible to determine the throughput for ALOHA using the Chase combining (ALOHA with CC).

We consider the system in saturation mode. Let the intensity of the input arrival rate $\lambda \leq T_{CC}$. Users transmit the message over the channel with probability $p = \frac{1}{M}$, where M is number of active users. The average SNR value is fixed. For the defined SNR, the optimum spectral efficiency R is selected according to

$$R < \log_2(1 + \text{SNR}). \tag{7}$$

We add a target user to the set of active users. Let us introduce a random value L, which determines the number of retransmissions of one message sent by target user. Let us define estimating of the average number of transmission attempts of the target user before SUCCESSFUL event as $E[L]$, one can find T_{CC} as

$$T_{CC} = \frac{1}{E[L]}, \tag{8}$$

where T_{CC} is throughput for ALOHA with CC.

At simultaneous transmission over the channel, user signals interfere. For the target user it is possible to calculate SINR for the j-th transmission. Using previously obtained formula (1) for the case with target user it can be written

$$\text{SINR}_j = \frac{\alpha_{*,j} E_*}{\sum_{i=1}^{N_j} \alpha_{i,j} E_i + N_0}, \tag{9}$$

where SINR_j is SINR ratio at the j-th transmission of the target user; N_j is the number of users transmitting with the target user at j-th transmission; $\alpha_{*,j}$ is the path loss of the labeled user for j-th transmission; $\alpha_{i,j}$ is the path loss of the i-th user for j-th transmission.

Since we are considering the system in saturation mode and users transmit with probability $p = \frac{1}{M}$, then N_j is distributed according to the Poisson distribution with parameter 1.

Based on the assumptions, it is assumed that the path loss and signals energy in the channel are such that the signals at the base station input for different users have the same energy equal to E, SINR can be defined as

$$\text{SINR}_j = \frac{E}{N_j E + N_0}. \tag{10}$$

Based on the article [15] and formula (5), the probability of sending a message by the target user for L attempts can be written as

$$\Pr\{L = i\} = \Pr\left\{\log_2\left(1 + \sum_{j=1}^{i} \frac{E}{N_j E + N_0}\right) > R\right\}. \tag{11}$$

The value under the sum can be divided by E then we get

$$\Pr\{L = i\} = \Pr\left\{\log_2\left(1 + \sum_{j=1}^{i} \frac{1}{N_j + \frac{N_0}{E}}\right) > R\right\}. \tag{12}$$

In formula (12) $\frac{N_0}{E}$ can be replaced by SNR^{-1} Then $E[L]$, according to the formula (12), can be defined as

$$E[L] = \sum_{i=1}^{\infty} i \Pr\left\{\log_2\left(1 + \sum_{j=1}^{i} \frac{1}{N_j + \text{SNR}^{-1}}\right) > R\right\}. \tag{13}$$

The value of $E[L]$ can be calculated with any given accuracy. As can be seen from formula (12), the throughput value depends on the SNR.

For system parameters defined as $E = 1$, $R = 1$, $\text{SNR}^{-1} = \frac{1}{N_0 - \epsilon}$, where $\epsilon \to 0$, $E[L] = 1.912$ was obtained to within a third sign. And as a result, for the ALOHA algorithm with CC, was calculated $T_{CC} = 0.523$. With increasing SNR, the throughput value decreases and tends to e^{-1}.

Simulation of the random multiple access system for the algorithm ALOHA with CC was carried out. As a result of the simulation, we obtained graphs of the dependence of the average delay on arrival rate. These graphs are shown in Fig. 1. The same figure shows a similar dependence for the basic Aloha algorithm, obtained numerically [20]. It should be noted that for Aloha with CC it is difficult to obtain this dependence, since the work of the system is described by a multidimensional Markov chain.

As can be seen from Fig. 1, the average delay for the basic Aloha model is finite for $\lambda < e^{-1} \approx 0.36$. For the Aloha algorithm with CC, the average delay is finite for $\lambda < 0.52$.

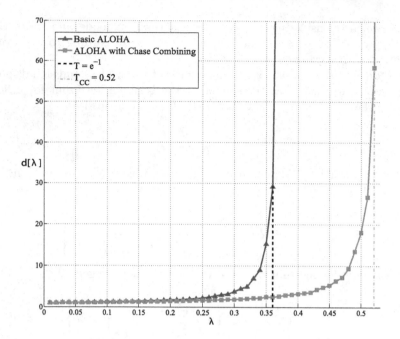

Fig. 1. Delay vs average arrival rate

4   Conclusion

In this paper the class of algorithms of random multiple access type of adaptive Aloha Chase combining was examined. We described the system model and introduced several assumptions. Most valuable ones are: base station knows the amount of active users, base station can differ the users regardless of success of their transmissions, and Chase combining for any amount of users can be performed. This allowed us to obtain an upper bound for throughput (0.523). This obtained estimation can be clarified by softening proposed assumptions (for example using adaptive estimation on the amount active users).

Acknowledgment. The authors Burkov and Matveev are supported by scientific project No. 8.8540.2017 "Development of data transmission algorithms in IoT systems with limitations on the devices complexity".

References

1. Lema, M.A., Laya, A., Mahmoodi, T., Cuevas, M., Sachs, J., Markendahl, J., Dohler, M.: Business case and technology analysis for 5G low latency applications. IEEE Access **5**, 5917–5935 (2017)
2. Andreev, S., Galinina, O., Pyattaev, A., Hosek, J., Masek, P., Yanikomeroglu, H., Koucheryavy, Y.: Exploring synergy between communications, caching, and computing in 5G-grade deployments. IEEE Commun. Mag. **54**(8), 60–69 (2016)

3. Orsino, A., Ometov, A., Fodor, G., Moltchanov, D., Militano, L., Andreev, S., Yilmaz, O.N., Tirronen, T., Torsner, J., Araniti, G., et al.: Effects of heterogeneous mobility on D2D-and drone-assisted mission-critical MTC in 5G. IEEE Commun. Mag. **55**, 79–87 (2017)
4. Ratasuk, R., Prasad, A., Li, Z., Ghosh, A., Uusitalo, M.A.: Recent advancements in M2M communications in 4G networks and evolution towards 5G. In: Proceedings of 18th International Conference on Intelligence in Next Generation Networks (ICIN), pp. 52–57. IEEE (2015)
5. Ometov, A.: Short-range communications within emerging wireless networks and architectures: A survey. In: Proceedings of 14th Conference of Open Innovations Association (FRUCT) (2013)
6. ITU: Recommendation ITU-R M.2083-0 (09/2015), IMT Vision - Framework and overall objectives of the future development of IMT for 2020 and beyond
7. Abramson, N.: THE ALOHA SYSTEM: another alternative for computer communications. In: Proceedings of the November 17–19, 1970, Fall Joint Computer Conference, pp. 281–285. ACM (1970)
8. Gerasimenko, M., Petrov, V., Galinina, O., Andreev, S., Koucheryavy, Y.: Energy and delay analysis of LTE-advanced RACH performance under MTC overload. In: Globecom Workshops (GC Wkshps), pp. 1632–1637. IEEE (2012)
9. Lien, S.Y., Chen, K.C., Lin, Y.: Toward ubiquitous massive accesses in 3GPP machine-to-machine communications. IEEE Commun. Mag. **49**(4), 66–74 (2011)
10. Ometov, A.: Fairness characterization in contemporary IEEE 802.11 deployments with saturated traffic load. In: Proceedings of 15th Conference of Open Innovations Association FRUCT, pp. 99–104. IEEE (2014)
11. METIS: ICT-317669 METIS project. Proposed solutions for new radio access. Deliverable D2.4, February 2015
12. Liva, G.: Graph-based analysis and optimization of contention resolution diversity slotted ALOHA. IEEE Trans. Commun. **59**, 477–487 (2011). ISSN 0090–6778
13. Larsson, P., Rasmussen, L.K., Skoglund, M.: Throughput analysis of hybrid-arq–a matrix exponential distribution approach. IEEE Trans. Commun. **64**(1), 416–428 (2016)
14. Ai, Y., Cheffena, M.: Performance analysis of hybrid-ARQ with chase combining over cooperative relay network with asymmetric fading channels. In: Proceedings of 84th Vehicular Technology Conference (VTC-Fall), pp. 1–6. IEEE (2016)
15. Caire, G., Tuninetti, D.: The throughput of hybrid-ARQ protocols for the Gaussian collision channel. IEEE Trans. Commun. **47**, 1971–1988 (2001). ISSN 1557–9654
16. Gallager, R.G.: A perspective on multiaccess channels. IEEE Trans. Commun. **31**, 124–142 (1985). ISSN: 1557–9654
17. Cover, T.M., Thomas, J.A.: Elements of Information Theory. Wiley, Hoboken (1991)
18. Krouk, E., Semenov, S.: Modulation and Coding Techniques in Wireless Communications. Wiley, Hoboken (2011)
19. Tsybakov, B.S.: Survey of USSR contributions to random multiple-access communications. IEEE Trans. Commun. **31**, 143–165 (1985). ISSN 1557–9654
20. Falin, G.I.: Performance evaluation for a class of algorithms of random multiple access to a radio-communication channel. Probl. Peredachi Inf. **18**, 85–90 (1982)

Characterizing Time-Dependent Variance and Coefficient of Variation of SIR in D2D Connectivity

Anastasia Ivchenko[1], Yuri Orlov[1], Andrey Samouylov[2,3(✉)],
Dmitri Molchanov[2,3], and Yuliya Gaidamaka[3,4]

[1] Department of Kinetic Equations, Keldysh Institute of Applied Mathematics,
4 Miusskaya Sq., 125047 Moscow, Russia
13.nightcrawler@li.ru, yuno@kiam.ru
[2] Department of Electronics and Communications Engineering,
Tampere University of Technology, 10 Korkeakoulunkatu,
33720 Tampere, Finland
[3] Department of Applied Probability and Informatics,
Peoples' Friendship University of Russia (RUDN University),
6 Miklukho-Maklaya St., 117198 Moscow, Russia
{samuylov_ak,molchanov_da,
gaydamaka_yuv}@rudn.university
[4] Institute of Informatics Problems, Federal Research Center
"Computer Science, and Control" of the RAS, 44-2 Vavilov St.,
119333 Moscow, Russia

Abstract. Attempting to build a uniform theory of mobility-dependent characterization of wireless communications systems, in this paper, we address time-dependent analysis of the signal-to-interference ratio (SIR) in device-to-device (D2D) communications scenario. We first introduce a general kinetic-based mobility model capable of representing the movement process of users with a wide range of mobility characteristics including conventional, fractal and even non-stationary ones. We then derive the time-dependent evolution of mean, variance and coefficient of variation of SIR metric. We demonstrate that under non-stationary mobility behavior of communicating entities the SIR may surprisingly exhibit stationary behavior.

Keywords: SIR · Mobility model · Time dependence · Fokker-Plank equation

1 Introduction

The conventional way to study the performance of wireless network is to the well-established set of methods of stochastic geometry [1, 2]. According to this approach, communicating entities are first modeled using spatial stochastic processes in two- or three-dimensional space. Then, recalling that metrics of interest pertaining to wireless networks, e.g., signal-to-interference ratio (SIR), are conventionally represented as a function of the distance between communicating and interfering entities, the sought metrics are obtained using direct geometrical probability arguments [3].

© Springer International Publishing AG 2017
O. Galinina et al. (Eds.): NEW2AN/ruSMART/NsCC 2017, LNCS 10531, pp. 526–535, 2017.
DOI: 10.1007/978-3-319-67380-6_49

Although the abovementioned approach does introduce realistic randomness into potential positions of the communicating entities, it inherently neglects another inherent property of wireless networks – mobility of terminals. The latter is known to severely affect the metrics of interest as the distance between communicating entities is gradually changing in time that may lead to periods of stable communications and intervals of outage, where no communications are feasible [4].

Over the last years, several attempts have been made to specify and analyze models, where users are allowed to freely move over the landscape. Unfortunately, no general analytical framework have been developed and most of those studies were performed in ad hoc approach using the system level simulation studies that often hide the principal trade-offs between involved parameters. One of the underlying reasons for the lack of general theory is that the family of mobility models used to represent movement of users is extremely wide having different natures and parameters, random waypoint mobility model (RWP, [5]), random direction model (RDM, [6]), Brownian motion (BM, [7]), Levy flights [8] to name a few. Such a zoo of models has led to the ad hoc nature of research in mobility-dependent characterization of wireless systems preventing from concentrating efforts on a single class of models.

One of the inherent features of the conventional mobility models is that they explicitly represent the location of the moving user at any instant of time t. In an attempt to develop a unified theory of mobility-dependent wireless networks in our recent studies [9–14] we have proposed to use the kinetic equations to model the movement process of users. The distinguishing feature of the proposed framework is that it represents the change of the coordinates' density of moving users instead of their locations at each t. Using this approach a wide variety of movement process can be modeled starting from the conventional BM process to more complex ones exhibiting fractal, long-range dependent and even non-stationary behavior. The reason behind this flexibility is that the model is parameterized by the probability density function (pdf) of the user coordinates at time t and the factors affecting its change in time.

In [9–12], a model for predicting a sample distribution function of an unsteady time series was developed. They described a numerical algorithm and software complex that generates an ensemble of time series trajectories, the distribution of which evolves in accordance with a certain kinetic equation. In these studies it was believed that the values of the random process are fixed at definite time intervals, i.e. the unit of time was the interval between consecutive events. The evolution equations for SIR and its ensemble-averaged values were derived in [13, 14].

In this paper, we continue the efforts on building the uniform theory of mobility dependent characterization of wireless system. Particularly, using the kinetic equation mobility model we analytically describe the time-dependent evolutions of the mean signal-to-interference ratio (SIR), the SIR variance and the coefficient of variation (CoV) of SIR for device-to-device (D2D) communications scenario. We address the general case when the random walk of the communicating entities is nonstationary. One of the important results is that under non-stationary movement SIR is statistically invariance in time having stationary distribution.

The rest of the paper is organized as follows. In Sect. 2, we briefly remind the basics of kinetic equation based mobility model. Then, in Sect. 3, we derive the time-dependent evolution of the mean, variance and CoV of SIR. Numerical results are given in Sect. 3. The paper is concluded in the last section.

2   Kinetic Approach for Modeling the Evolution of SIR

In this section, we first introduce the system model of the considered D2D scenario. Then, we briefly sketch the basic properties of the kinetic models. Finally, we provide a brief account of related work.

2.1   System Model

Consider the following D2D communications scenario. We assume that N D2D communicating pairs move in some closed subspace V of two-dimensional space. The transmitters and receivers of all the D2D partners are assumed to be identical and equipped with omnidirectional antennas. The emitted power of transmitters is assumed to be constant. All the D2D pairs are using the same frequency for communications. The only assumption imposed on the communications technology is that it is sensitive to interference created by simultaneous transmissions, e.g., LTE Rel. 13 in-band D2D. The specified scenario is typical for a number of realistic D2D use cases, e.g., crowd in the city center square, customers at the floor of the shopping mall.

The static time-averaged behavior of SIR metrics in such scenario has been considered in a number of studies [15–19]. In this paper, we target time-dependent evolution of SIR. Let $f(\mathbf{x}, t)$ be the probability density function (pdf) of the user location at time t. The movement of users is described by the kinetic model

$$\frac{\partial f(\mathbf{x}, t)}{\partial t} + div_{\mathbf{x}}(\mathbf{u}(\mathbf{x}, t)f(\mathbf{x}, t)) - \frac{\lambda(t)}{2}\Delta f(\mathbf{x}, t) = 0, \qquad (1)$$

where $\mathbf{u}(\mathbf{x}, t)$ is the drift parameter, $\lambda(t)$ is the diffusion coefficient. For simplicity, we assume that the null boundary conditions for the distribution function are given on the boundary of the domain V.

The principal difference between (1) and the conventional models proposed in the literature for modeling the moving patterns of users is that in (1) the input parameter is pdf of users' locations not the directions and velocity of individual users. At one hand, it creates additional problems as we first need to derive $f(\mathbf{x}, t)$ from the real data. However, at the same time, the model in terms of kinetic equation adds great flexibility in terms of the unified analytical description of a wide variety of mobility patterns imposing no restrictions on the form of $f(\mathbf{x}, t)$. Finally, the use of kinetic theory opens up a large set of tool available for analytical and numerical analysis.

As a long distance goal we are interesting in characterizing the communications link stability in terms of durations of the stable communications and outage periods durations and outage frequency. In this paper, the metrics of interest are the time-evolution of the variance and CoV for a non-stationary form of $f(\mathbf{x}, t)$.

2.2 Basic Properties of the Kinetic Models

Assume for a moment that $f(\mathbf{x}, t)$ is the probability density function (pdf) of some physical phenomena, not necessarily pdf of coordinates of communicating entities. Let $\mathbf{m}(t)$ be an average value of a random variable $\mathbf{x}$. Assuming that $\partial f(\mathbf{x}, t)/\partial t$ is continuous, using (1), the time derivative of $\mathbf{m}(t)$ is

$$\frac{d\mathbf{m}(t)}{dt} = \int_V \mathbf{x} \frac{\partial f(\mathbf{x}, t)}{\partial t} d\mathbf{x} = -\int_V \mathbf{x} \, div_{\mathbf{x}}(\mathbf{u}(\mathbf{x}, t) f(\mathbf{x}, t)) d\mathbf{x} + \frac{\lambda(t)}{2} \int_V \mathbf{x} \Delta f(\mathbf{x}, t) d\mathbf{x}, \quad (2)$$

Integrating (2) by parts we obtain

$$\frac{d\mathbf{m}(t)}{dt} = \int_V \mathbf{u}(\mathbf{x}, t) f(\mathbf{x}, t) d\mathbf{x} \equiv \mathbf{v}(t). \quad (3)$$

The result in (3) implies that the evolution of the average value of a random variable $\mathbf{x}$ whose density evolves according to (1) is determined by the average drift velocity only, $\mathbf{v}(t)$.

Applying the same technique the evolution of the variance and coefficient of variation we get

$$\frac{d\sigma^2(t)}{dt} = \int_V (\mathbf{x} - \mathbf{m}(t))^2 \frac{\partial f(\mathbf{x}, t)}{\partial t} d\mathbf{x} = \lambda(t) + 2\text{cov}_{\mathbf{x}, \mathbf{u}}(t), \frac{d\mu}{dt} = \frac{\mathbf{v}}{\sigma} - \frac{\mathbf{m}}{\sigma^3}\left(\frac{\lambda}{2} + \text{cov}_{\mathbf{x}, \mathbf{u}}\right). \quad (4)$$

When $f(\mathbf{x}, t)$ is interpreted the SIR pdf, without the loss of generality, we may assume that the link is stable when $m(t) > 1$, i.e. communications between D2D partners is feasible. Alternatively, when $m(t) < 1$ the link is in outage conditions. However, the mean value alone does not fully characterize the communications process alone evolving in time as the variance may severely affect the frequency and the duration of the outage and stable communications periods.

2.3 Related Work

Consider the evolution of the mean value of the SIR as an explicit function of the pdf of transmitter locations, $f(\mathbf{x}, t)$. Let N be the number of D2D transmitters in the area of interest V, and $\phi(\mathbf{x})$ is a signal attenuation that depends on the distance to the receiver. We can write

$$q(t) = \frac{1}{N} \int_V \frac{\phi(\mathbf{x})}{U(\mathbf{x}, t)} f(\mathbf{x}, t) d\mathbf{x}, \quad U(\mathbf{x}, t) = \int_V \phi(|\mathbf{x} - \mathbf{x}'|) f(\mathbf{x}', t) d\mathbf{x}', \quad (5)$$

The time evolution of the "average potential" of $U(\mathbf{x}, t)$ and mean SIR have been obtained in [14] in the following forms

$$\frac{\partial U}{\partial t} = \frac{\lambda}{2}\Delta U - div_{\mathbf{x}}\mathbf{J}, \quad \mathbf{J}(\mathbf{x},t) = \int\limits_V \phi(|\mathbf{x}-\mathbf{x}'|)\mathbf{u}(\mathbf{x}',t)f(\mathbf{x}',t)d\mathbf{x}',$$

$$N\frac{dq}{dt} = \int\limits_V \left(\mathbf{u}\nabla\left(\frac{\phi}{U}\right) + \frac{\phi}{U^2}div_{\mathbf{x}}\mathbf{J}\right)f(\mathbf{x},t)d\mathbf{x} + \frac{\lambda}{2}\int\limits_V \left(\Delta\left(\frac{\phi}{U}\right) - \frac{\phi}{U^2}\Delta U\right)f(\mathbf{x},t)d\mathbf{x}.$$

$$(6)$$

As one may observe, the evolution of the mean SIR value expressed as an explicit function of $f(\mathbf{x},t)$ is described by rather complex equation. Despite the complex structure, (6) still allows to make qualitative conclusions regarding the contributions of each component. The first integral is determined by the drift of the average potential, and the second is related to the diffusion of this potential. Note that the diffusion coefficient is usually a small value [13], so that, the evolution of the mean SIR is determined mainly by the first term (demolition effect). Therefore, if it turns out to be possible to introduce a control such that the first term is close to zero, then the behavior of SIR turns out to be quasi-stationary, which is of high practical importance.

3   Time-Dependent Variance and CoV of SIR

We now proceed deriving the time evolution of the variance and CoV of SIR as an explicit function of pdf of transmitter coordinates, $f(\mathbf{x},t)$. Observe that using (1) and interpreting $f(\mathbf{x},t)$ as pdf of the transmitter coordinates the SIR variance is defined as follows

$$\Sigma^2(t) = \frac{1}{N^2}\int\limits_V \left(\frac{\phi(\mathbf{x})}{U(\mathbf{x},t)} - \int\limits_V \frac{\phi(\mathbf{x}')}{U(\mathbf{x}',t)}f(\mathbf{x}',t)d\mathbf{x}'\right)^2 f(\mathbf{x},t)d\mathbf{x}. \qquad (7)$$

Expanding the power we get

$$N^2\frac{d\Sigma^2(t)}{dt} = \int\limits_V \left(\frac{\phi(\mathbf{x})}{U(\mathbf{x},t)} - Nq(t)\right)^2\frac{\partial f(\mathbf{x},t)}{\partial t}d\mathbf{x}$$

$$- 2\int\limits_V \left(\frac{\phi(\mathbf{x})}{U^2(\mathbf{x},t)}\frac{\partial U(\mathbf{x},t)}{\partial t} + N\frac{dq(t)}{dt}\right)\left(\frac{\phi(\mathbf{x})}{U(\mathbf{x},t)} - Nq(t)\right)f(\mathbf{x},t)d\mathbf{x}.$$

$$(8)$$

Observe that the first integral on the right side of (8) can be further expanded using (1). The second integral represents the averaging of the derivatives found earlier in (5) and (6). As a result, the time derivative on the left hand side of the equation can be expressed in terms of the derivatives of $\mathbf{x}$ on the right-hand side. For the first integral we may write

$$\int_V \left(\frac{\phi(\mathbf{x})}{U(\mathbf{x},t)} - Nq(t)\right)^2 \frac{\partial f(\mathbf{x},t)}{\partial t} d\mathbf{x}$$

$$= \int_V \left(\frac{\phi(\mathbf{x})}{U(\mathbf{x},t)} - Nq(t)\right)^2 \left(-div_x(\mathbf{u}(\mathbf{x},t)f(\mathbf{x},t)) + \frac{\lambda(t)}{2}\Delta f(\mathbf{x},t)\right) d\mathbf{x}$$

$$= 2\int_V \left(\frac{\phi(\mathbf{x})}{U(\mathbf{x},t)} - Nq(t)\right)\left(\mathbf{u}(\mathbf{x},t)\nabla\frac{\phi(\mathbf{x})}{U(\mathbf{x},t)}\right)f(\mathbf{x},t)d\mathbf{x}$$

$$+ \lambda(t)\int_V \left(\frac{\phi(\mathbf{x})}{U(\mathbf{x},t)} - Nq(t)\right)\left(\Delta\frac{\phi(\mathbf{x})}{U(\mathbf{x},t)}\right)f(\mathbf{x},t)d\mathbf{x} + \lambda(t)\int_V \left(\nabla\frac{\phi(\mathbf{x})}{U(\mathbf{x},t)}\right)^2 f(\mathbf{x},t)d\mathbf{x}.$$

(9)

Adding the second integral to the resulting expression and replacing the $\partial U/\partial t$ in it with (6), we obtain the following:

$$N^2\frac{d\Sigma^2(t)}{dt} = 2Nq(t)\int_V \frac{\phi(\mathbf{x})}{U^2(\mathbf{x},t)}\left(\frac{\lambda(t)}{2}\Delta U - div_x\mathbf{J}\right)f(\mathbf{x},t)d\mathbf{x}$$

$$- 2\int_V \frac{\phi^2(\mathbf{x})}{U^3(\mathbf{x},t)}\left(\frac{\lambda(t)}{2}\Delta U - div_x\mathbf{J}\right)f(\mathbf{x},t)d\mathbf{x} +$$

$$- 2Nq(t)\int_V \left(\mathbf{u}(\mathbf{x},t)\nabla\frac{\phi(\mathbf{x})}{U(\mathbf{x},t)}\right)f(\mathbf{x},t)d\mathbf{x} + \int_V (\mathbf{u}(\mathbf{x},t)\nabla)\left(\frac{\phi(\mathbf{x})}{U(\mathbf{x},t)}\right)^2 f(\mathbf{x},t)d\mathbf{x}$$

$$+ \lambda(t)\int_V \left(\frac{\phi(\mathbf{x})}{U(\mathbf{x},t)} - Nq(t)\right)\left(\Delta\frac{\phi(\mathbf{x})}{U(\mathbf{x},t)}\right)f(\mathbf{x},t)d\mathbf{x} + \lambda(t)\int_V \left(\nabla\frac{\phi(\mathbf{x})}{U(\mathbf{x},t)}\right)^2 f(\mathbf{x},t)d\mathbf{x}.$$

(10)

Now we group the terms in the resulting expression in accordance to their influence on the transfer of the SIR. First, these are the terms due to the transfer of the local SIR ϕ/U caused by the drift velocity in the Fokker-Planck Eq. (1). Secondly, there are the terms proportional to the product of the average SIR for the average demolition of the local SIR heterogeneity, i.e., quantities $Nq(t)$. Finally, there are terms related to diffusion in (1), proportional to $\lambda(t)$. As a result, we arrive to

$$N^2\frac{d\Sigma^2(t)}{dt} = 2\int_V \frac{\phi(\mathbf{x})}{U(\mathbf{x},t)}\left(\mathbf{u}(\mathbf{x},t)\nabla\frac{\phi(\mathbf{x})}{U(\mathbf{x},t)} + \frac{\phi(\mathbf{x})}{U^2(\mathbf{x},t)}div_x\mathbf{J}\right)f(\mathbf{x},t)d\mathbf{x}$$

$$- 2Nq(t)\int_V \left(\frac{\phi(\mathbf{x})}{U^2(\mathbf{x},t)}div_x\mathbf{J} + \mathbf{u}(\mathbf{x},t)\nabla\frac{\phi(\mathbf{x})}{U(\mathbf{x},t)}\right)f(\mathbf{x},t)d\mathbf{x}$$

$$+ Nq(t)\lambda(t)\int_V \left(\frac{\phi(\mathbf{x})}{U^2(\mathbf{x},t)}\Delta U - \Delta\frac{\phi(\mathbf{x})}{U(\mathbf{x},t)}\right)f(\mathbf{x},t)d\mathbf{x}$$

$$+ \lambda(t)\int_V \left(\frac{\phi(\mathbf{x})}{U(\mathbf{x},t)}\Delta\frac{\phi(\mathbf{x})}{U(\mathbf{x},t)} + \left(\nabla\frac{\phi(\mathbf{x})}{U(\mathbf{x},t)}\right)^2 - \frac{\phi^2(\mathbf{x})}{U^3(\mathbf{x},t)}\Delta U\right)f(\mathbf{x},t)d\mathbf{x}.$$

(11)

Having obtained the variance of SIR, the CoV is immediately available.

4   Reliability of D2D Connections

We now proceed illustrating the effect of time-dependent SIR variance on the reliability of D2D connection. Particularly, we present the results of modeling the mean SIR $q(t)$, the standard deviation of SIR $\Sigma(t)$, and the value $s(t) = q(t)/\Sigma(t)$ for the ensemble of trajectories obtained using kinetic mobility model in (1) using the procedure described in [14]. We consider two cases: (Case A) a relatively small average SIR value and $s(t)$ close to unity (Case B) large average SIR value with a small $s(t)$. These two cases are chosen to illustrate principally different behavior of SIR. In both case non-stationary movement was considered.

Figures 1 and 2 show standard deviation of SIR and CoV, respectively, calculated using the set of trajectories for N = 10 transmitting D2D devices for two considered

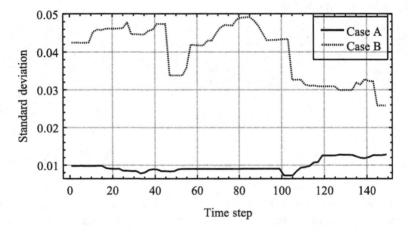

Fig. 1.   Standard deviation, $\Sigma(t)$, estimated using the ensemble of trajectories.

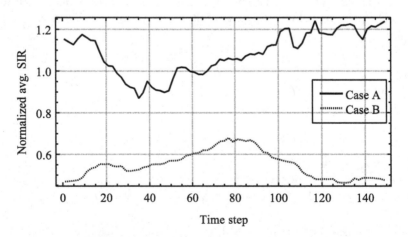

Fig. 2.   CoV estimated using the ensemble of trajectories

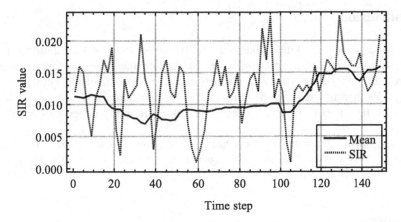

Fig. 3. The mean SIR and a single realization of SIR for Case A.

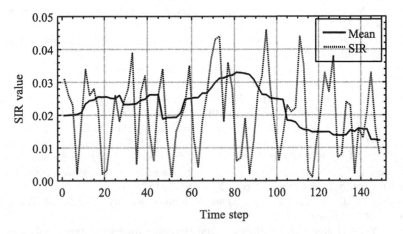

Fig. 4. The mean SIR and a single realization of SIR for Case B.

cases. Figures 3 and 4 illustrate the time-series of the mean SIR values and a single time-series of SIR over a non-stationary ensemble of trajectories for both cases.

Analyzing the presented results, one may observe that for a small SIR variance of Case A, the number of link outages for an arbitrarily chosen pair of D2D devices is less than for high average SIR of Case B characterized by equally high variance. In the provided examples, the number of outages at the level of reliability of the 0.01 is 7 and 13, respectively. Consequently, the introduced indicators of the stability of the connection between devices are of great practical importance, especially in cases of nonstationary random walk.

5 Conclusion

In this paper, using the kinetic based mobility model capable of capturing a wide variety of users' movement pattern, we have derived the analytical expressions for variance and CoV of the SIR of a link between two communicating entities in D2D scenario. We then investigated the effect of the variance and CoV on the link stability in terms of the number of outages for non-stationary mobility patterns of users.

Acknowledgement. The publication was financially supported by the Ministry of Education and Science of the Russian Federation (the Agreement number 02.a03.21.0008) and by RFBR (research projects Nos. 15-07-03051, 17-07-00845).

References

1. Baccelli, F., Błaszczyszyn, B.: Stochastic geometry and wireless networks volume II applications. Found. Trends® Netw. **4**(1–2), 1–312 (2010)
2. Haenggi, M.: Stochastic Geometry for Wireless Networks. Cambridge University Press, Cambridge (2012)
3. Klain, D.A., Rota, G.-C.: Introduction to Geometric Probability. Cambridge University Press, Cambridge (1997)
4. Rappaport, T.S.: Wireless Communications: Principles and Practice, vol. 2. Prentice Hall PTR, Upper Saddle River (1996)
5. Bettstetter, C., Hartenstein, H., Pérez-Costa, X.: Stochastic properties of the random waypoint mobility model. Wirel. Netw. **10**(5), 555–567 (2004)
6. Nain, P., et al.: Properties of random direction models. In: INFOCOM 2005, Proceedings IEEE 24th Annual Joint Conference of the IEEE Computer and Communications Societies, vol. 3. IEEE (2005)
7. Groenevelt, R., Altman, E., Nain, P.: relaying in mobile ad hoc networks: the brownian motion mobility model. Wirel. Netw. **12**(5), 561–571 (2006)
8. Rhee, I., et al.: On the levy-walk nature of human mobility. IEEE/ACM Trans. Netw. (TON) **19**(3), 630–643 (2011)
9. Orlov, Yu.N., Fedorov, S.L.: Modeling and statistical analysis of functionals defined on samples from a nonstationary time series, Preprints IPM. M.V. Keldysh. no. 43, 26 p. (2014)
10. Orlov, Yu.N.: Kinetic methods for investigating time-dependent time series, 276 p. MIPT, Moscow (2014)
11. Orlov, Yu.N., Fedorov, S.L.: Methods for numerical modeling of processes of a nonstationary random walk, 112 p. MIPT, Moscow (2016)
12. Orlov, Yu.N., Fedorov, S.L.: Modeling of distributions of functionals on the ensemble of trajectories of a nonstationary stochastic process, Pr. M.V. Keldysh, No. 101, 14 p. (2016)
13. Fedorov, S.L., Orlov, Yu.N., Samuylov, A., Moltchanov, D., Gaidamaka, Y., Samouylov, K., Shorgin, S.: Modeling of functional distribution for ensemble of trajectories, induced by the non-stationary stochastic process. In: proceedings of the 31th European Conference on Modelling and Simulation ECMS-2017, Budapest, Bulgaria, 23–26 May 2017, pp. 720–725 (2017)

14. Orlov, Yu.N., Zenyuk, D., Samuylov, A., Moltchanov, D., Andreev, S., Romashkova, O., Gaidamaka, Y., Samouylov, K.: Time-dependent SIR modeling for D2D communications in indoor deployments. In: proceedings of the 31th European Conference on Modelling and Simulation ECMS-2017, Budapest, Bulgaria, 23–26 May 2017, pp. 726–731 (2017)

15. Samuylov, D., Moltchanov, Y., Gaidamaka, V., Begishev, R., Kovalchukov, P., Abaev, S., Shorgin. SIR analysis in square-shaped indoor premises. In: proceedings of the 30th European Conference on Modelling and Simulation ECMS-2016, Regensburg, Germany May 31–June 03 2016, pp. 692–697 (2016). doi:10.7148/2016-0692

16. Samuylov, A., Gaidamaka, Y., Moltchanov, D., Andreev, S., Koucheryavy, Y.: Random triangle: a baseline model for interference analysis in heterogeneous networks. IEEE Trans. Veh. Technol. **65**(8), 6778–6782 (2016). doi:10.1109/TVT.2015.2481795

17. Begishev, V., Kovalchukov, R., Samuylov, A., Ometov, A., Moltchanov, D., Gaidamaka, Y., Andreev, S.: An analytical approach to SINR estimation in adjacent rectangular cells. In: Balandin, S., Andreev, S., Koucheryavy, Y. (eds.) ruSMART 2015. LNCS, vol. 9247, pp. 446–458. Springer, Cham (2015). doi:10.1007/978-3-319-23126-6_39

18. Etezov, S., Gaidamaka, Y., Samuylov, K., Moltchanov, D., Samuylov, A., Andreev, S., Koucheryavy, E.: On distribution of SIR in dense D2D deployments. In: 22nd European Wireless conference (EW 2016), 18–20 May 2016, Oulu, Finland, pp. 333–337 (2016)

19. Samuylov, A., Ometov, A., Begishev, V., Kovalchukov, R., Moltchanov, D., Gaidamaka, Y., Samouylov, K., Andreev, S., Koucheryavy, Y.: Analytical performance estimation of network-assisted D2D communications in urban scenarios with rectangular cells. Trans. Emerg. Telecommun. Technol. pp. 1–14 (2015). doi:10.1002/ett.2999

Analysis of Admission Control Schemes Models for Wireless Network Under Licensed Shared Access Framework

Ekaterina Markova[1], Dmitry Poluektov[1], Darya Ostrikova[1],
Irina Gudkova[1(✉)], Iliya Dzantiev[1], Konstantin Samouylov[1],
and Vsevolod Shorgin[2]

[1] Peoples' Friendship University of Russia (RUDN University),
6 Miklukho-Maklaya St, Moscow 117198, Russian Federation
{markova_ev, poluektov_ds, ostrikova_du, gudkova_ia,
samuylov_ke}@rudn.university
[2] Institute of Informatics Problems, Federal Research Center
"Computer Science and Control" of the Russian Academy of Sciences,
44-2 Vavilov St, Moscow 119333, Russian Federation
VShorgin@ipiran.ru

Abstract. Nowadays, mobile operators are faced with a problem of shortage of radio resources required for qualitative customer services. One of the possible solutions is the framework named LSA (Licensed Shared Access), which is developed with the assistance of ETSI. The LSA spectrum is shared between the owner (incumbent) and LSA licensee (e.g., mobile network operator). At any time, LSA spectrum could be used by incumbent or mobile network operator but not together at once. In this connection, if the incumbent needs its frequency, then LSA band becomes unavailable for mobile operator. This leads to service interruptions for mobile operator users. In this paper, we describe two possible Radio Admission Control (RAC) scheme models of the 3GPP LTE cellular network within LSA framework as finite queuing systems with reliable (single-tenant band) and unreliable (multi-tenant band) servers. Multi-tenant band is assumed to be intolerant to traffic delay. The formulas for calculating the performances measures – blocking probability, probability of service interruption and probability of service band changing – are proposed. The numerical analysis is provided for LSA example scenario of aeronautical telemetry.

Keywords: LSA · Cellular network · Shut-down policy · RAC scheme · Queuing system · Unreliable servers · Recursive algorithm · Interruption probability · Service band changing

The publication was financially supported by the Ministry of Education and Science of the Russian Federation (the Agreement number 02.a03.21.0008), RFBR according to the research project No. 16-37-00421 mol_a.

O. Galinina et al. (Eds.): NEW2AN/ruSMART/NsCC 2017, LNCS 10531, pp. 536–549, 2017.
DOI: 10.1007/978-3-319-67380-6_50

1 Introduction

The demand for mobile broadband services as well as the volume of traffic increases every year [1, 2]. A considerable amount of frequency resources is needed to provide to users services with a required level of quality of service (QoS) [3, 4]. The problem of resource shortage can be solved by means of the shared access to spectrum by several entities, implemented, for instance, by using the Licensed Shared Access (LSA) framework [5, 6], which is developed with the assistance of ETSI. LSA framework [7] can improve the efficiency of resource usage and ensure the access to a spectrum which otherwise would be underused. The spectrum is shared between the owners (incumbents) and a limited number of LSA licensees (e.g., mobile network operators). The LSA licensee has access to both bands – the single-tenant band assigned only to it and the multi-tenant band (LSA band) assigned also to the incumbent.

For the shared access, ETSI [8] proposes to use the spectrum allocated for aeronautical and terrestrial telemetry or specific applications including cordless cameras, portable video links, and mobile video links. Various policies of interference coordination between two entities could be considered. The authors of [3] propose three of them: the so-called ignore policy, shutdown policy [9–13], and limit power policy (used for aeronautical telemetry) [14–16]. According to shutdown policy, at any time, LSA spectrum could be used by incumbent or mobile network operator but not together at once. In this connection, if the incumbent needs its frequency, then LSA band becomes unavailable for mobile operator. This leads to service interruptions for mobile operator users. The limit power policy implies managing the user equipment (UE) power in uplink and eNodeB (eNB) power in downlink, so there is no service interruption of users due to the incumbent accessing spectrum.

In [9, 10], the models with a finite queue and tolerant to data traffic delays were discussed. There, the authors investigated the single- and multi-tenant bands, while the characterization of the LSA framework was completed by relying on the simulation tools and based on the queuing theory. The authors in [11, 12] focused on capturing the spatial user locations, which become a crucial factor in connection to the system performance. In the paper [13], capturing the inhomogeneous in time nature of rates of requests for access to spectrum and average times of spectrum use, the authors proposed a queuing model of binary access to spectrum as seen from the licensee's point of view. The queue is described by an inhomogeneous birth and death process with catastrophes and repairs. In [14] the limit power policy is described in the form of a queuing system with reliable (single-tenant band) and unreliable (multi-tenant band) servers. The authors of [15] proposed the mathematical models combined queuing theory and stochastic geometry for the LSA framework utilizing a multi-tenant band within the 3GPP LTE cellular system. The end-users are distributed uniformly and as the considered power control mechanism a simple limit power policy is implemented. The paper [16] contains experimental evaluation of dynamic LSA operation.

In this paper we consider shut-down policy of interference coordination between incumbent and mobile operator. Thus, we describe two novel Radio Admission Control (RAC) scheme models of the 3GPP LTE cellular network within LSA framework – without changing of service band and with priority processing on multi-tenant band – as finite queuing systems with reliable (single-tenant band) and unreliable (multi-tenant band) servers. Multi-tenant band is assumed to be intolerant to traffic delay. The numerical analysis of performance measures is provided for LSA example scenario of aeronautical telemetry.

The paper is organized as follows. In Sect. 2, we propose General Description of two RAC Schemes. In Sect. 3, we describe the mathematical models of two RAC Schemes: without changing of service band and with priority processing on multi-tenant band. In Sect. 4, we analyze numerically the performance measures: the blocking probability, the interruption probability, and the probability of service band changing. Section 5 concludes the paper.

2   General Description of RAC Schemes

2.1   Main Assumptions and Parameters

Let us consider a single cell of mobile network within LSA framework with two bands – single-tenant and multi-tenant. We suppose that the single-tenant band has the total capacity of C_1 bandwidth units (b.u.), the multi-tenant band has the total capacity of C_2 b.u. and has two modes s – operational ($s = 1$) and unavailable ($s = 0$). We assume that the multi-tenant band goes into unavailable mode with rate α and recovers into operational mode with rate β, i.e. the average time when multi-tenant band is available or unavailable is equal to α^{-1} or β^{-1} respectively. These intervals follow the exponential distribution.

Let users request only one service that generates streaming traffic with arrival rate λ, which is Poisson distributed. The average service time of one user's request processed on the single-tenant or multi-tenant band is exponentially distributed with mean μ_1^{-1} or μ_2^{-1} respectively. In accordance with the features of streaming traffic each user's request is served at the guaranteed bit rate (GBR) c. Without loss of generality, we assume c = 1 b.u. According to this the maximum number of single-tenant or multi-tenant band users is equal to C_1 or C_2 respectively.

All necessary further notations are given in Table 1.

Table 1. System parameters

Notation	Parameter description
C_1	Total capacity of the single-tenant band
C_2	Total capacity of the multi-tenant band
μ_1^{-1}	Average service time of one user's request processed on the single-tenant band
μ_2^{-1}	Average service time of one user's request processed on the multi-tenant band
s	State of the multi-tenant band, $s = 1$ if the band is operational and $s = 0$ if the band is unavailable
α^{-1}	Average time when multi-tenant band is available
β^{-1}	Average time when multi-tenant band is unavailable
c	Guaranteed bit rate
n_1	Number of single-tenant band users
n_2	Number of multi-tenant band users

2.2   Policy of Interference Coordination

In this paper we consider shut-down policy of interference coordination between the incumbent and the mobile operator. For a clearer understanding of the considered policy let us determine the rules for accepting requests of users. These rules are common to both RAC scheme models, which are discussed further in Sect. 3. When a new request arrives, the following scenarios are possible:

- The request will be accepted for service on the single-tenant band, which is possible if the request finds the number of single-tenant band users less than C_1.
- The request will be accepted for service on the multi-tenant band, which is possible if the number of single-tenant band users is equal to C_1, the multi-tenant band is operational, i.e. $s = 1$ and the request finds the number of multi-tenant band users less than C_2.
- Otherwise, the request will be blocked without any after-effect on the corresponding Poisson process.

If the incumbent needs resources, the multi-tenant band goes into unavailable mode, then we can consider two cases:

- The request processed on the multi-tenant band will continue its service on the single-tenant band, which is possible if the request finds the number of single-tenant band users less than C_1.
- Otherwise the request processed on the multi-tenant band will be interrupted.

Next, we turn to a detailed description of the features of both considered RAC schemes.

3   Mathematical Models of RAC Schemes Without and with Changing of Service Band

According to the considerations proposed in the Sect. 2, we can describe our systems by a Markov process $\mathbf{X}(t) = (N_1(t), N_2(t), S(t), t \geq 0)$, where $N_1(t)$ – the number of single-tenant band users, $N_2(t)$ – the number of multi-tenant band users, $S(t)$ – the state of the multi-tenant band.

Let us consider the first and simpler model with two bands within LSA framework. In this model the request is processed on the band to which it originally arrives and changes the service band, namely the multi-tenant band, only when it goes into unavailable mode. Then the state space of the system (Fig. 1) is defined as follows

$$\mathbf{X}_1 = \{(n_1, n_2, s) : s = 0, n_1 \in [0, C_1], n_2 = 0 \cup s = 1, n_1 \in [0, C_1], n_2 \in [0, C_2]\}. \quad (1)$$

Let us turn to the second model. Unlike the previous model, the request processed on the multi-tenant band arrives to the single-tenant band as soon as the number of the single-tenant band users becomes less than C_1. Thus, the change of the band is possible not only when the multi-tenant band goes into unavailable mode. The state space of the second model (Fig. 2) has the following form

$$\mathbf{X}_2 = \{(n_1, n_2, s) : s = 0, 1\, n_1 \in [0, C_1], n_2 = 0 \cup s = 1, n_1 = C_1, n_2 \in [0, C_2]\}. \quad (2)$$

3.1 Systems of Equilibrium Equations

The state transition diagram for the first model has the following form (Fig. 1).

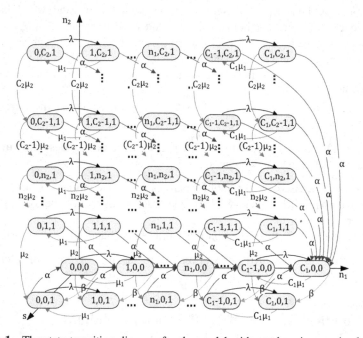

Fig. 1. The state transition diagram for the model without changing service band.

The corresponding Markov process representing the system states is described by the system of equilibrium equations

$$(\lambda + \beta)p_1(0,0,0) = \mu_1 p_1(1,0,0) + \alpha p_1(0,0,1);$$

$$(\lambda + \alpha)p_1(0,0,1) = \beta p_1(0,0,0) + \mu_1 p_1(1,0,1) + \mu_2 p_1(0,1,1);$$

$$(\lambda + n_1\mu_1 + \beta)p_1(n_1,0,0)$$
$$= \lambda p_1(n_1 - 1,0,0) + (n_1 + 1)\mu_1 p_1(n_1 + 1,0,0) + \alpha \sum_{m=0}^{n_1} p_1(n_1 - m, m, 1),$$
$$n_1 = 1, \ldots, C_1 - 1;$$

$$(\lambda + n_1\mu_1 + \alpha)p_1(n_1,0,1)$$
$$= \lambda p_1(n_1 - 1,0,1) + (n_1 + 1)\mu_1 p_1(n_1 + 1,0,1) + \beta p_1(n_1,0,0) + \mu_2 p_1(n_1,1,1),$$
$$n_1 = 1, \ldots, C_1 - 1;$$

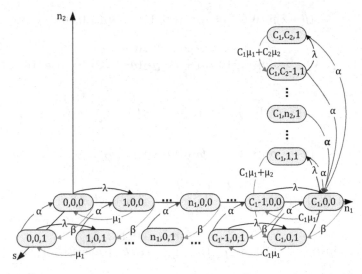

Fig. 2. The state transition diagram for the model with priority processing on multi-tenant band.

$$(\lambda + n_2\mu_2 + \alpha)p_1(0, n_2, 1) = \mu_1 p_1(1, n_2, 1) + (n_2 + 1)\mu_2 p_1(0, n_2 + 1, 1),$$
$$n_2 = 1, \ldots, C_2 - 1;$$

$$(\lambda + n_1\mu_1 + n_2\mu_2 + \alpha)p_1(n_1, n_2, 1)$$
$$= \lambda p_1(n_1 - 1, n_2, 1) + (n_1 + 1)\mu_1 p_1(n_1 + 1, n_2, 1) + (n_2 + 1)\mu_2 p_1(n_1, n_2 + 1, 1),$$
$$n_1 = 1, \ldots, C_1 - 1, n_2 = 1, \ldots, C_2 - 1$$

$$(C_1\mu_1 + \beta)p_1(C_1, 0, 0)$$
$$= \lambda p_1(C_1 - 1, 0, 0) + \alpha\left(p_1(C_1, 0, 1) + \sum_{m=1}^{C_2}(p_1(C_1, m, 1) + p_1(C_1 - m, m, 1))\right);$$

$$(\lambda + C_1\mu_1 + \alpha)p_1(C_1, 0, 1) = \lambda p_1(C_1 - 1, 0, 1) + \beta p_1(C_1, 0, 0) + \mu_2 p_1(C_1, 1, 1));$$

$$(\lambda + C_1\mu_1 + n_2\mu_2 + \alpha)p_1(C_1, n_2, 1)$$
$$= \lambda(p_1(C_1 - 1, n_2, 1) + p_1(C_1, n_2 - 1, 1)) + (n_2 + 1)\mu_2 p_1(C_1, n_2 + 1, 1),$$
$$n_2 = 1, \ldots, C_2 - 1;$$

$$(C_1\mu_1 + C_2\mu_2 + \alpha)p_1(C_1, C_2, 1) = \lambda(p_1(C_1 - 1, C_2, 1) + p_1(C_1, C_2 - 1, 1)),$$

where $p_1(n_1, n_2, s), (n_1, n_2, s) \in \mathbf{X}_1$ is the stationary probability distribution.

Let us denote $p_2(n_1, n_2, s), (n_1, n_2, s) \in \mathbf{X}_2$ the stationary probability distribution for the model with priority processing on multi-tenant band. Then using the state transition diagram (Fig. 2), we can represent the system of equilibrium equations for the second model:

$$(\beta + \lambda)p_2(0,0,0) = \alpha p_2(0,0,1) + \mu_1 p_2(1,0,0);$$

$$(\beta + n_1\mu_1 + \lambda)p_2(n_1,0,0)$$
$$= \lambda p_2(n_1 - 1,0,0) + \alpha p_2(n_1,0,1) + (n_1+1)\mu_1 p_2(n_1+1,0,0), n_1 = 1,\ldots, C_1 - 1;$$

$$(C_1\mu_1 + \beta)p_2(C_1,0,0) = \lambda p_2(C_1 - 1,0,0) + \alpha \sum_{m=0}^{C_2} p_2(C_1,m,1);$$

$$(\alpha + \lambda)p_2(0,0,1) = \beta p_2(0,0,0) + \mu_1 p_2(1,0,1);$$

$$(\alpha + n_1\mu_1 + \lambda)p_2(n_1,0,1)$$
$$= \lambda p_2(n_1 - 1,0,1) + \beta p_2(n_1,0,0) + (n_1+1)\mu_1 p_2(n_1+1,0,1), n_1 = 1,\ldots, C_1 - 1;$$

$$(\alpha + C_1\mu_1 + \lambda)p_2(C_1,0,1)$$
$$= \lambda p_2(C_1 - 1,0,1) + \beta p_2(C_1,0,0) + (C_1\mu_1 + \mu_2)p_2(C_1,1,1);$$

$$(\alpha + C_1\mu_1 + n_2\mu_2 + \lambda)p_2(C_1,n_2,1)$$
$$= \lambda p_2(C_1, n_2 - 1, 1) + (C_1\mu_1 + (n_2+1)\mu_2)p_2(C_1,n_2+1,1), n_2 = 1,\ldots, C_2 - 1;$$
$$(\alpha + C_1\mu_1 + C_2\mu_2)p_2(C_1,C_2,1) = \lambda p_2(C_1, C_2 - 1, 1)$$

3.2   Infinitesimal Generators

Let us denote the stationary probability distribution $p_i(n_1, n_2, s)$, $(n_1, n_2, s) \in \mathbf{X}_i$, as $\mathbf{p}_i$, $i = 1, 2$. We introduce the infinitesimal generator $\mathbf{A}_i = \left(a_i\left((n_1, n_2, s)(n_1', n_2', s')\right)\right)$, $i = 1, 2$ of the Markov process $\mathbf{X}(t)$ for the first and the second model respectively. In accordance with the state transition diagrams (Figs. 1 and 2) and the systems of equilibrium equations, which are presented above in the Sect. 3.1, the elements $a_i\left((n_1, n_2, s)(n_1', n_2', s')\right)$ of infinitesimal generators $\mathbf{A}_i, i = 1, 2$ are defined as follows:

- for the model without changing the service band

$$a_1\left((n_1, n_2, s)(n_1', n_2', s')\right) = \begin{cases} \alpha, & n_1' = \min(C_1, n_1 + n_2), n_2' = 0, s' = s - 1, \\ \beta, & n_1' = n_1, n_2' = n_2 = 0, s' = s + 1, \\ \lambda, & n_1' = n_1 + 1, n_2' = n_2, s' = s \text{ or} \\ & n_1' = n_1 = C_1, n_2' = n_2 + 1, s' = s = 1, \\ n_1\mu_1, & n_1' = n_1 - 1, n_2' = n_2, s' = s, \\ n_2\mu_2, & n_1' = n_1, n_2' = n_2 - 1, s' = s = 1, \\ *, & n_1' = n_1, n_2' = n_2, s' = s, \\ 0, & \text{otherwise}, \end{cases}$$
(3)

where $* = -(s\alpha + (1 - s)\beta + \lambda \cdot 1\{n_1 + sn_2 < C_1 + sC_2\} + n_1\mu_1 + n_2\mu_2);$

- for the model with changing the service band

$$a_1\big((n_1,n_2,s)(n_1',n_2',s')\big) = \begin{cases} \alpha, & n_1'=n_1, n_2'=0, s'=s-1, \\ \beta, & n_1'=n_1, n_2'=n_2=0, s'=s+1, \\ \lambda, & n_1'=n_1+1, n_2'=n_2=0, s'=s \text{ or} \\ & n_1'=n_1=C_1, n_2'=n_2+1, s'=s=1, \\ n_1\mu_1, & n_1'=n_1-1, n_2'=n_2, s'=s, \\ C_1\mu_2 & n_1'=n_1=C_1, n_2'=n_2-1, s'=s=1, \\ n_2\mu_2, & n_1'=n_1=C_1, n_2'=n_2-1, s'=s=1, \\ *, & n_1'=n_1, n_2'=n_2, s'=s, \\ 0, & \text{otherwise,} \end{cases} \quad (4)$$

where $* = -(s\alpha + (1-s)\beta + \lambda \cdot 1\{n_1 + sn_2 < C_1 + sC_2\} + n_1\mu_1 + n_2\mu_2)$.

The system probability distribution $p_i(n_1,n_2,s)$, $(n_1,n_2,s) \in X_i$ could be numerically computed as the solution of the system of equilibrium equations $\mathbf{p}_i\mathbf{A}_i = \mathbf{0}$, $\mathbf{p}_i\mathbf{1}^T = \mathbf{1}$, $i = 1,2$.

3.3 Recursive Algorithm

For the scheme model with priority processing on the multi-tenant band (the second model) the system probability distribution $p_2(n_1,n_2,s)$, $(n_1,n_2,s) \in X_2$ could be numerically computed not only as the solution of the system of equilibrium equations. For this RAC scheme model we propose a recursive algorithm.

Let us consider the determination of unnormalized probabilities $q(n_1,n_2,s)$, $(n_1,n_2,s) \in X_2$.

Lemma 1. (1) The values of unnormalized probabilities $q(n_1,n_2,s)$ are calculated by formulas

$$q(0,0,0) = 1, q(0,0,1) = x; \quad (5)$$

$$q(n_1,n_2,s) = \gamma_{n_1,n_2,s} + \delta_{n_1,n_2,s}x, \quad (n_1,n_2,s) \in X_2 : n_1 > 0; \quad (6)$$

$$x = \frac{(C_1\mu_1 + C_2\mu_2 + \alpha)\gamma_{C_1,C_2,1} - \lambda\gamma_{C_1,C_2-1,1}}{\lambda\delta_{C_1,C_2-1,1} - (C_1\mu_1 + C_2\mu_2 + \alpha)\delta_{C_1,C_2,1}}. \quad (7)$$

(2) The coefficients $\gamma_{n_1,n_2,s}$ and $\delta_{n_1,n_2,s}$ are calculated by recursive formulas

$$\gamma_{0,0,0} = 1, \delta_{0,0,0} = 0; \quad (8)$$

$$\gamma_{0,0,1} = 0, \delta_{0,0,1} = 1; \quad (9)$$

$$\gamma_{1,0,0} = \frac{\lambda + \beta}{\mu_1}, \delta_{1,0,0} = -\frac{\alpha}{\mu_1}; \quad (10)$$

$$\gamma_{1,0,1} = -\frac{\beta}{\mu_1}, \delta_{1,0,1} = \frac{\lambda + \alpha}{\mu_1}; \tag{11}$$

$$n_1 = 2, \ldots, C_1, \ n_2 = 0, s = 0$$
$$\gamma_{n_1,0,0} = \frac{(n_1 - 1 + \gamma_{1,0,0})}{n_1} \gamma_{n_1-1,0,0} + \frac{\delta_{1,0,0}}{n_1} \gamma_{n_1-1,0,1} - \frac{\lambda}{n_1\mu_1} \gamma_{n_1-2,0,0}, \tag{12}$$

$$\delta_{n_1,0,0} = \frac{(n_1 - 1 + \gamma_{1,0,0})}{n_1} \delta_{n_1-1,0,0} + \frac{\delta_{1,0,0}}{n_1} \delta_{n_1-1,0,1} - \frac{\lambda}{n_1\mu_1} \delta_{n_1-2,0,0}; \tag{13}$$

$$n_1 = 2, \ldots, C_1, \ n_2 = 0, s = 1$$
$$\gamma_{n_1,0,1} = \frac{(n_1 - 1 + \delta_{1,0,1})}{n_1} \gamma_{n_1-1,0,1} + \frac{\gamma_{1,0,1}}{n_1} \gamma_{n_1-1,0,0} - \frac{\lambda}{n_1\mu_1} \gamma_{n_1-2,0,1}, \tag{14}$$

$$\delta_{n_1,0,1} = \frac{(n_1 - 1 + \delta_{1,0,1})}{n_1} \delta_{n_1-1,0,1} + \frac{\gamma_{1,0,1}}{n_1} \delta_{n_1-1,0,0} - \frac{\lambda}{n_1\mu_1} \delta_{n_1-2,0,1}; \tag{15}$$

$$\gamma_{C_1,1,1} = \frac{\lambda + C_1\mu_1 + \alpha}{C_1\mu_1 + \mu_2} \gamma_{C_1,0,1} - \frac{\lambda}{C_1\mu_1 + \mu_2} \gamma_{C_1-1,0,1} + \frac{\beta}{C_1\mu_1 + \mu_2} \gamma_{C_1,0,0}, \tag{16}$$

$$\delta_{C_1,1,1} = \frac{\lambda + C_1\mu_1 + \alpha}{C_1\mu_1 + \mu_2} \delta_{C_1,0,1} - \frac{\lambda}{C_1\mu_1 + \mu_2} \delta_{C_1-1,0,1} + \frac{\beta}{C_1\mu_1 + \mu_2} \delta_{C_1,0,0}; \tag{17}$$
$$n_1 = C_1, \ n_2 = 2, \ldots, C_2, s = 1$$

$$\gamma_{C_1,n_2,1} = \frac{\lambda + C_1\mu_1 + (n_2 - 1)\mu_2 + \alpha}{C_1\mu_1 + n_2\mu_2} \gamma_{C_1,n_2-1,1} - \frac{\lambda}{C_1\mu_1 + n_2\mu_2} \gamma_{C_1,n_2-2,1}, \tag{18}$$

$$\delta_{C_1,n_2,1} = \frac{\lambda + C_1\mu_1 + (n_2 - 1)\mu_2 + \alpha}{C_1\mu_1 + n_2\mu_2} \delta_{C_1,n_2-1,1} - \frac{\lambda}{C_1\mu_1 + n_2\mu_2} \delta_{C_1,n_2-2,1} \tag{19}$$

Notice 1. The probability distribution $p_2(n_1, n_2, s)$ is calculated by formula

$$p_2(n_1, n_2, s) = \frac{q(n_1, n_2, s)}{\sum_{(i,j,k) \in \mathbf{X}_2} q(n_1, n_2, s)}, (n_1, n_2, s) \in \mathbf{X}_2. \tag{20}$$

3.4 Performance Measures

Having found the probability distribution $p_i(n_1, n_2, s), (n_1, n_2, s) \in \mathbf{X}_i$ one may compute the performance measures of the considered RAC schemes – the scheme without changing of service band ($i = 1$) and the scheme with priority processing on the multi-tenant band ($i = 2$) respectively:

- Probability $B_i, i = 1, 2$ that a target user's request will be blocked

$$B_i = p_i(C_1, C_2, 1) + p_i(C_1, 0, 0); \tag{21}$$

- Probability $I_i, i = 1, 2$ that for a target user's request processed on the multi-tenant band the service will be interrupted

$$
\begin{aligned}
I_1 = &\sum_{n_2=1}^{C_2} \sum_{n_1=C_1-n_2+1}^{C_1-1} p_1(n_1, n_2, 1) \frac{\alpha}{\alpha + n_1\mu_1 + n_2\mu_2 + \lambda} \cdot \frac{n_2 - C_1 + n_1}{n_2} \\
&+ \sum_{n_2=1}^{C_2} p_1(C_1, n_2, 1) \frac{\alpha}{\alpha + C_1\mu_1 + n_2\mu_2 + \lambda \cdot 1\{C_2 - n_2 > 0\}};
\end{aligned}
\tag{22}
$$

$$
I_2 = \sum_{n_2=1}^{C_2} p_2(C_1, n_2, 1) \frac{\alpha}{\alpha + C_1\mu_1 + n_2\mu_2 + \lambda \cdot 1\{C_2 - n_2 > 0\}};
\tag{23}
$$

- Probability $P_i, i = 1, 2$ that a target user's request processed on the multi-tenant band will continue its service on the single-tenant band

$$
\begin{aligned}
P_1 = &\sum_{n_2=1}^{C_2} \sum_{n_1=0}^{C_1-n_2} p_1(n_1, n_2, 1) \frac{\alpha}{\alpha + n_1\mu_1 + n_2\mu_2 + \lambda} \\
&+ \sum_{n_2=1}^{C_2} \sum_{n_1=C_1-n_2+1}^{C_1-1} p_1(n_1, n_2, 1) \frac{\alpha}{\alpha + n_1\mu_1 + n_2\mu_2 + \lambda} \cdot \frac{C_1 - n_1}{n_2},
\end{aligned}
\tag{24}
$$

i.e. the operator does not need to discontinue service for this user's request;

$$
P_2 = \sum_{n_2=1}^{C_2} p_2(C_1, n_2, 1) \frac{\mu_2}{\alpha + C_1\mu_1 + n_2\mu_2 + \lambda \cdot 1\{C_2 - n_2 > 0\}},
\tag{25}
$$

i.e. the operator redirects a target user's request for service on the single-tenant band though multi-tenant band is in available mode.

4 Numerical Analysis

In order to be more specific on the expected parameters of our LSA model, let us review the system under investigation. The LTE base station typically has 10 MHz main band that is always available (single-tenant band), and may also have 5 MHz band that is rented under the LSA license (multi-tenant band). Hence, the throughput of the frequency band (in bits per second) can be calculated by multiplying the width of this band by the spectral efficiency. That is, 4 by 10 MHz for the single-tenant band and 4 by 5 MHz for the multi-tenant band. This gives us 40 MBits per second and 20 MBits per second for the first and the second bands, respectively [17]. Let us assume that we have mobile LTE users that are trying to upload the short fragments of video of size 1 Mbyte to save battery. The bit rate for uploading video content is set to 1 Mbps for one user on single-tenant band and to 2 Mbps on multi-tenant band.

Thus, we can then determine that the capacity of the bands is $C_1 = 40$ users for the first band and $C_2 = 10$ users for the second band. The mean time of data uploading on single-tenant band is equal $\mu_1^{-1} = 1$ Mbyte/1 Mbps $= 8$ s and on multi-tenant band is equal $\mu_2^{-1} = 1$ Mbyte/2 Mbps $= 4$ s.

We consider a scenario of aeronautical telemetry with airplane flies [3], which do not depend on time of the day. Let us assume that the multi-tenant band goes into unavailable mode every three minutes (180 s) or every four minutes (240 s) or every five minutes (240 s) on average. The takeoff speed of all airplanes in the airport are the same and the average time during which an airplane is flying over one cell is equal to $\beta^{-1} = 1$ min.

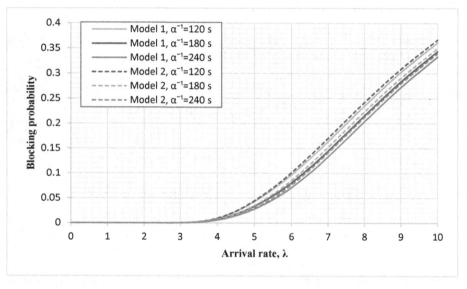

Fig. 3. Blocking probabilities B_1 and B_2 for different values of α^{-1}

The figures below show the behavior of each performance measure under examination – blocking probabilities B_1 and B_2 for model 1 (without changing of service band) and model 2 (with priority processing on multi-tenant band) (Fig. 3), probabilities of service interruption I_1 and I_2 (Fig. 4), and probabilities of service band changing P_1 (Fig. 5) and P_2 (Fig. 6) – for different values of α^{-1} (the average time when the multi-tenant band is available). First two figures show that the blocking probability and service interruption probability are lower for the model without changing of service band.

Among these characteristics is the probability of service band changing for one user, which is of particular interest as it is a nonmonotonic function with two monotone intervals and an extreme point. This can be explained as follows. At low loads, the LSA system mostly employs the servers of the single-tenant band. Accordingly, there are not

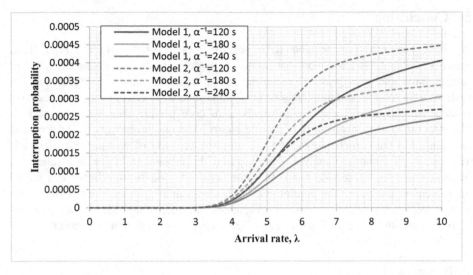

Fig. 4. Probabilities of service interruption I_1 and I_2 for different values of α^{-1}

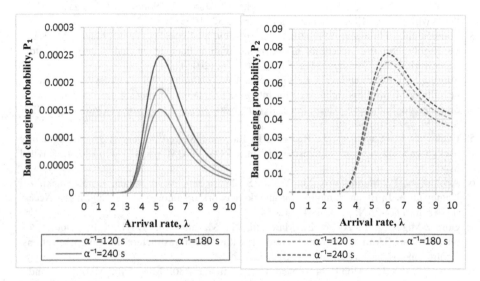

Fig. 5. Probability of service band changing P_1 for different values of α^{-1}

Fig. 6. Probability of service band changing P_2 for different values of α^{-1}

too many users served on the multi-tenant band. On the other hand, when the load is significant, there are not enough free resources on the single-tenant band. The extreme point on the plot represents the optimal load under which both bands would be utilized most efficiently.

5 Conclusion

We have presented two possible RAC scheme models of the 3GPP LTE cellular network within LSA framework for intolerant to delay traffic – without changing of service band and with priority processing on multi-tenant band – as finite queuing systems with reliable and unreliable servers. To compare the effectiveness of these two models we performed the numerical analysis of core performance measures such as blocking probability, probability of service interruption and probability of service band changing. For the numerical example, we used the scenario of aeronautical telemetry. The results of numerical analysis demonstrated the advantage of the first scheme – without changing of service band. In summary, our models may become a useful instrument in analyzing the emerging LSA systems and optimize their performance with respect to important operational parameters. An interesting task for future studies is to evaluate the quality of the schemes by involving the different cost criteria.

References

1. Cisco Visual Networking Index: Forecast and Methodology, 2015–2020 (2016)
2. Andrews, J., Buzzi, S., Choi, W., Hanly, S.V., Lozano, A., Soong, A.C.K., Zhang, J.C.: What will 5G be? IEEE J. Sel. Areas Commun. **32**, 1065–1082 (2014)
3. Ponomarenko-Timofeev, A., Pyattaev, A., Andreev, S., Koucheryavy, Y., Mueck, M., Karls, I.: Highly dynamic spectrum management within licensed shared access regulatory framework. IEEE Commun. Mag. **54**(3), 100–109 (2015)
4. Shorgin, S.Y., Samouylov, K.E., Gudkova, I.A., Galinina, O.S, Andreev, S.D.: On the benefits of 5G wireless technology for future mobile cloud computing. In: 1st International Science and Technology Conference Modern Networking Technologies (MoNeTec): SDN & NFV, pp. 151–154 (2014)
5. Buckwitz, K., Engelberg, J., Rausch, G.: Licensed Shared Access (LSA) – regulatory background and view of Administrations. In: CROWNCOM, pp. 413–416 (2014). (Invited paper)
6. Ahokangas, P., Matinmikko, M., Yrjola, S., Mustonen, M., Luttinen, E., Kivimäki, A., Kemppainen, J.: Business models for mobile network operators in Licensed Shared Access (LSA). In: DYSPAN, pp. 407–412 (2014)
7. Gomez-Miguelez, I., Avdic, E., Marchetti, N., Macaluso, I., Doyle, L.E.: Cloud-RAN platform for LSA in 5G networks – tradeoff within the infrastructure. In: Communications, Control and Signal Processing, pp. 522–525 (2014)
8. ETSI TS 103 113 Mobile broadband services in the 2300 MHz 2400 MHz band under Licensed Shared Access regime (2013)
9. Borodakiy, V.Y., Samouylov, K.E., Gudkova, I.A., Ostrikova, D.Y., Ponomarenko A.A., Turlikov, A.M., Andreev, S.D.: Modeling unreliable LSA operation in 3GPP LTE cellular networks. In: 6th International Congress on Ultra Modern Telecommunications and Control Systems, ICUMT 2014, pp. 490–496 (2014)
10. Gudkova, I.A., Samouylov, K.E., Ostrikova, D.Y., Mokrov, E.V., Ponomarenko-Timofeev, A.A., Andreev, S.D., Koucheryavy, Y.A.: Service failure and interruption probability analysis for Licensed Shared Access regulatory framework. In: 7th International Congress on Ultra Modern Telecommunications and Control Systems, ICUMT 2015, pp. 123–131 (2015)

11. Ahmadian, A., Galinina, O., Gudkova, I.A., Andreev, S., Shorgin, S., Samouylov, K.: On capturing spatial diversity of joint M2 M/H2H dynamic uplink transmissions in 3GPP LTE cellular system. In: 15th International Conference on Next Generation Wired/Wireless Advanced Networks and Systems, NEW2AN 2015, pp. 407–421 (2015)

12. Galinina, O., Andreev, S., Gerasimenko, M., Koucheryavy, Y., Himayat, N., Yeh, S.-P., Talwar, S.: Capturing spatial randomness of heterogeneous cellular WLAN deployments with dynamic traffic. IEEE J. Sel. Areas Commun. **32**(6), 1083–1099 (2014)

13. Gudkova, I., Korotysheva, A., Zeifman, A., Shilova, G., Korolev, V., Shorgin, S., Razumchik, R.: Modeling and analyzing licensed shared access operation for 5G network as an inhomogeneous queue with catastrophes. In: 8th International Congress on Ultra Modern Telecommunications and Control Systems and Workshops, ICUMT 2016, pp. 282–287 (2016)

14. Samouylov, K., Gudkova, I., Markova, E., Yarkina, N.: Queuing model with unreliable servers for limit power policy within Licensed Shared Access framework. In: Galinina, O., Balandin, S., Koucheryavy, Y. (eds.) NEW2AN/ruSMART 2016. LNCS, vol. 9870, pp. 404–413. Springer, Cham (2016). doi:10.1007/978-3-319-46301-8_34

15. Gudkova, I., Markova, E., Masek, P., Andreev, S., Hosek, J., Yarkina, N., Samouylov, K., Koucheryavy, Y.: Modeling the utilization of a multi-tenant band in 3GPP LTE system with Licensed Shared Access. In: 8th International Congress on Ultra Modern Telecommunications and Control Systems, ICUMT 2016, pp. 179–183 (2016)

16. Masek, P., Mokrov, E., Pyattaev, A., Zeman, K., Ponomarenko-Timofeev, A., Samuylov, A., Sopin, E., Hosek, J., Gudkova, I., Andreev, S., Novotny, V., Koucheryavy, Y., Samouylov, K.: Experimental evaluation of dynamic Licensed Shared Access operation in live 3GPP LTE system. In: 2016 IEEE Global Communications Conference, IEEE GLOBECOM 2016 (2016)

17. 3GPP TS 36.300 Evolved Universal Terrestrial Radio Access (E-UTRA) and Evolved Universal Terrestrial Radio Access Network (E-UTRAN); Overall description; Stage 2: Release 13 (2015)

Low-Complexity Iterative MIMO Detection Based on Turbo-MMSE Algorithm

Mikhail Bakulin[1], Vitaly Kreyndelin[1], Andrey Rog[2], Dmitry Petrov[3(✉)], and Sergei Melnik[4]

[1] IT Department, Moscow Technical University of Communications and Informatics (MTUSI), Moscow, Russian Federation
m.g.bakulin@gmail.com, vitkrend@gmail.com
[2] GlobalInformService, Moscow, Russian Federation
andhorn@mail.ru
[3] Peoples' Friendship University of Russia (RUDN University), Moscow, Russian Federation
dr.dmitry.petrov@ieee.org
[4] Central Scientific Research Institute of Communication, Moscow, Russian Federation
svmelnik@mail.ru

Abstract. Next generation Multiple Input Multiple Output (MIMO) communication systems should meet strict performance requirements and keep reasonable complexity. This paper presents a new low-complexity approach for iterative MIMO detection which is based on enhanced Turbo procedure. In the algorithm such components as linear Minimum Mean Square Error (MMSE) detection and soft symbol estimation based on the MMSE solution are utilized. A new original procedure of getting extrinsic data essentially improves the receiver performance and reduces its complexity. The results of the study are validated by link-level simulations.

Keywords: MIMO · Turbo decoding · Iterative detection · MMSE · ML

1 Introduction

The task of efficient MIMO detection for the large number of antennas raised up already in LTE networks. However, this problem is even more crucial for the emerging 5G networks where MIMO channels with 100 or even more transmitting and reviving antennas can be used [1,2].

The main role of MIMO detector (or MIMO equalizer) is to produce an estimation of probabilities of transmitted symbols and corresponding bits. It is an important part of MIMO receiver. Since the introduction of V-Blast [3,4], dozens of algorithms were proposed for MIMO detection. The algorithms differ in performance and complexity. Surveying this big set of detection approaches one can observe that Maximum Likelihood (ML) or Maximum a Posteriori (MAP)

© Springer International Publishing AG 2017
O. Galinina et al. (Eds.): NEW2AN/ruSMART/NsCC 2017, LNCS 10531, pp. 550–560, 2017.
DOI: 10.1007/978-3-319-67380-6_51

MIMO detector demonstrates the best (optimal) performance but at the cost of very high-complexity. The number of operations shows an exponential growth with regard to the number of transmitting antennas. High complexity makes ML hardly implementable especially for the case of a large number of transmitting and receiving antennas. However, ML detection represents only one class of algorithms. On the other hand, the complexity of Zero Forcing (ZF) and MMSE detectors is characterized by $O(N^3)$, where N is the rank of the channel matrix. Of course, complexity reduction is accompanied with performance degradation, which can achieve 5 dB and even more in the case of large-size MIMO systems. There are many sub-optimal algorithms proposed in the literature, such as Spherical Decoder (SD) [5], and M-algorithm combined with QR Decomposition (QRD-M) [6]. Their performance is comparable with ML but the complexity is significantly reduced. One can also find modifications of the MMSE algorithm, such as MMSE with Sorted QR Decomposition (MMSE SQRD) [7], and ordered layers cancellation algorithms [8–11]. Their performance is improved compared with MMSE at the expense of relatively small complexity growth. Still, there is performance gap (a few dBs) between MMSE based algorithms and ML or sub-ML. In [12] it was proposed to use the combination of ZF Ordered Successive Interference Cancellation (ZF OSIC) scheme and ML. ZF OSIC is applied to strong layers while ML is used for remaining weak layers. Such approach is intended to provide the proper performance/complexity trade-off by combining the schemes with different characteristics. Another efficient approach for Parallel Interference Cansellation (PIC) together with MMSE detection is proposed in [13].

To avoid misunderstanding and possible criticism about the way we have classified the algorithms by their complexity and performance we have to note that in some conditions, such as high signal-to-noise ratio (SNR), the complexity of sub-ML algorithms (e.g. SD or QRD-M) looks to be comparable with MMSE. However, if we consider all possible SNRs, one can observe that as SNR becomes smaller and the spherical radius (or parameter M in QRD-M) should be increased in order to keep the performance to be close to ML. Thus, the complexity increases. On the other hand, when M is kept the same as for the high SNR, the performance of sub-ML algorithms experiences serious degradation and loses to MMSE.

In this paper, we present a new detection scheme which is based on the MMSE solution. We called the scheme Turbo-MMSE because it utilizes the approach of extrinsic information extraction like it is done in turbo decoding. It is shown in our study that the complexity of Turbo-MMSE is comparable with MMSE but the performance can be much better than MMSE and its improved modifications [9–11]. Moreover, the new scheme assumes parallel layer processing resulting in the reduced calculation time.

The remaining parts of this paper are organized as follows. In Sect. 2, we briefly present the classical model of the MIMO-OFDM system. In Sect. 3, we review the most efficient MMSE-based schemes, which were presented before. In Sect. 4, we introduce the new Turbo-MMSE scheme and describe its difference with other MMSE-based schemes. In Sect. 5, we present simulation results and compare the performance of different MMSE based schemes. Finally, we conclude this paper in Sect. 6.

2   System Description

The typical multi-carrier MIMO system is described by matrix equation in frequency domain:

$$\mathbf{y} = \mathbf{Hx} + \boldsymbol{\eta}, \tag{1}$$

where $\mathbf{x} = (x_1, \ldots, x_M)^T$ is the transmitted signal vector, $\mathbf{y} = (y_1, \ldots, y_N)^T$ is received signal vector, $\boldsymbol{\eta} = (\eta_1, \ldots, \eta_N)^T$ is additive noise vector, $\mathbf{H}$ is channel matrix of size $N \times M$, all variables are complex. Equation (1) describes signal transmission through flat MIMO channel, which is typical case for most MIMO-OFDM systems utilized in most broadband wireless standards. In the simulations we will use MIMO transmitting schemes defined for LTE-A (see [14–16]) with channel models defined in [17].

3   MMSE-Based Layer Cancellation MIMO Detectors

MMSE-based Ordered Successive Interference Cancellation detector (MMSE OSIC) was introduced in [8]. Further, its improved version with soft output was defined in [9,11]. MMSE OSIC makes interference cancellation layer-by-layer in the order defined by layer ordering procedure. Layer ordering is done according to maximal post-detection SNR or to maximal signal power.

Layer cancellation means subtraction of signal transmitted on a layer from the received signal $\mathbf{y}$. After cancellation of k layers, the received signal contains the data transmitted on unprocessed layers, additive noise $\boldsymbol{\eta}$ and also extra noise caused by improperly detected symbols on canceled layers. Received signal with subtracted layers is defined by equation

$$\tilde{\mathbf{y}} = \mathbf{h}_{k+1} x_{k+1} + \underbrace{\sum_{i=k+2}^{M} \mathbf{h}_i x_i}_{\mathbf{I}_u} + \underbrace{\sum_{i=1}^{k} \mathbf{h}_i (x_i - \breve{x}_i)}_{\mathbf{I}_c} + \boldsymbol{\eta}, \tag{2}$$

where $\mathbf{I}_u$ defines the interference from undetected layers, $\mathbf{I}_c$ defines the interference from detected and canceled layers due to decision errors, x_i is a true symbol transmitted on layer i, while $\breve{x}_i$ is the estimation of x_i. For simplicity, we assume that Eq. (2) corresponds to already sorted layers, i.e. matrix columns $\mathbf{h}_1, \ldots, \mathbf{h}_M$ represent sorted layers in the order from best to worst. Solution x_{k+1} for the current layer $k+1$ is defined by MMSE filter

$$\mathbf{g} = \mathbf{h}_{k+1}^H \left[\mathbf{H}_{k+2:M} \mathbf{H}_{k+2:M}^H + \frac{1}{\sigma_s^2} \mathbf{R}_{I_D} + \frac{\sigma_\eta^2}{\sigma_s^2} \mathbf{I}_N \right]^{-1}, \tag{3}$$

where matrix $\mathbf{H}_{k+2:M}$ is the sub-matrix consisting of columns corresponding to undetected layers, $(\cdot)^H$ is the sign of Hermitean conjugation, $\mathbf{R}_{I_D}$ is covariance matrix defined by decision errors of previously processed layers, $\mathbf{I}_N$ is identity matrix, σ_s^2 and σ_η^2 are variances characterizing signal and noise powers, respectively. Covariance matrix $\mathbf{R}_{I_D}$ is defined by equation

$$\mathbf{R}_{I_D} = \mathbf{H}_{1:k}\mathbf{Q}_{D_e}\mathbf{H}_{1:k}^H, \tag{4}$$

where $\mathbf{H}_{1:k}$ is sub-matrix consisting of columns corresponding to already detected layers, $\mathbf{Q}_{D_e}$ is covariance matrix characterizing detected symbols errors (i.e. the difference between true symbols and symbol estimations obtained within detection). The elements of $\mathbf{Q}_{D_e}$ are defined as follows:

$$q_{u,v} = \begin{cases} \mathbb{E}\left[\left|x_u - \breve{x}_u\right|^2 \Big| \breve{x}_u\right], \, u = v \\ \mathbb{E}\left[(x_u - \breve{x}_u)\big|\breve{x}_u\right]\mathbb{E}\left[(x_v - \breve{x}_v)\big|\breve{x}_v\right]^*, \, u \neq v \end{cases}$$

$$u, v \in \{1, \ldots, k\}. \tag{5}$$

Estimation $\breve{x}_i$ is defined based on detected symbol probability in accordance to soft symbol detection procedure:

$$\breve{x}_i = \sum_{p=1}^{N_c} S_p \mathrm{Pr}(x_i = S_p|\hat{x}_i, \mu_i), \tag{6}$$

where N_c is the number of points in QAM constellation, $\mathrm{Pr}(x_i = S_p|\hat{x}_i, \mu_i)$ is probability that symbol x_i transmitted on layer i belongs to constellation point S_p. The probability is defined by unbiased MMSE estimation characterized by equivalent Gaussian channel approximation:

$$\hat{x}_i = \mathbf{g}\tilde{y} = \mu_i x_i + \eta_i \tag{7}$$

where wight coefficient $\mu_i = \mathbf{gh}_i$, and η_i has Gaussian distribution $\mathbb{N}(0, \sigma_{\eta_i}^2)$ with 0 mean and variance $\sigma_{\eta_i}^2 = (\mu_i - \mu_i^2)\sigma_s^2$.

Note that instead of soft symbol $\breve{x}_i$ it is possible to use the most probable symbol in the constellation. The elements $q_{u,v}$ of the covariance matrix are also defined by the probability model (7).

Matrix inversion in (3) is necessary to get the MMSE filter for each transmission layer. A simplified version of the matrix $\mathbf{Q}_{D_e}$ was proposed in [10]. It has all elements equal to zero except the main diagonal. As it was proven by simulations, such simplification leads to practically negligible performance degradation compared to full covariance matrix defined by (5).

It should be noted that in [9,10] the authors compare their low-complexity scheme of matrix inversion to high-quality but rather complex scheme based on Singular Value Decomposition (SVD). Therefore, the conclusion about very small complexity growth regarding the total complexity of the MMSE algorithm is made. If we compare, however, proposed matrix inversion scheme with more calculation efficient approaches, for example, with the inverse of partitioned matrices (see [18]), we can find out that the complexity of simplified MMSE OSIC is about 2–3 times the complexity of MMSE. For sure it is just a rough estimation and one can make more accurate calculations taking into account extra operations necessary for layers sorting, soft symbol detection and layers cancellation.

Finalizing the observation of MMSE OSIC, we can conclude that described algorithm provides improved BLock Error Rate (BLER) performance compared with MMSE while its complexity increases not more than 2–3 times. However, the improved performance is still rather far away from optimal ML. In the next section, we will present new iteration algorithm which demonstrates improved performance while its complexity is just 2–2.5 times more than MMSE.

4   New Interleave MIMO Detector Scheme with MMSE Kernel

The advantage of the proposed algorithm is in Parallel Interference Cancellation (PIC), which is successfully performed on each iteration for all transmitted layers. Before starting with the algorithm itself, we will introduce the procedure for getting extrinsic information.

Let's suppose we estimate some parameter x, which has Gaussian distribution and for which we have a priori information about its mean value $\bar{x}_{pr}$ and variance σ_{pr}^2, i.e. $x \sim \mathbb{N}(\bar{x}_{pr}, \sigma_{pr}^2)$. Suppose additionally that we obtained a new measurement and defined a new estimation $\bar{x}_{ms}$:

$$\bar{x}_{ms} = x + \eta, \tag{8}$$

where $\eta \sim \mathbb{N}(0, \sigma_{ms}^2)$ is estimation error. Based on these data we can get posterior Probability Density Function (PDF) for x:

$$\mathrm{Pr}_{pos}(x) = C_1 \exp\left(-\frac{(x - \bar{x}_{pr})^2}{2\sigma_{pr}^2} - \frac{(x - \bar{x}_{ms})^2}{2\sigma_{ms}^2}\right), \tag{9}$$

where C_1 is normalization constant. Transforming the expression inside the exponent above we get:

$$\mathrm{Pr}_{pos}(x) = C_1 \exp\left(-\frac{x^2 - 2x\frac{\bar{x}_{pr}\sigma_{ms}^2 + \bar{x}_{ms}\sigma_{pr}^2}{\sigma_{ms}^2 + \sigma_{pr}^2}}{2\frac{\sigma_{pr}^2\sigma_{ms}^2}{\sigma_{ms}^2 + \sigma_{pr}^2}}\right) \exp\left(-\frac{\bar{x}_{pr}^2\sigma_{ms}^2 + \bar{x}_{ms}^2\sigma_{pr}^2}{2\sigma_{ms}^2\sigma_{pr}^2}\right). \tag{10}$$

It is easy to see that (10) is still Gaussian PDF with variance and mean value defined by (11):

$$\sigma_{pos}^2 = \frac{\sigma_{pr}^2\sigma_{ms}^2}{\sigma_{pr}^2 + \sigma_{ms}^2},$$

$$\bar{x}_{pos} = \bar{x}_{pr}\frac{\sigma_{ms}^2}{\sigma_{pr}^2 + \sigma_{ms}^2} + \bar{x}_{ms}\frac{\sigma_{pr}^2}{\sigma_{pr}^2 + \sigma_{ms}^2}. \tag{11}$$

Therefore,

$$\Pr_{pos}(x) = C_1 \exp\left(-\frac{(x - \bar{x}_{pos})^2}{2\sigma_{pos}^2}\right) \exp\left(-\frac{\bar{x}_{pr}^2\sigma_{ms}^2 + \bar{x}_{ms}^2\sigma_{pr}^2}{2\sigma_{pr}^2\sigma_{ms}^2}\right) \times$$

$$\exp\left(\frac{(\bar{x}_{pr}\sigma_{ms}^2 + \bar{x}_{ms}\sigma_{pr}^2)^2}{2\sigma_{pr}^2\sigma_{ms}^2(\sigma_{pr}^2 + \sigma_{ms}^2)}\right) = C_2 \exp\left(-\frac{(x - \bar{x}_{pos})^2}{2\sigma_{pos}^2}\right). \quad (12)$$

Correspondingly, in the case when we are solving backward problem of getting extrinsic information from the posterior PDF, we obtain:

$$\sigma_{ms}^2 = \frac{\sigma_{pr}^2\sigma_{pos}^2}{\sigma_{pr}^2 - \sigma_{pos}^2},$$

$$\bar{x}_{ms} = -\bar{x}_{pr}\frac{\sigma_{pos}^2}{\sigma_{pr}^2 - \sigma_{pos}^2} + \bar{x}_{pos}\frac{\sigma_{pr}^2}{\sigma_{pr}^2 - \sigma_{pos}^2}. \quad (13)$$

From this point, we can continue with the description of iterative detection. The block diagram of the proposed algorithm is shown in Fig. 1.

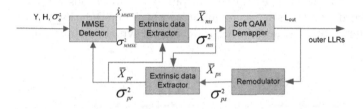

Fig. 1. Iterative detection using MMSE kernel.

MMSE detector estimates $\hat{\mathbf{x}}_{\mathrm{MMSE}}$ and solution variance σ_{MMSE}^2 based on a priori data obtained on the previous iteration, where a priori data are defined as mean value $\bar{\mathbf{x}}_{pr}$ and variance σ_{pr}^2. MMSE detector produces estimation in accordance to equations:

$$\hat{\mathbf{x}}_{\mathrm{MMSE}} = \bar{\mathbf{x}}_{pr} + \mathbf{g}_{\mathrm{MMSE\_Pr}}\left(\mathbf{Y} - \mathbf{H}\bar{\mathbf{x}}_{pr}\right),$$

$$\mathbf{g}_{\mathrm{MMSE\_Pr}} = \mathbf{V}_{pr}\mathbf{H}^H\left(\mathbf{H}\mathbf{V}_{pr}\mathbf{H}^H + \sigma_\eta^2\mathbf{I}\right)^{-1},$$

$$\mathbf{V}_{pr} = diag(\sigma_{pr}^2), \quad (14)$$

$$\sigma_{\mathrm{MMSE}}^2 = diag(\mathbf{V}_{\mathrm{MMSE}}) = diag(\mathbf{V}_{pr} - \mathbf{G}_{\mathrm{MMSE\_Pr}}\mathbf{H}\mathbf{V}_{pr}).$$

Note that in (14) σ_{MMSE}^2, σ_{pr}^2 are vectors, $\mathbf{V}_{pr}$ is a diagonal matrix containing variances σ_{pr}^2 for each layer of estimated vector $\mathbf{x}$, correspondingly σ_{MMSE}^2 is the vector of variances obtained from the diagonal of matrix $\mathbf{V}_{\mathrm{MMSE}}$. Considering MMSE estimation as posterior PDF, which is assumed to be Gaussian, and recalling conclusions of (13), we can extract extrinsic data. It will be used as a new observation on the next processing step. Thus, following (13) we define each component m of vectors $\bar{\mathbf{x}}_{ms}$, $\boldsymbol{\sigma}^2_{ms}$:

$$
\begin{aligned}
\bar{x}_{ms,m} &= -\frac{\sigma^2_{\text{MMSE},m}}{\sigma^2_{pr,m} - \sigma^2_{\text{MMSE},m}}\bar{x}_{pr,m} + \frac{\sigma^2_{pr,m}}{\sigma^2_{pr,m} - \sigma^2_{\text{MMSE},m}}\hat{x}_{\text{MMSE},m}, \\
\sigma^2_{ms,m} &= \frac{\sigma^2_{pr,m}\sigma^2_{\text{MMSE},m}}{\sigma^2_{pr,m} - \sigma^2_{\text{MMSE},m}}.
\end{aligned}
\tag{15}
$$

One can observe that (15) is equivalent to unbiased MMSE estimation defined by (7). Soft QAM demapper produces Log-Likelihood Ratio (LLR) estimations based on obtained $\bar{\mathbf{x}}_{ms}$, $\boldsymbol{\sigma}^2_{ms}$:

$$
L(b_{m,l}) = \ln\left(\frac{\sum\limits_{s_1 \in s(b_{m,l}=1)} \exp\left(-\frac{(\bar{x}_{ms,m}-s_1)^2}{2\sigma^2_{ms,m}}\right)}{\sum\limits_{s_0 \in s(b_{m,l}=0)} \exp\left(-\frac{(\bar{x}_{ms,m}-s_0)^2}{2\sigma^2_{ms,m}}\right)}\right),
\tag{16}
$$

where $s(b_{m,l} = 1,0)$ is the set of all symbols from QAM constellation S of order L with bit $b_{m,l} = 1,0$ in the l-th position of m-th symbol, and $l \in (1,\dots,L)$, $m \in (1,\dots,M)$.

These data are directed to the output when iterations loop is over. Otherwise, the data are used as an input to the re-modulation (or re-mapping) processing module. Re-modulator produces soft symbol estimations based on bit probabilities:

$$
\begin{aligned}
\bar{x}_{ps,m} &= \sum_S s(b_{m,1},\dots,b_{m,L})\prod_{t=1}^{L}\Pr(b_{m,t}), \\
\sigma^2_{ps,m} &= \sum_S |s(b_{m,1},\dots,b_{m,L})|^2 \prod_{t=1}^{L}\Pr(b_{m,t}) - |\bar{x}_{ps,m}|^2,
\end{aligned}
\tag{17}
$$

where $s(b_{m,1},\dots,b_{m,L})$ is m-th QAM symbol and

$$
\Pr(b_{m,t}) = \frac{\exp\left((-1)^{b_{m,t}+1}\frac{L_{m,t}}{2}\right)}{\exp\left(\frac{L_{m,t}}{2}\right) + \exp\left(\frac{-L_{m,t}}{2}\right)}.
$$

Summation in (17) is done for all constellation points (all bit combinations) of layer m. Note that (17) is equivalent to (5)–(6). If it is not the last iteration, we can calculate symbol probabilities instead of bit probabilities in the same way as it was done in OSIC. It is important that (17) as well as (5) can provide different variances for in-phase and quadrature components of soft QAM symbols. Due to this reason, MMSE solutions at all iterations, except the first one, must be obtained for real-valued signal equation. Therefore, complex matrix $\mathbf{H}$ is substituted by its real-valued representation:

$$
\mathbf{H} \to \begin{bmatrix}\Re(\mathbf{H}) & -\Im(\mathbf{H}) \\ \Im(\mathbf{H}) & \Re(\mathbf{H})\end{bmatrix}, \quad \mathbf{x} \to \begin{bmatrix}\Re(\mathbf{x}) \\ \Im(\mathbf{x})\end{bmatrix}, \quad \mathbf{y} \to \begin{bmatrix}\Re(\mathbf{y}) \\ \Im(\mathbf{y})\end{bmatrix}.
$$

The same procedure should be done in MMSE OSIC algorithm after cancellation of layer number 1.

Considering $\bar{\mathbf{x}}_{ps}$ and $\boldsymbol{\sigma}_{ps}^2$ as posterior PDF with a priori data $\bar{\mathbf{x}}_{ms}$, $\boldsymbol{\sigma}_{ms}^2$, we can extract extrinsic data applying (13):

$$\bar{x}_{pr,m} = -\frac{\sigma_{ps,m}^2}{\sigma_{ms,m}^2 - \sigma_{ps,m}^2}\bar{x}_{ms,m} + \frac{\sigma_{ms,m}^2}{\sigma_{ms,m}^2 - \sigma_{ps,m}^2}\bar{x}_{ps,m},$$
$$\sigma_{pr,m}^2 = \frac{\sigma_{ps}^2\sigma_{ms,m}^2}{\sigma_{ms,m}^2 - \sigma_{ps,m}^2}. \tag{18}$$

These data are used in MMSE detection in the next iteration cycle. In the first iteration it is assumed that $\sigma_{pr,m}^2 = \sigma_{s_m}^2$, $x_{pr,m} = 0$, so MMSE filter works in the same way as in the conventional MMSE detector.

To avoid negative variances in (18), it is necessary to check the condition $\sigma_{ps,m}^2 < \sigma_{ms,m}^2$. If it is not true, then iteration for that particular layer does not bring accuracy improvement, and this result should be discharged, i.e. on the next iteration a priori data for the layer m are the same as in the iteration before. In most cases the proposed method, that we called "Turbo MMSE", requires just two iterations. Further iterations do not bring noticeable performance improvement. Each iteration of Turbo MMSE consists of the following steps: getting MMSE filter, obtaining MMSE estimation, extracting unbiased MMSE estimation, making re-modulation and extracting a priori data, subtract a priori data from received vector. Taking into account that the first 3 steps are the same as in conventional MMSE, while the last two are quite simple and performed only once, we conclude that complexity of proposed scheme is 2–2.5 times higher the complexity of MMSE.

5   Simulation Results

Figures 2, 3, and 4 demonstrate the performance of developed algorithm. Simulations were performed for LTE-A downlink model 8×8 MIMO configuration. Three Modulation and Coding Schemes (MCS) are presented: MCS7 (4QAM, Coding Rate (CR) 1/2), MCS15 (16QAM, CR 1/2) and MCS21 (64QAM, CR 1/2). We use Extended Pedestrian A (EPA) channel model [17] with low correlation between channels and assume perfect channel knowledge at the receiver. 1000 channel realizations are used in simulation. Each realization has duration equal to one sub-frame. Transmitted data are distributed over 100 resource blocks and system bandwidth is 20 MHz (1200 sub-carriers). The number of iterations in turbo-decoder is 4. Transmission scheme assumes no automatic repeat request procedures.

For comparison we provide results of few reference algorithms: MMSE, ML, MMSE OSIC soft output (using full covariance matrix (5)). ML algorithm was realized as QRD-M scheme [6] with very big number of symbol-candidates (parameter $M = 1000$). It can be seen that proposed iterative procedure exceeds MMSE OSIC and both schemes outperform MMSE. There is expected gap with

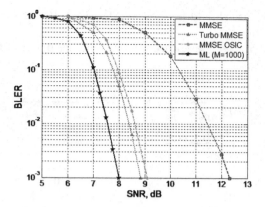

Fig. 2. BLER Performance in 8 × 8 LTE downlink, MCS7 (4QAM).

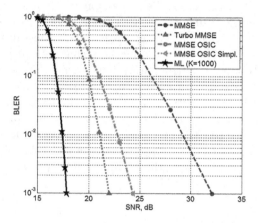

Fig. 3. BLER Performance in 8 × 8 LTE downlink, MCS15 (16QAM).

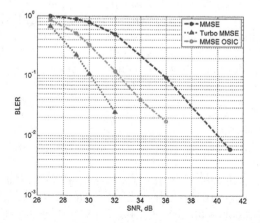

Fig. 4. BLER Performance in 8 × 8 LTE downlink, MCS21 (64QAM).

ML algorithm, but taking into account that proposed scheme has complexity comparable with MMSE and assumes parallel layer processing, it can be considered as an attractive performance-complexity trade-off.

6   Conclusion

In the paper the new iterative detection approach based on MMSE kernel and parallel interference cancellation is presented. Developed algorithm uses the principle of Turbo processing in its core. Unlike other known Turbo applications, however, we do not need to include decoder into the feedback loop. Additional information is obtained through non-linear processing in the demodulator. We show that proposed method demonstrates improved performance compared with conventional MMSE and MMSE based OSIC methods, while its complexity is only 2–2.5 times higher than MMSE.

Acknowledgment. The publication was financially supported by the Ministry of Education and Science of the Russian Federation (the Agreement number 02.a03.21.0008).

References

1. Vannithamby, R., Talwar, S.: Towards 5G: Applications, Requirements and Candidate Technologies. Wiley, Chichester (2017)
2. Luo, F.-L., Zhang, C.: Signal Processing for 5G: Algorithms and Implementations. Wiley, Chichester (2016)
3. Wolniansky, P.W., Foschini, G.J., Golden, G.D., Valenzuela, R.A.: V-BLAST: an architecture for realizing very high data rates over the rich-scattering wireless channel. In: International Symposium on Signals, Systems, and Electronics (ISSSE 1998), Pisa, Italy (1998)
4. Golden, G.D., Foschini, G.J., Valenzuela, R.A., Wolniansky, P.W.: Detection algorithm and initial laboratory results using V-BLAST space-time communication architecture. Electron. Lett. **35**(1), 14–16 (1999)
5. Hassibi, B., Vikalo, H.: On the sphere-decoding algorithm I. Expected complexity. IEEE Trans. Sig. Process. **53**(8), 2806–2818 (2005)
6. Dai, Y., Sun, S., Lei, Z.: A comparative study of QRD-M detection and sphere decoding for MIMO-OFDM systems. In: Personal, Indoor and Mobile Radio Communications (PIMRC), pp. 186–190, 11–14 September 2005
7. Wubben, D., Bonke, R., Kuhn, V., Kammeyer, K.-D.: MMSE extension of V-BLAST based on sorted QR decomposition. In: IEEE 58th Vehicular Technology Conference (VTC-Fall), vol. 1, pp. 508–512 (2003)
8. Lee, H., Lee, I.: New approach for coded layered space-time OFDM systems. In: IEEE International Conference on Communications (ICC), vol. 1, pp. 608–612, May 2005
9. Wang, J., Li, S.: Soft versus hard interference cancellation in MMSE OSIC MIMO detector: a comparative study. In: 4th International Symposium on Wireless Communication Systems (ISWCS), pp. 642–646, October 2007

10. Wang, J., Wen, O.Y., Li, S.: Soft-output MMSE-OSIC MIMO detector with reduced-complexity approximations. In: IEEE 8th Workshop on Signal Processing Advances in Wireless Communications (SPAWC), pp. 1–5, June 2007

11. Rog, A., Goreinov, S., Wu, D.S., Kim, D.: Method of MIMO signal decoding. Patent RU 2007114146, 16 April 2007

12. Iqbal, A., Kabir, M.H., Kwak, K.S.: Enhanced zero forcing ordered successive interference cancellation scheme for MIMO system. In: 2013 International Conference on ICT Convergence (ICTC), pp. 979–980, October 2013

13. Studer, C., Fateh, S., Seethaler, D.: ASIC implementation of soft-input soft-output MIMO detection using MMSE parallel interference cancellation. IEEE J. Solid-State Circ. **46**(7), 1754–1765 (2011)

14. 3GPP TS 36.211 v12.2.0, Evolved Universal Terrestrial Radio Access (E-UTRA); Physical Channels and Modulation (Release 12)

15. 3GPP TS 36.212 v12.1.0, Evolved Universal Terrestrial Radio Access (E-UTRA); Multiplexing and channel coding (Release 12)

16. 3GPP TS 36.213 v12.2.0, Evolved Universal Terrestrial Radio Access (E-UTRA); Physical layer procedures (Release 12)

17. 3GPP TS 36.104 v12.4.0, Evolved Universal Terrestrial Radio Access (E-UTRA); Base Station (BS) radio transmission and reception (Release 12)

18. Gentle, J.E.: Matrix Algebra: Theory, Computations, and Applications in Statistics, p. 95. Springer, New York (2007)

Rubidium Atomic Clock with Improved Metrological Characteristics for Satellite Communication System

Alexander A. Petrov[1(✉)], Vadim V. Davydov[1,2], Nikita S. Myazin[1], and Vladislav E. Kaganovskiy[2]

[1] Peter the Great Saint-Petersburg Polytechnical University,
St. Petersburg, Russia
Alexandrpetrov.spb@yandex.ru
[2] The Bonch-Bruevich Saint-Petersburg State University of
Telecommunications, St. Petersburg, Russia

Abstract. One of the essential elements of satellite communication system is an atomic clock. The proper operation of this system depends on the performance of atomic clock. In this paper one of the directions of modernization of the rubidium atomic clocks is considered. A new design of the frequency synthesis circuitry is presented. The new design of frequency synthesizer is based on method of direct digital synthesis. Theoretical calculations and experimental research showed a decrease in the step frequency tuning by several orders and significant improvement in the spectral characteristics of the output signal of frequency converter. A range of generated output frequencies is expanded. A summary of performance results of rubidium atomic clock, in particularly Allan variance, achieved to date is also provided.

Keywords: Quantum frequency standard · Rubidium atomic clocks · Frequency synthesizer · Direct digital synthesis · Allan variance

1 Introduction

Global navigation satellite constellations such as the European Galileo, Russian GLONASS, and the USA Global Positioning Systems (GPS) use atomic frequency standards for precision time-keeping and stable frequency generation [1–5].

Operation of measuring time systems, various electronic equipment, as well as, communication and information data transmission devices is also impossible without reliable operation of precision devices.

A necessary condition for reliable operation of info communication systems is coordinated work of the primary generator and receiver, so that the receiving unit can correctly interpret the digital signal. The difference in the synchronization of various units in a network, may lead to missing or to reread the information by receiving unit.

In order to solve this problem atomic clocks also called quantum frequency standards are used. Quantum frequency standards are devices that generate signals different

© Springer International Publishing AG 2017
O. Galinina et al. (Eds.): NEW2AN/ruSMART/NsCC 2017, LNCS 10531, pp. 561–568, 2017.
DOI: 10.1007/978-3-319-67380-6_52

types and frequencies. Thus, the frequency standards can synchronize operation of many sophisticated devices and instruments [1, 2, 5–7].

Standard frequencies on the atoms of rubidium or cesium are used as a clock generator in the communications equipment and in a data transmission devices, applied in the satellite navigation systems GLONASS and GPS as a clock generator and in the various metrological services. Also these standards perform a role of the reference signals with high precision and stability in radio equipment [1, 5, 6, 8–10].

With the development of scientific - technical progress operating conditions of rubidium atomic clocks are constantly changing. Therefore new requirements for measurement accuracy, reliability and weight - dimensional characteristics are produced to frequency standard. This leads constantly upgrade existing and develop new models of rubidium atomic clocks.

Development and commissioning of new atomic clock are very long and costly process, that in most cases hasn't enough funds and time. Therefore, in most cases, for specific tasks related to the operating conditions of frequency standards modernization is occur.

The process of modernization of frequency standards includes various directions: changing the weight and dimensions, reducing energy consumption, improvement metrological characteristics. And for frequency standards characterized by the fact that modernization may not be for the whole construction and may be for individual units or blocks.

In present work one of the directions of modernization of the rubidium atomic clock is considered. A new implementation of a digital frequency synthesizer for atomic clocks is presented. Improved output signal parameters frequency synthesizer leads to improvement characteristics of quantum frequency standard in particular the frequency stability.

2 The Optically Pumped Closed-Cell Rb Atomic Clock

In this section, the general principles of operation of a double resonance optically pumped sealed-cell atomic frequency standard are reviewed. The classical approach used in this field for the implementation of such a frequency standard has been based on the use of the ^{87}Rb isotope and is shown in Fig. 1 [1, 2, 5, 6].

In this setup light emitted by the spectral lamp excited at about 100 MHz is filtered by a cell containing ^{85}Rb. The fortuitous coincidence of the atomic transitions is such that the spectral line corresponding to the transition $S_{1/2}$, F = 2 $\leftrightarrow$ P emitted by the lamp is partly absorbed by this so-called hyperfine filter placed in the path of the radiation as illustrated. Consequently, the radiation reaching the resonance cell consists mainly of the spectral line $S_{1/2}$, F = 1 $\leftrightarrow$ P. This radiation causes optical pumping altering the population of the two hyperfine levels and making the cell transparent to the incident radiation. The cell is placed in a microwave cavity resonant at the ground state hyperfine frequency of ^{87}Rb. When energy is fed to the cavity at that frequency, the field created within the cell causes transitions between the two hyperfine energy levels and tend to equalize their population. The cell becomes absorbing again at the transition $S_{1/2}$, F = 1 $\leftrightarrow$ P.

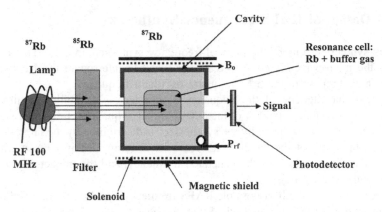

Fig. 1. General approach used in the implementation of a double resonance classical [87]Rb frequency standard using a [87]Rb spectral lamp and a [85]Rb filter.

A resonance line is observed in the light reaching the photodetector, and its characteristics reflect the properties of the ground state hyperfine transition: the line has a frequency relatively independent of the environment and is rather narrow, its width being determined largely by relaxation taking place in the $S_{1/2}$ ground state. An inert buffer gas is used in the cell in order to prevent relaxation by collisions of the atoms with the walls of the containing cell. In the case that the microwave is applied as a traveling wave without a cavity, Doppler broadening is avoided with the buffer gas through Dicke effect. In the present paper we will assume the presence of a cavity and of a standing wave. The relaxation rate is mainly controlled by spin exchange inter-action, collisions with the buffer gas atoms, diffusion to the container walls, and interaction of the atoms with the pumping light itself. In this approach, it is possible to obtain line widths of a few hundred Hz giving a line Q of the order of 10^7 to 10^8. In the arrangement shown in Fig. 1, the resonance cell and cavity are placed inside a solenoid creating a small magnetic field required to provide a quantization axis, and inside a magnetic shield preventing fluctuations in the environmental magnetic field from affecting the atoms resonant frequency.

For implementing a rubidium atomic frequency standard (RAFS), the microwave generator used to feed the cavity and excite the hyperfine transition of 87Rb is locked by means of a frequency control loop to the frequency of the resonance line as observed at the photodetector. In that approach, the microwave generator frequency is modulated at a low frequency of the order of one hundred Hz. The signal at the photodetector or analog error signal is digitized, synchronously detected, further processed digitally, then converted back to an analog voltage in order to steer the voltage controlled quartz oscillator.

This signal can be used to lock in frequency the microwave generator to the center of the resonance line. In practice the microwave generator consists of a quartz crystal oscillator synthesized numerically to the desired frequency.

3 New Design of RAFS Frequency Synthesizer

Frequency synthesizer one of the parts of frequency synthesis circuitry. This circuitry generates the microwave signal at ~ 6.8 GHz (used to interrogate the Rb hyperfine resonance transition) from the 5 MHz quartz oscillator frequency.

The main characteristic of the frequency synthesizer is an ability to impact the characteristic of frequency stability of the output signal of quantum frequency standard. Frequency instability introduced by the synthesizer is determined by the lateral discrete spectrum components of the signal that occurs in dividing, multiplying, mixing frequency signals, the accuracy of the generated frequency, and the impact on the signal of natural and technical noise.

In order to provide the best possible RAFS frequency stability, it is crucial that the microwave signal which interrogates the Rb atoms be as "clean" as possible; that is, free of unwanted sidebands and spurious signals which can cause Bloch-Siegert frequency shifts. In addition, it is very important that the microwave signal amplitude fluctuations and phase noise are small enough that the RAFS frequency stability is not impaired.

Experimental study showed that the present method of generating the frequency synthesizer output signal needs to increase the accuracy. The large resolution of step frequency is necessary. New scheme of the frequency synthesizer is designed by using method of direct digital synthesis (DDS - Direct Digital Synthesis). This method allows to generate the output signal of the synthesizer with accuracy about 10–5 Hz. Step frequency tuning ΔF_{out} calculated by the formula below:

$$\Delta F_{out} = \frac{F_{clk}}{2^N}, \tag{1}$$

where F_{clk} is the clock frequency, N is the capacity of accumulator.

In our scheme the clock frequency is equal $F_{clk} = 25$ MHz, the capacity of accumulator is equal $N = 40$. Step frequency tuning is equal $\Delta F_{out} = 2, 27 \cdot 10^{-5}$ Hz.

When we calculate step frequency tuning relatively resonant frequency of the microwave transitions of rubidium atoms, using formula (1), we get relative step frequency tuning $\Delta F_{out\,rel}$:

$$\Delta F_{out\,rel} = \frac{\Delta F_{out}}{6.834\,GHz} = \frac{2.27^{-5}\,Hz}{6.834\,GHz} = 3.32 * 10^{-15} \tag{2}$$

The application of direct digital synthesis gave the possibility of obtaining the generated frequencies in a wide range (0–5 MHz), in contrast to previous schemes, where this feature was absent. Range of generated output frequencies may be calculated by the formula below:

$$F_{out} = \frac{M*F_{clk}}{2^N}, \tag{3}$$

where M is the frequency code in decimal, F_{clk} is the clock frequency, N is the capacity of accumulator [1, 5, 6].

The implementation of our proposed method enables us to control the frequency of the output signal frequency synthesizer in real time. Time hopping t_h is the sum of the time adjustment the digital part of the system and the delay time of the low-pass filter (LPF). Time adjustment digital part in synthesizer is three clock cycles. The delay time of the LPF is inversely proportional to the low-pass filter bandwidth. If the cut-off frequency is equal $F_{cut} = F_{clk}/4$ then time delay is about four clock periods. It turns out that time hopping is equal $t_h = 7/F_{clk}$. The clock frequency F_{clk} is equal 25 MHz, then the time hopping is equal $t_h = 0.28$ µs. This ensures a high rate of frequency tuning, which makes it possible to more efficiently adjust the crystal oscillator frequency in contrast the previously used scheme.

To meet the requirements for spectral purity of output signal 10 bit DAC was used. It is possible to obtain the suppression of lateral amplitude components in the spectrum of the output signal is not worse than −60 dB.

In Fig. 2 a new design of the frequency synthesizer is presented.

Fig. 2. A new design of the frequency synthesizer

4 Experimental Results of New Frequency Synthesizer Design

Experimental research of the output signal parameters of a new scheme frequency synthesizer confirmed the simulation results and showed great advantages in comparison with previously used scheme.

In Figs. 3 and 4 as an example, oscillograms measured in the band of 6 kHz and 600 kHz of the output signal of a new design (a) and a previously used (b) of the frequency synthesizer are presented.

The experimental results show that the suppression of lateral components in the spectrum of microwave-excitation signal in the band of 6 kHz is improved on 29 dB and in the band of 600 kHz is improved about two times. Now value of suppression of lateral components equals 97 dB and 77 dB in the band of 6 kHz and 600 kHz consequently.

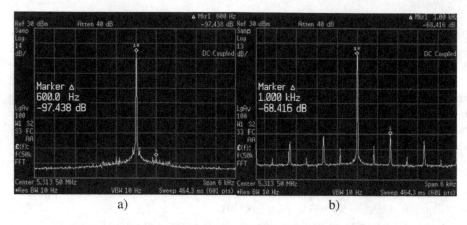

Fig. 3. Suppression of the lateral components in the band of 6 kHz.

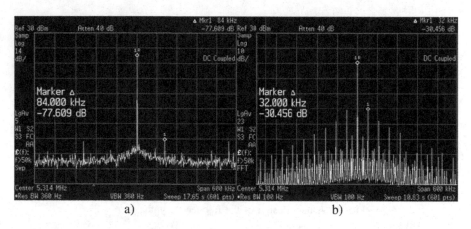

Fig. 4. Suppression of the lateral components in the band of 600 kHz.

With a decrease of lateral components more fine-tuning on the center of the resonance line is occur. This leads to a more accurate determination of the value of the nominal output frequency of frequency standard and, consequently, improves the long-term frequency stability of a rubidium atomic clock.

5 Experimental Research of Quantum Frequency Standard

Experimental research of quantum frequency standard consisted of the measure output frequency and calculation of the Allan variance, which allows to evaluate the frequency instability.

The formula for calculation the Allan variance is as follows:

$$\sigma_y^2 = \frac{\sum_i^n \sigma_{0i}^2}{2(n-1)}, \tag{4}$$

where $\sigma_{0i} = \frac{f_{i+1} - f_i}{f}$, i-th relative frequency variation, n- number of variations.

In Fig. 5 the Allan variance for previous design (1) and new design of frequency synthesizer is presented.

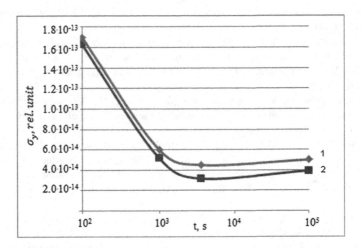

Fig. 5. The Allan variance for previous design (1) and new design of frequency synthesizer.

From these results it is clear that the use of new design of the frequency synthesizer makes it possible to obtain a better frequency stability of quantum frequency standard. It is achieved through an improved spectrum of the output signal of the frequency synthesizer.

6 Conclusion

Experimental research of frequency synthesizer showed improved parameters of the microwave-excitation signal, such as the step of frequency tuning, time of the frequency tuning, range of generated frequencies and spectral characteristics. Experimental research of the metrological characteristics of the quantum frequency standard on the atoms of rubidium with new design of frequency synthesizer showed improvement in long-term frequency stability on about 10%.

References

1. Riechle, F.: Frequency Standards: Basics and Applications. Wiley-VCH, Weinheim (2004). 526 p.
2. Semenov, V.V., Nikiforov, N.F., Ermak, S.V., Davydov, V.V.: Calculation of stationary magnetic resonance signal in optically oriented atoms induced by a sequence of radio pulses. Radiotekhnika I Elektronika (J. Commun. Technol. Electron.) **35**(10), 2179–2183 (1990)
3. Nazarov, L.E., Golovkin, I.V.: Symbol reception of signals corresponding to high-speed super-accurate codes and turbo codes based on them. J. Commun. Technol. Electron. **52**(10), 1125–1129 (2007)
4. Pavel'ev, A.G., Matyugov, S.S., Yakovlev, A.I.: J. Commun. Technol. Electron. **53**(9), 1021–1028 (2008)
5. Petrov, A.A., Davydov, V.V.: Improvement frequency stability of caesium atomic clock for satellite communication system. In: Balandin, S., Andreev, S., Koucheryavy, Y. (eds.) ruSMART 2015. LNCS, vol. 9247, pp. 739–744. Springer, Cham (2015). doi:10.1007/978-3-319-23126-6_68
6. Petrov, A.A., Vologdin, V.A., Davydov, V.V., Zalyotov, D.V.: Dependence of microwave – excitation signal parameters on frequency stability caesium atomic clock. J. Phys.: Conf. Ser. **643**(1), 012087 (2015)
7. Karaulanov, T.S., Graf, M.T., English, D.P., Rochester, S.M., Rosen, Y.J., Tsigutkin, K.F., Budker, D.T., Alexandrov, E.B., Balabas, M.V., Kimball, D.F.J., Narducci, F.A., Pustelny, S.E., Yashchuk, V.V.: Controlling atomic vapor density in paraffin-coated cells using induced atomic desorption. Phys. Rev. A – At. Mol. Opt. Phys. **79**(1), 012902–012914 (2009)
8. Petrov, A.A., Davydov, V.V.: Digital frequency synthesizer for ^{133}Cs-vapor atomic clock. J. Commun. Technol. Electron. **62**(3), 289–293 (2017)
9. Pakhomov, A.A.: Fast digital image processing of artificial earth satellites. J. Commun. Technol. Electron. **52**(10), 1114–1118 (2007)
10. Efimov, A.I., Łukanina, L.A., Samoznaev, L.N., Chashei, I.V., Bird, M.K.: Intensity of fluctuations in the frequency of radio signals of spacecraft in the near-solar plasma. J. Commun. Technol. Electron. **55**(11), 1253–1262 (2010)

Provision of Connectivity for (Heterogeneous) Self-organizing Network Using UAVs

Alexander Paramonov[1,2]($\boxtimes$), Ilhom Nurilloev[1],
and Andrey Koucheryavy[1]

[1] The Bonch-Bruevich State University of Telecommunications,
22 Pr. Bolshevikov, St. Petersburg, Russian Federation
alex-in-spb@yandex.ru
[2] The Peoples' Friendship University of Russia (RUDN University),
6 Miklukho-Maklaya st., Moscow 117198, Russian Federation

Abstract. This paper presents results of wireless sensor networks connectivity analysis. Authors propose the algorithm for WSN connectivity recovering by using of additional nodes which may be placed on the UAVs. Proposed algorithm based on the graph theory and data of network nodes positions. This algorithm allows selection of a minimal number of additional nodes and their positions.

Keywords: Wireless sensor network · Connectivity · Connectivity providing method · UAV network graph model

1 Introduction

This paper presents results of analysis of providing and recovering of connectivity in wireless self-organizing network, using unmanned aerial vehicles (UAV). The paper suggests the use of UAV positioning algorithm to organize connectivity between unconnected segments of network. The proposed algorithm is based on information of terrestrial nodes coordinates.

Using wireless communication technology allows to organize cooperation elements of automated systems, which has a certain freedom of movement in space. The development of UAV designing technology and robotic devices prompted the need of developing used methods in such systems, for providing elements cooperation [1, 2]. For example, to resolve a given row of tasks is recommended to use some robots or UAVs. To resolve tasks such of centralized control and interaction between devices maybe used a self-organizing network, that consists of transceivers, placed on these devices [7–9]. The structure of these networks depends on relative position of devices and it can change in time when relative position of the system's elements has changed. The change in network structure provoke the necessity of data service exchange between devices and route searching for data delivery. The variability or stability of network structure [3, 4] primary depend on object's motion character and at last depend on system's purpose and solved it's problem [5]. On designing homogeneous system, when all system's objects have equal possibility and movement character, parameters of network will be chosen based on these characteristics. When choosing the parameters,

© Springer International Publishing AG 2017
O. Galinina et al. (Eds.): NEW2AN/ruSMART/NsCC 2017, LNCS 10531, pp. 569–576, 2017.
DOI: 10.1007/978-3-319-67380-6_53

usually aimed to obtain the most stable network structure i.e. minimize network reorganization work [5]. In the most of practical applications, this purpose is achievable when there is a presence of certain limit on relative movement of system's elements. In UAV and terrestrial networks relative movement of system's elements can be limited by defining certain rules of their motion. These rules can be referred to all elements of the system. However, when interacting two or more systems i.e. in heterogeneous system, for example when UVA interacts with terrestrial system, there are problems in building an integrated network, because differences of characteristic of their movements. To ensure such interaction, it's necessary to provide enough stability of heterogeneous network's structure. The solution to this problem can be achieved by choosing rules for relative movement of elements of different networks.

2   Related Works

Due to the high importance of connectivity of WSN, a lot of work is devoted to this issue. We have considered some of them [1–7]. Some of the works propose using of specific nodes with long communication range [11].

3   Problem Statement

Interaction of UAV groups with terrestrial network segment can result in more different task of. In this article we focus only on the task of network connectivity of terrestrial network's segment. The nodes of terrestrial segment can be deployed in different ways. But due to various reasons, terrestrial segment may not provide full connectivity i.e. availability of each node to each nodes. Violation of network connectivity can occur because of a number of other reasons related to the consumption of energy reserves, technical failures of nodes, overload of communication channels, radio interferences, etc. As a result, the terrestrial segment of the network represents a set of disjointed subnets. The use of UAV allows to quickly provide connectivity between these subnets. Let us consider two cases:

The first case when nodes in UAV work autonomously i.e. they can perform traffic transit function between nodes of terrestrial segments. The second case, when the nodes in UAV can perform traffic transit via the network, that formed by UAV in a maintenance area i.e. UAV is a connected network. The two cases are shown in Fig. 1a and b respectively.

Traffic transit between disjointed terrestrial subnets allows to provide connectivity of terrestrial network. In the first case UAV works as a mobile transit node which can be quickly deployed at a certain point in the maintenance area. In the second case traffic transit can be performed via UAV network. In both cases, in order to provide transit traffic, it is necessary to select UAV placement points to provide a balance between the number of UAVs used and the quality of traffic servicing.

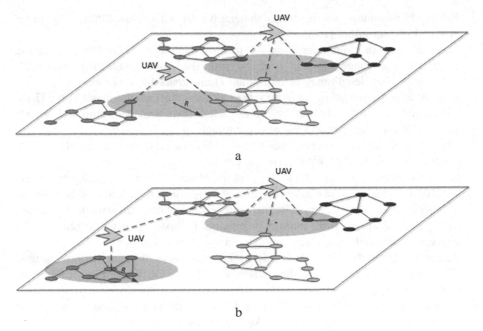

Fig. 1. Network connectivity provision using UAV.

4   Connectivity Provision of Terrestrial Segment

We are conceder connectivity provision of terrestrial segment as traffic transit between disjointed segments of terrestrial network with nodes placed on UAV. We assume that communication range of UAV node with terrestrial segment's nodes represents a circle of radius R.

This task is solved if node's coordinates of network's terrestrial segment is defined. We will describe the network as model of a fully connected undirected graph, where each vertex corresponds to each node of the considered network. We assume that the coordinates of terrestrial segment's nodes are defined, based on them can be calculated the matrix of distances between nodes of terrestrial segment $D = \{d_{ij}\}$, $i, j = 1 \ldots n$. In this matrix for flat model (2D) each of the elements is equal

$$d_{ij} = \sqrt{\left(x_i - x_j\right)^2 + \left(y_i - y_j\right)^2} \tag{1}$$

where x_i, y_i – coordinates of network's nodes, $i = 1 \ldots n$, n – number of nodes.

Matrix D describes lengths of graph edges. We assume that communication range of each node represent a circle of radius r.

To resolve the task of finding UAV deployment point is suggested the following algorithm.

1. Description of the network as model of undirected graph by the matrix of distances between vertexes D.

2. Finding of minimum spanning tree of the graph [10]. (Can be used Prim's algorithm or Kruskal's algorithm [10]).

3. Deletion from the minimum spanning tree of the graph edges longer than the communication range of node r, apart from the edges connecting node with UAV-node (their length may be more than r) and mark these vertices as v_1.

4. If the number of deleted edges is zero, then all the points of placement of the UAV are identified and the connectivity of the network is provided, **stop the search process**. If not, then perform the next paragraph (p. 5).

5. Clustering of marked vertices (v_1) using the FOREL algorithm when the cluster size equal to R, and the determination of the cluster centers.

6. Check the number of marked nodes (v_1) in the clusters that were found, mark these vertices as v_2 and remove clusters containing only one vertex. If after removing the number of clusters containing more than one vertex is more than zero, then go to the next paragraph (p. 7). If the number of remaining clusters is equal to zero (all the clusters contain only one vertex are marked v_2), then go to p. 8.

7. Add to the graph the vertices coincident with the centers of the remaining clusters. These vertices correspond to the locations of the UAV, performing the role of transit nodes, go to p. 2.

8. Choose the first vertex marked v_2, in the matrix of distances to find the nearest to her vertex with the same marker v_2. Add to the graph m vertices

$$m = \left\lceil \frac{L}{R} \right\rceil - 1 \qquad (2)$$

at a distance from the first vertex

$$d_i = i \frac{L}{\left\lceil \frac{L}{R} \right\rceil}, \quad i = 1 \ldots n \qquad (3)$$

where the sign $\lceil x \rceil$ means the ceiling of x, i.e. the smallest integer greater than or equal to x, and i is the number of added vertices. Unmark v_2 from considered vertices. The coordinates of the added vertices correspond to the coordinates of the UAV, that ensure the connectivity of the considered vertices.

Perform the same operation for all vertices marked v_2.

Go to p. 2.

Figure 2 shows a possible deployment plan of the UAV to ensure the connectivity of the terrestrial segment of the network.

If the UAV network is connected throughout the service area and can produce traffic transit of the nodes in terrestrial segment, this algorithm can be simplified by the deletion of paragraphs 6 and 8.

The result of the above algorithm is the set of points (coordinates) and location in which the nodes that are part of UAV, when they are sufficient quantity, provides full network connectivity.

The idea of proposed algorithm is that the network should be described by full-connected undirected graph model. First, necessary to find components of the graph (disjoint subnets). For this aim it is proposed to use finding the minimum spanning tree algorithm. Further, from found tree excludes edges whose length exceeds the radius of a node in terrestrial segment [6]. As a result, if presence such edges,

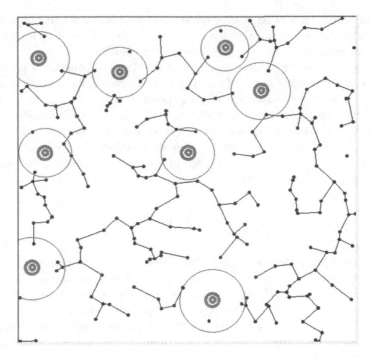

Fig. 2. Connectivity restoration of network's terrestrial segment using UAVs.

appear some components of the graph. The nodes that connected by excluded edges taken for the joint of the graph. Necessary, to provide the connectivity of these nodes with least of UAVs. Connection between joints, depending on their location, may be provided by one or more UAVs. It is desirable to minimize the number of UAVs. For the choice of their placement points, the position which provides communication between multiple joints, uses a clustering algorithm FOREL. Using this algorithm to try to find the centers of possible clusters (groups, two or more), connectivity of which can be provided by one UAV. If there are clusters containing only one vertex, suppose distance between them exceeds the radius of UAV's communication range. In this case, necessary to add enough number of vertices, that can provide connectivity of these vertices. After determining these points and the cluster centers and adding appropriate nodes to the graph, generated modification of the distance matrix D. This algorithm is repeated until achieving the state when after finding next minimum spanning tree there is not one edge that is greater than the radius of node's communication range, besides edges that connected with UAV.

5 Solution to the Problem with Limited Resources

When a number of UAVs limited, the probability of connectivity can be less than one, i.e., may remain disconnected components on the graph. In this case, it is advisable to introduce quality estimation of problem solution, herewith quality should be understood as degree of achievement of the desired result. It is obvious that at equal importance of all nodes in the network, this estimation can be probability of connectivity, i.e. probability of availability of all nodes to each other in the network.

As one of the possible options, consider the problem of connectivity providing in the network at various importance of nodes. Ascribe to each node of network a weight coefficient that illustrate the degree of its importance k_i, $i = 1 \dots n$. Then when choosing the placement for additional nodes (UAVs) to provide connectivity of graph's component should take into account the degree of node's importance of each component. For this, for each component of graph we introduce the coefficient of importance $.s$, where $s = 1 \dots C$, C - number of graph's components, obtained after compliance p. 3 of above algorithm.

$$\eta_s = \sum_{i=1}^{g} k_i \qquad (4)$$

where g – number of nodes, belonging to the i-th component of graph.

Further, we ascribe this coefficient to each joints of this component.

In this case, p. 5 of mentioned algorithm should be performed taking into account the coefficients of importance. When clustering algorithm is performing, these coefficients must be taken into account when calculating the mass centers of clusters. This makes it possible, primarily to provide connectivity for components, which are great importance for the network.

Implementation and use of the given algorithm allows to provide or restore the connectivity of terrestrial self-organizing network that contains quite large number of nodes.

6 Conclusions

1. Connectivity of wireless self-organizing network can be provided by introduction of additional nodes in the network. Rapid solution of this problem can be obtained by using the UAV as a carrier or delivery vehicle of these nodes. In some cases, using of UAVs allows to increase the communication range of node, thereby increasing the efficiency of its use.
2. Suggested determining method of coordinates for location UAVs (nodes), which provided connectivity between unconnected network segments based on usage information of network structure and geographical location of its nodes. In application of UAV this information can be obtained through research of network coverage by one or more UAVs.
3. The proposed algorithm allows to obtain one possible solution (near optimal solution), allowing to provide connectivity of disconnected network segments. Inasmuch used in this method clustering algorithm gives one possible solution, depending on the initial conditions, proposed method also allows, at certain conditions, to obtain a few different solutions.
4. Modification of proposed algorithm, taking into account the degree of node's significance in the network can be obtained solution for providing connectivity at

limited amount of resources (number of UAVs). In this case connectivity foremost must be provided between the most significant segments of network.

Acknowledgement. The publication was financially supported by the Ministry of Education and Science of the Russian Federation (the Agreement number 02.a03.21.0008).

References

1. Nurilloev, I., Paramonov, A., Koucheryavy, A.: Connectivity estimation in wireless sensor networks. In: Galinina, O., Balandin, S., Koucheryavy, Y. (eds.) NEW2AN/ruSMART - 2016. LNCS, vol. 9870, pp. 269–277. Springer, Cham (2016). doi:10.1007/978-3-319-46301-8_22
2. Muthanna, A., Masek, P., Hosek, J., Fujdiak, R., Hussein, O., Paramonov, A., Koucheryavy, A.: Analytical evaluation of D2D connectivity potential in 5G wireless systems. In: Galinina, O., Balandin, S., Koucheryavy, Y. (eds.) NEW2AN/ruSMART-2016. LNCS, vol. 9870, pp. 395–403. Springer, Cham (2016). doi:10.1007/978-3-319-46301-8_33
3. Dao, N., Koucheryavy, A., Paramonov, A.: Analysis of routes in the network based on a swarm of UAVs. In: Kim, K., Joukov, N. (eds.) Information Science and Applications (ICISA) 2016. LNEE, vol. 376, pp. 1261–1271. Springer, Singapore (2016). doi:10.1007/978-981-10-0557-2_119
4. Kirichek, R., Paramonov, A., Vareldzhyan, K.: Optimization of the UAV-P's motion trajectory in public flying ubiquitous sensor networks (FUSN-P). In: Balandin, S., Andreev, S., Koucheryavy, Y. (eds.) ruSMART 2015. LNCS, vol. 9247, pp. 352–366. Springer, Cham (2015). doi:10.1007/978-3-319-23126-6_32
5. Futahp, A., Koucheryavy, A., Paramonov, A., Prokopiev, A.: Ubiquitous sensor networks in the heterogeneous LTE network. In: 17th International Conference on Advanced Communications Technology (ICACT), pp. 28–32 (2015)
6. Shilin, P., Kirichek, R., Paramonov, A., Koucheryavy, A.: Connectivity of VANET segments using UAVs. In: Galinina, O., Balandin, S., Koucheryavy, Y. (eds.) NEW2AN/ruSMART-2016. LNCS, vol. 9870, pp. 492–500. Springer, Cham (2016). doi:10.1007/978-3-319-46301-8_41
7. Kirichek, R., Paramonov, A., Koucheryavy, A.: Swarm of public unmanned aerial vehicles as a queuing network. In: Vishnevsky, V., Kozyrev, D. (eds.) DCCN 2015. CCIS, vol. 601, pp. 111–120. Springer, Cham (2016). doi:10.1007/978-3-319-30843-2_12
8. Kirichek, R., Paramonov, A., Koucheryavy, A.: Flying ubiquitous sensor networks as a queueing system. In: 17th International Conference on Advanced Communications Technology (ICACT), pp. 127–132 (2015)
9. Muthanna, A., Paramonov, A., Koucheryavy, A., Prokopiev, A.: Comparison of protocols for ubiquitous wireless sensor network. In: International Congress on Ultra Modern Telecommunications and Control Systems and Workshops 6. Cep. 2014 6th International Congress on Ultra Modern Telecommunications and Control Systems and Workshops, ICUMT 2014, pp. 334–337 (2015)
10. Christofides, N.: Graph Theory. An Algorithmic Approach. Academic Press, Inc., Orlando (1975)
11. Yang, C., Shi, Y., Wang, Z., Lan, H.: Connectivity of wireless sensor networks for plant growth in greenhouse. **9**(1). Open Access http://www.ijabe.org

12. Akbari, A., Beikmahdavi, N., Khosrozadeh, A., Panah, O., Yadollahi, M., Jalali, S.V.: A survey cluster-based and cellular approach to fault detection and recovery in wireless sensor networks. World Appl. Sci. J. **8**(1), 76–85 (2010)
13. Akbari, A., Dana, A., Khademzadeh, A., Beikmahdavi, N.: Fault detection and recovery in wireless sensor network using clustering. Int. J. Wireless Mob. Netw. (IJWMN) **3**(1), 130–138 (2011)
14. Ghosh, A., Das, S.K.: Coverage and connectivity issues in wireless sensor networks. Pervasive Mob. Comput. **4**(3), 303–334 (2008)
15. Li, J., Andrew, L.L., Foh, C.H., Zukerman, M., Chen, H.-H.: Connectivity, coverage and placement in wireless sensor networks. Sensors **9**, 7664–7693 (2009)
16. Khoufi, I., Minet, P., Laouiti, A., Mahfoudh, S.: Survey of deployment algorithms in wireless sensor networks: coverage and connectivity issues and challenges. HAL Id: hal-01095749, submitted on 16 December 2014
17. Yu, J., Chen, Y., Ma, L., Huang, B., Cheng, X.: On connected target k-coverage in heterogeneous wireless sensor networks. Sensors **16**, 104 (2016)
18. Wang, Y.-C., Hu, C.-C., Tseng, Y.-C.: Efficient deployment algorithms for ensuring coverage and connectivity of wireless sensor networks. In: Proceedings of the First International Conference on Date of Conference on Wireless Internet, 10–15 July 2005
19. Sharma, L., Agnihotri, S.: Connectivity and coverage preserving schemes for surveillance applications in WSN. IJCSNS Int. J. Comput. Sci. Netw. Secur. **13**(6), 87 (2013)
20. Marks, M.: A survey of multi-objective deployment in wireless sensor networks. J. Telecommun. Inf. Technol. (2010)

Performance Evaluation of COPE-like Network Coding in Flying Ad Hoc Networks: Simulation-Based Study

Danil S. Vasiliev[✉], Irina A. Kaysina, and Albert Abilov

Department of Communication Networks and Systems,
Izhevsk State Technical University, ul. Studencheskaya, 7, 426069 Izhevsk, Russia
{danil.s.vasilyev,irina.kaysina,albert.abilov}@istu.ru
http://www.istu.ru/

Abstract. This paper investigates the performance of network coding in Flying Ad Hoc Networks (FANETs) with help of NS-3 simulation tool. We compared the efficiency of COPE-like network coding for "copter", "fixed-wing drone", and "swarm" scenarios by quality of service (QoS) metrics. We use 802.11g standard at the data-link layer and AODV routing protocol at the network layer of OSI model. Average QoS metrics were estimated in 50 simulation runs for each parameter set. Results show that using of network coding in "copter" and "fixed-wing" scenarios improves PDR and goodput metrics. In "swarm" scenario the efficiency of network coding degrades for swarms with low density of nodes.

Keywords: Mobile Ad Hoc networks · Unmanned Aerial Vehicles · Quality of service · Network coding · Routing protocols · Computer simulation

1 Introduction

Today, small Unmanned Aerial Vehicles (UAVs or drones) with on-board video cameras gain popularity: many of them are used in surveillance missions and for building monitoring. They depend on Wi-Fi standard for wireless transmission and could send data to ground stations and other drones during the flight. Networks of drones (Flying Ad hoc Networks, FANETs) consist of mobile nodes and wireless routes between them [1–3].

High throughput is essential to transmit large amount of video content through unsteady radio channels. A group of small drones could be used as a one swarm with help of special routing protocols (e.g. AODV, OLSR [12,13]). With evolving from point-to-point connections into mesh networks, it is imperative that the quality of service in wireless networks of flying robots be enhanced using new approaches. In FANET each node is not only a destination or source of information, but also a relay. Repeated messages create additional load in a network, and nodes need higher bandwidth to deliver messages to receivers flawless.

© Springer International Publishing AG 2017
O. Galinina et al. (Eds.): NEW2AN/ruSMART/NsCC 2017, LNCS 10531, pp. 577–586, 2017.
DOI: 10.1007/978-3-319-67380-6_54

For instance, one of possible solutions to improve the quality of video transmission in FANETs is network coding. Saving in bandwidth from this technique could be achieved using spatial redundancy of ad hoc networks [4,5]. Wireless network coding is a relatively new method for optimizing the transmission of data over a network. The method is based on the broadcast nature of wireless communication.

The paper is devoted to simulation-based performance evaluation of network coding in FANETs. We try to understand how mobility could influence on QoS (Quality of Service) parameters in an ad hoc network with enabled network coding. We analyse mobility scenarios to assess gain from practical network coding in highly mobile environment.

The remainder of this paper is organized as follows: Sect. 2, overview of network coding in FANET; Sect. 3, simulation scenarios; Sect. 4, QoS metrics; Sect. 5, results; Sect. 6, conclusion.

2   Network Coding in FANET

We use inter-session network coding implementation (COPE) to improve QoS metrics in FANETs. Each node in FANET is a router, which broadcasts data to neighbours and stores incoming packets for a certain period of time. The process of COPE-like network coding includes 3 stages: listening, encoding, and decoding.

In listening stage nodes store the received packets in their buffers. Each node reports its neighbours about stored packets using special header fields. Thus, each node in a network has information about packets stored in buffers of its neighbours.

In COPE-like network coding data flows are encoded in opportunistic manner. A node encodes two or more packets into one using simple XOR cipher (exclusive disjunction operation).

Each node has a buffer with received and overheard packets. When a node receives encoded packet, it compares it with packets in its buffer. After that, a node decodes the packet using control information from the header and gets the payload if possible.

Network coding can improve quality of service for high speed data streams in ad hoc networks [6–8]. We use simulation model of COPE in NS-3 created by Chi [10,11]. Also, the implementation of COPE scheme for Wi-Fi exists – CATWOMAN protocol [9]. This implementation could be used for further experimental research in simple FANET topologies.

3   Simulation Scenarios

Simulated FANET consists of two laptops A and B and a variable number of flying drones. AODV routing protocol guides packets inside the network. Laptops are located in points $(0, d/2, 0)$ and $(d, d/2, 0)$, correspondingly. All simulated nodes use 802.11g on 2.4 GHz and 12 Mbps bandwidth. Each node transmits signals with maximal range of 1200 m limited by the Friis Signal Propagation Model. During the simulation, nodes A and B transmit 1250 bytes UDP datagrams with data rates from 1 to 4 Mbps to each other. Intermediate nodes relay data using AODV routing protocol. We consider "copter", "fixed-wing drone", and "swarm" mobility scenarios.

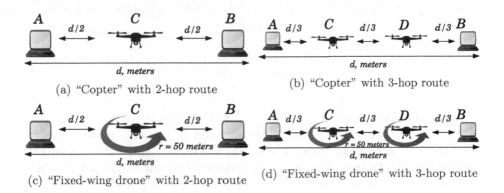

(a) "Copter" with 2-hop route

(b) "Copter" with 3-hop route

(c) "Fixed-wing drone" with 2-hop route

(d) "Fixed-wing drone" with 3-hop route

Fig. 1. Simulation scenarios "Copter" and "Fixed-wing drone"

In "copter" scenario A, B, and C nodes are stationary (Fig. 1a). The distance d between A and B is changing from 1300 to 2300 m. Node C is located between nodes A and B in the point with coordinates $(d/2, d/2,0)$.

The node A sends 5000 packets to the node B, and the node B sends 5000 packets to the node A. Each packet is 1250 bytes long. Packets are encoded at the node C. All packets use 2-hop route between nodes.

We also consider another case of this scenario with additional node D (Fig. 1b). This node makes 3-hop route between nodes A and B. The node D is located between nodes C and B. The distance between $A - C$ is equal to the distances $C - D$ and $D - B$. Coordinates of nodes C and D are $(d/3, d/2,0)$ and $(2*d/3, d/2,0)$ respectively. In this scenario we simulated different values of the distance d: from 2000 to 3200 m. The node D relays packets between nodes C and B. Data could be transmitted for higher distance using additional hop in the AODV route. But longer routes lead to low results for Quality of Service metrics. This is typical Alice-Bob scenario, that must show effectiveness of network coding approach.

In "fixed-wing drone" scenario intermediate nodes between nodes A and B rotate with radius $r = 50$ m (Fig. 1c). These nodes simulate fixed-wing drones. We also consider two cases of this scenario: with 2-hop route and 3-hop route between nodes A and B. In first case the node C moves in circle with the centre $(d/2, d/2,0)$. The distance d between nodes A and B is 2000 m in this case. In second case nodes C and D circle around the points $(d/3, d/2,0)$ and $(2*d/3, d/2,0)$ respectively (Fig. 1d). We also choose higher distance $d - 3000$ m. This scenario shows the influence of node mobility on QoS metrics and network coding efficiency.

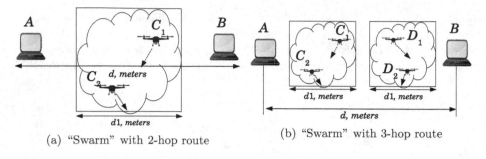

(a) "Swarm" with 2-hop route

(b) "Swarm" with 3-hop route

Fig. 2. Simulation scenario "Swarm"

In "swarm" scenario we simulate another type of node mobility (Fig. 2a). We use the swarm of randomly moving nodes to challenge network coding efficiency in 2-hop and 3-hop transmissions. The Gauss-Markov Mobility Model is used to simulate the UAV behavior in a swarm. In this scenario each intermediate node is choosing from swarms of 5 and 10 nodes. Swarm nodes are moving in imaginary box with the side $d1$. We also consider two different cases for 2-hop and 3-hop routes. For 2-hop route we consider $d1 = \{200, 1000, 2000\}$. The distance d is 2000 m. Coordinates x and y of each intermediate node could change from 900 to 1100 m for $d1 = 200$, $[500, 1500]$ for $d1 = 1000$, and $[0, 2000]$ for $d1 = 2000$. For 3-hop route we simulate two similar boxes with randomly moving nodes (Fig. 2b). Each box has the side $d1 = \{200, 400, 600\}$. The distance between nodes A and B is 3000 m. The center of first box is located in the point $(d/3, d/2, 0)$, and the center of the second box – $(2*d/3, d/2, 0)$. For example, if $d1 = 200$, coordinates x and y are in interval $[900, 1100]$ for nodes in the first box and $[1900, 2100]$ for nodes in the second box. We consider 1, 5, and 10 nodes in each box. That's why we consider swarm sizes $n = \{1, 5, 10\}$ for 2-hop case and $n = \{2, 10, 20\}$ for 3-hop case. Random locations of nodes were determined using different random seeds in NS-3. The velocity of each flying node is 50 m/s. These nodes are bound by the box and reflect from its borders without any speed reduction. Gauss-Markov Mobility Model is the model with memory: it defines coordinates of nodes using values of its previous coordinates [14].

In all scenarios nodes A and B run applications (OnOffApplication) that stream 5000 packets to each other. AODV routing protocol constructs routes between them using intermediate nodes. Node A starts the transmission at 1 s of the simulation, and node B – at 2 s. The simulation is 200 s long. The transmission finishes in around 50 s for 1 Mbps data rate. Network coding uses ARQ mechanism and could generate retransmission when main stream is finished. We also set $slrc = 1$ to turn off data link layer ARQ mechanism for 802.11g. We analyse the efficiency of network coding in all mobility scenarios using Quality of Service metrics.

4   Quality of Service Metrics

We evaluated the efficiency of network coding in FANETs using quality of service (QoS) metrics: packet delivery ratio and a goodput.

Average packet delivery ratio (PDR_{ave}) is the ratio between the number of received packets to the number of transmitted packets. This metric is measured based on sequence numbers generated by OnOffApplication objects in NS-3. PDR_{ave} is calculated by the formula:

$$PDR_{ave} = \frac{Rx}{Tx},\tag{1}$$

where Rx – a number of received packets, Tx – a number of transmitted packets.

A number of received packets is equal to a number of packets in the receiving buffer of OnOffApplication object at the end of the simulation run. Thus, we can calculate PDR_{ave} as follows:

$$PDR_{ave} = \frac{buf}{1250*5000},\tag{2}$$

where buf – the size of receiving buffer in bytes, 1250 – packet size in bytes, 5000 – the number of transmitted packets during the simulation.

$Goodput_{ave}$ is a payload received by the destination during the simulation divided by the transmission time:

$$Goodput_{ave} = \frac{buf*8}{T2-T1},\tag{3}$$

where buf – the size of receiving buffer in bytes, $T1$ – the start time of transmission in seconds, $T2$ – the last packet arrival time in seconds.

We calculated QoS metrics for each case separately: without network coding and with it. We measured these metrics for data rates from 1 mbps to 4 mbps and swarm sizes from 1 to 20 nodes. Fifty simulations were performed, and the plots show the average values for both streams.

We calculated the PDR gain as $G_{PDR} = PDR_{withNC}/PDR_{withoutNC}$ and the goodput gain as $G_{Goodput} = Goodput_{withNC}/Goodput_{withoutNC}$.

5   Results

We calculated PDR_{ave} and $Goodput_{ave}$ metrics for "copter", "fixed-wing drone", and "swarm" scenarios.

In "copter" scenario the node C is stationary. It broadcasts encoded packets to A and B nodes. The node C uses AODV routing protocol to relay packets between its neighbours. In this scenario the best results for PDR_{ave} metric were for distances from 1300 to 1600 m. When d is more than 1600 m, simulations with network coding show higher PDR_{ave} for all points (Fig. 3a). For example, for d=2000 packet delivery ratio gain G_{PDR} will be 1.6. We turned off data link ARQ algorithm ($slrc$=1) to show this advantage of network coding. This gain

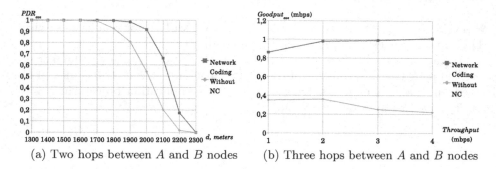

Fig. 3. PDR_{ave}-distance d(a) and goodput-throughput (b) dependences in "copter" scenario

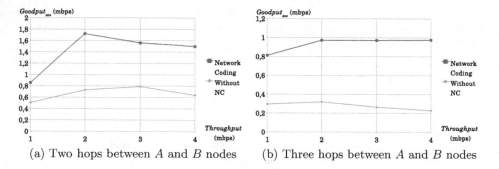

Fig. 4. $Goodput_{ave}$-throughput dependence in "fixed-wing drone" scenario

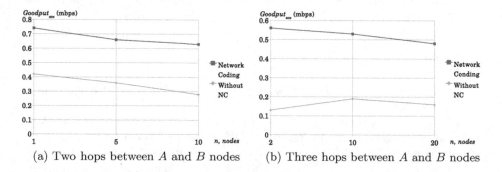

Fig. 5. $Goodput_{ave}$-nodes dependence for "swarm" scenario

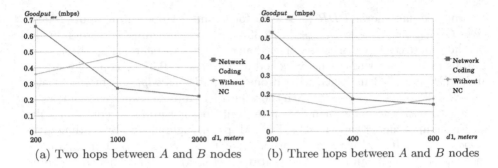

(a) Two hops between A and B nodes   (b) Three hops between A and B nodes

Fig. 6. $Goodput_{ave}$-distance $d1$ dependence for "swarm" scenario

G_{PDR} is achieved with help of ARQ mechanism that is enabled inside COPE-like network coding implementation. Collected statistics in log files showed that all packets transmitted by the node C were non-coded. Thus, even without encoding this algorithm demonstrates $G_{PDR} > 1.5$ using ARQ.

$Goodput_{ave}$ metric was measured in this scenario for $d = 2000$ m. The throughput was from 1 to 4 Mbps. Network coding demonstrated the maximal $Goodput_{ave} = 0.85$ Mbps with the throughput of 1 Mbps for A-B and B-A streams. Higher throughput caused higher losses and lower $Goodput_{ave}$ metric. For example, for the throughput of 4 Mbps only 0.45 Mbps reached nodes A and B. $Goodput_{ave}$ values for network coding were always higher due to higher PDR_{ave} results for all measured points. We can conclude that with increasing data transfer speed, the PDR_{ave} metric falls in both cases (with network coding and without). But due to network coding, the PDR_{ave} values are slightly better.

If we increase the number of hops between nodes A and B, it will cause negative impact on QoS metrics. We consider 3-hop route in "copter" scenario with $d = 3000$ and additional node D (Fig. 3b). Intermediate nodes C and D broadcast encoded packets. Log files show that more than 400 encoded packets were transmitted during each simulation run. We calculated $Goodput_{ave}$ metric to show the gain of network coding in this case. $Goodput_{ave}$ metric is enhanced by network coding for data rates (throughputs) from 1 to 4 Mbps. Maximal gain $G_{Goodput} = 4.6$ is measured for 4 Mbps throughput. Results demonstrate that the gain is higher for higher data rates. Network coding enhances both metrics: it uses retransmission technique to improve PDR_{ave} and XOR-encoding to improve $Goodput_{ave}$. But goodput is always lower than throughput for all points. $Goodput_{ave}$ degrades with each additional hop in FANET. That is why we got $Goodput_{ave} = 1$ Mbps for throughputs from 2 to 4 Mbps.

In "fixed-wing drone" scenario intermediate nodes are mobile. We calculate QoS metric for 2-hop and 3-hop routes and data rates from 1 to 4 Mbps. In both cases network coding shows higher PDR_{ave} and $Goodput_{ave}$ metrics. Average values of QoS metrics are slightly lower than in "copter" scenario.

When 2-hop route is used, maximal G_{PDR} is 2.56 for 4 Mbps throughput (Fig. 4a). $Goodput_{ave}$ is only 1.5 Mbps in this case due to low PDR_{ave} value.

Network coding shows $PDR_{ave} > 0.85$ for 1 and 2 Mbps throughput. It also gives $G_{Goodput} = 1.7$ and $G_{Goodput} = 2.35$ for these data rates, correspondingly. In 3-hop route case QoS metrics worser than in 'copter" scenario (Fig. 4b), e.g. for 1 Mbps throughput without network coding $PDR_{ave} = 0.35$ in 'copter" scenario and $PDR_{ave} = 0.3$ in "fixed-wing drone" scenario. $Goodput_{ave}$ metric is also lesser in all points: it does not reach the value of 1 Mbps. Network coding tries to cope with packet loss and high data rates, but does not grant acceptable results for throughput higher than 1 Mbps. For 1 Mbps throughput network coding gives $Goodput_{ave} = 0.81$ Mbps that is proportinal to $PDR_{ave} = 0.81$ in this case. In "copter" (Fig. 3) and "fixed-wing drone" (Fig. 4) scenarios network coding improves $Goodput_{ave}$ metric for throughput values from 1 to 4 Mbps and for 2-hop and 3-hop routes. Embedded ARQ mechanism grants higher PDR_{ave} and higher $Goodput_{ave}$. Even in mobile scenario nodes could transmit streams with higher data rate using this technique.

In "swarm" scenario we consider swarms of mobile nodes, nodes A and B stream data with throughput of 1 Mbps. In 2-hop case AODV protocol chooses one node from the swarm between nodes A and B (Fig. 5), in 3-hop case it chooses two nodes from two different swarms (Fig. 6). This scenario propose the most difficult challenge for network coding algorithm, because intemerdiate nodes C and D are not predefined. Relaying nodes are randomly chosen from swarms. All nodes in swarms are moving in the space limited by the side $d1$ of imaginary box. We consider $d1 = 200$ m and swarms of 1, 5, and 10 nodes. Results for PDR_{ave} and $Goodput_{ave}$ show that bigger swarms do not improve these metrics. This behaviour is caused by reactive nature of AODV protocol. Both metrics are higher for network coding, e.g. $G_{Goodput} = 1.76$ for $n = 1$ and $G_{Goodput} = 2.25$ for $n = 10$. We also consider 3-hop routes in "swarm" scenario. Each swarm is limited by $d1 = 200$ m and n values are 2, 10, and 20 nodes. Intermediate nodes are chosen from different swarms. In this case network coding demonstrated higher PDR_{ave} for all points. $Goodput_{ave}$ with network coding drops from 0.56 to 0.48 for number of nodes from 2 to 20. Values of both QoS metrics are very bad for all swarm sizes ($Goodput_{ave} < 0.2$) without network coding. Different values of distance $d1$ were considered. When we increase $d1$, values of PDR_{ave} and $Goodput_{ave}$ for network coding are lesser than without it. For example, in 2-hop case we have $Goodput_{ave} = 0.66$ Mbps for $d1 = 200$ m and $Goodput_{ave} = 0.27$ Mbps for $d1 = 1000$ m. Even without network coding AODV protocol could show $Goodput_{ave} = 0.47$ Mbps for $d1 = 1000$ m. The same is right in 3-hop case for $d1 = 600$ m: $Goodput_{ave} = 0.17$ Mbps without network coding and $Goodput_{ave} = 0.14$ Mbps with it. But in 3-hop case network coding algorithm generates encoded packets and demonsrate better results than pure AODV streaming for $d1 = 400$ m.

6   Conclusion

In the paper we investigated the efficiency of network coding technique in FANETs using NS-3. We calculated PDR_{ave} and $Goodput_{ave}$ metrics to estimate quality of service in simulated networks. In all scenarios intermediate nodes relay packets using AODV routing protocol between nodes A and B. Network coding showed better results for all measured points in "copter" and "fixed-wing drone" scenarios, e.g. $G_{Goodput} > 1.6$ for all cases. In "swarm" scenario the distance $d1$ grows, the density of nodes in swarms drops down. The gain of network coding for both metrics decreases for bigger distance $d1$ for 2-hop and 3-hop routes, because AODV routing protocol is not coding-aware.

FANETs propose many challenges: due to high mobility of nodes and unstable topologies such networks have low quality of service. Network coding is a new coding scheme that improves QoS metrics in ad hoc networks with stationary nodes. Results of our simulation-based study demonstrate that swarm of nodes does not strengthen network coding algorithm. To overcome node mobility new coding-aware algorithms are needed. These algorithms could be implemented as coding-aware routing protocols and metrics [10] or application layer software that will encode packets, collect neighbour packets in a buffer, and guide them between flying nodes using changing network topology as an advantage.

Acknowledgment. The reported study was funded by RFBR according to the research project No. 16-07-01201 a.

References

1. Bekmezci, I., Sahingoz, O.K., Temel, S.: Flying Ad-Hoc networks (FANETs): a survey. Ad Hoc Netw. **11**(3), 1254–1270 (2013)
2. Koucheryavy, A., Vladyko, A., Kirichek, R.: State of the art and research challenges for public flying ubiquitous sensor networks. In: Balandin, S., Andreev, S., Koucheryavy, Y. (eds.) ruSMART 2015. LNCS, vol. 9247, pp. 299–308. Springer, Cham (2015). doi:10.1007/978-3-319-23126-6_27
3. Kirichek, R., Vladyko, A., Paramonov, A., Koucheryavy, A.: Software-defined architecture for flying ubiquitous sensor networking. In: 2017 19th International Conference on Advanced Communication Technology (ICACT), Seoul, South Korea (2017)
4. Katti, S., Rahul, H., Hu, W., Katabi, D., Médard, M., Crowcroft, J.: XORs in the air: practical wireless network coding. IEEE/ACM Trans. Netw. (ToN) **16**(3), 497–510 (2006)
5. Chou, P., Wu, T.: Network coding for the internet and wireless networks. IEEE Sig. Process. Mag. **24**(5), 77–85 (2007)
6. Liu, Q., Wang, G., Feng, G., Guo, Y., Yao, Q.: Maximizing network coding opportunity through buffer management in wireless networks. In: 2016 IEEE International Conference on Internet of Things (iThings) and IEEE Green Computing and Communications (GreenCom) and IEEE Cyber, Physical and Social Computing (CPSCom) and IEEE Smart Data (SmartData), Chengdu, China (2016)
7. Singh, A., Nagaraju, A.: Network coding: ABC based COPE in wireless sensor and Mesh network. In: 2014 International Conference on Contemporary Computing and Informatics (IC3I), Mysore, India (2014)
8. Seferoglu, H., Markopoulou, A., Médard, M.: NCAPQ: network coding-aware priority queueing for UDP flows over COPE. In: 2011 International Symposium on Network Coding (NetCod), Beijing, China (2011)

9. Zhao, F., Médard, M., Hundebøll, M., Ledet-Pedersen, J., Rein, S., Fitzek, F.: Comparison of analytical and measured performance results on network coding in IEEE 802.11 Ad-Hoc networks. In: 2012 International Symposium on Network Coding (NetCod), Cambridge, MA, USA (2012)

10. Chi, Y., Agrawal, D.: HyCare: hybrid coding-aware routing with ETOX metric in multi-hop wireless networks. In: 2013 IEEE 10th International Conference on Mobile Ad-Hoc and Sensor Systems (MASS), Hangzhou, China (2013)

11. Chi, Y., Agrawal, D.: Network locality in wireless networks. In: 2013 ACS International Conference on Computer Systems and Applications (AICCSA), Ifrane, Morocco (2013)

12. Rosati, S., Kruzelecki, K., Heitz, G., Floreano, D., Rimoldi, B.: Dynamic routing for flying Ad-Hoc networks. IEEE Trans. Veh. Technol. $65(3)$, 1690–1700 (2016)

13. Vasiliev, D.S., Meitis, D.S., Abilov, A.: Simulation-based comparison of AODV, OLSR and HWMP protocols for flying Ad Hoc networks. In: Balandin, S., Andreev, S., Koucheryavy, Y. (eds.) NEW2AN 2014. LNCS, vol. 8638, pp. 245–252. Springer, Cham (2014). doi:10.1007/978-3-319-10353-2_21

14. Broyles, D., Jabbar, A., Sterbenz, J.: Design and analysis of a 3-D Gauss-Markov mobility model for highly dynamic airborne networks. In: Proceedings of the International Telemetering Conference 2010, San Diego, CA (2010)

Communication Technologies in the Space Experiment "Kontur-2"

Vladimir Muliukha[1,2(✉)], Vladimir Zaborovsky[1,2],
Alexander Ilyashenko[1], and Yuri Podgurski[1]

[1] Peter the Great Saint-Petersburg Polytechnic University,
Saint-Petersburg, Russia
vladimir@mail.neva.ru
[2] The Russian State Scientific Center for Robotics and Technical Cybernetics,
Saint-Petersburg, Russia

Abstract. The issues on using of the communication technologies for the remote control of on-ground robots from the Russian Segment of the International Space Station (ISS RS) in the frame of "Kontur-2" space experiment are presented. Characteristics of communication channels used in this control system ensure elements of telepresence for an operator using visual and tactile feedback. Results of space sessions are presented.

Keywords: Space experiment · Telepresence · Communication channels · Time delay · Satellite channel · Teleoperation

1 Introduction

"Kontur-2" is a joint space experiment for the in-flight verification of force feedback and telepresence technologies between the Russian Segment of the International Space Station (ISS RS), the German Aerospace Center (DLR) and the Russian State Scientific Center for Robotics and Technical Cybernetics (RTC). This mission follows after successful previous space experiments [1, 2]. The main objectives of "Kontur-2" space experiment are the implementation and evaluation of telerobotics and telepresence technologies for space telemanipulation applications using S-Band channel and Internet.

In August 2015 a force feedback joystick was developed and constructed by DLR. The joystick was designed according to a set of performance requirements and space related specifications needed for inflight operation in the ISS. Its materials, thermal design, motors and electronics have been developed to fulfill the required space qualification. The software for it was realized by combined team of programmers form RTC and Peter the Great Saint-Petersburg Polytechnic University (SPbPU). The joystick was delivered and installed onboard the ISS RS and a bilateral controller was designed to cope with the communication latencies between ISS, ground station on Earth at DLR and robot at RTC [3]. Both, DLR and RTC conducted several experiments during 2015 in which two cosmonauts successfully teleoperated on-ground robots at DLR and RTC from the ISS RS. Figure 1 shows the general mission setup of

© Springer International Publishing AG 2017
O. Galinina et al. (Eds.): NEW2AN/ruSMART/NsCC 2017, LNCS 10531, pp. 587–597, 2017.
DOI: 10.1007/978-3-319-67380-6_55

the space experiment "Kontur-2", including the infrastructure used for the cosmonaut training at Gagarin Research and Test Cosmonaut Training Center (GCTC).

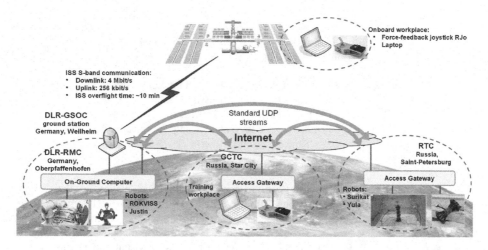

Fig. 1. "Kontur-2" Space experiment setup

The S-Band channel was implemented for real-time communication used in the ROKVISS experiment [1] and was used to connect force-feedback joystick onboard ISS RS and robots on Earth. The types of data transmitted between both, space operator (cosmonaut) and the robot on Earth are shown in Fig. 2. Haptic, visual and voice signals helped the cosmonaut to feel telepresence in the robotics system.

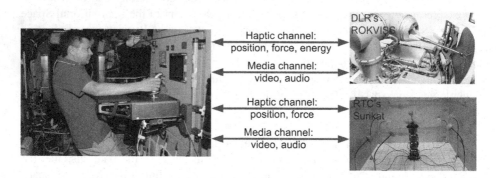

Fig. 2. Telepresence data streams

2 Communication and Control Requirements

It is well known in robotics, and especially in the control field, that closed loops systems that include non-negligible time delays can produce negative effects on system stability. This issue is magnified in those closed loop systems that are characterized by tight couplings, that is, where high frequency control actions are required to capture a reasonable spectrum of the dynamics of the controlled system. For instance, in order to capture the interaction of a robot's end-effector while contacting a hard surface, a well-established control-loop frequency is about 1000 Hz. In particular, the control loop frequency of the joystick is 3000 Hz. Our space experiment consisted of two parts: DLR and RTC sessions. Due to the limitations of the communication link, the transmission frequency through the S-Band link is 500 Hz for DLR mission. The control-loop frequency for the robot in RTC was even lower and not stable due to the use of Internet network channels between DLR and RTC. The purpose of this paper is to present the results of sessions of the space experiment "Kontur-2" within the framework of which the telepresence method was implemented, even with communication channels with random delays in the feedback loop.

The lower transmission frequency limits the overall control loop bandwidth. Nevertheless, while keeping higher local control loop frequencies at the joystick or robot sides has some control and performance benefits [5]. For instance, higher virtual damping values can be achieved or more accurate passivity observers resulting in better bilateral control performance. On the other hand, coping with latencies between master and slave systems can be challenging, especially when delay is non-constant and jitter and data losses are present.

Figure 3 shows round trip delay measurement in both scenarios. As it can be seen, the nature of these two links is quite different in terms of time delay, data losses and jitter. The time delay for DLR sessions varied from 20 to 30 ms (corresponding to azimuth and horizon points) with mean negligible data losses. RTC sessions has a mean delay of 65 ms between RTC in Saint-Petersburg and Oberpfaffenhofen (where the DLR is located) and highly oscillating package loss ratio, from 5% to 15%, due to the used UDP protocol. Though more limited in bandwidth, the ISS link is higher in performance. However, shadowing can occur resulting in signal attenuation and in turn higher package loss ratios or even communication blackouts. On the other hand, the Internet link measurements confirm a typical UDP behavior.

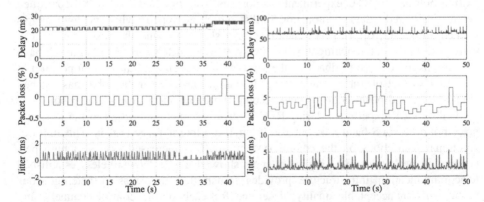

Fig. 3. DLR session through S-band link (left) and RTC session through Internet (right) round delay measurements

3 DLR Experiment

During the DLR sessions, cosmonauts controlled the ROKVISS arm, located in one of the laboratories in Oberpfaffenhofen. The approach for the bilateral controller in the force-feedback joystick is based on a 4-Channels architecture, whose stability in the presence of time delay, jitter and data losses is addressed through the Time Domain Passivity Control Approach (TDPA) [3] and the Time Delay Power Network (TDPN) representation. In this architecture, position and force signals are sent from the joystick to the ROKVISS robot, and computed and measured force signals are sent in the other direction. Both systems, the joystick and ROKVISS, are impedance controlled, i.e. the commanded signals to the joystick and to the robot are forces, and their outputs are positions. One of the most interesting features of this controller is its adaptability to any communication conditions, including different delay values, jitters and package losses.

The experiments with the joystick on board the ISS proved that all hard- and software components of joystick as well as the telepresence system are functioning reliably. The participating cosmonauts, Oleg Kononenko and Sergey Volkov, were able to perform the experimental tasks with the ROKVISS robot, located in the DLR in Oberpfaffenhofen from the ISS RS. Force-feedback and latency compensation technologies for bilateral control were successfully evaluated [4]. The cosmonauts reported that the tasks were easy to perform with the force feedback joystick. Different telepresence approaches were compared in terms of system and operator performance and the results from terrestrial and space sessions were compared to better understand the effects of microgravity on sensorimotor performance while controlling a telerobotic system. Preliminary analyses revealed that positional accuracy is degraded in microgravity compared to terrestrial conditions.

4 RTC Experiment

During first part of RTC experiment cosmonauts from onboard the RS ISS controlled cinematically redundant manipulator Surikat which equipped stylus and light targets located around [5, 6]. Targets lighted up for a short time and the cosmonaut had to extinguish the target, touching it by the stylus. Unlike DLR's robot Surikat didn't have torque sensors so force feedback on the joystick handle reflected the current position of the robot so that operator cannot move the handle faster than the robot has reached already specified position taking into account the delay and inertia of the robot.

Unlike DLR experiment during RTC control sessions connection to the ISS took place not only via S-band link, but also through the Internet (see Fig. 1). This imposes additional restrictions on the control system, such as increased delays, a high percentage of packet loss and higher jitter as showed above. Nevertheless established telecommunication infrastructure provides transmitted transmission heterogeneous traffic with an acceptable quality of service for each of the control channels, and feedback. The most critical to latency and bandwidth bi-directional control channel provided total transmission delay for the control and feedback loop (round trip time - RTT) about 85 ms. Figure 4 shows a graph of RTT for RTC session. The graph shows the change of RTT during the sessions which is connected to change in distance between ISS and ground station.

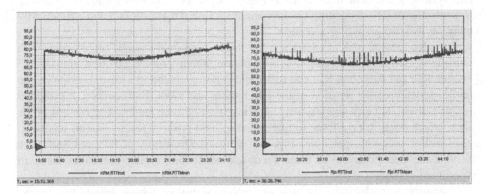

Fig. 4. RTT between the ISS and RTC during two different sessions

Established control system provided manual teleoperation from the ISS RS of robots located on Earth. Operator onboard the ISS RS receives feedback from the robot through several channels:

- Tactile feedback provided by transfer of information between robot and joystick drives controllers. Minimum delay for this type of feedback is provided by operation of controllers in real time and transmit this information flow via priority channel in S-band link. This type of feedback provided to the operator a tactile information about the inertia of the system in the absence of a torque sensor on the robot;
- Visual feedback, obtained by translation of the video from a camera mounted on the robot, or from the observation camera. For this type of feedback it was defined the balance between the quality of the transmitted image, the frequency of its update and delay in delivery when there are significant limitations on the bandwidth of the communication channel (256 kbit/s for upload). In connection with the need to use data compression algorithms, this kind of feedback provided an appreciable delay due to data processing at the transmitting and receiving sides;
- Visual feedback, obtained by visualization of the robot's 3D-model motion on the laptop. During control of Surikat for animation of 3D-model it was used telemetric information from the robot. This information was delivered in the priority channel, which provided low delays in display. Compactness of transmission data allowed for a low bandwidth to provide higher image refresh rate (compared to video), which significantly improves the dynamic picture representations of the surrounding area.

The effectiveness of combination of several types of feedback – haptic (force in the handle of RJo) and visual (video and 3D - model) – complement each other to provide a virtual "immersion" of operator to the environment of the robot operation, confirmed the successful implementation of manipulation and locomotion tasks carried out during teleoperation sessions.

Force feedback joystick was used in the experiment as part of different teleoperation systems for control heterogeneous robots. Software architecture allowed choosing the controller of RJo depending on the current session (DLR or RTC) and dynamically changing its settings during a session.

For teleoperation of RTC Surikat robot it was used dual-channel architecture with controllers, designed in RTC: the information about current position transmitted between RJo and the robot. RJo controller allowed implementing a virtual spring on the handle, which bind the position of the RJo handle and robot. Furthermore, it provided sensitive of virtual mass and viscosity. The insertion of force feedback to control of the robot, which doesn't have force-torque sensors, yield positive results like an increased speed and accuracy of operations. Figure 5 shows the trajectory of RJo and Surikat in the absence and presence of force feedback while performing the same tasks in experiment session.

When the force-feedback is used to remote control of robots, in addition to the requirements for communication channels, in microgravity conditions there are also requirements to the workplace of operator-cosmonaut. Figure 6 shows a photo of cosmonaut Oleg Novitsky that was made during the control of mobile RTC robot in December 2016. During the control sessions, cosmonauts have to hold a special hard handle with their left hand so not to lose the balance under the force impact from the joystick.

5   Sessions of RTC Mobile Robot Control

In 2016, additional sessions of the space experiment "Kontur-2" were organized to realize operations for remote control of the mobile robot from the ISS RS. To conduct these sessions, a specialized mobile robot and a three-dimensional polygon were created. The experiment used a model-based feedback. The essence of this type of feedback is that the force-torque actions on the handle of the joystick onboard the ISS are calculated in the program based on the current position of the obstacles, the robot, its speed and direction of movement. Thus, the model-based feedback increases demands on the accuracy and speed of the robot's positioning system [7].

To conduct the sessions of the space experiment "Kontur-2" there was used a Vicon positioning system, which gives very high accuracy of positioning of markers installed on the mobile robot. The data obtained from the Vicon system were used to calculate the force-torque effect on the handle of the joystick, as well as to form a three-dimensional model of the polygon on the astronaut's on-board computer using the Unity program. Figure 7 shows the appearance of the astronaut's interface consisting of a three-dimensional model of the environment formed in Unity program based on the robot's position data, as well as the polygon map (a top left corner of the GUI) and camera images (top right corner). In total, during the space experiment, three cameras were used: the first one is an overview camera, which allows to get an image of the whole polygon from the top, the second one is a camera on a mobile robot, and the last one is a camera used to organize videoconferences between developers and a cosmonaut after the control session of the experiment (Fig. 7).

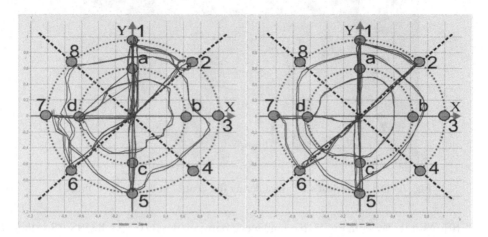

Fig. 5. Trajectory of RJo's handle (Master) and Surikat (Slave) in the absence (left) and presence (right) of force feedback

Fig. 6. Cosmonaut Oleg Novitsky controls the mobile RTC robot

All information flows in the space experiment "Kontur-2" are divided into priority and background. Priority flows include robot control commands and robot position data from the Vicon system. These data are transmitted onboard the ISS using the synchronous part of the S-Band channel. Background streams that include the transfer of video and audio information are transmitted over an asynchronous channel, if possible. Various algorithms for working with priority traffic were used in our experiment

Fig. 7. The cosmonaut's interface during the control of mobile RTC robot

[8–10]. As it is shown in Fig. 2, the bottleneck between the RTC and the ISS RS is the S-Band channel, whose uplink speed is limited to 256 Kbps. The result of conducting sessions of space experiment "Kontur-2" in 2015 was the necessity to develop new software to manage the quality and volume of the video stream. Figure 8 shows the appearance of the program that was developed to control the quality and volume of the video stream. This program allows to set the necessary video parameters, such as frame rate, resolution, chrominance, and also realize the picture-in-picture function, sending one video stream to the ISS instead of two. Also, the program automatically adjusts the video parameters for the given bandwidth of the communication channel, reducing the controlled parameters, which allows reacting quickly to changes in the communication environment to minimize the impact on the priority data stream.

The use of this video server in the sessions of the experiment in 2016 allowed to successfully compensate the negative changes in the characteristics of the telecommunication environment between the RTC and the ISS RS (Fig. 8).

The specificity of the space experiment "Kontur-2" in 2016 from the point of view of the used communication channels was the replacement of the physical synchronous communication channel between Weilheim (Germany), where is the receiving and transmitting S-Band antenna, and Oberpfaffenhofen (Germany), where is DLR equipment used in the experiment. Instead of the dedicated physical channel, was used the digital service of SDH E3 over MPLS, which led to a significant increase in the mean of data transfer delay when controlling the mobile RTC robot from the ISS RS. In 2015, RTT between the ISS and RTC during the space experiment varies from 70 to 85 ms. In this case, the dynamics of the delay is explained by the change in distance during the flight of the ISS over the antenna on Earth. RTT between DLR and RTC almost did not change during all the experiments and was about 60 ms. In 2016, the RTT between DLR and RTC did not change, but the total time for data exchange with the ISS increased to 110–140 ms, which is explained by the use of a new logical protocol in the data link. It should be noted that even such a sudden change in the parameters of the communication channel did not interfere with the software of the

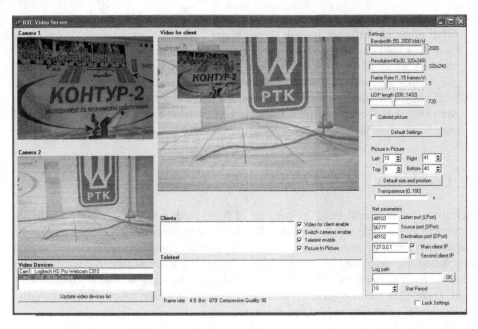

Fig. 8. The video server interface that controls the quality and volume of the video stream

force-feedback joystick on board the ISS RS to function successfully within the sessions of the "Kontur-2" experiment.

6 Control Loop

The specialized control loop in the Space Experiment "Kontur-2" makes it possible to significantly reduce the effect of random network delays on the control of the on-surface robot. The scheme itself was already presented by the authors in the framework of the preparation and creation of scientific equipment and software for the Space Experiment. Currently, the implemented version of the control loop is not significantly different from the previously announced [7, 11]. To implement the force sensing, as before, the feedback loop is divided into several sections:

- Local control loop in the Joystick, which operates at a frequency of 3 kHz and allows the operator to receive adequate tactile sensations. DLR research data showed that the frequency of feedback loop operation should be more than 500 Hz for telecontrol operation, the more the better. The frequency of 3 kHz was chosen based on the characteristics of the hardware platform created for use on the ISS;
- Local control circuit of a mobile robot, operating at a frequency of about 100 Hz. The developed software can process information much faster, but in our work as a sensor system is used a positioning system "Vicon". Thus, the frequency of operation is limited by the speed of updating the data received from Vicon and in our case it was about 100 Hz.

- The data transmission network loop "Transporter", functioning on the basis of the UDP protocol. Software module "Transporter" provides isochronous communication for local controllers: vector control samples that are uniformly received from the Joystick should be transferred to the robot. Similarly, in the opposite direction, the vector data from the telemetry system of the robot is transferred to Joystick. The frequency of sending network packets is 200 packets per second, while the packets are divided into even and odd. Even packets contain control commands, and odd ones contain service information to support the control system's operation.

7 Conclusion

Scientific and technical tasks of creating of force-feedback teleoperation system for control of on-ground robots from the ISS have been successfully solved within the frame of "Kontur-2" experiment. The network infrastructure for verification of such a distributed teleoperation systems has been created. For the first time the teleoperation sessions from onboard the ISS of on-ground robots using force-feedback joystick were held. Developed technologies are intended for using in space robotic systems during realization of planetary and orbital missions.

The special software was developed and created to implement the telepresence of the operator-cosmonaut during the remote control of the robot, which allows the implementation of force-feedback control using communication channels with high delays.

Acknowledgments. This research was supported by the RFBR grant 15-29-07131 ofi-m.

References

1. Albu-Schaeffer, A., Bertleff, W., Rebele, B., Schaefer, B., Landzettel, K., Hirzinger, G.: ROKVISS – robotics component verification on ISS current experimental results on parameter identification. In: IEEE International Conference Robotics and Automation, pp. 3879–3885 (2006)
2. Landzettel, K., Zaborovsky, V., Babkin, E., Belyaev, M., Kondratiev, A., Silinenko, A.: "KONTUR" Experiment on Russian Segment of the ISS. Scientific Readings in Memory of K.E. Tsiolkovsky, Kaluga (2009)
3. Artigas, J., Ryu, J.H., Preusche, C., Hirzinger, G.: Network representation and passivity of delayed teleoperation systems. In: 2011 IEEE/RSJ International Conference on Intelligent Robots and Systems, San Francisco, CA, pp. 177–183 2011. doi:10.1109/IROS.2011.6094919
4. Artigas, J., et al.: KONTUR-2: force-feedback teleoperation from the international space station. In: 2016 IEEE International Conference on Robotics and Automation (ICRA), Stockholm, pp. 1166–1173 (2016). doi:10.1109/ICRA.2016.7487246
5. Guk, M., Silinenko, A.: "Kontur-2" – joint German-Russian space experiment on force-feedback control of on-ground robots from onboard the ISS. In: Proceedings of International Scientific and Technical Conference «Extreme Robotics», 1–2 October 2014, Saint-Petersburg, "Polytechnica-service", pp. 59–64 (2014)

6. Zaborovskiy, V., Guk, M., Mulyukha, V., Silinenko, A., Volkov, O., Belyaev, M.: Teleoperation of on-ground robots from onboard the ISS in frame of "Kontur-2" space experiment. In: Proceedings of International Scientific and Practice Conference «Scientific Exploration and Experiments on the ISS», 9–11 April 2015, Moscow, IKI RAS, pp. 64–65 (2015)

7. Zaborovsky, V., Muliukha, V., Ilyashenko, A.: Cyber-physical approach in a series of space experiments "Kontur". In: Balandin, S., Andreev, S., Koucheryavy, Y. (eds.) ruSMART 2015. LNCS, vol. 9247, pp. 745–758. Springer, Cham (2015). doi:10.1007/978-3-319-23126-6_69

8. Muliukha, V., Ilyashenko, A., Zayats, O., Zaborovsky, V.: Preemptive queuing system with randomized push-out mechanism. Commun. Nonlinear Sci. Numer. Simul. **21**(1–3), 147–158 (2015). doi:10.1016/j.cnsns.2014.08.020

9. Ilyashenko, A., Zayats, O., Muliukha, V., Laboshin, L.: Further investigations of the priority queuing system with preemptive priority and randomized push-out mechanism. In: Balandin, S., Andreev, S., Koucheryavy, Y. (eds.) NEW2AN 2014. LNCS, vol. 8638, pp. 433–443. Springer, Cham (2014). doi:10.1007/978-3-319-10353-2_38

10. Ilyashenko, A., Zayats, O., Muliukha, V., Lukashin, A.: Alternating priorities queueing system with randomized push-out mechanism. In: Balandin, S., Andreev, S., Koucheryavy, Y. (eds.) ruSMART 2015. LNCS, vol. 9247, pp. 436–445. Springer, Cham (2015). doi:10.1007/978-3-319-23126-6_38

11. Zaborovsky, V., Guk, M., Muliukha, V., Ilyashenko, A.: Cyber-physical approach to the network-centric robot control problems. In: Balandin, S., Andreev, S., Koucheryavy, Y. (eds.) NEW2AN 2014. LNCS, vol. 8638, pp. 619–629. Springer, Cham (2014). doi:10.1007/978-3-319-10353-2_57

Accuracy of Secondary Surveillance Radar System Remote Analysis Station

Igor A. Tsikin and Ekaterina S. Poklonskaya[(✉)]

St. Petersburg Polytechnic University, 29 Politechnicheskaya st.,
St. Petersburg 195251, Russia
tsikin@mail.spbstu.ru, catherine091@mail.ru

Abstract. The Distance Analysis System (DAS) for a Secondary Surveillance
Radar System (SSRS) analysis is considered. The DAS assumes the usage of the
Remote Analysis Station (RAS) for the aircraft position determination and the
SSRS and DAS results comparison in order to detect the analyzed system errors.
The DAS error sources are considered. The RAS errors for the different aircraft
positions and different initial errors are estimated. The DAS and the analyzed
both monostatic and bistatic SSRS accuracies are compared for different aircraft
positions related to the RAS and the Secondary Surveillance Radar (SSR) locations. The recommendations about the RAS location provided the highest
DAS accuracy are given.

Keywords: Secondary Surveillance Radar · Distance analysis station · Aircraft
position determination

1 Introduction

The problem of Secondary Surveillance Radar Systems (SSRS) monitoring is critically
important for aircraft flight safety, especially when the abnormal intrasystem errors are
probable. Special "Parrot" transponders are a classical solution to the mentioned
problem [1]. SSRS errors are identified by comparing the determined position of the
"Parrot" to the real one. Nevertheless, the SSRS errors are detected only for certain
positions of the transponder, while the SSRS accuracy definition is substantial for any
aircraft location.

The problem of SSRS accuracy analysis can be solved by using a third-party radio
source (RC) for aircraft position determination with the help of a special remote
analysis station (RAS) simultaneously with SSRS [2–5]. In this regard both the signals
from RC (base cell stations, television and radio signal sources, etc.) and the signals
reflected from the aircraft are analyzed at the RAS. However, such systems can work
only when the RC is available in the line of sight. On the other hand, it is possible to
use the analyzed SSRS radar (SSR) as the RC [6–9] using the aircraft transponder
signals instead of passive reflected signals. However, this kind of distance analysis
system (DAS) accuracy is reduced because of the SSR location and working mode
general absence of a priori information.

O. Galinina et al. (Eds.): NEW2AN/ruSMART/NsCC 2017, LNCS 10531, pp. 598–606, 2017.
DOI: 10.1007/978-3-319-67380-6_56

In this paper, the influence of errors in the determination of the SSR antenna rotation speed and the request signals pulse repetition interval (PRI) is considered, assuming a known SSR location and the presence of an Additive White Gaussian Noise (AWGN) in the channels. It is also crucially to compare the DAS and the analyzed SSRS accuracies. At the same time, a monostatic SSRS accuracy is never lower than the accuracy of the DAS which is related to the class of bistatic systems [10], while the accuracy of the bistatic SSRS generally can be either lower or higher than the accuracy of the DAS depending on the aircraft location related to the SSR, SSRS receiver (SSRR) and RAS positions [11, 12]. This leads to the importance of the DAS and analyzed SSRS accuracies comparison for different locations of the aircraft and SSR, RAS, SSRR.

2 Aircraft Position Determination Algorithm

It is necessary to consider the DAS algorithm to define aircraft location determination error sources. This algorithm is similar to the classic aircraft position determination algorithm of bistatic SSRS but needs the preliminary SSR working mode evaluation. The aircraft position determination in any bistatic radar system, particularly in SSRS, is based on the path difference $L = R_1 + R_2 - b$ and the angle φ (Fig. 1) determination. In SSRS and also in DAS, the values of these parameters are being obtained from the SSR request signals and the transponder answer signals arriving time measurements [7, 8].

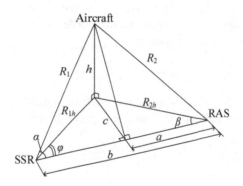

Fig. 1. The DAS scheme.

In Fig. 2 on the top axe the SSR request t_{ri} and the transponder answer $t_{ai}(i \geq K)$ signals emission time moments are represented. On the bottom one the requested t'_{ri} and answer $t'_{ai}(i \geq K)$ signals' arriving moments at the RAS are represented. The answer signal emitted at the moment t_{ai} is guaranteed as the transponder answer on the request signal emitted at the moment t_{ri}.

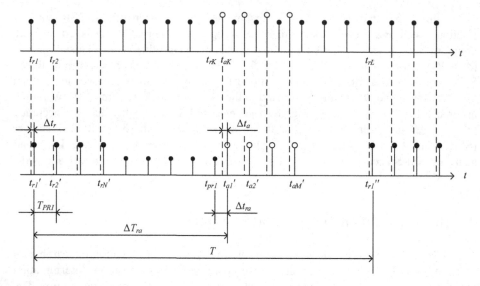

Fig. 2. SSR request and transponder answer signals emission (top axe) and receiving at RAS (bottom axe) moments.

First of all, we will need to consider a situation when SSR antenna period T and PRI T_{PRI} are known. As such, it is possible to get the predicted time moments t_{pri} (short lines on the bottom axe) of the assumed requested signals arriving in case of an isotopic SSR antenna. These signals are predicted up to the answer signals arriving time moments. Thus, it is possible to estimate the time difference $\Delta t_{ra} = t'_{ai} - t_{pri}$. The path difference can be found by this equation:

$$L = c \cdot (\Delta t_{ra} - t_h) \tag{1}$$

Here t_h – the time interval spent on the request SSR signal handle and answer signal formation at RAS, c – a light speed. The angle φ can be found using the equation:

$$\varphi = 2\pi \frac{\Delta T_{ra}}{T} \tag{2}$$

Thereby the aircraft position determination accuracy depends on the precision of the path difference L and the angle φ values determination which is influenced by the SSR antenna rotation period T, and the values of time intervals ΔT_{ra} and Δt_{ra} determination accuracy. This accuracy dependent on the answer and request signals time of arrival measurement precision is defined by AWGN influence. There are also "system" errors in SSRS. Such errors are connected to the SSR antenna rotation period and clock generator instabilities that inevitably affect the ΔT_{ra} and T parameter values determination accuracy.

3 Analysis of Different Error Sources Influences

It is obvious that all mentioned error sources influence the DAS accuracy in different ways depending on the aircraft location related to RAS and SSR positions. This impact also depends on the initial error source values such as standard deviations (SD) of request δt_r and answer δt_a signals errors in measured time of arrival (TOA) at RAS. These initial error source values also include the SD of SSR antenna rotation error δv_i and SD of SSR PRI error δT_{ci} at the moment of i-th request signal emission. The RAS accuracy under the impact of each error separately and for different aircraft positions is considered below. The path difference L value is assumed to be constant and the angle γ value and, respectively, the angle φ value are considered to be different. In this situation the aircraft location is determined by points situated at the ellipse with focuses in RAS and SSR positions (Fig. 3).

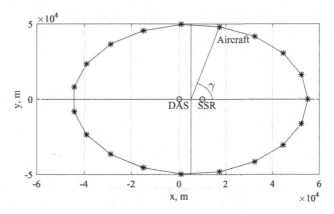

Fig. 3. Aircraft positions related to the RAS and SSR locations.

RAS accuracy in situations illustrated at Fig. 3 is examined using the MATLAB model. Modelling results are presented at Fig. 4. It illustrates the aircraft location error δR SD $\sigma_{\delta R}$ dependences on the aircraft actual locations for different signal-to-noise ratio (SNR) d^2 at the request signal processing device (Fig. 4a), PRI error δT_{PRIi} SD $\sigma_{\delta T_{PRIi}}$ (Fig. 4b) and antenna rotation error δv_i SD $\sigma_{\delta v_i}$ (Fig. 4c). The modelling refers to the case when the SSR working mode analysis time is equal to $N = 5$ SSR antenna rotations. SNR value influences on the SD $\sigma_{\delta t_r}$ and $\sigma_{\delta t_a}$ of time moment errors δt_r and δt_a respectively. The SD values $\sigma_{\delta t_r}$ and $\sigma_{\delta t_a}$ for different SNR values d^2 can be determined by an additional modelling where d^2 is the model input, and the model defines the $\sigma_{\delta t_r}$ and $\sigma_{\delta t_a}$ values statistically.

The modelling results presented at Fig. 4 show that the DAS accuracy has an order of 100 m. All considered error sources lead to the minimal aircraft location error when angle γ is close to π. The SD $\sigma_{\delta R}$ increases smoothly with the SNR value d^2 growth and quickly with the SD $\sigma_{\delta T_{PRIi}}$ and $\sigma_{\delta v_i}$ growth in the considered conditions.

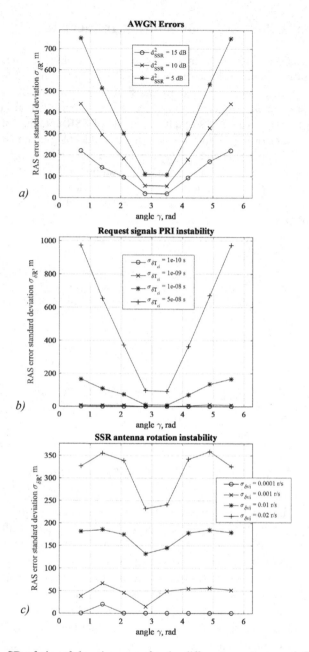

Fig. 4. SD of aircraft location error for the different error sources influence.

4 DAS and Analyzed SSRS Accuracies Comparison

Due to the fact that SSRS accuracy is generally close to 100 m [13] DAS accuracy must also have the same order. Considering the modelling results (Fig. 4), it is clear that allowable SSR error levels are $\sigma_{\delta T_{PRIi}} = 10^{-8}$ s and $\sigma_{\delta v_i} = 0,01$ r/s. Nevertheless, the important issues are the analysis of simultaneous influence of all considered errors on DAS and the analyzed SSRS (both monostatic and bistaic) and DAS accuracies comparison.

The DAS and analyzed SSRS accuracies dependent on the different angle γ values are considered. The most interesting situation is the worst one when the SSR contains no system for suppressing the answer signals transmission to the request signals received by the SSR antenna side lobes or such system is out of order [14]. In this situation, it is not possible to define the request signals emission time moments at DAS in any other way besides the prediction based on the defined request signals PRI value T_{PRI}. A lack of this information necessarily reduces the bistatic SSRS accuracy, but in the opposite situation, the DAS and analyzed SSRS accuracies are close for the same values of γ, and their comparison is not interesting. The modelling takes place in a condition of the lack of a priori information about the SSR working mode in DAS and the presence of the information about the SSR antenna rotation value in the analyzed bistatic SSRS. The time interval of the parameters T_{PRI} and T values definition at DAS is equal to $N = 5$ antenna rotations. The distance between RAS and SSR is considered to be equal to the distance between SSRR and SSR.

The dependences of SD $\sigma_{\delta R}$ values on the angle γ values are presented in Fig. 5 for different SNR d^2 values for DAS and monostatic (Fig. 5a) and bistatic (Fig. 5b, c) SSR and for error SD values $\sigma_{\delta T_{PRIi}} = 10^{-8}$ s, $\sigma_{\delta vi} = 0,01$ r/s. Figure 5b and c show the results for SSRS respectively with and without information about request signals transmission time. It is shown that the DAS accuracy is approximately equal to 100 m in a condition of all considered error sources influence. The DAS accuracy is always lower than the monostatic SSR accuracy for all considered conditions, but it can be equal or even higher than the bistatic SSR accuracy (especially in the situation shown in Fig. 5c). Thus, if the aircraft location provides angle values $\gamma = \pi$ in DAS and simultaneously $\gamma = 2\pi$ in bistatic SSR, the DAS accuracy is higher than the analyzed system accuracy. At the same time, for each γ value the DAS accuracy is always lower than the bistatic accuracy for the same γ value, and its difference grows with the SNR value decreasing and the value of γ tends to 2π.

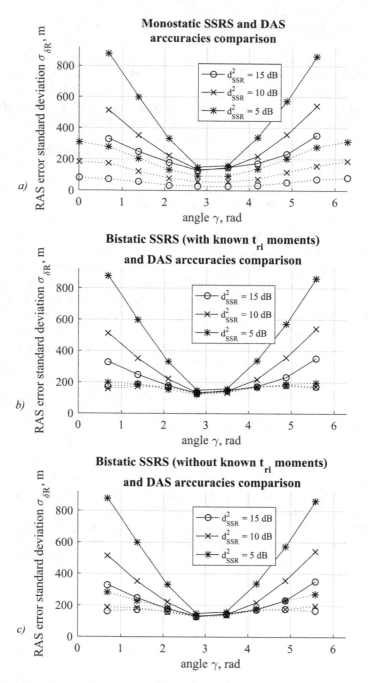

Fig. 5. Aircraft position determination error SD in the DAS (solid lines) and in the analyzed monostatic (a) and bistatic (b, c) SSRS (dashed lines).

5 Conclusion

DAS accuracy strongly depends on the SSR working mode instabilities. Thus, the aircraft location error SD $\sigma_{\delta R}$ value reaches approximately 50 m with PRI error SD value $\sigma_{\delta T_{PRi}} = 10^{-8}$ s and grows by an order of magnitude with $\sigma_{\delta T_{PRi}} = 10^{-7}$ s. Similarly the $\sigma_{\delta R}$ value increases twice with SSR antenna speed error SD value $\sigma_{\delta v_i}$ growth from 0.01 r/s to 0.02 r/s. Related to the AWGN errors which depend on the SNR value d^2 also affect the DAS accuracy. Thus, with $d^2 = 5$ dB DAS aircraft location error is up to $\sigma_{\delta R} = 200$ m, with $d^2 = 15$ dB this error increases approximately 4 times.

The worst situation for DAS is an absence of any request signals emission time moments determination methods except the prediction. An example of the situation is – the absence of additional undesirable answer signals transmission suppression antenna in SSR. In this situation the DAS and the analyzed bistatic SSRS accuracy difference is up to several hundred meters in the considered conditions ($\sigma_{\delta T_{PRi}} = 10^{-8}$ s, $\sigma_{\delta vi} = 0,01$ r/s, $d^2 = 5, 10, 15$ dB). Thus, this difference is up to 600 m with SNR value $d^2 = 5$ dB. The biggest difference takes place when angle values $\gamma = 2\pi$ for the DAS and $\gamma = \pi$ for the analyzed SSRS simultaneously. On the contrary, the least difference is when $\gamma = \pi$ for the DAS and analyzed SSRS (both DAS and SSRS accuracies are approximately 150 m for all considered SNR values). Moreover, if $\gamma \neq \pi$ for the analyzed bistatic SSRS simultaneously, the SSRS accuracy can be lower than the DAS accuracy. At the same time, the monostatic SSRS accuracy is always higher than the DAS accuracy (i.e., aircraft position determination error varies from 20 to 100 m for $\gamma = \pi$ and different SNR values, at the same time DAS accuracy for this situation is close to 150 m). Nevertheless, situation when $\gamma = \pi$ both in DAS and in analyzed monostatic SSRS also leads to the least accuracy difference. Thereby, it is important to locate RAS provided the highest probability of aircraft presence with $\gamma = \pi$ for DAS. It is easy to ensure given the availability of aircrafts tracks information.

References

1. Jacobs, M.H., Alon, Y., Geiger, J.H.: U.S. Patent No. 8,362,943. U.S. Patent and Trademark Office, Washington, DC (2013)
2. Macera, A., Bongioanni, C., Colone, F., Palmarini, C., Martelli, T., Pastina, D., Lombardo, P.: FM-based passive bistatic radar in ARGUS 3D: experimental results. In: GTTI 2013 – Session on Sensing (2013)
3. Pisane, J.: Automatic target recognition using passive bistatic radar signals. Doctoral dissertation, Supélec (2013)
4. Barott, W.C., Butka, B.: A passive bistatic radar for detection of aircraft using spaceborne transmitters. In: IEEE/AIAA 30th Digital Avionics Systems Conference (DASC) on Closing the Generation Gap - Increasing Capability for Flight Operations among Legacy, Modern and Uninhabited Aircraft. IEEE, New York (2011)
5. Malanowski, M.: Detection and parameter estimation of manoeuvring targets with passive bistatic radar. IET Radar Sonar Navig. **6**, 739–745 (2012)

6. Litchford, G.B.: U.S. Patent No. 4,115,771. U.S. Patent and Trademark Office, Washington, DC (1978)
7. Otsuyama, T., Honda, J., Shiomi, K., Minorikawa, G., Hamanaka, Y.: Performance evaluation of passive secondary surveillance radar for small aircraft surveillance. In: 12th European Radar Conference (EuRAD), pp. 505–508. IEEE, New York (2015)
8. Shiomi, K., Senoguchi, A., Aoyama, S.: Development of mobile passive secondary surveillance radar. In: Proceedings of 28th International Congress of the Aeronautical Sciences, Brisbane (2012)
9. Kim, E., Sivits, K.: Blended secondary surveillance radar solutions to improve air traffic surveillance. Aerosp. Sci. Technol. 45, 203–208 (2015)
10. Averyanov, V.Y.: Separated radiolocation stations and systems. Technika, Minsk (1978)
11. Willis, N.J., Griffiths, H.D. (eds.): Advances in Bistatic Radar, vol. 2. SciTech Publishing, Raleigh (2007)
12. Kovalyov, F.N.: The target location precision in bistatic radiolocation system. Radioengineering 8, 56–59 (2013)
13. Zhang, J., Liu, W., Zhu, Y.B.: Study of ADS-B data evaluation. Chin. J. Aeronaut. 24, 461–466 (2011)
14. Annex 10 to the Convention on International Civil Aviation. Surveillance Radar and Collision Avoidance Systems, vol. IV. ICAO (2014)

Spectral and Energy Efficiency of Optimal Signals with Increased Duration, Providing Overcoming "Nyquist Barrier"

Anna S. Ovsyannikova[1(⊠)], Sergey V. Zavjalov[1],
Sergey B. Makarov[1], Sergey V. Volvenko[1], and Trinh Luong Quang[2]

[1] Peter the Great St. Petersburg Polytechnic University, St. Petersburg, Russia
anny-ov97@mail.ru, zavyalov_sv@spbstu.ru,
{makarov,volk}@cee.spbstu.ru
[2] Binh Duong University, Ho Chi Minh, Vietnam
rusvibdu@gmail.com

Abstract. Optimal signals with increased duration, providing overcoming "Nyquist Barrier", are considered. Optimization task of searching envelopes of signals is presented and numerically solved. Solutions of optimization task for different signal durations and correlation coefficients are presented. Corresponding energy spectra are also checked. The dependence of occupied frequency bandwidth for different level of energy spectra is obtained. BER performance of proposed optimal signals was checked by simulation. It is shown, that energy losses can be reduced by reducing correlation coefficient in optimization task and in resulting optimal signals. Application of optimal signals with increased duration allows to be closer to Shannon boundary and provides realization of data transfer with overcoming "Nyquist Barrier". It was done by increasing spectral efficiency for different level of energy spectra.

Keywords: Spectral efficiency · Energy efficiency · Optimal signals · BER performance · Out-of-band emissions

1 Introduction

Development of modern wireless broadband 4G and 5G networks is associated with application of spectrally efficient signals. These signals have duration $T_c = T$ and different amplitude and phase trajectories [1–4]. Efficiency of data transfer can be increased by application of smoothed envelopes $a(t)$ with increased duration $T_c = LT$ ($L > 1$) [5–7]. This way leads to reducing of occupied frequency bandwidth ΔF. In addition, data transfer can be realized for the case of overcoming "Nyquist barrier" [8–10].

The term "Nyquist barrier" determine conditions for potential BER performance of orthogonal binary signals with symbol rate R_N in AWGN channel with frequency bandwidth ΔF_N. For binary signals with envelope $\sin(\pi t/T)/(\pi t/T)$ for $(-\infty < t < +\infty)$ symbol rate is equal to $R_N = 1/T$ and frequency bandwidth is equal to $\Delta F_N = 1/T$. Then "Nyquist barrier" in frequency domain is equal to $R_N/\Delta F_N = 1$ (bit/s)/Hz.

© Springer International Publishing AG 2017
O. Galinina et al. (Eds.): NEW2AN/ruSMART/NsCC 2017, LNCS 10531, pp. 607–618, 2017.
DOI: 10.1007/978-3-319-67380-6_57

Overcoming "Nyquist Barrier" is discussed in [11]. Faster-Than-Nyquist (FTN) signaling provide increase of symbol rate R with different forms of envelopes. Symbol rate of FTN system can be increased up to 22% with insignificant energy losses (Mazo's limit [10]).

Note, that for considered signals significant intersymbol ineterference ISI appears for $T_c > T$ and constant data rate $R = 1/T$. Significant ISI causes BER performance degradation for known forms of envelopes [12]. In other words, spectral efficiency $R/\Delta F$ can be increased, but energy efficiency degrades: required value E/N_0 increases for fixed error probability, where E – energy of the transmitted signal, and N_0 is average spectral density of AWGN.

It is known that function $a(t)$ for envelope can be obtained as solution of optimization task [5, 12–16]. This approach can provide minimization of ISI and required ΔF or reduction rate of the out-of-band emissions (OOBE). The output of solution of optimization task is optimal signals. These signals allow us to reach Shannon boundary for channel with additive white Gaussian noise and transfer capacity $C = \Delta F \log_2(1 + P_s/P_n)$, where P_s – average signal power, P_n – average noise power in occupied bandwidth ΔF [11]. So there is ambiguity for capacity estimation: value ΔF is different for different level of energy spectrum of random sequence of signals. These facts do not allow us to correctly determine the degree of approximation to the Shannon boundary [8, 9].

Note, that optimization task has next restrictions: restrictions on ΔF, which determine spectral efficiency $R/\Delta F$, and restriction on ISI, which determines BER performance and energy efficiency E/N_0.

Realization of receiver becomes more complex with the increasing L. We should save many already received symbols or construct more complex algorithms to obtain potential BER performance [17]. However, receiver realization can be simple if we apply optimal signals with controlled ISI. Therefore, optimization task should have restrictions on ISI.

The Objective of this Paper: Obtain an estimation of spectral and energy efficiency of optimal signals with increased duration. We will consider different ΔF for different level of energy spectrum of random sequence of signals and will compare efficiency of proposed signals with Shannon boundary for channel with additive white Gaussian noise.

2   Random Sequence of Optimal Signals with Increased Duration and Corresponding Energy Spectrum

In general terms, infinite random sequence of one-frequency signals with duration $T_c = LT$ ($L \geq 1$), carrier frequency f_0, arbitrary form of envelope $a(t)$, amplitude A_0 can be written as follows:

$$s(t) = A_0 \sum_{k=-\infty}^{+\infty} a(t - kT)d^{(k)}\cos\left(2\pi f_0 t + \varphi^{(k)}\right), \tag{1}$$

where $d^{(k)}$ – k-th modulation symbol, $\varphi^{(k)}$ – initial phase of k-th signal in the order. For example, $d^{(k)} = \pm 1$, $\varphi^{(k)} = 0$ for BPSK signals.

The expression of truncated j-th realization of a random sequence of signals (1) on time interval $[0; T_A]$ has next form:

$$\varsigma_j(t) = A_0 \sum_{k=0}^{Q} a(t - kT)d^{(k)}cos\left(2\pi f_0 t + \varphi^{(k)}\right), \tag{2}$$

where $T_A = QT + T_c$. In this article we will consider the case $\varphi^{(k)} = 0$.

Example of separate optimal signals is shown on Fig. 1a. We can see forms of envelopes for duration $T_c = 16T$. Resulting truncated j-th realization of a random sequence of signals (2) from Fig. 1a for $Q = 10$ is shown on Fig. 1b. Increasing of duration of signals results in reducing of occupied frequency bandwidth ΔF. Nevertheless, increasing of duration of signals also results in significant ISI. High level of ISI leads to dramatically degradation of BER performance for non-optimal signals.

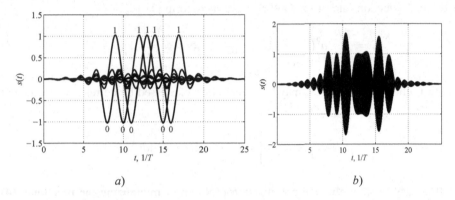

a)   b)

Fig. 1. Example of separate optimal signals (a), example of resulting truncated j-th realization of a random sequence of signals (b).

Energy spectrum $|S(f)|^2$ of random sequence of signals (2) for statistically independent modulation symbols is defined in area of positive frequencies by function a (t) and constant value Z [18, 19]:

$$|S(f)|^2 = \lim_{Q \to \infty} \frac{1}{QT_c} m_1\left\{|S_j(f)|^2\right\} = (Z/T_c)\left| \int_{-T_c/2}^{T_c/2} a(t) \exp(-j2\pi ft)dt \right|^2, \tag{3}$$

where value Z depends on signal constellation, $S_j(f)$ – spectrum of random sequence of signals (2), $m_1\{\ \}$ – mathematical expectation.

As can be seen from (3), function $a(t)$ defines form of energy spectrum $|S(f)|^2$ in area of occupied frequency bandwidth and in area of out-of-band emissions. In other words, function $a(t)$ defines reduction rate of the OOBE.

Optimization task should be solved to find function $a(t)$. Let's consider solution of this optimization task.

3 Optimization Task

Optimization task can be formulated as task of searching for the function $a(t)$, which provides minimization of functional J [5, 7, 16]:

$$arg\left(\min_{a(t)}\{J\}\right), \text{ where } J = \int_{-\infty}^{+\infty} g(f)\left|\int_{-\infty}^{+\infty} a(t)\exp(-j2\pi ft)dt\right|^2 df. \quad (4)$$

Note, that this task has restrictions on signal energy, reduction rate of the OOBE, level of ISI. Weighing function $g(f) = f^{2n}$ in expression (4) determines resulting occupied frequency bandwidth and reduction rate of the OOBE. Resulting occupied frequency bandwidth will be measured by different level of $|S(f)|^2$ relatively to the maximum value of energy spectrum: $\Delta F_{-30\ dB}$, $\Delta F_{-40\ dB}$, $\Delta F_{-50\ dB}$, $\Delta F_{-60\ dB}$.

It can be shown, that restriction on signal energy and boundary conditions, which determine reduction rate of the OOBE, have next form [5, 16]:

$$E = \int_{-T_c/2}^{T_c/2} a^2(t)dt = 1; \ a^{(k)}\big|_{t=\pm T_c/2} = 0, k = 1\ldots(n-1) \quad (5)$$

Restriction on level of ISI may be numerically expressed by correlation coefficient [5, 7, 16]:

$$K = \max_{k=1\ldots(L-1)}\left\{\int_{0}^{(L-k)T} a(t)a(t-kT)dt\right\}. \quad (6)$$

It is easy to show that the optimization problem of minimizing the functional (4) can be simplified to the problem of finding the expansion coefficients, which minimize function of many variables [5, 7, 16]:

$$J\left(\{a_k\}_{k=1}^m\right) = \frac{T_c}{2}\sum_{k=1}^m a_k^2(2\pi k/T_c)^{2n}. \quad (7)$$

Note, that restrictions and boundary conditions (5) and (6) must be modified [5, 7, 16].

Optimal functions $a(t)$ and corresponding energy spectra $|S(f)|^2$ of random sequence of optimal signals are shown on Figs. 2, 3, 4, 5, 6, 7, 8 and 9. Reduction rate of the OOBE for optimal signals is equal to $1/f^4$. Correlation coefficient K is equal to 0.1 on Figs. 2, 4, 6, 8 and equal to 0.01 on Figs. 3, 5, 7 and 9. Resulting expansion coefficients are presented in Table A.1.

Let's consider optimal signals and corresponding energy spectra. Increasing of T_c leads to transformation of $|S(f)|^2$ in area of OOBE. Firstly, values of $f^{(0)} = \min\{|S(f)|^2 = 0\}$ will be reduced for fixed values K. As we can see from Figs. 2 and 8, transition from $T_c = 10T$ to $T_c = 16T$ provides reducing of $f^{(0)}$ from $0.57/T$ to $0.50/T$ for $K = 0.1$. Application of optimal signals with $K = 0.01$ leads to increasing values of $f^{(0)}$.

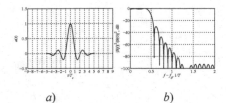

a)   b)

Fig. 2. $T_c = 10T$, $K_0 = 0.1$

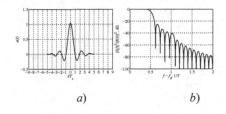

a)   b)

Fig. 3. $T_c = 10T$, $K_0 = 0.01$

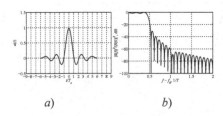

a)   b)

Fig. 4. $T_c = 12T$, $K_0 = 0.1$

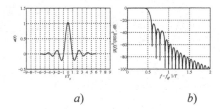

a)   b)

Fig. 5. $T_c = 12T$, $K_0 = 0.01$

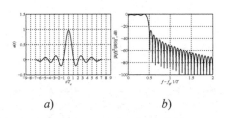

a)   b)

Fig. 6. $T_c = 14T$, $K_0 = 0.1$

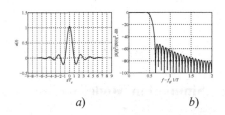

a)   b)

Fig. 7. $T_c = 14T$, $K_0 = 0.01$

For example, $f^{(0)} = 0.62/T$ for $T_c = 10T$ and $f^{(0)} = 0.55/T$ for $T_c = 18T$. Secondly, transition from $K = 0.1$ to $K = 0.01$ leads to increasing maximum value of function $a(t)$. Thirdly, reduction rate of the OOBE is almost constant and equal to $1/f^4$.

Increasing of T_c leads to reducing of ΔF. Note, that we consider doubled frequency bandwidth in comparison with Figs. 2, 3, 4, 5, 6, 7, 8 and 9. This is typical for signals transmitted on a carrier frequency. However, dependence $\Delta F(T_c/T)$ will be different for different level of energy spectra (Fig. 10): −30 dB, −40 dB, −50 dB, −60 dB.

It can be shown (Fig. 10a), that:

$$\Delta F_{-30\,\text{dB}}(T_c = 2T)/\Delta F_{-30\,\text{dB}}(T_c = 18T) = 2.0 \text{ for } K = 0.10;$$
$$\Delta F_{-60\,\text{dB}}(T_c = 2T)/\Delta F_{-60\,\text{dB}}(T_c = 18T) = 2.7 \text{ for } K = 0.10;$$
$$\Delta F_{-30\,\text{dB}}(T_c = 2T)/\Delta F_{-30\,\text{dB}}(T_c = 18T) = 2.6 \text{ for } K = 0.01;$$
$$\Delta F_{-60\,\text{dB}}(T_c = 2T)/\Delta F_{-60\,\text{dB}}(T_c = 18T) = 4.7 \text{ for } K = 0.01.$$

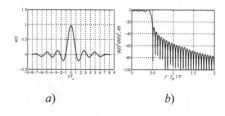

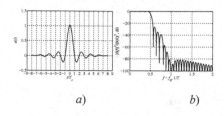

a)   *b)*   *a)*   *b)*

Fig. 8. $T_c = 16T$, $K_0 = 0.1$ **Fig. 9.** $T_c = 16T$, $K_0 = 0.01$

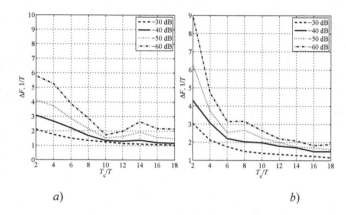

a)   *b)*

Fig. 10. Values ΔF for different level of energy spectra $K = 0.1$ (a), $K = 0.01$ (b).

4   Simulation Model

Simulation model was developed in Matlab. This model is presented on Fig. 11. Block "Initialization of the simulation parameters" performs functions of determination of a (t), the value of T_c/T, signal-to-noise ratio E/N_0. Then, the model proceeds to generation of random information bits. Information bits go to input of block "modulator". In our model we use BPSK modulation. Output signals are transferred through AWGN channel. The mixture of signals and AWGN is transferred to the input of the block "demodulator".

Coherent detection bit-by-bit algorithm [20] is used in our model. This algorithm is simple in realization and provides delay minimization for signal proceeding. At least 10^6 information bits were transmitted to check BER performance at each signal-to-noise ratio in developed simulation model.

Note, that there is block of computing $|S(f)|^2$. Energy spectrum is obtained by averaging on various realizations of a random sequence of signals (2) with $Q = 100$. After computing $|S(f)|^2$ we can calculate ΔF for different level of energy spectra.

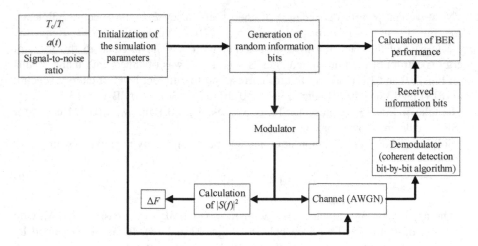

Fig. 11. Block diagram of simulation model.

5   Results and Discussions

Energy efficiency can be numerically estimated by value $\Delta(E/N_0)$. This value presents the difference between BER performance of optimal signals and theoretical BER performance of "classical" BPSK signals in area of error probability $p_{err} = 10^{-3}$. Dependence of $\Delta(E/N_0)$ on duration of optimal signals T_c is shown on Fig. 12.

As we can see, BER performance dramatically degrades if duration of optimal signals T_c increases for the case $K = 0.1$. This fact can be explained by high level of ISI, which increases with increasing of T_c. Transition to $K = 0.01$ leads to reducing of level of ISI and resulting $\Delta(E/N_0)$ is not more than 0.2 dB.

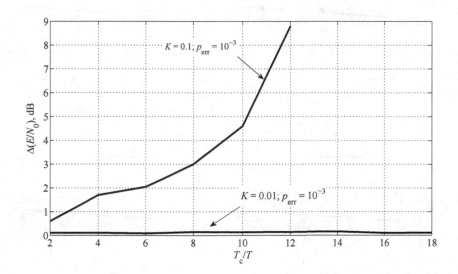

Fig. 12. Dependence of $\Delta(E/N_0)$ on duration of optimal signals T_c.

Now we can present dependence of energy efficiency E/N_0 on spectral efficiency $R/\Delta F$ (Fig. 13) in area of error probability $p_{err} = 10^{-3}$. We assume that the frequency characteristic of the data transmission channel with AWGN is equal to the form of energy spectrum of the transmitted signals. Figure 13 was obtained with using results from Figs. 10 and 12. Different figures correspond to different level of energy spectra: -30 dB (Fig. 13a), -40 dB (Fig. 13b), -50 dB (Fig. 13c), -60 dB Fig. 13d.

Results on Fig. 13 are presented for two cases: $K_0 = 0.1$ and $K_0 = 0.01$. Results for BPSK signals are shown for comparison.

Solid line shows the Shannon limit and can be described by next expression [8, 9]:

$$E/N_0 = \left(2^{(R/\Delta F)} - 1\right)/(R/\Delta F). \tag{8}$$

The application of optimal signals with $K_0 = 0.1$ leads to increasing of E/N_0 with increasing of T_c for fixed error probability (Figs. 12 and 13). It can be explained by application of restriction (6) in optimization task. This restriction does not provide full search on all realizations of signals in analyze interval at receiving. We need to reduce K_0 for BER performance improving, i.e. for E/N_0 reduction for fixed error probability.

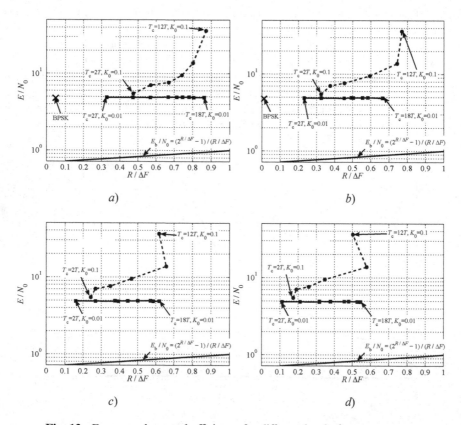

a)   b)

c)   d)

Fig. 13. Energy and spectral efficiency for different level of energy spectra.

Existing ambiguity of ΔF for different level of energy spectra effects on calculation of spectral efficiency. Note, that increasing of signal duration T_c allows to be closer to Shannon boundary (for $K_0 = 0.01$). However, this process slows down with increasing of T_c.

Let's consider spectral efficiency of optimal signals and "Nyquist barrier". Numerical estimation of "Nyquist barrier" is equal to $R_N/\Delta F_N = 1$ (bit/s)/Hz. For comparison of signals we can use conditional "Nyquist barrier". Conditional "Nyquist barrier" is determined for signals with envelope like $\sin(x)/x$ with limited duration. So these signals have specific values of frequency bandwidth for different level of energy spectra. Term "conditional" is associated with ambiguity of ΔF. Occupied frequency bandwidth ΔF for optimal signals and signals with envelope like $\sin(x)/x$ with limited duration is presented in Table 1.

These values are obtained for symbol rate $R = 1/T$. As we can see, overcoming of conditional "Nyquist barrier" for signal duration $T_c = 18T$ is equal to 10–23% for −30 dB level of energy spectra and 53–58% for −60 dB level of energy spectra.

Note, that BER performance of optimal signals with $K_0 = 0.01$ is equal to BER performance of classical BPSK signals. Obtained results shows dependence of possibility of overcoming "Nyquist barrier" on ambiguity of frequency bandwidth.

Table 1. Occupied frequency bandwidth

T_c	$\Delta F_{-30\ dB}$			$\Delta F_{-40\ dB}$		
	$K_0 = 0.01$	$K_0 = 0.1$	$\sin(x)/x$	$K_0 = 0.01$	$K_0 = 0.1$	$\sin(x)/x$
$2T$	$3.04/T$	$2.11/T$	$2.35/T$	$4.31/T$	$3.08/T$	$4.40/T$
$4T$	$2.21/T$	$1.75/T$	$1.95/T$	$3.08/T$	$2.67/T$	$3.20/T$
$6T$	$1.76/T$	$1.49/T$	$1.70/T$	$2.21/T$	$2.20/T$	$2.60/T$
$8T$	$1.50/T$	$1.35/T$	$1.56/T$	$2.03/T$	$1.68/T$	$2.34/T$
$10T$	$1.40/T$	$1.25/T$	$1.46/T$	$1.98/T$	$1.34/T$	$2.14/T$
$12T$	$1.34/T$	$1.15/T$	$1.40/T$	$1.80/T$	$1.30/T$	$2.05/T$
$14T$	$1.28/T$	$1.11/T$	$1.32/T$	$1.71/T$	$1.37/T$	$1.92/T$
$16T$	$1.24/T$	$1.07/T$	$1.30/T$	$1.51/T$	$1.21/T$	$1.80/T$
$18T$	$1.16/T$	$1.04/T$	$1.28/T$	$1.50/T$	$1.16/T$	$1.74/T$

T_c	$\Delta F_{-50\ dB}$			$\Delta F_{-60\ dB}$		
	$K_0 = 0.01$	$K_0 = 0.1$	$\sin(x)/x$	$K_0 = 0.01$	$K_0 = 0.1$	$\sin(x)/x$
$2T$	$6.27/T$	$4.12/T$	$7.38/T$	$8.96/T$	$5.80/T$	$13.05/T$
$4T$	$3.73/T$	$3.72/T$	$5.35/T$	$4.74/T$	$5.25/T$	$9.58/T$
$6T$	$2.54/T$	$2.86/T$	$4.50/T$	$3.16/T$	$3.88/T$	$7.55/T$
$8T$	$2.67/T$	$2.15/T$	$3.84/T$	$3.16/T$	$2.88/T$	$6.80/T$
$10T$	$2.25/T$	$1.52/T$	$3.56/T$	$2.65/T$	$1.73/T$	$6.10/T$
$12T$	$2.00/T$	$1.62/T$	$3.26/T$	$2.20/T$	$1.99/T$	$5.60/T$
$14T$	$1.75/T$	$1.92/T$	$3.06/T$	$2.09/T$	$2.65/T$	$5.24/T$
$16T$	$1.71/T$	$1.57/T$	$2.90/T$	$1.83/T$	$2.17/T$	$4.90/T$
$18T$	$1.62/T$	$1.48/T$	$2.74/T$	$1.89/T$	$2.16/T$	$4.60/T$

6 Conclusions

Ambiguity of frequency bandwidth for real channels with AWGN leads to impossibility of obtain accurate evaluation of overcoming "Nyquist barrier". Conditional overcoming "Nyquist barrier" is equal up to 58% and depends on signals duration and level of energy spectra. Note, that this process is carried out with energy efficiency comparable to energy efficiency of classical BPSK signals.

It is shown that we can be closer to Shannon boundary by application of optimal signals with increased duration. Joint use of optimal signals and large base of the alphabet of modulation symbols leads to effective data rate increasing in limited frequency bandwidth.

The results of the work were obtained under the State contract № 8.2880.2017 with Ministry of Education and Science of the Russian Federation and used computational resources of Peter the Great Saint-Petersburg Polytechnic University Supercomputing Center (http://www.scc.spbstu.ru).

Appendix A

The expansion coefficients of the envelope $a(t)$

Form of envelope $a(t)$ depends on number of expansion coefficients m of the finite Fourier series. We choose m by next expression:

$$\varepsilon(m) = \sqrt{\int_{-T/2}^{T/2} (a_m(t) - a_{m-1}(t))^2 dt} \leq 0.01,$$

Table A.1. The expansion coefficients of the envelope $a(t)$ for $n = 2$.

	$K = 0.1$			$K = 0.01$		
	$T_c = 2T$	$T_c = 10T$	$T_c = 18T$	$T_c = 2T$	$T_c = 10T$	$T_c = 18T$
a_0	+1.0915	+0.2012	+0.1263	+0.9876	+0.2013	+0.1128
a_1	+0.6323	+0.2181	+0.1068	+0.6990	+0.2005	+0.1113
a_2	+0.0654	+0.2014	+0.1244	+0.1489	+0.2014	+0.1122
a_3	−0.0142	+0.2173	+0.1111	−0.0369	+0.1995	+0.1108
a_4	+0.0043	+0.2015	+0.1160	+0.0108	+0.2027	+0.1128
a_5	−0.0018	+0.0615	+0.1236	−0.0045	+0.1339	+0.1094
a_6	0.0009	−0.0065	+0.0990	+0.0022	+0.0008	+0.1125
a_7	0	0	+0.1412	−0.0012	+0.0002	+0.1127
a_8	0	0	+0.0737	+0.0007	+0.0094	+0.1072
a_9		0	−0.0025	0	−0.0063	+0.0753
a_{10}		0	+0.0013	0	0.00380	+0.0166

where $a_m(t)$ and $a_{m-1}(t)$ – functions, which have m and $(m - 1)$ expansion coefficients of the finite Fourier series.

Results of solutions of optimization task are presented in form of expansion coefficients of the finite Fourier series in Table A.1.

References

1. Rodrigues, M., Darwazeh, I.: A spectrally efficient frequency division multiplexing based communication system. In: 8th International OFDM-Workshop, Hamburg, Germany, September 2003, pp. 70–74 (2003)
2. Gelgor, A., Gorlov, A., Nguyen, V.P.: The design and performance of SEFDM with the Sinc-to-RRC modification of subcarriers spectrums. In: 2016 International Conference on Advanced Technologies for Communications (ATC), Hanoi, pp. 65–69 (2016)
3. Xu, T., Darwazeh, I.: Spectrally efficient FDM: spectrum saving technique for 5G? In: 1st International Conference on 5G for Ubiquitous Connectivity, Akaslompolo, pp. 273–278 (2014)
4. Anderson, J.B., Rusek, F., Öwall, V.: Faster-than-nyquist signaling. Proc. IEEE **101**(8), 1817–1830 (2013)
5. Zavjalov, S.V., Volvenko, S.V., Makarov, S.B.: A method for increasing the spectral and energy efficiency SEFDM signals. IEEE Commun. Lett. **20**(12), 2382–2385 (2016)
6. Xu, T., Darwazeh, I.: Nyquist-SEFDM: pulse shaped multicarrier communication with sub-carrier spacing below the symbol rate. In: 2016 10th International Symposium on Communication Systems, Networks and Digital Signal Processing (CSNDSP), Prague, pp. 1–6 (2016)
7. Zavjalov, S.V., Makarov, S.B., Volvenko, S.V.: Duration of nonorthogonal multifrequency signals in the presence of controlled intersymbol interference. In: 2015 7th International Congress on Ultra Modern Telecommunications and Control Systems and Workshops (ICUMT), Brno, pp. 49–52 (2015)
8. Gattami, A., Ringh, E., Karlsson, J.: Time localization and capacity of faster-than-nyquist signaling. In: 2015 IEEE Global Communications Conference (GLOBECOM), pp. 1–7 (2015)
9. Zhou, J., Li, D., Wang, X.: Generalized faster-than-Nyquist signaling. In: 2012 IEEE International Symposium on Information Theory Proceedings (ISIT), pp. 1478–1482 (2012)
10. Mazo, J.E.: Faster-than-Nyquist signaling. Bell Syst. Tech. J. **54**, 1451–1462 (1975)
11. Fan, J., Guo, S., Zhou, X., Ren, Y., Li, G.Y., Chen, X.: Faster-than-Nyquist signaling: an overview. IEEE Access **5**, 1925–1940 (2017)
12. Zavjalov, S.V., Makarov, S.B., Volvenko, S.V.: Reduction of energy losses under conditions of overcoming "Nyquist Barrier" by optimal signal selection. In: Galinina, O., Balandin, S., Koucheryavy, Y. (eds.) NEW2AN/ruSMART -2016. LNCS, vol. 9870, pp. 620–627. Springer, Cham (2016). doi:10.1007/978-3-319-46301-8_53
13. Zavjalov, S.V., Makarov, S.B., Volvenko, S.V.: Application of optimal spectrally efficient signals in systems with frequency division multiplexing. In: Balandin, S., Andreev, S., Koucheryavy, Y. (eds.) NEW2AN 2014. LNCS, vol. 8638, pp. 676–685. Springer, Cham (2014). doi:10.1007/978-3-319-10353-2_63
14. Xue, W., Ma, W.Q., Chen, B.C.: Research on a realization method of the optimized efficient spectrum signals using legendre series. In: 2010 IEEE International Conference on Wireless Communications, Networking and Information Security (WCNIS), vol. 1, pp. 155–159 (2010)

15. Xue, W., Ma, W.Q., Chen, B.C.: A realization method of the optimized efficient spectrum signals using fourier series. In: 2010 6th International Conference on Wireless Communications Networking and Mobile Computing (WICOM) (2010)
16. Zavjalov, S.V., Makarov, S.B., Volvenko, S.V., Xue, W.: Waveform optimization of SEFDM signals with constraints on bandwidth and an out-of-band emission level. In: Balandin, S., Andreev, S., Koucheryavy, Y. (eds.) ruSMART 2015. LNCS, vol. 9247, pp. 636–646. Springer, Cham (2015). doi:10.1007/978-3-319-23126-6_57
17. Rashich, A., Kislitsyn, A., Fadeev, D., Ngoc Nguyen, T.: FFT-based trellis receiver for SEFDM signals. In: 2016 IEEE Global Communications Conference (GLOBECOM), Washington, DC, pp. 1–6 (2016)
18. Gelgor, A., Gorlov, A., Popov, E.: Multicomponent signals for bandwidth-efficient single-carrier modulation. In: 2015 IEEE International Black Sea Conference on Communications and Networking, BlackSeaCom 2015, Article no. 7185078, pp. 19–23 (2015)
19. Gelgor, A., Gorlov, A., Popov, E.: On the synthesis of optimal finite pulses for bandwidth and energy efficient single-carrier modulation. In: Balandin, S., Andreev, S., Koucheryavy, Y. (eds.) ruSMART 2015. LNCS, vol. 9247, pp. 655–668. Springer, Cham (2015). doi:10.1007/978-3-319-23126-6_59
20. Zavjalov, S.V., Makarov, S.B., Volvenko, S.V.: Nonlinear coherent detection algorithms of nonorthogonal multifrequency signals. In: Balandin, S., Andreev, S., Koucheryavy, Y. (eds.) NEW2AN 2014. LNCS, vol. 8638, pp. 703–713. Springer, Cham (2014). doi:10.1007/978-3-319-10353-2_66

Choosing Parameters of Optimal Signals with Restriction on Correlation Coefficient

Anna S. Ovsyannikova, Sergey V. Zavjalov[✉], Sergey B. Makarov, and Sergey V. Volvenko

Peter the Great St. Petersburg Polytechnic University, St. Petersburg, Russia
anny-ov97@mail.ru, zavyalov_sv@spbstu.ru,
{makarov,volk}@cee.spbstu.ru

Abstract. Optimal signals with smoothed envelopes and increased duration are considered. Smoothed envelopes were obtained by optimization task with restriction on the energy of signal, boundary conditions and restriction on intersymbol interference. Solutions of optimization task for different values of restrictions are considered. It is shown that increasing of duration of optimal signals leads to increasing of the maximum value of the envelope and nonlinear change of the energy spectrum's form. We can assume that as correlation coefficient becomes less, the form of envelope gets closer and closer to the form of the function $\sin(x)/x$ and the maximum value of the envelope becomes larger. Energy loss relatively to the theoretical BER performance of BPSK signals is not more than 0.2 dB for correlation coefficient 0.01. Increasing of correlation coefficient is accompanied by degradation of BER performance. This effect is especially evident for high values of signal duration. So for multifrequency signals choosing small values of correlation coefficient more than 0.1 may be reasonable because of reducing occupied frequency bandwidth. Energy losses which occur for small values of correlation coefficient, may be compensated by redundancy of transmitted data. For one-frequency signals it is better to choose small values of correlation coefficient (less than 0.1).

Keywords: Optimal signals · BER performance · Out-of-band emissions · Restrictions · Correlation coefficient

1 Introduction

Modern wireless broadband 4G and 5G networks are developing sweepingly in the direction of using multifrequency spectrally efficient signals which have dense spectrum $S(f)$ and limited bandwidth [1, 2]. Classical version [3, 4] supposes applying of SEFDM signals with duration T_c, rectangular envelope form and nonorthogonal frequency separation between subcarriers $\Delta f < 1/T$. Such signals allow to reduce occupied frequency bandwidth ΔF of SEFDM signals, but it leads to significant degradation of BER performance, especially for small values of Δf [5–7].

Extra reducing of ΔF may be achieved due to application of smoothed envelopes and signals with increased duration $T_c = LT$ ($L > 1$) [8–10]. Increasing of signal

© Springer International Publishing AG 2017
O. Galinina et al. (Eds.): NEW2AN/ruSMART/NsCC 2017, LNCS 10531, pp. 619–628, 2017.
DOI: 10.1007/978-3-319-67380-6_58

duration results in decreasing of the occupied frequency bandwidth defined for the specified level of out-of-band emissions (OOBE), e.g. for the level of −30 dB ($\Delta F_{-30\ \text{dB}}$).

Reducing Δf causes interference between signals transmitted on the neighbouring subcarriers. Consider that subcarriers are located in the points corresponding to the minimum value of solution of the equation $|S(f)|^2 = 0$ (Fig. 1). Frequency separation between subcarriers should be chosen according to this fact to minimize the influence of the interference. On Fig. 1 along the ordinate axis values of energy spectrum are plotted. These spectra belong to a random sequence of OFDM signals and multifrequency signals with optimal form of the envelope for the number of subcarriers $N = 7$. The central subcarrier on Fig. 1 is located at f_0. The bandwidth gain $\varepsilon_{\Delta F}$ is defined for the level of −30 dB and depends on the form of envelope $a(t)$ of signal [6, 10–12].

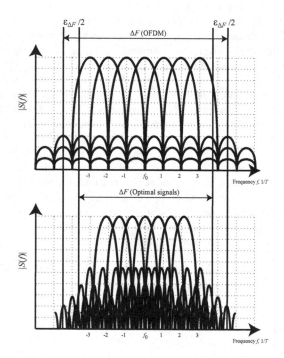

Fig. 1. Spectrum of random sequence of multifrequency signals.

Subcarriers' location (Fig. 1) allows to minimize mutual interference of signals transmitted on the neighbouring subcarrier frequencies [8–10, 13]. Therefore optimal envelope forms of signals should be chosen [8, 10, 14, 15]. These forms may be found as a solution of optimization task which has restrictions on the rate of OOBE (occupied frequency bandwidth) and BER performance [8, 10]. Transmitting data in such conditions happens with overcoming "Nyquist Barrier". In this work we try to determine quantitative estimations of parameters, specifically energy losses which are associated with degradation of BER performance, and bandwidth gains.

2 Optimal Signals and Optimization Task

In general terms, multifrequency signal with duration $T_c = LT$ ($L \geq 1$), arbitrary form of envelope $a(t)$ and amplitude A_0 may be written as follows:

$$s_j^{(k)}(t) = A_0 \sum_{n=-N/2}^{(N-1)/2} \sum_{k=-\infty}^{+\infty} a(t - kT)d_n^{(k)}cos(2\pi(f_0 + n\Delta f)t). \tag{1}$$

where $d^{(k)}$ – k-th modulation symbol on n-th subcarrier frequency. For example, $d^{(k)} = \pm 1$ for BPSK signals.

The truncated j-th realization of a random sequence of signals (1) on time interval $[0; T_A]$ has next form:

$$\varsigma_j(t) = A_0 \sum_{n=-N/2}^{(N-1)/2} \sum_{k=0}^{Q} a(t - kT)d_n^{(k)}cos(2\pi(f_0 + n\Delta f)t). \tag{2}$$

where $T_A = QT + T_c$.

Energy spectrum $|S(f)|^2$ of a random sequence of signals (2) for statistically independent modulation symbols is defined in the area of positive frequencies by function $a(t)$ and constant value Z [11]:

$$|S(f)|^2 = \sum_{n=-N/2}^{(N-1)/2} \lim_{Q \to \infty} \frac{1}{QT_c} m_1 \left\{ |S_j(f_0 + n\Delta f)|^2 \right\}$$

$$= \sum_{n=-N/2}^{(N-1)/2} (Z/T_c) \left| \int_{-T_c/2}^{T_c/2} a(t) \exp(-j2\pi(f_0 + n\Delta f)t)dt \right|^2, \tag{3}$$

where Z depends on signal constellation, $S_j(f)$ – spectrum of a random sequence of signals (2), $m_1\{\ \}$ – mathematical expectation.

As can be seen from (3), function $a(t)$ defines form of energy spectrum $|S(f)|^2$ both in the area of occupied frequency bandwidth and in the area of OOBE. We have to solve optimization task to find $a(t)$. Let's consider this solution.

Optimization functional is determined by next expression [10]:

$$arg\left(\min_{a(t)} \left\{ g(f) \left| \int_{-\infty}^{+\infty} a(t)\exp(-j2\pi ft)dt \right|^2 df \right\} \right). \tag{4}$$

Here g(f) = f 2n (n = 1, 2, 3...) [8, 10] is a weighing function. Assigned optimization task has a list of restrictions: restriction on the energy of signal, boundary conditions [8–10, 15]:

$$E = \int\limits_{-T_c/2}^{T_c/2} a^2(t)dt = 1, a^{(k)}\big|_{t=\pm T_c/2} = 0, k = 1\ldots(n-1) \tag{5}$$

The restriction on the level of ISI of signals transmitted at each subcarrier frequency may be set numerically by the coefficient of mutual correlation [8, 10]:

$$\max_{k=1\ldots(L-1)} \left\{ \int\limits_{0}^{(L-k)T} a(t)a(t-kT)dt \right\} < K_0. \tag{6}$$

To solve optimization task numerically we will use well known method [8, 10, 15]. This task transforms into the task of searching for the coefficients of expansion in a limited Fourier series which provide minimization of the function of several variables. Thus, forms of optimal signals may be obtained.

3 Optimal Envelope Forms of Signals with Increased Duration

Let's consider solution of optimization task (4) for different values of T_c and K_0, taking restrictions (5) and (6) into account. Adding restriction on the level of ISI leads to degradation of the rational envelope form of signals transmitted at each subcarrier frequency. Energy spectra (3) of random sequences of such signals degrades too. On Fig. 2a and c you can see envelope forms of optimal signals with duration $T_c = 10T$ and $T_c = 18T$ correspondingly. The rate of OOBE was chosen as $1/f^2$ during optimization procedure. Forms of energy spectra belonging to random sequences of optimal signals are shown on Fig. 2b and d.

We can see that increasing of duration of optimal signals leads to increasing of the maximum value of the envelope $\max\{a(t)\}$ and nonlinear change of the energy spectrum's form. The restriction on correlation coefficient (6) influences on the spectrum significantly. Comparing Fig. 2b and d, we can say that there is considerable dependence of Δf value on the choice of correlation coefficient (Table1). For $T_c = 10T$ and $K = 0.01$ the value Δf increases 3.3 times relatively to the one, obtained without restriction on the level of ISI (Table 1, Fig. 2b). For $T_c = 18T$ value Δf increases 6.7 times (Table 1, Fig. 2d). Occupied frequency bandwidth for the level $\Delta F_{-30\text{ dB}}$ increases similarly: up to 2.8 times for $T_c = 10T$ and up to 4.7 times for $T_c = 18T$.

To sum up, we can assume that as correlation coefficient becomes less, the form of rational envelope gets closer and closer to the form of the function $\sin(x)/x$ and the maximum value of the envelope becomes larger.

The dependencies of $\Delta F_{-30\text{ dB}}$ and Δf from ratio T_c/T for $K = 0.01$ and different values of the rate of OOBE n are shown on Fig. 3. The dependence $\Delta f(T_c/T)$ is almost the same for each n. Spectrum widening caused by decreasing of correlation coefficient becomes less notable when signal duration T_c increases. So we can suppose that Δf tends to the value $0.5/T$, and $\Delta F_{-30\text{ dB}}$ tends to the value $1/T$ as signal duration T_c increases (despite the value of n).

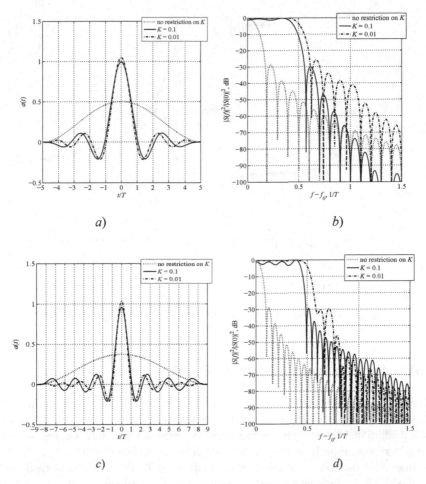

Fig. 2. Optimal envelope forms and energy spectra for $T_c = 10T$ (a, b) and $T_c = 18T$ (c, d).

Table 1. Values of Δf and $\Delta F_{-30\text{dB}}$

	$T_c = 10T$		$T_c = 18T$	
	Δf, $1/T$	$\Delta F_{-30\text{dB}}$, $1/T$	Δf, $1/T$	$\Delta F_{-30\text{dB}}$, $1/T$
No restriction on K	0,19	0,49	0,10	0,27
$K = 0.1$	0,57	1,23	0,50	1,04
$K = 0.01$	0,62	1,41	0,67	1,26

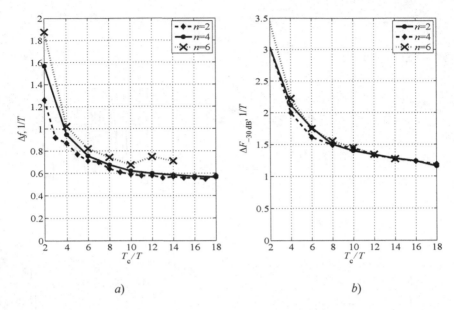

Fig. 3. Dependence of $\Delta F_{-30\ dB}$ and Δf from ratio T_c/T for $K = 0.01$.

4   BER Performance of Optimal Signals

BER performance of optimal signals was obtained by simulation modeling in Matlab system [16]. Error probabilities were calculated for the channel with constant parameters and additional Gaussian noise with power spectral density $N_0/2$. On Fig. 4 we can see graphs representing BER performance for different K. For $K = 0.01$ (Fig. 4a) BER performance almost does not depend on signal duration T_c. Energy loss relatively to the theoretical BER performance of BPSK signals is not more than 0.2 dB.

Increasing of correlation coefficient is accompanied by degradation of BER performance (Fig. 4b–d). This effect is especially evident for high values of signal duration T_c. So, for $T_c = 16T$ energy loss relatively to the theoretical BER performance is equal to 2.6 dB when $K = 0.05$. When $K = 0.2$, error probability is not less than 0.07 for $T_c = 16T$.

Resulting dependencies of error probability from correlation coefficient K are shown on Fig. 5. If duration of optimal signal increases, BER performance degrades. This fact conforms to the idea of "working over Nyquist Barrier". Numerically these losses may be estimated as an increase of error probability for specified signal-to-noise ratio. So, for $K = 0.1$ the conversion from $T_c = 2T$ to $T_c = 16T$ results in increasing of error probability 3.3 times if $E/N_0 = 6$ dB (Fig. 5a), 8.9 times if $E/N_0 = 8$ dB (Fig. 5b) and 44 times if $E/N_0 = 10$ dB (Fig. 5c).

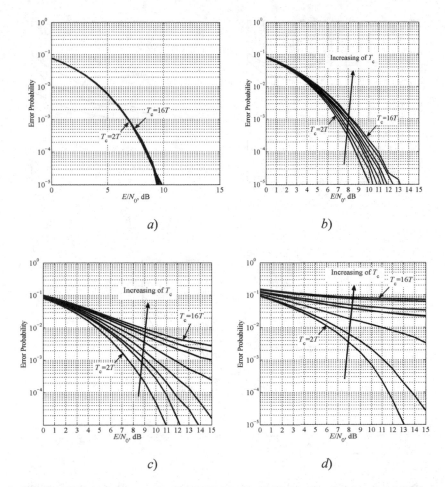

Fig. 4. BER performance for $K = 0.01$ (a), $K = 0.05$ (b), $K = 0.1$ (c), $K = 0.2$ (d).

Using the curves on Figs. 4 and 5, we can make next conclusion. For multifrequency signals choosing small values of correlation coefficient $K > 0.1$ may be reasonable because of reducing occupied frequency bandwidth (Fig. 1). Energy losses which occur for small values of K, may be compensated by redundancy of transmitted data. For one-frequency signals it is better to choose small values of $K < 0.1$.

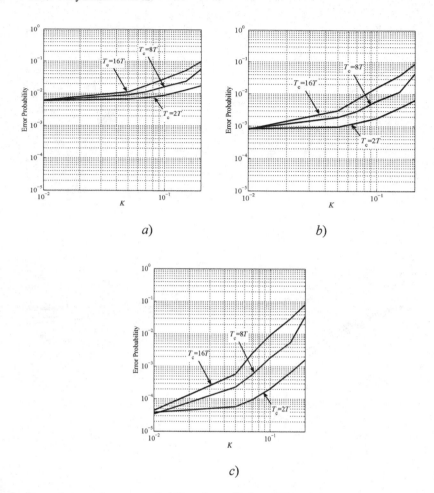

Fig. 5. Dependence of error probability from K for $E/N_0 = 6$ dB (a), $E/N_0 = 8$ dB (b), $E/N_0 = 10$ dB (c).

5 Conclusions

In the work we give quantitative estimates of energy losses for one-frequency optimal signals with increased duration under conditions of overcoming "Nyquist Barrier". It is shown that if correlation coefficient $K > 0.1$ is chosen as an optimization parameter, energy losses are equal to 7–8 dB. Therefore, the area of application of optimal signals is considerably limited.

For multifrequency optimal signals the choice of correlation coefficient $K > 0.1$ as an optimization parameter allows to reduce occupied frequency bandwidth significantly and improve spectral efficiency of data transmission.

Increase of duration of optimal signals with restriction on correlation coefficient $K < 0.1$ leads to the bias of the first zero of energy spectrum belonging to a random sequence of signals to the value $0.5/T$.

Variation of the parameter which determines the rate of OOBE of random sequence of signals almost does not influence on the value of energy spectrum's first zero and occupied frequency bandwidth.

The results of the work were obtained under the State contract № 8.2880.2017 with Ministry of Education and Science of the Russian Federation and used computational resources of Peter the Great Saint-Petersburg Polytechnic University Supercomputing Center (http://www.scc.spbstu.ru).

References

1. Xu, T., Darwazeh, I.: Spectrally efficient FDM: spectrum saving technique for 5G?. In: 1st International Conference on 5G for Ubiquitous Connectivity, Akaslompolo, pp. 273–278 (2014)

2. Andreev, S., Galinina, O., Pyattaev, A., Hosek, J., Masek, P., Yanikomeroglu, H., Koucheryavy, Y.: Exploring synergy between communications, caching, and computing in 5G-grade deployments. IEEE Commun. Mag. **54**(8), 60–69 (2016). Article no. 7537178

3. Rodrigues, M., Darwazeh, I.: A spectrally efficient frequency division multiplexing based communication system. In: 8th International OFDM-Workshop, Hamburg, Germany, pp. 70–74, September 2003

4. Mazo, J.E.: Faster-than-Nyquist Sig. Bell Syst. Tech. J. **54**, 1451–1462 (1975)

5. Rashich, A., Kislitsyn, A., Fadeev, D., Nguyen, T.N.: FFT-Based trellis receiver for sefdm signals. In: 2016 IEEE Global Communications Conference (GLOBECOM), Washington, DC, pp. 1–6 (2016)

6. Gelgor, A., Gorlov, A., Nguyen, V.P.: The design and performance of SEFDM with the Sinc-to-RRC modification of subcarriers spectrums. In: 2016 International Conference on Advanced Technologies for Communications (ATC), Hanoi, pp. 65–69 (2016)

7. Zavjalov, S.V., Makarov, S.B., Volvenko, S.V.: Nonlinear coherent detection algorithms of nonorthogonal multifrequency signals. In: Balandin, S., Andreev, S., Koucheryavy, Y. (eds.) NEW2AN 2014. LNCS, vol. 8638, pp. 703–713. Springer, Cham (2014). doi:10.1007/978-3-319-10353-2_66

8. Zavjalov, S.V., Makarov, S.B., Volvenko, S.V., Xue, W.: Waveform optimization of SEFDM signals with constraints on bandwidth and an out-of-band emission level. In: Balandin, S., Andreev, S., Koucheryavy, Y. (eds.) ruSMART 2015. LNCS, vol. 9247, pp. 636–646. Springer, Cham (2015). doi:10.1007/978-3-319-23126-6_57

9. Zavjalov, S.V., Makarov, S.B., Volvenko, S.V.: Duration of nonorthogonal multifrequency signals in the presence of controlled intersymbol interference. In: 2015 7th International Congress on Ultra Modern Telecommunications and Control Systems and Workshops (ICUMT), Brno, pp. 49–52 (2015)

10. Zavjalov, S.V., Volvenko, S.V., Makarov, S.B.: A Method for Increasing the spectral and energy efficiency SEFDM signals. IEEE Commun. Lett. **20**(12), 2382–2385 (2016)

11. Gelgor, A., Gorlov, A., Popov, E.: Multicomponent signals for bandwidth-efficient single-carrier modulation. In: 2015 IEEE International Black Sea Conference on Communications and Networking, BlackSeaCom 2015, Article no. 7185078, pp. 19–23 (2015)

12. Gelgor, A., Gorlov, A., Popov, E.: On the synthesis of optimal finite pulses for bandwidth and energy efficient single-carrier modulation. In: Balandin, S., Andreev, S., Koucheryavy, Y. (eds.) ruSMART 2015. LNCS, vol. 9247, pp. 655–668. Springer, Cham (2015). doi:10.1007/978-3-319-23126-6_59

13. Isam, S., Darwazeh, I.: Characterizing the intercarrier interference of non-orthogonal spectrally efficient FDM system. In: 2012 8th International Symposium on Communication Systems, Networks & Digital Signal Processing (CSNDSP), pp. 1–5, 18–20 July 2012
14. Xue, W., Ma, W.-Q., Chen, B.C.: Research on a realization method of the optimized efficient spectrum signals using Legendre series. Wireless Communications. In: 2010 IEEE International Conference on Networking and Information Security (WCNIS), pp. 155–159, 25–27 June 2010
15. Xue, W., Ma, W., Chen, B.: A realization method of the optimized efficient spectrum signals using Fourier series. In: 2010 6th International Conference on Wireless Communications Networking and Mobile Computing (WiCOM), pp. 1–4, 23–25 September 2010
16. Zavjalov, S.V., Makarov, S.B., Volvenko, S.V.: Reduction of Energy Losses Under Conditions of Overcoming "Nyquist Barrier" by Optimal Signal Selection. In: Galinina, O., Balandin, S., Koucheryavy, Y. (eds.) NEW2AN/ruSMART -2016. LNCS, vol. 9870, pp. 620–627. Springer, Cham (2016). doi:10.1007/978-3-319-46301-8_53

The Filtration of Composite Signals from Interference by the Maximum Likelihood Method

Kseniia V. Vlasova[1], Valerii A. Pakhotin[1], Evgenii V. Korotey[1],
and Sergey B. Makarov[2(✉)]

[1] Immanuel Kant Baltic Federal University, Kaliningrad, Russia
{p_ksenia, eugeny_korotey}@mail.ru,
VPakhotin@kantiana.ru
[2] Peter the Great St. Petersburg Polytechnic University, St. Petersburg, Russia
makarov@cee.spbstu.ru

Abstract. The technology of wideband signals filtering in the presence of non-orthogonal interference and the additive noise is presented in this work. It is shown that interference can be represented as a pulse with constant unknown parameters, which can be estimated by means of maximum likelihood method, and deleting from the received message. Presents the results of model calculations and experiments, confirming the main theoretical insights.

Keywords: Filtering · Noise immunity · Broadband signals · Likelihood functional

1 Introduction

Currently, there is a change from simple narrow band signals to complex wide band signals, and, respectively, the bandwidth of receiving devices becomes wider and can reach megahertz or more. In this context, the issues of noise immunity, the problems of filtering wide band signals from the interference are extremely relevant. Analogue and digital filters are widely known to provide consistent optimal signal filtering [3, 5], also the Wiener and the Kalman filters, which can be adapted to changing signal reception conditions [1, 2], can provide the same quality of filtering. However, such filters work effectively only with quasi-orthogonality of the signal and the interference. In the theory of optimal reception, interference is considered as a random process with the known laws for the distribution of its parameters [3, 5]. In this case, the likelihood function is statistically averaged. However, the efficiency of processing is drastically reduced.

This paper proposes a new technology of signal filtering from the interference by the maximum likelihood filters. The technology is based on the research results of the [4]. It allows you to efficiently filter the signal from the interference in the area of their non-orthogonality. The interference is considered random, but the correlation interval of its parameters is longer than the processing interval, i.e. duration of the signal. This assumption is fully justified for wide band signals with a duration of less than 100 μs. At such intervals, the interference may be represented as a radio impulse with constant

O. Galinina et al. (Eds.): NEW2AN/ruSMART/NsCC 2017, LNCS 10531, pp. 629–634, 2017.
DOI: 10.1007/978-3-319-67380-6_59

but unknown parameters. In this case, the received message contains a signal, interference and additive noise. Using the technology of direct minimization of the likelihood functional [4], it is possible to estimate unknown interference parameters and to exclude it from the received message. The capabilities of the presented technology are limited by the signal-to-noise ratio. However, during processing, the wide band signals significantly increase the signal-to-noise ratio, so the filtering process proves to be rather effective.

2   The Foundations of the Theory

Let describe the filtering technology of the composite signal from interference burst by the filters of maximum likelihood. The technology is based on maximum length sequence. Let assume that the received message includes composite signal and interference burst.

$$\hat{y}(t) = \hat{U}_C e^{i\omega_c(t-t_c)} \mu(t-t_c) + \hat{U}_\Pi e^{i\omega_\Pi(t-t_n)} + \hat{U}_{III}(t), \tag{1}$$

where $\hat{U}_C$ is the phasor of the composite signal, $\mu(t - t_c)$ is the modulating function of the composite signal, t_c is the time of reception of the composite signal, ω_c is the angular frequency of the signal, $\hat{U}_\Pi$, ω_Π are the phasor and the frequency of the interference.

The interference parameters $\hat{U}_\Pi$, ω_Π are constant over the signal duration interval. The interference burst can completely overlap the interval $[t_c; t_c + T]$. In this case, the duration of the interference is longer than the duration of the signal. The interference burst can partially overlap the interval $[t_c; t_c + T]$. These cases are taken into account automatically due to the coefficient of cross-correlation of the signal and interference $\hat{R}$.

Let write the logarithm of the likelihood function based on (1).

$$\ln\left(L\left(\bar{\lambda}'\right)\right) = -\frac{1}{2\sigma^2\tau_k} \int_{t_c'}^{t_\Pi'+T} \left|\hat{y}(t) - \hat{U}_C' e^{i\omega_c'(t-t_c')} \mu(t-t_c') - \hat{U}_\Pi' e^{i\omega_\Pi'(t-t_\Pi')}\right|^2 dt, \tag{2}$$

where $\bar{\lambda}'$ is the vector of the estimated signal and interference parameters;

T is the signal duration.

While differentiating (2) with respect to the amplitudes $\hat{U}_C'$ and $\hat{U}_\Pi'$, and equating differentials to zero, we obtain an equation system from which the functional dependences follow.

$$
\hat{U}'_c\left(t'_c, t'_\Pi, \omega'_C, \omega'_\Pi\right) = \frac{1}{T} \frac{\int\limits_{t'_c}^{t'_\Pi + T} \hat{y}(t)e^{i\omega'_c(t - t'_c)}\mu(t - t'_c)dt - \hat{R}\int\limits_{t'_c}^{t'_\Pi + T} \hat{y}(t)e^{i\omega'_\Pi(t - t'_\Pi)}dt}{1 - |\hat{R}|^2},
$$

$$
\hat{U}'_\Pi\left(t'_c, t'_\Pi, \omega'_C, \omega'_\Pi\right) = \frac{1}{T} \frac{\int\limits_{t'_c}^{t'_\Pi + T} \hat{y}(t)e^{i\omega'_\Pi(t - t'_\Pi)}dt - \hat{R}^*_i\int\limits_{t'_0}^{t'_0 + T} \hat{y}(t)e^{i\omega'_c(t - t'_c)}\mu(t - t'_c)dt}{1 - |\hat{R}|^2}, \qquad (3)
$$

where $\hat{R} = \frac{1}{T}\int\limits_{t'_c}^{t'_\Pi + T} e^{i\omega'_\Pi(t - t'_\Pi)}e^{-i\omega'_c(t - t'_c)}\mu(t - t'_c)dt$.

Inserting (3) into (2) the converted likelihood functional is received as

$$
\Delta\left(t'_c, t'_\Pi, \omega'_c, \omega'_\Pi\right) = \int\limits_{t'_c}^{t'_\Pi + T} |\hat{y}(t)|^2 dt - \hat{U}'_C \int\limits_{t'_c}^{t'_\Pi + T} \hat{y}^*(t)e^{i\omega_c(t - t'_c)}\mu(t - t'_c)dt
$$
$$
- \hat{U}'_\Pi \int\limits_{t'_c}^{t'_\Pi + T} \hat{y}^*(t)e^{i\omega_\Pi(t - t'_\Pi)}dt \qquad (4)
$$

The converted likelihood functional is a surface in the space of parameters $t'_0, t'_\Pi, \omega'_C, \omega'_\Pi$. The minimum of the surface determines estimated parameters of the signal and the interference $\hat{U}'_C, \hat{U}'_C, t'_0, \omega'_C, \omega'_\Pi$. The expected value of (4) determines the noise dispersion at the same point. Therefore, it is possible to completely estimate the interference parameters by the minimum of the converted likelihood functional. They allow to generate an interference signal $\hat{y}_\Pi(t) = \hat{U}'_\Pi \exp(i\omega'_\Pi(t - t'_\Pi))$ and subtract it from the received message. As a result, a new message $\hat{y}1(t) = \hat{y}(t) - \hat{y}_\Pi(t)$ is received. The accuracy of the estimates of the interference parameters in the received message determines the quality of the filtering.

3   The Results of the Model Studies

Let review the results of the model studies of filtering the composite signal from the frequency concentrated interference. Figure 1(a) shows the received message containing a composite signal based on the 16-positional maximum length sequence (signal position and its amplitude are marked by a line), and additive noise. The signal parameters: amplitude 1, time of reception 0.6 ms, signal duration 480 μs, duration of one positions 30 μs, initial phase 30°, frequency 50 kHz, SNR = −14 dB. The processing interval equals to the duration of the composite signal 480 μs. With such signal duration, the correlation filter significantly suppresses the noise component (Fig. 1(b)). At the same time the SNR increases by $\sqrt{N} = \sqrt{2400} \approx 50$ times (N is a number of uncorrelated noise in the processing interval). As a result, the signal estimates prove to be rather accurate. Figure 1(c) shows the spectrum of a composite signal.

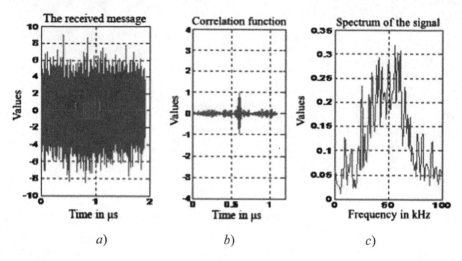

Fig. 1. The received message containing a composite signal and additive noise (left), a correlation function (center), and the spectrum of the received message. SNR = −14 dB. The correlation filter effectively suppresses the noise.

When adding the noise with parameters: amplitude 5, frequency 50 kHz, the received message, the correlation function, and the spectrum of the received message change (Fig. 2(a), (b), (c)).

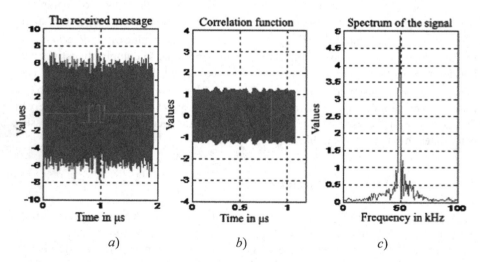

Fig. 2. The received message (left), containing a composite signal, additive noise and interference with an amplitude of 5 V. The SNR = 0 dB. The correlation filter cannot filter out this interference. The joint spectrum (right) consists of a wide signal spectrum and a narrow interference spectrum.

The correlation filter cannot filter out the interference (Fig. 2(b)). Only the orthogonal part of the interference is filtered.

Figure 3 shows the filtration results in accordance with the proposed processing technology.

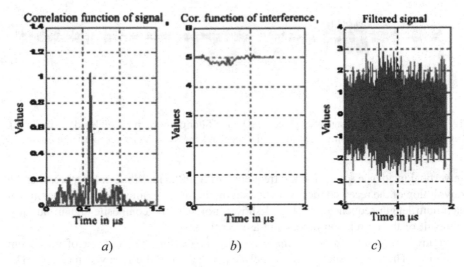

Fig. 3. The correlation functions of the signal (a) and interference (b), the received message with the excluded interference (c). SNR = 0 dB. The interference amplitude 5 V, the signal amplitude is 1 V. The frequencies of the interference and the signal are the same.

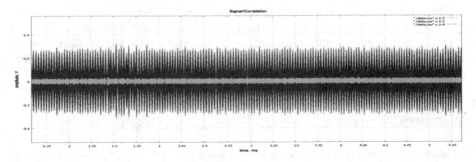

Fig. 4. The received message that contains the composite signal with a carrier frequency of 3500 kHz reflected from the ionosphere, and continuous interference with large amplitude. The correlation processing (white color) does not detect the signal in the noise.

The correlation function of the signal (Fig. 3(a)) provides a sufficiently accurate estimate of the signal parameters. The correlation function of the interference (Fig. 3 (b)) provides sufficiently accurate estimates of the interference parameters. As a result, the interference can be reproduced and excluded from the received message (Fig. 3(c)).

Resulting from the assumption about the structure of the interference in the processing interval, an experimental evidence of the existence of this kind of interference is required. Figure 4 shows the received message that contains the composite signal, interference, and additive noise. The received message was obtained with the help of

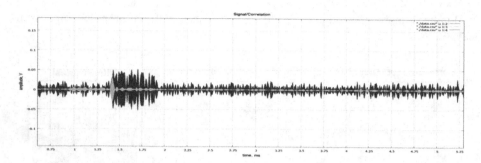

Fig. 5. The received message after processing by the filter of maximum likelihood (black color). The result of the correlation processing of the received message after filtering (white color).

the vertical incidence sounder at the frequency of 3500 kHz. White color represents the convolution of the received message with the image of the emitted signal (16-positional maximum length sequence). As Fig. 4 shows, not only the composite signal, but also the result of the correlation processing are hardly seen.

Figure 5 shows the received message after processing by the filter of maximum likelihood. The interference is completely excluded from the received message. The correlation processing after filtering allows to estimate the time of reception and the amplitude of the composite signal after reflecting from the ionosphere. The experiment fully confirms the possibility of filtering the signal from the interference in the assumption of the constancy of the interference parameters in the processing interval.

The presented filtration technology of simple and composite signals from concentrated interference and interference burst, both natural and intentional, allows to almost completely exclude the interference without distorting the signal information. It can be used in location, navigation and communication systems with digital processing of signal in the noise.

References

1. Kalman, R.E.: A new approach to linear filtering and predication problems. Trans. ASME Ser. D **8**, 35–45 (1960)
2. Winer, N.: The Interpolation, Extrapolation and Smoothing of Stationary Time Series, p. 162. Wiley, Hoboken (1949)
3. Perov A.I.: Statisticheskaya teoriya radiotekhnicheskih sistem [Statistical Theory of Radio Technical Systems], 400 p. Radiotekhnika Publ., Moscow (2003)
4. Vlasova, K.V., Pachotin, V.A., Klionskiy, D.M., Kaplun, D.I.: Estimation of radio impulse parameters using the maximum likelihood method. Mach. Learn. Data Anal. **1**(14), 1977–1990 (2015)
5. Shirman, Y.D.: Razreshenie i szhatie signalov [The Resolution and Compression of Signals]. Sov. Radio Publ., Moscow (1974)

Direct Signal Processing for GNSS Integrity Monitoring

Igor A. Tsikin$^{(\boxtimes)}$ and Antonina P. Melikhova

St. Petersburg Polytechnic University, 29 Politechnicheskaya St.,
St. Petersburg 195251, Russia
tsikin@mail.spbstu.ru, antonina_92@list.ru

Abstract. Angle-of-arrival (AOA) integrity monitoring method for global-navigation satellite systems (GNSS) is discussed. In accordance with the AOA method, the direct decision making procedure (DDMP) about integrity presence or absence is created on the basis of direct antenna array signal processing. The DDMP is statistically simulated, and thus the most important characteristics, such as false alarm and missed detection probabilities, are estimated for different planar antenna arrays. The results of simulations show the algorithms efficiency areas for different structures of antenna arrays, so that finally the relationship between the number of antenna elements and missed detection probability are achieved for different number of analyzing signals. Comparison of DDMP with the decision making procedure for the AOA method, based on direction finding results post-processing, shows efficiency of DDMP both in terms of probability based characteristics and required computational costs.

Keywords: Interference mitigation · Global navigation satellite systems · GNSS integrity monitoring · Generalized likelihood ratio test · Angle of arrival · Direction finding

1 Introduction

Global navigation satellite systems' (GNSS) users calculate their position on the basis of receiving and processing signals from several satellite vehicles (SV). The signal radiated by the transmitter of the SV number m can be described as [1]:

$$s_0^{(m)}(t) = A_0^{(m)}(t) \cdot \cos\left(\omega_0^{(m)} t + \varphi_m(t) + \psi_0^{(m)}\right), \tag{1}$$

where $\psi_0^{(m)}$, $\omega_0^{(m)}$, $A_0^{(m)}(t)$ and $\varphi_m(t)$ – respectively, initial phase, frequency, amplitude and phase modulation of the signal from the SV number m. As a result of M ($M > 4$) signals (1) processing user's equipment measures signal's time delay τ_m, coordinates x_m, y_m, z_m and ionospheric corrections $\Delta\tau_m$ for each SV. The set of these values x_m, y_m, z_m, $\Delta\tau_m$, τ_m and $\Delta\omega_0^{(m)}$ is used during user position estimating [2]. User's coordinate statistical estimations $\hat{x}, \hat{y}, \hat{z}$ can differ from user's position true values x, y, z. The situation when the differences significantly exceed ones which occur when signals are processed under the influence of additive noise is referred to as GNSS integrity

© Springer International Publishing AG 2017
O. Galinina et al. (Eds.): NEW2AN/ruSMART/NsCC 2017, LNCS 10531, pp. 635–643, 2017.
DOI: 10.1007/978-3-319-67380-6_60

failure. Such failures can be either unintentional (ionospheric, tropospheric or intrasystemic GNSS faults [3, 4]) or caused by the radiation of special interference sources [5–7]. Such integrity failure is caused by a single false navigation signal source (FNSS). The signals of the FNSS are identical to the legal GNSS signals, but have their own values for parameters $\Delta\tau'_m$, τ'_m and $\Delta\omega_0'^{(m)}$, that lead to incorrect $\hat{x}, \hat{y}, \hat{z}$ user's position estimation. It is very important to provide a user with actual integrity information which can be received during an integrity monitoring procedure [8, 9]. Methods based on antenna array (AA) processing are widespread. In particular in Angle-of-Arrival (AOA) method decision-making procedure about GNSS integrity presence or absence is realized by the post-processing of navigation signal sources direction finding (DF) results [10–13]. On the other hand, it is interesting to develop and analyze the integrity monitoring method based on direct decision making procedure (DDMP). This method would be optimized by the criteria of missed detection probability P_{MD} minimum when false alarm probability P_{FA} is fixed.

2 Decision Making Algorithm on the Base of Direct Signal Processing for GNSS Integrity Monitoring

When GNSS integrity failure is absent, an appropriate mathematical model of the signal transmitted by SV number m and received by the element (EAA) number l of AA can be described as [14, 15]:

$$
\begin{aligned}
s_l^{(m)}(t) &= \alpha_m A_0^{(m)}\left(t - \tau_l^{(m)}\right) \\
&\times \cos\left[\left(\omega_0^{(m)} + \Delta\omega_0^{(m)}\right)t - \left(\omega_0^{(m)} + \Delta\omega_0^{(m)}\right)\tau_l^{(m)} + \varphi_m\left(t - \tau_l^{(m)}\right) + \psi_0^{(m)}\right],
\end{aligned}
\tag{2}
$$

where $\tau_l^{(m)}$ – time delay, caused by signal's transmission from SV number m to EAA number l; $\Delta\omega_0^{(m)} = \omega_0^{(m)} v_R^{(m)} \big/ c$ – Doppler shift of the carrier frequency; $v_R^{(m)}$ – radial velocity component of the relative movement between user and SV number m. The value $\tau_l^{(m)}$ in (2) can be decomposed:

$$
\tau_l^{(m)} = \tau_1^{(m)} + \delta\tau_l(\mu_m, \eta_m),
\tag{3}
$$

where $\tau_1^{(m)}$ – time delay, caused by signal's transmission from SV number m to the « reference » element in AA; $\delta\tau_l(\mu_m, \eta_m)$ – signal's time of arrival difference between reference element and EAA number l; (μ_m, η_m) – SV number m direction (respectively azimuth and elevation) from the user's location. In the current situation of integrity failure absence directions (μ_m, η_m) are well known parameters because they can be calculated on basis of Ephemerise data information which is carried by navigation signal.

For simplicity, we shall assume that GNSS signals have constant envelopes $A_0^{(m)}$ and signals frequency Doppler shifts $\Delta\omega_0^{(m)}$ can be totally compensated by the phase locked loop [14]. Also, using (3) we can ignore the difference between slowly varying functions $\varphi_m\left(t - \tau_l^{(m)}\right)$ and $\varphi_m\left(t - \tau_1^{(m)}\right)$ for each l from 1 to L. In this case we may write $s_l^{(m)}(t)$:

$$s_l^{(m)}(t) = \alpha_m A_0^{(m)} \times \cos\left[\omega_0^{(m)} t - \delta\varphi_l(\mu_m, \eta_m) + \varphi_m\left(t - \tau_1^{(m)}\right) + \psi_m\right], \qquad (4)$$

where $\delta\varphi_l(\mu_m, \eta_m) = \omega_0^{(m)} \delta\tau_l(\mu_m, \eta_m)$ – phase difference between signals received by reference element and EAA number l; ψ_m – unknown initial phase of the signal $s_l^{(m)}(t)$ received by reference element of antenna array.

Using the definition in (4), total signal from M visible SV received by the EAA number l has a form:

$$x_l(t) = \sum_{m=1}^{M} s_l^{(m)}(t) + n_l(t), \qquad (5)$$

where $n_l(t)$ – additive white Gaussian noise (AWGN) component on EAA number l. In case of integrity failure absence (hypothesis H_0) likelihood function for signal (5) is proportional to the value

$$B_l(\psi, H_0) = \exp\left\{\frac{2}{N_0} \int\limits_{(T_a)} x_l(t) \sum_{m=1}^{M} s_l^{(m)}(t)dt\right\} \times \exp\left\{-\frac{1}{N_0} \int\limits_{(T_a)} \left(\sum_{m=1}^{M} s_l^{(m)}(t)\right)^2 dt\right\},$$

where $\psi = (\psi_1, \ldots \psi_M)^T$ – column vector whose components are random initial phases and $N_0/2$ – mean power spectral density of AWGN. Integration is performed on the interval of analysis T_a determined by the $s_l^{(m)}(t)$ signal's duration. Taking into account the orthogonality of the signals from various SV, we have:

$$B_l(\psi, H_0) = \exp\left\{\frac{2}{N_0} \int\limits_{(T_a)} x_l(t) \sum_{m=1}^{M} s_l^{(m)}(t)dt - \frac{1}{N_0} \sum_{m=1}^{M} \alpha_m^2 E_m\right\}, \qquad (6)$$

where $E_m = \int\limits_{(T_a)} (s_{0m}(t))^2 dt$ – SV number m signal's energy.

Let's introduce complex envelopes $F_l(t)$ and $F_0^{(l,m)}(t)$ for signals $x_l(t)$ and $s_l^{(m)}(t)$ respectively, where $F_0^{(l,m)}(t) = \alpha_m A_0^{(m)} e^{-j\delta\varphi_l(\mu_m, \eta_m)} e^{j\varphi_m\left(t - \tau_1^{(m)}\right)} e^{j\psi_m}$. We transform (6) to the form:

$$B_l(\psi, H_0) = \exp\left\{\frac{1}{N_0} \sum_{m=1}^{M} \text{Re}\left[\left|V_l^{(m)}\right| \alpha_m A_0^{(m)} e^{j\delta\varphi_l(\mu_m, \eta_m)} e^{j\left(\Phi_l^{(m)} - \psi_m\right)}\right] - \frac{1}{N_0} \sum_{m=1}^{M} \alpha_m^2 E_m\right\}, \qquad (7)$$

where $tg\left(\Phi_l^{(m)}\right) = \dfrac{\text{Im}(v_l^{(m)})}{\text{Re}(v_l^{(m)})}$, $V_l^{(m)} = \int\limits_{(T_a)} F_l(t) C_0^{(m)}(t)dt$ and $C_0^{(m)}(t) = e^{-j\varphi_m\left(t - \tau_1^{(m)}\right)}$.

Using (7) and taking into account the independence between AWGN components on different EAA, we can describe the likelihood function for column vector $\mathbf{x}(t) = (x_1(t), \ldots, x_l(t), \ldots, x_L(t))^T$, whose components are signals received by each EAA of antenna array, to be proportional:

$$
\begin{aligned}
B(\boldsymbol{\psi}, H_0) &= \prod_{l=1}^{L} B_l(\boldsymbol{\psi}, H_0) \\
&= \exp\left\{ \frac{1}{N_0} \sum_{l=1}^{L} \sum_{m=1}^{M} \alpha_m A_0^{(m)} \left| V_l^{(m)} \right| \mathrm{Re}\left[e^{j\delta\varphi_l(\mu_m, \eta_m)} e^{j\left(\Phi_l^{(m)} - \psi_m\right)} \right] - \frac{L}{N_0} \sum_{m=1}^{M} \alpha_m^2 E_m \right\}.
\end{aligned} \tag{8}
$$

In case of integrity failure caused by a single FNSS (hypothesis H_1) the total signal received by the EAA number l can be described as:

$$
x_l(t) = \sum_{m=1}^{M} s_l'^{(m)}(t) + s_l^{(m)}(t) + n_l(t), \tag{9}
$$

where $s_l^{(m)}(t)$ is described in (2), and signal $s_l'^{(m)}(t)$ is transmitted by FNSS and simulates the signal from SV number m. When this signal is received by EAA number l it can be described as:

$$
s_l'^{(m)}(t) = \alpha_m' A_0^{(m)} \cos\left[\omega_0^{(m)} t + \delta\varphi_l(\mu_0, \eta_0) + \varphi_m\left(t - \tau_1'^{(m)} \right) + \psi_m' \right], \tag{10}
$$

where $\delta\varphi_l(\mu_0, \eta_0)$ – phase difference between FNSS signals received by reference element and EAA number l. Signals time delays $\tau_1'^{(m)}$ are different to their respective values $\tau_1^{(m)}$ observed in case of integrity failure absence. It is assumed that FNSS has unknown and random azimuth μ_0 and elevation η_0.

There is a principal difference between the signals (4) and (10) analyzing on hypothesis H_0 and H_1 respectively. In first case (under H_0) phase differences $\delta\varphi_l(\mu_m, \eta_m)$ have known and different values for each SV. These values are determined by SV direction from user's location. In second case (under H_1) mentioned phase differences have unknown but equal values (which depend on FNSS direction from user's location) for each received signal.

We will assume the parameters α_m' and α_m have values providing significant FNSS signals power exceeding over the signals from SV ($\alpha_m' > > \alpha_m$). So, in the case of the signals (9) processing user's equipment detects FNSS signal components $s_l'^{(m)}(t)$. In this situation, the user obtains incorrect information about its own location. Under the hypotheses H_1 by analogy of (8), we can describe likelihood function for column vector $\mathbf{x}(t)$ as proportional to the value:

$$
B(\boldsymbol{\psi}, H_1) = \exp\left\{ \frac{1}{N_0} \sum_{l=1}^{L} \sum_{m=1}^{M} \alpha_m' A_0^{(m)} \left| V_l^{(m)} \right| \mathrm{Re}\left[e^{j\delta\varphi_l(\mu_0, \eta_0)} e^{j\left(\Phi_l^{(m)} - \psi_m\right)} \right] - \frac{L}{N_0} \sum_{m=1}^{M} \alpha_m'^2 E_m \right\}. \tag{11}
$$

In all of our discussions we assume that parameters $\tau_1^{(m)}$ and $\tau_1'^{(m)}$ ($l = 1...L$) have precisely known values, but column vectors $\boldsymbol{\psi} = (\psi_1,...\psi_M)^T$, $\boldsymbol{\psi}' = (\psi_1',...\psi_M')^T$, $\boldsymbol{\alpha} = (\alpha_1...\alpha_M)$, $\boldsymbol{\alpha}' = (\alpha_1'...\alpha_M')$ and FNSS direction (μ_0,η_0) are random and unknown. Applying (8), (11) and using maximum likelihood estimate values for these parameters, let's consider a generalized likelihood ratio test [15]:

$$
\frac{\max\limits_{\mu_0,\eta_0,\alpha',\psi'} W(\mathbf{x}(t)/H_1)}{\max\limits_{\alpha,\psi} W(\mathbf{x}(t)/H_0)} \underset{H_0}{\overset{H_1}{\underset{<}{>}}} \Lambda_0.
\tag{12}
$$

Likelihood ratio $\Lambda(H_1/H_0) = \dfrac{\max\limits_{\mu_0,\eta_0,\alpha',\psi'} W(\mathbf{x}(t)/H_1)}{\max\limits_{\alpha,\psi} W(\mathbf{x}(t)/H_0)}$ in the left part of (12) can be described as:

$$
\Lambda(H_1/H_0) = \exp\left\{\frac{1}{N_0}\max\limits_{\mu_0,\eta_0,}\left\{\sum_{m=1}^{M}\left(A_0^{(m)}\right)^2\left|\sum_{l=1}^{L}V_l^{(m)}e^{j\delta\varphi_l(\mu_0,\eta_0)}\right|^2\right\}\right\}
$$
$$
\times \exp\left\{-\frac{1}{N_0}\sum_{m=1}^{M}\left(A_0^{(m)}\right)^2\left|\sum_{l=1}^{L}V_l^{(m)}e^{j\delta\varphi_l(\mu_m,\eta_m)}\right|^2\right\}.
\tag{13}
$$

Turning to the consideration of log likelihood ratio (13) and assuming the amplitudes $A_0^{(m)}$ equality for all signals we could derive the structure of DDMP:

GNSS integrity failure is detected if

$$
\max\limits_{\mu_0,\eta_0}\left\{\sum_{m=1}^{M}\left|\sum_{l=1}^{L}V_l^{(m)}e^{j\delta\varphi_l(\mu_0,\eta_0)}\right|^2\right\} - \sum_{m=1}^{M}\left|\sum_{l=1}^{L}V_l^{(m)}e^{j\delta\varphi_l(\mu_m,\eta_m)}\right|^2 > \ln\Lambda_0',
\tag{14}
$$

otherwise, integrity presence is detected.

3 Probability Based Characteristics for Direct Decision Making Procedure

Let's obtain the DDMP (14) operating characteristics as relations between missed detection probability P_{MD} and false alarm probability P_{FA} for various values of L. For example, we consider the situation when the user located in St. Petersburg (Russia) observes GPS constellation in which dilution of precision is less than 3 [17]. In this case P_{MD} and P_{FA} for the typical h^2 value are illustrated on Figs. 1 and 2. These operating characteristics are obtained in the context of integrity monitoring system implementation on compact drones using antenna arrays with a small number of elements ($L = 2 ... 6$).

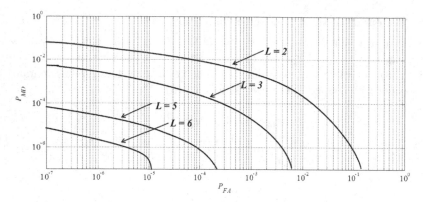

Fig. 1. Dependence of missed detection probability on false alarm probability for DDMP $(M = 3, h^2 = 10 \text{ dB})$

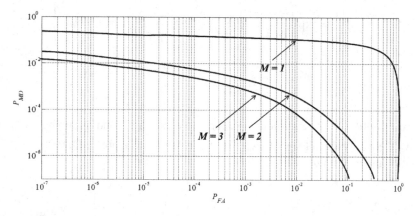

Fig. 2. Dependence of missed detection probability on false alarm probability for DDMP $(L = 2, h^2 = 10 \text{ dB})$

Let's compare DDMP (14) and decision-making algorithm of DF results post-processing from [11]. Performance comparison is graphically illustrated in Fig. 3. Typical operating characteristics for constellations observed in St. Petersburg shown in solid lines whereas dotted lines illustrate average characteristics for various constellations. Looking at Fig. 3 we can see that the DDMP (14) has a significant performance gain in comparison with the algorithm from [11] in both cases of interest.

4   On the Computational Complexity of Algorithms

The DDMP (14) assumes the following sequence of actions:

- measuring the complex envelope $F_l(t)$ values for signals $x_l(t)$ from each element of antenna array (complex envelop measurer, CEM in Fig. 4);

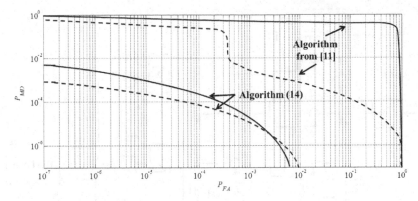

Fig. 3. Dependence of missed detection probability on false alarm probability for DDMP and post-processing algorithm from [11] ($M = 3$, $h^2 = 10$ dB)

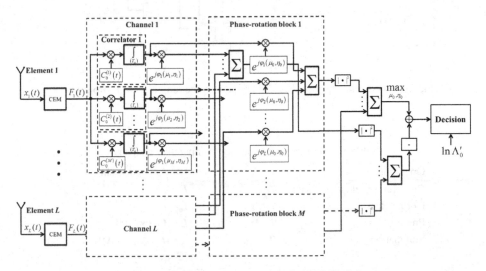

Fig. 4. The form of a processor for DDMP (14)

- multiplying each of the complex envelopes by the reference signal $C_0^{(m)}(t)$ and integrating with respect to time (Correlators in Fig. 4);
- generalized likelihood ratio (14) testing.

The decision making algorithm from [11] assumes the use of results at the correlators' outputs to implement DF procedure at first. The results $(\hat{\mu}_1, \hat{\eta}_1)$, $(\hat{\mu}_2, \hat{\eta}_2)$ ... $(\hat{\mu}_M, \hat{\eta}_M)$ of DF procedure are post-processed in accordance with the likelihood ratio test from [11]. The form of a processor to implement such DF post-processing decision-making procedure is shown in Fig. 5. Let's compare these forms with respect to the ease of computation. Table 1 contains the number of operations required for both comparing algorithms implementation. It is clear that post-processing algorithm from [11] is computationally tedious unlike DDMP.

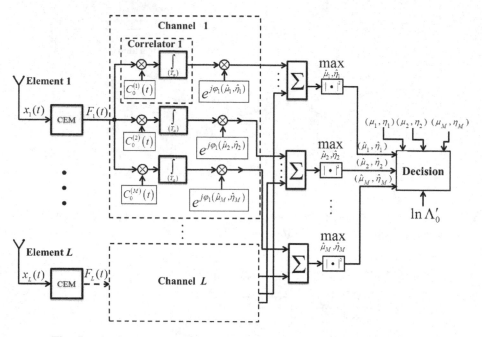

Fig. 5. The form of a processor for DF post-processing algorithm from [11]

Table 1. Algorithms complexity comparative analysis

Operation	DDMP	DF post-processing
Complex envelope estimation	L	L
Dot product	$L \cdot M$	$L \cdot M$
Phase rotation	$L \cdot M + 360 \cdot 90$	$L \cdot M \cdot 360 \cdot 90$
Summation	$2 \cdot L \cdot M$	$LM + 2M + 4MM + {}$ $+ 360 \cdot 90 \cdot (L-1)M$
Module	$2M$	$M + 360 \cdot 90 \cdot M$
Square	$2M$	$6M + M \cdot 360 \cdot 90$

5   Conclusions

In accordance with generalized likelihood ratio test, direct decision-making procedure (DDMP) for GNSS integrity monitoring system is derived. It is shown that DDMP has a significant performance gain in comparison with the DF post-processing algorithm from [11] in both cases of interest considered in the problems. In a typical situation of GPS constellation observation in range of interest false alarm probability values $(10^{-5} \ldots 10^{-7})$, DDMP provides missed detection probability values to be three orders less than ones provided by the DF post-processing algorithm from [11]. Besides, another advantage of a DDMP is a weak dependence of false alarm probability on

GNSS constellation geometry. So the DDMP operating characteristics depend first of all on the number of visible satellites and antenna array structure. In range of interest 10^{-5} ... 10^{-7} of false alarm probability values, DDMP performance improves monotonically as the number of elements increases. Thus, increasing the number of elements by one leads to the missed detection probability decreasing as much as one order.

References

1. Xu, G.: GPS: theory, algorithms and applications (2007)
2. Kaplan, E., Hegarty, C.: Understanding GPS: principles and applications. Artech house, Norwood (2005)
3. Milner, C.: Bayesian Inference of GNSS failures (2016)
4. Ahn, J.: Orbit ephemeris failure detection in a GNSS regional application (2015)
5. Sathyamoorthy, D.: Global navigation satellite system (GNSS) spoofing: a review of growing risks and mitigation steps. Defence S&T Tech. Bull. 6(1), 42–61 (2013)
6. Schmidt, D.: A survey and analysis of the GNSS spoofing threat and countermeasures (2016)
7. Tippenhauer, N.O.: On the requirements for successful GPS spoofing attacks, 2011\Tao, A recursive receiver autonomous integrity monitoring (Recursive-RAIM) Technique for GNSS Anti-Spoofing (2015)
8. Binjammaz, T., Al-Bayatti, A., Al-Hargan, A.: GPS integrity monitoring for an intelligent transport system. In: 2013 10th Workshop on Positioning Navigation and Communication (WPNC). IEEE (2013)
9. Bitner, T.L.: Detection and removal of erroneous GPS signals using angle of arrival, Dissertation. Auburn University (2013)
10. Melikhova, A., Tsikin, I.: Antenna array with a small number of elements for angle-of-arriving GNSS integrity monitoring. In: IEEE 39th International Conference on Telecommunications and Signal Processing (TSP), pp. 190–193 (2016)
11. Harms, J.K.: Synthetic aperture digital beamsteering array for global positioning system interference mitigation: a study on array topology. Dissertation (2014)
12. Irteza, S.: Compact antenna array receiver for robust satellite navigation systems. Int. J. Microwave Wirel. Technol. 7(06), 735–745 (2015)
13. Glonass. Principy postroeniya i funkcionirovaniya/ pod red. A. I. Pero-va, V.N. Harisova. Izd. 4-oe, pererab. i dop. – M.: Radiotekhnika, – 800 s (2010)
14. Navstar GPS Space Segment/Navigation User Segment Interfaces, IS-GPS-200. http://www.gps.gov
15. Van Trees, H.L.: Detection, Estimation, and Modulation Theory. Wiley, Hoboken (2004)
16. Balanis, C.: Antenna Theory: Analysis and Design. Wiley, Hoboken (2016)
17. Langley, Richard B.: Dilution of precision. GPS World 10(5), 52–59 (1999)

An Intentional Introduction of ISI Combined with Signal Constellation Size Increase for Extra Gain in Bandwidth Efficiency

Van Phe Nguyen, Anton Gorlov$^{(\boxtimes)}$, and Aleksandr Gelgor

Peter the Great St. Petersburg Polytechnic University, St. Petersburg, Russia
nvphe1905@gmail.com, anton.gorlov@yandex.ru,
a_gelgor@mail.ru

Abstract. In this paper, we investigate the possibility to obtain an additional bandwidth efficiency gain for signals with intentionally introduced inter symbol interference (ISI) by increasing the signal constellation size. We analyzed single-carrier signals with QPSK and 16-QAM signal constellations. The ISI is introduced by the optimal finite pulses obtained by the linear optimization. The criterion of the optimization problem was maximization of the free Euclidean distance under a fixed bandwidth comprising 99% of signal power and for the chosen constellation. We used the sphere decoding (SD) algorithm for the signal detection under assumption of additive white Gaussian noise channel. The simulation results have shown that the maximal increase of bandwidth efficiency is provided when the ISI introduction is utilized simultaneously with the increase of signal constellation size.

Keywords: Bandwidth efficiency · Inter symbol interference · Partial response signaling · Optimal pulse · Sphere decoding

1 Introduction

In recent years, there is an increasing interest to signals with controlled inter symbol interference (ISI). The possibility of improving spectral efficiency by using such signals was originally proposed by Mazo in [1] where he introduced the term FTN – Faster than Nyquist signaling. The symbol rate exceeded the Nyquist barrier in the proposed modulation scheme. However, the implementation of the FTN idea was impossible for a long time due to high computational complexity of the detection algorithm. In most communication systems, the bandwidth efficiency is usually improved by increasing the size of the signal constellation. At present, modern digital signal processors provide a computational power, which is sufficient to detect signals with controlled ISI. It motivates many researches to return to the idea of improving the spectral efficiency by intentional ISI introduction.

According to [2, 3] the main advantage of signals with controlled ISI is in the fact that they can provide a higher channel capacity (throughput) in comparison to the case of conventional orthogonal signals. Thus, the spectral and energy characteristics of

© Springer International Publishing AG 2017
O. Galinina et al. (Eds.): NEW2AN/ruSMART/NsCC 2017, LNCS 10531, pp. 644–652, 2017.
DOI: 10.1007/978-3-319-67380-6_61

modern communication systems potentially can be overcome by utilization of signals with controlled ISI.

Despite the called perspectives of signals with ISI, the development of efficient methods for their generation and detection represents an actual problem. Empirical methods for generation of single- and multicarrier signals with controlled ISI are represented in [1, 4], these methods are based on the idea of closer allocation of sinc-pulses. In [5] Liveris and Georghiades have extended the idea from [1] by transition from sinc- to RRC-pulses (Root Raised Cosine), and the same approach for multicarrier signals has been proposed in [6] thus extending the idea from [4]. The methods for generation of optimal single- and multicarrier signals with controlled ISI are represented in [7–9]. These works also demonstrate the significance of utilization of optimal pulses instead of sinc- or RRC-pulses for generation of signals with controlled ISI. In more detail, the utilization of optimal pulses provides an additional gain in energy and bandwidth efficiency with respect to the modes with RRC-pulses.

The Viterbi [10], BCJR [11] (Bahl, Cocke, Jelinek and Raviv) and SD [12] (Sphere Decoding) algorithms are usually utilized for detection of signals with ISI. The BCJR algorithm is the decoding algorithm based on the maximum a posteriori probability (MAP) criterion. It is usually used in systems with forward error correction because it provides a possibility for iterative decoding. The Viterbi and SD algorithms implement the maximum likelihood estimation of entire sequence of symbols (MLSE criterion). The computational complexity of the Viterbi and BCJR algorithms is approximately $O((N + L - 1)M^{L - 1})$, where N is the number of symbols in a frame, L is the pulse duration as a number of symbol transmission periods, M is the constellation size. If the pulse duration L is high and/or a large constellation ($M > 4$) is utilized, the implementation of the Viterbi and BCJR algorithms becomes practically impossible because of the high computational complexity. In contrast, the SD algorithm complexity does not depend on the pulse duration nor the constellation size. The SD has the polynomial complexity from the frame size N.

However, one drawback of signals with controlled ISI can be observed if they are used for significant increase of bandwidth efficiency without changing signal constellation. In these models, the empirical methods of ISI imposition [1, 4–6] lead to enormous energy losses and the signals with controlled ISI lose their practical meaning in comparison to orthogonal signals with enlarged constellations. If the ISI is introduced optimally [7–9], a gain in spectral efficiency can be obtained by increase of duration of pulses (in time or frequency domains). This also leads to increase of a computational complexity of detection algorithms. To sum up, a joint utilization of two methods should be considered to figure out if it is possible to obtain an additional gain of bandwidth efficiency or reduce a detection algorithm complexity.

In this work, we consider single-carrier signals with QPSK, 16-QAM constellations and optimal pulses obtained by the technique from [7]. The considered optimal pulses provide the maximal free Euclidean distance under a fixed bandwidth comprising 99% of signal power and for a chosen constellation. There is a high computational complexity of the Viterbi and the BCJR algorithms applied to detection of signals with controlled ISI and large signal constellations. Therefore, we will utilize the SD algorithm for detection of short frames with relatively low number of symbols with respect to frame lengths utilized in modern communication systems.

2 Optimal Pulses for Constellations QPSK and 16-QAM

It is possible to independently process in phase and quadrate components for the signals with QPSK and 16-QAM constellations. In this case, the computational complexity of detection algorithm is significantly reduced. However, the optimization problem can be simplified by searching solutions for 2-PAM and 4-PAM (Pulse Amplitude Modulation) constellations instead of QPSK and 16-QAM constellations respectively. In [7], authors proposed a formulation and solution of a linear optimization problem for searching a discrete autocorrelation function $g_b[n]$. The discrete-time pulses $b[n]$ can be obtained by analyzing roots of a polynomial, which is a z-transform of the optimal autocorrelation function. The optimization criterion is the maximal free Euclidean distance d_{free} for all possible signals sequences. This criterion provides the asymptotically minimal error probability when the signal-to-noise ratio is increasing. The additional constraint is the normalized bandwidth $W_{99\%}T$ comprising 99% of signal power, where $1/T$ is a symbol rate. If the energy of pulse is also normalized, the solution of the optimal problem is unique.

If we increase the length of pulse and the size of the signal constellation, the number of constraints is increased exponentially. Then, the computational complexity of the optimization problem is significantly increased. This is a crucial disadvantage of the technique from [7]. In such cases, the solution of the optimization problem can be obtained by iterative solving a simplified problem, which contains fewer constraints. The normalized energy of pulse is also used to ensure a solution uniqueness.

If we assume that the discrete-time pulse $b[n]$ contains L samples, then the continuous shaping filter impulse response $h(t)$ can be expressed by

$$h(t) = \sum_{n=0}^{L-1} b[n]\psi(t - nT), \tag{1}$$

where $\psi(t)$ is an interpolation pulse, e.g., one of the RRC-functions. The signal with ISI is a result of linear modulation:

$$y(t) = \sum_{n=0}^{N-1} u[n]h(t - nT), \tag{2}$$

where $u[n]$ is a discrete sequence of modulation symbols from the used signal constellation of size M (Fig. 1).

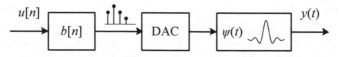

Fig. 1. Block diagram of generator of signal (2)

In this work, we study the optimal pulses with $L = 4, 8, 12, 14$ for each 2-PAM and 4-PAM constellations. For each value of L, the values of normalized bandwidth were varied from the $W_{99\%}T = 0.76$ down to the possible minimum value with a 0.04 step: to the $W_{99\%}T = 0.52$ for $L = 4$, to the $W_{99\%}T = 0.28$ for $L = 8$, to the $W_{99\%}T = 0.2$ for $L = 12$, to the $W_{99\%}T = 0.16$ for $L = 14$. The obtained maximal squared normalized values of d_{free} for each pair $(W_{99\%}T, L)$ and for two constellation types are represented in the Table 1.

Table 1. Values d_{free}^2 by values $W_{99\%}T$, L and type of constellation

2-PAM signal constellation					4-PAM signal constellation				
$W_{99\%}T$	L				$W_{99\%}T$	L			
	4	8	12	14		4	8	12	14
0.76	1.902	2.0	2.0	2.0	0.76	0.761	0.80	0.80	0.80
0.72	1.533	1.770	1.986	2.0	0.72	0.613	0.70	0.794	0.80
0.68	1.345	1.543	1.716	1.758	0.68	0.538	0.617	0.686	0.702
0.64	1.232	1.405	1.513	1.549	0.64	0.438	0.562	0.601	0.612
0.60	1.054	1.280	1.366	1.389	0.60	0.353	0.482	0.537	0.548
0.56	0.934	1.183	1.229	1.274	0.56	0.299	0.381	0.479	0.495
0.52	0.832	0.969	1.121	1.137	0.52	0.230	0.322	0.348	0.380
0.48	-	0.850	0.926	0.980	0.48	-	0.274	0.290	0.312
0.44	-	0.779	0.785	0.807	0.44	-	0.256	0.260	0.261
0.40	-	0.637	0.665	0.709	0.40	-	0.20	0.237	0.242
0.36	-	0.457	0.557	0.572	0.36	-	0.140	0.192	0.190
0.32	-	0.333	0.452	0.461	0.32	-	0.110	0.150	0.161
0.28	-	0.250	0.311	0.340	0.28	-	0.079	0.098	0.112
0.24	-	-	0.206	0.238	0.24	-	-	0.078	0.085
0.20	-	-	0.149	0.152	0.20	-	-	0.060	0.061
0.16	-	-	-	0.090	0.16	-	-	-	0.033

Figures 2 and 3 show the free distance losses (FDL) depending on the normalized bandwidths for 2-PAM and 4-PAM constellations respectively. The FDL are calculated from the free Euclidean distances which take place for the orthogonal pulses $(d_{free}^2 = 2.0$ and $d_{free}^2 = 0.8$ for 2-PAM and 4-PAM constellations respectively). It follows that an increase of the pulse length L leads to the narrower normalized bandwidth $W_{99\%}T$. An additional gain can be provided by the increase of L. The gain is either in a reduction of the normalized bandwidth $W_{99\%}T$ for a fixed FDL or in a reduction of FDL for a fixed value of the normalized bandwidth $W_{99\%}T$. However, this additional gain becomes smaller with increase of L. Furthermore, characteristics for $L = 12$ and for $L = 14$ almost do not differ.

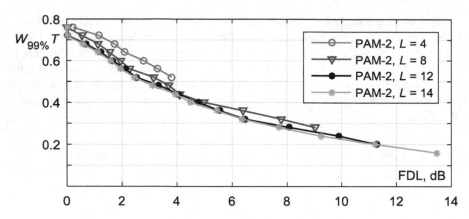

Fig. 2. Normalized bandwidth by free distance losses for 2-PAM constellation

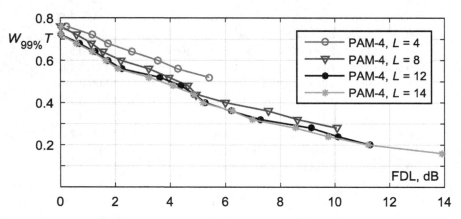

Fig. 3. Normalized bandwidth by free distance losses for 4-PAM constellation

3 Model Description

As it was said above, the generated signal (2) is passed through a channel with AWGN. The matched filtering is required for detection of this signal. In this case, the impulse response of the matched filter is $\psi(t)$. However, it is possible to omit the interpolation and the matched filtering during the simulation and process discrete sequences. The values $W_{99\%}T$ of all obtained pulses is less than 0.76. Thus, it is always possible to find $\psi(t)$ with constant spectral density in the bandwidth $W_{99\%}T$ (and even beyond this bandwidth), e.g. one of the RRC-pulses satisfies this requirement (Fig. 4). In this way, the interpolation during the generation and the matched filtering during the detection do not affect the signal spectrum. Therefore, these procedures are required only in practice when it is necessary to implement real detection and transmission units.

The generated signal is obtained as a result of linear filtering

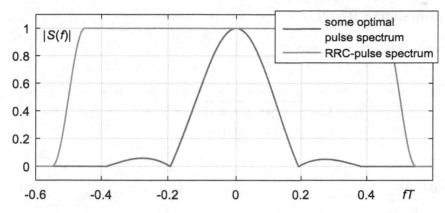

Fig. 4. Optimal pulse and some RRC-pulse spectrums

$$y[k] = \sum_{n=0}^{N-1} u[n]b[k-n]. \tag{3}$$

In other words, signals are obtained by choosing a finite number of points from an N-dimensional lattice in the Euclidean space Z^N. The procedure of searching the maximal likelihood sequence is equivalent to searching the lattice point closest to the point of the received signal. This finding is equivalent to minimizing the following metric:

$$\min_{\hat{u}[n] \in Z^N} \sum_k |\hat{y}[k] - r[k]|^2, \tag{4}$$

where $r[k] = y[k] + w[k]$ are samples of the received signal, $w[k]$ are samples of AWGN, $\hat{y}[k]$ are samples of the reference signal defined by sequence $\hat{u}[n]$.

The SD algorithm attempts to check all of the lattice points, which lie in a certain sphere of radius d. The center of this sphere is the point of the received signal. The complexity of the sphere decoding depends on the radius d. If d is too large, the sphere contains too many lattice points. On the other hand, if d is too small, the sphere may contain no lattice points. The value of radius d is usually chosen based on the values of SNR (Signal to Noise Ratio). If the sphere does not contain any lattice point, the radius d should be increased and the decoding process is repeated. In [12], the analysis of the computational complexity of the sphere algorithm has been studied, and in [13], the algorithm of choosing an initial radius d has been proposed.

In our simulations we used $N = 52$ as the length of the information sequence. On the one hand, it provides rather low computational complexity of the SD algorithm. On the other hand, an additional study of the Viterbi detection performance for signals with the small level of ISI ($L = 8$) and QPSK and 16-QAM signal constellations showed almost equal performance for $N = 52$ and $N = 1000$ at BER $< 10^{-2}$. Therefore, we can expect adequate results for $N = 52$ for signals with a high level of ISI and, hence, with high values of L.

4 Simulation Results

Figure 5 shows the bandwidth efficiency

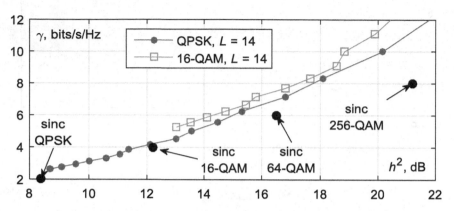

Fig. 5. Bandwidth efficiency by signal-to-noise ratio for $L = 14$

$$\gamma = \frac{R}{W_{99\%}} = \frac{\log_2(M)/T}{W_{99\%}} = \frac{\log_2(M)}{W_{99\%}T}, \tag{5}$$

depending on the signal-to-noise ratio SNR

$$h^2 = E_{\mathrm{bit}}/N_0. \tag{6}$$

The R in (5) is the information rate bit/s; in (6), E_{bit} is the bit energy, N_0 is the AWGN one-sided spectral power density. During the simulation we marked the SNR values corresponding to the error performance BER = 10^{-4}. The required signal-to-noise ratio values for signals with ISI were obtained from BER-curves analysis. The BER curves were obtained for pulses with $L = 14$ (the characteristics are represented in Table 1). We also considered the modes with RRC-pulses (full response signaling, i.e. orthogonal pulses) with roll-off factor $\beta = 0$ (marked *sinc* in Fig. 5).

From Fig. 5 it follows that, at first, the signals with ISI provide a significant gain with respect to the orthogonal signals. At second, the curves for the signals with ISI have common range of values γ and h^2. The curve of the 16-QAM signal constellation always lies higher than the corresponding curve of QPSK signal constellation. Therefore, it is possible to obtain a gain with the constant ISI level by increasing the size of the signal constellation. This gain is either up to 1 dB at the required value of signal-to-noise ratio for a fixed value of bandwidth efficiency, or up to 16% for bandwidth efficiency at a fixed value of the signal to noise ratio. Also, signals with ISI always outperform orthogonal signals.

According to Figs. 2, 3, the both curves in Fig. 5 are asymptotic for each constellation, i.e. with a chosen technique of optimal pulses design it is not possible to obtain better results. In this way, the introduction of ISI and the increase of signal constellation size should not be utilized separately for increasing of bandwidth efficiency. The best results can be achieved only when the both methods are utilized.

The computational complexity of the SD detection algorithm utilized for obtaining the points from Fig. 5 is approximately equal for all points. However, the SD algorithm does not allow efficient processing long signal sequences. If we use the Viterbi or BCJR algorithms, the computational complexity for 16-QAM signal constellation will be significantly larger than for QPSK signal constellation. Figure 6 shows curves for QPSK and 16-QAM constellations with the pulses of lengths $L = 14$ and $L = 4$ respectively. The Viterbi and BCJR algorithms have approximately equal computational complexity in these two modes. However, the curve for 16-QAM does not provide a gain with respect to the curve for QPSK. Therefore, when the Viterbi and BCJR algorithms are utilized, the bandwidth efficiency can be increased only with increase of the computational complexity.

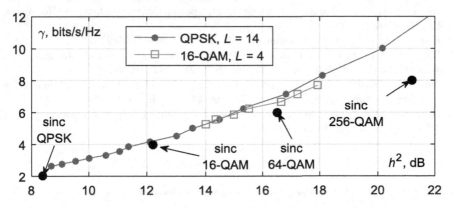

Fig. 6. Bandwidth efficiency by signal-to-noise ratio for QPSK, $L = 14$ and 16-QAM, $L = 4$

5 Conclusions

In this work, we have shown that the maximal increase of bandwidth and energy efficiencies can be reached by simultaneous introduction of the intentional ISI and increase of the signal constellation size. In this case, the computational complexity of the Viterbi and BCJR algorithms becomes practically impossible. The SD algorithm can be utilized only for detection of short frames. Thus, the developing an effective algorithm for detecting long frames of signals with high level of ISI remains as a subject for further work.

References

1. Mazo, J.E.: Faster-than-Nyquist signaling. Bell Syst. Tech. J. **54**(8), 1451–1462 (1975)
2. Rusek, F., Anderson, J.B.: Constrained capacities for faster-than-Nyquist signaling. IEEE Trans. Inf. Theory **55**(2), 764–775 (2009)

3. Rusek, F., Anderson, J.B.: The two dimensional Mazo limit. In: Proceedings of International Symposium on Information Theory, pp. 970–974 (2005)
4. Liveris, A.D., Georghiades, C.N.: Exploiting faster-than-Nyquist signaling. IEEE Trans. Commun. **51**(9), 1502–1511 (2003)
5. Kanaras, I., Chorti, A., Rodrigues, M.R.D., Darwazeh, I.: Spectrally efficient FDM signals: bandwidth gain at the expense of receiver complexity. In: International Conference on Communications ICC, pp. 1–6 (2009)
6. Gelgor, A., Gorlov, A., Nguyen, V.P.: The design and performance of SEFDM with the Sinc-to-RRC modification of subcarriers spectrums. In: International Conference on Advanced Technologies for Communications ATC, pp. 65–69 (2016)
7. Said, A., Anderson, J.B.: Bandwidth-efficient coded modulation with optimized linear partial-response signals. IEEE Trans. Inf. Theory **44**(2), 701–713 (1998)
8. Gelgor, A., Gorlov, A., Popov, E.: Multicomponent signals for bandwidth-efficient single-carrier modulation. In: Black Sea Conference on Communications and Networking BlackSeaCom, pp. 19–23 (2015)
9. Gelgor, A., Gorlov, A., Nguyen, V.P.: Performance analysis of SEFDM with optimal subcarriers spectrum shapes. In: Black Sea Conference on Communications and Networking BlackSeaCom (2017)
10. Forney, G.D.: The viterbi algorithm. Proc. IEEE **61**(3), 268–278 (1973)
11. Bahl, L., Cocke, J., Jelinek, F., Raviv, J.: Optimal decoding of linear codes for minimizing symbol error rate. IEEE Trans. Inf. Theory **20**(2), 284–287 (1974)
12. Fincke, U., Pohst, M.: Improved methods for calculating vectors of short length in a lattice, including a complexity analysis. Math. Comput. **44**(170), 463–471 (1985)
13. Hassibi, B., Vikalo, H.: On the sphere-decoding algorithm I. Expected complexity. IEEE Trans. Signal Process. **53**(8), 2806–2818 (2005)

Foreground Detection Using Region of Interest Analysis Based on Feature Points Processing

Nikita Ustyuzhanin[1,2(✉)] and Marat Gilmutdinov[1]

[1] Saint Petersburg State University of Aerospace Instrumentation (SUAI),
67, Bolshaya Morskaya str., Saint-Petersburg 190000, Russia
nikita.ustyuzhanin@skolkovotech.ru, mgilmutdinov@gmail.com
[2] Skolkovo Innovation Center, Skolkovo Institute of Science and Technology,
Building 3, Moscow 143026, Russia

Abstract. Analysis of regions of interest is a promising approach for performance improvement of many applications related to video signal processing and transmission. Many state-of-the-art methods face challenges like global luminance changing, objects camouflage, etc. This paper is dedicated to a new method of foreground detection employing region of interest analysis. The main idea of the proposed method is processing feature points located in the regions with object movement. The performance of the foreground detection was estimated using ground truth data and F1-Score criterion.

Keywords: Video surveillance · Computer vision · Region of interest · Feature point · SURF · SIFT

1 Introduction

According to the research and forecasts of Cisco [2], video surveillance traffic is envisioned to increase extremely each year [1], therefore, the efficient compression of video traffic becomes a crucial problem. At the same time, major improvements in computational capacity allow for execution of high-complexity analytic methods on mobile devices. Many computer vision-related tasks employ specific areas called *Region of Interest* (ROI). This paper is dedicated to detection of the Regions of Interest used in video compression and videoanalytic tasks. In video compression *Background Frames* may be used as additional reference data. Preliminary analysis of feasibility of background modeling methods in lossy compression system is presented in paper [4]. It is assumed that regions with moving objects would be sent to decoder that can estimate background frame in the same manner as encoder.

The remaining part of this paper is organized as follows. Next section contains a description of state-of-the-art methods considered in this investigation. Known challenges are described in Sect. 3. Proposed method for determination ROI is presented in Sect. 4. Section 5 is an experimental result analysis of proposed algorithm.

The research of Nikita Ustyuzhanin was partially supported by FASIE.

O. Galinina et al. (Eds.): NEW2AN/ruSMART/NsCC 2017, LNCS 10531, pp. 653–661, 2017.
DOI: 10.1007/978-3-319-67380-6_62

2   Related Works

Detection of Regions of Interest (ROI) is an important task for various applications in computer vision, e.g. face recognition, car license plate, object tracking and video compression. A typical approach for searching ROI is based on *Background Models*. The Background Model is presented as a frame containing only static scene information. The operation of generating a background frame is called *Background Modeling*. It is based on main features of video surveillance systems [3]:

- fixed position of camera view;
- static or slightly changing background scene;
- fixed values of capturing parameters, e.g. focus, exposure, white-balance.

Many background modeling methods are described in [3]. The task of foreground highlighting is solved by using background estimation results. The basic approach of foreground highlighting is called *Background Subtraction*:

$$F(x, y) = I_t(x, y) - B_t(x, y), \tag{1}$$

where I – video sequence frame, B – background frame, (x, y) – coordinates of pixel, t – instant of time.

After background subtraction, *Threshold Filtration* is usually be applied to obtain a binary *Foreground Mask* indicating:

$$Mask(x, y) = \begin{cases} 1, F(x, y) > Thres \\ 0, else \end{cases}, \tag{2}$$

where *Thres* is a value of threshold.

The example of background subtraction with threshold filtration is depicted in the Fig. 1. This method is low complexity, however, it is not accurate and has many artefacts. In addition, it is necessary to select parameter *Thres* which is sensitive to illumination value in the threshold filter. The flowchart of this method is depicted in the Fig. 2.

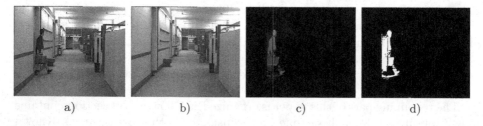

a)   b)   c)   d)

Fig. 1. Example of background subtraction based algorithm: (a) frame 10 of *Hall* sequence, (b) background frame, (c) frame difference, (d) foreground map

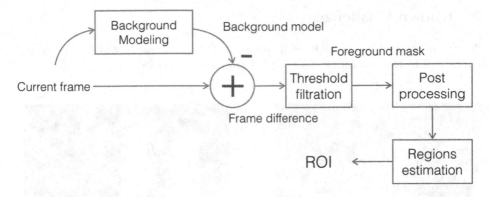

Fig. 2. Typical ROI search approach based on background subtraction

To improve the foreground mask, *erosion* and *dilatation* [15] algorithms are applied on *post processing* stage. Estimation of ROI is based on foreground information analysis. In the further figures this method is called *BS-based method*. Input of this algorithm is the current frame of video sequence and background frame. Output is the set of ROI coordinates.

Another class of ROI determination methods uses *Feature Points*. This approach is also used in ROI detection for classification applications [13,14]. The most popular methods of points of interests detection are based on SIFT [6] and SURF [7]. They are usually used in the following tasks:

- panoramas stitching [19]
- stereo-pairs adjusting [20];
- reconstruction of 3D-models using 2D-projection [7];
- object recognition on sample [6,18].

Figure 3 depicts an example of feature points found by algorithms SURF and SIFT. It should be observed that many points are common for both methods. In this work the union of closest points of both methods is used.

Fig. 3. Example of feature points on image: (a) SURF points, (b) SIFT points, (c) the union of closest points.

3 Known Challenges

Above-mentioned methods have a number of disadvantages which are known challenges in the computer vision. In [3] they are described as:

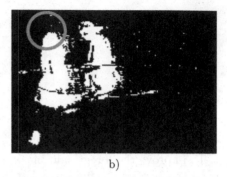

a)   b)

Fig. 4. Visual example of camouflage artifacts in background subtraction based algorithm: (a) frame 25 of *Traffic* sequence, (b) foreground mask.

1. **Camouflage.** A pixel of a foreground object may have the same color as the background in adjacent regions. In such a case the foreground and the background cannot be distinguished reliably within a single frame. This can take place at threshold filtration after background subtraction, for the foreground and background pixels having close values. The visual example of this challenge is depicted in the Fig. 4.
2. **Illumination changes.** This problem rises in case of scene modification due to local or global illumination change. Such a change can be gradual or sudden. The example of gradual illumination change is the change of relative location of the sun during the day in outdoor scenes. The example of sudden illumination change can be switching indoor lights on or off, or an outdoor change between cloudy and sunny conditions. The example of global sudden illumination change is depicted in the Fig. 5.
3. **Sleeping foreground object.** Foreground object may stop at a certain place and stay there for a long time. As a result, at some moment the system would include it in the background. This effect impacts the accuracy of video analytic and compression functions.
4. **Camera jitter.** The camera may sway back due to various conditions, such as the wind or a transport shake. It causes nominal motion but is undesirable in the sequence.
5. **Image noise.** The cause of image noise can be a camera with low video capturing performance or quantization noise in low bitrate video streams.

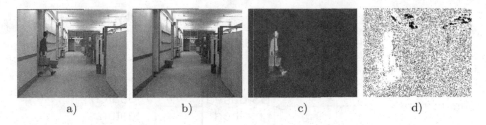

a)   b)   c)   d)

Fig. 5. Example of illumination change in background subtraction based algorithm: (a) frame 11 of *Hall* sequence, (b) background frame, (c) background subtraction, (d) foreground mask.

4 Proposed Method of Refining Regions of Interest

The background subtraction method which is described in Sect. 2 has many disadvantages, namely:

- Low accuracy in the estimation of foreground objects;
- Impossibility to distinguish objects by mask.

One of the alternatives of ROI detection improvement is feature points analysis. The similar approach is used in paper [13]. The main advantage of the proposed method from [12] is decreasing of computational complexity by performing search only in the areas where some movement was detected. Advantages of the proposed BS-based method is robustness to external factors, such as camera jitter, image noises, illumination changes, etc. The analysis of proposed approach is described in Sect. 5. Proposed method analyses foreground and background feature points. Similar points which belong to foreground and background are analyzed in neighboring regions to avoid camera jitter artifacts. The pseudocode of detecting areas operation is presented in the scheme Algorithm 1.

Algorithm 1. ROI detection using features estimation

---

procedure ROI Detection
    for each area with movement **do**
        areaPoints ← findFeatures(area, fg)
        backgroundPoints ← findFeatures(area, bm)
        bmPointsInArea ← $areaPoints - backgroundPoints$
        ratio ← $\frac{size(bmPointsInArea)}{size(areaPoints)}$
        if ratio ≥ 0.05 **then**
            return *background*
        else
            return *foreground*

The scheme of this method is illustrated in the Fig. 6. For object tracking it is necessary to add to the given scheme the features' list for objects and analyse move of points "cloud". Analysis of proposed method is presented in Sect. 5.

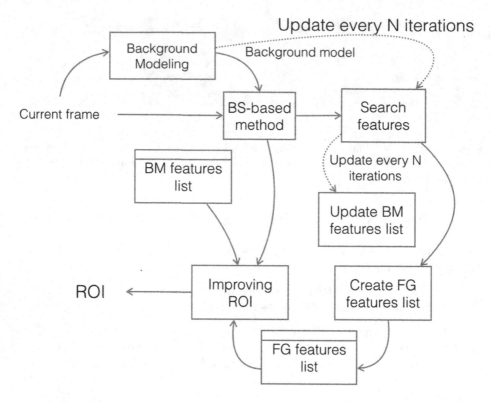

Fig. 6. Scheme of proposed method of refining ROI by feature points

5 Experimental Results Analysis

The proposed method is tested in some situations. For these cases, a special video test set was prepared which is described in Table 1.

A situation in which illumination change does not happen and there are no noises and camera jitter, is called *No affect*. For this case used [8,9] video sequences. Foreground masks corresponding to the sets were obtained and compared with ground truth by F1-Score [16]. The f1-score is the basic approach for estimation of binary classifiers. The estimation is calculated by following equation:

$$F1\text{-}score = \frac{2TP}{2TP + FP + FN}, \tag{3}$$

where TP – right detected foreground point (true positive), FP – wrong detected foreground point (false positive), FN – wrong detected background point (wrong negative).

The results are presented in the Table 2. In the Fig. 7 foreground masks (Fig. 7d, e, g) in the situation with illumination change are illustrated.

Table 1. Test set description

Sequence	Type	Link	Description
111	No affect	[9]	Synthetic
121	No affect	[8]	Synthetic
Light switch	Illumination changes	[10]	Real
Noisy night	Image noises	[11]	Real
Camera jitter	Camera jitter	[12]	Real

Table 2. Comparison of methods by F1-Score criteria [16]

Method	BS-based	Proposed
No affect	**0.8**	0.75
Illumination change (Fig. 7)	0.23	**0.64**
Noise	0.03	**0.25**
Camera jitter	0.06	**0.12**

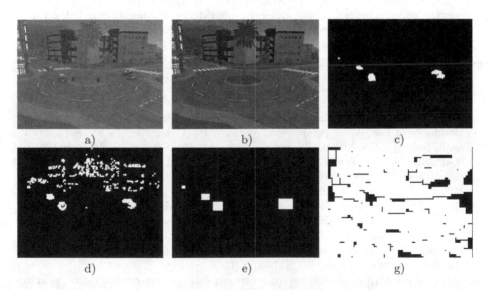

a)   b)   c)

d)   e)   g)

Fig. 7. Illumination change challenge: (a) 1059 frame of [8] sequence, (b) background frame generated by *Mixture of Gaussian* [17], (c) ground truth of 1059 frame, (d) ROI mask obtained by BS-based method, (e) ROI mask obtained by feature points method, (g) ROI mask obtained by global motion vectors method [5].

Proposed method solves successfully the problems of illumination change, random reflections, impulse noises and jitter. It should be noticed that regions in the ground truth sequences have a complex form, at the same time rectangular shapes in output of proposed method appear as they are most often used in the image and video processing. Nevertheless, obtained results demonstrate sensitivity to noise for cameras located in some high place (e.g. mast, roof of the house). In this case the size of small foreground objects is comparable to the size of noise artefacts.

6 Conclusion

Feature points analysis is a conventional approach used in computer vision. It can be used in many applications, e.g. video surveillance systems, online video streaming, etc. A new approach based on feature points counting allows for improving performance of region of interest detection. Unlike state-of-the-art methods employing background subtraction, proposed method enables avoiding detection faults caused by sudden luminance changing. This advantage is verified by using ground truth sequences and F1-score criterion.

References

1. The Zettabyte Era - Trends and Analysis - Cisco: White Paper. http://www.cisco.com/c/en/us/solutions/collateral/service-provider/visual-networking-index-vni/vni-hyperconnectivity-wp.html. Accessed 30 May 2017
2. Visual Networking Index – Cisco. http://www.cisco.com/c/en/us/solutions/service-provider/visual-networking-index-vni/index.html. Accessed 30 May 2017
3. Bouwmans, T., Porikli, F., Hoferlin, B., Vacavant, A.: Background Modeling and Foreground Detection for Video Surveillance. Taylor and Francis Group, LLC, Abingdon (2015)
4. Gilmutdinov, M., Ustyuzhanin, N.: Analysis of background modeling methods performance in lossy video compression systems. In: Vishnevskiy, V.M., Samouylov, K.E. (eds.) DCCN-2016. CCIS, vol. 628, pp. 481–489. Springer, Heidelberg (2016)
5. Gilmutdinov, M., Nemcev, N.: Approach for foreground detection based on parallax effect analysis, Saint Petersburg State University of Aerospace Instrumentation (2016)
6. Lowe, D.G.: Object recognition from local scale-invariant features. In: The proceedings of the seventh IEEE international conference on Computer Vision (1999)
7. Bay, H., Tuytelaars, T., Gool, L.: SURF: speeded up robust features. In: Leonardis, A., Bischof, H., Pinz, A. (eds.) ECCV 2006. LNCS, vol. 3951, pp. 404–417. Springer, Heidelberg (2006). doi:10.1007/11744023_32
8. 121 Dataset. http://bmc.iut-auvergne.com/wp-content/ressources/videos/learning/121_png.zip. Accessed 30 May 2017
9. 111 Dataset. http://bmc.iut-auvergne.com/wp-content/ressources/videos/learning/111_png.zip. Accessed 30 May 2017
10. Light Switch. http://www.vis.uni-stuttgart.de/hoeferbn/bse/dataset/SABS-LightSwitch.rar. Accessed 30 May 2017

11. Noisy   Night.   http://www.vis.uni-stuttgart.de/hoeferbn/bse/dataset/SABS-NoisyNight.rar. Accessed 30 May 2017
12. Camera Jitter. http://wordpress-jodoin.dmi.usherb.ca/static/dataset/cameraJitter/boulevard.zip. Accessed 30 May 2017
13. Olaode, A.A., Naghdy, G., Todd, C.A.: Unsupervised region of interest detection using fast and surf. Comput. Sci. Inf. Technol. **5**, 63–72 (2015)
14. Viola, P., Jones, M.: Rapid object detection using a boosted cascade of simple features. In: Proceedings of the 2001 IEEE Computer Society Conference on Computer Vision and Pattern Recognition, CVPR 2001. IEEE (2001)
15. Stevenson, R., Arce, G.: Morphological filters: statistics and further syntactic properties. IEEE Trans. Circ. Syst. **34**, 1292–1305 (1987)
16. Powers, D.M.: Evaluation: from precision, recall and F-measure to ROC, informedness, markedness and correlation. J. Mach. Learn. Technol. **2**, 37–63 (2011)
17. Stauffer, C., Grimson, W.: Adaptive background mixture models for real - time tracking. In: Proceedings of IEEE Conference on Computer Vision and Pattern Recognition (1999)
18. Lowe, D.G.: Distinctive image features from scale-invariant keypoints. Int. J. Comput. Vis. **60**, 91–110 (2004)
19. Brown, M., et al.: Recognising panoramas. In: ICCV (2003)
20. Delponte, E., et al.: SVD-matching using SIFT features. Graph. Models **68**(5), 415–431 (2006)

Modified EM-Algorithm for Motion Field Refinement in Motion Compensated Frame Interpoliation

Nikolay Nemcev[(✉)] and Marat Gilmutdinov

Saint Petersburg State University of Aerospace Instrumentation (SUAI),
67, Bolshaya Morskaya str., Saint Petersburg 190000, Russia
nicknemcev@gmail.com, mgilmutdinov@gmail.com

Abstract. Motion-compensated frame interpolation is required in many applications, e.g. packet-based video transmission error concealment, frame rate up-conversion, etc. One of the most important challenges of temporal interpolation is the accuracy of motion estimation. Several approaches for improving the motion estimation performance are proposed in this paper. Global motion vectors and motion segmentation allow for motion vectors field refinement. Performance of proposed motion-compensated temporal interpolation method is compared with several modern temporal interpolation methods.

Keywords: True motion estimation · True motion model · Optimization problem · Spatial-temporal interpolation of visual data · Frame rate up-conversation

1 Introduction

The procedure of temporal interpolation is one of the most important components of modern visual data processing systems. Temporal interpolation is widely used in many tasks, namely:

- frame rate up-convertion,
- making the slow-motion effect,
- recovery of damaged or lost frames of the video sequence (error concealment),
- temporal mode of scalable video coding (decompression of video data encoded at low bitrates with a reduction in the number of frames).

One of the most important methods used for temporal interpolation of frames is the estimation of the true motion of objects in the video sequence. It should be noted that the problem of the true motion estimation differs from the task of motion estimation for video compression. These approaches consider sligihtly different optimization tasks. One of the most important of the assumptions is piecewise smoothness of the vector field. The main reason is that the motion vectors corresponding to spatially close regions must be correlated [1]. From

© Springer International Publishing AG 2017
O. Galinina et al. (Eds.): NEW2AN/ruSMART/NsCC 2017, LNCS 10531, pp. 662–670, 2017.
DOI: 10.1007/978-3-319-67380-6_63

the theoretical point of view, the model of true motion is well formalized, and there are algorithms that explicitly take into account this model in motion estimation [1,2]. However, many assumptions used in the true motion model are not true for real frame sequences. That requires the development of heuristic algorithms oriented to improving the visual quality of interpolated frames. In this paper, we propose a motion estimation algorithm that uses improved true motion model to solve the problem of temporal frame interpolation. It is based on motion field segmentation that allows for improved estimation of accuracy in areas with complex motion of objects. This paper is organized as follows. In Sect. 2 the problem of the true motion estimation is formulated. In Sect. 3 the structure of the proposed algorithm is presented. In Subsect. 3.1, the proposed procedure of global motion estimation is introduced. Subsect. 3.2 is dedicated to proposed procedure of local motion estimation. Subsect. 3.3 is dedicated to proposed procedure of motion field refinement. Subsect. 3.4 is dedicated to procedure of motion-compensated frame interpolation. Section 4 presents the results of the performance evaluation of the proposed method and well-known existing implementations of algorithms for temporal frame interpolation.

2 Theoretical Background

The developed motion estimation algorithm is based on bidirectional motion estimation/compensation and uses a pair of adjacent frames of the original video sequence $\mathbf{F}_{i-1}$ and $\mathbf{F}_{i+1}$ to estimate the vector field $\mathbf{V}$, describing the movement of objects between these frames. Frames $\mathbf{F}_{i-1}$ and $\mathbf{F}_{i+1}$ will be called basic or reference frames. Frame $\mathbf{F}_i$ will be called an interpolated, or intermediate, frame. Model vector field for the problem of estimation of the true motion can be introduced as a result of the optimization of the following functions with fixed frames $\mathbf{F}_{i-1}$ and $\mathbf{F}_{i+1}$ [1,2]:

$$\mathbf{V}^* = argmin_{\mathbf{V} \in \Omega} \mathbf{E}\left(\mathbf{F}_{i-1}, \mathbf{F}_{i+1}, \mathbf{V}\right), \tag{1}$$

where Ω – is the space of all possible vector fields and

$$E\left(\mathbf{F}_{i-1}, \mathbf{F}_{i+1}, \mathbf{V}\right) = E_1\left(\mathbf{F}_{i-1}, \mathbf{F}_{i+1}, \mathbf{V}\right) + \alpha E_2(\mathbf{V}). \tag{2}$$

The term E_1 characterizes the energy of the difference frame obtained using the reference frames $\mathbf{F}_{i-1}$ and $\mathbf{F}_{i+1}$ and motion field $\mathbf{V}$, term E_2 characterizes the smoothness of a vector field $\mathbf{V}$, $\alpha \geq 0$ – regularization coefficient between these terms.

In this formulation E_2 is the inverse value to the smoothness, that is, the minimum value of the term E_2 corresponds to a maximum level of smoothness of the vector field $\mathbf{V}$.

It should be noted that there are algorithms that find the minima of function (2) explicitly [3]. Since the function (2) depends on a large number of parameters and is not unimodal, the founded minima are usually local. However, it should be noted that usage of formula (2) describes only the motion model, and the vector field corresponding to local and global minima of this model does not necessarily accurately reflect the true motion of the frame objects. For example, the model does not take into account that on the boundaries of objects a smooth vector field cannot accurately convey the true motion. In this regard, better results in terms of visual quality are achieved using suboptimal algorithms, considering the optimization of (1) implicitly. The basis of all true motion estimation methods is the procedure of optimal motion vector searching, which can be realized using either a complete search of all possible vectors (optimal search) or some suboptimal algorithm. Complete search guarantees the minimal energy of the difference frame (E_1), but its results do not match with the model of true motion, especially in texture regions [2]. In addition, the procedure of complete search has high computational complexity. Therefore, the optimal vector search in the true motion estimation problem is usually performed using suboptimal methods, for example, gradient descent [6]. As a rule, schemes based on motion prediction are used to calculate the true motion vector field [2,6,11]. In these algorithms, the vector for the new region is estimated using vectors generated earlier for the neighboring regions [7]. Developed motion estimation algorithm belongs to this class of algorithms.

3   Proposed Temporal Interpolation Pipeline

The scheme of developed algorithm is depicted in Fig. 1. The main structure is based on the sheme proposed in [2]. The task of motion estimation is divided into three stages:

- global motion estimation;
- local motion estimation;
- refinement of a vector field.

Information on global motion helps to significantly reduce the complexity of the subsequent stages of the motion estimation algorithm. It also allows for more accurate descriptions of the displacements of fast moving objects.

At the stage of local motion estimation, the field of motion vectors is generated using bilateral approach. First, the initial motion field using information about global motion is constructed. Thereafter, the elements of this field classified and possibly corrected using an iterative hierarchical prediction procedure similar to the approaches proposed in [6,7].

To refine motion vectors located in non-textured regions and regions with complex motion, an iterative procedure based on the principles of the EM algorithm [9] is proposed.

Motion-compensated frame interpolation is performed by bilateral algorithm.

3.1   Global Motion Estimation

The global motion estimation module searches for motion models and describes them using a set of global motion vectors $\mathbf{V}_g$. It should be noted that different

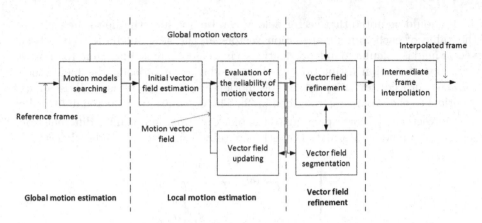

Fig. 1. Scheme of the developed algorithm.

areas of the frame can correspond to different global motion vectors. The number of global motion vectors depends on the frame motion characteristics.

In the paper it is proposed to use the random sampling method for estimation of the global motion (RANSAC, RANdom SAmple Consensus) [4]. The procedure of global motion searching can be described as the following iterative process:

1. randomly select a fixed number of pixels with coordinates $\mathcal{M} : \left\{ \begin{matrix} x^{(i)} \\ y^{(i)} \end{matrix} \right\}_{i=1}^{N}$;

2. find the shift $\left(\begin{matrix} \Delta x^* \\ \Delta y^* \end{matrix} \right)^k$, most accurately describing their cooperative shift (movement);

3. select the global vector $\mathbf{Vg}^{(m)} = \left(\begin{matrix} \Delta x^* \\ \Delta y^* \end{matrix} \right)^n$ of the current layer m, which is shift with the maximum amount of pixel, which movement this vector describes with a given accuracy;

4. remove from consideration pixels assigned to the current layer, after which the procedure can be repeated.

3.2   Local Motion Estimation

Global vectors $\mathbf{V}_g : \left\{ \begin{matrix} vx_g^{(i)} \\ vy_g^{(i)} \end{matrix} \right\}_{i=1}^{d}$ used as an initial information for local motion estimation step. Local motion estimation is performed using the gradient descent search [5] a initial vector field $\mathbf{V} : \left\{ \left(\begin{matrix} vx \\ vy \end{matrix} \right)_{i,j} \right\} \begin{matrix} i = 0,..,h-1 \\ j = 0,..,w-1 \end{matrix}$ are generated at first.

It should be noted that vector field $\mathbf{V}$ can be not smooth due to the fact that the value of each individual motion vectors at the stage of building the basic vector field is calculated independently. To increase of smoothness and to improve accuracy of motion estimation at the boundaries of objects, the hierarchical iterative procedure for updating the motion vector field is used. It is almost identical to the procedure described in [2]. The calculation of the reliability label for the vector $v_{i,j}^{(h)}$, corresponding to the block $g_{i,j}^{(h)}$, is significantly different from [2], basing on edge information which is obtained using Canny edge detector:

$$
r_{i,j}^{(h)} = \begin{cases}
1, & if\ e_{i,j}^{(h)} > errThr \\
2, & if\ C_{i,j}^{(h)} > Ca_{i,j}^{(h)}\ I_{i,j}^{(h)} = 0 \\
3, & if\ C_{i,j}^{(h)} > Ca_{i,j}^{(h)}\ I_{i,j}^{(h)} = 1 \\
4, & if\ C_{i,j}^{(h)} \leq Ca_{i,j}^{(h)}\ I_{i,j}^{(h)} = 0 \\
5, & else
\end{cases} , \tag{3}
$$

where $C_{i,j}^{(h)}$ and $Ca_{i,j}^{(h)}$ are local smoothness characteristics of the motion vector field and calculated according to [6,7], $e_{i,j}^{(h)}$ – block matching error (in this paper the SAD (sum of absolute differences [5]) is used), $I_{i,j}$ – function indicating that the block belongs to the area of the boundaries of objects.

3.3   Motion Field Refinement

To analyze the movement in regions with complex motion an iterative procedure based on EM algorithm (expectation-maximization) [8] is used. Motion segmentation algorithm used in refinement initially was proposed in [9].

The input parameters of the algorithm are: the information about the field of motion vectors $\mathbf{V}$, field of vectors reliability estimates $\mathbf{r}$, the characteristics of the motion models $\mathbf{V}_g$ and information about the pixels of the basic frames $\mathbf{F}_{i-1}$ and $\mathbf{F}_{i+1}$. The procedure output is the updated vector field $\mathbf{V}$.

At the preliminary stage of the procedure, the field of labels $\mathbf{L}$ is constructed to determine the vector belonging to a specific model of motion by evaluating the Euclidean distance between the vector of the field $\mathbf{V}$ and the vectors of the motion models $\mathbf{V}_g$.

At the first stage, for each pixel of the intermediate frame, the error function of belonging to a particular motion model is calculated:

$$
E_k(y, x) = w_k(y, x) D_k^2(y, x) + w_k(y, x) \log w_k(y, x)
$$
$$
- \mu \sum_{m,n \neq y,x} W(y, x; m, n) w_k(y, x) w_k(m, n) - \gamma r_k(y, x)
$$
$$
+ \tau \sum_{s \in N(L(y,x))} V_c(L(y, x), L(s)) \tag{4}
$$

where $E_k(y, x)$ – comparison error for pixel with the coordinates y, x to the motion model under the number $k \in [1, .., d]$, d – number of motion models,

μ, τ and γ – regularization coefficients, $w_k(y, x)$ – weighting coefficient of the EM-algorithm [8], $W(y, x; m, n)$ – quantity that takes into account the spatial similarity of the pixels (located next to similar pixels most likely relate to the same object and their movement can be described by one model of movement) described in [9], $r_k(y, x)$ – reliability label of the vector of the model k, placed in vector field $\mathbf{V}$ on position (y, x), obtained according to the expression (3), $V_c(L(y, x), \mathbf{L}(s))$ is used to take into account the label field L spatial structure, it is the so-called clique [10], $\mathbf{L}(y, x)$ – the current vector class label, $L(s)$ – class label of its neighboring pixel.

The calculation of the updated motion model label for the pixel $F_{i-1}(y, x)$, can be described using the following optimization problem:

$$L(y, x) = argmin_{k \in [1:d]} E_k(y, x).$$ (5)

After updating the label field $\mathbf{L}$, the vectors corresponding to the changed labels are updated. The update is performed using the gradient descent procedure. Gradient descent starts from the global vector corresponding to the current label.

After updating the vector field $\mathbf{V}$, in accordance with the implementation of the EM-algorithm described in [9] the new values of $\mathbf{V}_g$ are computed.

3.4   Motion-Compensated Frame Interpolation

Motion compensation is performed using bilateral approach, according to the expression:

$$\mathbf{F}_i(x, y) = \frac{1}{2}(\mathbf{F}_{i-1}\left(x - \frac{1}{2}vx(y, x), y - \frac{1}{2}vy(y, x)\right)$$
$$+ \mathbf{F}_{i+1}\left(x + \frac{1}{2}vx(y, x), y + \frac{1}{2}vy(y, x)\right))$$ (6)

where $F_i(x, y)$ – pixel of the intermediate frame, $vx, vy \in \mathbf{V}$.

4   Experimental Results

The evaluation of the interpolation performance was done in accordance with the following procedure:

- from the initial video sequence frames with odd numbers was removed;
- deleted frames were interpolated using remaining frames;
- average $Y - PSNR$ value was used for numerical comparison of proposed method and various exiting solutions. The $Y - PSNR$ is calculated between original frames, which were deleted, and interpoliation results.

During the evaluation of the interpolation efficiency, the arithmetic mean of PSNR, is calculated according to expression:

$$Y - PSNR_{avg} = \frac{\sum_{i=1}^{N} PSNR(\mathbf{Y}_i, \hat{\mathbf{Y}}_i)}{N}, \tag{7}$$

where $Y - PSNR(\mathbf{Y}_i, \hat{\mathbf{Y}}_i)$ – peak signal-to-noise ratio between a pair of frames, calculated as:

$$PSNR\left(\mathbf{Y}_i, \hat{\mathbf{Y}}_i\right) = 10lg\frac{w \cdot h \cdot \{max(\mathbf{Y})\}^2}{\sum_{y=1}^{h} \sum_{x=1}^{w} \left(Y_i(y, x) - \hat{Y}_i(y, x)\right)^2}, \tag{8}$$

where $\mathbf{Y}_i$ – the luminance component of original frame of the video sequence, $\hat{\mathbf{Y}}_i$ – component of the interpolated frame, N – number of interpolated frames.

In this paper, we compared the efficiency of the following algorithms for frame rate up-conversion:

- algorithm for temporal frame interpolation (MSU-FRC) [11];
- algorithm for temporal frame interpolation (Correlation-Based FRC), described in [6];
- algorithm for temporal frame interpolation (SUAI-FRC), described in [7];
- algorithm for temporal frame interpolation (MME-TE-FRC), described in [12];
- method used in MVTools Avisynth plugin (MVTools) [13];
- method based on the proposed procedure of motion estimation (PROP).

For all compared algorithms, the parameters corresponding to the maximum interpolation quality were used. The results are shown in Fig. 2 and Table 1.

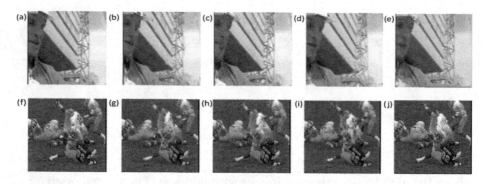

Fig. 2. A visual comparison of the Y component for the 186th frame of the Foreman sequence: original frame; b MSU-FRC, PSNR = 26,24; MME-TE-FRC, PSNR = 27,63; d Correlation-Based FRC, PSNR = 24,85; e PROP, PSNR = 27,9 and the 220th frame of the Football sequence: f original frame; g MSU-FRC, PSNR = 21,79; h MME-TE-FRC, PSNR = 22,25; i Correlation-Based FRC, PSNR = 22,21; j PROP, PSNR = 23,76.

Table 1. An objective comparison of the results of the interpolation

Method	$Y - PSNR_{avg}$, dB			
	Bus	Foreman	Football	Soccer
SUAI-FRC	27.10	35.58	23.63	30.50
MSU-FRC	28.18	35.08	22.52	29.94
MVTools	27,85	35,03	23.54	**30.83**
MME-TE-FRC	**28.32**	35.35	24,04	—
PROP without global motion estimation and vector field refinement	26.83	35.31	24.12	28.79
PROP without vector field refinement	26.87	35.41	25.17	30.23
PROP	26.98	**35.64**	**25.94**	30.63

The comparison results show that the usage of the global motion estimation procedure brings significant improvements in interpolation results for sequences containing fast moving objects and sequences taken from a moving camera (Soccer, Football). Nevertheless, it does not significantly increase the quality of interpolation for sequences with objects moving at a relatively low speed (Foreman, Bus). Usage of the proposed procedure of the motion vector field refinement to specify the motion in complex regions of frame improves the performance of interpolation for all test video sequences.

Proposed algorithm is not efficient for video sequences with a large number of small moving objects and complex texture regions (Bus), comparable with the best algorithms when processing sequences with relatively slow motion (Foreman,Soccer), and also significantly exceeds efficiency of the existing analogues in processing video sequences with fast and complex motion (Football).

5   Conclusion

The modification of EM-algorithm considered in this paper is used for refinement procedure in local motion estimation. Motion-compensated frame interpolation algorithm uses this refinement and several additional methods, e.g. global motion estimation and reliability classification of vectors, for analysis of regions with complex motion. Proposed method outperforms modern temporal interpolation methods, especially in high-motion test video sequences with fast-moving objects.

References

1. Rajala, S., Abdelqadar, I., Bilbro, G., Snyder, W.: Motion estimation optimization. In: Proceedings of IEEE Internernational Conference on Acoustics, Speech and Signal Processing (ICASSP 1992), San Francisco, 23–26 March 1992, pp. 253–256 (1992). doi:10.1109/ICASSP.1992.226203

2. Veselov, A., Gilmutdinov, M.: Algorithm of motion field optimization for temporal frame interpolation. Informatsionno-upravliaiushchie sistemy **4**(71), 25–28 (2014)
3. Boykov, Y., Kolmogorov, V.: An experimental comparison of min-cut/max-flow algorithms for energy minimization in vision. IEEE Trans. Pattern Anal. Mach. Intell. **26**(9), 1124–1137 (2004). doi:10.1109/TPAMI.2004.60
4. Fischler, M., Bolles, R.: Random sample consensus: a paradigm for model fitting with application to image analysis and automated cartography. Commun. Assoc. Comp. Mach. **24**, 381–395 (1981)
5. Haan, G.: Motion estimation and compensation. Ph.D. thesis, Delft University of Technology, 17 September 1992
6. Huang, A.-M., Nguyen, T.: Correlation-based motion vector processing with adaptive interpolation scheme for motion-compensated frame interpolation. IEEE Trans. Image Process. **18**(4), 740–752 (2009)
7. Veselov A., Gilmutdinov M.: Iterative hierarchical true motion estimation for temlioral frame interliolation. In: IEEE International Worksholi on Multimedia Signal Processing, MMSP (2014)
8. Dempster, A.P., Laird, N.M., Rubin, D.B.: Maximum likelihood from incomplete data via the EM algorithm. J. R. Stat. Soc. B **39**(1), 138 (1977)
9. Weiss, Y., Adelson, E.H.: Perceptually organized EM: a framework for motion segmentation that combines information about form and motion. In: M.I.T Media Laboratory Perceptual Computing Section Technical Report No. 315. ICCV 1998 (1998)
10. Wang, Q.: HMRF-EM-image: implementation of the Hidden Markov random field model and its expectation-maximization algorithm, 18 December 2012. arXiv:1207.3510v2 [cs.CV]
11. Frame rate conversion filter MSU. http://compression.graphicon.ru/video/motion_estimation/
12. Seong-Gyun, J., Chul, L., Chang-Su, K.: Motion-compensated frame interpolation based on multihypothesis motion estimation and texture optimization. IEEE Trans. Image Process. **22**(11), 4497–4509 (2013)
13. Frame rate conversion filter MVTools Avisynth plugin. http://avisynth.org.ru/mvtools/mvtools2.html

The Models of Moving Users and IoT Devices Density Investigation for Augmented Reality Applications

M. Makolkina[1,2(✉)], A. Koucheryavy[1], and A. Paramonov[1]

[1] The Bonch-Bruevich State University of Telecommunications,
22 Pr. Bolshevikov, St. Petersburg, Russian Federation
makolkina@list.ru, akouch@mail.ru,
alex-in-spb@yandex.ru
[2] Peoples' Friendship University of Russia (RUDN University),
6 Miklukho-Maklaya St, Moscow 117198, Russian Federation

Abstract. Applications of augmented reality penetrate into all spheres of human life. With the emergence of glasses of augmented reality, the introduction of this technology in VANET (Vehicular ad hoc network), etc. a number of interesting questions arise. For example, the amount of data that a user can perceive and understand the significance of the received content. The article develops a user perception model, which depends on the type of data, the amount of information and the significance of the data. The user is a queuing system object that receives various data from the surrounding objects. The data is ranked according to the priorities for which the transmission characteristics are determined. With the movement of the user, objects in his environment change. The user perception model defines the requirements for the service delivery model, which will allow the maximization of information that the user can perceive.

Keywords: Augmented reality · User behavior · Internet of Things · Service model

1 Introduction

Today, modern technologies provide to human a whole range of new opportunities that were not available decades ago, and the implementation of which seemed impossible [1–3]. However, now they are gradually being introduced into our everyday life and become an integral part of it [4–6]. From year to year, the development goes to a whole new level, making a lot of useful things. They are able to recognize a person's face, his voice, facial expressions and gestures, as well as determine emotions and moods [7]. For example, existing 3D scanning devices (so-called biometric scanning) used in various fields of medicine can scan a person's face and body using all its anatomical features, which is widely used for plastic surgery [8, 9].

Most of the technologies are oriented not only to the entertainment side of life, but also, first of all, to bringing benefits to people and simplifying their activities. One of them is the Augmented Reality. Recently, it is increasingly interested in the IT-area by experts, the creators of games and applications on the Android and iOS platforms, as

© Springer International Publishing AG 2017
O. Galinina et al. (Eds.): NEW2AN/ruSMART/NsCC 2017, LNCS 10531, pp. 671–682, 2017.
DOI: 10.1007/978-3-319-67380-6_64

well as marketers [10, 11]. This technology allows a person to receive information absolutely in new ways, combining both the virtual and the real, and does so in simple and clear ways. It converts real-world objects in a different quality, revealing them to the user on the other side.

The basis of augmented reality (AR) is not a completely artificial world, but the real life—what the user himself sees directly. People using AR-technology can get additional information about the object and its location, thereby interacting with virtual content in the real world. Therefore, virtual content is superimposed on the image of an object in the real world that the user sees, and it is not difficult for him to distinguish virtual objects from real ones. As well, the user can enable or disable the selected functions of the Augmented Reality (functions which can be associated with specific objects). For example, if the user is only interested in locating a specific shopping center, then he can configure the augmented reality options in his application on the device so that no other information except the selected one is displayed. It should also be noted that AR can not only add three-dimensional models, but also hide some objects from reality at the user's command. The word "augmented" in the definition may have the opposite meaning.

The user's interest in augmented reality services poses new challenges to operators that provide this type of service. They must maintain the network at the proper level and provide the required values of QoS (Quality of Services) and QoE (Quality of Experience) indicators. To implement such services, it is necessary to revise the approaches of communication networks construction. Thus, there are a number of works that consider a new type of communications – D2D (Device-to-Device) and a new principle of networking – SDN (Software-Defined Network) network [12–14].

2   Related Works

Today, a lot of research is devoted to different aspects of augmented reality technology. In addition to describing a large field of application, for example, real-time traffic flow prediction and preventing traffic congestion [15] or recognizing traffic signs in order to impair driving safety [16], there are a number of articles devoted to analyzing the network traffic of augmented reality services. In the paper [17] authors introduce a new approach to support the planning and design of road construction projects using AR technology. AR and wireless communications allow to create a new way of position and orientation of data transfer between moving vehicles and an AR animation. Today users can deploy a wide variety of different application. However, many applications require a great computing resources and high network performance. There are some solutions to provide the required parameters. For example, cloud computing services reduce central processing unit (CPU) usage. Thus, in [18] machine virtualization is investigated and its impact on network characteristics, such as throughput and delay variations. Authors in [19] proposed a method to increase wireless network capacity for mobile traffic. They investigated the interaction of a cellular network in licensed bands and a device-to-device network in unlicensed bands, which allows a significant increase in wireless capacity gains.

Another problem is the type and amount of information. For example, there are cloud computing service, a network which provides the necessary QoS and QoE parameters and a user. Evidently, the user can perceive a limited set of data, especially if he moves at high speed. In [20] authors research vehicle augmented reality traffic information system. The driver should easily and quickly understand the information to ensure safe driving. They propose a new approach for AR traffic signs recognition, which improve the accuracy of AR traffic information system in various driving situations and unfavorable weather conditions. In [21] the authors study the impact that augmented reality technology has on the behavior of old drivers. Much attention is paid to the traffic signs detection by drivers. The size of the sign, font, color, etc. are studied. The article [22] presents a review of researches in this field. Many AR applications are associated with video transfer. As an augmented scene is viewed in real time it is important to display the relevant information in timely manner. In [23] packetization issues and video rate control are considered and their influence on packet delay for AR video applications.

In this paper, we proposed the user perception model, which depends on the type of data, the amount of information and the significance of the data.

3 Service Model

Suppose a user walks down the street in glasses of the augmented reality. Glasses are connected to a smartphone, for example, via Bluetooth technology. When the user looks in any direction, the Augmented Reality glasses read the markers from the objects in the view field. EpsonMoverioBT 200 Glasses, for example, are managed by a special remote control MoverioBT 200 (controller) Android version 4.0.4, through which the user gives to glasses a command to visualize in front of him additional information about the selected object. The user's smartphone has an application for interacting with AR glasses. Thus, AR glasses are accessed through the smartphone from the database of the augmented reality server. The database stores all information about objects in different formats. After the object was successfully identified in the database, the server extracts information about it and sends it to the application via the network. After receiving an affirmative answer, it superimposes layer in text form describing the name and purpose of the object. The scheme for providing this service is shown in Fig. 1.

In addition to the scenario described above, objects within the user's review range can directly transmit information about themselves. This may be a Bus Schedule if the user has chosen the "bus stop" object or store opening hours. For this, various technologies can be used, such as Wi-Fi and D2D communications. It should also be noted that data can come from objects without user intervention.

The user does not select the object about which he wants to receive information, but simply looks around and the glasses read data from all surrounding objects. Anyway, the question that arises is how much information a user can receive for a certain period of time. The user can be in a place with a large density of augmented reality objects or move at high speed, and then person simply does not have time to read or understand some of received information.

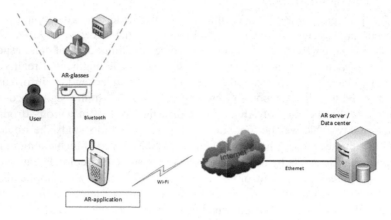

Fig. 1. Scheme for the provision of AR services.

4   The User Perception Model

The user of augmented reality service receives a certain capacity of additional information in the form of messages of various types: text, graphic, speech, sound, tactile and, possibly, other types. Messages are entered into the user's environment and are accessible to perception.

Let's make the following assumptions:

- The user is able to adequately perceive some quantity of information μ per unit of time (the intensity of information perception);
- The intensity of information perception is different for different types of messages (text, graphics, voice, sound, tactile sensations) μ_k, $k = 1...K$;
- Messages are in the user's environment for some, in general, limited time t;
- Messages of various types arise in the user's environment independently of each other at random times;
- The user is able to select the message according to the degree of its importance (priority);
- The user is able to interrupt the processing (perception of the current message) when there is a significant message on entrance.

Under the assumptions made, the model can be described by the queueing model with the selection of request according to their priorities (with relative priorities), Fig. 2.

Thus, the service time is random, and service time distribution is determined by a composition of distributions.

Let's estimate quantity of new objects in a surrounding during t as

$$f(x) = \sum_{i=1}^{k} \xi_i p_i(x) \tag{1}$$

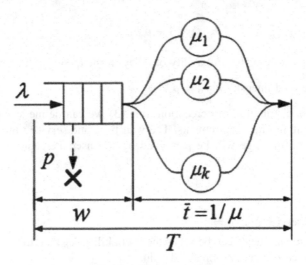

Fig. 2. AR user model.

where $p_i(x)$ – Service time probability density of the k-th type message;

$0 \le \xi_i \le 1, \quad \sum_{i=1}^{n} \xi_i = 1$ – The probability of a i-th type message being received.

As noted above, in general, the augmented reality information perception model can be described by a queuing system. For description, let's consider the use of G/G/1 models with priority asset. Taking into account the random character of the elements arrangement and the fact that the user is moving, we can admit a more particular case of M/G/1 [7], which under such conditions can be sufficiently close to the real situation. The average waiting time for the start of service of priority (type) p in this system can be defined as:

$$
w_p = \frac{\sum_{i=1}^{K} \lambda_i M(t_i^2)}{2(1 - \theta_p)(1 - \theta_{p-1})}
\tag{2}
$$

Where, λ_i – flow message intensity i-th priority (type);

μ_i – intensity of service (perception) of i-th messages priority (type);

$M(t_i^2)$ – the second initial moment of the distribution function of the message service time of the i-th priority (type) messages;

$$
\theta_p = \sum_{i=1}^{p} \frac{\lambda_i}{\mu_i}
$$

Approximation of the waiting time distribution of the message of the i-th priority type can be approximated by the expression [1]:

$$h_p(t) \approx 1 - p_{>0} e^{-\frac{p_{>0}}{w_p}} \tag{3}$$

For the simplest flow, the probability of delay in the queue is numerically equal to the value of the load intensity $p_{>0} = \rho = \sum_{i=1}^{K} \frac{\lambda_i}{\mu_i}$.

Knowing the distribution of perception time (3) we can define system parameters based on the waiting time requirements. The wait time and service time determine the time in which the message will be perceived by the user. Perception time of i-type message

$$T_i = w_i + \bar{t}_i \tag{4}$$

where, average service time $\bar{t}_i = \frac{1}{\mu_i}$.

The system requirements can be set as the probability Λ_{0i} that, the message waiting time does not exceed a certain threshold value T_{0i}

$$h_p(T_{oi}) = p(t < T_{oi}) = \Lambda_{0i} \tag{5}$$

The choice of the waiting time threshold T0i can be made from various considerations, Depending on the service features and the significance of the messages. As a characteristic approach, we take in consideration the fact that the waiting time should not exceed the staying time of the message in the user's environment, i.e. the time of its existence.

The staying time of the message in the user's environment is determined by the staying time of the corresponding object and depends on the user's behavior [23]. For example, if the user's environment is defined by an area bounded by a circle of radius R, then, if the user moves at rectilinear uniform motion, Fig. 3.

With a random distribution of objects, the distance traversed by the object in the perception area will be a random variable with a density probability.

$$g(d) = \frac{1}{\pi R \sqrt{1 - \left(\frac{d}{2R}\right)^2}} \tag{6}$$

And mean value will be defined as

$$M[d] = \bar{d} = \frac{4}{\pi} R \tag{7}$$

At v user's motion speed, the staying time of the object in the area will also be a random variable with a density probability:

$$q(t) = \frac{1}{\pi R \sqrt{1 - \left(\frac{v}{2R} t\right)^2}} \tag{8}$$

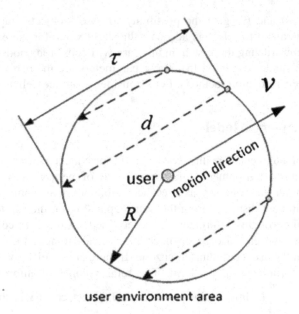

Fig. 3. User motion model

Then, mean value:

$$M[t] = \bar{\tau} = \frac{4R}{\pi v} \tag{9}$$

A graph of density probability (4) is shown in Fig. 4.

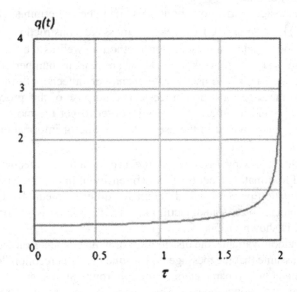

Fig. 4. Staying time of object in the area.

Expressions (8) and (9) give the possibility to assess objects staying time in a user-accessible area and to select an adequate threshold value for the expression (5).

The task of constructing the system in this case will consist in choosing the system parameters values under condition (5). These parameters are the part of each type of messages and their formation methods. Let's consider a service delivery model.

5   Service Delivery Model

The provision of augmented reality service is to provide the user with additional information about his environment in the form of a set of messages [24]. The provided information can refer to any objects as direct accessible or not available for observation and located at different distances from the user, depending on the service features. In general, the volume of this information is determined by the data contained in the service databases, and can also be obtained from local storages located directly on objects located in the user communication zone. If there are n objects with information in the user's environment Ij, $j = 1...n$. The total volume of information will be determined as $I = \sum_{j=1}^{n} I_j$. Information about each J object can be in the form of mj messages, which is formed according to some principles $m_j = R(I_j)$. Principles of a message formation include the manner in which the message is presented and its formation method. As presentation methods, for example, a representation in text form, graphics (pictograms), speech, sound, tactile and other presentation modes can be chosen. The method of messages formation involves the construction of messages based on the available information about the object and the message size requirements. For example, a text message can be detailed or reduced; a graphical message may contain a photograph, a plan or a short icon, etc.

Thus, the message formation principles can be determined as $R(I_j, U) = \{M_k(I_j, U), k(U)\}$. Where, k(U) defines the message construction method for the chosen display mode k and the presentation method as well as the formation method depend on the user's state data, U ($k = 1...K$, where K is the number of possible ways of presenting the message). The user's state (behavior) affects both the choice of the message type and message formation process. The purpose of this process is to select the type of message and its informativeness in order to get the most effect from the additional information provided to the user. The volume of information in the generated message is $I_k = M_k(I_j, U)$.

Let's make the following assumption: the user is able to adequately perceive a certain volume of information u per time unit (Intensity of information perception). The intensity of information perception is different for different methods of messages formation (text, graphics, voice, sound, tactile) $k = 1...K$. The message formation related to the AR object is shown on Fig. 5.

Thus, it is a random variable, the distribution of which depends on the message type distribution. We assume that the message m_j is in the user's perception domain for time τ. This time depends on the parameters of the environment area and the nature of the user's movement, as shown above. Thus, service delivery parameters, such as

m_j

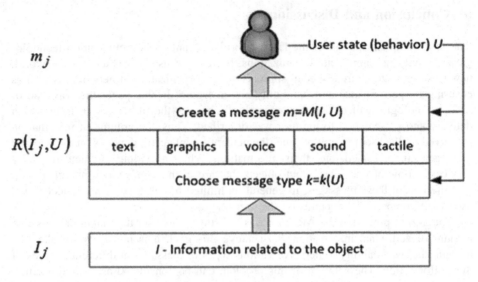

$R(I_j, U)$

I_j

Fig. 5. Message formation related to the AR object.

messages presentation and the volume of information in a message, affect the number of messages perceived by the user, therefore the volume of perceived information.

Let's assess the part of user-perceived messages through waiting time distribution (5). Then, choosing the value of T_0, proceeding from the characteristics of the user's movement (8) and (9), the following objective function can be formulated as

$$\{M, k\} = \arg\left(\max_{k,M}\left(h_p(T_{oi})I_p\right)\right) \tag{10}$$

$$k = 1\ldots K;$$

$$M = M(I, U);$$

where I_p – the volume of information in the message.

The task of getting the maximum of the objective function is to achieve the maximum volume of information perceived by the user in a given interval of time T0. This is achieved by selecting message types k = 1…K and their presentation method M (I, U). In fact, the message presentation method actually determines the volume of information in the message (detailed, reduced, short, etc.). The choice of the message presentation method, in general, requires a separate consideration. In this paper, we assume that M(I, U) is a discrete function whose value is a message that can be represented in three formats: detailed, reduced and short formats. The choice of one or another format is determined by solving the function (10).

In the given conditions, function (10) is a problem to solve for optimizing a discrete objective function. As a method of its solution, the method of dynamic programming can be used.

6   Conclusion and Discussion

Due to the implementation of the latest information ant telecommunication technologies, not only in commercial organizations, but also in users lives around the world, new services that require a special approach to the organization of network, as well as ensuring its performance and optimality are emerging. In this paper the way one of such technologies called "augmented reality" changes the principles of information delivery and presentation to the user was described. Thus, the volume of information perceived by the user per unit of time (the perception intensity) depends on the way the messages are presented, and their information. When providing augmented reality services, a flow of messages about objects that are in the user's environment occurs. The lifetime of these messages, in general, is limited by the time of the object in the user's environment and depends on the user's behavior.

The aim of providing the AR service is to bring to the user the maximum possible amount of additional information. To achieve this goal, it is necessary to take into account the user's ability to perceive various types of messages at different degrees of their information. The solution of this problem can be considered as an optimization problem, in which the variables are the types of provided messages and the degree of their information.

In the future work, we will investigate the effect of user moving motion on the quality of transmission when providing augmented reality services using AR glasses.

Acknowledgment. The publication was financially supported by the Ministry of Education and Science of the Russian Federation (the Agreement number 02.a03.21.0008), RFBR according to the research project No. 16-37-00209 mol_a "Development of the principles of integration the Real Sense technology and Internet of Things".

References

1. Kirichek, R., Vladyko, A., Zakharov, M., Koucheryavy, A.: Model networks for internet of things and SDN. In: 18th International Conference on Advanced Communication Technology (ICACT), pp. 76–79 (2016)
2. Al-Fuqaha, A., Guizani, M., Mohammadi, M., Aledhari, M., Ayyash, M.: Internet of things: a survey on enabling technologies, protocols, and applications. IEEE Commun. Surv. Tutor. **17**(4), 2347–2376 (2015)
3. Kirichek, R., Koucheryavy, A.: Internet of things laboratory test bed. In: Zeng, Q.-A. (ed.) Wireless Communications, Networking and Applications. LNEE, vol. 348, pp. 485–494. Springer, New Delhi (2016). doi:10.1007/978-81-322-2580-5_44
4. Broschart, D., Zeile, P.: Architecture: augmented reality in architecture and urban planning. In: Peer Reviewed Proceedings of Digital Landscape Architecture 2015 at Anhalt University of Applied Sciences, pp. 111–118 (2015)
5. Billinghurst, M., Clark, A., Lee, G.: A survey of augmented reality. Found. Trends Hum.-Comput. Inter. **8**(2–3), 73–272 (2015)

6. Park, H.S., Kim, K.-H.: AR-based vehicular safety information system for forward collision warning. In: Shumaker, R., Lackey, S. (eds.) VAMR 2014. LNCS, vol. 8526, pp. 435–442. Springer, Cham (2014). doi:10.1007/978-3-319-07464-1_40

7. Koucheryavy, A., Makolkina, M., Paramonov, A.: Applications of augmented reality traffic and quality requirements study and modeling. In: Vishnevskiy, V.M., Samouylov, K.E., Kozyrev, D.V. (eds.) DCCN 2016. CCIS, vol. 678, pp. 241–252. Springer, Cham (2016). doi:10.1007/978-3-319-51917-3_22

8. BusinessWire: Global Augmented Reality & Virtual Reality in Healthcare Industry Worth USD 641 Million by 2018 - Analysis, Technologies & Forecasts 2013–2018 - Key Vendors: HologicInc, Artificial Life Inc, Aruba Networks - Research and Markets

9. Konstantinova, J., Jiang, A., Althoefer, K., Dasgupta, P., Nanayakkara, T.: Implementation of tactile sensing for palpation in robot-assisted minimally invasive surgery: a review. IEEE Sens. J. **14**(8), 2490–2501 (2014). Bulling, A., Cakmakci, O., Kunze, K., Rehg, J.M.: Eyewear computing-augmented the humsn with head-mounted wearable assistants. Dagstuhl reports, vol. **6**(1), SchlossDagstuhl-Leibniz-Zentrumfuerinformatik (2016)

10. Hara, H., Kuwabara, H.: Innovation in On-site work using smart devices and augmented reality. Fujitsu Tech. J. **51**(2), 12–19 (2015)

11. 5G and e-Health. White paper, 5GPPP, October 2015

12. Pyattaev, A., Johnsson, K., Surak, A., Florea, R., Andreev, S., Koucheryavy, Y.: Network-assisted D2D communications: implementing a technology prototype for cellular traffic offloading. In: Wireless Communications and Networking Conference (WCNC), pp. 3266–3271. IEEE (2014)

13. Vladyko, A., Muthanna, A., Kirichek, R.: Comprehensive SDN testing based on model network. In: Galinina, O., Balandin, S., Koucheryavy, Y. (eds.) NEW2AN/ruSMART -2016. LNCS, vol. 9870, pp. 539–549. Springer, Cham (2016). doi:10.1007/978-3-319-46301-8_45

14. Zhang, M.: Real-time traffic flow prediction using augmented reality. University of Windsor Scholarship at UWindsor. Paper 5687 (2016)

15. Abdi, L., Meddeb, A., Abdallah, F.B.: Augmented reality based traffic sign recognition for improved driving safety. In: Kassab, M., Berbineau, M., Vinel, A., Jonsson, M., Garcia, F., Soler, J. (eds.) Nets4Cars/Nets4Trains/Nets4Aircraft 2015. LNCS, vol. 9066, pp. 94–102. Springer, Cham (2015). doi:10.1007/978-3-319-17765-6_9

16. Behzadan, A.H., Kamat, V.R.: Visualization of vehicular traffic in augmented reality for improved planning and analysis of road construction projects. In: Proceedings of 2009 NSF Engineering Research and Innovation Conference, Honolulu, Hawaii (2009)

17. Wang, G., Ng., T.S.E.: The impact of virtualization on network performance of amazon EC2 data center. In: Proceeding INFOCOM 2010 Proceedings of the 29th Conference on Information Communications, pp. 1163–1171 (2010)

18. Andreev, S., Galinina, O., Pyattaev, A., Johansson, K., Koucheryavy, Y.: Analyzing assisted offloading of cellular user sessions onto D2D links in unlicensed bands. IEEE J. Sel. Areas Commun. **33**(1), 67–80 (2014)

19. Abdi, L., Meddeb, A., Abdallah, F.B.: In-vehicle augmented reality traffic information system: a new type of communication between driver and vehicle. In: Proceedings of the International Conference on Advanced Wireless, Information, and Communication Technologies (AWICT) (2015)

20. Fu, W.T., Gasper, J., Kim, S.W.: Mixed and augmented reality (ISMAR). In: 2013 IEEE International Symposium, pp. 59–66. IEEE (2013)

21. Mogelmose, A., Trivedi, M.M., Moeslund, T.B.: Vision-based traffic sign detection and analysis for intelligent driver assistance systems: perspectives and survey. IEEE Trans. Intell. Transp. Syst. **13**(4), 1484–1497 (2012)

22. Razavi, R., Fleury, M., Ghanbari, M.: Low-delay video control in a personal area network for augmented reality. IET Image Proc. **2**(3), 150–162 (2008)
23. Iversen, V.: Teletraffic engineering and network planning, Department of Photonics Engineering, Technical University of Denmark. [http://www.fotonik.dtu.dk]
24. Futahi, A.: Wireless sensor networks with temporary cluster head nodes. In: Futahi, A., Paramonov, A., Koucheryavy, A. (eds.) 18th International Conference on Advanced Communication Technology (ICACT), pp. 283–288 (2016)

Quality of Experience Estimation for Video Service Delivery Based on SDN Core Network

Maria Makolkina[1,2], Ammar Muthanna[1(✉)], and Steve Manariyo[1]

[1] The Bonch-Bruevich State University of Telecommunications,
22 Pr. Bolshevikov, St. Petersburg, Russian Federation
makolkina@list.ru, ammarexpress@gmail.com,
mansteve06@mail.ru
[2] Peoples' Friendship, University of Russia, (RUDN University),
6 Miklukho-Maklaya St, Moscow 117198, Russian Federation

Abstract. Video traffic plays an increasingly important role in today's telecommunication networks. Most services are difficult to imagine without video stream transmission, and it is not only IPTV and OTT services, but also the services of augmented reality that are gaining popularity. At the same time, users' requirements for quality of experience provision are constantly tightened. Therefore, operators need to look for new ways to deliver video content to the user, which will allow the transfer of large amounts of traffic with the appropriate quality of experience. In this article, the possibilities of using SDN networks for the transmission of video traffic are investigated. For this purpose, we have created a laboratory testbed, which consists of a multimedia complex for the delivery of IPTV content and a SDN segment of the network. To estimate the quality of experience of the transmitted video traffic, we used, in addition to the generally accepted parameters, such as delays, losses, throughput, also the Hurst parameter.

Keywords: IPTV · Video traffic · Soft-Defined Network (SDN) · Hurst parameter · Quality of experience (QoE)

1 Introduction

In the struggle for the user operators constantly expand the range of services provided. Currently, the main trend in the development and maintenance of the competitiveness of telecommunications networks is the introduction of various types of services, including requiring an increasing share of the resource of IPTV services and others [1–3]. Most of these services are somehow connected with the transfer of video. With the development of the technology of augmented reality, the load from video in the network is constantly increasing. Thus, in the basis of providing augmented reality services also processing, delivery and storage of video. The employment of augmented reality applications has penetrated into all spheres of a person's daily life [4, 5]. The technology of augmented reality is actively used in conjunction with the concept of the Internet of Things [6]. There are many applications that collect information from surrounding objects and display to the user in a convenient way, for example, glasses of

O. Galinina et al. (Eds.): NEW2AN/ruSMART/NsCC 2017, LNCS 10531, pp. 683–692, 2017.
DOI: 10.1007/978-3-319-67380-6_65

augmented reality [7]. The user can see the timetable, the route of the bus, just by looking at the augmented reality glasses at a bus stop from afar. Augmented reality technology plays a special role in medicine [8], where high-resolution video transmission allows significantly improve and facilitate the work of doctors.

This type of services makes strict requirements to the quality of experience, forcing operators to upgrade existing communication networks, as well as approaches and methods for assessing transmission quality, taking into account the features of video traffic [9]. One of the characteristic features of video traffic is a high degree of self-similarity [10], which is estimated using the Hurst parameter. It is known that such network parameters as delays and losses impair with increasing degree of self-similarity. Thus, the evaluation of self-similar properties and their inclusion in the calculation of network parameters will allow more accurate description and reproduction of video sequences, which will positively affect the QoE estimates.

Existing networks that are characterized by a distributed structure are not suitable for implementing a large set of complex video services. While the SDN concept, in which data transmission and management are separated, are well suited for such tasks. SDN consists of the level of networking applications, the control level and the network infrastructure level [11, 12]. The level of networking applications is implementing different network management functions including traffic monitoring, security management, QoE management, management of data flows in the network, etc. The control level is responsible for the network topology and providing the unified interface with the higher layer. The network infrastructure level includes the main element of the SDN network – the controller. The SDN controller manages topology, network devices, applications and control of available server resources.

The increase in the volume of video traffic in networks necessitates the development of new ways of delivering video to the user and an integrated estimation of the quality of experience video services both on the user side and the operator. One of the main tools for providing a wide range of video services is the IPTV technology. As a basis for building a transport network, the technology SDN has been chosen, which today has a number of significant advantages when delivering content to the user's home. Our goal is to investigate the possibility of using SDN approach for efficient delivery of IPTV traffic with the appropriate quality of experience.

The remainder of this article is organized as follows: Sect. 2 analyzes the other works in this field of research. Section 3 shows the experimental investigation and laboratory testbed. Section 4 presents testing results. Finally, Sect. 5 concludes this experience paper.

2 Related Works

By now, several works on the delivery of IPTV services by using SDN have been known in the context of the increasing efficiency of video services, QoE estimations, traffic monitoring and control. Thus, in [13] authors investigated the control and routing IPTV traffic over SDN. The implementation of IPTV requires a multicast mode, which allows increasing the efficiency of video transmission by saving the network transport resource. The authors offer a new protocol developed by them and evaluate the efficiency of its work using OMNET ++ simulator. They show that the use of this protocol in SDN networks increases switch performance in time. In the paper [14] the transmission time of the first joint/receive packet to a client when using Dijkstra's and Prim's algorithms is researched. They deployed Dijkstra's and Prim's algorithms in the POX controller to calculate the shortest paths with delay time as the edge weight. They experimentally compared both algorithms and defined the edge weight as the distance between two adjacent nodes. Researches [15] investigated searching optimization algorithms and tools for managed video delivery networks. They offer the methodology for QoE estimation through different types of evaluation: content provider, operator and client. They use SDN controller to differentiate between two type of flows to satisfy QoE requirements from the three actors. In [16] experience of building an IPTV Software Defined Network testbed that can be used to develop and validate new approaches for service assurance in IPTV networks is described. They created a cost-effective prototype for research purposes which includes comprehensive description of new metrics and reconfiguration mechanisms implemented in OF and Flood-light. With the prototype, they conduct a number of experiments including network reconfiguration and routing experiment. Also the great interest is in the study of IPTV traffic patterns in SDN networks. In paper [17] is developed the evaluation model that allows examining how networks' performance will be affected as the traffic loads and network utilization change. This model for an OpenFlow SDN based on queueing theory, and resolves its average packet processing time.

3   Description of the Topology of the Built SDN Network and IPTV System

The infrastructure consists of IPTV ecosystem over SDN - software-defined networks (Fig. 1). The topology of the constructed network (testbed) is the connection of three Openflow switches, in the ring one openflow switch is logically divided into three switches for reservation in the event of an overload of one of the channels. This was taken into account since the SDN controller can allocate traffic route by default, in the network. For research facilities, an Ethernet switch is connected between openflow switches and the controller, whose purpose is creating an aggregation layer to SDN switches, thereby the possibility of network scaling, what increase the number of connected switches to one network controller.

As a controller of the developed testbed, the open source SDN controller Open-daylight Beryllium SR4 is chosen. The choice of this controller is based on its multifunctionality, the possibility of network scaling, and as already noted, this controller develops a whole community relevant to this project. Likewise, there are regular updates, improvements and correction of errors in the implementations of the controller and its modules. One of the important facts of choosing this controller is its architectural feature – modularity. Thus, if required to expand its functions, the installation of a new module does not take much time and this can be done without stopping the operation of the controller itself, respectively, the main network controlled by the controller will work without stopping. Another important fact of choosing this particular controller for building a network is its multifunctional API (Application

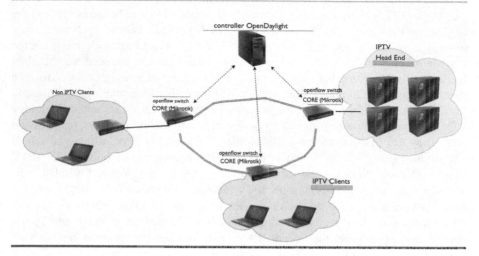

Fig. 1. Tesbed of a software-configurable network with an IPTV system

programming interface), Which allows working with each of its modules. At the same time, there can be used both the Java API and the RESTful API, which clearly provides ample opportunities to the development of controller applications, like those based on the same server where the controller is working or those remotely connected to the controller through an API in the authentication condition. Opendaylight controller is an open source and uses Java-Karaf technology, this allows you to develop your own Java protocol modules. Thus, the process of connecting its own developed modules is well reflected in the documentation for developers, which is also freely available on the official website of the OpenDaylight project.

IPTV system consists of an end-to-end solution designed by BeeSmart. BeeSmart is complete middleware solution for IPTV system. The system is composed of two subsystems, the IPTV headend end IPTV user's network.

The headend is composed of IPTV BeeSmart middleware, MSR-508 multiswitch, Cisco D9854 satellite receiver, Cisco D9900 transcoder and Cisco Ethernet switch C3560.

Video streams are received from digital satellite (Quattro LNB), converted by D9900 transcoder to multicast IP streams for further use by middleware server. The middleware task is to form a shell for the multicast streams for further view of these streams on the TV set-top boxes, has an administrative web interface for channel parameters management, user groups creation and control over TV set-top boxes.

Multicast Streams arrive on the Ethernet switch, and then are sent to IPTV client through SDN network. The control is implemented by BeeSmart middleware server, which interacts with multicast streams on the network and provides user preferences and content for TV set-top boxes.

Equipment used in building a testbed

Table 1 shows the description of the equipment used to build the testbed of the programmable network. Displayed also are the tasks for which this or that equipment was Supplied.

Table 1. Description of the testbed equipment

	Denotation	Technical description	Executable tasks
1.	OpenDaylight sdn controller	CPU: Intel Xeon(R) E3-1220V2 RAM: 16 Гб OS: Linux Ubuntu 14.04 LTS Version: OpendaylightBeryllium SR4	Programmable network controller
2.	Openflow switch	Mikrotik RB 201 1UI AS-RM Version: openflow 1.0	Logically divided into three switches, as SDN switches
3.	Aggregation switch	Cisco Catalyst 3750G series PoE-24	Aggregation level switch
4.	Aggregation switch	Cisco Catalyst 3750G series PoE-24	Aggregation level switch
5.	MSR-508 multiswitch	Radio spectrum: 950–2400 MHz Signal: 0–20 dB	Power distribution of satellite inverter
6.	Cisco D9854 satellite receiver	Demodulation: DVB-S QPSK, DVB-S2 QPSK, 8PSK Signal decoding: MPGE-4 AVC, MPEG-2 1080i and 720p	Decoder
7.	Cisco D9900 transcoder	Bit rate: 0.1–213 Mbps Supports: SPTS and MPTS	Content manager
8.	Aggregation switch	Cisco Catalyst WS-3560X-24T-S	Aggregation level switch
8.	Dune HD TV-101	Supports: 1080p (Full HD) MKV, DivX, XviD, APE, FLAC HDMI interface Ethernet network connection	TV set-top boxes

4 SDN Network vs Traditional Network

The experiment was conducted to analyze the impact of SDN network over video delivery.

Packet delay distribution was analyzed for video traffic based on SDN core network, another based on traditional core network without SDN controller for comparison. "Russia 24" channel was chosen in HD format during experiment.

When an IP packet traverse the network the following thinks usually happen.

After reaching the network the packet is mapped into a Logical link control (LLC) frame at the LLC layer. In the next step the encapsulated LLC frame is handled to the Physical Layer for Delivery. The medium Access Control (MAC) layer then convert the Frame and transmitted the signals.

The reverse procedure is applied at the receiver side.

In SDN network physical and network layers function are virtualized through openflow protocol in switches, which makes faster packets encapsulation and too many other process packet sending included.

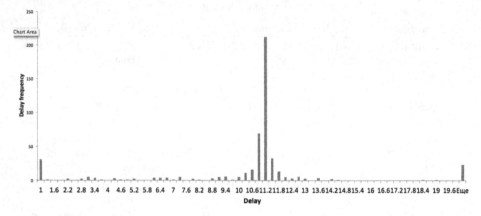

Fig. 2. Packet delay distribution over a no SDN network

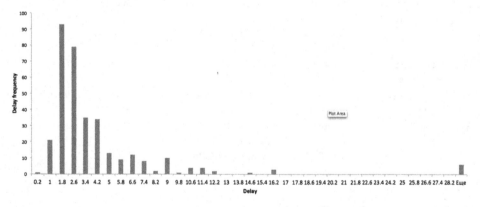

Fig. 3. Packet delay distribution based on SDN network

Analysis of the two graphics (see Figs. 2 and 3) show that the basic density of packets comes with a smaller delay over SDN network.

SDN functions are better for video traffic, and this can be explained by the following SDN features:

- Data transmission and management processes are separated;
- Network management is centralized through unified software tools

Flow properties for networks with packet switching don't correspond to those of the simplest flow. Studies have shown that these flows often have properties described by the theory of fractals. So often additional features are introduced to take into account their specific features, such as autocorrelation (autocovariance) function, Hurst coefficient, fractality coefficient. The Hurst parameter can be used as an objective metric for assessing the video quality of experience by analogy with the R-factor used to assess the quality speech transmission in the Recommendations of the International Telecommunication Union [18]. Such flows are often called self-similar. Self-similar flow is a random flow requests for which the Hurst coefficient is between 0.5 and 1. The Hurst coefficient is widely used in the analysis of time series due to its stability.

There are several methods of Hurst coefficient calculation. In this work, the analysis of packets delays distribution was used to calculate Hurst parameter, which can be used for video quality estimation in future works.

In practical terms, self-similarity estimation and calculation of the Hurst parameter are complex problems. In real conditions, they operate with certain data sets, so there is no way to check the traffic path for self-similarity. Therefore, it becomes necessary to investigate the various properties of self-similarity in real measured traffic, without having full information about all scales. To date, a number of methods for estimating self-similarity are known.

As the main method for estimating the values of the Hurst parameter, the method of analysis of the dispersion variation graph was chosen, which allows estimating the degree of self-similarity of the aggregated flow. The method consists in investigating the slowly decaying dispersion of a self-similar aggregated process.

For the Hurst coefficient estimation, the following expression is used:

$$RS = (aN)H, \tag{1}$$

therefore

$$H = \log RS \log aN, \tag{2}$$

where;
H   – Hurst coefficient;
S   – Standard deviation of a number of observations;
R   – Deviation range;
N   – Number of observation;
a   – constant.

In this paper we Estimate the Hurst parameter in two different scenarios: with SDN controller and without SDN controller.

In all scenarios Hurst parameter value is greater than 0.5 what is considered to be a sufficient basis for the recognition of self-similar process. It should be noted that the value of Hurst parameter closes to one, may mean that the process is deterministic, i.e. not random.

Estimation of the Hurst parameter was conducted by the analysis of the distribution of delays (see Table 2), which can be used for video quality estimation in future works.

The average value of Hurst coefficient is 0.78 in case with SDN controller and 0.56 without SDN controller.

Table 2. Hurst exponent in two different scenarios

With SDN controller	Without SDN controller
H = 0.77	H = 0.58
H = 0.81	H = 0.54
H = 0.78	H = 0.56
H = 0.69	H = 0.57
H = 0.72	H = 0.55

5   Conclusion

The above study allows us assessing the ability of the VOS network to transmit video with the proper quality. In the course of the study, a testbed was created that allows evaluating the functions of the SDN elements, such as traffic monitoring, management of data flows in the network, security and QoE management, that makes service deployment and management easier and allows for better utilization of network resources.

During the investigation we used equipment to provide IPTV services to create a real traffic load on the network. A comparison between SDN and no SDN core network for IPTV video was made considering packet delay distribution and Hurst parameter was calculated. The Hurst parameter, which determines the degree of self-similarity of traffic, can be used as an objective metric for assessing the video quality of experience. The comparison represents a practical approach, since packet delay may critically affect the video quality while Hurst parameter may improve video quality assessment. The result shows how SDN network improve video delivery due to its features of virtualizing physical network resources and network optimization. SDN technology is convenient for dynamic traffic management. There is a great improvement of network control and monitoring when using Software-Defined Networking policies.

Further authors will develop an SDN controller application for dynamic distribution of video traffic witch definitely will improve the video Quality of Experience. Authors will develop model for assessing video quality by adding Virtual Reality Headsets Glasses on client side then associate the video quality with Hurst parameter.

Acknowledgment. The publication was financially supported by the Ministry of Education and Science of the Russian Federation (the Agreement number 02.a03.21.0008), RFBR according to the research project No. 16-37-00209 mol_a "Development of the principles of integration the Real Sense technology and Internet of Things".

References

1. Bergenti, F., Gotta, D.: Augmented reality for field maintenance of large telecommunication networks. In: Conference and Exhibition of the European Association of Virtual and Augmented Reality (2014)

2. Joo, H.J., Hong, B.H., Lee, E.S., Choi, H.K.: Analysis of IPTV service quality applying real-time QoE measurement technology. In: Park, J., Leung, V., Wang, C.L., Shon, T. (eds.) Future Information Technology, Application, and Service. Lecture Notes in Electrical Engineering, vol. 179, pp. 103–109. Springer, Dordrecht (2012). doi:10.1007/978-94-007-5064-7_15
3. Schatz, R., Hossfeld, T., Casas, P.: Passive youtube QoE monitoring for ISPs. In: 2012 Sixth International Conference on Innovative Mobile and Internet Services in Ubiquitous Computing (IMIS), pp. 358–364, July 2012
4. Abdi, L., Abdallah, F.B., Meddev, A.: In-vehicle augmented reality traffic information system: a new type of communication between driver and vehicle. Procedia Comput. Sci. **73**, 242–249 (2015)
5. Zollmann, S., Hoppe, C., Langlotz, T., Reitmayr, G.: FlyAR: augmented reality supported micro aerial vehicle navigation. IEEE Trans. Vis. Comput. Graph. **20**, 560–568 (2014)
6. Makolkina, M., Kirichek, R., Teltevskaya, V., Surodeeva, E.: Research of interaction between applications of augmented reality and control methods of UAVs. In: Koucheryavy, Y., Mamatas, L., Matta, I., Ometov, A., Papadimitriou, P. (eds.) WWIC 2017. LNCS, vol. 10372, pp. 186–193. Springer, Cham (2017). doi:10.1007/978-3-319-61382-6_15
7. Billinghurst, M., Clark, A., Lee, G.: A survey of augmented reality. Found. Trends Hum.-Comput. Interact. **8**(2–3), 73–272 (2015)
8. Putnam, K.: Interconnected wearable devices streamline care delivery. **103**(6), P10–P12 (2016). Doi:10.1016/S0001-2092(16)30220-4
9. Lucrezia, F., Marchetto, G., Risso, F.: In-network support for over-the-top video quality of experience. In: The Sixth International Conference on Advances in Future Internet, AFIN 2014, pp. 72–78 (2014)
10. Makolkina, M., Prokopiev, A., Paramonov, A., Koucheryavy, A.: The quality of experience subjective estimations and the hurst parameters values interdependence. In: Balandin, S., Andreev, S., Koucheryavy, Y. (eds.) NEW2AN 2014. LNCS, vol. 8638, pp. 311–318. Springer, Cham (2014). doi:10.1007/978-3-319-10353-2_27
11. Vladyko, A., Muthanna, A., Kirichek, R.: Comprehensive SDN testing based on model network. In: Galinina, O., Balandin, S., Koucheryavy, Y. (eds.) NEW2AN/ruSMART -2016. LNCS, vol. 9870, pp. 539–549. Springer, Cham (2016). doi:10.1007/978-3-319-46301-8_45
12. Volkov, A., Khakimov, A., Muthanna, A., Kirichek, R., Vladyko, A., Koucheryavy, A.: Interaction of the IoT traffic generated by a smart city segment with SDN core network. In: Koucheryavy, Y., Mamatas, L., Matta, I., Ometov, A., Papadimitriou, P. (eds.) WWIC 2017. LNCS, vol. 10372, pp. 115–126. Springer, Cham (2017). doi:10.1007/978-3-319-61382-6_10
13. Ushakov, Y., Polezhaev, P., Legashev, L., Bolodurina, I., Shukhman, A., Bakhareva, N.: Increasing the efficiency of IPTV by using software-defined networks. In: Galinina, O., Balandin, S., Koucheryavy, Y. (eds.) NEW2AN/ruSMART -2016. LNCS, vol. 9870, pp. 550–560. Springer, Cham (2016). doi:10.1007/978-3-319-46301-8_46
14. Rattanawadee, P., Ruengsakulrach, N., Saivichit, C.: The transmission time analysis of IPTV multicast service in SDN/openflow environments. In: 12th International Conference on Electrical Engineering/Electronics, Computer, Telecommunications and Information Technology (ECTI-CON), pp. 1–5 (2015)
15. Diallo, M.T., Fieau, F., Abd-Elrahmane, E., Afifi, H.: Utility-based approach for video service delivery optimization. In: International Conference on Systems and Network Communication, ICSNC 2014, pp. 5–10 (2014)
16. Thorpe, C., Olariu, C., Hava, F., McDonagh, P.: Experience of developing an openflow SDN prototype for managing IPTV networks. In: IFIP/IEEE International Symposium on Integrated Network Management (IM), pp. 966–971, May 2015

17. Muhizi, S., Shamshin, G., Muthanna, A., Kirichek, R., Vladyko, A., Koucheryavy, A.: Analysis and performance evaluation of SDN queue model. In: Koucheryavy, Y., Mamatas, L., Matta, I., Ometov, A., Papadimitriou, P. (eds.) WWIC 2017. LNCS, vol. 10372, pp. 26–37. Springer, Cham (2017). doi:10.1007/978-3-319-61382-6_3
18. Makolkina, M., Prokopiev, A., Paramonov, A., Koucheryavy, A.: The quality of experience subjective estimations and the hurst parameters values interdependence. In: Balandin, S., Andreev, S., Koucheryavy, Y. (eds.) NEW2AN 2014. LNCS, vol. 8638, pp. 311–318. Springer, Cham (2014). doi:10.1007/978-3-319-10353-2_27

LTE MCS Cell Range and Downlink Throughput Measurement and Analysis in Urban Area

Yi Hua Chen$^{(\boxtimes)}$, Kai Jen Chen, and Jyun Jhih Yang

Institute of Information and Communication Engineering,
Oriental Institute of Technology, New Taipei City, Taiwan
ff011@mail.oit.edu.tw,
s100109243@gmail.com, gn02090800@gmail.com

Abstract. The measurement in this study was conducted using a TEMS Pocket engineering model test phone, through which the cell coverage and cell capacity of LTE were analyzed using the KPI parameters of packet switch (PS) between the 4G EnodeB base stations (BSs) and user equipment (UE). Of the 60 EnodeB BSs that were measured, 46 supported the 700 MHz band, 35 supported the 1800 MHz band, and 6 supported the 2600 MHz band. The analysis results for the measurement data showed that the modulation code scheme (MCS) cell range means for MCS (0)–MCS (28) in Banqiao District were approximately 0.2–0.91 km, with an overall MCS average of 0.45 km. Cell capacity was presented using downlink (DL) throughput; the analysis results illustrated that the average DL throughputs were roughly 5.37 Mb/s–49.71 Mb/s, and the overall MCS average throughput was 18.87 Mb/s. The signal-to-interference-plus-noise ratio (SINR) means for MCS (0)–MCS (28) were approximately −5.33–20.38 dB, and the overall MCS averaged SINR was approximately 6.04 dB. The reference signal received power (RSRP) means for MCS (0)–MCS (28) were approximately −111.86−−73.83 dBm, and the overall MCS averaged RSRP was approximately −94.09 dBm. Finally, the analysis and comparison of the theoretical values and actual measurement illustrated that the data curves for this study were consistent with the 3GPP specifications, with SINR and the theoretical curve achieving a correlation of up to 0.936, and the DL throughput and the theoretical curve achieving a correlation of up to 0.933.

Keywords: LTE · MCS · SINR · Cell range · Downlink throughput

1 Introduction

This study mainly focused on the practical 4G network use of Internet users in the densely populated Banqiao District in New Taipei City. In 2015, Chen et al. [1, 2] used an engineering model mobile phone equipped with Nemo Handy to analyze small scale fading and fast Ricean fading; moreover, they examined the BS traffic volume changes through big data in Kenting, Pingtung County [3].

In their study on the relationship between users and BS using a smart phone in 2016, Lehne et al. [4] noted that a site is divided into three sectors; however, their measurement

© Springer International Publishing AG 2017
O. Galinina et al. (Eds.): NEW2AN/ruSMART/NsCC 2017, LNCS 10531, pp. 693–707, 2017.
DOI: 10.1007/978-3-319-67380-6_66

data demonstrated that the distribution of users differs according to the BSs of the distinct bands they are connected to. The LTE 800 MHz users in this study were mainly distributed between the cell-centered range of −110°–50°, whereas those using 1800 and 2100 MHz were distributed between the cell-centered range of −60°–60°. From the perspective of spatial distribution, the density of users' distribution evidently declined as their distance from the BSs increased. A lower-frequency signal was capable of supporting a longer distance, with 800 MHz capable of reaching 1.62 km, whereas the distance was reduced to 580 m at 2100 MHz. In general, 90% of the users in the two case studies were located within a distance of 1.35–3.47 km from the BS.

In 2012, Sainju [5] conducted a study at Singerjärvi (66.069731°, 29.034217°) and Kumpuvaara (66.086234°, 28.481345°) in Kuusamo, Finland. Noting that Kuusamo belongs to a rural area, the study measured the three bands of LTE 800, LTE 1800, and UMTS 900 MHz from two BSs. The coverage area of the LTE 800 MHz BS at the Singerjärvi area had a maximum distance of up to 13.7 km, with a mean downlink (DL) throughput of 13.8 Mb/s, whereas a throughput of more than 10 Mb/s was achieved at a distance of 7.1 km. At the distance of roughly 11.8 km, the throughput further declined to 1 Mb/s. The coverage area for the LTE 800 MHz base area in the Kumpuvaara area had a maximum distance of 17.3 km with a mean DL throughput of 9.2 Mb/s. The throughput at 9.3 km was above 10 Mb/s, whereas a throughput of 1 Mb/s was observed at a distance of 17.3 km. The analyzed data revealed that lower-band signals were capable of greater coverage distances and had superior throughput.

The LTE signal strength can be directly determined through measurement; the enhancement or attenuation of reference signal receive power (RSRP) and signal-to-interference-plus-noise ratio (SINR) are correlated with channel quality indicator (CQI) and modulation code scheme (MCS) level [6]. The limits of cell edge and the possibility of improvement were also understood through the SINR and DL throughput in the measurement data.

The remainder of this study is organized as follows: Sect. 2.1 describes the measuring tools, methods, and routes; Sect. 2.2 describes the use of latitude and longitude as the conversion basis for the straight line distance between base station and user equipment (UE); and Sect. 2.3 introduces the calculation of the average KPI signal measured. Section 3.1 describes the distribution of the measured samples and the operation process between MCS level and CQI; Sect. 3.2 presents the overall data measurement analysis for all bands; and Sect. 3.3 illustrates the measurement data analysis results for each band (700, 1800, and 2600 MHz). Section 4 compares the measurement results with 3GPP specifications, and the results are illustrated in tables. Section 5 organizes the measurement results.

2 Measurement and Analysis Principle

2.1 Measurement Planning

This study used an engineering model phone equipped with the engineering model test software TEMS Pocket to measure the obtained the log files, and the parameter sampling and signal quality analysis were conducted using the Actix Analyzer software.

MCS, RSRP, SINR, information bandwidth (IB), and throughput were analyzed to provide a complete presentation of correlation analysis and to compare the theoretical and actual values for signal coverage.

Figure 1 shows the measurement route for Banqiao in New Taipei City. Lock mode was adopted for the network measurement. Thus, the 4G LTE 700-, 1800-, and 2600-MHz bands were locked, although specific BS were not locked during measurement.

Fig. 1. Measurement route for Banqiao, New Taipei City

2.2 GIS Distance Calculation

This study used a geographic information system (GIS), which is a digitized operating system capable of integrating relevant geographic information. The straight line distance between two points was calculated by combining geography and cartography. Because the calculation was conducted using the latitude and longitude of the Earth, spherical rather than planar coordinates were used. Latitude and longitude were used to calculate the distance between two points on a sphere, also known as the great circle distance, and its equation is as follows.

If the latitude and longitude for two points x and y are given by $(\emptyset_1, \lambda_1)$ and $(\emptyset_2, \lambda_2)$, the straight line distance d between the two points is calculated as follows (1):

$$d = R \times arccos(\sin \phi_1 \times \sin \phi_2 \\ + \cos \phi_1 \times \cos \phi_2 \times \cos (\lambda_2 - \lambda_1)) \tag{1}$$

where R represents the radius of the Earth (6,378 km).

2.3 MCS Weighting Average Distance, SINR and DL Throughput Calculation

MCS samples as a weighting factor was included when calculating the essential parameter average for MCS levels, SINR, and DL throughput using (2), (3), and (4), in which the distribution function of MCS changes in the channel is considered.

(1) Average Distance:

$$D = \frac{\sum_{i=0}^{28} d_i \times n_i}{N} \tag{2}$$

where d_i represents the straight line distance corresponding to each MCS between the UE and EnodeB, n_i represents the corresponding sample for each MCS when the UE successfully connected to EnodeB, N is the total sample count of the successfully connect radio resource control, and D represents the average MCS distance.

(2) Average SINR:

$$SINR = \frac{\sum_{i=0}^{28} SINR_i \times n_i}{N} \tag{3}$$

where $SINR_i$ represents the corresponding SINR of each MCS between the UE and EnodeB, n_i represents the corresponding sample for each SINR when the UE was successfully connected to EnodeB, N represents the total sample count of the successfully connected RRC, and $SINR$ represents the average SINR of the MCS weighting.

(3) Average DL Throughput:

$$Tp = \frac{\sum_{i=0}^{28} Tp_i \times n_i}{N} \tag{4}$$

where Tp_i is the corresponding DL throughput for each MCS between the UE and EnodeB, n_i represents the corresponding sample for each MCS when the UE was successfully connected to EnodeB, N represents the total sample count of the successfully connected RRC, and Tp represents the average DL throughput for the calculated MCS level.

3 KPI Signal Analysis

3.1 MCS Index Profile

This chapter describes the analysis results for the measurement in Banqiao District in New Taipei City, Taiwan. The GIS theory mentioned in Sect. 2 was employed to calculate the distance between EnodeB and the UE. The MCS, SINR, DL throughput, RSRP, and relationships between the essential parameters of the LTE network, were presented with a series of broken line graphs. The corresponding MCS cell range, SINR, DL throughput, and RSRP were observed at different MCS levels. The measured data were then compared with the 3GPP theoretical values in Sect. 4.

The sample numbers determine the accuracy of the data and the actual situation of sample distribution during the data analysis and comparison. More than 110 000 samples were measured in this study, with each band including tens of thousands of test samples. Of the 60 BS that were measured, 46 supported the 700 MHz channel, 35 supported the 1800 MHz channel, and 6 supported the 2600 MHz channel. The sample distribution function indicates that the sample rose rapidly from MCS (6) and that most samples fell between MCS (6) and MCS (19) (more than 4000 samples for each level). The spectral efficiency was in the range of 0.877–3.029. The figure indicates that by constructing 4G LTE BSs, network operators enabled most users to use modulation technology above MCS (6). The test was conducted in an intense city.

3.2 All Band Signal Analysis

Figure 2 displays the measured SINR against the MCS cell range; the horizontal axis represents the MCS level, the blue line represents the corresponding principle coordinates for the MCS cell range values, and the red line represents the corresponding sub coordinates for the SINR values (unit: dB). The cell range means of MCS (0) to MCS (28) were in the range of 0.91–0.19 km, and the mean overall MCS was 0.45 km. The corresponding mean SINRs for each MCS from MCS (0) to MCS (28) were in the range of −5.33–20.38 dB, whereas the mean overall MCS SINR was approximately 6.04 dB. The curve changes of measured SINR and MCS cell range were almost linear with a correlation of −0.886, indicating that a more satisfactory MCS level was obtained with an improvement in SINR following a shorter distance to the BSs [7]. The average SINR value of MCS (0) was superior to those of MCS (1) and MCS (2). Because fewer samples were lower than MCS (4), the accuracy differed from the theoretical values.

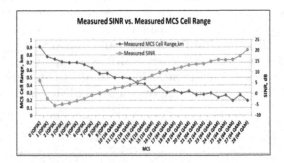

Fig. 2. Overall measurement of SINR for MCS (0)–MCS (28) vs. MCS cell range for each MCS level in Banqiao District (Color figure online).

Figure 3 shows the measured DL throughput against the MCS cell range. The horizontal axis represents the MCS level, the blue line is the corresponding principle coordinates for the MCS cell range values (unit: km), and the green line is the corresponding sub coordinates for the measured DL throughput values (unit: Mb/s). The average DL throughputs for each MCS from MCS (0) to MCS (28) were in the range of 22.91–49.71 Mb/s, with an average total MCS DL throughput of 18.87 Mb/s. The curves for the corresponding distance of each MCS and the measured DL throughput

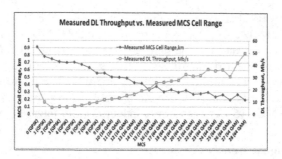

Fig. 3. Overall measurement of DL throughput for each MCS level from MCS (0) to MCS (28) vs. the MCS level in Banqiao District. The sub coordinates represent the DL throughput (Mb/s) measured in Banqiao District. The curves for the corresponding distance (Blue line) of each MCS and the measured DL throughput (Green line) had a correlation of −0.814 (Color figure online).

had a correlation of −0.814, revealing a significant negative correlation. The data indicated that a shorter distance to the BS led to a rise in SINR. This yielded a more satisfactory modulation method with a superior transmission efficiency, which resulted in an increase in the number of bits transferred per second and also enhanced the DL throughput. Figure 3 shows that the DL throughput performance of MCS (0) reached 22.91 Mb/s, which was higher than those of MCS (1) and (2), possibly because fewer samples were lower than MCS (4), thus causing the accuracy to differ from the theoretical values.

In addition, Fig. 4 compares the RSRP and SINR. The horizontal axis represents the MCS level, the purple line represents the corresponding principle coordinates for the RSRP values (unit: dBm), and the red line represents the sub coordinates for the SINR values (unit: dB). The RSRPs for MCS (0)–MCS (28) were in the range of −106.9 to −73.83 dBm, and the average RSRP was approximately −94.09 dBm. The SINR means for each MCS from MCS (0) to MCS (28) were in the range of −106.9 to −73.83 dBm, and the average overall MCS SINR was 6.04 dB.

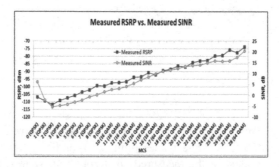

Fig. 4. Overall measurement of RSRP for each MCS level from MCS (0) to MCS (28) vs. SINR in Banqiao District. The sub coordinates represent the mean SINR (dBm) for the MCS level. The data revealed a positive correlation between RSRP (Purple line) and SINR (Red line) of up to 0.962 (Color figure online).

The data revealed a positive correlation between RSRP and SINR of up to 0.962. The performances of RSRP and SINR for MCS (0) were superior to those for MCS (1) and MCS (2), possibly because fewer samples were lower than MCS (4), causing the accuracy to differ from the theoretical values.

3.3 Individual Band Signal Analysis

Figure 5 illustrates the variations of SINR and MCS cell range. The horizontal axis represents the MCS level, the red line represents the corresponding principle coordinates for the MCS cell range (unit: km), and the blue line represents the corresponding sub coordinates for SINR (unit: dB).

(a) The corresponding cell range means for each MCS from MCS (0) to MCS (28) under the 700-MHz band were approximately 0.38–0.12 km, and the overall MCS average was 0.31 km. The corresponding SINR means for each MCS between MCS (0) and MCS (28) were in the range of 0.50–20.06 dB, and the overall MCS SINR mean was 5.04 dB. The correlation between the changes in the corresponding distance of each MCS and SINR was −0.981, displaying a significant negative correlation.

(b) The corresponding cell range means for each MCS from MCS (0) to MCS (28) under the 1800 MHz band were in the range of 0.3 km to 0.25 km, and the overall MCS average was 0.26 km. The corresponding SINR means for each MCS between MCS (0) and MCS (28) were in the range of −3.88–21.81 dB, and the overall MCS SINR mean was 6.19 dB. The changes of the corresponding distance for each MCS and SINR had a correlation of-0.840, indicating a significant negative correlation.

(c) The corresponding cell range means for each MCS from MCS (0) to MCS (28) under the 2600 MHz band were in the range of 1.17–0.52 km, and the MCS average was 0.96 km. The corresponding SINR means for each MCS between MCS (0) and MCS (28) were between 7.00 and 20.88 dB, and the overall MCS SINR mean was 5.71 dB. The correlation between the changes of the corresponding distance for each MCS and SINR was −0.967, demonstrating a significant negative correlation.

The average MCS distances for the 700-, 1800-, and 2600-MHz bands in this measurement data analysis were approximately 0.31, 0.26, and 0.96 km, respectively. According electromagnetic theory, a lower frequency for the 700-MHz band yields lower attenuation, which should result in greater coverage. However, the measurement data indicated that the coverage distances for the 700-, 1800-, and 2600-MHz bands did not match the theoretical estimation, possibly because the BSs of the 700- and 1800-MHz bands were more densely deployed in the urban area than those of the 2600-MHz band, thus reducing the coverage distance. The 2600-MHz band is a newly deployed band with fewer BSs in the urban area. Table 1 lists the number of BSs for each band deployed in the urban area.

Moreover, the measurement data indicated that the MCS level and SINR had a significantly negative correlation. Thus, a shorter distance to the BS yielded a higher SINR, a larger MCS level used by UE, and a higher transmission efficiency, and vice versa.

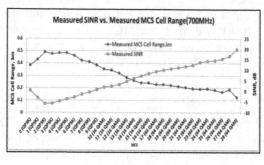

(a) LTE 700MHz

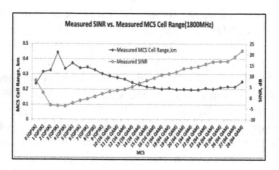

(b) LTE 1800MHz

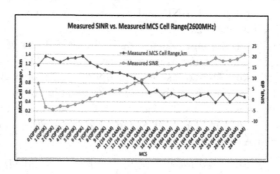

(c) LTE 2600MHz

Fig. 5. Overall measurement of SINR for MCS (0)–MCS (28) vs. MCS cell range corresponding to the MCS levels of each band in Banqiao District. The sub coordinates represent the mean SINR (dB) for each MCS level. (a) LTE 700 MHz, (b) LTE 1800 MHz, and (c) LTE 2600 MHz. The correlation between the changes in the corresponding distance of each MCS and SINR was −0.981 (a), -0.840 (b), and −0.967 (c) (Color figure online).

Figure 6 shows the cell capacity DL throughput against the MCS cell range. The horizontal axis represents the MCS level, the blue line represents the corresponding principle coordinates for the MCS cell range (unit: km), and the green line represents the corresponding sub coordinates for the DL throughput (unit: Mb/s).

Table 1. Summary table of measured KPI signal analysis for each band in the urban areas of Banqiao District

Band	700 MHz	1800 MHz	2600 MHz	Total
EnodeB	46	35	6	60
Cell range (Km)	0.31	0.26	0.96	0.45
SINR (dB)	5.04	6.19	5.71	6.04
DL throughput (Mb/s)	11.34	17.37	28.3	18.87
RSRP (dBm)	−87.22	−87.19	−115.7	−94.09

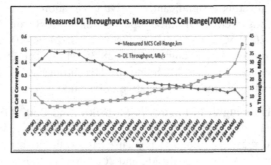

(a) LTE 700MHz

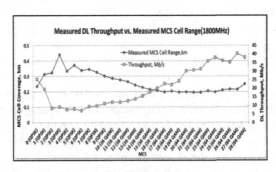

(b) LTE 1800MHz

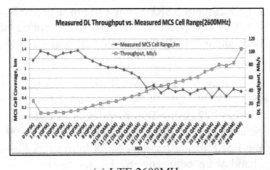

(c) LTE 2600MHz

Fig. 6. Overall measurement of DL throughput for MCS (0)–MCS (28) vs. MCS cell range for MCS (0)–MCS (28) corresponding to the MCS levels for each band in Banqiao District. (a) LTE 700 MHz, (b) LTE 1800 MHz, and (c) LTE 2600 MHz (Color figure online).

(a) The corresponding average DL throughput for each MCS from MCS (0) to MCS (28) under the 700-MHz band were in the range of 11.32–40.18 Mb/s, with an overall MCS average DL throughput of 11.34 Mb/s. The changes in the corresponding distances for each MCS and the DL throughput had a correlation of −0.854, displaying a significant negative correlation.

(b) The corresponding average DL throughputs for each MCS from MCS (0) to MCS (38) under the 1800-MHz band were in the range of 25.38–38.09 Mb/s, and the overall MCS average DL throughput was 173.37 Mb/s. The correlation between the changes in the corresponding distances for each MCS and the DL throughput was −0.747, indicating a significant negative correlation.

(c) The corresponding average DL throughputs for each MCS from MCS (0) to MCS (38) under the 2600-MHz band were in the range of 24.94–104.28 Mb/s, and the overall MCS average DL throughput was 28.30 Mb/s. The changes in the corresponding distances for each MCS and the DL throughput had a correlation of −0.9, demonstrating a significant negative correlation.

The average DL throughputs for the 700-, 1800-, and 2600-MHz bands in this measurement data analysis were approximately 11.34, 17.37, and 28.30 Mb/s respectively. According to the data, although LTE uses the orthogonal frequency-division multiple access technology, the 700-, 1800-, and 2600-MHz bands possess distinct IBs, which strongly influence DL throughput, with the wider bandwidth yielding a higher throughput. The bandwidths for both 700- and 1800-MHz bands measured in this study were 10 MHz, whereas the 2600-MHz band had a bandwidth of 20 MHz. Because a wider bandwidth results in a higher throughput, as shown in the data, the DL throughput of the 2600-MHz band was up to 28.30 Mb/s, which was greater than those of the 700- (11.34 Mb/s) and 1800-MHz (17.37 Mb/s) bands.

These data also indicated a negative correlation between DL throughput and distance, with SINR as the major influencing factor for DL throughput; thus, a shorter distance to the BS yielded a higher SINR. The UE obtained a higher MCS level, leading to a higher transmission rate, which increased the number of bits per second transferred, thereby enhancing the transmission of DL throughput in the same transmission bandwidth. Conversely, a greater distance away from the BS yielded a lower DL throughput.

Figure 7 presents the measured RSRP against the measured SINR. The horizontal axis represents the MCS level, the purple line represents the corresponding coordinates of the RSRP (unit: dBm), and the red line represents the corresponding sub coordinates for the SINR (unit: dB).

(a) The RSRP means for each MCS from MCS (0) to MCS (28) under the 700-MHz band were between −113.68 and −90.86 dBm, and the overall MCS RSRP mean was approximately −87.22 dBm. The correlation between the changes in the corresponding distance for each MCS and the RSRP was 0.994.

(b) The corresponding RSRP means for each MCS from MCS (0) to MCS (28) under the 1800-MHz band were between −87.13 and −71.3 dBm, and the overall MCS RSRP mean was 87.19 dBm. The correlation between the changes in the corresponding distance for each MCS and the RSRP was 0.979.

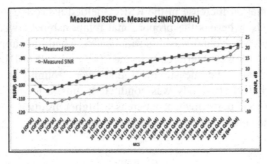

(a) LTE 700MHz

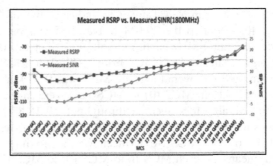

(b) LTE 1800MHz

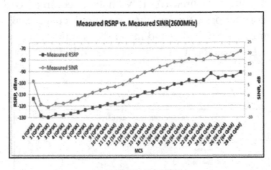

(c) LTE 2600MHz

Fig. 7. RSRP for MCS(0)–MCS(28) vs. SINR corresponding to the MCS levels of each band measured in Banqiao District. (a) LTE 700 MHz, (b) LTE 1800 MHz, and (c) LTE 2600 MHz (Color figure online).

(c) The RSRP means for each MCS from MCS (0) to MCS (28) under the 2600-MHz band were between −115 and −90.86 dBm, and the overall MCS RSRP mean was approximately −115.70 dBm. The correlation between the changes in the corresponding distance for each MCS and RSRP was 0.994.

The resulting means for the 700-, 1800-, and 2600-MHz bands obtained from the average RSRP in this analysis were −87.22, −87.19, and −115.70 dBm, respectively.

The data indicated that an increase in the RSRP yielded a stronger SINR for the 700- and 2600-MHz bands, whereas the proportional relationship between the RSRP and SINR under the 1800-MHz band was less satisfactory when the strength of the former increased, signifying a lower interference in the 700- and 2600-MHz bands compared to the 1800-MHz band. The measurement data were roughly consistent with the original estimation before the study was conducted; thus, a lower frequency yields a more satisfactory RSRP performance, whereas a higher frequency results in a more rapid attenuation, causing an inferior RSRP performance.

4 Discussion

The results for Banqiao District in New Taipei City indicated that of the 60 linked BSs that were measured, 46 supported the 700-MHz band, 35 supported the 1800-MHz band, and 6 supported the 2600-MHz band. In total, more than 110 000 samples were obtained. The cell range means for each MCS from MCS (0) to MCS (28) were in the range of 0.91–0.19 km, and the overall average MCS was 0.93 km. The average DL throughputs for each MCS from MCS (0) to MCS (28) were in the range of 22.91–49.71 Mb/s, with an overall MCS average throughput of 18.87 Mb/s. The corresponding SINR means for each MCS from MCS (0) to MCS (28) were between −5.33 and 20.38 dB, and the overall MCS SINR mean was approximately 6.04 dB. The corresponding RSRP means for each MCS from MCS (0) to MCS (28) were between −106.9 and −73.83 dBm, with an overall MCS RSRP of approximately −94.09 dBm.

Figure 8 compares the measured SINR with that required for 3GPP. The figure reveals a positive correlation between the measured curve and the 3GPP curve. According to the figure, the SINR demand increased following a rise in the MCS level, and the measured DL throughput and required 3GPP DL throughput displayed a correlation of up to 0.930.

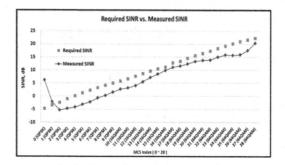

Fig. 8. Comparison of the theoretically required SINR for MCS (0)–MCS (28) and the measured SINR in Banqiao District.

The cell capacity DL throughput was also compared with the 3GPP with MIMO 2 × 2 DL throughput in Fig. 9.

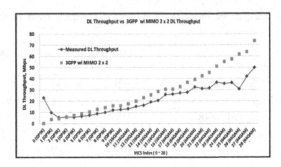

Fig. 9. Comparison of the overall measured theoretical DL throughput for each MCS level from MCS (0)–MCS (28) and the measured DL throughput in Banqiao District.

The measured curve on Fig. 9 was consistent with the 3GPP DL throughput curve; the theoretical and the measured DL throughputs demonstrated a correlation of up to 0.93. The data indicated that the superior modulation technology (QPSK/16QAM/64QAM) increased the number of bits transmitted per second, thereby enhancing the DL throughput under the same bandwidth.

Table 1 lists the measurement data for Banqiao District. As mentioned in Sect. 3.2, electromagnetic theory states that a lower frequency in the 700-MHz band leads to a smaller attenuation, which should result in a greater coverage distance.

However, the measurement data illustrated that the coverage distances for the 700-, 1800-, and 2600-MHz bands were inconsistent with the theoretical estimation. The average distances for the 700-, 1800-, and 2600-MHz bands were 0.31, 0.26, and 0.96 km, respectively. The 700- and 1800-MHz bands shared the characteristics of a shorter coverage distance following a rise in frequency. However, the characteristics displayed by the 2600-MHz band were inconsistent with the description according to electromagnetic theory, whereby a higher frequency leads to a greater attenuation and a shorter coverage distance.

Although the measurement in this study locked on separately to the 700-, 1800-, and 2600-MHz bands, the study did not lock on to a specific BS. Because the bases stations supporting the 700- and 1800-MHz bands were more densely deployed than those supporting the 2600-MHz band, mobile phones determine the current signal strength when the UE is farther away from the BSs during the mobile test. Because the BSs supporting the 2600-MHz band were deployed farther apart, the mobile phones were unable to locate neighboring BSs that with a satisfactory signal to support the 2600-MHz band even when their signal strengths were extremely weak, thus causing the data to indicate that the 2600-MHz band possessed a greater coverage distance.

Table 2 shows that the correlation analysis of the measured KPI signals for each band in the urban areas of Banqiao District. The measurement data showed a significant negative correlation between the MCS cell range and the RSRP, SINR, and DL throughput of each band, indicating that a shorter distance to the BSs yielded a more satisfactory RSRP, and a higher SINR was obtained. The rise in SINR led to a higher MCS level used by the UE and a superior transmission rate, thereby achieving a higher throughput. The comparison between the measured KPI and the required 3GPP

Table 2. Correlation analysis of measured KPI signals for each band in the urban areas of Banqiao District

Band	700 MHz	1800 MHz	2600 MHz	All band
Measured SINR vs. MCS cell range	−0.981	−0.840	−0.967	−0.886
Measured DL throughput vs. MCS cell range	−0.854	−0.747	−0.900	−0.814
Measured RSRP vs. measured SINR	0.994	0.979	0.994	0.962
Required SINR vs. measured SINR	0.968	0.911	0.941	0.936
Required DL throughput vs. measured DL throughput	0.943	0.889	0.982	0.930

indicated that they had a correlation higher than 0.9. The measured cell range and required DL throughput exhibited a correlation of 0.889. The measured numerical curves in this study are thus consistent with the required 3GPP.

5   Conclusion

The measurement data showed that the SINR and DL throughput of the 700-, 1800-, and 2600-MHz bands were negatively correlated to the MCS cell range, whereas SINR and RSRP had a significantly positive correlation. The data also indicated that a shorter distance to the BS enhanced the SINR, thus obtaining a superior modulation method that improved the transmission rate and enhanced the DL throughput. A higher band under the same transmission bandwidth also possessed a higher throughput. The data illustrated an enhancement in SINR following an increase in the RSRP of the 700- and 2600-MHz bands, whereas the positive correlation between RSRP and SINR under the 1800-MHz band was less satisfactory following an enhancement in the former. This result revealed a lower interference in the 700- and 2600-MHz bands compared with the 1800-MHz band.

Finally, the measurement data were compared with the 3GPP theory; a high correlation was observed between them. The cell range calculation in this study will be an important step for coverage probability estimation of the on-site EnodeB in the future.

References

1. Chen, Y.H., Chu, C.L., Chen Jr., Y., Yang, J.J.: Small scale fading signal measurement and analysis in lantern festival hot zone. In: 2015 International Conference on Computer Science and Information Engineering (CSIE2015) (2015)
2. Chen, Y.H., Jou, C.C., Chen Jr., Y.: Fast ricean fading signal measurement and analysis in ken-ting spring scream festival hot zone. In: World Conference on Control, Electronics and Electrical Engineering (WCEE 2015) (2015)
3. Chen, Y.H., Su, M.L., Yang, J.J., Chen Jr., Y.: Relational analysis of the parametric model of daily traffic and customer behavior. In: 2015 International Conference on Computer Science and Information Engineering (CSIE2015) (2015)

4. Lehne, P.H., Glazunov, A.A., Karlsson, K.: Finding the distribution of users in a cell from smart phone based measurements. In: 2016 International Symposium on Wireless Communication Systems (ISWCS), pp. 538–542 (2016)
5. Sainju, P.M.: LTE performance analysis on 800 and 1800 MHz bands, Tapere University of Technology, Master of Science Thesis, Computing and Electrical Engineering on 7th May 2012
6. Afroz, F., Subramanian, R., Heidary, R., Sandrasegaran, K., Ahmed, S.: SINR, RSRP, RSSI and RSRQ measurements in long term evolution networks. Int. J. Wirel. Mob. Netw. (IJWMN) 7(4), 113–123 (2015)
7. Sevindik, V., Wang, J., Bayat, O., Weitzen, J.: Performance evaluation of a real long term evolution (LTE) network. In: 2012 IEEE 37th Conference on Local Computer Networks Workshops (LCN Workshops), pp. 679–685 (2012)

Transfer of Multimedia Data via LoRa

Ruslan Kirichek[1,2(✉)], Van-Dai Pham[1], Aleksey Kolechkin[1],
Mahmood Al-Bahri[1], and Alexander Paramonov[1]

[1] State University of Telecommunication, St. Petersburg, Russian Federation
kirichek@sut.ru, daipham93@gmail.com, clearbox204@gmail.com,
albahri.89@hotmail.com, alex-in-spb@yandex.ru
[2] Peoples' Friendship University of Russia (RUDN University),
6 Miklukho-Maklaya St., Moscow, Russian Federation

Abstract. This article provides the results of the multimedia data
transmission parameters research by LoRa using, in particular the results
of images and voice transmission using the fragment of the model network in the Internet of Things Laboratory SPbSUT. During the series of
experiments there was noticed the LoRa radio modules performance variation of different parameters (Bandwidth, Spreading Factor and Coding
Rate), which have affected on the time and quality of image transmission. For image compression were used JPEG and JPEG 2000 methods, which have allowed to achieve an acceptable compression and image
reconstruction while transmitting in the low-speed network. In the course
of the experiment, the images were transferred from a camera mounted
on a quadrocopter at distance of several kilometers. We considered such
parameters as the time of data transfer, packet loss, estimation of the
images quality obtained on the basis of subjective and objective methods. For voice compression, the A-law method was used, which allowed
to compress the frame size by 4 times. Experiments of real-time speech
transmission were conducted in different languages and evaluated by the
experts. During the results analyzing there were defined the lower subjective score for Arabic, and the higher scores for English and Vietnamese.
In conclusion, this article provides the results of the quantitative and
qualitative dimensions evaluation and presents directions for the further
research.

Keywords: LoRa · Internet of Things · Quality of experience · Images ·
Voice

1 Introduction

Currently, the development of telecommunications is intimately associated with
the concept of the Internet of Things (IoT) [1,2]. Whereas, the development
of telecommunications and the Internet has gave humanity infinite possibilities of knowledge and information exchange, forever changed the world around
us. In the last 20 years there variety of wired and wireless data transmission
technologies have occurred that are meeting the requirements of the constantly

© Springer International Publishing AG 2017
O. Galinina et al. (Eds.): NEW2AN/ruSMART/NsCC 2017, LNCS 10531, pp. 708–720, 2017.
DOI: 10.1007/978-3-319-67380-6_67

growing amount of data transfer between people. With the advent of the concept of Internet of things the key element of internetworking became devices - the Internet-things. In this regard, over the past decade, there are transmission technologies appeared that are directed at low battery supplied power. Examples may include such technologies and protocols as ZigBee, 6LowPAN, BLE, LoRa, LoRaWAN, Sigfox, Weightless P, etc. By reason of the necessity of data delivery from devices with low power consumption directly to the general communications network there has appeared the new LPWAN class of networks. The most sought-after technologies for this class are: Sigfox, Weightless P, NB-IoT LoRa, LoRaWAN, which to any extent are competing technologies. In the [3, 4] there is made comparison and conducted of the LPWAN technologies. There are also given the examples of these technologies usage for the various IoT-applications. The LoRa is noticed as the most widespread technology in terms of open structure and unlicensed band. For the purpose of technology supporting and development there was incorporated LoRa Alliance [5].

The application of the LoRa technology is widespread. Technology is used in the sea-craft monitoring system [6], in the farming sector [7], in the medicine for human health monitoring [8, 9], and as applications for the Smart cities as well [10].

The transmission of images and voice in Personal Area Network study for the ZigBee Protocol is described in [11, 12]. Based on these results there was set a task of research conduction of JPEG and JPEG 2000 image transmission using the network "point-to-point" fragment for the LoRa technology. The LoRa module can be installed on UAV (unmanned aerial vehicle), which collects data from sensor nodes in a remote area [13]. Due to the ability to transfer data using LoRa technology over long distances (in line of sight conditions), it is possible to deliver the combined data to the public communication network with a low delay. In some cases, it is necessary to perform visual control of the flying territory (Fig. 1). Examples of such situations may be natural disasters, fires, earthquakes, equipment failures, etc.

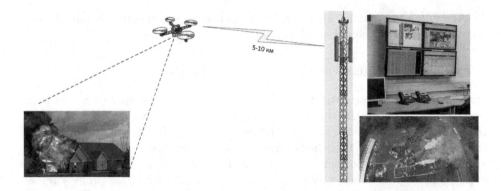

Fig. 1. Example of transferring images from a UAV to a monitoring center

Obviously, the main purpose of LoRa technology is the transfer of small telemetric information from the Internet of things, but the results obtained on the basis of the existing infrastructure can be used in the planning of visual control systems in emergency situations.

2   LoRa Technology

The LoRa technology [14, 15] was developed by Semtech Company and it stands for Long Range. This technology is the unique of the existing technical solutions used for the LPWAN design. LoRa technology combines the LoRa modulation technique in wireless networks LPWAN and open protocol LoRaWAN.

The method of LoRa modulation is based on the spread spectrum modulation, and chirp spread spectrum (CSS) with integrated forward error correction (FEC), by which data are encoded by broadband pulses with the frequency that is increasing or decreasing at a certain time interval.

This solution, as the opposed to the technology of direct spectrum expansion, makes the receiver robust to frequency departures from the nominal value and simplifies the requirements for clock signal generator which allows the usage of inexpensive quartz-crystal resonator in LoRa modules. LoRa technology operates in unlicensed frequency bands: 868 MHz in Europe, 915 MHz in the USA and 433 MHz in Asia [4].

In practice, besides the frequency range specifying possibility, in LoRa technology it is also possible to configure the nodes parameters: bandwidth (BW), spreading factor (SF) and coding rate (CR). By varying these parameters it is possible to determine the optimal mode of the operation for different data transmission conditions between the transmitter and the receiver. In the used modules, the bandwidth values can be 125 kHz and 250 kHz, the spreading factor is in the range from 6 to 12. The parameter of coding rate has a fixed value of 4/5. The relationship between these parameters is determined by the formula (1):

$$DR(Kbps) = \frac{BW \cdot SF \cdot CR}{2^{SF}} \tag{1}$$

According to formula (1) is possible to calculate the theoretical data transmission rate, the results of which are presented in the Table 1.

3   Transfer Images and Voice over LoRa

Due to the fact of popularization of low power wireless technologies and long propagation distance the complete coverage for LoRa networks it is currently realized in the major cities and the surrounding areas. The possibility of images transmission based of these networks opens up new horizons in the security applications, observation and control in farming and production sector. However, to develop an effective system of multimedia data transmission in wireless long-range networks, there are many issues caused by the bandwidth resources limitation, power consumption and the data transmission tolerance to the channel errors.

Table 1. The parameters of the tested LoRa module

DR type	BW (MHz)	SF	CR	Data rate (Kbps)
DR 0	125	12	4/5	0.29
DR 1	125	11	4/5	0.53
DR 2	125	10	4/5	0.97
DR 3	125	9	4/5	1.75
DR 4	125	8	4/5	3.12
DR 5	125	7	4/5	5.46
DR 6	250	7	4/5	10.93

3.1   Restriction of LoRa Technology

The LoRa technology provides throughput 50 bps to 37 kbps under a variety of different parameters of modulation [5,14,15]. The data link layer supports transmission of the small size frames, that are limited by 250 bytes, including the valuable and control information. Function of fragmentation and collection of the valuable information is executed at the level of application. In this regard, for images transmission with the use of LoRa modules it is not required to apply the hardware methods of information compression.

3.2   Methods of JPEG and JPEG 2000 Compression

Methods of JPEG and JPEG 2000 compression can be distinguished by different parameters, beginning from efficiency and complexity of compression to scalability [16]. The format of JPEG images is the most widespread algorithm of compression with losses, which means that the compressed image is very close to initial, but is not absolutely coincided with him. The value of losses and quality of the received image are set by the parameter q (Quality) lying within 1 to 100. The compression algorithm has been developed by the Joint Photographic Expert Group that consists of experts of two leading organizations in the standardization sphere: ITU (International Telecommunication Union) and ISO (International Organization for Standardization). Other format of the compressed images storage is JPEG 2000. The algorithm of compression possesses has a number of advantages: higher compression index in comparison with JPEG, the mode of compression and restoration "without losses", gradual display of images during the loading, scaling, errors correction, etc. In case of JPEG format the Huffman's coding is used, but JPEG 2000 format assumes the usage of an algorithm of pyramidal Wavelet-transform. In this regard JPEG 2000 has higher complexity of coding and, as a result, efficiency higher complexity of compression. The key parameter in JPEG 2000 standard is k-level of compression that the compressed image size to the initial size ratio and is ranging from 0 to 1. On the 2 there are two images that are compressed according to JPEG and

JPEG 2000 algorithms with compression index of 1:15 respectively and an initial image size of 20.3 kB. Comparing two images subjectively it is obvious that the right image, compressed by the JPEG 2000 algorithm has higher quality. In networks with a low capacity, it is obvious that the usage of the JPEG 2000 format is more expedient, therefore it is possible to provide the best quality of the transferred image (Fig. 2).

(a)   (b)

Fig. 2. Images were compressed in JPEG (a) and JPEG 2000 (b)

3.3   Methods of Voice Compression

In the modern digital networks of telephony ITU-T has standardized the G.711 standard for audio data transmission systems [17]. According to the G.711 standard there are two versions of audio data compression: A-law and U-law. A-law is the algorithm of the signal compression – relaxation that is in Europe telecommunication systems for the purpose of the systems optimization used, for example, by analog signal dynamic range reduction for the signal quantization and binary coding - after that transmission is made. A-law is used for the E1 standard. U-law is similar to the A-law algorithm and is used in North America and Japan. Distinction consists in the method of an analog signal selection. The signal is not selected from both diagrams linearly, but logarithmic. A-law provides more dynamic range in comparison with the U-law version. As a result because of smaller quantizing errors of an analog signal the sound is less distorted. The audio data transfer within low speed system will require compression on the part of the transmitter and decompression on the part of the receiver.

To transfer audio data in a low-speed system, then compression on the transmitter side and decompression at the receiver side will be required. After two methods of voice compression - A-law and U-law – were considered, for a further research A-law algorithm of voice compression has been chosen.

4   Description of Experiment

The experimental study was performed in the Internet of Things Laboratory at the Bonch-Bruevich State University of Telecommunications [18]. In the experiment were used Lora modules, based on the radio chip SX1276 and the microcontroller STM32L, which are developed by the "Unwired Device" company [19].

The low-level software of the module allows to configure LoRa technology parameters, such as DR type which represents a combination of parameters (BW, SF, CR). The DR type is established at the beginning of the configuration and defines an operating mode of the module. The available DR types with calculation speed are presented in the Table 1. According to values in this table, for researches were chosen several types - DR 4, DR 5, and DR 6. In this experiment were used two pairs of LoRa modules, which were located at the distance of 5000 m in the conditions of direct visibility (Table 1). Schemes of the experimental stand are presented in Fig. 3.

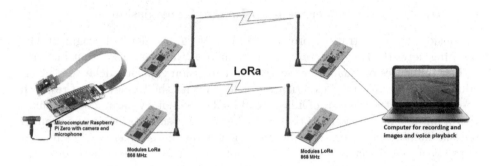

Fig. 3. The scheme of the experimental stand

4.1 Hardware

Two LoRa Unwired Range radio modules (running at 868 MHz) with UMDK-RF adapters were used as transmitters. The radio module consists of the SX1267chip, the microcontroller STM32L and coordinating elements. The radio modules were connected to the single-board microcomputer Raspberry Pi Zero via the UART serial ports using adapter CP2102. In order to transfer the image, the camera, which was connected to the microcomputer, was used to capture images with a resolution of 160 × 120 pixels and converting the color image to grayscale. The captured image was compressed using JPEG and JPEG 2000 formats and split into packets of 220 bytes for subsequent transfer to the receiver side. For voice transmission was used the microphone, which was connected to the microcomputer, recording the voice with the following parameters:

- Sample type: 16-bit format;
- Flow rate in Hz: 2500;
- Number of input channels and number of output channels: 2;
- Frame size: 220 bytes.

One frame of 16-bit format with one input channel has a size of 220 bytes. According to this in our case we have every time $220 \times 2 \times 2 = 880$ bytes of voice information. The transmitter uses the A-law voice compression method, which compresses from 880 bytes to 220 bytes. The generated packet that is 220 bytes-size is added to the buffer memory for the subsequent transfer to the receiving side.

4.2 The Images and Voice Reception on the Basis of LoRa Technology

The same radio modules (running at 868 MHz), which were connected to a personal computer via serial UART interfaces, were used as receivers. The receiver caught a sequence of incoming packets, which were recorded after the decompression in a file and thus an image was formed.

At the input of voice receiver, the receiver caught a sequence of incoming packets, which are reproduced using the A-law method after decompression.

4.3 Results of the Experiment on Image Transmission

To consider the data transfer process based on the LoRa network fragment, the following network characteristics were identified as the most important in terms of their influences on qualities of service: transmission capacity, delay and jitter, value losses and quality of experience (QoE). Within the framework of research were performed the analysis of influence of LoRa technology parameters on qualities of image transmission, the time of image transmission over the radio channel and the number of losses. The obtained images on the receiver were estimated on the basis of subjective methods of perception. Results of the experiment were analyzed for images, which are compressed in JPEG format (file extension *.jpg) and JPEG 2000 (file extension *.jp2) depending on the parameters of the LoRa technology transmission. Experimental results of the packets transmission and time of packets transmission over the radio channel are presented in Tables 2 and 3.

Table 2. The results of transmission for images with the extension *.jp2

Type DR	Number of packets loss	Total number of packets	Rate of packet loss (%)	Time (ms) of packet transmission over the radio channel		
				Minimum	Maximum	Average
DR 4	84	700	12.00	570	582	576
DR 5	110	700	15.70	328	336	332
DR 6	69	700	9.86	167.4	170.6	169

Because of that fragmentation is performed at the application layer, as a result of the image transmission if one or more packets are lost, this image will not open. In this case, we can assume that the entire image is lost. The experimental results of the number of image losses are shown in Table 4.

Further we considered the dependence of the peak signal-to-noise ratio (PSNR) for accepted images. PSNR can be defined through the mean squared error (MSE) by the formula (2). With the size of the transmitted images

Table 3. The results of transmission for an images with the extension *.jpg

Type DR	Number of packets loss	Total number of packets	Rate of packet loss (%)	Time (ms) of packet transmission over the radio channel		
				Minimum	Maximum	Average
DR 4	128	700	18.29	572	581	576.5
DR 5	98	700	14.00	326	340	330
DR 6	93	700	13.29	165.2	170.5	167.8

Table 4. Results of the number of image losses

Type DR	Number of transmitted images	Number of image losses		Positive results of image delivery	
		*.jp2	*.jpg	*.jp2	*.jpg
DR 4	100	10	16	90%	84%
DR 5	100	14	12	84%	88%
DR 6	100	8	12	92%	88%

$(m \times n) = (160 \times 120)$, we assume that the pixel value of the transmitted image at (i, j) location is denoted by I(i, j), and the accepted image - by K(i, j).

$$MSE = \frac{1}{mn} \sum_{i=0}^{m-1} \sum_{j=1}^{n-1} [I(i,j) - K(i,j)]^2 \qquad (2)$$

In this case, PSNR is defined by the formula (3):

$$PSNR = 10 \log_{10}(\frac{MAX_I{}^2}{MSE})[\text{dB}] \qquad (3)$$

where MAX_I is the maximum value received by the image pixel. In our research, each pixel has 8 bits, which means that $MAX_I = 255$.

The comparison between the transmitted and received images is performed in the GNU Octave using the formulas (2) and (3) [20]. The results of the calculations show that after recording on the receiving side the JPEG 2000 received images remained identical with the faithful images, i.e. $MSE = 0$. Subjective comparison between the received JPEG images and the original images is shown in Fig. 4.

When transmitting images in JPEG format, subjectively there was a difference between the accepted images and the original ones.

The PSNR index for transmitting JPEG images is shown in Fig. 5. Most of the received images have index $PSNR = 23$ dB.

After receiving a series of JPEG-images, they were combined into a sequence for the formation of underclocked video timelapseJPG.avi and images of JPEG

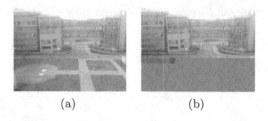

(a)   (b)

Fig. 4. Comparison between the transmitted images (a) and the received images (b)

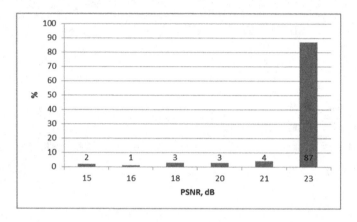

Fig. 5. Index PSNR of received JPEG images

2000 format were combined in timelapseJP2.avi, respectively. Video clips were played within the speed of 10 frames per second. The expert group evaluated the underclocked video using subjective methods. Five experiments were conducted, as a result of which the experts have watched several times the received underclocked video and have given the estimates. Expert evaluations were defined in accordance with the following five-point scale: 5 - excellent, 4 - good, 3 - acceptable, 2 - bad, 1 - unacceptable. Table 5 shows the evaluation results for each video. In order to analyze the obtained evaluations, the confidential interval is calculated according to the formula (4), which is obtained from the standard evaluations deviation (5) and the size of each sample is received from Recommendation ITU BT.500-13 [21,22].

$$\delta_{jkr} = 1.96 \frac{S_{jkr}}{\sqrt{N}} \tag{4}$$

$$S_{jkr} = \sqrt{\sum_{i=1}^{N} \frac{(\overline{u}_{jkr} - u_{jkr})^2}{N-1}} \tag{5}$$

As it is shown the results of subjective evaluation of video clips for the compression method JPEG 2000 image, underclocked video is more enjoyable for

Table 5. Results of subjective video evaluation by a group of experts

Type	Expert 1	Expert 2	Expert 3	Expert 4	Expert 5	Average	Confidence interval
JPEG	2	2.5	2	2.5	2	2.2	0.24
JP2000	3.5	3	3.2	3	3.5	3.24	0.22

watching than with the JPEG compression method. According to evaluations of the experts, the video clip timelapseJP2.avi is acceptable for watching. So we can summarize that the transmission of images and video clips, which was formed from several images in JPEG 2000 format, is higher quality than from images in JPEG format.

4.4   Results of the Experiment of Voice Transmission

In our experiment, the parameters were evaluated such as the number of losses, the degree of speech recognition in different languages, using the DR 6 type within the speed of 10.938 kbps. Five experts were selected to evaluate the received voice from the transmitter. In total about 5 experiments on each considered language were conducted. These experimental conditions, which were available for each language, were presented different. Subjective evaluation can differ from one language to another, because languages differ in sound, some are more melodic, some are more clear and easy for recognition and understanding, etc. Therefore, our experiments were conducted in following languages: Russian, English, Arabic, French and Vietnamese. In order to reduce probability of exposure of lower evaluation by the experts for the foreign voice, the expert group consisted of people with knowing at least two languages. Expert evaluations were defined in accordance with the following five-point scale: 5 - excellent, 4 - good, 3 - acceptable, 2 - bad, 1 - unacceptable.

In order to evaluate the quality of voice transmission, there was selected methodology, proposed in ITU-T Rec. P.913 [23]. As a subjective evaluation method, we chose the ACR (Absolute Category Rating) method, which is described in ITU-T Rec. P.910 [22].

This method uses categorical evaluations. The advantage of the ACR method is the ability to evaluate only the obtained data on the receiver without the etalons, which allows to approach real network operating conditions and to evaluate end users.

In order to analyze the results, we calculated the average score of each expert by formula (6):

$$\overline{u}_{jkr} = \frac{1}{N} \sum_{i=1}^{N} u_{jkr} \tag{6}$$

For further processing of the results, we calculate the confidential interval according to formulas (4) and (5) for all languages. The final results of the calculations are combined in Table 6.

Table 6. Results of the voice transmission in all languages

Language	Evaluation	Losses (%)	Confidence interval
Russia	3.41	13.82	0.03
English	3.51	12.33	0.03
Arabic	3.38	11.69	0.02
French	3.4	10.78	0.03
Vietnamese	3.53	10.15	0.02
Average	3.44	11.75	0.03

According to the results, the voice transmissions in the English are higher than in the Russian with approximately the same transmission conditions. This is due to the subjective perception of information by experts. All experts have knowledge of the English language, and also in the group of experts there were foreign students who are more familiar with English. Therefore, evaluations in the voice transmission in English were overestimated. When comparing the results, we realized that the voice transmission in Arabic had a lowest score because the sound of the Arabic language was difficult.

5   Summary

This article provides the experimental study results of images and voice transmission on the basis of LoRa technology. For image transfer implementation within wireless low-speed networks there have been chosen the JPEG and JPEG 2000 compression methods. According to images transmission results within the network, and the video-clips assessment which are from several images in the JPEG 2000 format created, it is shown that the accepted images are not distorted, and the majority of JPEG images are transferred with distortion and the received *PSNR index* = 23 dB in comparison with the transferred images. Thus the format of JPEG 2000 compression is suitable for image compression while transmission on the basis of LoRa technology and as a result within the LPWAN networks.

For voice transfer implementation there have been chosen the A-law compression method which gives the possibility of the voice transmission in different languages while using LoRa technology with the DR6 parameter. During the data analysis we have received lower grade when using Arabic and higher grade for English and Vietnamese by subjective evaluation.

Further research will be devoted to images transmission within the LoRaWAN and SigFox networks with different images permissions and distances between transmitter and base station. For this research was developed the model of an images transmission system on the basis of LoRaWAN technology.

Acknowledgements. The publication was financially supported by the Ministry of Education and Science of the Russian Federation (the Agreement number 02.a03.21.0008) and RFBR according to the research project No. 16-37-00215 mol_a "Biodriver".

References

1. ITU-T Recommendation Y. 2060. Overview of Internet of Things, Geneva, February 2012
2. Kirichek, R., Kulik, V., Koucheryavy, A.: False clouds for Internet of Things and methods of protection. In: IEEE 18th International Conference on Advanced Communication Technology: ICACT 2016, Phoenix Park, Korea, pp. 201–205 (2016)
3. Raza, U., Kulkarni, P., Sooriyabandara, M.: Low power wide area networks: an overview. IEEE Commun. Surv. Tutor. **19**, 855–873 (2017)
4. Kirichek, R., Kulik, V.: Long-range data transmission on Flying Ubiquitous Sensor Networks (FUSN) by using LPWAN protocols. In: Vishnevskiy, V.M., Samouylov, K.E., Kozyrev, D.V. (eds.) DCCN 2016. CCIS, vol. 678, pp. 442–453. Springer, Cham (2016). doi:10.1007/978-3-319-51917-3_39
5. https://www.lora-alliance.org
6. Li, L., Ren, J., Zhu, Q.: On the application of LoRa LPWAN technology in sailing monitoring system. In: 13th Annual Conference on Wireless On-demand Network Systems and Services, WONS 2017, pp. 77–80 (2017)
7. Pham, C., Rahim, A., Cousin, P.: Low-cost, long-range open IoT for smarter rural African villages. In: IEEE International Smart Cities Conference, ISC2 2016 (2016)
8. Kirichek, R.: The model of data delivery from the wireless body area network to the cloud server with the use of unmanned aerial vehicles. In: Proceedings - 30th European Conference on Modeling and Simulation, ECMS 2016, pp. 603–606 (2016)
9. Petäjäjärvi, J., Mikhaylov, K., Hämäläinen, M., Iinatti, J.: Evaluation of LoRa LPWAN technology for remote health and wellbeing monitoring. In: 10th International Symposium Medical Information and Communication Technology, ISMICT 2016 (2016)
10. Cesana, M., Redondi, A.E.C.: IoT communication technologies for smart cities. In: Angelakis, V., Tragos, E., Pöhls, H.C., Kapovits, A., Bassi, A. (eds.) Designing, Developing, and Facilitating Smart Cities, pp. 139–162. Springer, Cham (2017). doi:10.1007/978-3-319-44924-1_8
11. Kirichek, R., Makolkina, M., Sene, J., Takhtuev, V.: Estimation quality parameters of transferring image and voice data over ZigBee in transparent mode. In: Vishnevsky, V., Kozyrev, D. (eds.) DCCN 2015. CCIS, vol. 601, pp. 260–267. Springer, Cham (2016). doi:10.1007/978-3-319-30843-2_27
12. Pekhteryev, G., Sahinoglu, Z., Orlik, P., Bhatti, G.: Image transmission over IEEE 802.15.4 and ZigBee networks. In: IEEE International Symposium on Circuits and Systems 2005, vol. 4, pp. 3539–3542 (2005)
13. Kirichek, R., Vladyko, A., Paramonov, A., Koucheryavy, A.: Software-defined architecture for flying ubiquitous sensor networking. In: International Conference on Advanced Communication Technology, ICACT 2017, pp. 158–162 (2017)
14. Semtech White Paper. Smart Cities Transformed Using LoRa Technology, November 2016
15. SX1272/3/6/7/8: LoRa Modem Designers Guide. Semtech Co., Camarillo, AN1200.13 (2013)

16. Taubman, D., Marcellin, M.: JPEG 2000 Image Compression Fundamentals Standards and Practice. The Springer International Series in Engineering and Computer Science, vol. 642, 773 p. Springer, Boston (2002)
17. ITU-T Recommendation G.711 (1998). Amd. 1, August 2009
18. Kirichek, R., Koucheryavy, A.: Internet of Things laboratory test bed. In: Zeng, Q.-A. (ed.) Wireless Communications, Networking and Applications. LNEE, vol. 348, pp. 485–494. Springer, New Delhi (2016). doi:10.1007/978-81-322-2580-5_44
19. https://www.unwireddevices.com
20. https://www.gnu.org/software/octave
21. ITU-R Recommendation BT.500-13: Methodology for the subjective assessment of the quality of television pictures, Geneva, January 2012
22. ITU-T Recommendation P.910: Subjective video quality assessment methods for multimedia applications, Geneva, April 2008
23. ITU-T Recommendation P.913: Methods for the subjective assessment of video quality, audio quality and audiovisual quality of Internet video and distribution quality television in any environment, January 2014

Practical Results of WLAN Traffic Analysis

A. Paramonov[1,2(✉)], A. Vikulov[1], and S. Scherbakov[3]

[1] The Bonch-Bruevich State University of Telecommunications,
22 Pr. Bolshevikov, St. Petersburg, Russian Federation
alex-in-spb@yandex.ru
[2] The Peoples' Friendship University of Russia (RUDN University),
6 Miklukho-Maklaya St., Moscow 117198, Russian Federation
[3] Beltel, ul. Mayakovskogo 3BA, St. Petersburg 191025, Russian Federation
serg@beltel.ru

Abstract. This paper presents some results of the WLAN traffic measurements and analysis. The main idea of the work is to describe some common parameters of the originated traffic which may be useful in the network planning and analysis tasks. Authors have considered such parameters as session duration, user activity, packet size, traffic intensity, and some others.

Keywords: Traffic · WLAN · Statistical data · Measurements · Session duration · Packet size · Distribution · Access point · Access network

1 Introduction

The modern communication network presents heterogeneous networks based on the different technologies at different network architecture layers. In particular different wideband access technologies may be used at the subscriber access layer. The technology progress leads to the traffic and quality of service growth. This is manifested in growth of the network capabilities, services development and implementation processes. The network planning and maintenance processes require technical solutions which provide appropriate balance between subscriber originated traffic, quality of service and amount of the network resources (throughput, number of channels, bandwidth etc.). However, only the amount of network resources can be easily assessed. Quality of service estimations have to be based on the user requirements and may be obtained by different methods, expert evaluation, for example. The user traffic characteristics are the most important elements in the initial data necessary for the network planning and maintenance tasks. A complete enough description of the traffic makes it possible to select or construct an adequate model of the service system (communication network) and consequently to solve the task of selecting the required network resources with its help. The difficulty of obtaining data about subscriber's traffic is in the need of conducting observations (monitoring) in the existing communication network, which results should correspond to the conditions where these results are intended to be used. The range of services provided in modern communication networks is very wide, with every service having it's traffic features. The selection of traffic of various services in monitoring tasks is quite a complex process and is often

© Springer International Publishing AG 2017
O. Galinina et al. (Eds.): NEW2AN/ruSMART/NsCC 2017, LNCS 10531, pp. 721–733, 2017.
DOI: 10.1007/978-3-319-67380-6_68

not justified when collecting descriptive statistical material. Therefore, there is the task of determining possibly minimal set of characteristics and traffic parameters, which will allow describing the user traffic quite accurately.

The present work is devoted to the analysis of traffic of existing broadband wireless access networks. In the selected WLAN with its quite large geometric size we have been concentrated on the parameters of traffic, which in our opinion can be considered as generally relevant for the majority of users, regardless of the departmental affiliation of the network and its other features.

2 Related Works

For today there are many works dedicated to the WLAN characteristics, quality of service and traffic parameters. These issues are interrelated, therefore they are often considered together. Characteristics of the originated traffic use may heavily depend on user, location (office, town, country), field of activity, application etc. Therefore it is very difficult to describe traffic in common, i.e. to give description which may be useful for network planning and analysis in any arbitrary place and application. Thus knowledge of the traffic is very important for practical purposes.

Most of the known works are related to network analysis (throughput, capacity, QoS etc.), traffic pattern analysis (flow characteristics, selfsimilarity, etc.), statistical characteristics of the WLAN traffic for any specific network or application. Therefore we split all related works in three categories: network analysis, flow characteristics and user traffic analysis.

A lot of works are dedicated to the network characteristics such as throughput, capacity, losses and delay parameters. These works often consider the WLAN MAC layer as a queuing system and propose analytical models for estimation of the delay or losses parameters based on traffic characteristics. Here we will mention some of them. The first well-known analytical service model was proposed in Bianchi work [1]. Many other works proposed different analytical models and simulation results, for example [2–5]. Importance of these works is in the network model description, which is necessarily based on traffic parameters.

Some works are dedicated to the flow characteristics of the WLAN traffic. In the [12] authors present results of modeling and simulation of the traffic and consider the traffic characteristics from the point of view of flow characteristics. In the [10] authors propose the signal processing methods use for the WLAN traffic analysis. Author of the [16] consider traffic selfsimilarity based on the simulation results. In the [7] authors present results of analysis of the statistical measurement data based on flow characteristics. Considered works describe WLAN traffic as a flow with selfsimilarity properties.

Works [6, 8, 9, 11, 13–15] are dedicated to the traffic analysis based on the measurement data. In the [6] authors present results of measurement of WLAN traffic in specific places (buildings of the Stony Brook University). They present traffic patterns per day for different buildings, average throughput for different access points, traffic between access points, amount of data, control and management traffic.

Authors of the [8] present analysis of the unwanted traffic in the wireless access network. They present results of measurements of the WLAN installed in the specific building. Authors show that clients associate several times with the APs. They analyze the traffic to understand when handover occurs and whether the handovers were preferred or should have been avoided.

In the [9] authors present measurement results for the WLAN with a large amount of access points distributed in large amount of buildings for a long time period. They consider total amount of traffic by days per week, by hours per day, originated in different buildings etc. Work [11] presents results of measurement and analysis of traffic in the large scale WLAN. Authors present such parameters as traffic pattern per days of week, distribution of traffic per access points, session duration, number of handover per session. Authors of the [13] study network performance in conditions of security protocols use. They present results of traffic measurement and modeling during experiments. Presented results show impact of security protocols use on the network performance. Works [14, 15] presents some results of WLAN measurements.

All considered related work is dedicated to the traffic measurements and analysis is aimed to solve any specific problem for specific WLAN. Authors of considered works present many useful results which may be compared to any other specific networks or applications. The main idea of this paper is to get results for any common parameters of the WLAN traffic, which may be useful for network planning and analysis.

3 General WLAN Information and Experiment Procedure

In order to extend the analytics of mentioned network characteristics, the large hotel WLAN has been monitored.

The WLAN monitoring information has been kindly furnished by the "BELTEL" system integrator after the complex network audit having been accomplished. It was carried out in order check the RF condition, coverage, client structure and user experience of customer's WLAN. All end user data have been analyzed in impersonal form.

During network audit, the traffic information has been gathered during 28 days since 13 Feb 2017 till 13 Mar 2017. The monitored WLAN includes: 369 dual radio access points with IEEE802.11ac support, two redundant WLAN controllers, wired LAN infrastructure and monitoring system.

The Hotel has 10 floors with guest rooms and 2 floors with public areas. 10 guest room floors are covered by 32 APs each. Ground floor and first floor are covered by 49 APs that provide Wi-Fi coverage for hotel lobby, restaurants, conference halls, night club and hotel administration rooms.

The Wi-Fi coverage is provided in both bands (2.4 GHz and 5 GHz) with minimum signal strength of −65 dBm. The handover requirement is 75 dBm signal strength for 2 APs. The maximum transmission power for APs is 50 mW (17 dBm), which is needed in order to provide balance in both uplink and downlink frame flow, because of typical client mobile device transmission power limitation at 30 mW.

RF coverage measurement has also been carried out, but its results are beyond the scope of this research which is focusing on traffic characteristics and user experience.

Here we will just state, that physical coverage characteristics show that PHY layer of WLAN is in very good condition.

The WLAN coverage area on two lower floors is an approximate rectangle of 18000 m^2. The upper floors have complex form with total space of approximately 10000 m^2 each. The average WLAN coverage area for every AP is approximately 300 m^2 which results in wireless network coverage, dense and solid enough to neglect the user to AP distance effects.

It is important to mention that during monitoring period all APs had aggregate downtime due to restarts, power failures or reconfiguration issues for less than 5 min per every AP. This means that both wired and wireless infrastructure of the Hotel is in very good condition and is well serviced.

4 Results and Statistics Analysis

In this section the monitored parameters will be analyzed. During monitoring period 518492 formal user sessions have been established by 8473 unique user devices. Under unique user device we understand unique MAC address as it was registered by WLC during client device association process. Under formal client session we understand the Client-AP data transfer session. Under real client session we understand a sequence of continuous formal client sessions, when association time of the next formal session equals the end of association time of the previous one, i.e. the handover process took place. Thus client device handover can unite several formal sessions into one real client session.

Among 226994 real client sessions, 56726 sessions had zero quantity of packets sent or received by AP radio, which gives us 170268 user sessions with non-zero amount of data sent in both uplink or downlink direction. All user experience metrics have been calculated for these 170268 real user sessions.

It is important to claim, that the network is not overloaded with client traffic, i.e. the airtime and RF resources are at clients' disposal and do not limit client requests to network services. The general parameters of WLAN utilization are shown in Table 1.

Table 1. WLAN utilization.

WLAN parameter		Value
Average downlink throughput	a_{DL}, Mbps	3.15
Average uplink throughput	a_{UL}, Mbps	3.03
Peak traffic intensity per AP	$a_{RM\ max}$, Mbps	36.23
Maximum TX airtime usage	$\rho_{TX\ max}$, %	30
Maximum RX airtime usage	$\rho_{RX\ max}$, %	29

4.1 Band and Standard Revision

Table 2 shows all real client sessions according to IEEE 802.11 standard revision. It shows that a major number of client sessions have been established in 2.4 GHz band in IEEE802.11n standard.

Table 2. IEEE 802.11 standard revision per session.

Latest IEEE 802.11 revision	Sessions quantity	Sessions, %
802.11a	949	0.4
802.11b	0	0.0
802.11g	2709	1.2
802.11n (2.4 GHz)	186525	82.2
802.11a (5 GHz)	15100	6.6
802.11ac	21711	9.6
Total	226994	100.0

4.2   Client Session Duration

Analysis of the session duration statistical data shows that it has complex distribution which presumably consists of a number of different sessions, but the accuracy of the measurements (minimum 5 min interval) does not allow a sufficiently detailed analysis. Figure 1 shows empirical distribution represented by the histogram and analytical model. The analytical model represented by the Gamma distribution with probability density function

$$f(x) = \begin{cases} \frac{\alpha^k}{\Gamma(k)} x^{k-1} e^{-\alpha x}, & x \geq 0 \\ 0, & x < 0 \end{cases} \tag{1}$$

where α and k – are distribution parameters.

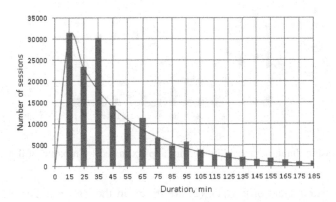

Fig. 1. Session duration probability density.

The probability density function (pdf) is given by the histogram at Fig. 1.

Gamma distribution was chosen due to the relatively good fit the histogram represented the empirical data. The accuracy of fit of the empirical data is not very good due to the some peaks at the 35, 65, 95, 125, 155, 185 min duration.

For detailed analysis of the session distribution we have to consider reasons of the session close. The following main reasons are: manual disassociation by a user, moving a user device away from the network and automatically closing by timeout (device sleep or hibernation mode). The first two reasons cause random session duration defined by a user behavior. The third reason causes session length defined by user devices turnings (defaults as usually). Thus we can propose that the empirical distribution may be described by a mixture of a number of random processes. The first is a process caused by the first and the second reasons. It may be depicted by continuous infinite distribution. The second, third and so on are processes caused by timeouts (possible user settings). As seen from the picture there are 35, 65, 95, 125, 155, 185 min. For pdf construction they may be depicted as discrete function. As we assume the correspondent timeouts are 5, 30, 60 and 90 min (session close after timeout, at the next 5th minutes interval).

$$f(x) = \begin{cases} f(x) & x \neq x_i, \ i = 1\ldots n \\ f(x) + r_i & x = x_i, \ i = 1\ldots n \end{cases} \tag{2}$$

The first process pdf (1) and empirical distribution represented by histogram are depicted at the Fig. 1. The mixed flow pdf (2) and empirical distribution represented by histogram are depicted at the Fig. 2.

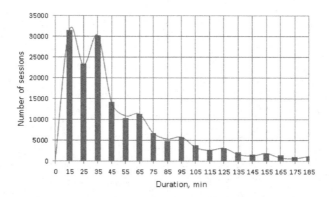

Fig. 2. Session duration probability density approximated by mixture of distributions.

Values of parameters of pdf (1), (2) are given in the Table 3. The Kolmogorov–Smirnov test shows that the hypothesis of pdf (2) is not rejected with confidential probability 0.95. So we can conclude that the proposed analytical model (2) is good description of the session duration distribution of the real traffic.

Following the hypothesis of the timeout duration, we can suppose that in the real case the user's settings can be completely arbitrary, but our study shows that it is very probably that they have some most commonly used values.

Table 3. Parameters of the probability density function.

Parameter	Value
α	2633.570
k	0.930
m, s	2448.785
σ, s	2539.499
r_1	0.006
r_2	0.079
r_3	0.017
r_4	0.010
r_5	0.007
r_6	0.005
r_7	0.004

4.3 Client Activity

To describe client activity we measure the part of time when a user terminal was in the active state, i.e. in the active session state. This value is equal to the ratio of the total session time to the total measurement time. So this value reflects activity of the user. The empirical distribution (pdf) of the total work time is depicted at the Fig. 3.

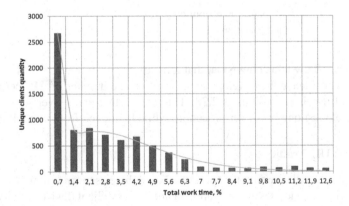

Fig. 3. Session duration probability density approximated by mixture of distributions.

$$f(x) = \eta_1 f_1(x) + \eta_2 f_2(x) \tag{3}$$

$$f_{1,2}(x) = \begin{cases} \dfrac{\alpha^k}{\Gamma(k)} x^{k-1} e^{-\alpha x} & , \quad x \geq 0 \\ 0 & , \quad x < 0 \end{cases} \tag{4}$$

With distribution parameters α and k. Values of parameters of pdf (3), (4) are given in the Table 4. The pdf of the mixed flow is given by

Table 4. Parameters of the probability density function.

Parameter	$f_1(x)$	$f_2(x)$
η_i	0.50	0.50
α_i	0.38	3.30
k_i	0.02	0.01
m_i (%)	0.92	3.96

4.4   Session Close Reason

Another important characteristic of client behavior is how client device disassociates to WLAN. For each real client session it's end reason as it was recorded by controller have been recorded. Table 5 shows session end reason. "Disassociation detected" means that client device has sent a disassociation frame to AP, "No longer seen from controller" means that client device has left the coverage area with no further handover to another AP.

Table 5. Session end reason.

Reason	Sessions, %
No longer seen from controller	86.7
Disassociation detected	13.3

Thus we can draw to a conclusion, that most of clients do not formally disassociate to WLAN and just leave the coverage area. The average number of APs per real client session for all real user sessions is 2,28. This means that every average client in the network had at least one handover procedure while he moved from one AP to another.

4.5   DL and UL Traffic

In this section we will examine the total number of bytes sent in uplink and downlink direction by client device session. Figures 4a and b show client device activity from the point of view of user data sent in uplink (UL) and downlink (DL) direction respectively. For both data flow directions, sessions with amount of data transferred under 10 MB have been analyzed. With the selected maximum value 88.14% and 95.32% of sessions have been analyzed for DL and UL data flows respectively.

Downlink and uplink amount of traffic is approximated with Pareto distribution. Table 6 shows the distribution parameters for both directions of data flow per user session. We can conclude that the amount of transferred data in DL close to UL and both links has very close distribution of sessions per amount of transferred data.

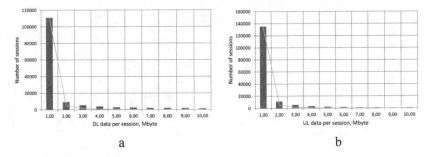

a   b

Fig. 4. Downlink (a) and uplink (b) bytes per session distribution and approximation by Pareto distribution.

Table 6. DL and UL bytes per client session distribution parameters.

Data flow direction	Selected max value, MB	Empirical mean, MB	Expected value, MB	k	x_m
DL	10	0.98	0.97	1.098	0.086
UL	10	0.67	0.68	1.155	0.091

4.6   Packet Size

Analysis of the packet size statistical data has shown that the empirical distribution can't be described by any analytical model with a good accuracy. So we propose that the packet flow may consist of number of statistically different flows. With the use of cluster analysis we found two statistically different groups of the data. We assume that the packet flow is mixed flow which consists of two independent flows. Each of these flows (groups of data) can be described by own distribution function.

Taking in account the nature of the object under study we have to note that the real packet size distribution function must be discrete and finite. But by doing this we can significantly complicate the model and its usage. Therefore, to describe the distribution function, we use models of infinite continuous distribution. Figure 5a shows empirical distribution represented by the histogram and two analytical models. Both of them are represented by the Gamma distribution with probability density function

$$f_i(x) = \begin{cases} \dfrac{\alpha_i^{k_i}}{\Gamma(k_i)} x^{k_i-1} e^{-\alpha_i x} & , \quad x \geq 0 \\ 0 & , \quad x < 0 \end{cases} \tag{5}$$

where α_i and k_i – are distribution parameters. The pdf of the mixed flow is given by

$$f(x) = \eta_1 f_1(x) + \eta_2 f_2(x) \tag{6}$$

The mixed flow pdf and empirical distribution represented by histogram are depicted at the Fig. 5b.

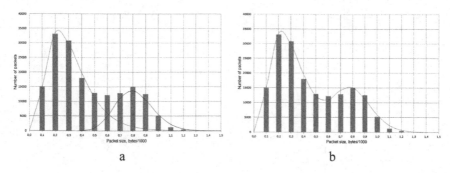

a   b

Fig. 5. Packet size probability density.

Values of parameters of pdf (6) are given in the Table 7. The Kolmogorov–Smirnov test show that the hypothesis of pdf (6) is not rejected with confidential probability 0.95. So we can conclude that the proposed analytical model (6) is good description of the of the packet size distribution of the real traffic.

The real traffic consists of many flows originated by different. The results obtained show that these flows may be divided by two groups: relatively "short" and "long" packets. We can suppose that the first group of packets is formed by such network services as text messages and voice data. The second group is generated by WEB-surfing, video streaming services and file transmit services. Relatively large sample size allows assuming that the obtained packet size model is specific not only for this case. In some special case traffic may differ according to different network services but the model may be applicable by using appropriate numerical values of parameters.

4.7   Traffic Intensity Per User Session

In this section we will analyze the average traffic intensity in both directions per user session. The average traffic intensity is calculated as

$$\bar{\lambda} = \frac{A_{TX} + A_{RX}}{t} \text{ bit/s} \tag{7}$$

where A_{TX} and A_{RX} - number of bytes sent in each direction of traffic flow (bits), t – is a session duration (s). Figure 6 shows average traffic intensity per user session distribution. This parameter can be quite accurately approximated by Pareto distribution

$$f_1(x) = \begin{cases} \dfrac{k\, x_m^k}{x^{k+1}} & , \quad x \geq x_m \\ 0 & , \quad x < x_m \end{cases} \tag{8}$$

where x_m and k are parameters of the pdf.

Table 8 shows the distribution parameters for average throughput per user session. Sessions with average throughput under 20 KB/s were analyzed. With the selected maximum value 96.11% of client sessions have been analyzed.

Table 7. Parameters of the probability density function.

Parameter	$f_1(x)$	$f_2(x)$
η_i	0.71	0.29
α_i	2.61	27.75
k_i	103.67	27.58
m_i (bytes)	270	765
σ_i (bytes)	167	145

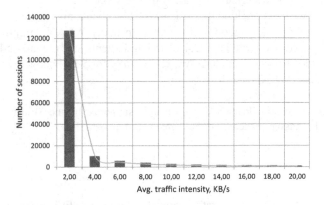

Fig. 6. Average traffic intensity per session.

Obtained data shows that the average traffic intensity per session is much lower than the network throughput. It means that distribution of traffic is very uneven during the session.

Table 8. Traffic intensity per user session distribution parameters

Selected max value, KB/s	Empirical mean, KB/s	Expected value, KB/s	k	x_m
20	1.61	1.60	1.054	0.082

5   Conclusions

Presented results of the WLAN traffic analysis show some important parameters such as: session duration, user activity, DL and UL traffic value per session, packet size and average traffic intensity.

Study of session duration has shown that it may described by Gamma distribution function with difference of existence of a number of fixed duration sessions caused by user equipment settings (timeout). The mean duration of the session is equal to 36.64 min. Study of the activity of a user has shown that the average active state of a user is equal to 3,6%. Distribution of users by activity may be described by mixture distribution consisting of two gamma-distributed independent random values (low and high activity users).

The values of traffic per session in DL and UL are equal statistically. They have close mean values and identical distribution functions which are well described by the Pareto distribution. The average traffic rate through DL a UP is 0.98 and 0.67 Mbit/s respectively.

Packet size may be described by mixture distribution and consists of two independent random values: "short" (mean 270 bytes) and "long" (mean 765 bytes) packets. Both of these values may be described by gamma distributions.

Traffic intensity per session is not uniformly distributed. Distribution of sessions by traffic intensity may be described by Pareto distribution.

Additionally we can conclude that IEEE 802.11b clients no longer can be found in real WLANs. This fact is in good correlation with many WLAN equipment vendors' recommendations, for example [17]. 2.4 GHz band clients still prevail and though it is often recommended to design modern WLANs mainly for 5 GHz coverage [18], legacy users are still main WLAN services target.

Acknowledgement. The publication was financially supported by the Ministry of Education and Science of the Russian Federation (the Agreement number 02.a03.21.0008).

References

1. Bianchi, G.: IEEE 802.11—saturation throughput analysis. IEEE Commun. Lett. **2**(12), 318–320 (1998)
2. Bianchi, G., Di Stefano, A., Giaconia, C., Scalia, L., Terrazzino, G., Tinnirello, I.: Experimental assessment of the backoff behavior of commercial IEEE 802.11b network cards. In: IEEE Xplore, 29 May 2000
3. Han, C., Dianati, M., Tafazolli, R., Kernchen, R.: Throughput analysis of the IEEE 802.11p enhanced distributed channel access function in vehicular environment. In: IEEE Xplore, 04 October 2010
4. Challa, R.K., Chakrabarti, S., Datta, D.: An improved analytical model for IEEE 802.11 distributed coordination function under finite load. Int. J. Commun. Netw. Syst. Sci. **3**, 169–247 (2009)
5. Camps-Mur, D., Garcia-Saavedra, A., Serrano, P.: Device to device communications with WiFi Direct: overview and experimentation. IEEE Wirel. Commun. **20**(3), 96–104 (2013)
6. Rajesh, G., Srikanth, K.: WiFi Traffic Analysis Project Report. http://www3.cs.stonybrook.edu/~rgolani/docs/Wireless_Report.pdf. Accessed 12 Mar 2015
7. Phillips, C., Singh, S.: Analysis of WLAN Traffic in the wild. In: International Conference on Research in Networking, NETWORKING 2007. Ad Hoc and Sensor Networks, Wireless Networks, Next Generation Internet, pp. 1173–1178 (2007)
8. Raghavendra, R., Belding, E.M., Papagiannaki, K., Almeroth, K.C.: Unwanted link layer traffic in large IEEE 802.11 wireless networks. IEEE Trans. Mob. Comput. **9**(9), 1212–1225 (2010)
9. Kotz, D., Essien, K.: Analysis of a campus-wide wireless network. Wirel. Netw. **11**(1), 115–133 (2005)
10. Partridge, C., Cousins, D., Jackson, A.W., Krishnan, R., Saxena, T., Timothy Strayer, W.: Using signal processing to analyze wireless data traffic. In: WiSE 2002 – Proceedings of 1st ACM Workshop on Wireless Security Atlanta, GA, USA, pp. 67–76, 28–28 September 2002

11. Lee, J., Choi, S., Jung, H.: Analysis of user behavior and traffic pattern in a large-scale 802.11a/b network. https://www.researchgate.net/publication/228627517_Analysis_of_ user_behavior_and_traffic_pattern_in_a_large-scale_80211_ab_network
12. Yamkhin, D., Won, Y.: Modeling and analysis of wireless LAN traffic. J. Inf. Sci. Eng. **25**, 1783–1801 (2009)
13. Ochang, P.A., Irving, P.: Performance analysis of wireless network throughput and security protocol integration. Int. J. Future Gener. Commun. Netw. **9**(1), 71–78 (2016)
14. Yeo, J., Youssef, M., Agrawala, A.: Characterizing the IEEE 802.11 traffic: the wireless side. CS-TR-4570 and UMIA CS-TR-2004–15, University of Maryland, 1 March 2004
15. Viecco, C.H., Gupta, M.: A dense wireless LAN case study. Technical report TR640, Indiana University, December 2006
16. Jiang, M.Z.: Analysis of wireless data network traffic. Thesis submitted in partial fulfillment of the requirements for the degree of master of applied science, Simon Fraser University, April 2000
17. Cisco Systems: Cisco Wireless LAN Controller Configuration Best Practices. Cisco Press (2015)
18. Ruckus Wireless: Deploying Very High Density Wi-Fi. Best Practice Design Guide (2012)

Wi-Fi Based Indoor Positioning System Using Inertial Measurements

Mstislav Sivers[1], Grigoriy Fokin[2], Pavel Dmitriev[2],
Artem Kireev[2]([⊠]), Dmitry Volgushev[2],
and Al-odhari Abdulwahab Hussein Ali[2]

[1] St. Petersburg State Polytechnical University, St. Petersburg, Russia
m.sivers@mail.ru
[2] The Bonch-Bruevich St. Petersburg State University of Telecommunications,
St. Petersburg, Russia
grihafokin@gmail.com, {p-dmitriev,kireyev}@list.ru,
{d.volgushev,abdwru2011}@yandex.ru

Abstract. Recent time Wi-Fi-based positioning leads in non-GNSS methods for indoor positioning. Globally existing infrastructure is the main advantage of Wi-Fi-based methods. Traditional methods or gathering data for RSSI Fingerprint offline mode require costly site survey. This paper presents a method of RSSI data acquisition connected with inertial navigation system (INS) that takes less time without loss of positioning accuracy. Proposed algorithm gives opportunities to detect location for users without MEMS-platform on their mobile devices. Furthermore, Kalman filter is developed to decrease measurement errors. We evaluated our system in real conditions.

Keywords: Wi-Fi · RSSI · Radiomap · INS · Kalman filtering

1 Introduction

The problem of navigating indoors becomes more and more urgent. New buildings get voluminous and rather complex structures, only those who visit such places constantly can orient themselves, and for unprepared person navigation in such places turns into torture. Furthermore, accurate indoor localization systems find their use in significant areas of modern applications such as augmented reality, check-in and geo-fencing services, guided excursions, location-based advertising and others. Due to big commercial potential big companies like Apple, Alphabet, Qualcomm already have some projects in this area.

Traditional Global Navigation Satellite Systems (GNSS) such as Global Positioning System (GPS) and GLObal NAvigation Satellite System (GLONASS) provide rather accurate position estimates outdoor, however for indoor localization this is not the case because of non-line of sight (NLOS) conditions as a result of reduced satellite visibility and its weak signal inside buildings.

Available solutions for indoor location include special wireless technologies such as Subpos [1], IndoorAtlas [2], Accuware [3] and NanoLOC [4] which are dedicated for positioning and give accurate estimates, for example, IndoorAtlas provide accuracy

O. Galinina et al. (Eds.): NEW2AN/ruSMART/NsCC 2017, LNCS 10531, pp. 734–744, 2017.
DOI: 10.1007/978-3-319-67380-6_69

less than 1 m and use radio signals, geomagnetic fields and barometric pressure. However all these systems need extra expenses for infrastructure customization and suffer from dynamically changing surroundings. That's why our group decided to use existing infrastructure without possibility to influence on its normal functionality.

Wireless networks provide several positioning techniques, which can be classified by the primary measurements: Received Signal Strength Indication (RSSI), Angle of Arrival (AoA), Time of Arrival (TOA), Time Difference of Arrival (TDOA) [5, 6]. AoA methods are not often used because of complexity and cost of transceivers and antenna modules [7] Keeping track of our experience [8] in creating comprehensive data maps for RSSI Fingerprinting, this method has been evaluated as most attractive.

Modern smartphones have some sensors that can be used to improve performance of localization such as accelerometer and gyroscope. Combination of these sensors call inertial navigation system (INS). INS can estimate user's position with a high update speed but accumulates integration drift, which makes impossible standalone appliance of INS. That's why integration of RSSI and INS appears to be quite perspective.

The aim of this paper is to implement algorithm of optimization RSSI data map building and method of combining RSSI and INS for positioning indoors. Results of implementing will be checked experimentally.

The material in the paper organized as follows. Indoor positioning techniques by INS and RSSI with algorithms, protocols and related works are presented in the second part. Experimental analysis including scenarios, flowcharts, implementation features and obtained results are presented in the third part. Finally, we draw the conclusions in the fourth part.

2   Indoor Positioning Techniques

Indoor positioning techniques experimentally validated during investigation include RSSI Fingerprint method and INS localization system.

2.1   Inertial Navigation System

Inertial navigation system (INS) is a localization system that use micro-machined electromechanical systems (MEMS) which provide velocity and the bearing angle of an object. Movement of an object is controlled by six parameters conducted by accelerometer and gyroscope. Accelerometer provides translational linear acceleration in x, y, and z directions and gyroscope computes angular velocity in cartesian coordinates. Coordinate systems according to smartphone body are shown in Fig. 1.

The most important source of error of an accelerometer is the bias. The bias of an accelerometer is the offset of its output signal from the true value, in m/s^2. A constant bias error of ε, when double integrated, causes an error in position which grows quadratically with time. The accumulated error in position is [9]:

$$s(t) = \varepsilon * \frac{t^2}{2} \qquad (1)$$

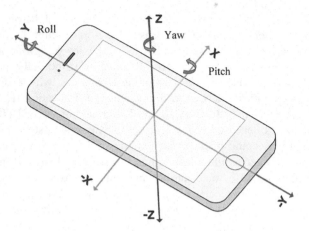

Fig. 1. Coordinate system on phone frame.

where t is the time of the integration.

It is possible to estimate the bias by measuring the long term average of the accelerometer's output when it is not undergoing any acceleration. Uncorrected bias errors are typically the error sources that limit the performance of the device [9].

The bias of a rate gyro is the average output from the gyroscope when it is not undergoing any rotation. A constant bias error of ε, when integrated, causes an angular error which grows linearly with time:

$$\theta(t) = \varepsilon * t \tag{2}$$

The constant bias error of a rate gyro can be estimated by taking a long term average of the gyro's output whilst it is not undergoing any rotation. Once the bias is known, it is trivial to compensate for it by simply subtracting the bias from the output.

Mathematical integration of the acceleration a(t) computes the velocity v(t), double integration determines the distance from the starting point s(t):

$$v(t) = v(0) + \int_0^t a(t)dt \tag{3}$$

$$s(t) = s(0) + \int_0^t v(t)dt \tag{4}$$

Integrating the angular velocity ω(t) gives orientation φ(t):

$$\Delta\varphi(t) = \int_0^t \omega(t)\, dt \tag{5}$$

In order to minimize cumulative integration drift, there are some known techniques as Zero Velocity Updates (ZUPTs) [10] and Step-Detection [11]. ZUPT exploits an internal characteristic of pedestrian walking: static contact with the floor results in zero acceleration and velocity during of each step taken. This algorithm effectively applied in pedestrian systems, but demands extra hardware.

Instead of performing double integration, Step-Detection operates patterns in variations of the acceleration by detecting the peak points of the curve (Fig. 2). Properly aligned orientation sensor determines the heading orientation.

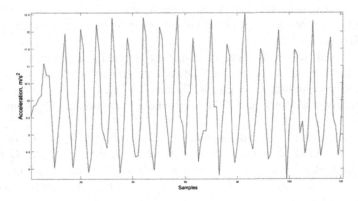

Fig. 2. Typical pattern of full acceleration for step detection

Step-detection tracking is based on step length estimation and counting detected steps:

$$X_k = X_{k-1} + S_k \qquad (6)$$

where X_k and X_{k-1} estimated location vectors after and before k^{th} step. S_k is the estimated displacement vector for the kth step:

$$S_k = [l_k \cos \theta_k, l_k \sin \theta_k]^T, \qquad (7)$$

where l and θ are the step length and heading orientation for the kth step:

However, step length varies from step to step even for the same pedestrian. That's why we use the model for dynamic step length estimation [12]

$$l_k \approx \sqrt[4]{a_{max} - a_{min}} * K, \qquad (8)$$

where a_{max} maximum of full acceleration measured in a single step, a_{min} minimum of full acceleration measured in a single step, K is a constant for unit conversion that can be determined experimentally. Full acceleration is calculated:

$$Acc_{full} = \sqrt{a_x^2 + a_y^2 + a_z^2}, \qquad (9)$$

where a_x, a_y, a_z are the output values of the triaxial accelerometer.

It is worth to note that Step-detection algorithm and heading orientation is the primary part of user equipment position estimation in the proposed scheme.

2.2   Fingerprint Method

Measuring the RSSI from all accessible Wi-Fi access points (AP) in a certain number of points is called Fingerprinting method. Collected measurements result in database called "radiomap" [13]. Mobile object location is calculated by comparing the current RSSI measurements with the data stored in radiomap. In general, procedure of determining the location of the target object consists of two phases: the offline acquisition and the online positioning.

Offline phase is following: first, we divide the targeted positioning area into grids, usually the grid is 1 m * 1 m; the coordinate of each grid is regarded as reference point (RP). Each RP corresponds to a vector R. Typical $R = (r_1, r_2, \ldots, r_N)$ consists of RSSI values from N APs. Combination of RP levels and grid coordinates results in radiomap database.

Online phase is the location determination phase. In this phase, RSSI levels at target object location should be compared with each vector R from database. The location with best match of RSSI in radiomap database should be the estimated location.

However, creating radiomap takes a lot of time and human resources especially if the number of RP is large.

3   Experimental Analysis

In this section, we describe experimental analysis including scenarios, flowcharts, implementation features and obtained results. Experiments was conducted with initial conditions: St. Petersburg State University of Telecommunications in the territory of 400 m^2 with 124 Wi-Fi APs. As we use RSSI fingerprint method Wi-Fi APs location is unnecessary to be identified. Primary RSSI and INS measurements were collected by Iphone 6 smartphone, which equipped with 6-axis Inertial measurement unit Inven-Sense MP67B [14]. For correct azimuth angle estimation smartphone orientation was fixed horizontally and y-axis points in walking direction. Data from Wi-Fi and INS were structured by software package Matlab mobile [15].

Our group conducted several experiments with different length and trajectories within one floor. This section describes one of tests that demonstrates results in full measure.

Positioning accuracy was evaluated in terms of Root Mean Square Error (RMSE) position estimate in meters with respect to true position in Cartesian coordinates:

$$\Delta = \frac{1}{N} \left(\sum_{p=1}^{N} \sqrt{(x_{est} - x_a)^2 + (y_{est} - y_a)^2} \right), \tag{10}$$

where N – number of measurements for one estimated location, x_a and y_a – actual position coordinates.

To evaluate algorithm of RSSI radiomap building and method of combining RSSI and INS measurements, processing and visualization we developed Matlab software.

Figure 3 shows the algorithm for the mobile object location calculation. Main part of this scheme is a database, where INS and Wi-Fi measurements are inputs. Wi-Fi measurement represented by RSSI data include MAC addresses, APs signal levels and timestamps for each reference point in radiomap. INS measurements consist of accelerometer, azimuth data, timestamps and represents walking track calculated by Step-Detection algorithm. RSSI data and walking track matched by timestamp reference points to form full database. In this way radiomap reference points (RSSI) are associated with points of walking track (INS). In practice, this part of algorithm is presented in offline phase.

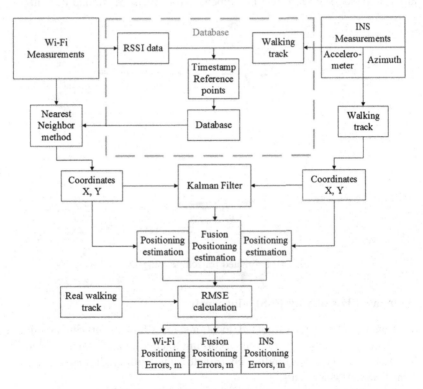

Fig. 3. Coordinate estimation algorithm

Nearest Neighbor Method (NNM) for instant RSSI measurements is applied in online phase to estimate (x, y) coordinates by searching nearest location point from database. Step-Detection algorithm is also used in online phase to calculate mobile object position.

Combining two coordinates evaluations from NNM and INS tracking by Kalman filter results in Fusion Position Estimation. RMSE calculation is performed by comparing estimated position with real walking track in three cases.

3.1   Estimated Position with INS

Real walking track and estimated INS track presented in Fig. 4. Real walking track has 175 m length. INS trajectory calculated by Step-Detection algorithm looks similar and has 172.5 m length. It was observed that error of 2.5 m is a result of acceleration measurements deviation.

Inertial navigation system RMSE in this case is equal to 0.76 m (Fig. 5). Most of errors placed in interval from 0.5 to 1 m. Such small error caused by initial conditions. Moreover in some cases (Table 1) RMSE became bigger because of integration drift influence. Since, this method is not suitable for ordinary users because smartphone orientation is fixed. Therefore, we implement it for precise radiomap building.

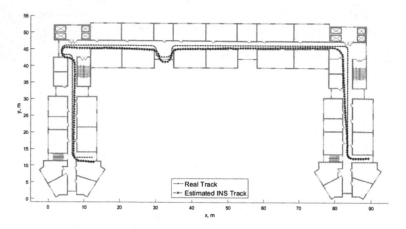

Fig. 4. Estimated trajectory using INS

3.2   Estimated Position by RSSI Fingerprint

Visualization of actual and estimated locations for RSSI fingerprint positioning in Wi-Fi networks is illustrated in Fig. 6. According to coordinate estimation algorithm in this case NNM selects corresponding coordinate values from radiomap closest relative to current RSSI measurements.

RMSE of RSSI Fingerprint tests is equal to 0.97 m (Fig. 7). Most of errors placed in interval from 0.9 to 1.1 m. In comparison with INS location error raised since in this case smartphone gets RSSI measurements less frequently. That's why user position estimation indoors can't be estimated accurately especially in case of continuous movement. To solve this problem Kalman filtering was developed.

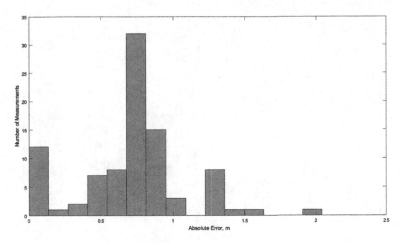

Fig. 5. INS error distribution

Table 1. Algorithm accuracy results

Test number	Mean error (INS)	Mean error (fingerprinting)	Mean error (Kalman filter)
1	0.73	1.06	0.80
2	0.79	0.97	0.78
3	0.79	1.03	0.87
4	7.00	0.91	0.70
5	1.78	0.90	0.69
6	1.73	1.00	0.97
7	0.90	0.91	0.85
8	2.87	1.01	0.86
Average Δ	2.07	0.97	0.81

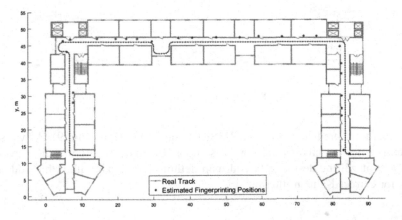

Fig. 6. Estimated locations using RSSI

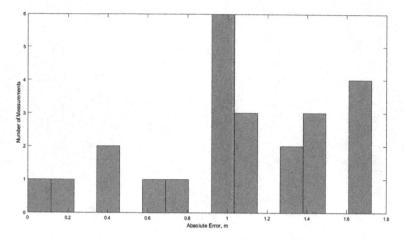

Fig. 7. RSSI fingerprint error distribution

3.3 Kalman Filtering

We employed Kalman filter to fuse and predict RSSI Fingerprint localization results. In this case, we can observe full user trajectory (Fig. 8).

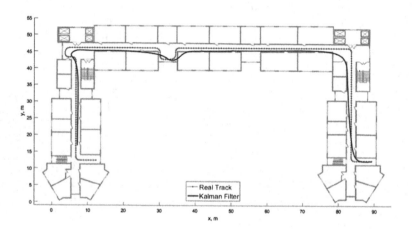

Fig. 8. Kalman filtering trajectory

Kalman filter prediction average RMSE is equal to 0.81 m (Fig. 9). As we see in some turns trajectory crosses walls and it's impossible to define user position precisely. That's why it will be desirable to add map defining to our algorithm that adds more factors for correct Kalman filtering.

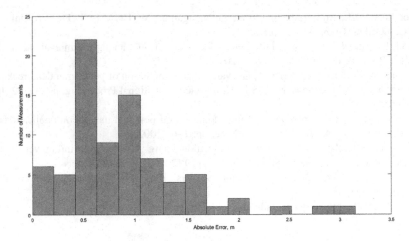

Fig. 9. Kalman filter error distribution

4   Conclusion

In this paper we evaluated the algorithm of collecting data for radiomap merged with INS. Simultaneous INS and RSSI radiomap building takes much less time than traditional methods. Employment of structured INS and RSSI database gave 1 m accuracy but in case of dynamically changing user position algorithm is not reliable. Kalman filter employment resolved this problem with minimal loss of accuracy. Performed experiment confirms algorithm efficiency in buildings with Wi-Fi infrastructure. In future we plan to add map defining and create mobile application for ordinary mobile user test inside university.

References

1. SubPos a «Dataless» Open-Source WiFi Positioning System. http://www.subpos.org/
2. IndoorAtlas. http://www.indooratlas.com/
3. Accuware Indoor Location Technology. https://www.accuware.com
4. Nanoloc Real Time Location System. http://nanotron.com
5. Galov, A., Moschevikin, A., Voronov, R: Combination of RSS localization and ToF ranging for increasing positioning accuracy indoors. In: 11th International Conference on ITS Telecommunications (ITST), pp. 299–304. IEEE (2011)
6. Dmitriev, P., Pisarev, S., Sivers, M.: Analysis of methods and algorithms of positioning. In: WiFi Networks, vol. 10, p. 44. Vestnik Sviazy (2015). ISSN 0320-8141
7. Kireev, A., Fokin, G.: Radio direction finding of LTE emissions using mobile spectrum monitoring station with circular antenna array. Trudy Nauchno-issledovatel'skogo instituta radio, vol. 2, pp. 68–71 (2015). ISSN 0134-5583
8. Sivers, M., Fokin, G., Dmitriev, P., Kireev, A., Volgushev, D., Hussein Ali, A.-O.A: Indoor positioning in WiFi and NanoLOC networks. In: Galinina, O., Balandin, S., Koucheryavy, Y. (eds.) NEW2AN/ruSMART -2016. LNCS, vol. 9870, pp. 465–476. Springer, Cham (2016). doi:10.1007/978-3-319-46301-8_39

9. Woodman, O.J.: An introduction to inertial navigation. №. UCAM-CL-TR-696, University of Cambridge, Computer Laboratory (2007)
10. Skog, I., Handel, P., Nilsson, J.O.: Zero-velocity detection—an algorithm evaluation. IEEE Trans. Biomed. Eng. **57**, 2657–2666 (2010)
11. Jimenez, A.R., Seco, F., Prieto, C., Guevara, J.: A comparison of pedestrian dead-reckoning algorithms using a low-cost MEMS IMU. In: Intelligent Signal Processing, pp. 37–42. IEEE (2009)
12. Weinberg, H.: Using the ADXL202 in pedometer and personal navigation applications. In: Analog Devices AN-602 Application Note, pp. 1–6 (2002)
13. Moka, E., Retscher, G.: Location determination using WiFi fingerprinting versus WiFi trilateration. J. Locat. Based Serv. **1**(2), 145–159 (2007). Taylor&Francis
14. InvenSense's 6-Axis Family of Motion Sensors. https://www.invensense.com/products/motion-tracking/6-axis/
15. Matlab Mobile. https://www.mathworks.com/products/matlab-mobile.html

Power Allocation in Cognitive Radio with Distributed Antenna System

Jerzy Martyna[✉]

Institute of Computer Science, Faculty of Mathematics and Computer Science,
Jagiellonian University, ul. Prof. S. Lojasiewicza 6, 30-348 Cracow, Poland
martyna@ii.uj.edu.pl

Abstract. In this paper, the power allocation (PA) for a cognitive radio (CR) system with a distributed antenna system (DAS) is investigated. In the DAS, remote antennas are separated and connected to a central unit (CU) via optical fibre. In this article, the problem of power allocation for CR with DAS is formulated as an optimisation problem. Based on the Karush-Kuhn-Tucker (KTT) conditions for this optimisation problem, a suboptimal PA scheme is developed. The proposed algorithm maximises energy-efficiency (EE) while maintaining proportional fairness. Finally, simulation examples are presented to verify the proposed method. It is shown that the proposed algorithm is effective in power allocation for a CR system with DAS implementation.

Keywords: Distributed antenna system (DAS) · Cognitive radio system · Power allocation · Convex programming

1 Introduction

The cognitive radio system (CR), introduced by Mitola [10–12], is one of the most promising technologies developed in dozens of research centres. It is anticipated that these systems will have new functionality, including learning, decision making, etc., which will allow for flexible detection user needs, create smart networks and utilize available radio resources in the best way. Furthermore, according to the definition of the CR system, given by Haykin [5], they will vary their performance parameters, including the frequency carrier, modulation type, with a view to securing high quality communication and the use of spectral resources. This will be possible by building knowledge through these devices and tuning into radio environment changes.

The use of distributed antenna systems (DASs) is a very promising technology in wireless networks [19]. Thanks to their use, transmission efficiency improves without increasing bandwidth. Unlike systems with conventional, central antennas, which are located in the middle of the cell, DAS systems have antennas located on its periphery. Then the signal is delivered from the central unit (CU) using fibre to all distributed antennas [13,19]. The central unit only processes signals, while the DAS is responsible for sending and receiving signals.

© Springer International Publishing AG 2017
O. Galinina et al. (Eds.): NEW2AN/ruSMART/NsCC 2017, LNCS 10531, pp. 745–754, 2017.
DOI: 10.1007/978-3-319-67380-6_70

The additional benefit of using this system is the statistical distance reduction between users and access point. This increases the signal strength and also decreases interchannel interference, which greatly improves system performance.

In radiocommunication systems, distributed antenna systems have been the subject of extensive research and analysis. Among others, in the paper [4] the capacity of the distributed antenna system in the decay channels is given. Resource allocation in such systems for various methods of subcarrier allocation are given in [18]. He et al. [6] provided an optimal allocation scheme resources in distributed antennaed OFDM systems. In turn, [7] presents an analysis of the distributed system large-scale distributed antenna system (L-DAS). In the paper by Bulu *et al.* [3] an attempt was made to increase the system performance of the DAS system by designing antenna locations and antenna selection. Zhao et al. in the paper [16] presented the concept of a coordinated spectral distribution in a distributed antenna system. The authors in [8] proposed bit allocation and paring methods for multiuser DAS with limited feedback. In the paper by Yu *et al.* [15] the channel capacity of the DAS over a multi-path frequency-selective fading channel was investigated. The application of an auction algorithm for a cognitive distributed antenna system is shown by Zhao *et al.* [17].

The purpose of this paper is to develop a new power allocation scheme in cognitive radio systems with distributed antenna systems. For the adopted model, such a system has formulated an optimisation problem and then developed its suboptimal solution. By this method, an algorithm for suboptimal power allocation (PA) in cognitive radio systems with distributed antenna systems is developed. The performance of the proposed approach is evaluated through a number of simulation scenarios. The results obtained demonstrated the effectiveness of proposed energy-efficiency (EE) power allocation for CR system with distributed antennas.

Section 2 shows the accepted model of cognitive radio system with distributed antennas. Section 3 provides a suboptimal scheme of power allocation in a cognitive radio system with distributed antennas. The simulation results are presented in Sect. 4. Section 5 summarizes the work.

2   System Model

It is assumed that the DAS distributed antenna system, consisting of M antennas, is located in a single cell (Fig. 1) which contains K secondary users and L mobile users (MU). The primary base station (PBS) is located outside the cell and transmits data to the original users, including primary users (PUs) located inside the cell. Primary users have access to licensed frequency bands. The cell is equipped with a secondary system, which also consists of a secondary base station (SBS). It consists of a central unit (CU) and a distributed antenna system. It is assumed here that each secondary user (SU) and mobile user (MU) has a single transceiver antenna. Furthermore, it is assumed that all DAS antennas are uniformly distributed inside the cell and are connected to the CU through optical fibre. The mobile user is also a secondary user and may use free frequencies intended for the PUs both for transmitting and receiving their data.

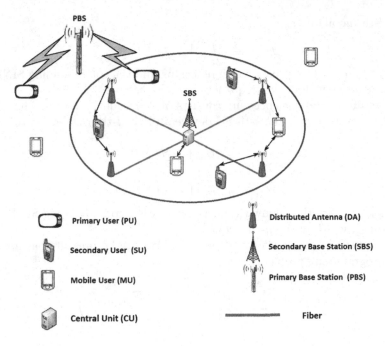

Fig. 1. Distributed antenna system (DAS) located in a single cell which contains cognitive radio system.

The signal-to-noise ratio (SINR) for the k-th user SU is equal to

$$SINR_k = \frac{s_{k,i}p \mid h_{k,i} \mid^2}{\sum_{j=1,j \neq i}^{M} s_{k,j}p \mid h_{k,j} \mid^2 + \sum_{l=1,l \neq i}^{L} s_{k,l}p \mid h_{k,l} \mid^2 + p' \mid g_k \mid^2 + \sigma_k^2} \quad (1)$$

where $s_{k,i}$ corresponds to the large scale fading of the signal sent from i-th antenna and received in k-th antenna and can be expressed in the form

$$s_{k,i} = \left(\frac{d_{k,i}}{d_0}\right)^{-\alpha} \cdot 10^{\frac{\mu}{10}} \quad (2)$$

$D_{k,i}$ defines the distance between k-th and i-th antenna, d_0 is the reference distance, α path loss exponent, μ represents a random variable with a mean of σ^2. P is the signal strength of i-th antenna, $h_{k,i}$ is the channel between i-th and k-th antenna, p' is PU transmission power, g_k is a channel between PSB and k-th SU device.

The SINR value for i-th mobile user is given by

$$SINR_l = \frac{s_{l,i}p \mid h_{l,i} \mid^2}{\sum_{j=1,j \neq i}^{M} s_{l,j}p \mid h_{l,j} \mid^2 + \sum_{k=1,l \neq k}^{K} s_{l,k}p \mid h_{l,k} \mid^2 + p' \mid g_l \mid^2 + \sigma_l^2} \quad (3)$$

The total achievable data transfer rate for k-th SU device is equal to

$$R_k = \log_2(1 + SINR_k) \quad (4)$$

and for l-th mobile user

$$\nu_{l,n} = \gamma_{l,n} B_n (1 + SINR_l) \tag{5}$$

where $\gamma_{k,n}$, $gamma_{l,n}$ are bandwidth functions and the corresponding SINR, B_n is the bandwidth n-th of the frequency.

It is assumed that the spectrum consists of N frequency bands. Then a portion of the spectrum per the k-th SU device and the l-th mobile user are equal, respectively:

$$\nu_{k,n} = \gamma_{k,n} B_n (1 + SINR_k) \tag{6}$$

$$\nu_{l,n} = \gamma_{l,n} B_n (1 + SINR_l) \tag{7}$$

where $\gamma_{k,n}$, $\gamma_{l,n}$ are the bandwith functions and the corresponding SINR, B_n is the bandwith width of the n-th frequency.

The problem of power allocation optimisation can be presented as nonlinear integer programming problem

$$Z = \max \sum_{n=1}^{N} \left(\sum_{k=1}^{K} x_{k,n} \nu_{k,n} + \sum_{l=1}^{L} x_{l,n} \nu_{l,n} \right) \tag{8}$$

subject to

$$\text{for } \forall k \ \sum_{n=1}^{N} x_{k,n} \leq 1 \tag{9}$$

$$\text{for } \forall l \ \sum_{n=1}^{N} x_{l,n} \leq 1 \tag{10}$$

$$x_{k,n} = \{0,1\}, \quad k = 1, 2, \ldots, K, \ n = 1, 2, \ldots, N \tag{11}$$

$$x_{l,n} = \{0,1\}, \quad l = 1, 2, \ldots, L, \ N = 1, 2, \ldots, N \tag{12}$$

$$x_{k,n} + x_{l,n} \leq 1, \quad \text{if } c_{k,l} = 1 \tag{13}$$

where $c_{k,l}$ is an element of the neighborhood matrix $C_{k,l} = \{c_{k,l}\}_{K \times L}$, where $c_{k,l} = 1$ indicates that there is interference between k-th SU and l-th mobile user, $c_{k,l} = 0$ otherwise.

The lower constraint on the value of the objective function of the problem given by Eq. (8) can be obtained thanks resolving the task of Karusha-Kuhn-Tucker (KKT) [2] using the Lagrange multiplier method. The created Lagrange function has the form:

$$L(\overline{\lambda}, x_{k,n}, x_{l,n}) = \sum_{n=1}^{N} \left(\sum_{k=1}^{K} x_{k,n} \nu_{k,n} + \sum_{l=1}^{L} x_{l,n} \nu_{l,n} \right) + \sum_{k=1}^{K} \lambda_k \left(1 - \sum_{n=1}^{N} x_{l,n} \right)$$

$$+ \sum_{l=1}^{L} \lambda_l \left(1 - \sum_{l=1}^{L} x_{l,n} \right) \tag{14}$$

With the constraints given by Eqs. (9), (10) and (13), where the vector contains Lagrange multipliers. The conditions necessary for the existence of a saddle point of Lagrange functions $L(\overline{\lambda}, x_{k,n}, x_{l,n})$, which must be met simultaneously, are as follows:

$$\frac{\partial L}{\partial \lambda_1} = \sum_{k=1}^{K} x_{k,n}(\nu_{k,n} - \lambda_k) = 0 \tag{15}$$

$$\frac{\partial L}{\partial \lambda_2} = \sum_{l=1}^{L} x_{l,n}(\nu_{l,n} - \lambda_l) = 0 \tag{16}$$

with the constraints given by Eqs. (9), (10), (13).

Solving the system of Eqs. (15), (16) and taking into considerations the constraints given by Eqs. (9), (10) and (13), we can find the lower limit on value of the optimal solution target function for allocation power in the CR system with distributed antennas.

3   Sub-optimal Algorithm for Power Allocation in CR System with DAS

From the conditions of the optimal solution to the problem of power allocation in a CR system for $\forall n \in N$ can be obtained:

(a) if $\displaystyle\sum_{k=1}^{K}(\nu_{k,n} - \lambda_k) \geq 1$, then $x_{k,n} = 1$

(b) if $\displaystyle\sum_{k=1}^{K}(\nu_{k,n} - \lambda_k) < 1$, then $x_{k,n} = 0$

(c) if $\displaystyle\sum_{l=1}^{L}(\nu_{l,n} - \lambda_l) \geq 1$, then $x_{l,n} = 1$

(d) if $\displaystyle\sum_{l=1}^{L}(\nu_{l,n} - \lambda_l) < 1$, then $x_{l,n} = 0$

Under these conditions we can obtain suboptimal results for the power allocation algorithm for a cognitive radio system with scattered antennas that consists of two procedures (see Fig. 2). The first procedure checks sequentially all user devices of secondary (SUs) and mobile users (MUs) inside the cell, whether the above conditions are met and there is no interference. If so, it follows call a second procedure, which for each antenna separately depending on interference between

---

Algorithm 1 Power allocation in CR system with DAS

1: **procedure** DAS IN SINGLE CELL(a, b)
2: Each individual SU and MU compute their local PU signal sensing bits;
3:     **while** SU and MU actions are allowed **do**
4:         **for** $j \leftarrow 1, K$ **do**
5:             **if** $\sum_{k=1}^{K}(v_{k,n} - \lambda_k) \geq 0$ **then**
6:                 $x_{k,n} \leftarrow 1$
7:             **else**
8:                 $x_{k,n} \leftarrow 0$
9:             **end if**
10:        **end for**
11:        **for** $l \leftarrow 1, L$ **do**
12:            **if** $\sum_{l=1}^{L}(v_{l,n} - \lambda_l) \geq 0$ **then**
13:                $x_{l,n} \leftarrow 1$
14:            **else**
15:                $x_{k,n} \leftarrow 0$
16:            **end if**
17:        **end for**
18:        **if** $(C_{k,l} = 1$ and $v_{k,n} > v_{l,n})$ **then**
19:            $x_{k,n} \leftarrow 1;\ x_{l,n} \leftarrow 0$
20:        **end if**
21:        **for** $m \leftarrow 1, M$ **do**
22:            Power Allocation$(m, x_{k,n}, x_{l,n})$
23:        **end for**
24:    **end while**
25: **end procedure**

26: **procedure** POWER ALLOCATION$(m, x_{k,n}, x_{l,n})$
27:    **if** $(C_{k,j} = 0$ and $x_{k,n} = 1)$ **then**
28:        $p_m \leftarrow p_m + \Delta p_m$
29:    **else**
30:        **if** $(x_{k,n} = 1$ and $x_{l,n} = 0)$ or $(x_{k,n} = 0$ and $x_{l,n} = 1)$ **then**
31:            $p_m \leftarrow p_m - \Delta p_m$
32:        **end if**
33:    **end if**
34: **end procedure**

Fig. 2. Algorithm for the power allocation in a cognitive radio system with DAS.

devices increases or decreases the assigned power antennas with power Δp_m. The value of Δp_m can be determined based on following dependence:

$$\Delta p_m = \frac{\max\left(L, L(\overline{\lambda}, x_{k,n}, x_{l,n})\right) - L(\overline{\lambda}, x_{k,n}, x_{l,n})}{L(\overline{\lambda}, x_{k,n}, x_{l,n})} \tag{17}$$

4   Simulation Results

In the simulation, the energy efficiency of the distributed antenna system and the single, central antenna at cognitive radio system. The system structure was adopted as a single cell with a radius of 1000 m, where 9 antennas of the distributed antenna system (DAS) were arranged. For comparison it was assumed that a central antenna system could be located at the centre of the cell, which is considered in the case of a single central antenna system (CAS). In addition, it is assumed that there are only one PBS and several primary users (PUs) and several mobile users (MUs) within the cell. In addition, It is assumed that the number of SUs within the cell can vary from 5 to 60. The assumed noise level is N_0 is equal to $-110\,\text{dBm}$, and the transmission power of the PBS base station is 0.1 W, while the transmission power of a single SU is equal to 0.05 W.

In the first simulation scenario, energy efficiency was compared for a single system, central antenna and distributed antenna system for different values of emitted energy by distributed antennas, varying from 0.1 to 1.4 W. Figure 3 shows the energy efficiency (EE) for cognitive radio system varying with emitted power of antenna system. Energy efficiency is defined here according to Miao *et al.* [9] as the overall number of bits per unit energy. The results of the energy efficiency simulation obtained for the distributed antenna system (DAS) and the single central antenna (CAS) are comparable to those given by He *et al.* [6] for

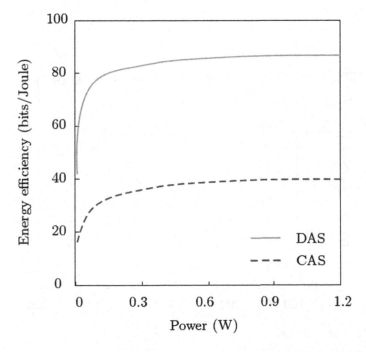

Fig. 3. Energy efficiency versus emitted power of antenna system for DAS and CAS systems.

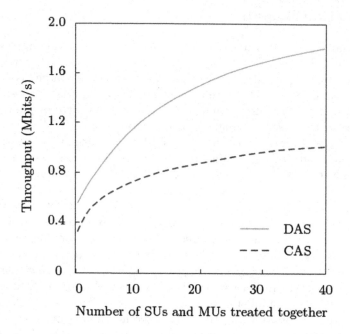

Fig. 4. Throughput system for the DAS and the CAS systems.

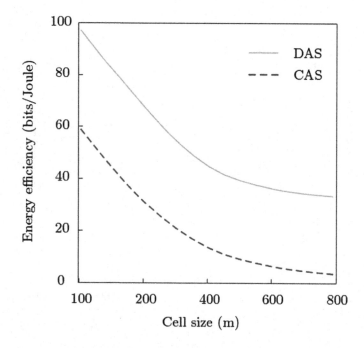

Fig. 5. Energy efficiency versus size of cell for the DAS and the CAS systems.

the optimal solution to this problem. Based on the results obtained, it is stated that the energy efficiency of the distributed antenna system (DAS) increased in comparison with a single, central antenna (CAS) increased by approximately four times.

In the second simulation scenario, the system throughput in dependence on the number of SUs and MUs users treated together was taken into account. The results obtained are shown in Fig. 4. From the graphs, it can be observed that the system throughput is ca. 25% larger for DAS system compared with CAS system.

Figure 5 shows the energy efficiency versus the cell size for both systems: CAS and DAS. It can be observed that the energy efficiency is almost double for the DAS system compared with the CAS system for the same size of cell.

5   Conclusion

The use distributed antenna systems in cognitive radio systems increases the energy efficiency of the antenna system. In addition, this makes it practical to green communication by reducing the total energy emitted by the antennas. The problem of optimal allocation has been formulated as resources in a cognitive radio system with scattered antennas. The algorithm presented makes it possible to improve spectrum efficiency in a cognitive radio system with scattered antennas. The results of the simulation confirmed the correctness of the proposed solution.

References

1. Akyildiz, I.F., Lee, W.-Y., Vuran, M.C., Mohanty, S.: NeXt generation/dynamic spectrum access/cognitive radio wireless networks: a survey. Comput. Netw. **50**(13), 2127–2159 (2006)
2. Boyd, S., Vandenberghe, L.: Convex Optimization. Cambridge University Press, Cambridge (2004)
3. Bulu, G., Ahmad, T., Gohary, R., Yanikomeroglu, H., Toker, C.: Generalized coordinated port selection in a multi-cell distributed antenna system using semidefinite relaxation. In: Proceedings of the Twentyfourth IEEE International Symposium on Personal Indoor and Mobile Radio Communications (PIMRC), London, September 2013
4. Chen, H.-M., Chen, M.: Capacity of the distributed antenna systems over shadowed fading channels. In: Vehicular Technology Conference, VTC Spring, pp. 1–4 (2009)
5. Haykin, S.: Cognitive radio: brain-empowered wireless communications. IEEE J. Sel. Areas Commun. **23**(2), 201–220 (2005)
6. He, C., Li, G.Y., Zheng, F.-C., You, X.: Energy-efficient resource allocation in OFDM systems with distributed antennas. IEEE Trans. Veh. Technol. **63**(3), 1223–1231 (2014)
7. Joung, J., Chia, Y.K., Sun, S.: Energy-efficient, large-scale distributed-antenna system (L-DAS) for multiple users. IEEE J. Sel. Top. Signal Process. **8**(5), 954–965 (2014)

8. Lee, H., Park, E., Park, H., Lee, I.: Bit allocation and pairing methods for multi-user distributed antenna systems with limited feedback. IEEE Trans. Commun. **62**(8), 2905–2915 (2014)

9. Miao, G., Himayat, N., Li, G.Y.: Energy-efficient link adaptation in frequency-selective channels. IEEE Trans. Commun. **58**(2), 545–554 (2019)

10. Mitola, J., Maguire, G.Q.: Cognitive radio: making software radios more personal. IEEE Pers. Commun. **6**(4), 13–18 (1999)

11. Mitola, J.: Cognitive radio – an integrated agent architecture for software defined radio (2000)

12. Mitola, J.: Cognitive Radio Architecture: The Engineering Foundation of Radio XML. Wiley, Hoboken (2006)

13. Park, E., Kim, H., Park, H., Lee, I.: Feedback bit allocation schemes for multi-user distributed antenna systems. IEEE Commun. Lett. **17**(1), 99–102 (2013)

14. Thomas, R.W., Friend, D.H., DaSilva, L.A., MacKenzie, A.B.: Cognitive networks. In: Arslan, H. (ed.) Cognitive Radio, Software Defined Radio, and Adaptive Wireless Systems, pp. 17–41. Springer, Dordrecht (2007)

15. Yu, Q., Li, Y., Xiang, W., Meng, W., Tang, W.: Power allocation for distributed antenna systems in frequency-selective fading channels. IEEE Trans. Commun. **64**(1), 212–222 (2016)

16. Zhao, J., Feng, W., Shi, R., Wang, J., Zhao, M.: Coordinated multi-user transmission in distributed antenna-based spectrum-sharing systems. In: 2015 International Conference on Wireless Communications and Signal Processing (WCSP), pp. 1–5. IEEE (2015)

17. Zhao, F., Ji, S., Chen, H.: A spectrum auction algorithm for cognitive distributed antenna systems. Ad Hoc Netw. **58**, 269–277 (2016)

18. Zhu, H., Karachontzitis, S., Toumpakaris, D.: Low-complexity resource allocation and its application to distributed antenna systems [Coordinated and Distributed MIMO]. IEEE Wirel. Commun. **17**(3), 44–50 (2010)

19. Huiling, Z., Wang, J.: Radio resource allocation in multiuser distributed antenna systems. IEEE J. Sel. Areas Commun. **31**(10), 2058–2066 (2013)

Multi-level Cluster Based Device-to-Device (D2D) Communication Protocol for the Base Station Failure Situation

Abdelhamied A. Ateya[1,2], Ammar Muthanna[1(✉)],
Anastasia Vybornova[1], and Andrey Koucheryavy[1]

[1] State University of Telecommunication,
Pr. Bolshevikov 22, St. Petersburg, Russia
a_ashraf@zu.edu.eg, ammarexpress@gmail.com,
a.vybornova@gmail.com, akouch@mail.ru
[2] Electronics and Communications Engineering,
Zagazig University, Zagazig, Egypt

Abstract. With the massive increase of wireless devices and the challenges associated, device-to-device (D2D) communication becomes a vital solution. Employing D2D communication achieves higher throughput, better cell coverage, increases the spectrum efficiency and other valuable benefits for the cellular networks. In this paper, D2D communication is used to overcome the problems occurred in the situation of Base station (BS) failure and maintain the communication inside the cell without BS. An energy efficient multi-level cluster based D2D communication protocol is introduced to maintain the cell operation even if the Base station is out of use. The protocol employs D2D communication and multi-level clustering to provide communication paths through the neighbouring cells and BSs. The proposed work also efficient if there is a catastrophic situation.

Keywords: D2D · Clustering · BS · Cellular networks · Multi-level · 5G · Wi-Fi Direct

1 Introduction

By 2020 it is expected that the number of wireless devices will be of order of trillions [1]. With the massive increase of wireless devices and the challenges associated, D2D communication becomes a vital solution. Unlike the existing cellular networks and earlier generations (1G...4G), the future fifth generation (5G) cellular system will employ D2D communication as an enabling technology [2]. Based on the 3GPP report, D2D communication provides a flexible efficient communication paradigm for the future 5G cellular system [3]. Recent research activities in LTE-A also consider to employ D2D communication in the current existing 4G networks [4, 5].

Recently, cellular network operators promote their users to use D2D communication technology to make use of the benefits of the new technology [3]. This results in an increase of the wireless devices support D2D communication. Wireless certified

© Springer International Publishing AG 2017
O. Galinina et al. (Eds.): NEW2AN/ruSMART/NsCC 2017, LNCS 10531, pp. 755–765, 2017.
DOI: 10.1007/978-3-319-67380-6_71

devices support D2D communication reach to 4 million devices in 2017 and are predicted to reach to 10 million in 2018 based on the Wi-Fi alliance watch report [6].

D2D communication enables direct communication between proximate devices without relaying information to the BS [7]. Employing D2D communication in cellular networks achieves lots of benefits, including [8, 26]:

1. D2D communication provides an offloading way for the cellular network,
2. Increases spectrum efficiency,
3. Achieves higher throughput,
4. Reduces the traffic congestion, and
5. Achieves better cell coverage.

D2D communication has two main classifications as illustrated in Fig. 1. In the first classification, D2D communication is categorized based on the spectrum used in the direct communication to two main categories In-band and Out-band D2D communication [7, 27]. In In-band D2D communication, the direct communication is run over the licensed spectrum of cellular system. In another word, the cellular spectrum is used for both cellular communication and D2D communication. The In-band D2D communication can be viewed as underlay or overlay. If D2D communication uses the same resources as cellular system it is referred to as underlay In-band D2D communication. However, in Overlaying a part of cellular spectrum is devoted to the D2D communication. In Out-band D2D communication, the links is initiated over unlicensed spectrum (referred to as a radio interface) using an appropriate wireless technology such as Bluetooth, Zigbee, IEEE 802.11 family [9] and Wi-Fi Direct [10]. If the radio resources are managed and coordinated by the users themselves, the D2D communication is Out-band autonomous. In the other hand, if the BS manages and controls the unlicensed radio resources, the D2D communication is an Out-band controlled [11].

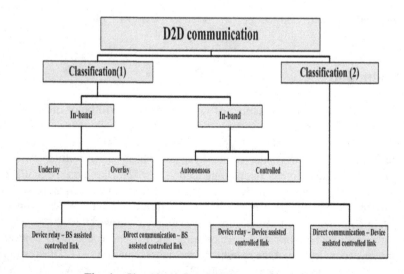

Fig. 1. Classifications of D2D communications.

The second classification is based on the degree of involvement of the BS [12]. If the BS controls the communication link, the D2D is either device relaying with base station assisted controlled link or direct communication between devices with base station assisted controlled link. While in the other two types the devices control the communication link themselves.

There are some challenges associated with the implementation of D2D communication technology. The most important challenges are the resource allocation, the interference management and the delay [13, 14]. In this work, a multi-level cluster based autonomous D2D communication protocol is introduced to maintain the in cell communication without the BS. The rest of the work is organized as the following: in Sect. 2 the literature and related work is presented and in Sect. 3 the proposed work is introduced.

2 Related Work

Recently, the research in the area of D2D communications becomes more attractive. D2D communication represents the key solution for current and future cellular systems. Clustering represents the efficient mechanism to employ D2D communication with higher benefits. There are a lot of literatures in clustering and D2D communication, we provide a brief description for the most related works to our proposed work.

In [15], authors concern with the cluster formation and the selection of the cluster heads. The main purpose of the work is to introduce an energy efficient clustering algorithm that reduces the energy consumption through the direct communication process. Thus authors provide a new strategy for changing cluster heads based on their energy. BS is responsible for the selection of the cluster heads and monitoring them. When the energy of a CH goes below a threshold level estimated by the BS, the BS takes the responsibility to select a newer CH with higher energy. The main advantage of the work is the high energy efficiency introduced while in the other hand; the introduced clustering protocol put a great load on the BS as it is mainly dependent on the BS. Another point of view, the work doesn't consider a certain type of D2D communication and claims that the clustering protocol can be used with all types of D2D. This work must be used in the existence and proper work of the BS unlike our proposed work, also it applies clustering only one time with different mechanism in cluster formation.

In [16], a multi-cast clustering D2D communication protocol is introduced. The authors concern with the cluster head selection, in a way that reduces the blocking probability and increases the power efficiency. The BS controls the clustering inside the cell and decides the decisions of CHs. Simulation results illustrates that the protocol is efficient in terms of blocking probability and energy efficiency. The work makes no sense of D2D communication types and the clustering depends mainly on the BS which remains the main disadvantage of the system.

In [17], a multi-cast clustering D2D communication protocol is introduced such as in [16] but the mechanism of clustering is different. The cluster head selection depends on the data distribution and thus the users who store a multimedia data serve as CHs. The declaration of the CHs is held by the BS.

In [18], a new approach of clustering is introduced. Authors provide a self clustering algorithm based on D2D communication. The cluster formation and CH selection is done using Reed-Solomon (RS) coding and frequency hopping (FH) mechanism. The contribution provides a distribution and illustration for the new idea but without performance metrics. The BS doesn't involve in clustering process.

In [19], authors introduce an architecture and mathematical model approach for Out-band D2D communication. The interface technology used for direct communication is the Wi-Fi Direct and the cluster formation is done based on the specifications of Wi-Fi Direct. The D2D clustering protocol gives a good simulation results in terms of throughput. Also, authors provides a modification approach for changing CHs based on their energy through the BS in order to achieve a higher energy efficiency. The main advantage of the work is that the BS doesn't involve in the cluster formation and thus the clustering protocol offloads the cellular load and reduces traffic congestion.

The novelty of the proposed work is that it solves the problem of the BS failure through multi-level clustering. There are no previous work deals with this problem. Also, it is the first time to introduce Multi-level clustering with D2D communication.

3 Proposed Work

The proposed protocol is used to maintain the cell communication in case of the BS failure (due to either a technical problem or a catastrophic situation), thus the vision is to maintain the cell operation without the BS. If the cell users lose the control of the BS they can switch to the D2D proposed protocol. The protocol employs the short range D2D communication as an alternative method to put the hole cell in communication with the neighbouring cells. Beside D2D communication, the cell nodes should be clustered more than once to enable reaching to the neighbouring cells. The main reason for applying clustering multi-times is to make all nodes be able to communicate with the boundary nodes. The boundary nodes are nodes that located at the cell boundary and can communicate directly with nodes in the neighbouring cells. Thus, in order to maintain the cell operation without BS, the proposed protocol employs multi-level cluster based D2D procedures. The idea behind multi-level clustering is to perform clustering between cluster heads of the first level and select some of them to be a second level cluster heads then the second level cluster heads selects some of them to be a third level cluster heads and the procedure goes on until the final level cluster heads [20].

Cluster Formation

Acluster is a group of several users with a CH that acts as the leader of the group and CMs. In the proposed protocol, cluster formation is achieved by users as it is assumed that there is no BS. Multi-level clustering protocol employs clustering more than once based on the cell capacity and the distribution of the nodes inside the cell. First, the first level of clustering is formed and then the levels are formed in descending order. Assume that the protocol performs N levels of clustering, so the first level (L_1) is formed first then the last level (L_N) and then $L_{(N-1)}$, $L_{(N-2)}$.....2. In each clustering stage there will be CHs and CMs for different clusters. Figure 2 illustrates the relation between different levels CHs.

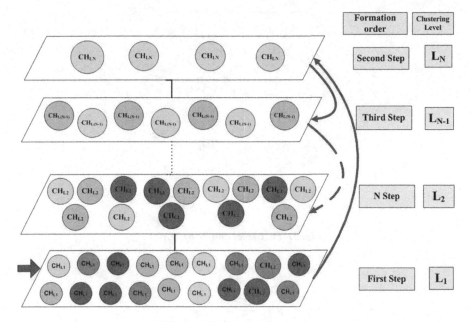

Fig. 2. CHs of different levels

First Level Clustering

As the clusters are formed inside the cell, they are refereed as intra-cell clusters. All intra-cluster communications run over a WLAN in an unlicensed spectrum. The cell is divided into clusters with a CH (CH_{L1}) and CMs for each cluster; and this is the first level of clustering. The communication inside clusters is Out-band autonomous D2D type as CMs communicate with their CH using Wi-Fi direct.

The first level of clustering is formed based on the announced Wi-Fi Direct specifications [6]. Each Wi-Fi Direct group represents a cluster and the group owner is the head of the cluster. Each CH receives the basic information of its CMs. This information contains the Temporary Mobile Subscribe Identity (M/S-TMSI) and International Mobile Subscriber Identity (IMSI). As a response, each CH forms a schedule that contains all member nodes in its cluster and their details. The schedule data contains the member node identification (CM-ID) allocated by the CH and other received data identifying the member nodes. CHs share their schedules and the downlink traffic with their CMs through a broadcasting message. Figure 3 illustrates the first stage of cell clustering and the cluster heads of the first level are illustrated in Fig. 4.

Wi-Fi Direct

Wi-Fi Direct is a recent Wi-Fi technology that enables the short range direct communication between devices [10]. Wi-Fi Direct certified devices are assigned a software access point that enables multi-hop communication [21]. Wi-Fi Direct has the following benefits [22, 23]:

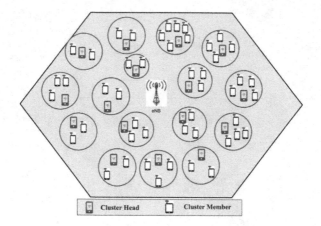

Fig. 3. First level clustering (L_1)

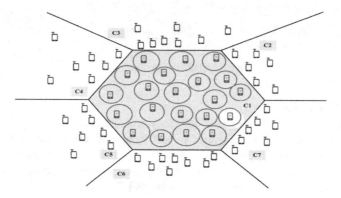

Fig. 4. First level cluster heads (CH_{L1}) with the neighbouring cells.

1. High Mobility, as the Wi-Fi Direct enabled devices can communicate without Wi-Fi router or access point,
2. Ease of use, due to recent advances in discovery features, users can discover the available devices and services before connection set up,
3. High security,
4. Support the third-party application,
5. Ability to form group work, and
6. Higher throughput due to short range distance.

The proposed work assumes that all devices support Wi-Fi Direct technology. Wi-Fi Direct enabled devices support a mandatory Discovery and power efficient mechanisms and thus, these devices are able of forming Groups or clusters with a group owner that acts as the head of the group (CH). The first stage in forming groups is the device discovery stage. Device discovery is the mechanism used by the devises to find, identify other devices and set up connections with them. The user has the ability to go on the discovery mode at any time (for our work the user will turn automatically to the discovery mode since he lose the BS signal). The group is formed automatically once a connection is established between two devices or more and the members of the group are negotiated to select the CH. Further, details about Wi-Fi Direct cluster formation are presented in [24].

Higher Level Clustering

To set up connections with the neighbouring cells, the first level cluster heads should found their way to communicate directly with nodes in the neighbouring cells. Not all CHs can communicate directly with the neighbouring cells because of the distance and thus higher levels of clustering must be formed. After forming first level of clustering, the final level (level N) of clustering is then formed by the following procedure.

The first step to build the final level of clusters LN is to select the final level cluster heads CH_{LN}. CHs (CH_{L1}) go on a discovery stage to discover which ones are able to reach and communicate with the neighbouring cells. All CHs broadcast a message and wait for a response from any nodes in the neighbouring cells [26]. CHs able to reach to the neighbouring cells, announce themselves as a boundary CHs (BCHs) and those are the cluster heads of the last clustering stage (N) (CH_{LN}). Figure 5 illustrates the CH_{LN}s marked with the yellow colour.

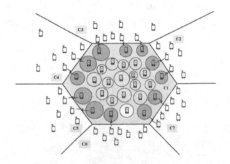

Fig. 5. N-level cluster heads (CH_{LN}) (Color figure online)

BCHs form their clusters by broadcasting discovery messages to other first level CHs. CHs of the first level able to receive the message of BCHs respond to the message and announce themselves as cluster members of the last level of clustering CM_{LN} (some of them will be cluster heads of the clustering stage N − 1 ($CH_{L(N-1)}$)). The response is in the form of a responding message contains identification data including M/S-TMSI, IMSI and the number of first level member nodes associated with each one. Each BCH receives respond message from one or more nodes build a table with the identification data of these nodes and give them identifications as CM_{LN}. Thus, the final level (N) of clusters is formed with CH_{LN}s that provides a communication path for them and for CM_{LN}s with the neighbouring cells. Figure 6 illustrates the N-level clusters, CH_{LN}s and CM_{LN}s.

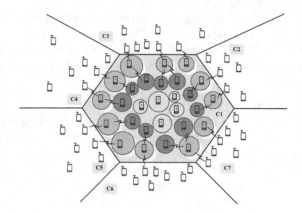

Fig. 6. N-level clusters

With the same mechanism, the level $N-1$ of clustering is formed. Cluster members of the last clustering level (CM_{LN}) go on a discovery stage and broadcast an advertisement message to the rest of CHL1. CHL1s able to receive the advertisement message of CM_{LN} respond with a message contains the identification data and announce themselves as $CM_{L(N-1)}$. CM_{LNs} that receive a response message, announce themselves as $CH_{L(N-1)s}$ and the procedure continues until the $(N-1)$ clustering level is formed. Figure 7 illustrates the clustering level $N-1$ with $CH_{L(N-1)s}$ and $CM_{L(N-1)s}$.

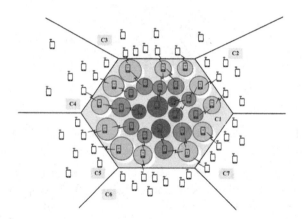

Fig. 7. $(N-1)$ clustering level

The procedure of forming clusters goes on until all CHL1s can reach to BCHs and thus to the neighbouring cells. The number of clustering levels depends on the cell capacity and the distribution of nodes in the cell. The clustering goes to an end only when all CHL1s form parts of a higher level clusters. At this time all nodes in the cell have a communication path to the neighbouring cells and their BSs and thus they can perform all their communication through this way with a proper way and without their cell's BS. Figure 8 shows an example of a 3-level clustering with a cluster head labelled.

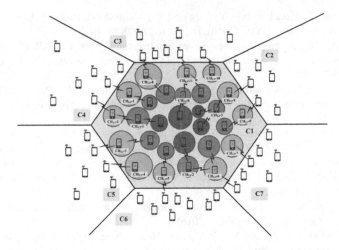

Fig. 8. Example of (3-level) clustering with cluster head labelled

4 Conclusion and Future Work

D2Dcommunication achieves many benefits to the cellular system and thus it must be embedded to the existing 4G and the future 5G cellular systems. Behind the benefits, D2D communication can also be used to solve some problems associated with the cellular system; one of them is the BS failure. Employing multi-level clustering with D2D communication is an efficient way to maintain the inter-cell communication without BS. Multi-level clustering gives all nodes inside the cell, communication paths to neighbouring cells and thus can perform all required communication processes. In case of catastrophic situation, the proposed protocol is an efficient way to preserve the cell communication. The worst case is the failure of the BS and those of all neighbouring cells, in this situation the proposed protocol not help. The best case for employing the proposed protocol is the BS failure due to a technical problem or a regional catastrophic situation with all BSs of neighbouring cells are in work.

The future vision is to build a mathematical model for the proposed algorithm and implement it using FPGA to check the performance.

References

1. Mohammed, A., Nordin, R.: Evolution towards fifth generation (5G) wireless networks: current trends and challenges in the deployment of millimetre wave, massive MIMO, and small cells. Telecommun. Syst. **64**(4), 617–637 (2017)
2. Mansoor, S., Molisch, A., Smith, P., Haustein, T., Zhu, P., De Silva, P., Tufvesson, F., Benjebbour, A., Wunder, G.: 5G: a tutorial overview of standards, trials, challenges, deployment and practice. IEEE J. Sel. Areas Commun. **35**, 1201–1221 (2017)
3. Muthanna, A., Masek, P., Hosek, J., Fujdiak, R., Hussein, O., Paramonov, A., Koucheryavy, A.: Analytical evaluation of D2D connectivity potential in 5G wireless systems. In: Galinina,

O., Balandin, S., Koucheryavy, Y. (eds.) NEW2AN/ruSMART -2016. LNCS, vol. 9870, pp. 395–403. Springer, Cham (2016). doi:10.1007/978-3-319-46301-8_33

4. Chongdeuk, L.: A collaborative power control and resources allocation for D2D (device-to-device) communication underlaying LTE cellular networks. Clust. Comput. **20** (1), 559–567 (2017)

5. Ometov, A., Orsino, A., Militano, L., Araniti, G., Moltchanov, D., Andreev, S.: A novel security-centric framework for D2D connectivity based on spatial and social proximity. Comput. Netw. **107**(2), 327–338 (2016)

6. Wi-Fi Alliance: Wi-Fi certified Wi-Fi direct: personal, portable Wi-Fi (2014)

7. Andreev, S., Moltchanov, D., Galinina, O., Pyattaev, A., Ometov, A., Koucheryavy, Y.: Network-assisted device-to-device connectivity: contemporary vision and open challenges. In: Proceedings of the 2015 21th European Wireless Conference, Budapest, Hungary, 20–22 May 2015, pp. 1–8 (2015)

8. Pimmy, G., Kumar, R.: Device-to-device communication in cellular networks: a survey. J. Netw. Comput. Appl. **71**, 99–117 (2016)

9. Boris, B., Bononi, L., Bruno, R., Kassler, A.: Next generation IEEE 802.11 wireless local area networks: current status, future directions and open challenges. Comput. Commun. **75**, 1–25 (2016)

10. Wi-Fi Alliance: http://www.wi-fi.org/discover-wi-fi/wi-fi-direct. Accessed April 2017

11. Gábor, F., Roger, S., Rajatheva, N., Ben Slimane, S., Svensson, T., Popovski, P., Silva, J., Ali, S.: An overview of device-to-device communications technology components in METIS. IEEE Access **4**, 3288–3299 (2016)

12. Asadi, A., Wang, Q., Mancuso, V., Clerk, J.: A survey on device-to-device communication in cellular networks. IEEE J. Commun. Surv. Tutor. **16**, 1801–1819 (2014)

13. Araniti, G., Raschellà, A., Orsino, A., Militano, L., Condoluci, M.: Device-to-device communications over 5G systems: standardization, challenges and open issues. In: Xiang, W., Zheng, K., Shen, X. (eds.) 5G Mobile Communications, pp. 337–360. Springer, Cham (2017). doi:10.1007/978-3-319-34208-5_12

14. Shuo, Y., Ejaz, W., Guan, L., Anpalagan, A.: Resource allocation schemes in D2D communications: overview, classification, and challenges. Wirel. Pers. Commun. 1–20 (2017)

15. Bhaskara, N., Fahmi, A., Syihabuddin, B., Julia, A.: Cluster head rotation: a proposed method for energy efficiency in D2D communication. In: IEEE International Conference on Communication, Networks and Satellite (COMNESTAT), pp. 89–90, December 2015

16. Pushpalatha, M., Ramesh, T., Giriraja, C., Kumar, S.: Power efficient multicast routing protocol for dynamic intra cluster device to device communication. In: International Conference on Electrical, Electronics, and Optimization Techniques (ICEEOT), pp. 1853–1856. IEEE, March 2016

17. Peng, B., Peng, T., Liu, Z., Yang, Y., Hu, C.: Cluster-based multicast transmission for device-to-device (D2D) communication. In: Vehicular Technology Conference (VTC Fall), pp. 1–5. IEEE, September 2013

18. Tofar, C., Wei, C., Hsu, M., Lin, C., Yu, T.: Distributed clustering and spectrum-based proximity device discovery in a wireless network. In: IEEE International Symposium on Broadband Multimedia Systems and Broadcasting (BMSB), pp. 1–4, June 2016

19. Arash, A., Mancuso, V.: Network-assisted Outband D2D-clustering in 5G cellular networks: theory and practice. IEEE Trans. Mob. Comput. (October 2016)

20. Ateya, A., Sayed, M., Abdalla, M.: Multilevel hierarchical clustering protocol for wireless sensor networks. In: IEEE International Conference Engineering and Technology (ICET), pp. 1–6, April 2014

21. Colin, F., Tapparello, C., Heinzelman, W.: Enabling multi-hop ad hoc networks through WiFi direct multi-group networking. In: IEEE International Conference on Computing, Networking and Communications (ICNC), pp. 491–497, January 2017

22. Wenlong, S., Yin, B., Cao, X., Cai, L., Cheng, Y.: Secure device-to-device communications over WiFi direct. IEEE Netw. **30**(5), 4–9 (2016)

23. Pyattaev, A., Galinina, O., Johnsson, K., Surak, A., Florea, R., Andreev, S., Koucheryavy, Y.: Network-assisted D2D over WiFi direct. In: Mumtaz, S., Rodriguez, J. (eds.) Smart Device to Smart Device Communication, pp. 165–218. Springer, Cham (2014). doi:10.1007/978-3-319-04963-2_7

24. Karim, J., Farhat, O., Al-Jurdi, G., Sharafeddine, S.: Optimized group owner selection in WiFi direct networks. In: IEEE 24th International Conference on Software, Telecommunications and Computer Networks (SoftCOM), pp. 1–5, September 2016

25. Athul, P., Kunz, A., Velev, G., Samdanis, K., Song, J.: Energy-efficient D2D discovery for proximity services in 3GPP LTE-advanced networks: ProSe discovery mechanisms. IEEE Veh. Technol. Mag. **9**(4), 40–50 (2014)

26. Masek, P., Zeman, K., Hosek, J., Tinka, Z., Makhlouf, N., Muthanna, A., Novotny, V.: User performance gains by data offloading of LTE mobile traffic onto unlicensed IEEE 802.11 links. In: Proceedings of the 38th International Conference on Telecommunication and Signal Processing, TSP 2015, pp. 1–5. Asszisztencia Szervezo Kft., Prague (2015). ISBN 978-1-4799-8497

27. Masek, P., Muthanna, A., Hosek, J.: Suitability of MANET routing protocols for the next-generation national security and public safety systems. In: Balandin, S., Andreev, S., Koucheryavy, Y. (eds.) ruSMART 2015. LNCS, vol. 9247, pp. 242–253. Springer, Cham (2015). doi:10.1007/978-3-319-23126-6_22

Author Index

Abilov, Albert 577
Akishin, Vladimir 375
Akulich, Natalya V. 227
Al-Bahri, Mahmood 708
Al-Mistarihi, Mamoun F. 258, 270
Anagnostopoulos, Theodoros 151, 163
Andreeva, S.L. 337
Andrey, Shorov 97
Apanasenko, Nikolay 510
Apatova, N.V. 326
Araniti, Giuseppe 395
Asghar, Muhammad Zeeshan 132
Ateya, Abdelhamied A. 755

Bakulin, Mikhail 550
Belyakhina, Tamara 75
Beschastnyi, Vitalii 395
Bezzateev, Sergey 292
Blinnikov, Mikhail 196
Borodulin, Roman U. 455
Boychenko, O.V. 326
Burkov, Artem 519
Bushehrian, Omid 247
Büther, Florian 214

Campolo, Claudia 395
Chen, Kai Jen 693
Chen, Yi Hua 693
Chernogorskiy, Sergey A. 351

da Silva, Catarina Pires 109
David, Gil 280
Davydov, Roman V. 177
Davydov, Vadim V. 177, 227, 561
Devos, Mathieu 40
Dmitriev, Pavel 734
Dyubo, Dmitrii 206
Dzantiev, Iliya 395, 536

Ebers, Sebastian 214
Efimova, Maria 97
Egorkin, Vladimir I. 490

Fedchenkov, Petr 151, 163
Filimonov, Alexey V. 490
Fokin, Grigoriy 734

Gaidamaka, Yuliya 526
Galimova, A.I. 337
Galimova, M.P. 13
Garnaev, Andrey 382
Gelgor, Aleksandr 644
Gileva, T.A. 13
Gilmutdinov, Marat 653, 662
Glukhov, V.V. 13, 315
Glushakov, Ruslan 196
Goldstein, Alex 375, 441
Goldstein, Boris 375
Gorlov, Anton 644
Grishunin, Sergei 300
Gudkova, Irina 395, 536

Habib, M. Ahsan 132
Hämäläinen, Timo 132, 235, 280
Hussein Ali, Al-odhari Abdulwahab 734

Ilyashenko, Alexander 432, 587
Ismagilova, L.A. 13
Ivannikova, Elena 235
Ivanov, Sergey I. 482
Ivchenko, Anastasia 526

Jayaraman, Prem Prakash 53, 75

Kaganovskiy, Vladislav E. 561
Kaysina, Irina A. 577
Kholod, Ivan 97, 498
Kiesewetter, Dmitry V. 3
Kim, Jinup 473
Kireev, Artem 734
Kirichek, Ruslan 66, 196, 708
Klinkov, Victor A. 184
Kolechkin, Aleksey 708
Komarek, Ales 121

Korenevskaya, Maria   432
Korotey, Evgenii V.   629
Korotkov, Konstantin G.   3
Koucheryavy, A.   671
Koucheryavy, Andrey   196, 569, 755
Kreyndelin, Vitaly   550
Kuropatkin, Nikolay   519

Lau, Florian-Lennert   214
Lavrov, Alexander P.   466, 482
Lee, Seung-Que   473
Lenets, Vladimir A.   177, 227
Leventsov, Valery   142
Lima, Solange Rito   109
Linkov, A.J.   315

Magableh, Amer M.   258, 270
Makarov, Sergey B.   455, 607, 619, 629
Makolkina, M.   671
Makolkina, Maria   683
Malkov, S.V.   326
Malyugin, Victor I.   3
Manariyo, Steve   683
Maresova, Petra   87
Markova, Ekaterina   536
Martyna, Jerzy   745
Masek, Pavel   40
Maslevtsov, Andrey V.   490
Matveev, Nikolay   510, 519
Medvedev, Alexey   151, 163
Melikhova, Antonina P.   635
Melnik, Sergei   550
Mercl, Lubos   121
Mitra, Karan   53, 75
Mohammadi Asl, Reza   420
Molchanov, Dmitri   526
Mukhanova, Natalya V.   337
Muliukha, Vladimir   432, 587
Muthanna, Ammar   196, 683, 755
Myazin, Nikita S.   561

Naumov, Valeriy   395
Nejad, Shayeste Esmail   247
Nekrasova, Tatyana P.   326
Nemcev, Nikolay   662
Nguyen, Van Phe   644
Nikolaevskiy, Nikolay   142
Nuojua, Viivi   280
Nurilloev, Ilhom   569

Orlov, Yu.N.   408
Orlov, Yuri   526
Ostrikova, Darya   536
Ovsyannikova, Anna S.   607, 619

Pakhotin, Valerii A.   629
Paramonov, A.   671, 721
Paramonov, Alexander   569, 708
Parfenova, Yu.A.   408
Parra, Octavio José Salcedo   365
Pavlik, Jakub   121
Petropulu, Athina   382
Petrov, Alexander A.   561
Petrov, Dmitry   550
Pham, Van-Dai   708
Pirmagomedov, Rustam   196
Plokhoi, Nikolai   498
Podgurski, Yuri   587
Poklonskaya, Ekaterina S.   598
Poluektov, Dmitry   536
Pourabdollah, Elham   420

Quang, Trinh Luong   607

Radaev, Anton   142
Rico, Miguel José Espitia   365
Rog, Andrey   550
Rud', Vasily Yu.   177
Rukavitsyn, Andrey   97

Sadov, Oleg   151, 163
Saenko, Igor I.   482
Saguna, Saguna   53, 75
Samouylov, Andrey   526
Samouylov, Konstantin   395, 536
Saveleiv, Ivan K.   177
Savosin, Sergey   29
Scherbakov, S.   721
Scopelliti, Pasquale   395
Semencha, Alexandr V.   184
Semenov, Konstantin P.   3
Shilov, Nikolay   29
Shorgin, Vsevolod   536
Shorov, Andrey   498
Shvetsov, K.V.   351
Silva, João Marco   109
Sivers, Mstislav   734
Smirnov, Alexander   29
Sobeslav, Vladimir   87, 121

Sosunov, Boris V.   455
Sosunova, Inna   151, 163
Stelzner, Marc   214
Suloeva, Svetlana   300

Tanash, Islam M.   258, 270
Tarasenko, Margarita Yu.   177, 227
Teslya, Nikolay   29
Trappe, Wade   382
Tsikin, Igor A.   598, 635
Tsiligiridis, Theodore   420
Tsimerman, Evgenia A.   184
Tsybin, Oleg Yu.   206
Turlikov, Andrey   510

Ustyuzhanin, Nikita   653

Vainshtein, Sergey N.   490
Vargas, Mónica Pineda   365
Vasilenko, N.V.   315
Vasiliev, Danil S.   577
Vekshin, Yuriy V.   466

Vikulov, A.   721
Vinogradova, Ye. Yu.   337
Vladimirov, Sergey   66
Vlasova, Kseniia V.   629
Volgushev, Dmitry   734
Voloshina, Natalia   292
Volvenko, Sergey V.   607, 619
Vybornova, Anastasia   755

Wurz, Marc Christopher   490

Yalunina, Tatiana R.   177
Yalunina, Tatyana R.   227
Yang, Jyun Jhih   693

Zaborovsky, Vladimir   587
Zaslavsky, Arkady   53, 75, 151, 163
Zavjalov, Sergey V.   607, 619
Zayats, Oleg   432
Zemlyakov, Valery E.   490
Zhalgasbekova, Aigerim   53
Zolotukhin, Mikhail   235

Printed in the United States
By Bookmasters